CRITICAL METALS HANDBOOK

Critical Metals Handbook

Edited by

Gus Gunn
British Geological Survey
Keyworth
Nottingham
UK

Published in collaboration with the British Geological Survey

This work is a co-publication between the American Geophysical Union and Wiley

British Geological Survey
NATURAL ENVIRONMENT RESEARCH COUNCIL
AGU American Geophysical Union WILEY

This edition first published 2014 © 2014 by John Wiley & Sons, Ltd
This work is a co-publication between the American Geophysical Union and Wiley

Registered Office
John Wiley & Sons, Ltd, The Atrium, Southern Gate, Chichester, West Sussex, PO19 8SQ, UK

Editorial Offices
9600 Garsington Road, Oxford, OX4 2DQ, UK
The Atrium, Southern Gate, Chichester, West Sussex, PO19 8SQ, UK
111 River Street, Hoboken, NJ 07030-5774, USA

For details of our global editorial offices, for customer services and for information about how to apply for permission to reuse the copyright material in this book please see our website at www.wiley.com/wiley-blackwell.

Library of Congress Cataloging-in-Publication Data

Critical metals handbook/edited by Gus Gunn.
pages cm
Includes bibliographical references and index.
ISBN 978-0-470-67171-9 (cloth)
1. Metals–Handbooks, manuals, etc. I. Gunn, Gus, 1951-
TA459.C75 2014
669–dc23

2013022393

A catalogue record for this book is available from the British Library.

Wiley also publishes its books in a variety of electronic formats. Some content that appears in print may not be available in electronic books.

Cover image: The Spor Mountain open-pit beryllium mine in Utah operated by Materion Brush Natural Resources Inc. (Courtesy of Materion Corp.)
Cover design by Steve Thompson

Set in 9/11.5pt Trump Mediaeval by SPi Publisher Services, Pondicherry, India

1 2014

Contents

List of Contributors, xi
Acknowledgements, xiii

1 Metal resources, use and criticality, 1
T.E. Graedel, Gus Gunn and Luis Tercero Espinoza
The geology and technology of metals, 1
Key concepts, 1
Definitions and terminology, 3
Will we run out of minerals?, 5
Geological assessment, 6
Considerations of supply and demand, 6
Recycling and reuse of metals, 9
The concept of criticality, 10
Assessments of criticality, 11
Improving criticality assessment, 14
Implications of criticality for corporate and governmental policy, 16
Outlining this book, 16
Acknowledgements, 17
Note, 18
References, 18

2 The mining industry and the supply of critical minerals, 20
David Humphreys
Suppliers of minerals – miners and explorers, 21
Industry dynamics, 23
Constraints on mineral supply response, 27
Natural constraints, 27
Economic constraints, 29
Institutional constraints, 31
Critical minerals and the role of China, 34
Policy issues, 38
Notes, 39
References, 39

3 Recycling of (critical) metals, 41
Christian Hagelüken
Rationale and benefits, 41
The urban mine, 41
Recycling benefits, 43
Status and challenges of recycling critical metals, 45
The metals life cycle, 45
Waste and resource legislation, 47
The recycling value chain, 47
Recycling challenges, 48
The seven conditions for effective recycling, 50
Recycling technologies, 51
Collection and pre-processing, 52
Metallurgical recovery, 54
Status of recycling of the EU critical metals, 57
The significance of life-cycle structures, 58
Case study 1: Industrial PGM applications, 59
Case study 2: Automotive PGM applications, 60
Case study 3: Electronic PGM applications, 60
Global flows of old products, 60
Differences in recycling rates and pathways for improvement, 61
Conclusion and the way forward, 62

Innovation needs, 62
Resource security as a societal driver for recycling, 64
Mining and recycling as complementary systems, 64
Conclusions, 66
Notes, 66
References, 67

4 Antimony, 70
Ulrich Schwarz-Schampera
Introduction, 70
Definitions and characteristics, 70
Abundance in the Earth, 71
Mineralogy, 71
Major deposit classes, 72
Gold–antimony (epithermal) deposits, 74
Greenstone-hosted quartz-carbonate vein and carbonate replacement deposits, 77
Reduced magmatic gold systems, 78
Extraction methods and processing, 78
Mining, 78
Ore processing, beneficiation and conversion to metal, 79
Specifications, 82
Uses, 82
Antimony trioxide, 84
Sodium antimonate, 84
Other non-metallurgical uses, 85
Antimony metal, 85
Recycling, 85
Substitution, 86
Resources and reserves, 86
Production, 87
Projects under development, 90
World trade, 91
Prices, 92
Environmental aspects, 94
Outlook, 95
References, 96

5 Beryllium, 99
David L. Trueman and Phillip Sabey
Introduction, 99
Properties of beryllium, 99
Distribution and abundance in the Earth's crust, 100
Uses of beryllium, 100
Alloys containing less than 2% beryllium, especially copper–beryllium, 101
Pure beryllium metal and alloys containing over 60% beryllium, 102
Beryllia (BeO) ceramics, 103
World production, 103
World trade, 105
World resources, 106
Mineralogy of beryllium, 106
Beryllium deposits, 107
Pegmatite deposits, 107
Hydrothermal deposits, 110
Mining and processing of beryllium, 110
Beryl ores, 110
Bertrandite ores, 110
Processing of beryl and bertrandite to beryllium hydroxide, 111
Production of metal and alloys from beryllium hydroxide, 113
Production of beryllium oxide from beryllium hydroxide, 113
Recycling, 115
Substitution, 116
Environmental aspects, 116
Prices, 118
Outlook, 118
Note, 119
References, 119

6 Cobalt, 122
Stephen Roberts and Gus Gunn
Introduction, 122
Physical and chemical properties, 122
Distribution and abundance in the Earth, 122
Mineralogy, 122
Deposit types, 123
Hydrothermal deposits, 123
Magmatic deposits, 129
Laterites, 130
Manganese nodules and cobalt-rich ferromanganese crusts on the seafloor, 132
Extraction, processing and refining, 134
Cobalt from nickel sulfide ores, 134

Cobalt from nickel laterite ores, 134
Cobalt from copper–cobalt ores in DRC and Zambia, 135
Other sources of cobalt, 136
World production and trade, 138
Resources and reserves, 139
Uses, 140
Recycling, 142
Substitution, 142
Environmental issues, 143
Prices, 144
Outlook, 144
Acknowledgements, 146
Notes, 146
References, 146

7 Gallium, 150
Thomas Butcher and Teresa Brown
Introduction, 150
Physical and chemical properties, 150
Mineralogy and distribution, 150
Sources of gallium, 151
Bauxite, 151
Sphalerite (ZnS), 151
Other geological settings, 152
Recovery methods and refining, 152
Primary recovery, 152
Secondary recovery, 153
Refining and purification, 155
Gallium in GaAs semiconductors, 155
Specifications and uses, 157
Gallium metal, 157
Gallium antimonide, 157
Gallium arsenide, 157
Gallium chemicals, 159
Gallium nitride, 160
Gallium phosphide, 162
Photovoltaics, 162
Substitution, 163
Environmental aspects, 163
World resources and production, 164
Production in 2010, 164
Future supplies, 166
World trade, 167
Prices, 167
Outlook, 170
Acknowledgements, 171
References, 172

8 Germanium, 177
Frank Melcher and Peter Buchholz
Introduction, 177
Physical and chemical properties, 177
Distribution and abundance in the Earth, 177
Mineralogy, 178
Deposit types, 179
Accumulation of germanium in sulfide deposits, 181
Enrichment of germanium in lignite and coal, 185
Extraction methods, processing and beneficiation, 186
Extraction, 186
Processing, 186
Specifications, 188
Germanium tetrachloride, $GeCl_4$, 188
Germanium dioxide, GeO_2, 188
First reduction metal, 188
Production of zone-refined metal ('intrinsic' metal), 188
Single crystals, 188
Uses, 189
Recycling, re-use and resource efficiency, 189
Substitution, 191
Environmental aspects of the life cycle of germanium and its products, 192
Resources and reserves, 192
Production, 194
Future supplies, 196
World trade, 197
Prices, 197
Outlook, 198
Supply challenges, 198
Demand drivers, 199
Supply and demand scenario, 200
Acknowledgments, 200
Notes, 200
References, 200

9 Indium, 204
Ulrich Schwarz-Schampera
Introduction, 204
Physical and chemical properties, 204
Abundance in the Earth's crust, 205
Mineralogy, 205
Major deposit classes, 206
Base-metal sulfide deposits, 209
Polymetallic vein-type deposits, 209
Base-metal-rich tin–tungsten and skarn deposits, 210
Base-metal-rich epithermal deposits, 210
Extraction methods and processing, 210
Mining, 210
Processing, beneficiation and conversion to metal, 212
Indium production from copper ores, 213
Indium production from tin ores, 214
Indium recovery from secondary sources, 214
Specifications and uses, 214
Indium–tin oxide (ITO), 215
Alloys and solders, 215
Semiconductors, 216
Others, 216
Resources and reserves, 217
Production, 218
Production from residues and scrap, 220
Projects under development, 221
Abandoned production, 221
World trade, 222
Prices, 223
Recycling and substitution, 224
Environmental aspects, 225
Outlook, 226
References, 227

10 Lithium, 230
Keith Evans
Introduction, 230
Properties and abundance in the Earth, 230
Mineralogy and deposit types, 230
Pegmatites, 232
Continental brines, 232
Geothermal brines, 234
Oilfield brines, 234
Hectorite, 234
Jadarite, 235
Extraction methods and processing, 236
Specification and uses, 238
Recycling, 240
Substitution, 240
Environmental factors, 241
World resources and production, 241
Reserves and resources, 241
Production, 244
Current producers, 245
Production costs, 248
Future supplies, 249
Pegmatite-based projects, 249
Continental brines, 250
Geothermal brine, 251
Oilfield brine, 251
Hectorite, 252
Jadarite, 253
World trade, 253
Prices, 254
Outlook, 255
Acknowledgements, 258
Notes, 258
References, 258

11 Magnesium, 261
Neale R. Neelameggham and Bob Brown
Introduction, 261
Physical and chemical properties, 261
Distribution and abundance in the Earth, 262
Mineralogy, 262
Deposit types, 263
Extraction methods, processing and beneficiation, 263
Nineteenth-century magnesium production processes, 266
Commercial magnesium production processes of the twentieth century, 266
Specifications and uses, 267
Recycling, re-use and resource efficiency, 269
Substitution, 271
Environmental aspects, 272
Non-greenhouse-gas regulations – electrolytic magnesium production, 272
Non-greenhouse-gas regulations – thermal magnesium, 273
Greenhouse-gas emission studies, 273

World resources and production, 275
Future supplies, 277
World trade, 277
Prices, 277
Outlook, 279
References, 281

12 Platinum-group metals, 284
Gus Gunn
Introduction, 284
Properties and abundance in the Earth, 284
Mineralogy, 285
Major deposit classes, 285
PGM-dominant deposits, 286
Nickel–copper-dominant deposits, 292
Other deposit types, 293
Extraction and processing, 294
Extraction methods, 294
Processing, 294
Specifications and uses, 297
Uses of platinum, palladium and rhodium, 297
Uses of ruthenium, iridium and osmium, 300
Recycling, re-use and resource efficiency, 300
Substitution, 301
Environmental issues, 301
World resources and production, 302
Resources and reserves, 302
Production, 302
World trade, 304
Prices, 306
Outlook, 306
Acknowledgements, 309
Note, 309
References, 310

13 Rare earth elements, 312
Frances Wall
Introduction, 312
Physical and chemical properties, 312
Distribution and abundance in the Earth's crust, 313
Mineralogy, 315
Deposit types, 317
Carbonatite-related REE deposits, 319
Alkaline igneous rocks, 323
Other hydrothermal veins, 324
Iron oxide–apatite deposits, including iron-oxide–copper–gold (IOCG) deposits, 324
Placer deposits (mineral sands), 324
Ion adsorption deposits, 324
Seafloor deposits, 325
By-products, co-products and waste products, 325
Extraction methods, processing and beneficiation, 325
Mining, 325
Beneficiation, 325
Extraction and separation of the REE, 327
Specifications and uses, 328
Recycling, re-use and resource efficiency, 328
Substitution, 330
Environmental aspects, 330
World resources and production, 331
Future supplies, 332
World trade, 333
Prices, 334
Outlook, 336
Note, 337
References, 337

14 Rhenium, 340
Tom A. Millensifer, Dave Sinclair, Ian Jonasson and Anthony Lipmann
Introduction, 340
Physical and chemical properties, 340
Distribution and abundance, 341
Mineralogy, 341
Deposit types, 342
Porphyry deposits, 342
Vein deposits, 345
Sediment-hosted copper deposits, 345
Uranium deposits, 346
Magmatic nickel–copper–platinum-group element (PGE) deposits, 346
World resources and production, 346
Future supplies, 348
Extraction methods, processing and beneficiation, 350
Specifications and uses, 352
Recycling and re-use, 354

Catalysts, 354
Superalloys, 355
Substitution, 355
Environmental issues, 356
World trade, 356
Prices, 357
Outlook, 358
References, 359

15 Tantalum and niobium, 361
Robert Linnen, David L. Trueman and Richard Burt
Introduction, 361
Physical and chemical properties, 361
Distribution and abundance in the Earth, 361
Mineralogy, 362
Deposit types, 363
Carbonatite deposits, 363
Alkaline to peralkaline granites and syenites, 367
Peraluminous pegmatites, 368
Peraluminous granites, 370
Extraction methods and processing, 371
Specifications and uses, 374
Recycling, re-use and resource efficiency, 375
Substitution, 375
Environmental aspects of niobium and tantalum, 376
Geopolitical aspects, 376
World resources and production, 377
Future supplies, 379
Prices, 380
Outlook, 381
Note, 382
References, 382

16 Tungsten, 385
Teresa Brown and Peter Pitfield
Introduction, 385
Physical and chemical properties, 385
Distribution and abundance in the Earth's crust, 385
Mineralogy, 386
Deposit types, 386
Vein/stockwork deposits, 387
Skarn deposits, 389
Disseminated or greisen deposits, 390
Porphyry deposits, 390
Breccia deposits, 391
Stratabound deposits, 391
Pegmatite deposits, 392
Pipe deposits, 392
Hot-spring deposits, 392
Placer deposits, 392
Brine/evaporite deposits, 392
Extraction methods, processing and beneficiation, 392
Extraction, 392
Processing, 393
Specifications and uses, 395
Specifications, 395
Uses, 396
Recycling, re-use and resource efficiency, 398
Old scrap, 398
New scrap, 398
Unrecovered scrap, 399
Recycling methods, 399
Substitution, 399
Environmental aspects of the life cycle of the metal and its products, 399
World resources and production, 400
Resources and reserves, 400
Production, 401
Future supplies, 402
World trade, 404
Prices, 406
Outlook, 406
Acknowledgements, 409
References, 409

Appendices, 414
Glossary of technical terms, 419
Index, 431

Contributors

Bob Brown
Publisher
Magnesium Monthly Review
Prattville
Alabama
USA

Teresa Brown
British Geological Survey
Keyworth
Nottingham
UK

Richard Burt
GraviTa Inc.
Elora
Ontario
Canada

Thomas Butcher
Independent Consultant
New York
USA

Peter Buchholz
Mineral Resources Agency (DERA) at the Federal Institute for Geosciences and Natural Resources (BGR)
Dienstbereich Berlin
Wilhelmstraße 25-30
13593 Berlin-Spandau
Germany

Keith Evans
Independent Consultant
San Diego
California
USA

T.E. Graedel
Center for Industrial Ecology
Yale University
New Haven
Connecticut
USA

Gus Gunn
British Geological Survey
Keyworth
Nottingham
UK

Christian Hagelüken
Director EU Government Affairs
Umicore AG & Co. KG
Hanau-Wolfgang
Germany

David Humphreys
Independent Consultant
London
UK

Ian Jonasson
Formerly research scientist at Geological Survey of Canada

Ottawa
Ontario
Canada

Robert Linnen
Robert W. Hodder Chair in Economic Geology
Department of Earth Sciences
University of Western Ontario
London
Ontario
Canada

Anthony Lipmann
Managing Director
Lipmann Walton & Co Ltd
Walton on Thames
Surrey
UK

Frank Melcher
Federal Institute for Geosciences and Natural Resources (BGR)
Stilleweg
Hannover
Germany

Tom A. Millensifer
Executive Vice President and Technical Director of Powmet, Inc.
Rockford
Illinois
USA

Neale R. Neelameggham
'Guru'
Ind LLC
9859 Dream Circle
South Jordan
Utah
USA

Peter Pitfield
British Geological Survey
Keyworth
Nottingham
UK

Stephen Roberts
School of Ocean and Earth Science
National Oceanography Centre
University of Southampton
Southampton
UK

Phillip Sabey
Manager
Technology and Quality
Materion Natural Resources
Delta
Utah
USA

Ulrich Schwarz-Schampera
Federal Institute for Geosciences and Natural Resources (BGR)
Stilleweg
Hannover
Germany

Dave Sinclair
Formerly research scientist at Geological Survey of Canada
Ottawa
Ontario
Canada

Luis Tercero Espinoza
Fraunhofer Institute for Systems and Innovation Research ISI
Karlsruhe
Germany

David L. Trueman
Consulting Geologist
Richmond
British Columbia
Canada

Frances Wall
Head of Camborne School of Mines and Associate Professor of Applied Mineralogy
Camborne School of Mines
University of Exeter
Penryn
UK

Acknowledgements

I would like to thank the authors and reviewers of each chapter who worked hard to deliver high-quality content suitable for the intended non-specialist readership of this book. I am particularly grateful to colleagues at the British Geological Survey for their expert contributions: Teresa Brown for many contributions to editing, map preparation, provision of statistical data, and compilation of appendices and the glossary; Debbie Rayner for preparing all the diagrams and tables; Ellie Evans for formatting text and references; and Chris Wardle for assisting with the cover design. I would also like to express my gratitude to the Natural Environment Research Council (NERC) UK for provision of funding through a knowledge exchange grant that allowed me to work on this project. Finally, I would like to thank my wife, Barbara, for her patience, understanding and support throughout the preparation of this book.

Gus Gunn

British Geological Survey
Keyworth, Nottingham, UK
April 2013

1. Metal resources, use and criticality

T.E. GRAEDEL[1], GUS GUNN[2] AND LUIS TERCERO ESPINOZA[3]

[1]*Center for Industrial Ecology, Yale University, New Haven, Connecticut, USA*
[2]*British Geological Survey, Keyworth, Nottingham, UK*
[3]*Fraunhofer Institute for Systems and Innovation Research ISI, Karlsruhe, Germany*

The geology and technology of metals

Key concepts

In a book such as this, which is intended for a broad audience, it is important to discuss some key concepts and terminology relating to minerals and metals which, although widely used, are seldom defined. In some cases the meaning may be obvious, while in others they are anything but obvious. To avoid confusion and misuse, and to minimise the risks of misunderstanding, we define in the first part of this chapter certain fundamental terms that will provide a foundation for the chapters which follow.

Minerals are essential for economic development, for the functioning of society, and for maintaining our quality of life. Everything we have or use is ultimately derived from the Earth, produced either by agricultural activities or by the extraction of minerals from the crust. Unlike crops, which are grown for the essential purpose of maintaining life by providing the nutrients we need to survive, mankind does not generally need the minerals themselves. Rather, minerals are extracted for the particular physical and chemical properties their constituents possess and which are utilised for specific purposes in a huge range of goods and products. Following some form of processing and purification, a mineral, often in combination with certain other minerals, is incorporated into a component which is used in a product. It is the need or desire for the products that generates a demand for minerals, rather than demand for the mineral itself. As a result, there is always the possibility of finding an alternative material to provide the required functionality. The only exceptions to this possibility are nitrogen, phosphate and potash, which are essential to life itself and cannot be substituted.

The term 'mineral' is used to describe any naturally occurring, but non-living, material found in, or on, the Earth's crust for which a use can be found.[1] Four principal groups of minerals may be distinguished according to their main uses:

1. Construction minerals – these comprise bulk minerals such as sand and gravel, crushed rock and clay, which are used for making concrete and bricks to provide foundations and strength in buildings, roads and other infrastructure. They are produced in large quantities at low cost from extensive deposits that are widely distributed at shallow depths in the Earth's crust.
2. Industrial minerals – these are non-metallic minerals that, by virtue of specific chemical or physical properties, are used for particular applications in a wide range of industrial and consumer products. There are numerous industrial minerals

Critical Metals Handbook, First Edition. Edited by Gus Gunn.

but the most widely used include salt, gypsum, fluorspar, and kaolin. They tend to occur in large quantities but only at relatively few locations. They generally require specialist processing in their production and consequently they are relatively expensive.

3. Energy minerals – these are minerals such as oil, gas and coal that are used to generate energy that is captured when they are burned. They are used in the production of electricity, in fuels for transportation and heating, and also in the manufacture of plastics. Coal is relatively easy to find and cheap to extract; in contrast, oil and gas are generally difficult to find and extract and, therefore, command high prices.

4. Metals – metals are distinguished by distinctive chemical and physical properties, such as high electrical and thermal conductivity, malleability, ductility and the ability to form alloys. They are exploited for a multitude of purposes and some, such as iron, aluminium and copper, are used in huge quantities. Other metals with fewer or more specialised applications, such as platinum, indium and cobalt, are used in much smaller quantities, ranging from tens to hundreds or thousands of tonnes per year. Economic deposits of metals are rare and difficult to locate. The metal-bearing ores are expensive to mine and to process, and consequently metals command a high price.

Another term in common usage is 'mineral commodity' which is used to refer to any mineral raw material that can currently be extracted from the Earth for a profit.

The abundance of individual metals in the Earth's crust varies greatly (Figure 1.1) and influences the costs involved in locating, mining and preparing the metals for use. Some of the major industrial metals, like iron, aluminium and calcium, have crustal abundances similar to the main rock-forming elements, such as oxygen, silicon and calcium, and are several orders of magnitude more abundant than many of the widely used base metals such as copper, lead and zinc. Many others, such as the precious metals gold and platinum, are considerably rarer. However, crustal abundance is only one factor that influences production costs. Some metals that are common in the crust, such as magnesium, aluminium and titanium, occur in forms that need a high input of energy to separate them from their ores, thus making them relatively expensive. It is also important to note that the localised concentrations of metals that can be exploited economically result from unusual geological processes. Consequently, the distribution of economic deposits in the Earth's crust is highly dispersed, with some regions richly endowed in metals and others largely devoid of them. Furthermore, our knowledge of the processes that lead to the concentration of particular metals in the Earth's crust varies widely. For metals that are used in large quantities, such as copper and zinc, we have a reasonably good idea of where and how to locate new deposits. However, for many of the scarcer metals, especially those that have been brought into wide use relatively recently, information on their occurrence, concentration and processing is generally very limited.

It is a complex and expensive process to prove economic viability once an unusual enrichment of a potentially useful mineral or assemblage of minerals, commonly referred to as a 'mineral occurrence', is discovered. This involves determination of the quantity of mineral present and the assessment of the optimum methods for mining and processing the ore. Apart from geological processes that determine the physical availability of a metal there are a host of other factors that influence access to the resources in the ground – cheap labour or cheap power may confer a competitive advantage to a particular country or region while, on the other hand, government regulation, fiscal and administrative requirements, or social and cultural constraints may restrict or prevent access to potentially valuable deposits.

The timescale from discovery of a mineral occurrence to mine production is generally a long one. It commonly takes more than ten years to evaluate the mineral resource in the ground, to raise the funds to build a mine, to acquire the necessary regulatory approvals and to secure the trust and cooperation of the local communities. Once these are in place, and provided that favourable economic conditions prevail, the mine and

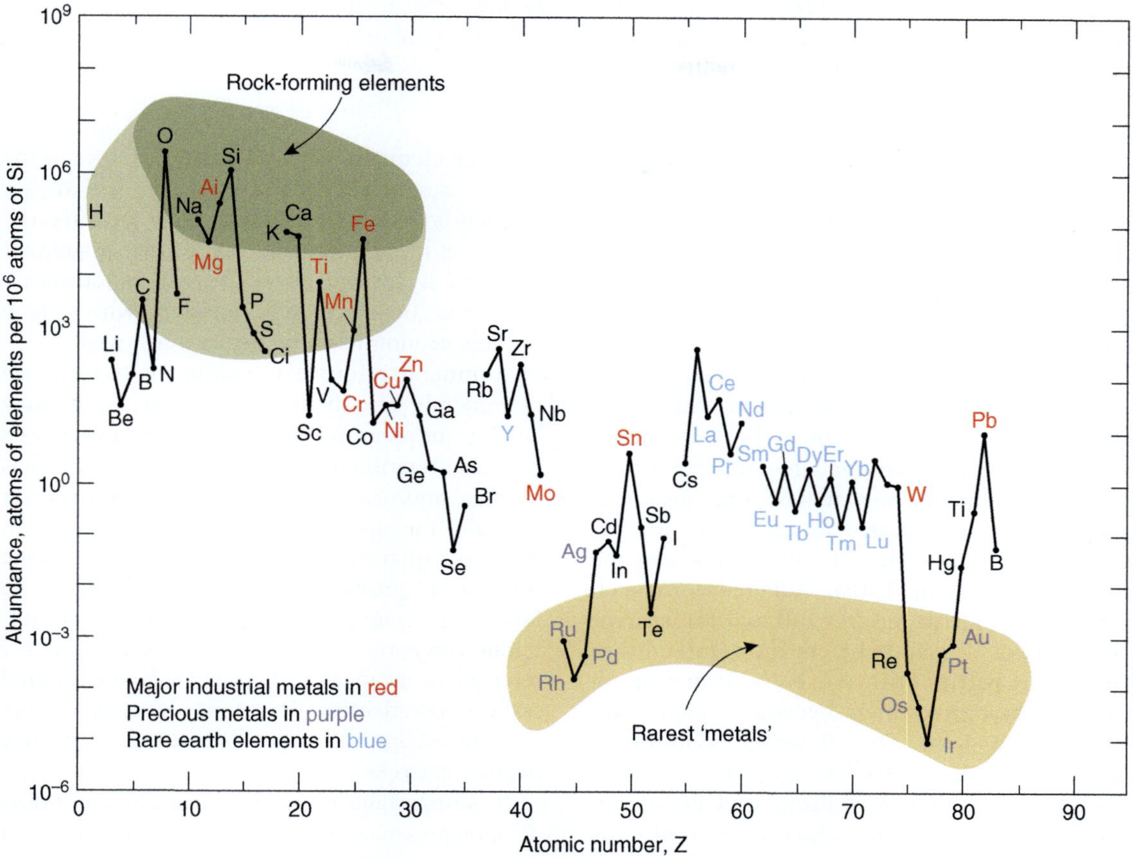

Figure 1.1 The abundance of the chemical elements in the Earth's upper continental crust as a function of atomic number. Many of the elements may be classified into partially overlapping categories. (Modified from USGS, 2002.)

supporting infrastructure can be built and mineral extraction can commence.

Definitions and terminology

The costs involved in bringing a new mine into production today commonly amount to hundreds of millions of dollars or, in the case of a large new mine on a greenfield site, more than a billion dollars. A metal mine typically operates for a minimum period of a decade although, depending on economic and other circumstances, it may continue for more than 100 years. Given the size and duration of these investments it is essential that all parties – the mining company, investors, local communities, governments and regulators – 'speak the same language' and fully understand their obligations and expectations throughout the life of the mine, from construction to operation, closure and site rehabilitation. Without effective communication, based on clear unambiguous terminology, such understanding can never be attained and problems may well arise at some stage.

The first steps in determining the economic viability of a mineral deposit are the exploration and resource assessment stages which involve drilling and detailed sampling to determine the quantity of material present and its quality – or, in the case of a metallic mineral deposit, its grade, which is the percentage of metal that the rock contains. The consistent and correct use of terminology is essential for the reporting and assessment of exploration results and to underpin sound decision making. Without this, discrimination between genuinely economic deposits and those of marginal or unproven economic significance is impossible.

The assessment is, therefore, based on a system of resource classification the main objective of which is to establish the quantities of minerals likely to be available in the future. Many governments now require that resources and reserves are reported according to internationally accepted codes in countries where the company's stock is listed. Adherence to such reporting standards ensures full and transparent disclosure of all material facts and is intended to provide all parties with reliable information on which to base investment decisions. Such codes include the Joint Ore Reserves Committee (JORC) code in Australia and the Canadian Institute of Mining, Metallurgy and Petroleum (CIM) reporting standard which is referred to as National Instrument (NI) 43-101. Following an era of industry self-regulation, these codes were developed in response to scandals in Australia and Canada where many people were misled by speculation and rumour leading to unfounded spectacular rises in share prices and, soon after, rapid falls. In the short term these led to huge financial losses and, in the longer term and more significantly, to a prolonged loss of investor confidence in the mining industry. Accordingly these, and other codes, were developed to set minimum standards of reporting of exploration results, mineral resources and ore reserves. They provide a mandatory system of classification of tonnage and grade estimates according to geological confidence and technical/economic considerations. They require public reports to be prepared by appropriately qualified persons and provide guidance on the criteria to be used when preparing reports on exploration results, mineral resources and ore reserves.

Resources and reserves

The key elements of the reporting codes are the terms 'resources' and 'reserves', which are frequently confused and/or used incorrectly. They are, in fact, fundamental to the distinction between a mineral deposit that is currently economic (reserves) and another which may become economic in the future (resources).

A mineral 'resource' is a natural concentration of minerals or a body of rock that is, or may become, of potential economic interest as a basis for the extraction of a commodity. A resource has physical and/or chemical properties that makes it suitable for specific uses and is present in sufficient quantities to be of intrinsic economic interest. To provide more information about the level of assurance, resources are divided into different categories which, in the JORC code, are referred to as measured, indicated and inferred resources, reflecting decreasing level of geological knowledge and hence decreasing confidence in their existence.

It is important to note that identified resources do not represent all the mineral resources present in the Earth, a quantity that is sometimes referred to as the 'resource base.' In addition to identified resources, there are resources that are undiscovered or unidentified (Figure 1.2). Undiscovered resources may be divided into hypothetical and speculative categories. Hypothetical resources are those which may reasonably be expected to occur in deposits similar to those known in a particular area under similar geological conditions. Speculative resources are those which may be present either in known deposit types in areas with favourable geological settings but where no discoveries have yet been made or in new types of deposit whose economic potential has not yet been recognised.

A mineral 'reserve' is that part of a mineral resource that has been fully geologically evaluated and is commercially and legally mineable. Mineral reserves are divided in order of increasing confidence into probable and proved categories.

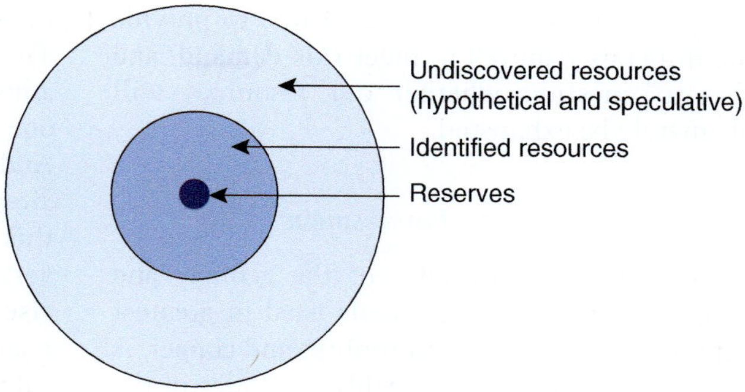

Figure 1.2 Schematic representation of the relative size of the quantities represented by the terms resources and reserves. Reserves generally represent only a tiny fraction of resources.

The ultimate fate of a mineral reserve is either to be physically worked out or to be made non-viable, either temporarily or permanently, by a change in circumstances (most often economic, regulatory or social). So-called 'modifying factors' (economic, mining, metallurgical, marketing, social, environmental, legal and governmental) contribute to the viability of a mineral deposit and determine whether or not it will be exploited.

Figure 1.2 is a simple graphical depiction of the relative sizes of the quantities represented by the terms undiscovered and identified resources and reserves. If this figure were drawn to scale the circle representing the reserves would be very small relative to the resources because reserves are only a tiny fraction of the resources of any mineral.

The term 'reserve base' was also formerly used when discussing mineral resources and mineral availability. This term, introduced by the United States Geological Survey (USGS) and the United States Bureau of Mines (USBM) in 1980, was used as an estimate of the size of the mineral reserve and those parts of the resources that had reasonable potential for becoming economic within planning horizons beyond those that assume proven technology and current economics. However, the reserve base estimates were generally based on expert opinion rather than on data and were not readily defensible, especially at times of rapid growth in mineral demand and consequent massive increases in exploration expenditure, as happened during much of the first decade of the 21st century. Consequently, the USGS abandoned use of the reserve base category in 2010 (USGS, 2010).

Will we run out of minerals?

We are using minerals and metals in greater quantities than ever before. Since 1900 the mine production of many metals has grown by one, two, or even three orders of magnitude (Graedel and Erdmann, 2012). For some metals, especially those used in high-tech applications, the rate of use has increased particularly strongly in recent decades, with more than 80 per cent of the total global cumulative production of platinum-group metals (PGM), indium, gallium and rare earth elements (REE) having taken place since 1980 (Hagelüken et al., 2012). We are also using a greater variety of metals than ever before. For example, turbine blade alloys and coatings make use of more than a dozen metals and high-level technological products, such as those used in medicine, incorporate more than 70 metals. In the quest for improved performance, microchips now use about 60 metals, whereas in the 1980s and 1990s only about 20 were commonly incorporated into these devices.

The main reasons for these changes are increased global population and the spread of prosperity across the world. New technologies, such as those needed for modern communication and computing and to produce clean energy, also require considerable quantities of numerous metals. In the light of these trends it has become important to ask if we can continue to provide the minerals required to meet this demand, and also to question whether our resources will ultimately be exhausted.

Geological assessment

In general, our knowledge of the geology and industrial uses of those metals used in greatest amounts, such as iron, aluminium and copper, is extensive. There is a reasonably good idea of the geological processes responsible for the formation of economic deposits of these metals, and consequently how to identify the best places to look for additional resources. Experience over many decades and centuries has taught geologists and mining engineers how to find, extract and process these metals to provide the goods and services we need. As a result it has been possible to find new deposits to replace those that are worked out, and economic development has not been constrained by metal scarcity.

However, reliable estimates of the total amount of any metal that may be available in the Earth's crust are not in place. Various authors have calculated the maximum quantities present based on estimates of mean elemental crustal concentrations and have concluded that the amounts potentially available are huge (e.g. Cathles, 2010). Although these estimates provide upper limits to availability, they have little real practical value because they take no account of the costs, economic, environmental or social, that would be involved in extracting metals from these sources. Some researchers have adopted a different, 'bottom up' approach based on probabilistic estimates of the crustal endowment of particular metals in specific deposit types. Perhaps the best known and largest study of this type is the United States Geological Survey's Global Mineral Resource Assessment Project, which is being undertaken to assess the world's undiscovered non-fuel mineral resources. One of the first studies completed was a quantitative mineral resource assessment of copper, molybdenum, gold and silver in undiscovered porphyry deposits of the Andean mountain belt in South America (Cunningham et al., 2008). This study concluded that there may be a huge amount of copper to be discovered to a depth of one kilometre below the Earth's surface in the Andes, equivalent to 1.3 times as much as has already been found in porphyry copper deposits in this region. Estimates derived in this way are very useful, not only to mining companies but also to planners, economists, governments and regulators. The approach also has real practical value because it assesses the availability of resources of a type that are well known and can be mined and processed economically with current technology. However, this method is dependent on the availability of high-quality geological data and on a sound understanding of the target mineral deposit class. Unfortunately, such geological information is not generally available and knowledge of many mineral deposit classes that may contribute to global metal production is poor. Consequently, this approach is not likely to yield reliable estimates of global metal availability in the near future; rather, its application will be restricted to a particular deposit type within specific areas. Of course, rather than having accurate estimates of what might ultimately be available to us, what really matters is how can we be sure that we have enough metal to meet our needs and that we will not run out in the future as demand grows.

Considerations of supply and demand

Much of the recent debate has focused on the adequacy of mineral deposits to meet future demand rather than on the political and economic barriers. Several authors have concluded that mineral scarcity and, ultimately, depletion are unavoidable (Ragnarsdottir, 2008; Cohen, 2007). Some have made alarmist forecasts that suggest that for some minerals and metals depletion may occur over relatively short timescales of a few decades or even years. However, these predictions are based on 'static lifetimes' derived from existing known resources or reserves divided by current or projected future demand (Cohen, 2007; Gilbert, 2009; Sverdrup et al., 2009). These forecasts fail to recognise that resources

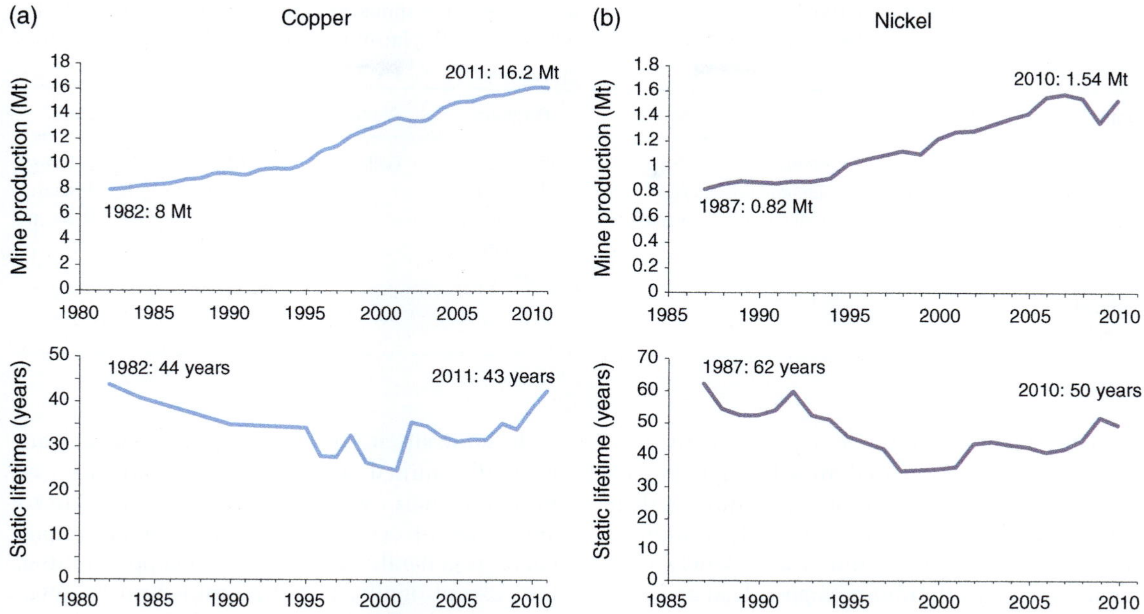

Figure 1.3 Despite escalating global production of metals, reserves have continually been replenished. These graphs show that static lifetimes (number of years' supply remaining equals reserves divided by annual production), in this case of (a) copper and (b) nickel, are extended ahead of production (Mt, million tonnes, metal content). (Mine production data from BGS World Mineral Statistics Database; reserve data from USGS Mineral Commodity Summaries, 2012 and earlier editions.)

and reserves are neither well known nor fixed. Reserves are economic entities that depend on scientific knowledge of minerals and on the price of the target metal or mineral. As our scientific understanding has improved, reserves have continually been replenished through new discoveries, by improved mining and processing technology, and by improved access to deposits. Furthermore, market mechanisms help to overcome supply shortages for major metals – if prices rise, then reserves will extend to include lower-grade ore; if prices fall then they will contract to include higher-grade material. High prices will also stimulate increased substitution, recycling and resource efficiency and thus will contribute to improved security of supply.

Crowson (2011) has discussed changes in reserve levels of some major industrial metals since 1930. He showed that, despite escalating production, reserve levels have actually grown over time and outpaced production. For example, global copper reserves in the early 1930s were reported to be about 100 million tonnes, thought at the time to be sufficient for about 80 years. However, in 2010 the USGS reported copper reserves of 540 million tonnes (USGS, 2010) and in 2011 the estimate was again revised upwards to 630 million tonnes, an increase of more than 16 per cent in a single year (USGS, 2011). Similar trends can be seen in the global reserve levels for some minor metals. For example, tungsten reserves grew by more than 50 per cent between 2000 and 2011, while reserves of REE grew by 25 per cent between 2008 and 2011. It is clear, therefore, that reserve estimates are unreliable indicators of the long-term availability of metals as their definition depends on current science, technology and economics (Figure 1.3).

A type of scarcity referred to as 'technical scarcity' or 'structural scarcity' presents a particular challenge and may be difficult and expensive to

Table 1.1 By-product metals derived from the production of selected major industrial metals (top row, bold). Those metals shown in italics may also be produced from their own ores. (PGM, platinum-group metals; REE, rare earth elements.)

Copper	**Zinc**	**Tin**	**Nickel**	**Platinum**	**Aluminium**	**Iron**	**Lead**
Cobalt	Indium	*Niobium*	*Cobalt*	Palladium	Gallium	*REE*	*Antimony*
Molybdenum	Germanium	*Tantalum*	*PGM*	Rhodium		*Niobium*	Bismuth
PGM	Cadmium	Indium	Scandium	Ruthenium		Vanadium	Thallium
Rhenium				Osmium			
Tellurium				Iridium			
Selenium							
Arsenic							

alleviate. Technical scarcity applies chiefly to a range of rare metals used mostly in high-tech applications. Many of these are not mined on their own; rather they are by-products of the mining of the ores of the more common and widely used metals, such as aluminium, copper, lead and zinc (Table 1.1). These by-product or companion metals are present as trace constituents in the ores of the host metals and, under favourable economic conditions, they may be extracted from these ores, or from concentrates and slags derived from them. For example, indium and germanium are chiefly by-products from zinc production, while tellurium is mainly a by-product of copper mining. However, the low concentration of the companion metal in the host ores means that there is little economic incentive to increase production at times of shortage. For example, only about 25–30 per cent of the 1000 tonnes of indium that is potentially available globally each year from mining indium-rich zinc ores is actually recovered. The rest ends up in wastes because it is not economic to install the additional indium extraction capacity at zinc refineries or because the efficiency of the indium recovery is poor (Mikolajczak and Harrower, 2012). It is therefore difficult to predict the capacity of the supply chain to meet increased demand for the by-product. If the high level of by-product demand is expected to be sustained, for example because of a particular well-established technological requirement such as indium in flat-panel displays and portable electronic devices, then a good economic case for increased indium production can be made.

In some situations certain elements which are normally mined as by-products may also be mined in their own right if their concentrations and mode of occurrence allow it. For example, cobalt is generally a by-product of copper mining, but, exceptionally, it can be mined on its own. Similarly, the PGM are commonly by-products of nickel mining but most production is from PGM-only mines in South Africa.

In some instances groups of metals have to be produced together as coupled elements because they are chemically very similar and cannot be easily separated from the minerals in which they occur. The best examples of coupled elements are the platinum-group metals (PGM: rhodium, ruthenium, palladium, osmium, iridium and platinum) and the rare earth elements (REE comprising 15 lanthanides, scandium, and yttrium). In these cases there is no major carrier metal, but normally one or two of the group determines production levels and the economic viability of the extractive operations. In the case of the PGM, platinum is commonly the main driver for production, with palladium, iridium and ruthenium derived as by-products.

The petroleum industry's debate about 'peak oil' has been extended to the non-fuel minerals industry. The peak concept was developed from the work of oil geologist Hubbert in the 1950s who predicted, on the basis of the existence of a well-known 'ultimately recoverable reserve', that oil production in the USA would peak about 1970 and then enter a terminal decline (Hubbert, 1956). Others extended this approach to predict that global oil production would peak in 2000. These predictions proved largely correct, although global oil production peaked a few years later than forecast. Hubbert's model is based on symmetrical (bell-shaped) curves, with the production peak occurring when approximately half of the extractable resource has been extracted. More recently various authors have advocated 'peak metals' as a tool for understanding future trends in the production of metals (Bardi and Pagani, 2007; Giurco et al., 2010). Bardi and Pagani (2007) examined global production data for 57 minerals and concluded that 11 of these had clearly peaked and several others were approaching peak production.

The application of the peak concept to metals production has been criticised by various authors who have questioned both the validity of the assumptions underlying the model when applied to metals and also the failure to address the real causes of variations in production and consumption in the mineral markets (Crowson, 2011; Ericsson and Söderholm, 2012). Records from the last 200 years show that the prices of major metals are cyclical, with intermittent peaks and troughs closely linked to economic cycles. Declining production is generally driven by falling demand rather than by declining resources or lack of resource discovery. At times of increasing scarcity the price of minerals will increase, which, in turn, will tend to stimulate increased substitution and recycling and encourage investments in new capacity and more exploration. High prices may also lead to more focus on improving current exploration and production technologies. Historically, technological innovation has often succeeded in developing new lower-cost methods for finding and extracting mineral commodities.

It is concluded, therefore, that the peak concept is not valid for modelling mineral resource depletion and cannot provide a reliable guide to future metal production trends. Furthermore, estimates of reserves and resources, and the static lifetime of mineral raw materials calculated from them, should not be used in the assessment of future mineral availability as they are highly likely to give rise to erroneous conclusions with potentially serious implications for policy making and investment decisions.

Recycling and reuse of metals

Modern technology is largely designed around the use of virgin materials extracted from geological sources. It is increasingly apparent, however, that materials that have been incorporated into products no longer in use (secondary materials, scrap) can provide a valuable supplement to virgin stocks. This reuse will generally require that the secondary materials are comparable in quality to those generated from the virgin stocks.

Primary metals are produced through a sequence of actions following their discovery and evaluation: mining the ore, milling it (crushing the rock and separating the metal-containing minerals from the waste material), smelting (to transform the metal oxides and sulfides into impure metal), and refining (to purify the smelted material). None of these processes is perfect, so metal is lost at each stage. The sequence for secondary metals has some of the same characteristics. It begins with collection of the discards, separation of the metals in the discards, sorting of the separated metals, and smelting or similar metallurgical processes to transform the results of the previous processes into metals pure enough for reuse. As with primary processes, metal is lost at each stage.

In a world of increasing resource use, secondary supplies of metals will, however, be insufficient to meet overall demand. Even if all the metals incorporated into products were collected and recycled with 100 per cent efficiency at the end of their useful life, there would inevitably be a shortfall in supply which would have to be filled through production from primary resources (Figure 1.4).

Nonetheless, secondary supplies provide a resource supplement that generally requires less energy than primary metals (often much less), and has generally lower environmental impacts. Through recycling activities, most metals have the potential for reuse over and over again, but only if product designers enable recycling by judicious choice of metal combinations and assembly

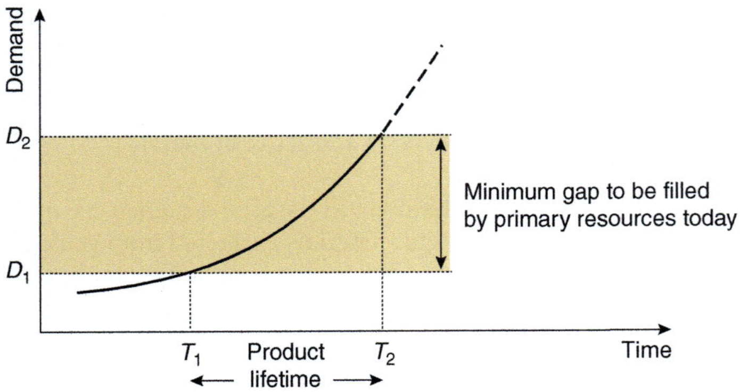

Figure 1.4 When demand for a commodity increases over time recycling alone cannot meet the higher demand. At the beginning of the lifetime of a product, T_1, demand is at a level D_1. At the end of its lifetime, T_2, demand has risen to D_2 but the amount potentially available from recycling will be D_1. The gap in supply (D_2–D_1) can only be met from primary resources.

practices, if governments and individuals optimise product collection at end of life, and if recycling technology is able to produce secondary material whose quality is sufficiently high to enable reuse without downgrading. Certain elements in specific applications are used in a highly dispersed state and cannot be recovered. For example, potassium, phosphate and nitrogen in fertilisers are dissipated in use, as are metals like zinc and magnesium, which are also used for agricultural purposes. Other unrecoverable losses of metals include titanium in paint pigments, and platinum and ruthenium used in very thin layers in hard-disc drives. A wide range of other metals is also lost due to wear and corrosion in use.

Recycling of metals and minerals and the challenges associated with improving its uptake and efficiency are discussed in more detail in Chapter 3 of this book.

The concept of criticality

Without minerals we would not enjoy the lifestyle that we enjoy in the West and to which many others aspire. Without the continued development in the twentieth century of technology for mineral exploration, processing and manufacturing we would not benefit from cheap and reliable products ranging from aeroplanes and cars, to computers, mobile phones and a panoply of other portable personal electronic products that are currently proliferating, such as tablet computers.

This book deals with certain metals that have become increasingly important in recent years for a variety of purposes and for which demand is rapidly increasing. For example, as technology has progressed so new markets for metals, which were previously little used, have arisen or, in some instances, greatly expanded in response to society's needs. Of particular importance are so-called 'green' technologies, especially as the major world economies attempt to shift from carbon-based energy systems.

What is meant by the 'criticality' of metals? Dictionary definitions (e.g. "the quality, state, or degree of being of highest importance") suggest that the term relates to 'essential' or nearly so. In the first few years of the 21st century the label was applied to metals, and particularly to the possibility that some metals might become scarce enough to cease being routinely available to technology. This is more than an idle concern: there have been a number of instances in the past few decades when war, technological change or geopolitical decisions have resulted in temporary shortages. We ask a more fundamental question here, however: might some metals be particularly susceptible to long-term scarcity regardless of the reason or reasons? If we entertain this possibility, could we forecast this situation far enough in advance to mitigate some of its most challenging implications? Or, to simplify, can we determine a metal's criticality and turn that knowledge to use?

The first complexity to point out is that criticality is a matter of degree, not of state. Figure 1.5 makes this point graphically: criticality is not the

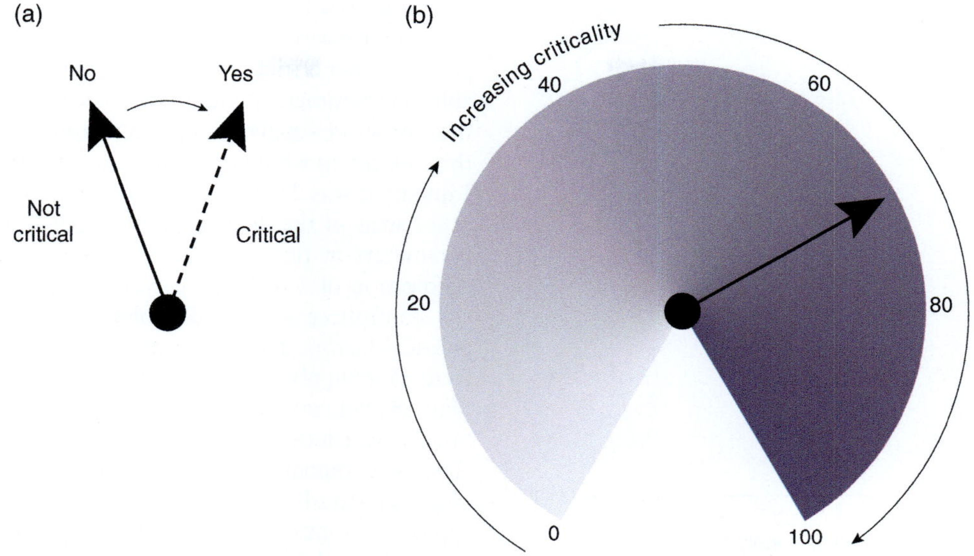

Figure 1.5 Criticality is not simply a designation of 'critical' or 'not critical' as indicated in (a); rather it is a matter of degree, as indicated in (b) where an arbitrary 'criticality level' (here 70) is defined.

position of a switch, such that a metal is either critical or non-critical (Figure 1.5a), but rather a position on a dial where any position above a certain level could arbitrarily be designated as the dividing line between critical or not. The next complexity concerns the metric itself: what is the dial measuring? As we will see, methodologies for determining degrees of criticality can be very complex and are generally multi-dimensional, so the arrow in Figure 1.5b points to a location in two-dimensional or three-dimensional space. This reflects the fact that scarcity may be a consequence of geological factors, economic factors, technology evolution, potential for substitutes, environmental impacts, and many more. This complexity has spawned a variety of analytical approaches and, unfortunately for those wishing to employ the information from those studies, a variety of results.

It is also important to point out that criticality is not a property whose determination is identical to all potential users. For a company whose business is making electrical cables, copper is essential. For a maker of fine jewellery, gold is essential. However, the cable-maker's business does not utilise gold, nor the jeweller's copper (i.e. for those users, either gold or copper cannot be deemed a critical metal). In sum, the degree of criticality of a metal is related to the physical and chemical properties of the metal itself, to a number of factors influencing supply and demand, and to the questioners themselves.

Assessments of criticality

As mentioned earlier in this chapter, concerns about the possible scarcity of natural resources are a recurring theme in history. The main focus has been on the potential impacts of supply disruptions to the economy, especially where it is dependent on imported materials. In the minerals industry finding rapid solutions is particularly challenging because of the high costs and long lead times required to make new mineral supplies available. Buijs and Sievers (2012) noted that the criticality studies conducted in the USA and EU in the 1970s and 1980s adopted basically similar approaches to those used today to identify critical raw materials. Nevertheless, the critical minerals

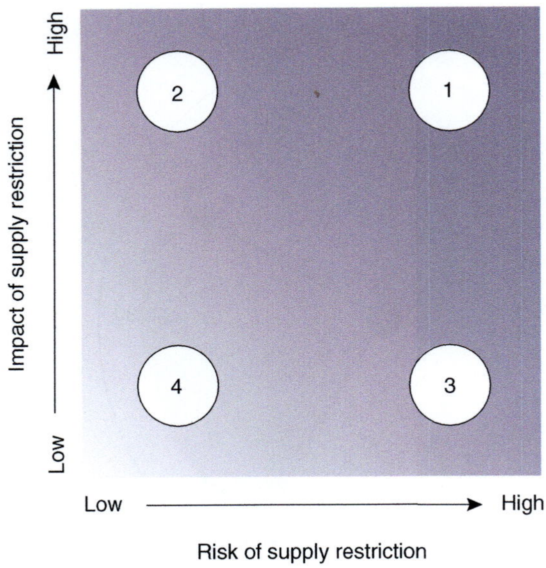

Figure 1.6 The criticality matrix originated by the U.S. National Research Council (2008), as revised by Graedel and Allenby (2010). Metals falling in sector 1 are more critical than those falling in sectors 2 or 3, and much more critical than those falling in sector 4.

identified in those earlier assessments differ from those now classified as critical, thus highlighting that such studies provide only a 'snapshot' of a dynamic system and have little predictive value. However, Buijs and Sievers also observe that the analysis conducted in the earlier studies and the solutions proposed at that time are similar to those of today. Then, as now, it was concluded that, although geological scarcity was highly unlikely, the main supply risks were companion/host relationships, import dependence, the concentration of production in a small number of politically unstable countries, and increased resource nationalism in various forms as the governments in producing countries seek to derive greater benefits from the exploitation of indigenous resources. The measures proposed to alleviate future supply shortages include stockpiling of raw materials, establishment of long-term supply contracts and exploitation of indigenous resources.

The first recent attempt to define metal criticality and suggest metrics that might be employed to assess it was that of a committee of the US National Research Council (2008). The committee proposed that criticality was a two-parameter variable, one parameter being supply risk and the other the impact of supply disruption. Figure 1.6 shows the concept, in which an element falling in the area 1 quadrant was deemed more critical than those in other areas of the diagram. Further, each of those parameters in turn was regarded as some sort of aggregation of a number of contributory metrics: the committee suggested geological availability, political factors, technological capacity and other factors for Supply Risk, and substitutability, importance of applications and other factors for Impact of Supply Restriction. The committee did not select specific components nor delineate the methodology in detail, but did make rough criticality approximations for 11 metals and groups of metals. Those showed the most critical to be rhodium, the least copper, and the others at various locations in between. The committee emphasised that the evaluations were largely to demonstrate the concept, not in any way to be definitive.

A second important evaluation was initiated by the European Commission (EC) in 2009, with a report published in the following year (European Commission, 2010). The EC working group retained the two-axis concept, with supply risk being one of the parameters, but defined the second axis on the basis of the potential economic impact of supply disruption on European industry. Supply risk was further defined as an aggregate of three parameters: the political stability of the producing countries, the potential to substitute the metal being evaluated, and the extent to which the metals are recycled. The evaluation also included environmental risks as a separate concern, and the classification 'critical' was assigned to a raw material if a certain threshold for both economic importance and at least one of the complementary metrics was exceeded. In practice, the metals ranked of most environmental concern were already designated as critical based on other factors.

The EC working group evaluated forty one metals and minerals. The result is shown in Figure 1.7. Arbitrarily drawing lines of

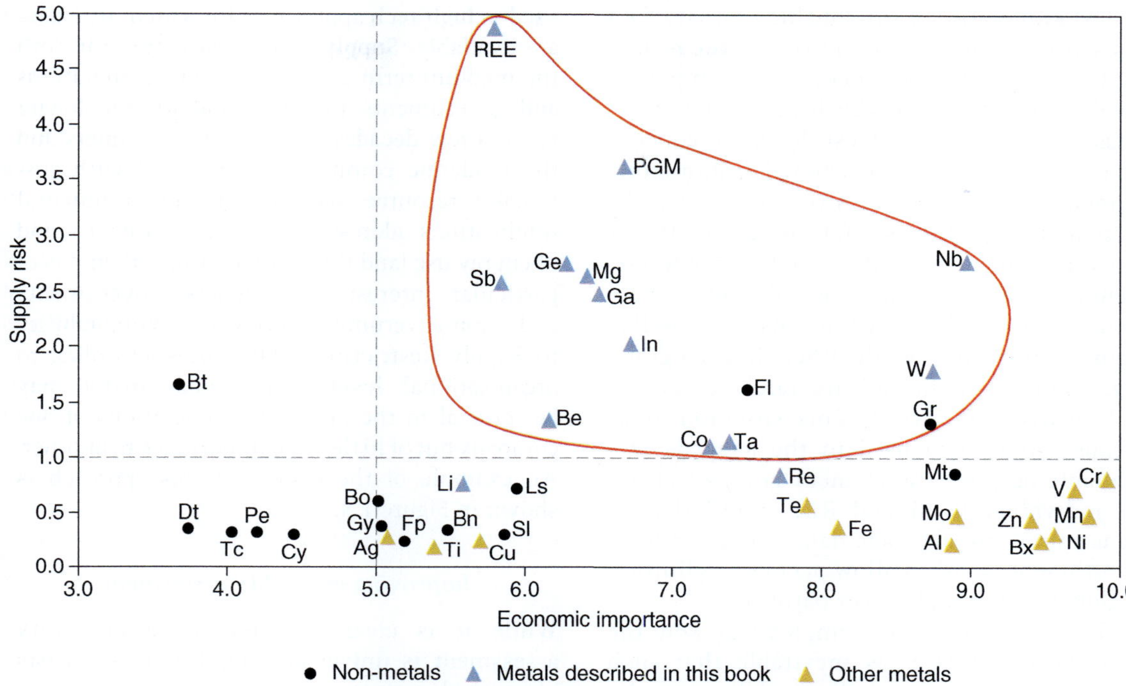

Ag, silver; Al, aluminium; Be, beryllium; Bt, barytes; Bx, bauxite; Bn, bentonite; Bo, borate; Co, cobalt; Cr, chromium; Cu, copper; Cy, clays; Dt, diatomite; Fe, iron; Fp, feldspar; Fl, fluorspar; Ga, gallium; Ge, Germanium; Gr, graphite; Gy, gypsum; In, indium; Li, lithium; Ls, limestone; Mg, magnesium; Mn, manganese; Mo, molybdenum; Mt, magnesite; Nb, niobium; Ni, nickel; Pe, perlite; PGM, platinum-group metals; Re, rhenium; REE, rare earth elements; Sb, antimony; Sl, silica; Ta, tantalum; Tc, talc; Te, tellurium; Ti, titanium; V, vanadium; W, tungsten; Zn, zinc.

Figure 1.7 The criticality matrix of the European Commission (2010). The horizontal axis reflects the economic impact of supply restriction on a broad group of European industries; supply risk constitutes the vertical axis. The 14 raw materials falling within the top-right cluster are regarded as critical to the European Union. (Modified from European Commission, 2010.)

demarcation, the working group designated ten metals as critical: antimony, beryllium, cobalt, gallium, germanium, indium, magnesium, niobium, tantalum and tungsten, as well as two groups of metals, the rare earth elements and the platinum-group metals.

There have been other efforts to designate metals as critical, including those of Morley and Etherley (2008), the U.S. Department of Energy (2010 and 2011) and the Joint Research Council of the EC (JRC, 2011). These, together with the National Research Council and European Commission studies and others, have been reviewed by Erdmann and Graedel (2011) and Buijs and Sievers (2012). They found that the great differences in methodology, the sets of metals reviewed, and selection criteria render it less than convincing at present to single out some metals for special attention while neglecting others, as distinctions between critical and non-critical metals are too complex to be easily resolved. It is clear that, although this topic is generating a high level of interest from governments and corporations throughout the world, the methodology is immature and the results are not necessarily helpful to all parties whose ultimate aim is to secure future supplies of minerals (Buijs et al., 2012).

The availability of suitable high-quality data is a serious issue that can impact on the results of the criticality assessment. For example, in the EU study (EC, 2010) the diagram (Figure 1.7) suggested that the highest level of concern should be for the rare earth and platinum-group elements. These groupings turn out not to be particularly helpful so far as criticality is concerned, in view of the fact that some elements in each group (e.g. platinum, neodymium) are widely used and have a possible claim to criticality, while others in each group (e.g. osmium, holmium) are rarely employed and clearly not critical. This situation arose because some data used in the analysis was available only for the element groups and not for individual PGM and REE. Similarly, for some minor metals trade data is not available in sufficient detail to allow accurate definition of global import and export patterns.

Given the inherent complexities and the data shortcomings it is inevitable that such criticality assessments will not deliver results of universal application, and also that they may fail to identify potential problems. They may suggest that certain materials are at risk when, in fact, market forces may be able to solve the problems in the short or medium term. They may also produce false negatives whereby supplies of some materials are incorrectly identified as secure. However, as these limitations have come to be appreciated and while interest in criticality remains at a high level, so there have been continual refinements of the methodology, adapting it for particular purposes, different organisational levels (corporate, national and global), and over different timescales.

More recently, Graedel and co-workers at Yale University have proposed a comprehensive and flexible methodology for the determination of metal criticality by enhancing the US National Research Council approach (Graedel et al., 2012; NRC, 2008). This method involves three dimensions: Supply Risk, Environmental Implications and Vulnerability to Supply Restriction. It uses a combination of data and expert judgement, the latter especially important for speciality metals used in high-tech application for which little data are available. Supply risk is estimated for both the medium term (5–10 years, with corporations and governments in mind) and for the longer term (a few decades, of interest to planners and the academic community concerned with sustainable resource management). Environmental Implications address both issues of toxicity and of energy use (and thus climate impact), and is of particular interest to designers, governments and non-government agencies. Vulnerability to Supply Restriction (VSR) varies according to organisational level: a particular metal may be crucial to the products or operations of one company but of little or no importance to another. An example of the results of this approach is shown in Figure 1.8.

Improving criticality assessment

While it is clear that no single criticality assessment is universally applicable, shortlists of critical raw materials have an important role to play in warning decision makers in government and industry about current issues of concern and possible impacts on security of supply in the short term. Development of a longer-term capacity to explore potential supply issues is the ultimate goal of such assessments, but there are many intricacies to address before this can be achieved. Key requirements include the necessity to analyse individual metals and underlying issues in more detail, to acquire better data, and to analyse trends and patterns of future demand.

One of the challenges of providing perspective on the long-term supply and demand of metals is that their uses evolve in ways not always predictable. Nonetheless, various studies have attempted to consider technology scenarios considering how wind power, photovoltaic solar power, automotive fuel cells, and other technologies could develop in the next few decades (e.g. European Commission, 2003; IEA, 2008; Shell, 2008). In a typical study, Kleijn and van der Voet (2010) evaluated the resource requirements needed to meet several technology

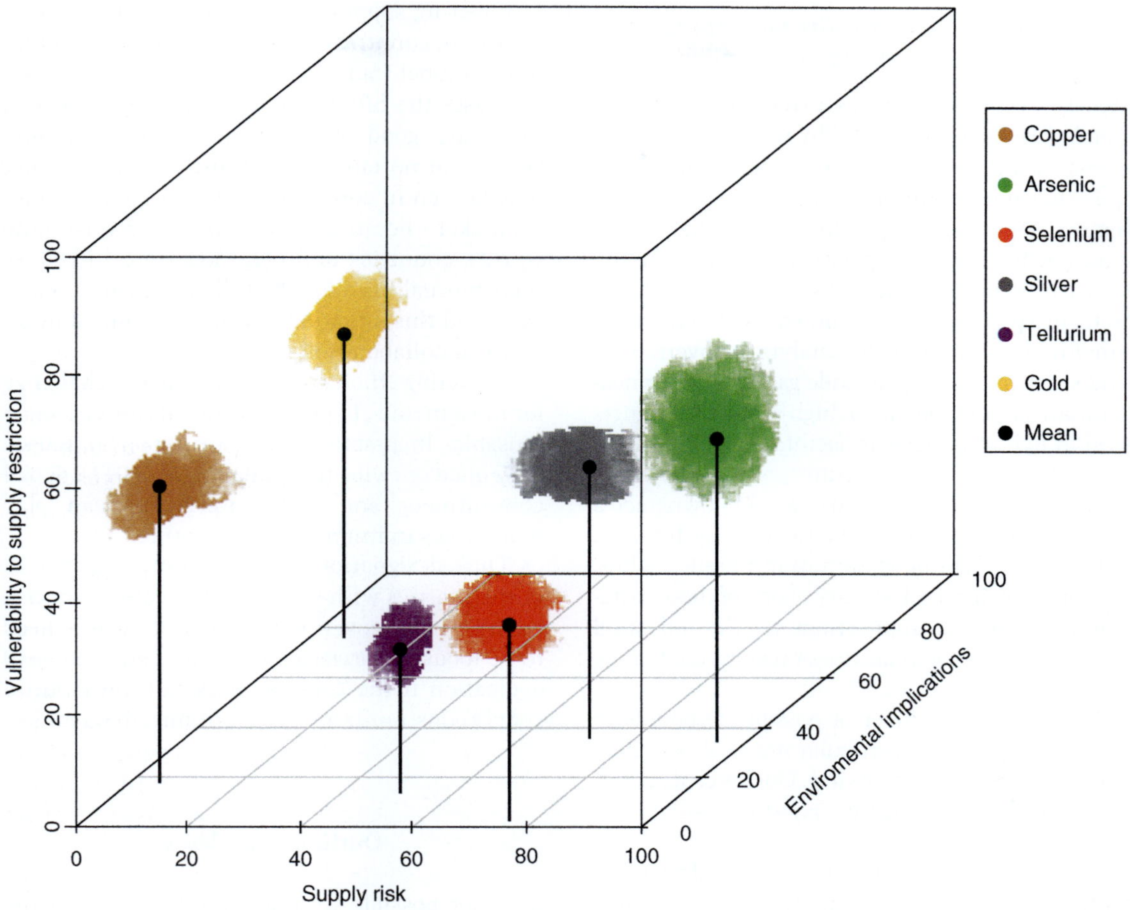

Figure 1.8 The criticality of the geological copper group of metals as determined by the Yale University methodology. (After Nassar et al., 2012.)

projections. They found that substantial deployment of wind turbines, photovoltaic solar cells, hybrid vehicles, enhanced transmission grids, among others, have a strong potential to be restricted because of the large quantities of metal that would be required. Their study indicates that future technology planning will need to have at its centre an assessment of the impacts on metal demand, especially for the scarce metals that are acquired as by-products.

Very few studies have attempted to predict demand for a broad spectrum of technologies (e.g. Angerer et al., 2009) and most have focused on material requirements for the clean energy sector (e.g. U.S. Department of Energy, 2010 and 2011; JRC, 2011). In general, the inclusion of projections in criticality assessment will be a step forward because it will reduce reliance on the future validity of indicators compiled from historic and current data. However, projections inevitably represent a present view of future market states and, though useful for orientation, cannot be relied upon to provide accurate assessments of future demand.

Implications of criticality for corporate and governmental policy

Modern technology makes extensive use of the metals designated as critical by the various assessments discussed above. In virtually all cases, these uses result in improved product performance: faster computers, sharper images on the display screen, wider ranges of operating temperatures, etc. Sometimes no suitable substitute for a critical metal in a particular use is known, as with rhodium (employed in automobile catalytic converters to oxidise harmful nitrogen oxide gases, NO_x), or neodymium (a component of high-strength magnets used in hybrid vehicles to facilitate electric motor performance). In other circumstances a substitute might be available, but its use would downgrade a product's utility, as would be the case for hafnium in computer chips or samarium in missiles. Thus, the potential or actual scarcity of one of these materials has dramatic implications for the industrial using sectors, or for countries or regions containing those sectors.

There exist a number of possible responses to the realisation that a particular material is or may be critical. For corporations (e.g. Duclos et al., 2010):

- vigorously investigate possible substitute materials;
- improve material utilisation in manufacturing;
- redesign products to eliminate or reduce critical material use;
- investigate the potential for recycled materials to replace or supplement virgin material supplies;
- consider entering into long-term contracts or creating stockpiles to ensure supplies for future manufacturing activities.

For governments:

- support geological research to locate new mineral deposits and to better evaluate known deposits;
- support research into improved technologies for recycling;
- consider voluntary programmes or legislation to improve rates of collection and appropriate processing of discarded products containing recyclable materials.

Ensuring supplies of critical materials to corporations, countries or regions inevitably involves international trade, because no country or region possesses the full palette of materials – one area may have good platinum-group metal deposits but few or no rare earth deposits, while another may be rich in copper deposits but lacking those of nickel. Because metal use is diverse, the world's countries and continents are linked by their mutual need for the full spectrum of materials, and this situation requires continued international collaboration.

Recycling efficiency remains a major challenge for most metals. In principle, metals are endlessly reusable. In practice, they are typically reused only once or twice (Eckelman et al., 2011). Social commitment and policy initiatives can play major roles in improving this picture.

Thus, designation of metals or metal groups as critical carries with it policy implications for corporations and governments. The responses need to be focused, forward-looking and pursued with dedication if the consequences of critical metal supply constraints are to be minimised or avoided.

Outlining this book

It is not possible in a single book to cover the entire range of potentially critical metals, nor to unambiguously select those that might be of most concern. As a practical and reasonable choice, however, we address those deemed critical by the European Union working group (2010): antimony, beryllium, cobalt, gallium, germanium, indium, magnesium, niobium, the platinum-group metals, the rare earth elements, tantalum and tungsten. Lithium is included as well, on account of its increasing importance in battery technology and current concerns over its long-term availability. A chapter on rhenium has also been added.

Following this first chapter, two chapters address topics generic to all the metals. The first treats the mining industry, explaining its nature and how it responds to changing demand. The second is on

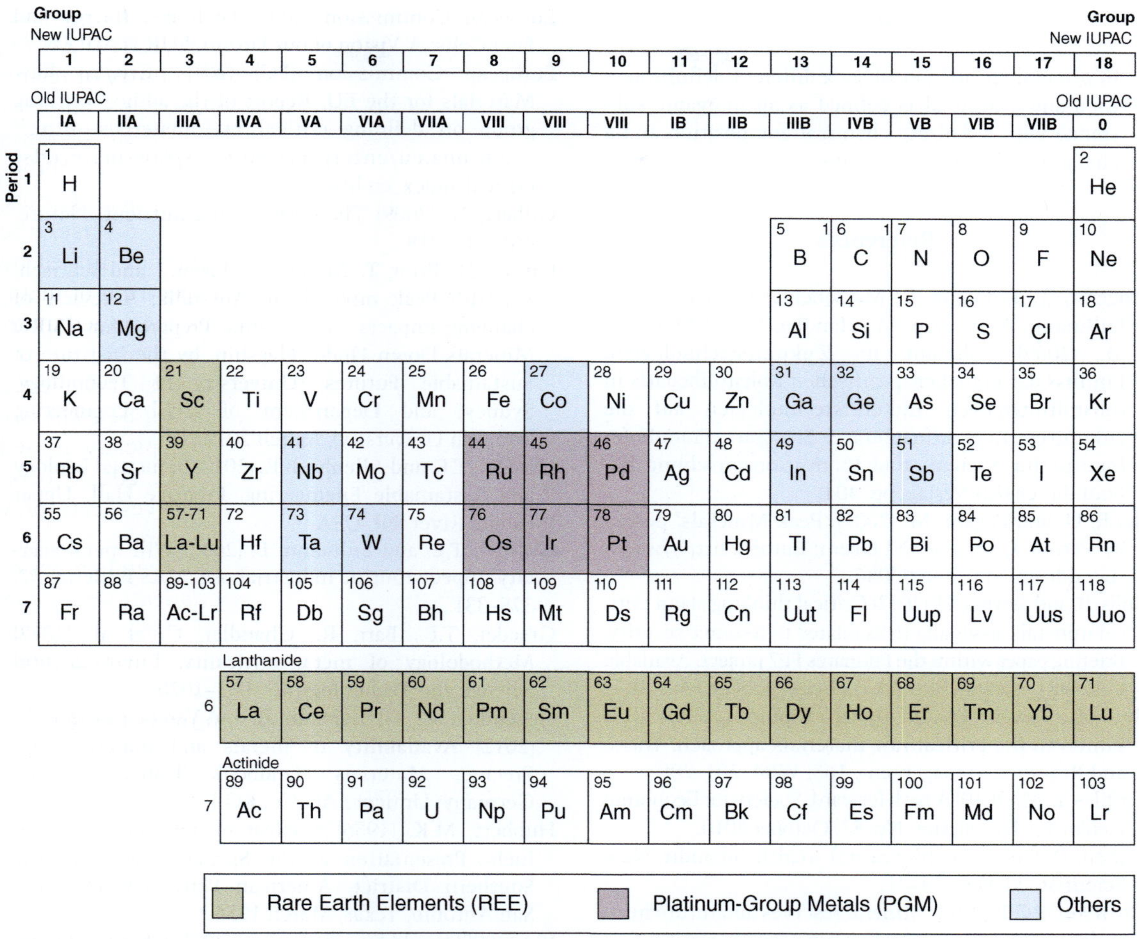

Figure 1.9 The Periodic Table of the chemical elements highlighting the metals described in this book. The new IUPAC (International Union of Pure and Applied Chemistry) naming system for the groups is used in this book. The old IUPAC system is shown for comparison.

recycling, and provides the basis for an understanding of recycling prospects and limitations.

Each of the individual metals or metal groups listed above is then given its own chapter, which provides a summary of appropriate information, including physical and chemical properties, geology, production, trade, recycling and future outlook. While not exhaustive, this information constitutes a basic understanding of the element or element group's criticality aspects and challenges, as well as a perspective on its supply, demand and prospects. These metals and the metal groups covered are shown in the Periodic Table (Figure 1.9).

Acknowledgements

Gus Gunn publishes with the permission of the Executive Director of the British Geological Survey.

Note

1. In mineralogy and petrology a different definition is used and a mineral is defined as an inorganic substance with a definite chemical composition and a characteristic crystal structure.

References

Angerer, G., Erdmann, L., Marscheider-Weidemann, F., Lüllmann, A., Scharp, M., Handke, V. and Marwede M. (2009) Rohstoffe für Zukunftstechnologien. Einfluss des branchenspezifischen Rohstoffbedarfs in rohstoffintensiven Zukunftstechnologien auf die zukünftige Rohstoffnachfrage. Stuttgart: Fraunhofer Institut für System- und Innovationsforschung ISI; Fraunhofer-IRB-Verlag, pp. 404.

Bardi, U. and Pagani, M. (2007) 'Peak Minerals' posted by Vernon C. in The Oil Drum: Europe. http://www.theoildrum.com/node/3086.

Buijs, B. and Sievers, H. (2012) Critical thinking about critical minerals: assessing risks related to resource security. Briefing paper within the Polinares FP7 project. Available from: http://www.clingendael.nl/ciep/publications/2011/

Buijs, B., Sievers, H. and Tercero Espinoza, L.A. (2012) Limits to the critical raw materials approach. Waste and Resource Management, 165, WR4, 201–208.

Cathles, L. M. (2010) A path forward. Society for Economic Geologists Newsletter, No. 83, October 2010.

Cohen, D. (2007) Earth's natural wealth: an audit. New Scientist, 23 May, 34–41.

Crowson, P.C.F. (2011) Mineral reserves and future minerals availability. Mineral Economics 24, 1–6.

Cunningham, C.G., Zappettini, E.O., Waldo Vivallo S. et al. (2008) Quantitative mineral resource assessment of copper, molybdenum, silver and gold in undiscovered porphyry copper deposits in the Andes mountains of South America. USGS Open-File Report 2008–1253.

Duclos, S. J., Otto, J.P. and Konitzer, D.G. (2010) Design in an era of constrained resources. Mechanical Engineering 132/9, 36–40.

Eckelman, M., Reck, B.K. and Graedel, T.E. (2011) Exploring the global journey of nickel with Markov models. Journal of Industrial Ecology 16/3, 334–342.

Erdmann, L. and Graedel, T.E. (2011) The criticality of non-fuel minerals: A review of major approaches and analyses. Environmental Science and Technology 45, 7620–7630.

Ericsson, M and Söderholm, P. (2012) Mineral depletion and peak production. Polinares, Working Paper No. 7, September 2010.

European Commission (2003) Hydrogen Energy and Fuel Cells: A Vision of our Future. EUR 20719 EN.

European Commission (EC) (2010) Critical Raw Materials for the EU. Report of the ad-hoc working group on defining critical raw materials. http://ec.europa.eu/enterprise/policies/raw-materials/critical/index_en.htm.

Gilbert, N. (2009) The disappearing nutrient. Nature, 461, 716–718.

Giurco, D., Prior, T., Mudd, G., Mason, L and Behrisch, J. (2010) Peak minerals in Australia: A review of changing impacts and benefits. Prepared for CSIRO Minerals Down Under Flagship, by the Institute of Sustainable Futures (University of Technology, Sydney) and Department of Civil Engineering (Monash University), March 2012.

Graedel, T.E. and Allenby, B.R. (2010) Industrial Ecology and Sustainable Engineering, Prentice Hall, Upper Saddle River, NJ, USA.

Graedel, T.E. and Erdmann, L. (2012) Will metal scarcity impede routine industrial use. MRS Bulletin, 37, 325–331.

Graedel, T.E., Barr, R., Chandler, C. et al. (2012) Methodology of metal criticality. Environmental Science and Technology 46 1063–1070.

Hagelüken, C., Drielsmann, R. and Ven den Broeck, K. (2012) Availability of metals and materials. In: Precious Materials Handbook (Hanau-Wolfgang, Germany: Umicore AG Co. KG.)

Hubbert, M.K. (1956) Nuclear energy and the fossil fuels. Presentation at the Spring Meeting of the Southern District, American Petroleum Institute, San Antonio, Texas, March 1956.

IEA (2008) World Energy Outlook 2007. China and India Insights. OECD/IEA 2007, Paris, France.

Kleijn, R. and van der Voet, E. (2010) Resource constraints in a hydrogen economy based on renewable energy sources: An exploration. Renewable and Sustainable Energy Reviews 14, 2784–2795.

Joint Research Council (2011) Critical metals in strategic energy technologies. (Luxembourg: Publications Office of the European Union.)

Mikolajczak, C. and Harrower, M. (2012) Indium Sources and Applications. Minor Metals Conference, February 2012.

Morley, N. and Etherley, D. (2008) Material Security: Ensuring Resource Availability to the UK Economy. Oakedene Hollins: C-Tech Innovation Ltd: Chester, UK.

Nassar, N.T., Barr, R., Browning, M. et al., (2012) Criticality of the geological copper group. Environmental Science and Technology 46, 1071–1076.

National Research Council (2008) Minerals, Critical Minerals, and the U.S. Economy. Washington, DC: National Academy Press.

Ragnarsdóttir, K.V. (2008) Rare metals getting rarer. Nature Geoscience. 1, 720–721.

Shell (2008) Shell Energy Scenarios to 2050. Shell International BV.

Sverdrup, H., Koca, D. and Robert, K.H. (2009) Towards a world of limits: The issue of human resource follies. Goldschmidt Conference Abstracts 2009.

U.S. Department of Energy (2010) Critical Materials Strategy. Washington, DC.

U.S. Department of Energy (2011) Critical Materials Strategy. Washington, DC.

USGS. (2002) Rare Earth Elements – Critical Resources for High Technology. U.S. Geological Survey Fact Sheet 087-02.

USGS. (2010) Mineral Commodity Summaries 2010.

USGS. (2011) Mineral Commodity Summaries: Copper.

2. The mining industry and the supply of critical minerals

DAVID HUMPHREYS

Independent Consultant, London, UK

Mineral products are bought for their utility, this utility being reflected in the price which consumers are prepared to pay for them. Properly functioning markets should ensure that an appropriate supply of such products is available to meet consumer demand. A shortage of the sought-after mineral serves to push prices up and stimulate companies to invest in new production capacity. A surfeit of supply leads to a fall in price and a curtailment of output.

The issue of a mineral's 'criticality' enters into the equation because the global economy is composed not just of companies and consumers but also of nations, and nations have strategic interests. Within the broader, strategic, context, mineral products are viewed not only as having utility to consumers but also in terms of the contribution they make to national projects, such as raising the living standards of the nation's citizens, maintaining a capability to produce certain important industrial goods, or ensuring that the nation has the ability to defend itself militarily. In making the transition from being simply 'useful' to being 'critical', minerals and their supply become not just matters for the market but also matters of national security. The process of transition is thus often referred to as 'securitisation'.

The role played by the mining industry in meeting the demand for minerals is subject to a similar duality. The economic function of mining companies is to respond to the requirements of the market, as expressed through mineral prices. For the most part, the industry does this quite effectively. The industry has always had a strong enterprise culture and rising mineral prices can usually be relied upon to prompt mining and exploration companies to develop mines and search for new mineral deposits.

As with mineral consumers, producers operate in a national setting. National authorities are responsible for establishing the legal, fiscal and environmental parameters within which mining companies work. However, like consuming nations, producing nations have strategic objectives. In this context, mining may be perceived as a vehicle for the promotion of broader objectives such as economic development, the reduction of poverty or the assertion of national self-determination. In a direct parallel with the process of securitisation in consuming countries, the assertion of these strategic priorities results in the politicisation of the mineral products and conditions the ability of the mining industry to respond to market signals and thus to supply the minerals that consumers require.

This chapter is divided into five sections. The first looks at the mining industry and its major corporate components, the miners and explorers. The second discusses how the mining industry

Critical Metals Handbook, First Edition. Edited by Gus Gunn.

Table 2.1 World's largest mining companies by market capitalisation, mid-March 2013. (Data from author's estimates based on web sources.)

Rank	Company	Country	Market Cap $bn
1	BHP Billiton	Australia	190
2	Rio Tinto	UK	92
3	Vale	Brazil	90
4	Xstrata	Switzerland	51
5	Anglo American	UK	39
6	Freeport McMoRan	USA	34
7	Grupo Mexico	Mexico	32
8	Norilsk Nickel	Russia	32
9	Barrick Gold	Canada	29
10	Goldcorp	Canada	26
11	Newmont Mining	USA	20
12	Newcrest Mining	Australia	18
13	Teck Resources	Canada	17
14	Antofagasta	UK	16
15	Fresnillo	UK	16
16	AngloGold Ashanti	South Africa	13
17	Fortescue Metals Group	Australia	13
18	Yamana Gold	Canada	11
19	Impala Platinum	South Africa	9
20	Kinross Gold	Canada	9

responds to the demand for minerals and to changes in the level of demand. The third examines the factors which inhibit the mining industry's responses to changes in demand. The fourth looks at some of the specific issues posed for miners by the minerals currently deemed 'critical' and at the role of China in mineral markets. The fifth considers some of the things that governments of consuming countries can do to promote the supply responsiveness of the mining industry.

Suppliers of minerals – miners and explorers

The mining industry exists to meet the mineral requirements of consumers and, in so doing, make profits for shareholders. Although not on the scale of the oil and gas industries, the mining industry is, nevertheless, a very large industry. The enterprise value[1] of the global mining industry in 2010 is estimated to have been around US$2100 billion (Citi, 2011a). London lies right at the heart of this industry, and is host to the headquarters of several of the world's largest mining companies. As of March 2013, there were thirteen mining and metals companies in the FTSE 100 having a combined market capitalisation of US$340 billion, 12.7 per cent of the total value of the FTSE100 (FTSE, 2013). Seven years earlier, the share was six per cent.

The structure of the global mining industry today is the product of a long and complex history. The largest and most publicly visible companies are the so-called 'global diversified miners', or mining 'majors'. These are, by any standards, large companies, operating across many geographies and minerals. Following a period of consolidation during the first decade of the century, this group currently comprises BHP Billiton, Vale, Rio Tinto, Anglo American and Xstrata.[2] The market capitalisation of the world's largest mining companies is shown in Table 2.1. The country indicated is the country of the company's primary stock market listing. The table, it should be noted, excludes aluminium companies, this because most of the value of aluminium, like steel, is created through metallurgical processing rather than through mining.

At the next level down in terms of scale, companies tend to be more focused with respect to either commodity or country. Freeport McMoRan, Grupo Mexico and Antofagasta, for example, are focused on copper, while Barrick Gold, Goldcorp and AngloGold Ashanti are, as their names suggest, focused on the production of gold. Companies which produce a variety of products, but which operate predominantly in one country, include several from the former Soviet Bloc, most notably Norilsk Nickel, but also Kazakhmys and ENRC (Eurasian Natural Resources Corp.) which fall just outside the top twenty companies listed.

Most of the world's largest miners, and all of those in Table 2.1, are public companies, quoted on stock markets (from which their market capitalisations are derived). There are, in addition, a few mining companies comparable in the scale of their mineral output to those listed in the table which are either wholly or predominantly owned by the state. These include the world's largest copper producer, Codelco, which is owned by the state of Chile, and a handful of Chinese companies such as China Shenhua, Yanzhou Coal, China Minmetals Corporation (Minmetals), Chinalco, Metallurgical Corporation of China, (MCC) China Nonferrous Metal Mining Corp. (CNMC) and the Jinchuan Group. Although production from state-owned enterprises is significant and growing, the extent of state ownership in mining is still very much less than is the case with oil and gas.

Beyond the larger and mid-sized mining companies, there are huge numbers of smaller miners, ranging from quoted companies with two or three mines to small family enterprises. Some produce for international markets and some just for local markets. The nature of the mineral product and the form of its occurrence play an important part in determining what products such producers focus on. Small miners do not generally try to compete in mineral markets where producers need scale economies and correspondingly large capital outlays, like iron ore. They can, however, operate in markets where demand is small or where ore deposits can be worked on a relatively small scale, like precious metals or semi-precious stones. At the extreme end of this part of the industry are the artisanal miners. These are very small, maybe even part-time, operators, recovering minerals that can be easily mined near surface (such as alluvial gold, tin, tantalum and diamonds) using very little capital. Such production activity is commonly lightly regulated or indeed wholly unregulated, with miners operating under very basic, and often unsafe and environmentally unsound, conditions. Artisanal mines do, nonetheless make a significant, if not always terribly reliable, contribution to the supply of several critical minerals.

The other key players in the mineral supply equation are exploration companies. This is the entrepreneurial end of the business – the equivalent of technology start-ups – the end where small companies go out to find mineral deposits in the hope either of being able to mine them themselves or else (and more often) sell them on at a good profit to a larger company for development. Since exploration can create enormous value for shareholders, turning what might otherwise be a fairly worthless piece of land into a profitable business opportunity, exploration companies have a strong pioneering quality. The highest rewards typically go to those with innovative ideas about ore genesis (an example might be those which uncovered significant diamond resources in Canada) or which are prepared to go looking in remote and difficult places. By the same token, exploration is also an extremely high-risk activity, and much exploration ends in failure and in investors losing their money.

Accordingly, exploration companies have their own particular economics and their own specialist investors. Banks, which might well be interested in helping a mining company with proven mineral reserves to finance the construction of a mine, are not generally interested in financing exploration. Exploration companies therefore tend to have to rely on equity (i.e. stock market) financing for their activities or on the support of large private

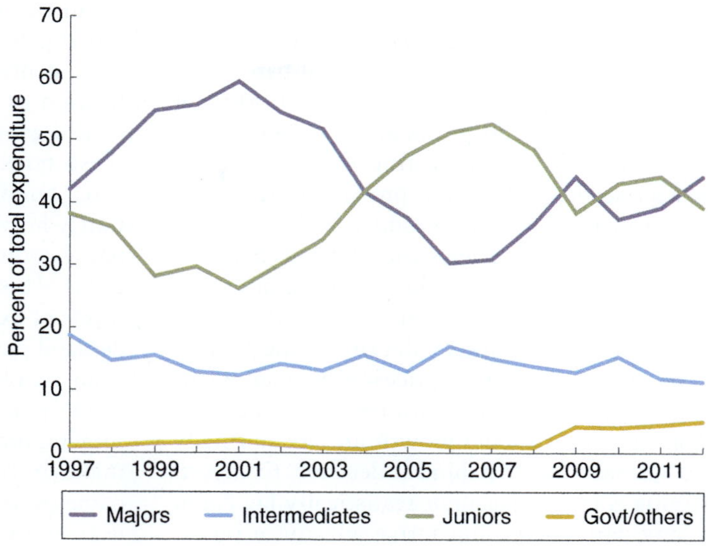

Figure 2.1 Worldwide exploration by company type: per cent shares, 1997–2012. (Data from MEG, 2012.)

investors. Some stock markets specialise in the provision of this sort of financing, notably the Toronto stock exchange (TSX) the Australian stock exchange (ASX) and the alternative investment market (AIM) of the London stock exchange (LSE). Because of the nature of its activities and of its financing, this is much the most responsive part of the mining industry and the part that is quickest to adjust to changes in market perceptions.

Metals Economics Group (MEG) has, for many years, compiled data on global exploration spending. For 2012, it estimated that expenditure was at a record level of US$21.5 billion (MEG 2012). Figure 2.1 shows the distribution of exploration expenditure in recent years split between that undertaken by mining majors, by intermediates, by juniors and by government or other organisations. Two points are apparent from this figure. First, spending by the juniors was much more responsive to rising prices during the course of the metal price boom in 2004–2007 and more responsive also to the falling off of prices in 2008–2009. Secondly, despite the small size of the companies in this sector, the juniors collectively account for a very large proportion of total exploration, this share rising to over 50 per cent of total spend in 2006 and 2007. A high proportion of exploration spending by juniors is accounted for by gold, the small scale of many gold deposits combined with the easy saleability of the product making this metal the target of choice for many juniors. A final point to note is that MEG data is focused on private-sector exploration and accordingly does not take full account of exploration by state companies and other state organisations. In light of the fast growth of state-funded exploration in countries such as Russia, India and, above all, China, in recent years, Raw Materials Group of Sweden considers that MEG's data understate the total exploration spend (Ericsson, 2011a).

Industry dynamics

The larger mining companies do not generally give much thought to a mineral's perceived criticality when evaluating an investment. Their role is to produce minerals for which there is a proven market and to make a profit by so doing.

It is certainly the case that part of the assessment of whether something can be mined profitably resides in a miner's judgement about the strength of demand for the mineral in question and the price that consumers will be prepared to pay for it. However, for the most part these cannot be very accurately determined. Mineral demand and mineral prices are functions of the economic cycle, the forecasting of which is a very inexact science. Moreover, proving up resources and bringing them into production is a process that can take several years and a lot can change in the condition of markets during that time. Thus, while a miner must have some general level of confidence that a market will exist for the product to be produced and that prices will be sufficient to generate a positive return on capital, detailed projections of demand growth are not normally the primary factor behind a decision to invest. Mining companies cannot realistically lay claim to any particular comparative advantage in the art of economic forecasting and will generally, and rightly, be sceptical about the claims which appear in the popular press from time to time about the glittering prospects of this or that exotic-sounding mineral.

The situation with junior miners and exploration companies is a little different. As already noted, these companies are generally dependent on equity markets for their financing. Their survival thus depends on their ability to spark and to sustain interest amongst investors. Accordingly, they tend to be rather more sensitive to market perceptions about the desirability of different minerals than are large mining companies and will often creatively talk up the prospects for the products which they are hoping to find and to mine.

This being the case, exploration companies and junior miners are that much more likely than larger, well-established, mining companies to be responsive to the notion of a mineral's criticality. A project becomes easier to promote if the product it is expected to recover is viewed as having an exciting growth prospect, or is used in new and exotic applications; especially when this is reflected in strongly rising prices. It may not be that the mineral in question is suffering from insufficient investment, or even that there is a realistic prospect of getting a mine into operation in time to relieve any shortage, it is simply that funding is more readily available at such times. The identification of rare earth elements and lithium as critical minerals in recent years has helped generate huge interest in exploration for these minerals. There are believed to be some three hundred rare earth deposits under evaluation (Chegwidden and Kingsnorth, 2011) and over one hundred lithium projects (Mining Journal, 2011). This gold rush mentality – wherein high levels of exploration feed expectations about the demand prospects for a mineral, and vice versa – is an age-old feature of the mining industry.

Only a very few of the many thousands of mineral prospects that are explored ever actually make it through to production. And when it comes to the determination of whether a mineral deposit is to be developed, then judgements about the outlook for demand may well take second place to judgements about the economics of production. After all, if too many companies are pursuing the same growth segment for a given mineral, then there is always the risk that the market will at some point tip over into serious oversupply, at which point the relative competitiveness of producers becomes rather important. Many large mining companies, it might be noted, talk about their strategic objective as being to secure and operate low-cost, long-life, mines without reference to any particular mineral or its demand outlook.

In order for a prospect to be developed, a mining company will generally want to be sure that the resource is of a scale, quality and consistency to support production long enough to permit the recovery of the initial capital investment. It will need to be sure that the conditions of the rock are such as to permit safe and efficient mining. It will need to be sure that power and water are available to the project and that transport exists to get the product to market. In essence, what this will all ultimately boil down to is that the company will want to be confident, or as confident as it is possible in business to be, that it will be able to produce at costs which will make it profitable over the long term. This will, of course, depend in part on its assessment of the long-run price of the product to be produced. However, because of the uncertainties attaching to forecasting long-run commodity prices, the mining company will also be seeking the comfort of knowing that its costs are competitive relative to those of others in the industry.

Such comparative cost assessments play an absolutely vital part in mining company decision making. Having low production costs is not in itself sufficient to justify investment in a project. The object of the exercise, after all, is to make money, and while low operating costs are clearly better than high operating costs, if low production costs can only be achieved through very high levels of spending on capital, it may still be that a project does not merit development. However, comparative assessments of operating costs remain important to miners for several reasons. They permit companies to benchmark themselves against others in the industry and to determine how efficiently they are producing. They provide information about where prices might trend longer term, in as far as prices in a competitive market tend towards the cost of the marginal producer; that is, the highest cost producer required in production to meet prevailing demand. And, by the same token, they will provide information about which companies are likely to be making positive cash returns throughout the cycle, something in which those financing a mine's development, be they shareholders or banks, will be particularly interested.

For the most part, the mining industry has been successful in responding to the changing requirements of the market, and indeed it is organised and incentivised so to do. Rising prices provide a signal of actual or impending shortage and companies accordingly respond by increasing output from existing operations and by instigating searches for new resources. Rising prices also have a stimulus effect on financial markets and facilitate the raising of debt and equity funding by miners and explorers. Reflecting these factors, global mining investment in 2011 soared to US$175 billion, the highest level ever recorded (UBS, 2011) and investment in 2012 is believed to have been higher still. For minerals currently deemed critical, and whose demand is expected to grow rapidly in coming years, there is no reason in principle why the industry should not be able to respond to the challenge and to match increases in demand.

Aside from encouraging the development of known high-quality deposits, higher mineral prices have the effect of converting what were once marginally economic resources into mineable reserves, and may similarly convert waste dumps from earlier workings into sources of recoverable product. Sustained interest in these minerals will also likely stimulate an interest in investing in, and improving the technologies used for, the recovery of these minerals. It is interesting to note that Molycorp Minerals's Mountain Pass rare earths mine in California, which has re-opened in response to higher rare earth element prices, has been wholly reconfigured since it was closed in 1995 following a thorough review of all aspects of its technical and environmental performance. The new mine will have process recoveries of 95 per cent as against 60–65 per cent at the old mine, will use 30 per cent less reagents and only four per cent of the fresh water. Accordingly, it will be able to produce rare earth elements at much lower costs than it could when it operated previously.

Mining is, however, a capital-intensive industry with long lead times from discovery to production and its responses are necessarily lagged. For new mine production to flow, mining companies have to be convinced that they have a viable project and then secure funding for it. Combined with the need to assess the environmental impacts of a mine, to acquire permits to mine and, frequently, forge agreements with local communities, this process can take several years. And then there is the not insignificant matter of building the mine itself. This will require ground preparation, the construction of plant, the acquisition of specialised equipment and the creation of facilities for mine waste. Not uncommonly, it will also require the building of railways, ports and power stations.

Accordingly, while it may be the case that eventually miners will catch up with imbalances in supply and demand for minerals, there can

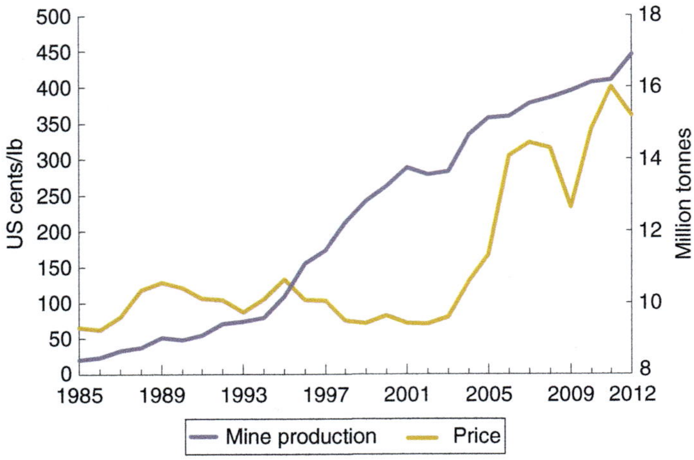

Figure 2.2 Copper prices and mine production, 1985–2012. (Price data are annual averages calculated from London Metal Exchange monthly averages; production data compiled from World Metal Statistics Yearbook, published annually by World Bureau of Metal Statistics.)

nonetheless be periods of shortage while they are bringing on new supply, these periods potentially lasting several years. Such long lags are one of the defining characteristics of the mining industry and the basis of its sticky supply responses. The point can be illustrated at a high level with the example of copper. Figure 2.2 shows a plot of world mine production against the world copper price. A close examination of this figure reveals that there was an eight-year gap between the uptick in prices which took place in 1987 and the acceleration of mine production growth in 1995. A similar delayed supply response is evident in the cyclical downswing. The decline in prices which occurred following the price peak of 1995 did not result in a visible reduction in global production until 2002. In the most recent cyclical upswing, the price increases which began in 2004 finally resulted in an acceleration in mine production in 2012. Like the proverbial oil tanker, mineral production can take a long time to turn around.

During these lengthy periods of supply adjustment, prices are required to take the strain of forcing supply and demand into alignment by rising to levels which choke off the portion of demand which cannot be satisfied. This is sometimes referred to as demand rationing or demand destruction.

Demand rationing works in several ways. First and foremost, high prices encourage consumers to use less of a product or to use it more efficiently. Thus, high copper prices have resulted in the thin-walling of copper pipes and in gauge reduction. High prices of nickel have encouraged a shift towards the production of stainless steels which use less, or indeed, no, nickel. High prices also encourage consumers who can do so to switch to using cheaper alternative materials. Thus, high platinum prices have led jewellers to substitute white gold and palladium for platinum, while high copper prices have encouraged the substitution of plastic plumbing pipe for brass pipe. Finally, high prices result in a reduced call on mined materials by encouraging recycling (see Chapter 3). A key element in the economics of recycling is played by the cost of collection and separation. Although the effects tend to be highly mineral specific, as mineral prices rise so generally they provide an incentive for the collection of old scrap and an increase in the rate of recycling.

These effects of higher prices are part of the normal mechanism of adjustment in mineral markets. Indeed, some of the changes brought about by price-induced changes in demand have beneficial long-run effects in terms of increasing the efficiency of materials use and promoting advances in technology. Although it is not so hard to point to examples where mineral markets have suffered sustained shortfalls in supply – the shortages of cobalt arising from civil war in Zaire in the late 1970s, for example – it is quite difficult to think of peacetime examples where shortages of minerals have had serious adverse long-run effects on an economy. Normally, markets adjust in the short term through price-induced demand rationing and in the longer run through increased supply.

Constraints on mineral supply response

While it may generally be the case that properly functioning markets will provide solutions to mineral shortages, there is a variety of natural, economic and institutional factors which in practice can inhibit the responses of miners and explorers to shortages and thereby prolong the period of supply adjustment. These are important in assessing the likely future availability of critical minerals. The next three sections look at these factors in turn.

Natural constraints

One of the defining characteristics of the minerals industry is that mineral resources are depleted through exploitation. (For a fuller discussion on resource definitions and related issues, see Chapter 1.) This gives rise to a common perception that physical availability may become a constraint on future mineral supplies. In point of fact, physical availability of minerals in the ground is scarcely, if ever, a constraint on mineral supply. While some minerals are inevitably easier to find and develop than others, most minerals mined commercially are available in quantities which are adequate for very many years to come.

A simple way of looking at the adequacy of mineral reserves is to divide through the reported reserves of a mineral by annual production to get a 'static life index'; that is, an estimate of the expected life of the remaining reserves expressed in years. Such estimates are, unfortunately, commonly subject to misrepresentation in that casual users of these data have a tendency to overlook the fact that reserves is a dynamic concept. For many of the most commercially important minerals, including copper, nickel, lead, tin and zinc, reserves life tends to fall into a range of twenty to fifty years (USGS, 2012). However, it should be noted that these reserves lives have not much changed in many years (Crowson, 2011). As production has risen, so reserves have gone up. The commercial incentives simply do not exist for companies to go out and prove up reserves which will be required more than fifty years out. Moreover, it should also not be forgotten that most mineral materials are not destroyed by use and that in addition to reserves in the ground there are substantial amounts of above-ground materials available for re-use and recycling.

For the minor metals which are the focus of concern in the debate on critical minerals, the picture on reserves life is more varied. Some, such as the rare earths, lithium and tantalum, have reported reserves which are very large indeed relative to current levels of production, stretching out in the case of the first two to several hundred years. For others, because they are recovered as by-products, there are no meaningful estimates of reserves so a calculation of reserves life cannot be made.

If the physical availability of minerals is not a major constraint on the supply response of the minerals industry, the geographic concentration of mineral resources can be. Such a concentration exists in the case of the platinum group metals (PGMs).[3] (A more detailed discussion on the PGMs is to be found in Chapter 12.) According to the US Geological Survey, some 95 per cent of the world's reserves of PGMs are located in South Africa, in the Bushveld Complex (USGS, 2012). Such concentration of reserves does not in itself represent a constraint on supply, but it does make supply more susceptible to constraint.

Thus, while South Africa's reserves are sufficient to supply world markets with PGMs for many years to come, several factors have restricted access to the reserves.

A large part of the better reserves are tied up by one of the existing major producers and are therefore not available to industry newcomers. Those that are available often present serious challenges with respect to mining conditions, permitting and access to smelting and refining facilities. Meanwhile, all mining companies operating in South Africa, including the majors, face tough challenges associated with ownership requirements (including the strictures of black economic empowerment (BEE)), health and safety standards, power availability, rising labour costs and a strong Rand. It is largely for these reasons that, despite the existence of strong demand for platinum over the past decade, driven by demand for autocatalysts, investment in the industry has been constrained and production has grown scarcely at all (see Figure 12.5), forcing up prices and, as noted above, resulting in a substantial displacement of jewellery demand for platinum.

Other cases where a high concentration of reserves represents a potential constraint on supply are phosphates in Morocco and cobalt in DR Congo. In the case of Morocco, which accounts for three-quarters of the world's reserves of phosphate rock (USGS, 2012), access to the industry is restricted by the fact that the industry is wholly state-owned. With respect to cobalt, the DR Congo accounts for around half the world's reserves of cobalt and around half its production (USGS, 2012). The challenging investment climate in DR Congo, and the reluctance of large international companies to operate there, has long made this a matter of concern for consuming nations. The invasion of the Katanga province in DR Congo in the 1970s (or the Shaba province and Zaire as they were respectively known at the time), and the disruptions to supply that this caused, were a key event in triggering the 'resource war' concerns of the 1980s. In addition to the concentration of reserves by country, corporate concentration can sometimes be an issue in supply vulnerability as the case of the PGMs also illustrated. Examples of minerals where corporate concentration of production is unusually high are niobium (where CBMM is the dominant producer) and tantalum (Talison Minerals).

Another natural constraint on the ability of the mining industry to respond to shortages is the fact that some metals are produced predominantly as by-products of other, economically more important, metals. (For more on the subject, see Chapter 1.) This is not to say that they cannot be recovered in their own right, only that the cost of doing so will be very much higher, perhaps prohibitively so. Thus, cobalt is produced largely as a by-product of copper and nickel. Cadmium, indium and germanium are produced as by-products of zinc production. Germanium is also recovered from coal fly ash. Gallium is produced as a by-product of bauxite. Molybdenum, rhenium, selenium and tellurium are all produced as by-products of copper.

The problem for these metals is that their supply is largely dependent on the production economics of the metal of which they are a by-product. If production of the principal metal ceases to be economically viable then the by-product will cease to be recovered too. It is for this reason that the supply of by-products tends to be insensitive to changes in the level of demand and why the prices of mineral recovered as by-products can fluctuate wildly. The price of molybdenum (Figure 2.3) shows a fairly classic by-product profile, which is to say, long periods of low and stable prices (reflecting times when markets are well supplied) interspersed by explosive price peaks (reflecting times when supply is unable to respond to increased demand). Rhenium, another by-product of copper which is discussed in Chapter 14, shows a similar price pattern to molybdenum. Quite a few of the minerals deemed critical in the US and the EU fall into this by-product category and accordingly display the rather erratic supply and price behaviour of such metals.

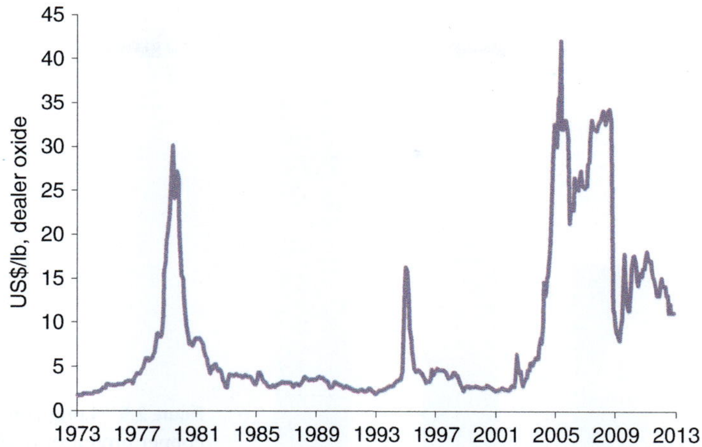

Figure 2.3 Molybdenum price, 1973–2012. (Price data are quarterly averages from Metal Bulletin.)

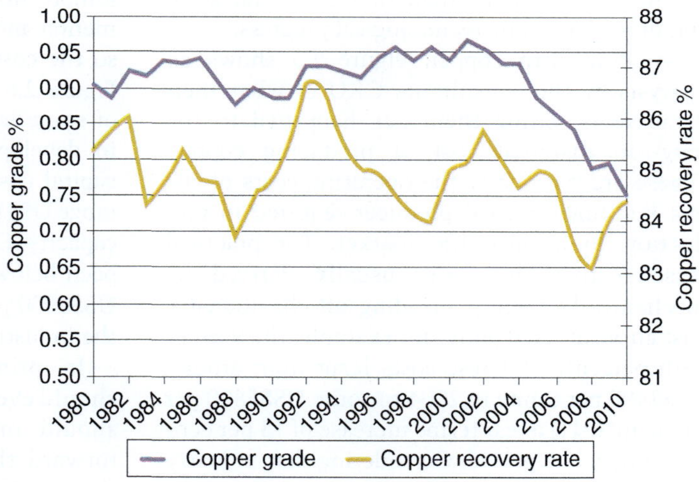

Figure 2.4 Copper mine grades and recoveries, 1980–2010. (After Citi, 2011b.)

Economic constraints

A second set of constraints on mineral development – referred to here as economic constraints – are a product of the fact that the quality of mineral resources has a tendency to deteriorate over time. Ores become lower in grade or more difficult to treat, while ore deposits are found at greater depth or in more difficult locations. As an illustration of this, Figure 2.4 shows the recent declining trend in copper ore grades and in recoveries from those ores.

To some degree, the upward pressure on industry costs which results from these trends can be offset – or even more than offset – by cost-reducing improvements in technology, and historically this has been the general experience of the industry. However, there is no law which says that this has to be the case and, for a number of mineral commodities, it would appear that the declining quality of reserves, combined with other factors such as higher energy prices, water availability and tougher environmental requirements,

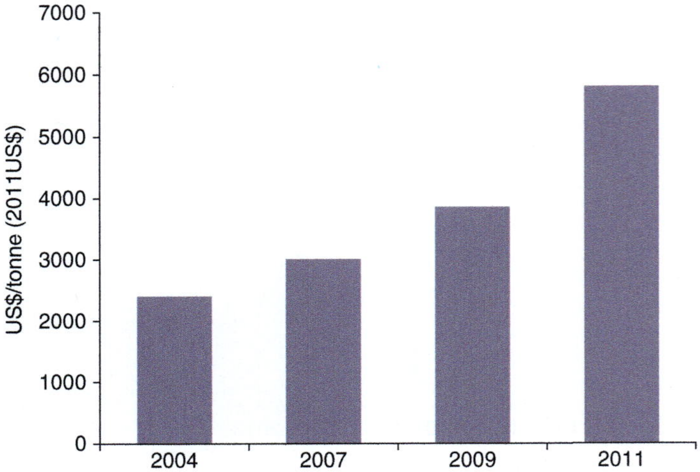

Figure 2.5 Long-run marginal cost of copper production. (Adapted from Robinson, 2011.)

are pushing up net production costs, notwithstanding continuing technological progress.

Sticking with copper, Figure 2.5 shows an analysis by the consultants CRU of what their database is telling them has happened to the long-run marginal cost of producing copper. (These are notionally the operating costs of the last (i.e. highest cost) producer required in production to balance the market. For practical reasons, they are more usually derived by mechanically taking a reading off the industry cost curve at fixed point, for example, the 90th or 95th percentile.) These costs leapt from around US$2400 per tonne in 2004 to some US$5800 per tonne in 2011; a real terms increase of 13 per cent a year over the period. Considering that industry operating costs declined for the twenty-five years prior to this the increase is extraordinary. Partly, of course, the effects are cyclical, but it seems probable that, underlying these cyclical influences, a structural shift is taking place. Moreover, this experience is not exclusive to copper. Similar evidence of deteriorating quality of ore resources and rising production costs can be adduced for nickel, PGMs and gold. At the same time, it should be noted, evidence of declining ore quality is less evident in other cases, for example, in iron ore, coal and bauxite.

What applies to operating costs applies also to capital costs. As mines become deeper and more remote from infrastructure, and as the environmental and political challenges of mining mount, so the cost of building mines has escalated too. Figure 2.6 shows estimates of the capital costs of some large greenfield copper mines currently in development or undergoing evaluation. The capital costs of these mines typically fall in the range US$10,000–20,000 per tonne of annual mine capacity. Capital costs historically have generally been below US$7500 per tonne of capacity, with US$5000 per tonne for a long time being used by the industry as a rough rule of thumb.

In principle, higher costs of production should eventually result in higher prices, which should in turn contribute towards bringing forward the necessary investment to balance the market. However, there are lags in the system. The long-run prices used by companies in the evaluation of their projects have been rising, but companies, and the banks providing them with finance, have to be absolutely convinced that prices are going to stay substantially higher on a sustainable basis before risking a commitment to large long-life projects. It also might be noted that exchange rates can be as important as mineral prices in determining a mining project's viability. The emergence of an increasingly multi-polar global economy, and the associated decline in the role of the US dollar, is likely to bring with it

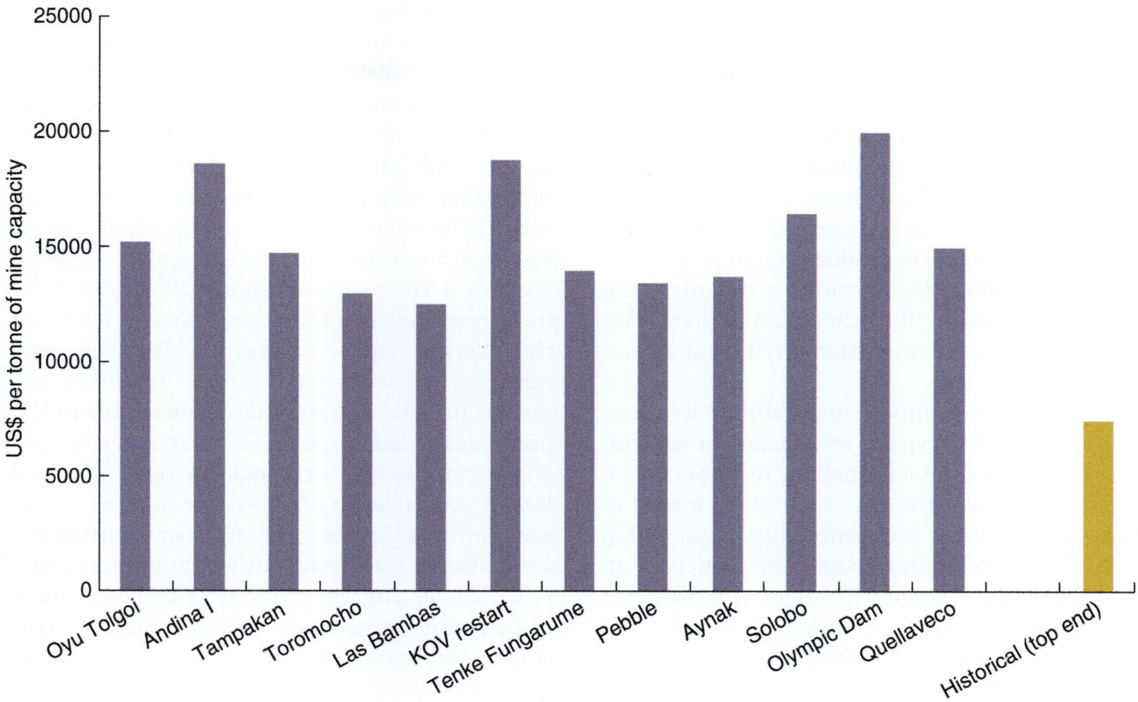

Figure 2.6 Capital cost of green-field copper mines. (Data from various industry sources.)

increased currency instability, adding a further layer of complexity and risk to mine project evaluation. In short, while companies may be investing heavily in new capacity, they are having to overcome higher economic barriers, and assume greater exposure to risk, to do so.

Institutional constraints

A third set of constraints on mineral supply involve institutional factors. For the purposes of the section, these are taken to include the laws and taxes to which mining companies are subject wherever they operate, through to intermittent geopolitical interventions in the industry.

Although a widespread perception exists in the more economically developed countries that issues of institutional risk is largely a matter for emerging and developing nations, this is far from the case where the mining industry is concerned. Pressures on land for housing, agriculture and leisure use, and concerns about impacts of mining on the environment and on local communities, make many developed countries highly ambivalent about the industry. On one level, this simply reflects the widespread perception that mining is a dirty and unsightly business and the fact that developed countries have large numbers of articulate people with the leisure to fight mineral projects. Such opposition to mining activities is often underpinned by the view that developed countries can make their living in other, cleaner, ways and import the mineral raw materials they want from elsewhere.

While many mining companies will want to persist with mining in the more developed countries because of the political stability and legal protections which such countries typically offer, mineral projects in developed countries often confront extremely demanding and lengthy permitting procedures and very tight restrictions on emissions, noise, visibility, effluents and transportation. Moreover, these restrictions are getting tougher with time. Although the provisions in themselves may be entirely reasonable, their cumulative effect can sometimes render a project marginal and encourage miners to go where they feel more appreciated and the wealth and employment they create are more highly valued. It does, however, leave developed countries more heavily dependent on imports than they might otherwise be and in a morally weak position to demand that others supply them with products they are demonstrably reluctant to produce themselves.

A growing impediment to the ability of miners and explorers to respond to changes in mineral demand is resource nationalism in mineral-rich countries. As was the case with the commodity boom of the 1970s, the commodity boom which started in 2004 helped stoke up a debate in mineral-producing countries over whether host nations were receiving a sufficient share of the proceeds from mining's growing success. Countries throughout the world have taken the opportunity to increase taxes and royalties on the industry. These are factors which mining companies have to take into account when assessing the likely returns to shareholders from an investment and, at the margin, can be an important factor influencing the decision whether or not to proceed with an investment. The threat of a wide-ranging Resource Super Profits Tax in Australia in the first half in 2010 resulted in many mining companies pointedly cancelling and deferring projects. (A revised Minerals Resource Rent Tax was subsequently introduced in July 2012.)

However, the concerns of resource nationalism are not confined to the distribution of income. They stretch also into the ownership and control of the industry. As noted in the introduction to this chapter, this politicisation of minerals in mineral-producing countries is a direct parallel to the securitisation of minerals in consuming nations. Minerals viewed within this broader political context become not just the basis of wealth-generating economic activity but a potential component in a project of national economic and social development, and a symbol of a country's sovereign right of self-determination. As the Washington consensus gives ground to the Beijing consensus at the level of the global economy, so the emphasis on the role of the nation and of the state is becoming more prominent within the confines of the resources sector. The forces for economic liberalism, as represented, for example, by the attempt to complete the World Trade Organisation's Doha Round, are in retreat and in international institutions like the World Trade Organisation the sovereign rights of nations over natural resources are increasingly being asserted in opposition to the principles of economic efficiency which underpin and legitimise the free trade system. Less and less, it seems, will mineral-rich countries accept the idea that other countries, or multilateral institutions, have the authority to determine how they develop their resources and how much of their mine production they must make available on international markets.

In the course of the recent minerals boom, a number of countries have come to the view that their national interests are best served by insisting that the state has a stake in mining operations on their home soil, or else that mine developments are undertaken by domestic private companies. Bolivia, for example, embarked on a programme of nationalisation for the mining industry in 2005. Mongolia has insisted on a major holding for the state in the large Oyu Tolgoi copper mine. Zimbabwe passed an Act in 2008 to promote the 51 per cent 'indigenisation' of mining companies operating within its borders. The Government of Guinea stripped Rio Tinto of some of its permits to mine iron ore in 2008, on the grounds that the company was not advancing the projects quickly enough, and now requires a substantial direct holding in all large new mining projects undertaken within the country. In DR Congo, the government in 2010 expropriated two mines belonging to TSX-quoted First Quantum Minerals.

Nor should it be supposed that such interventionism is confined to developing countries. The Australian government blocked the purchase

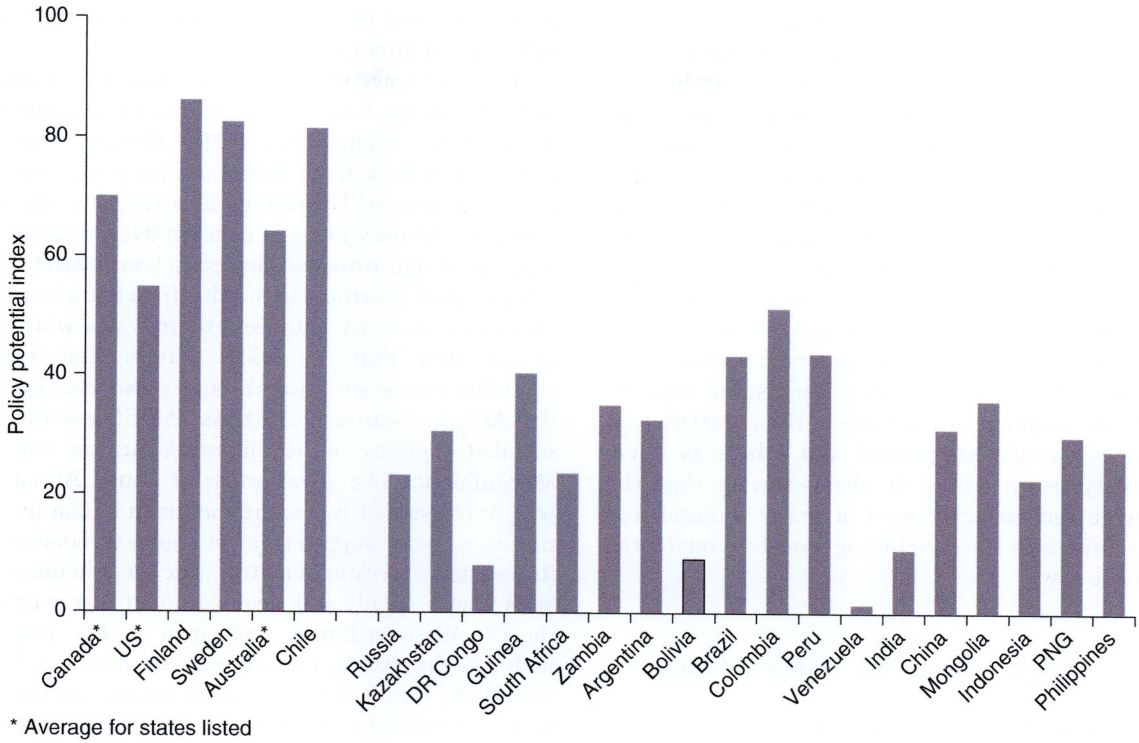

Figure 2.7 Policy attractiveness of large mining countries. (Data from Fraser Institute, 2011.)

of the Prominent Hill copper mine by China Minmetals in 2009, while the Canadian government blocked BHP Billiton's proposed take-over of the Potash Corporation of Saskatchewan in 2010 on the grounds that it was not in the national interest.

State interventionism inevitably adds another layer of uncertainty to investment decision making by mining companies. Political risk assessment is difficult and unreliable and there are only so many things that companies can do to mitigate risk. Many available strategies for risk mitigation, such as bringing in partners or buying political risk cover, result in reduced control over projects and/or increased costs. Despite this, on the basis that mining companies have to go where the minerals are found (and presumably also because mineral prices remain high), companies are continuing to commit to invest in what might be regarded as 'difficult' countries such as DR Congo, Guinea and Mongolia in the hope and expectation that they can manage the geopolitical risks involved and not become victims several years down the road of the 'obsolescing bargain' (the situation in which the investment has been made and the rules are changed). Time will tell whether this confidence is justified. The experience of the oil sector, it must be said, which is now wholly dominated by state firms, provides a somewhat discouraging example.

An objective assessment of the nature and scale of geopolitical risk across the industry poses obvious problems and, in the last resort, it is the geopolitics of the particular country in which a miner is thinking of investing which matter. However, attempts are routinely made to try and provide some comparative context for the assessment of this issue. Figure 2.7 shows the results of an investor perceptions survey carried out annually by the Fraser Institute of Canada. Amongst other things, this survey seeks to capture the mining industry's perception of the relative attractiveness of mining policy in a variety of mineral-rich countries, taking account of political risk alongside a range of other factors such as taxation, the administration, interpretation and enforcement of mining laws, and environmental regulation. While such analyses have their limitations, Figure 2.7 serves to make the point that perceptions of policy attractiveness vary significantly across different countries, with some regimes (those on the left-hand side of the chart) viewed as essentially supportive and others as effectively no-go zones. It also suggests that the perceived attractiveness of many mineral-rich countries in the developing world is considered quite low.

Critical minerals and the role of China

The specific minerals which are the object of concern for mineral-consuming countries vary through time, as also do the countries viewed as unreliable sources of supply. This in turn has implications for the producers whose role it is to seek to ensure that adequate supplies of the minerals are forthcoming.

The concerns which arose in the immediate post-WWII era over mineral supply were focused largely on the military requirements of having to fight a sustained conventional (i.e. non-nuclear) war. Efforts to address the threat in the US – which included the creation of a large materials stockpile – were therefore focused on a lot of relatively basic industrial raw materials, particularly those which were not abundant in the US, such as bauxite, manganese, zinc, lead, nickel, chromium and tin (Anderson and Anderson, 1998). With respect to the threat of nuclear confrontation, there were parallel concerns relating to the availability of uranium. The USSR, the source of the presumed threat to the USA in the context of the Cold War, would have had similar concerns, although the perception in the West at the time was that the USSR was broadly self-sufficient in minerals.

A second wave of concern over the supply of critical minerals followed (not coincidentally) the commodities boom of the 1970s. Although the Cold War was still on-going at that time, and the USSR was still perceived as a threat to the West, the primary focus of concern over mineral supplies at that time had shifted to South Africa. The policy of apartheid in South Africa had alienated many western states and there was a widespread view that the USSR was seeking to capitalise on the situation through its support for the African National Congress (ANC) and the socialist regimes of neighbouring Angola and Mozambique. The government of South Africa was, at the same time, using the threat of disruption to mineral supplies to the region to bolster its position in western capitals. The specific minerals whose supply was deemed under threat by the US, Western Europe and Japan at this time included the PGMs, manganese, chromium and vanadium, for all of which South Africa was the western world's leading supplier (House of Lords, 1982 and Maull, 1986). Because of Zaire's[4] proximity to South Africa and its dependence on South Africa's transport routes, supplies of cobalt were also considered vulnerable to unfolding events in South Africa (this quite apart from issues related to Zaire's own political instability).

The concern over mineral supplies at this time was less to do with the threat to the military capabilities of mineral consuming countries and more to do with the threat of economic disruption. Manganese, chromium, vanadium and cobalt were used in the production of high-performance steels, such as stainless steels and high-strength low-alloy (HLSA) steels, as well as superalloys, which were in turn used for the manufacture of important high-technology products such as petrochemical plant, oil pipelines and jet and gas turbines. The fear was that curtailment of supplies of these metals from South Africa would cause serious dislocation in strategically important industrial sectors, from energy production to aerospace. These concerns gave rise in a number of countries, including

Japan, the Republic of Korea and the UK, to strategic stockpiling of the threatened metals and to the adoption of schemes to incentivise mineral exploration.

In the most recent manifestation of concern over the supply of critical minerals, the focus of concern has once more shifted. With respect to the specific minerals which are deemed under threat, the focus has shifted to a range of specialised, low-volume metals used in the production of technologically advanced consumer electronics, green energy products and defence applications. Many of these are discussed in detail elsewhere in this book.

The sophisticated nature of the products in which minerals now designated as critical are used and the growing complexity of linkages between different sectors of modern economies (as well as the blurring of the distinction between commercial and military products), makes the threat which their non-supply would pose rather harder to evaluate than was the case in earlier era (Anderson and Anderson, 1998). While there is a clear sense in consuming countries that these metals are important for certain cutting-edge applications, it is evident from studies published in the US and EU that the authors of these studies have struggled with the matter of how to assess the relative importance of different end-uses of the minerals designated as critical and the likely economic impact of their non-availability (NRC, 2008 and European Commission, 2010). While the approaches adopted in the US and EU reports differ – with the US study resting more on expert judgment and the EU study adopting a more quantitative approach – both generate some rather counter-intuitive results. Thus, for example, the US study determines that the economic impact of restrictions on the supply of rhodium would be greater than those on copper, while in the EU study the economic importance of tellurium and rhenium is rated as higher than that of copper and the PGMs.

The geographic focus of concern has also shifted since the 1980s. It has shifted towards China. This follows from the facts that China is a major producer and supplier of many high-technology minerals, and that western consumers became heavily dependent on supplies from China during years when China was offering these minerals at substantially lower prices than were available from suppliers elsewhere. Of the fourteen minerals judged most critical by the European Commission – antimony, beryllium, fluorspar, gallium, germanium, graphite, indium, magnesium, rare earth elements, tungsten, niobium, PGMs, cobalt and tantalum – no less than ten (the first ten minerals listed) are sourced by the EU substantially from China (European Commission, 2010). The concern is that China's own growing domestic use of these minerals is reducing the supplies being made available for export, creating increased competition for supplies amongst other users of these minerals and putting upward pressure on prices.

Not surprisingly, given that it is the world's largest and fastest growing market for minerals, China shares many of the concerns of the US and the EU about minerals availability. In fact, because of the need for a good supply of raw materials to support the rapid industrialisation and urbanisation of the country, and because the legitimacy of China's leadership depends in no small part on its ability to sustain high growth rates, China takes the matter of mineral supplies very seriously indeed (FT.com, 2011 and Ericsson, 2011b). For those minerals which it can source internally, the Chinese government has generally encouraged local mine development. Recently, however, this objective has awkwardly become conflicted with another policy objective, namely the need to regulate the mining industry more tightly so as to improve its environmental performance and conserve resources, and led the Chinese government to seek to restrict the export of certain minerals considered important to national economic development.

There is, however, a long list of minerals which China cannot source wholly from domestic sources. As of 2010, China had to import 100 per cent of its PGMs, 85 per cent of its copper and nickel, and 70 per cent of its iron ore. For these minerals, China has had to turn to international markets and its purchases of these and many other minerals have been a major factor driving global markets and mineral investment in recent years. In addition, since 2004, and the promulgation of its 'go out' policy, the Chinese Government has been actively encouraging its companies to invest in mining overseas as a means to secure supplies for its domestic metallurgical operations. Chinese companies have been particularly active in the pursuit of iron-ore investment opportunities overseas, notably in Australia, but they have also invested in other mineral projects such as those for copper, nickel and coal. In addition to China's direct investment in foreign mining projects, Citi analysts have identified 217 M&A (mergers & acquisition) deals involving Chinese companies in recent years, totalling almost US$50 billion (Citi, 2011a). The Metallurgical Miners' Association of China (MMAC) has said it would like to see 40 per cent of imports of iron ore coming from Chinese-invested mines by 2015 (China Economic Net, 2011a). Also, as a means to secure adequate supplies of mineral, the Chinese Government operates a strategic stockpile, the State Reserves Bureau (SRB), to hold and manage supplies of metals it deems critical to its industrial development such as aluminium, copper, nickel and zinc.

The particular range of the minerals designated as critical in the US and the EU has important implications for the nature of the response from the world's miners. For the most part, these are not minerals of any great interest to the major miners. They are simply too small in terms of their market size. Minerals do not, it should be noted, attract public interest in direct proportion to the scale of their markets. Raw Materials Group of Stockholm has calculated that global mine output of metals plus diamonds and uranium was worth around US$386 billion at the mine in 2009. Its analysis is shown in Figure 2.8. (Note that it excludes coal, which would add very substantially to the total.) Two thirds of the total was accounted for by iron ore, copper and gold. By way of comparison, the value of mined cobalt in that year was around US$2 billion, the value of rare earths was around US$1 billion, the value of antimony was somewhat less than US$1 billion, while the markets for gallium, germanium, indium and tantalum combined amounted to less than US$1 billion. The value of rare earths production, it might be noted, was around one per cent the value of iron ore production, revealing a striking difference between the importance accorded these minerals by policy makers and the importance accorded them by the industry.

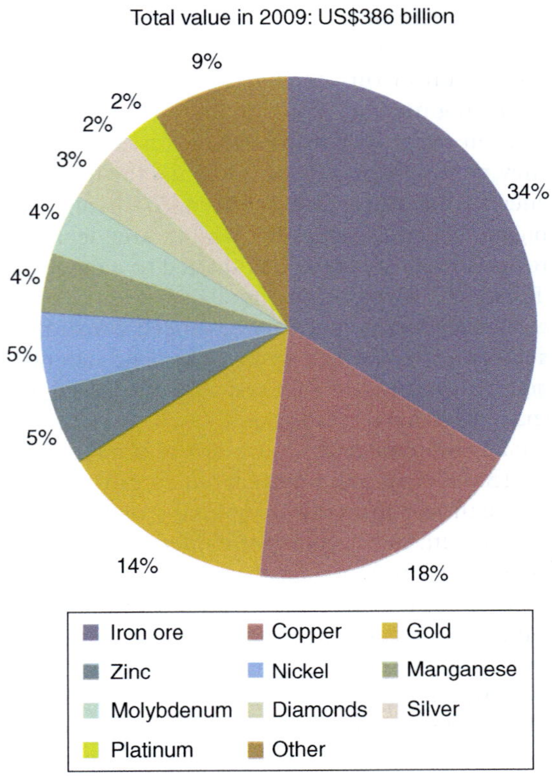

Figure 2.8 Metals value at the mine in 2009 (includes diamonds and uranium). (Data from Raw Materials Group, Stockholm, 2010.)

The large mining companies, having revenues measured in tens of billions of dollars a year, naturally like to focus their financial resources and management time on commodities which can make a material contribution to their businesses. As a result of this, the development of projects producing many of the minerals deemed critical is often left to smaller companies which, while they may be enterprising, often lack the experience, political clout and financial muscle of the big companies, making the route from discovery to production lengthier and more uncertain. These smaller companies also face the challenge that the markets for many minor metals lack transparency. The absence of exchange pricing and forward markets for these metals inevitably makes some potential financial backers nervous about investment.

That said, smaller miners have been extremely active in pursuing projects targeted on these critical minerals and on others facing declining supplies from China. Thus, for example, in the case of the rare earth elements, Molycorp Minerals has re-opened the Mountain Pass mine in California, Lynas Corporation has re-opened the Mount Weld mine in Western Australia, Great Western Minerals Group (GWMG) is re-opening the Steenkampskraal mine in South Africa's Western Cape, Toyota is planning to open a mine at Dong Pao in Vietnam with Sojitz and the Vietnamese Government, while there is a raft of other projects in Canada, Australia and elsewhere undergoing exploration and evaluation. In the case of tungsten, North American Tungsten Corp. has re-opened its Cantung Mine in the Northwest Territories of Canada and is evaluating the Mactung deposit in Yukon. There are also advanced plans in Australia to develop the King Island Scheelite mine on Tasmania and the Molyhil project in Northern Territory. Woulfe Mining Corp. is hoping to re-open the Shangdong tungsten mine in South Korea. In the UK, Wolf Minerals has conducted a feasibility study on the Hemerdon tungsten deposit in Devon and has raised funds to re-open the mine there.

Much the same goes for fluorspar, another mineral on the EU's list of critical minerals. In Mexico, there are expansions planned at Mexichem Fluor SA de CV (the world's largest producer) and at Fluorita de Mexico SA de CV, while in Mongolia, Monros is expanding its operations. In South Africa, Sephaku is expanding its operations by developing a new mine at Nokeng and ENRC is planning a mine at Doornhoek. In Canada, Canada Fluorspar is re-activating the St Lawrence fluorspar mines in Newfoundland. In the US, the Klondike II fluorspar mine in Livingston County, Kentucky, has been permitted for re-opening. In Vietnam, Dragon Capital Vietnam Resource Investments is building the Nui Phao tungsten-fluorspar project in North Vietnam while in Thailand, SC Mining Co is developing the Doi Ngom deposit in the north of the country.

All of which serves to make the point that resources of many of the minerals currently sourced from China are, at a price, available elsewhere in the world. It is just that China's preparedness in the past to supply these commodities at low prices made it uneconomic for many producers elsewhere to do so. Consumers, who themselves operate in competitive markets, were opportunistically led towards buying cheap Chinese minerals, in doing so creating a degree of dependency on China that was, in retrospect, perhaps unwise. As supplies from China have diminished and prices have increased, so miners in these other countries have been granted the opportunity to start, or re-start, production. The same applies in the case of the critical materials sourced significantly from DR Congo, namely cobalt and tantalum. Substantial resources of these metals exist outside DR Congo (for tantalum in Brazil and Australia, for example). It is just that in the past there was insufficient economic incentive for producers in these other regions to grow their output. A diminution of supply out of DR Congo, or heightened concern over political risk in the country, would provide the required incentive.

For minerals such as gallium, germanium and indium which are recovered as by-products and which are sourced from China as a result of China's rapid development as the world's largest processor of metals, there is very little the mining industry can do to relieve supply shortages.

Generally, it will not be economic for miners to pursue production of these minerals in their own right, and the issue of a supply response rests rather with metals processors outside China and on the question of whether it is profitable for them to add recovery circuits to existing plants to produce the relevant metals. In the longer term, it may depend on an ability to find new resources of these metals or to work different types of ores containing them.

With respect to large-scale, more basic mineral products, China represents a rather different challenge for western states. Certainly it is the case that China's growing demand for these products has resulted in a tightening of global supplies and increased prices, much as it has for more specialist metals. However, China's attempts to source an increasing amount of minerals from their own overseas mines – which may serve to ease the pressure on global supplies in all regions – is an additional competitive pressure on western mining companies. Just as western consumers are becoming more conscious of competition from Chinese manufacturers and purchases of minerals so western miners are feeling similar pressures in their business. As of the moment, this threat is relatively modest (Ericsson, 2011b) but it is one that is likely to increase with time. (Humphreys, 2011).

Policy issues

In the main, mineral markets work and deliver an appropriate level of supply to mineral users. Mining is a highly adaptable and enterprising industry and miners are constantly on the lookout for opportunities to make money by identifying gaps in the market and filling them. Although concerns over the availability of mineral supplies in consuming nations are understandable when markets are tight, there are few examples one can point to in history where the non-availability of mineral supplies has resulted in serious economic trauma. The shortages which have given rise to concerns over critical minerals in recent years are largely transitional and result from the lagged supply responses which are an unavoidable feature of the mining industry. High mineral prices are part of the mechanism for transition; they force supply and demand into alignment, in the short run by choking off demand and in the longer run by stimulating new supply by encouraging increased exploration and technical innovation.

Policy makers, whose time horizons, being politically determined, are generally shorter than are those of the mining industry, need to be aware of the underlying reasons why the industry suffers from lagged responses and why therefore the adjustment to imbalances in supply and demand takes time. This is normal, if frustrating. They also need to be aware of the characteristics of the individual mineral products under threat and understand better where and how they are produced, especially those which come as by-products of other minerals, or flow from the processing of imported ores. Europe, for example, is a major producer of cobalt but the cobalt it produces all comes as a by-product from the processing of imported copper and nickel ores. Without these metal processing activities in Europe, there would be no cobalt produced.

For miners and explorers to perform their functions effectively, markets must be allowed to operate. Price signals must be reliable and companies must be allowed – or, better, encouraged – to respond to these signals. The following comment, which was made at a time when governments were busy disengaging themselves from involvement in the minerals sector, still seems relevant to today's challenges.

> "Fair and efficient markets are man-made constructs. Their effective functioning depends not on the absence of policy but on a particular type of policy. They do not create or maintain themselves. Those who decide to trust to the markets for their minerals therefore need to accompany this decision with a commitment to ensure that mineral products are able to flow without undue hindrance from tariffs, subsidies or spurious environmental conditions, and that investment in the industry can similarly flow to where it can most productively be employed." (Humphreys, 1995.)

With respect to the last point relating to the free flow of investment, here there do appear to be some significant challenges facing the industry. While there are few physical impediments to investment in new mineral supply, there are some significant economic and institutional constraints and these appear to be getting more severe with time. In particular, there are the growing pressures from resource nationalism and from the growing involvement of the state in the mining sector in many countries. These are matters about which the mining industry can do relatively little but which should be of interest to policy makers.

While current and recent concerns over critical minerals naturally lead governments of consuming countries to want to do something to prevent a recurrence of supply shortages, it carries the risk of fighting the last war. As has been shown, problems of minerals supply do not always come back in the same form or apply to the same minerals. Policies such as stockpiling of critical minerals have a superficial appeal but they are cumbersome and costly and have not in the past proven very effective; in addition, the buying of minerals for a strategic stockpile always risks aggravating supply problems by pushing up prices and distorting markets.

There are, nonetheless, things that policy makers can usefully do to assist with the adjustment to supply shortages and to support future industry supply responses. However, these tend to be more long term and structural in nature. Governments of consuming countries can, for example, help consumers adjust to mineral shortages and accompanying high prices by encouraging R&D in materials technologies and facilitating recycling.

With regard to supporting future industry supply responses, governments of mineral-consuming regions can promote local mine production, where this could be viable. They could also do more to promote the development of new technologies for mining and mineral processing. Neither of these are things for which the authorities in the US and the EU have shown much enthusiasm in recent years. For minerals which cannot be supplied locally, there is a need to fight for open and competitive mineral markets both within multilateral forums and through bilateral agreements. It may be (and this would represent a departure from past and present practices) that governments should stand up more prominently for their companies where their legal rights are being flouted by host countries or where they are subject to unfair competition by state-owned enterprises. They could also support and encourage institutions which provide finance for exploration and mining, including providing guarantees for companies making investments in higher-risk countries, and provide more support for educational establishments which are training up the next generation of mining industry personnel. The best protections against sustained mineral shortages are efficiently working markets and free-flowing investment. Although the point has a tendency to get lost in the world of political cut and thrust, the US, the EU and China are all major mineral consuming and importing regions and share a common interest in healthily supplied global markets.

Notes

1. Enterprise value (EV) is calculated as market capitalisation plus debt, minority interests and preferred shares minus total cash and cash equivalents.
2. In late 2012, Glencore and Xstrata agreed terms for the merger of the two companies, to take place in 2013.
3. These comprise platinum, palladium, rhodium, ruthenium, iridium and osmium.
4. Since 1997, Zaire has become the Democratic Republic of the Congo.

References

Anderson, E.W. and Anderson, L.D. (1998) Strategic Minerals: Resource Geopolitics and Global Geo-Economics. John Wiley and Sons, Chichester, UK.

Chegwidden, J. and Kingsnorth, D. (2011) Rare earths – an evaluation of current and future supply. Presentation to the Institute for the Analysis of Global Security, 2011.

China Economic Net. (2011) http://en.ce.cn/Industries/EnergyandMining/201104/19/t20110419_22371368.shtml
Citi (2011a) The Changing Face of Global Mining Markets. Citibank Global Markets.
Citi (2011b) Good Assets or Good Management. Citibank Global Markets.
Crowson, P.C.F. (2011). Mineral reserves and future minerals availability. Mineral Economics 24, (1), 1–6.
Ericsson, M. (2011a) Private communication.
Ericsson, M. (2011b) Mineral Supply from Africa: China's Investment Inroads into the African Mineral Resource Sector. The Journal of The Southern African Institute of Mining and Metallurgy 111 (July).
European Commission (2010) Critical Raw Materials for the EU. Report of the Ad-hoc Working Group on defining critical raw materials.
Fraser Institute (2011) Survey of Mining Companies: 2010–2011,
FT.com (2011) Minmetals warns of a gulf in resources.
FTSE. (2013) Available at http://www.ftse.com/
House of Lords (1982) Strategic Minerals, 20th report of the House of Lords Select Committee on the European Communities, Session 1981–82, HMSO.
Humphreys, D. (1995) Whatever happened to security of supply? Minerals policy in the post-Cold War world. Resources Policy 21 (2), 91–97.
Humphreys, D. (2011) Emerging Miners and their Growing Competitiveness. Mineral Economics 24 (1), 7–14.
Maull, H.W. (1986) South Africa's Minerals: The Achilles Heel of Western Economic Security? International Affairs, RIIA 62 (4).
MEG. (2012) Corporate Exploration Strategies. SNL Metals Economics Group, Halifax, Nova Scotia.
Mining Journal (2011) Answering the call. Lithium Focus, Mining Journal.
NRC. (2008) Minerals, Critical Minerals, and the US Economy. National Research Council, The National Academies Press, Washington D.C.
Robinson, P. (2011) Copper – Returning to Normality? Presentation to CRU Copper Conference, Santiago, Chile.
UBS. (2011) Mining and Steel Primer. UBS Investment Research.
U.S. Geological Survey, USGS. (2012) Mineral Commodity Summaries. US Geological Survey.

3. Recycling of (critical) metals

CHRISTIAN HAGELÜKEN

Director EU Government Affairs, Umicore AG & Co. KG, Hanau-Wolfgang, Germany

Rationale and benefits

The world faces major societal challenges including climate change, energy supply and availability of critical resources. Metals are essential to the economy as a whole and to the products and technologies that will allow transition towards a society that is able to address these challenges. Technology metals,[1] which include the critical metals, play a key role in this context not only for the European Union but for all industrialised regions in the world. Moving towards a competitive and sustainable economy requires access to these materials in adequate quantities and at competitive costs. It is increasingly recognised that, among others, one important use area for technology metals is energy generation (Achzet et al., 2011; Graedel, 2011; and Resnick Institute, 2011). As especially Europe is highly dependent on imports of metals and metal concentrates, and primary resources from European deposits are not available in sufficient amounts for most metals, more recycling is a cornerstone in any strategy for securing long-term supply and becoming more independent of geopolitical constraints, as described in Chapter 2.

The urban mine

Unlike energy raw materials, metals are, in principle, not consumed but can be kept in an 'eternal' life cycle. After initial extraction from mineral deposits in the 'geosphere' metals are used to construct infrastructure and to manufacture products. The use of metals, particularly of technology metals in industrial and consumer products, has grown rapidly in recent decades. For example, more than 80 per cent of the global mine production of the platinum-group metals (PGM), the rare earth elements, indium and gallium since 1900 took place in the last three decades (Hagelüken and Meskers, 2010), and many of these metals are still bound in the 'technosphere' or 'anthroposphere'. As a result, products such as cars, electronics, batteries and industrial catalysts have evolved into a potential 'renewable' metal resource for the future, often referred to as an 'urban' or 'above ground mine', that must not be wasted. Moreover, compared to natural ores, metal concentrations in many products are relatively high. For example, a typical primary gold mine will yield around five grams of gold per tonne (g/t) of ore. In electronic scrap, this figure rises to 200–250 g/t for computer circuit

Critical Metals Handbook, First Edition. Edited by Gus Gunn.

Table 3.1 Average content of precious metals, copper and cobalt in mobile phones and computers, and resulting metals demand from global sales in 2010, compared with world mine production.

	a) Mobile phones		b) PCs and laptop computers			
	1600 million units/year, each with a lithium-ion battery		350 million units/year, of which ca 180 million have a lithium-ion battery		a + b = Urban mine	
Metal	Unit metal content	Total metal content	Unit metal content	Total metal content	Global mine production (2010)	Share a + b of global mine production
Silver	250 mg	400 t	1000 mg	350 t	22,900 t	3%
Gold	24 mg	38 t	220 mg	77 t	2650 t	4%
Palladium	9 mg	14 t	80 mg	28 t	225 t	19%
Copper	9 g	14,000 t	500 g	175,000 t	18 Mt	<1%
Cobalt	3.8 g	6100 t	65 g	11,700 t	88,000 t	20%

t, tonnes; Mt, million tonnes; g, grams; mg, milligrams

boards and as high as 300–350 g/t for mobile phone handsets, grades which are very uncommon in primary gold deposits. An autocatalyst contains some 2000 g/t of PGMs in the ceramic brick, compared to average PGM concentrations of less than 10 g/t in most mines which produce PGMs. Taking the high environmental impact of primary production of precious metals due to low ore concentrations, difficult mining conditions and other factors into account, scrap recycling becomes even more beneficial from an environment protection viewpoint. Furthermore, if state-of-the-art technologies are used, the environmental benefits of the recycling of scrap metals are further enhanced (Hagelüken and Corti, 2010).

The challenge, however, is that most consumer products are widely distributed and often difficult to trace around the planet and that, although metal concentrations are high, the absolute precious metal content in a single device is often very low. Accordingly, in order to economically exploit these urban mines, it is first necessary to gather sufficient quantities of these dispersed products to create a true above-ground deposit that will fulfil the second basic criterion for an economically viable 'ore body' apart from its concentration, that is sufficient volume. This is illustrated in Table 3.1 using the example of mobile phones and computers. Mobile phones contain over 40 different chemical elements including base metals such as copper (Cu), nickel (Ni) and tin (Sn), speciality metals including cobalt (Co), indium (In), and antimony (Sb), and the precious metals silver (Ag), gold (Au) and palladium (Pd). Metals, mostly copper, make up about one quarter of the weight of such a telephone.

In one tonne of mobile phones (equivalent to about 13,000 units without batteries) this adds up to an average of 3.5 kg Ag, 340 g of Au, 130 g of Pd, and as much as 130 kg of Cu. The value of these metals can exceed €10,000 per tonne, with more than 80 per cent of the total attributed to the precious metals. For a single unit, however, the precious-metal content is only of the order of milligrams: 250 mg Ag, 24 mg Au, 9 mg Pd on average, while Cu is not more than nine grams and the rechargeable battery contains about four grams of Co (Hagelüken and Meskers, 2008). Thus, the net metal value of a single mobile phone is only about one euro even at the high metal prices prevailing in 2011, and so does not per se provide a real economic incentive for recycling. It is the sheer number of mobile phones in use that makes this metal resource significant: about 1.6 billion units were sold worldwide in 2010 alone (Gartner, 2011), and, by the end of 2010, total global sales had reached around 10 billion units, roughly equivalent to 1.5 telephones for every human being currently living on the planet. They contain a total of 2400 tonnes of Ag, 230 tonnes of Au, and 90 tonnes of Pd. The gold and silver content of the combined 2010 sales volumes of mobile phones and computers are equivalent to approximately four per cent of the global mine production of silver and gold and 20 per cent of palladium and cobalt (Table 3.1). But how much of this 'urban mine' will eventually be recycled and what is required to do this effectively?

The use of metals in cars provides another example of metals potentially available from the 'urban mine'. In cars, the catalyst in the exhaust system has accounted for well over 50 per cent of the mine production of PGMs (in particular platinum, palladium and rhodium) for many years and the recent proliferation of electronic components in road vehicles (such that a car is effectively a 'computer on wheels') will have an increasing impact on the demand for gold, silver, and a number of critical metals. Given the anticipated boom in the use of electric vehicles, even more technology metals will be required. These will include more PGM for fuel-cell stacks, cobalt, lithium and rare earths for batteries and electric motors, and bismuth and semiconductor metals, such as antimony, tellurium, germanium, for thermo-electric applications in which waste heat is used to generate electric power. Figure 3.1 shows annual demand figures for PGM in automotive catalysts, both for the whole world and separately for Europe. Cumulative gross demand (5430 tonnes PGM globally) and recycling volumes (730 tonnes) up to 2010 are also shown. In Europe, the large increase in demand for PGM in this application started in the mid-1990s. As a result, in view of the relatively long lifetimes of cars (10–15 years), this means that some 1100 tonnes of PGM are still in use in European cars and up to 3000 tonnes globally. Most car catalysts that are recycled today are from the late 1990s, so in the future the potential for recycling PGM from cars on the road today is huge. However, there are major deficiencies in the recycling of automotive catalysts, mainly related to the failure to collect vehicles at the end of their lives. If these structural problems of this 'open cycle' (see Section 'The significance of life-cycle structures') cannot be overcome, the potential of this urban mine will be significantly undermined.

Based on an in-depth analysis of PGM flows from automotive catalysts (GFMS et al., 2005) and ongoing industrial experience in the autocatalyst recycling market, I estimate that the cumulated wastage from car catalysts already add up to some 900 tonnes of PGM globally, of which Europe accounts for about 200 tonnes. PGM losses occur at all stages of the life cycle, including dissipative driving losses 'through the exhaust pipe' (if road conditions are bad or the car is not well maintained), losses from cars that were not recycled, losses from catalysts that were not removed from a car before shredding, losses during (inappropriate) handling and mechanical pre-treatment of car catalysts, and losses during metallurgical catalyst processing.

Recycling benefits

There is a broad consensus today that recycling offers significant benefits, from an environmental perspective as well as by increasing the supply security of metals. It is generally acknowledged that recycling:

- reduces the environmental burden that would otherwise occur (by preventing emissions from discarded products and landfills into soil, water and air, and by reducing the use of land needed to accommodate waste materials);
- mitigates the environmental impact of mining (by reducing the amount of mining required and thus decreasing energy demand, carbon dioxide emissions, land and water use, and impacts on the biosphere, e.g. in rain forests, Arctic regions, ocean floors);
- extends the lifetime of and preserves primary geological resources (and thus contributes to buying time to develop improved mining and processing techniques that might facilitate the extraction of metals from low-grade or deep ore bodies);
- reduces geopolitical dependencies arising where critical metal resources are concentrated in a few mining countries and/or are in

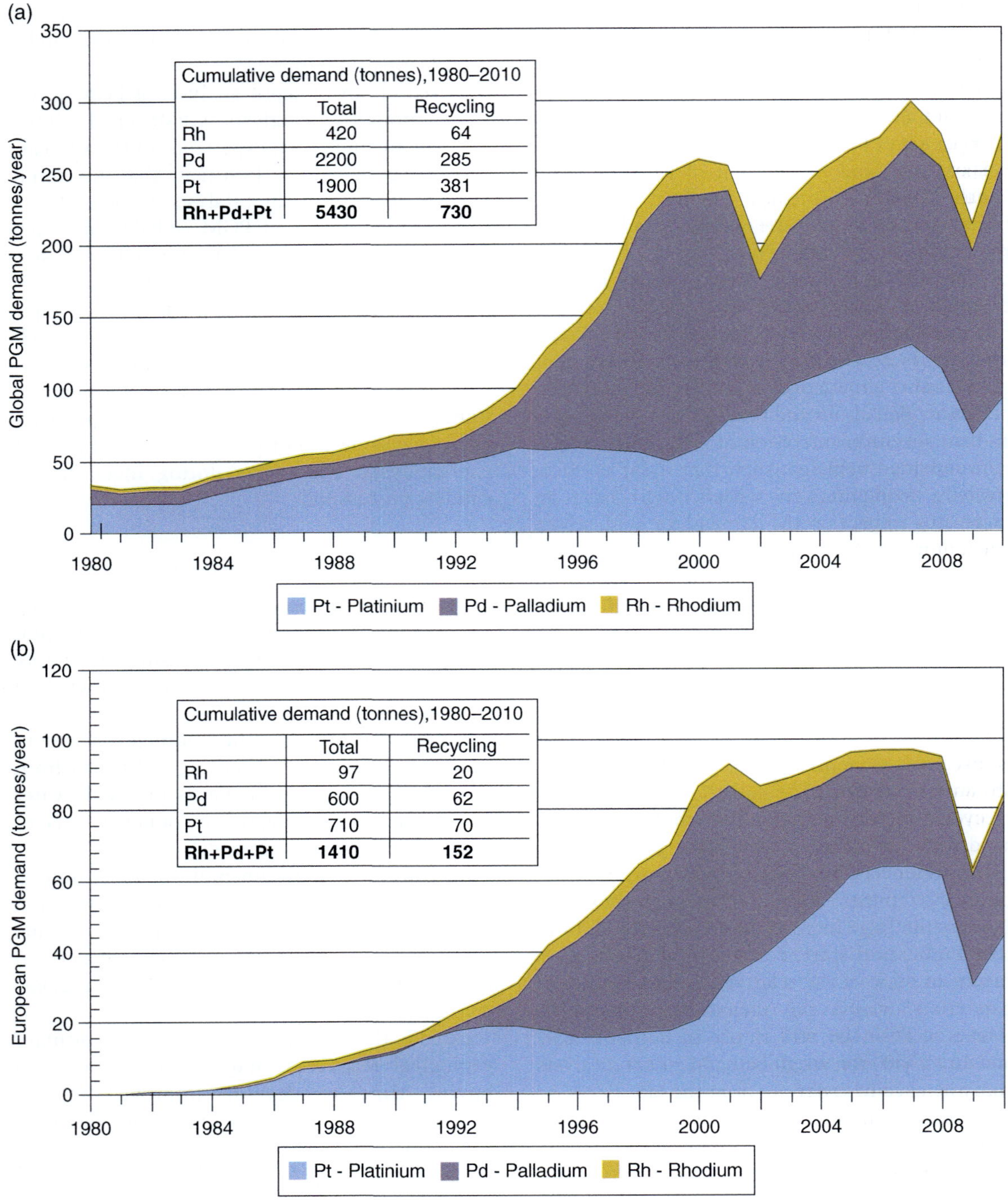

Figure 3.1 (a) Gross global demand for platinum, palladium and rhodium for automotive catalysts (annual and cumulative) and cumulative recycling volumes. (b) Gross European demand for platinum, palladium and rhodium for automotive catalysts (annual and cumulative) and cumulative recycling volumes. (Data from Johnson Matthey, 2011.)

the hands of a small number of companies (recycling therefore effectively creates a significant local resource);

- contributes to the supply security of minor and critical metals by partially decoupling their production from the primary production of the associated major carrier metals (the strong link between the mine production of carrier metals and by-product minor metals limits the primary supply of important technology metals – see Chapters 1 and 2 for further discussion);
- supports ethical sourcing of raw materials (through improved transparency of the supply chain, ensuring that no 'conflict metals' are used that originate from war regions in Africa or elsewhere);
- dampens metal price fluctuations by improving the demand–supply balance and limiting speculation by broadening the supply base;
- creates significant employment potential, including high-technology jobs and infrastructure.

Status and challenges of recycling critical metals

The metals life cycle

Fully effective recycling could, in principle, lead to an infinite cycle of metal use. The quality of recycled metals matches that of primary metals and they are traded accordingly at identical prices. Hence, once metals are recycled there is no need to incentivise their use. In practice, however, some quantity of metal is inevitably lost from product life cycles because they cease to be accessible for recovery for various reasons. Some materials, like plastics, often face a 'downcycling' issue because the recycled material is of lower quality than the virgin material and hence can only be used in lower-value applications.

The key challenge, therefore, is to minimise metal losses. As Figure 3.2 shows, the metals life cycle starts with exploration and mining which bring materials from the geosphere into the technosphere. Thereafter, the basic concept for sustainable metals use is simple in principle:

- Extract metals in an efficient manner from the ores.
- Use them as efficiently as possible in the manufacturing of products.
- Avoid dissipation during use and at their end-of-life (EoL).
- Minimise losses of metals into residues during all phases of the life cycle. Such residues may include tailings and slags from primary production, production scrap from manufacturing, products discarded during their useful lifetime, non-recycled streams, and slags, effluents and final waste at the end of life.

Where such residues cannot be avoided they need to be recycled with high efficiencies. In this way metals can be re-used at a high rate in successive life cycles. This implies that both scrap generated during production and derived from end-of-life products are new sources of raw materials. It is therefore desirable to reduce unrecycled residues derived from all stages of the life cycle. This demands 'intelligent' product design (durable products, appropriate material combinations, ease of disassembly, etc.) and the use of appropriate and innovative processes throughout the value chain as described later in this chapter. Historic wastes should also be regarded as additional future resources in addition to primary geological resources. Moreover, substitution in product manufacturing may also help mitigate raw materials supply constraints, but it could lead to new challenges for product recyclability or for sourcing the substitutes. The encouragement of increased re-use and extended lifetimes of products and materials is also beneficial, providing it ultimately leads to recycling at the end of the extended lifetime. In this sense re-use is not an alternative option to recycling, rather it delays the time until recycling should take place (Eurometaux, 2011).

Primary metal production is not usually considered when recycling is discussed. This misses the fact that improved treatment of tailings, slag, or other side streams from mining, smelting and refining can contribute significantly to resource efficiency and to the supply of technology metals. Together, improved

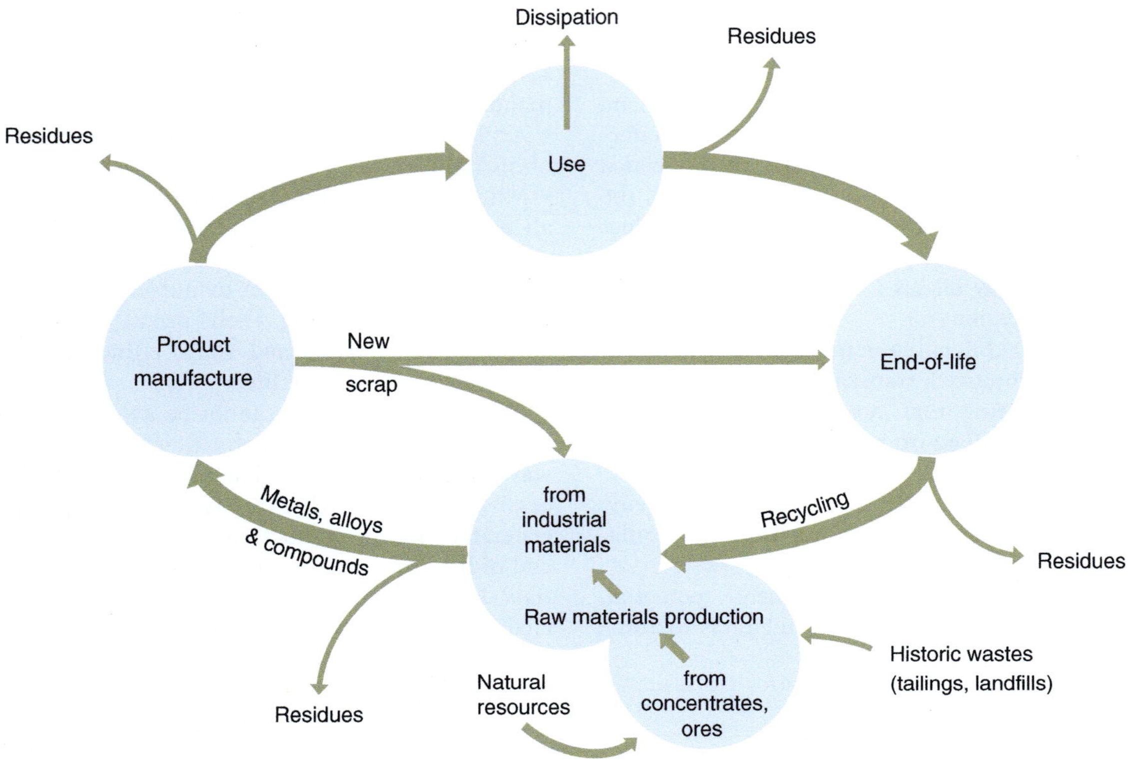

Figure 3.2 Sustainable use of metals along product life cycles – efficient recycling of residues generated at all stages can minimise use of primary raw materials. (Adapted from Meskers, 2008.)

efficiency in primary production and reworking of historic primary wastes, such as mine tailings and smelter slags, comprise a large and accessible potential additional source for most by-product metals, such as indium, germanium, molybdenum, rhenium and gallium. However, for most precious metals, such as the PGM, these inefficiencies in the primary supply chain have already been largely overcome due to their high value.

In manufacturing, recycling production scrap usually becomes more challenging when moving along the production process. For example, while recycling indium from sputter target manufacturing and spent ITO (indium–tin oxide) targets is relatively simple, reclaiming the metal from sputtering chamber scrapings, from broken or out-of-spec liquid crystal display (LCD) glass or finally from entire out-of spec or obsolete LCD monitors or photovoltaic modules becomes increasingly difficult. In the early stages of the manufacturing process there is significant recycling potential for many technology metals. In the case of indium, germanium and ruthenium, these have been realised to a growing extent in recent years, again driven by rising prices. Metal losses occurring during the use of a product are not generally recoverable due to their mostly dissipative nature (Hagelüken and Meskers, 2010). Recycling EoL products will be key to achieving sustainable use of metals. This has been recognised by governmental bodies such as the European Commission (EU-COM, 2012), the USA (US NRC, 2008; US DOE, 2010) and Japan seeking to curb waste generation and to use waste as a resource.

Waste and resource legislation

EU legislation like the Directive on End-of-life Vehicles (ELV) of September 2000 (EU-ELV, 2000) and the Directive on Waste Electrical and Electronic Equipment (WEEE) of January 2003 (EU-WEEE, 2003) demonstrates the approach in Europe. However, this legislation was developed in the 1990s and its main motivation was environmental protection through the avoidance of hazardous emissions and the recovery of mass materials, such as steel and plastics. The recovery of critical metals was hardly considered at that time. Hence, these directives define only mass-based recycling targets (e.g. 80 per cent of the weight of a car needs to be recycled) and the recovery of critical and technology metals is not specifically addressed.[2] The effect of this has, in fact, been counter-productive with the recycling technologies applied to recover the mass materials actually leading in some case to losses of technology/critical metals.[3]

The current calculation method for the EU recycling rates has two additional deficiencies: (1) it considers only the EoL products which are recycled within a country, i.e. the major exports of old cars or electronics from some countries, such as Germany, are not reflected in the reported recycling rate; and (2) it ignores the real recycling efficiency of the final (metallurgical) recycling process, since for this step only input data are considered (regardless of which output of a specific metal is achieved). In other words, the difference in efficiency attained in a state-of-the-art metallurgical process and a primitive 'backstreet' operation are not taken into account. Although for steel, copper and precious metals losses are small *if* state-of-the-art processes are used, they can be significant for many technology metals. Hence, the high recycling rates reported in the EU for WEEE or ELV create a false impression that the recycling systems are uniformly advanced and effective across the board. To obtain a true circular economy for technology metals, new approaches are needed, which demand better product design as well as innovative business models, well-organised and efficient recycling systems, and high-performance technical processes over the entire value chain.

The recycling value chain

As shown in Figure 3.3 recycling needs to be considered as a system comprising a number of key parts. It goes beyond process technology and requires a complete chain that begins with collecting, sorting, and dismantling/pre-processing in order to separate components containing valuable metals or to upgrade relevant fractions prior to final metallurgical processing.

Each of the main steps involves different stakeholders and companies. The process complexity, the skills and infrastructure requirements, and the investment volumes rise significantly along the process chain:

- *Collection*: this can be organised by national or municipal take back schemes, original equipment manufacturer (OEM) systems, commercial or charity organisations and involves logistical companies. Collection generally takes place on a local or regional level and commonly involves small and medium enterprises (SMEs).
- *Dismantling and pre-processing*: this uses manual and/or mechanical processes, often in combination. It takes place on a local, regional and inter-regional level, again involving many SMEs, but also some large waste management companies.
- *Metallurgical metals recovery*: comprises a combination of pyrometallurgical (smelting) and hydrometallurgical (chemical) processes, followed by a final step of metal refining. The key players are usually large metallurgical companies, such as Aurubis, Boliden, Dowa and Umicore, many of which are also active in processing primary materials. The plant feed is sourced globally and technological sophistication and economies of scale are crucial for success.

The resulting overall recycling efficiency is the product of the efficiency of each stage – recycling is only as good as its weakest link.

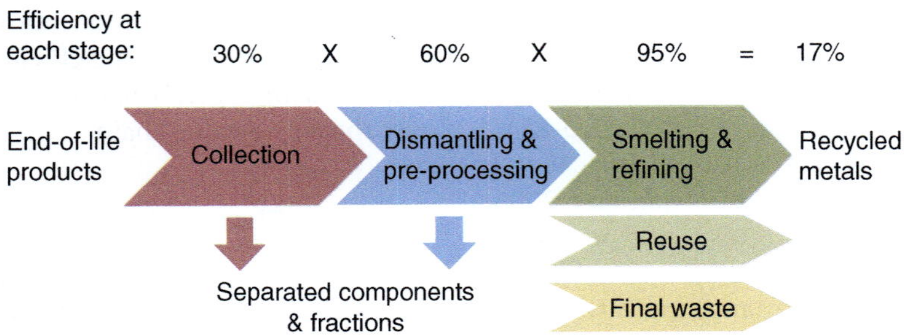

Figure 3.3 The main stages in a recycling chain for consumer products. The total recycling efficiency results from the product of the single stage efficiencies – the example shown is for average gold recovery from electronic scrap.

For example, the overall efficiency of gold and palladium recycling from WEEE in Europe is estimated to be below 20 per cent (Hagelüken and Corti, 2010). This reflects the combined efficiency of collection (30 per cent), pre-processing (60 per cent) and metallurgical recovery (over 95 per cent) (Figure 3.3).

While recycling of mass metals works relatively well and further improvements may be possible, there are significant constraints and deficits for many technology metals, including some identified as critical. The main deficits usually occur at the collection stage (either because the products are not collected at all or are exported for sub-standard treatment after collection) and at the pre-processing stage (Meskers et al., 2009). In contrast, for many, but not all, metals the metallurgical smelting and refining stages are already very efficient. In order to improve metal recycling overall it is necessary to develop specialised, systematic approaches along the chain for specific products and metal combinations.

Recycling challenges

Metal combinations in products usually differ from those in primary deposits, which results in new technological challenges for their efficient recovery. In many cases, very low concentrations or 'inappropriate' combinations of technology metals in certain products will set economic and technical limits to their reclamation. Complex products such as vehicles and computers pose particular technical challenges (Reuter et al., 2005; Van Schaik and Reuter, 2012).

Recycling technologies for precious metals and base metals, such as copper or lead, have been developed over centuries and are mostly capable of achieving high recovery yields today when it comes to the final metallurgical step. Recycling of speciality metals is often connected to base and precious metal metallurgy and still offers significant potential for optimisation in many cases. Metals that are dissipated during their path along the recycling chain, for example into dust fractions, or that are diverted into an inappropriate final process by unintended co-separation are lost. Examples are precious metals which end up in steel mills or aluminium smelters.

For speciality metals, the situation is usually worse because it is far more difficult to recycle technology metals from complex products than to recycle mass materials from waste streams such as bottle glass or steel scrap. While the latter deals with 'mono-substance' materials largely without hazardous ingredients, the former has to cope with 'poly-substance' compositions, including hazardous materials, that are present in complex components within complex products. In the one case, the focus is on volume and cost optimisation, while environmentally sound value recovery from materials present in low concentrations is of key importance in the other case. It is evident that

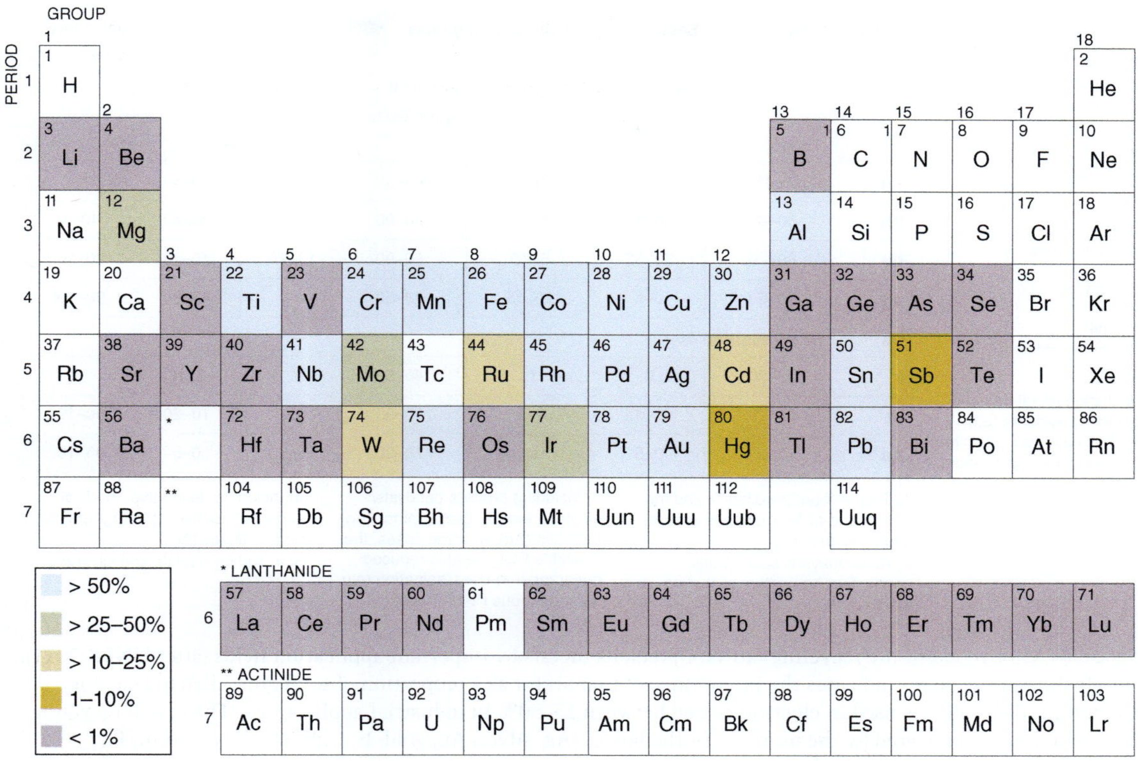

Figure 3.4 Global end-of-life recycling rates for metals in metallic applications. (Modified after UNEP, 2011.)

recycling of technology and critical metals requires a fundamentally different recycling chain setup and that a 'mass-materials approach' will inevitably lead to significant losses of trace materials.

These challenges are underlined in the report "Recycling Rates of Metals" published by the UNEP Resource Panel (UNEP, 2011). It shows that, while for most base and precious metals end-of-life recycling rates of over 50 per cent are achieved on global average, recycling rates for most speciality metals are extremely low, less than one per cent (Figure 3.4).

Even for the valuable precious metals a division of recycling rates by main application fields reveals significant deficits in most consumer applications (Figure 3.5).

The high average precious metals recycling rates are attributed to the very low level of losses in jewellery, coinage and investment ingot uses, as well as the highly efficient (>90 per cent) industrial life cycles of PGMs, e.g. in (petro) chemical process catalysts. PGM recycling rates from automotive catalysts are only 50–60 per cent, while for gold, silver and PGM in electronic applications they fall below 15 per cent. As will be shown later in this chapter recycling technology is not the main reason for these losses.

The recycling potential or recyclability of a product or material depends on various technical, economic, structural and organisational factors:

- *The intrinsic metal value* of the base material. This depends on its absolute metal content and the metal price, and determines the economic attractiveness of recycling. It sets a benchmark for the recycling technology and the overall recycling costs. Materials containing precious metals often make recycling quite attractive.

Importance of end use sector for each metal listed (% of total gross metal demand)*

- > 50%
- > 25–50%
- > 10–25%
- < 10%
- nil

* Including metal demand for closed systems (e.g. process catalysts, glass and other industrial applications)

	EOL recycling rates 1)	Sector-specific EOL recycling rates: Vehicles 2)	Electronics	Industrial applications 3)	Dental	Others 4)	Jewellery, coins 5)
Ru	5–15		0–5	40–50		0–5	
Rh	50–60	45–50	5–10	80–90		30–50	40–50
Pd	60–70	50–55	5–10	80–90	15–20	15–20	90–100
Ag	30–50	0–5	10–15	40–60		40–60	90–100
Os	no relevant end use						
Ir	20–30	0	0	40–50		5–10	
Pt	60–70	50–55	0–5	80–90	120	10–20	90–100
Au	15–20	0–5	10–15	70–90	15–20	0–5	90–100

1) Total without jewellery, coins (no typical end-of-life managements for these products).
2) Autocatalysts, spark plugs, conductive Ag-pastes, excluding car electronics.
3) Including process catalysts/ electrochemical, glass, mirror (Ag), batteries (Ag). In some cases, the available EOL metal is reduced due to prior in-use dissipation (e.g. homogeneous Pt-catalysts).
4) Including decorative, medical, sensors, crucibles, photographic (Ag) photovoltaics (Ag).
5) Including medals and silverware.

Figure 3.5 Global end-of-life recycling rates for precious metals by important application fields (after UNEP, 2011). The shading in the boxes indicates the importance of that sector as a proportion of total gross demand e.g. for ruthenium (Ru), >50% is used in electronics and between 25–50% in industrial applications. The actual recycling rates achieved are indicated by the numbers in the boxes. (Ag, silver; Au, gold; Ir, iridium; Os, osmium; Pd, palladium; Pt, platinum; Rh, rhodium; Ru, ruthenium.)

- *The material composition*: this goes beyond the chemical composition to include physical characteristics such as shape, size and the type of connection between materials or components. It influences the choice of the technical recovery methods used for sorting, pre-treatment and refining. The product or material composition usually has a significant impact on recovery costs as well as on the technically achievable recovery yields.
- *The application field of a product and how it is used*: the former refers to the area of use in consumer or business markets and the latter to new or re-use products, user behaviour, risk of dissipation, product mobility, country of usage, etc. These factors together determine the probability of particular products and metals entering into an appropriate recycling channel at their end-of-life. (see section 'The significance of life cycle structures' below) (Hagelüken and Meskers, 2010).

The seven conditions for effective recycling

For effective recycling of a product, material or metal, seven basic conditions must be met:

1. **Technical recyclability** of the material or metal combination. For example, all precious metals, most base metals and a number of speciality metals can be recovered from a printed circuit board if state-of-the-art processes are used, but some speciality metals such as gallium, germanium or rare earth elements contained in this mix are not recoverable (see Section 'Metallurgical recovery').
2. **Accessibility** of the relevant components. For example, an underfloor automotive catalyst, a lead-acid car battery, or a personal computer (PC) motherboard is easily accessible for dismantling, whereas a circuit board used in car electronics (e.g. in the engine management system) usually is not. As long as such components are dismantled *before*

the car is put through the shredder, most metals they contain are technically recyclable (condition 1; see Section 'Collection and pre-processing').

3. **Economic viability**, whether intrinsically (triggered by a positive net value) or externally (by authorities, retailers, etc. to compensate a negative net value) created. Dismantled automotive catalysts or PC motherboards have a positive net value, therefore recycling is economically viable by itself. In contrast, a dismantled ultra-thin PGM-coated PC hard disk or a LCD screen coated with indium–tin oxide (ITO) usually has a negative net value due to the cost of processing it. Recovering the critical metals from such products would currently not be economically viable unless paid for externally or subsidised. However, it should be noted that such external creation of economic viability is not unusual in recycling, e.g. in the case of treatment of household waste (paid by consumer fees) or in deposit fund systems for empty beer bottles or aluminium cans (see also condition 4).

4. **Collection mechanisms** to ensure the product is available for recycling. If collection mechanisms are not in place, items such as old PCs or mobile phones may end up being stored in households or discarded into the waste bin for landfill or municipal incineration. The metals they contain would effectively be lost to the recycling chain.

5. **Entry into the recycling chain** and remaining therein up to the final step. Items such as PC motherboards, mobile phones or cars containing catalysts are often sent (either legally or illegally) to countries without the proper infrastructure for recycling at their end of life. The same applies for components or fractions from EoL products which escape at a later stage from a state-of-the-art recycling chain. This usually results in technology metals being lost to the recycling chain.

6. **Optimal technical and organisational** setup of this recycling chain. Comprehensive recycling chains exist within Europe and in other industrialised regions, though it is important that certain items, such as PCs and mobile phones, are not mixed with other low-grade electronic waste and channelled into a shredder process without prior removal of the precious metal-containing circuit boards. The same applies to the PGM-containing catalyst in a car or fuel cell.

7. **Sufficient capacity** along the entire chain to make comprehensive recycling happen. Once conditions 1–6 are met, the only requirement is to ensure that there is sufficient capacity to process the volume of material available for recycling. Metal refiners are willing to invest in building up such capacities provided there is sufficient security of feed later on. Conditions 4 to 6 are thus crucial to trigger timely investments in the growing market for metals recycling.

Particular end-of-life products and materials will satisfy certain conditions in this sequence; the further they get, the easier it will be to find appropriate measures to make use of this recycling potential. Figure 3.10, later in this chapter, shows a schematic representation of a life cycle for a typical critical-metals-containing product, highlighting the various points at which losses of metal can occur. Each of these provides an opportunity for making an improvement to the process or for adding incentives that encourage recycling to take place.

Recycling technologies

Before the actual recovery of the technology metals can take place, the recycling material needs to be conditioned in most cases as shown in Figures 3.3 and 3.6. Examples of such conditioning include dismantling and/or mechanical pre-processing (e.g. by shredding and sorting) of electronic scrap, de-canning car catalysts (extracting the catalyst monolith from the steel case) and burning oil refining catalysts contaminated with carbon. Whatever pre-processing is employed, it should be conducted in a way that its output fractions provide an optimal fit to the subsequent metallurgical recovery processes and that losses of valuable substances during pre-processing are minimised.

Different metals require different final pathways for metallurgical processing. Three main

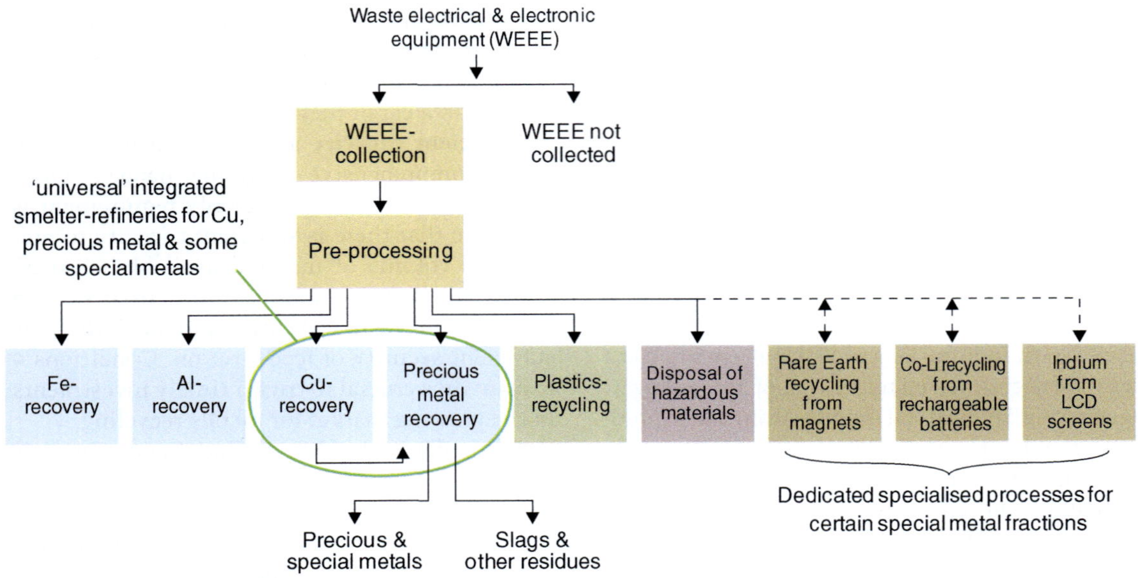

Figure 3.6 Dismantling and pre-processing are crucial to channel substances into the most appropriate metallurgical recovery facilities for final processing. This is illustrated here for waste electrical and electronic equipment. (Al, aluminium; Co, cobalt; Fe, iron; Li, lithium.)

routes and a large number of dedicated processes are currently used (Figure 3.6):

- Ferrous metals recovery takes place in steel plants.
- Aluminium recovery is conducted in aluminium refineries and re-melters.
- Copper, lead, nickel and precious metals are recovered in modern integrated smelter-refineries. A number of speciality metals, such as tellurium, selenium and antimony, also fit in this pathway and can be recovered in sophisticated 'universal processes' as by-products.

However, a number of other speciality metals are lost in such universal flowsheets and require dedicated processes, as shown in Figure 3.6. Some of the critical metals from the EU list (EC, 2010), such as PGM, antimony and some indium, can be recovered efficiently via the integrated smelter-refinery pathway. The other critical metals require dedicated processes, although for all of them metallurgical solutions do exist (for specific challenges see Section 'Metallurgical recovery').

It is clear that even the best metallurgical recovery process is useless if the material fractions that contain the target metals do not reach the appropriate final process. Hence, the organisation of the recycling chain and the management of the interfaces between the main steps are crucial for overall recycling success (Hagelüken, 2006a). Furthermore any system should be designed for a specific type of material (Van Schaik and Reuter, 2010).

Collection and pre-processing

Effective collection systems are a prerequisite for reclaiming metals and the infrastructure used for this purpose must be suited to local conditions. Collected EoL products are usually sorted into several categories, which, in the EU, are prescribed by legislation at the country level. In order to simplify logistics, attempts are often made to reduce the number of categories. However, too much reduction will result in a heterogeneous mixture of material types, which tends to reduce the effectiveness of the subsequent pre-processing and recovery processes. A balance must be struck between too many and too few categories to maximise the overall recovery rates of technology metals. A separate collection category for small ICT equipment, such as mobile phones, digital cameras and USB memory sticks, is sensible because they contain significant concentrations of technology metals that would be largely lost if mixed with other small low-grade electric household appliances such as razors, tooth brushes and toasters.

Pre-processing is required for most complex products in order to prepare for subsequent effective recovery of the contained metals. It must be able to cope with a material feed that changes over time and includes many different models and types of equipment (Reuter et al., 2005). While this works quite well for base metals, it is much more difficult to achieve for technology metals. Losses during pre-processing for the various recycling routes and the impact of different material combinations (product design) need to be quantified to evaluate the efficiency of the processes and the potential for improvement, mainly in the technical interface between (mechanical) pre-processing and metallurgy. The complexity of high technology products leads to incomplete extraction of materials at EoL, as these are strongly interlinked. For example, precious metals contained in circuit boards are associated with other metals in contacts, connectors, solders, etc.; with ceramics in multi-layer capacitors, integrated circuits (ICs), hybrid ceramics, etc.; with plastics in circuit board tracks, interboard layers, ICs, etc. Small-size material connections, coatings and alloys cannot be separated sufficiently by shredding. Hence, incomplete liberation and subsequent incorrect sorting result in losses of technology metals to side streams, including dust, from which they cannot be recovered by metallurgical treatment (Hagelüken, 2006a; Meskers et al., 2009; Van Schaik and Reuter, 2010 and 2012).

An industrial test with mixed IT electronic scrap treated in a modern shredder without prior dismantling of circuit boards revealed that the percentages of silver, gold and palladium ending up in fractions from which they could be recovered (circuit board and copper fractions) were only 12 per cent, 26 per cent, and 26 per cent, respectively (Chancerel et al., 2009; Chancerel, 2010). Automotive catalysts, batteries, high-grade circuit boards, mobile phones and MP3 players thus need to be separated prior to mechanical pre-processing to prevent irrecoverable losses. These components and devices can be fed into a smelter-refinery process directly and most metals recovered with an efficiency of greater than 90 per cent.

For larger and for low-grade electronic scrap, such as small domestic appliances and most audio-visual equipment, direct feeding to a smelter is usually not applicable and some degree of mechanical pre-processing is required. Instead of intensely shredding the material, a coarse size reduction, followed by manual or automated removal of circuit board fractions, can be a viable alternative. Trained workers are often able to remove certain complex target components more selectively than automated sorting. Wherever trained manual labour is available and affordable, it can be a viable alternative to dismantle, sort and remove critical fractions, such as circuit boards or batteries, by hand if this is combined subsequently with state-of-the-art industrial metal recovery processes (UNEP-StEP, 2009; Wang et al., 2012). Often, better results can be achieved by a 'negative selection': rather than removing clearly identified items or components of a specific type (e.g. positive selection of circuit boards) it can be more effective to remove everything which is clearly free of these items. Compared to a direct positive selection here the output stream with target material is still partly diluted with remaining non-target material, hence the concentration of target material after sorting is lower. However, it avoids the critical slip of value into the wrong fraction. With the reduced amount of the remaining less heterogeneous material then a second, more sophisticated (and costly) sorting step becomes affordable. A common misconception in mechanical pre-processing is that the aim should be to generate highly concentrated ('pure') output streams, since this can usually only be achieved at the cost of significantly reduced recovery rates. To mitigate this 'concentration dilemma' it is necessary to find the right sorting intensity, which enables sufficient separation and concentration without causing counter-productive effects on metal yields. For specific material mixes and separation technologies this can be expressed by the so-called grade-recovery curve, in the same way that separation efficiency is measured in the beneficiation of primary metallic ores (Hagelüken, 2006a; Van Schaik and Reuter, 2010 and 2012).

For improving the recycling efficiency of rare earth elements (REE), high-strength magnets, such as those used in electric motors and hard disk drives, are a potentially important source as they account for a significant share of total REE demand and the concentrations of REE within them are relatively high. However, these magnets are usually very brittle, so a shredder process would break them into many tiny still highly magnetic pieces which would stick to any iron surface and would not be accessible for REE recovery. Consequently, these magnets must be removed prior to any shredding process.

Overall, the optimisation of product design, combined with the use of appropriate sorting methods and depth, and routing of the various fractions produced, can lead to a substantial increase in yields, especially for technology metals. New developments in products continually lead to new recycling challenges. An example is the modern car. Such 'computers on wheels' include numerous electronic components, containing significant amounts of technology metals, which are scattered throughout the vehicle. This prevents current shredder technology from recovering them, while cost-effective manual dismantling is not currently feasible.

Metallurgical recovery

Although in principle, every metal can be recovered on its own, the mix of substances in a material stream creates challenges for metallurgical recovery. The more complex this mix the more challenging is the metallurgy and the greater are the inevitable metal losses. Complex material mixes are best recovered by initial pyrometallurgy ('smelting') followed by combined hydro ('leaching') and pyrometallurgical processes. Chemically noble metals (such as copper, PGM, selenium and tellurium) can be concentrated into a metallic phase, for example by alloying with copper, lead or nickel. Volatile elements (such as mercury, cadmium, zinc and rhenium) tend to go to the offgas stream, while metals that oxidise easily are transferred to the slag phase (REE, lithium, aluminium, silicon, magnesium, tantalum and germanium). While most metals are extracted via the metallic phase, some can be recovered from offgas or slags if concentrations are high enough and/or a 'property hook' (thermodynamic properties, vapour pressure, density, magnetic properties, etc.) is available to enable the process to be economic. Traces of metal oxides embedded in large volumes of slag volumes comprising, for example, mostly ceramics or silica, are not economic to extract.

Consequently, high-technology metallurgical processes are required to recover technology metals from complex components efficiently. For example, Umicore's integrated smelter refinery in Antwerp recovers in its universal flowsheet fourteen different precious and speciality metals together with the base metals copper, lead and nickel, which are used as metallurgical collectors (Figure 3.7).

For precious metals from circuit boards or catalysts, yields of close to 100 per cent are achieved in spite of their low concentrations, while, at the same time, tin, lead, copper, bismuth, antimony, indium, selenium and others are reclaimed (Hagelüken, 2006b). Other integrated smelter processes are operated by Aurubis, Boliden, Dowa and Xstrata (see website information under references). Furthermore, in dedicated processes Umicore recovers cobalt, nickel and copper from rechargeable batteries (Dewulf et al., 2010) (Figure 3.8), indium and gallium from photovoltaic sputter chamber scrapings (Meskers et al., 2010), germanium from wafer production scrap, and indium from indium–tin oxide (ITO) sputtering targets.

Research is underway to extend the range of feed materials and recover additional special

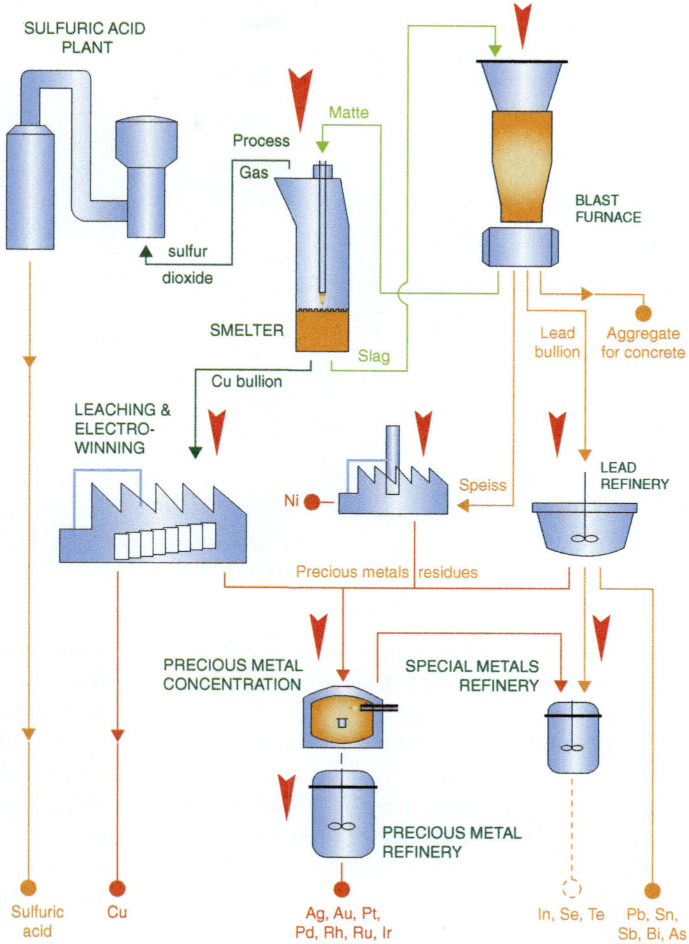

Figure 3.7 Metal combinations that fit thermodynamically can be recovered in a sophisticated integrated smelter-refinery process where copper, lead and nickel act as collectors for precious and some speciality metals. The example shown here is the Hoboken universal process of Umicore. The large red arrows indicate where recycling materials can be fed into the process (depending on concentration and properties). The main feed stream goes into the smelter (upper left arrow). (Ag, silver; As, arsenic; Au, gold; Bi, bismuth; In, indium; Ir, iridium; Ni, nickel; Pd, palladium; Pt, platinum; Rh, rhodium; Ru, ruthenium; Sb, antimony; Se, selenium; Sn, tin; Te, tellurium.)

metals, such as REE and lithium (Umicore-Rhodia, 2011). As described in the metal-specific chapters of this book, further dedicated processes exist for a number of other materials, like tantalum recovery from capacitor scrap or tungsten recovery from drilling tools.

For metals that already follow other metal streams or can be separated from offgas or effluents, economic recovery might be achieved through adjustments to the flow sheet and/or by developing appropriate after-treatment steps. In contrast, recovering metals economically that oxidise easily and are dispersed as a low-grade slag constituent can be extremely difficult or even thermodynamically impossible. Hence product design, mechanical pre-processing and metallurgical processing are linked in a highly interdependent system. The laws of thermodynamics determine which metals might ultimately be recovered under optimal metallurgical conditions (Habashi, 1997; Nakajima et al., 2011). It is important that product design avoids – if product functionality permits – incompatible combinations and ensures that materials which cannot be recovered in a common final treatment process can be separated during pre-processing ('design for disassembly'). Sophisticated modelling of

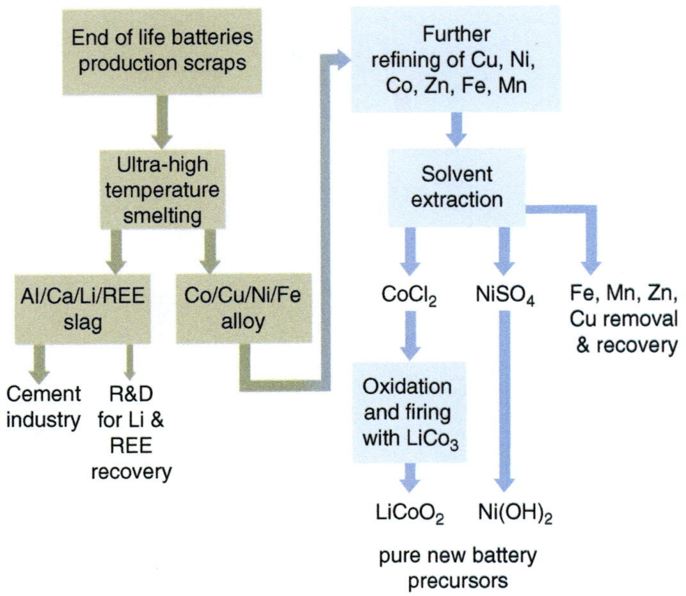

Figure 3.8 Certain metal combinations need dedicated processes. The Umicore process for lithium-ion and nickel-metal hydride batteries is shown here. (Al, aluminium; Ca, calcium; Co, cobalt; Cu, copper; Fe, iron; Li, lithium; Mn, manganese; Ni, nickel; REE, rare earth elements; Zn, zinc. $CoCl_2$, cobalt chloride; $LiCoO_2$, lithium cobalt oxide; $Ni(OH)_2$, nickel hydroxide; $NiSO_4$, nickel sulfate.)

this interlinked system has been undertaken by Reuter and Van Schaik (Van Schaik and Reuter, 2010 and 2012).

The combination of metals with toxic and/or organic substances with halogens in many EoL products requires specialised equipment and processes. Considerable investment is required for offgas and effluent management to ensure environmentally sound operations, preventing heavy metal and dioxin emissions. In practice, at the present time many recovery plants are not adequately equipped. In particular in the emerging Asian economies, electronic scrap is often 'industrially' treated in non-compliant smelters or leached with strong acids in hydrometallurgical plants with untested or ineffective effluent management, and with a primary focus on recovering gold and copper. Most electronic scrap in these countries is handled in the informal sector in thousands of backyard recycling facilities. This includes open-air incineration to remove plastics, 'cooking' of circuit boards over a torch for desoldering, cyanide leaching and mercury amalgamation (Puckett et al., 2002 and 2005; Kuper and Hojsik, 2008; Sepúlveda et al., 2010; Chi et al., 2011). Not only do these practices have very serious effects on health and environment, but their efficiency is also very low. An investigation in Bangalore, India revealed that only 25 per cent of the gold contained in circuit boards was recovered, compared to over 95 per cent in integrated smelters (Keller, 2006; Rochat et al., 2007). A UNEP report provides a comprehensive overview on the situation in developing countries (UNEP-StEP, 2009).

Recovering metals from combinations that do not occur in nature presents a particular challenge. Most metallurgical recovery processes were developed over centuries around the combinations of metal families and gangue minerals, as shown in the Web of Metals (Reuter et al., 2005). Some processes have been adjusted to secondary materials but the same laws of chemistry and thermodynamics apply. Most primary concentrates fit 'automatically' into one of the established metallurgical recovery routes. This is not the case for EoL products that contain man-made combinations, which are commonly very different from those found in nature. Once precious metals and speciality metals enter a steel plant or aluminium smelter it is almost impossible to recover them.

In most cases, however, metallurgical technology itself is not the barrier preventing good recycling rates. Appropriate processes for new and difficult materials (e.g. mobile phones, lithium-ion batteries, new process catalysts, diesel particulate filters and fuel cells) will be developed as long as there is an economic incentive to do so. The complex interplay between metallurgy and other technical and non-technical impact factors on metal recycling rates is described in depth in the report 'Metal Recycling – Opportunities, Limits, Infrastructure' (UNEP, 2013) which was prepared by a group of international recycling and metallurgy experts in the context of the International Resource Panel.

Status of recycling of the EU critical metals

With respect to technical and economical recyclability the critical metals defined by the EU working group (EC, 2010) can be divided into three main groups:

(a) Critical metals combined with precious metals that can be recycled economically using existing metallurgical processes.

The *PGM*s are the best performers in this group because their high price drives them into recycling. Technically, metallurgical yields of well over 95 per cent can be achieved for platinum and palladium. For rhodium, ruthenium and iridium metallurgical yields are also high but somewhat lower than for platinum and palladium. The most important PGM-containing products for recycling are car and process catalysts, equipment used in the glass industry, jewellery, circuit boards and mobile phones. In the latter two uses PGM (mainly palladium) are used in combination with other critical metals and with gold and silver. Some sophisticated integrated smelter-refinery processes can co-recover *antimony* and *indium* from such complex metal combinations (Hagelüken, 2006b). This co-recovery is triggered by the presence of the PGM and/or gold as 'paying metals', without which the recovery of low concentrations of indium or antimony would not be economically viable. In the future, platinum-containing fuel cell stacks may become an important source for PGM recycling. If end-of-life products are properly collected the main reason for metal losses in this group is inappropriate pre-processing.

(b) Critical metals, without associated 'paying' precious metals, that can be technically recycled using (new) dedicated processes

As for the first group, technical recyclability is a basic requirement. This is affected by material composition and complexity as well by the concentration of these metals and can be technically more challenging than for precious metals. Further differences occur here in the (current) economic viability. Recycling is most economically attractive for relative high-grade concentrates or components which contain critical metals. These include: *indium* from spent ITO targets; *indium* and *gallium* from high-grade photovoltaic (or other) production scrap; *germanium* from Ge wafers; *beryllium* from certain high grade production scraps; *tantalum* from capacitor production scrap; *tungsten*, *cobalt*, *tantalum* and *niobium* from machine tools, superalloys and alloyed steel (if used in steel alloys, recovery of these metals generally takes place by alloy recycling, without separating and purifying the individual element); and *cobalt* from certain rechargeable batteries or process catalysts.

It is generally not currently economic to recycle critical metals where their concentrations are low or for which new large-scale processes are still under development. These include: *REE* from lamp phosphors, batteries, magnets;[4] *indium* from LCD screens or photovoltaic panels; *cobalt* from rechargeable batteries with relative low cobalt content or high dismantling/pre-treatment requirements. Whether new processes will eventually become economically viable depends on many factors, including future metal prices, available quantities and associated economies of scale, R&D efforts and technical progress to improve metal yields and to reduce process costs. Effective pre-processing of the critical metal-containing components and materials is an essential prerequisite to ensure they are directed to the most appropriate metallurgical processes. For example, without removal of the REE-containing magnets from electric motors recycling of the REE will be impossible. Similarly, batteries must be removed from electrical devices and ITO coatings from glass surfaces to create an indium concentrate that is amenable to existing indium recovery processes (Böni and Widmer, 2011).

(c) Critical metals as part of complex material mixes with thermodynamic constraints

Electronic circuit boards are a good example of this group. Although *antimony*, and to a lesser extent *indium*, thermodynamically fit with the recovery of copper and precious metals, other critical metals present in circuit boards do not. *Tantalum, gallium, germanium* and the *REE* are likely to oxidise in the existing recycling processes and thus would end up highly diluted in the smelter slag. Recovery from this slag would not be economic. Hydrometallurgical processes for the treatment of circuit boards do not offer a better option, since the range and yields of recovered metals would become worse and treatment of strong leaching effluents is challenging and costly. Furthermore, sole hydrometallurgical treatment of complex products will lead to numerous other output streams and residues that need to be dealt with, ranging from precipitates and unclean resins to vapours or sludges. Shifting the focus from the precious-metals recovery to tantalum, REE or gallium does not make economic sense either since it would lead to high losses of the valuable precious metals. There are a number of other examples where incompatible material combinations will inevitably lead to the loss of one critical metal or another (e.g. PGM and REE in automotive catalysts). Hence, it is almost inevitable that in complex mixes technical conflicts of interest will exist. Consequently, 100 per cent recycling of critical metals will never be possible and priority choices need to be made (see also UNEP, 2013).

Such prioritisation has also to be made between the target materials for recovery of critical metals. For reasons given above, e.g. tantalum recovery from EoL circuit boards will not be possible without sacrificing other valuable and critical metals although tantalum can be recovered efficiently from many alloys or capacitor production scrap. Research on REE recovery should focus on magnets, lamp phosphors and rechargeable batteries which account for a major part of current total demand (Oakdene Hollins, 2010; Schüler et al., 2011). Research on the recovery of REE from car catalysts is unlikely to be worthwhile because of the likelihood of associated negative impacts on the efficiency of PGM recycling from these sources.

The significance of life-cycle structures

The largest challenges for recycling are to overcome the low levels of collection of consumer goods and their inefficient handling within the recycling chain. The life-cycle structure for consumer goods differs fundamentally from that of industrial products. This was investigated in a comprehensive research project on the life cycle of PGMs (GFMS et al., 2005). Good examples of efficient industrial cycles (business-to-business, B2B) are PGM-containing process catalysts. For the PGM contained therein, ownership usually remains with the industrial user, the product location is well known, and handling throughout the life cycle is conducted in a professional, transparent way. This is known as a 'closed-loop' recycling system. In contrast, ownership of consumer items (business-to-consumer, B2C) tends to shift frequently, goods such as mobile phones and cars are moved around the globe, manufacturers lose track of their devices and the flow of products becomes almost impossible to trace. This forms an 'open loop' in which recycling cannot be guaranteed. Even after an item reaches the recycling chain, the first steps in particular are not always handled by reputable agencies (Hagelüken et al., 2009).

The structural factors for PGM recycling may also be extended to other metals in industrial and consumer products. Figure 3.9 illustrates the difference between closed and open loop systems. The structure of the latter is considerably more

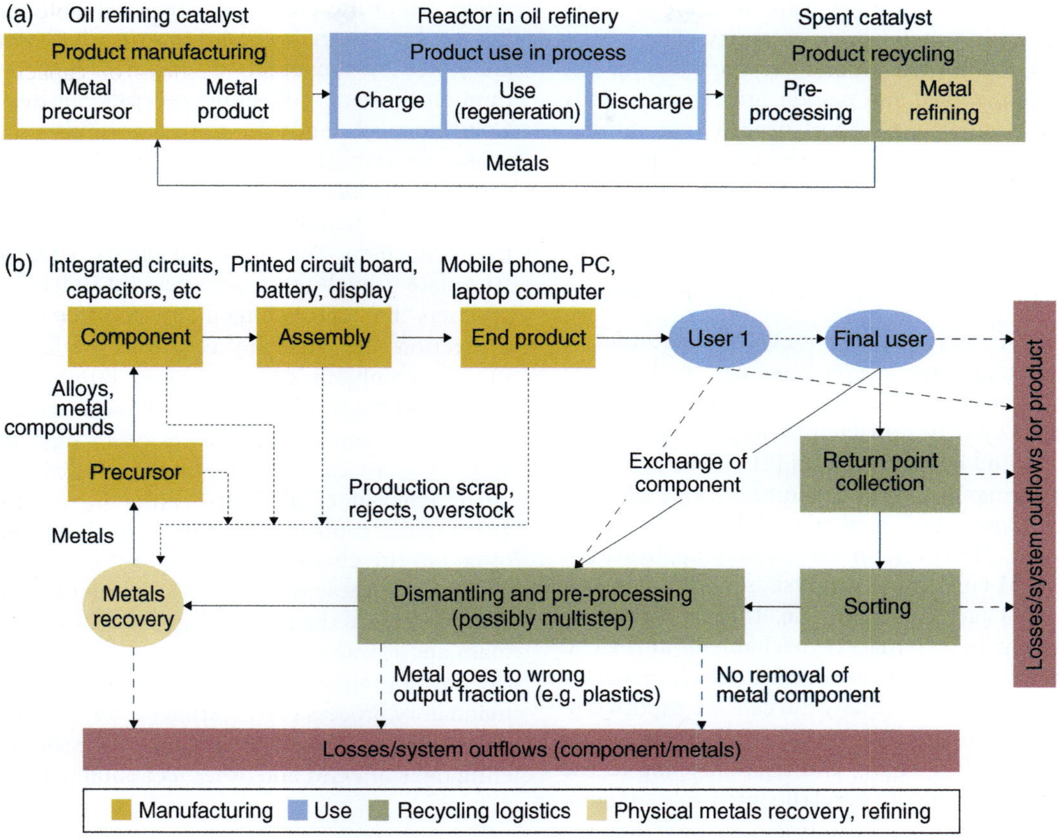

Figure 3.9 (a) Closed-loop recycling systems for industrial applications, illustrated by process catalysts. (b) Open-loop recycling systems for consumer goods, illustrated by electronic items.

complex, hence the probability of effective recycling of technology metals in these systems is much less. The following case studies illustrate the current recycling rates for PGM materials from industrial, automotive and consumer electronic applications (Hagelüken, 2012). Those of the seven conditions for effective recycling, as described above, which are not met are also indicated. All numbers refer to global averages (see also Figure 3.4).

Case study 1: Industrial PGM applications

Industrial applications, such as catalysts used in the production of fine chemicals and in petrochemical processing, and in PGM equipment used in the glass industry, are the current benchmark for PGM recycling, with end-of-life recycling rates comfortably exceeding 80 per cent. Recycling in this case is solely market driven and is an integral part of the product life cycle. Each of the conditions 1–7 for effective recycling are met, strongly supported by the high value of the PGM materials involved. This limits the need for primary-metal supply for these industries to covering the small life-cycle losses and to keeping up with market growth and new applications. As most industrial users own the PGM contained in their products and subcontract to specialists the PGM recovery from spent catalysts and the manufacture of fresh catalysts, they are also less vulnerable to metal price volatility. The metals are largely turned around on the basis of the weights involved and market prices impact only on the small gaps that need to be closed. Furthermore, the closed-loop system offers full transparency on metal sources, thus avoiding the use of 'conflict metals' in these applications.

Case study 2: Automotive PGM applications

End-of-life recycling rates for PGM in automotive applications have a global average in the range 50–60 per cent. Automotive PGM recycling in Europe is partly impacted by legislation such as the EU End of Life Vehicles (ELV) Directive, but the dominant driver is economic. As for all PGM applications, technical recyclability is not a problem. However, in automotive applications, the PGM-containing catalyst is only a sub-system of a larger product, i.e. the car, which is driven by its own market mechanisms at the end of its life. European car catalyst recycling mainly fails at condition 5. Many old cars are exported to countries outside Europe which lack an appropriate recycling chain, and it is only due to the excellent recyclability and the intrinsic economic value of automotive catalyst recycling that the PGM losses here are not even higher. Better enforcement of international waste shipment rules to limit the export of genuine scrap cars could enhance recycling rates within Europe.

Case study 3: Electronic PGM applications

Recycling rates for PGM in electronic applications are currently only about 5–10 per cent. The main driver for recycling in Europe is legislation such as the EU Waste Electrical and Electronic Equipment (WEEE) Directive. The largest use of PGM in electronics is palladium used in circuit boards, and market mechanisms at the end of life of products such as PCs, TVs, mobile phones and car electronics play an important role in the rates of collection and recycling of the circuit boards they contain. As in the case of automotive catalysts, a major challenge occurs at condition 5 when end-of-life electronics are exported out of Europe, but failure at condition 4, in which items are stored by consumers or disposed of through municipal waste collection, is also significant. Conditions 2 and 3, the accessibility and economic viability of components for recycling, can also be an issue in some cases, as can condition 6, inappropriate handling within the recycling chain. The EU WEEE Directive helps to stimulate recycling of electrical and electronic products but its enforcement is often weak. Collection rates in many countries are low and illegal or dubious exports out of Europe are a serious problem, especially for IT and telecommunication equipment, which are of high relevance for critical metals (Sander and Schilling, 2010; Prakash et al., 2010; Basel Convention, 2011). More transparency and better monitoring of end-of-life chains would improve the rates of recycling of these products. Most importantly, a shift of focus of the current legislation away from mass and towards a more pragmatic approach to the introduction of treatment standards and a mandatory process certification system along the recycling chain would help to increase the recycling rates of PGM and other technology metals.

Global flows of old products

Trading of old equipment and donations to charity have led to a steady but hardly transparent flow of EU devices to eastern Europe, Africa and Asia (Buchert et al., 2007; Sander and Schilling, 2010). Whether a product is ready for recycling or will be re-used largely depends on where it happens to be – what is regarded as waste in Europe or elsewhere in the developed world may well be re-usable in Africa. Dealers take advantage of this by 'exporting for re-use' and a fair amount of these exports evade the Basel Convention waste-export provisions. The Basel Convention restricts the exports of hazardous wastes, which includes most electronics scrap, and hence is only allowed to be exported to OECD countries under a notification procedure. However, a 'product' (i.e. not a waste) does not fall under the Basel regime, so declaring old equipment as a re-usable product is a way of by-passing the regulation.

In fact, used products collected in good faith for recycling or re-use are known to vanish in a dubious way only to resurface in primitive landfills or substandard backyard 'recycling' facilities in developing countries (Sander and Schilling, 2010). Insufficient cooperation within the life cycle and recycling chain, together with poor tracking of product and material streams along the entire chain, can explain why inefficient open cycles continue to exist.

Mobile phones provide a striking example (Nokia, 2008): although the technology is in place to recycle them economically and with high efficiency, the actual global recycling volume (excluding re-use) is less than 2000 tonnes per annum. This is less than 5 per cent of the global total of 80,000 tonnes per annum which is potentially available for recycling. Most of the old telephones that are collected are exported to developing countries for re-use (partly justified, partly 'fake re-use'), where efficient recycling does not take place at their final end of life. In many cases, collected phones are exported in bulk, without prior testing, mainly to Hong Kong. There specialised companies select re-usable and repairable phones from the rest. While the re-usable phones are traded within Asia and to Africa, the non-reusable ones mostly go to local backyard recyclers. In order to overcome this practice mandatory testing of such equipment before export is required to ensure that all non-reusable phones are identified as scrap. These should remain in Europe and be directed into legitimate recycling channels.

If mobile phones are truly exported for re-use they can help improve the communication infrastructure in developing countries at affordable prices. On the other hand, the current lack of any effective global recycling infrastructure inevitably leads to the significant loss of valuable raw materials. It is therefore advisable to set up global structures which promote the shipment of critical fractions derived from EoL products from developing and transition countries (when they are lacking in appropriate metallurgical recovery facilities) to state-of-the-art facilities in industrialised countries. Beyond a significant reduction of hazardous emissions and an increase in the amount of recycled metals, this approach also generates better net revenues in the developing/transition countries. This "Best of 2 Worlds" concept is elaborated in (Rochat et al., 2008; Wang et al., 2012).

Differences in recycling rates and pathways for improvement

The main reason for the differing recycling rates in the case studies is less a question of market and legislative drivers, and more connected to the step a product reaches in the sequence of conditions listed above. It is important to note that in all applications technical recyclability is not an issue: for example, yields of well over 95 per cent can be achieved if the PGM-containing material reaches a state-of-the-art precious metals refinery. Hence, neither PGM product manufacturers nor PGM refiners can act alone to improve the situation; it is the life-cycle system as a whole and the interactions of the stakeholders within that system which will make the difference.

What has been shown here for PGMs is basically true for most other critical metals. However, PGMs benefit from a very good metallurgical recyclability even from low-grade and complex products as well from a high economic attractiveness due to their value. As described earlier in this chapter, for metals contained in complex mixes with thermodynamical constraints even under an optimal life cycle system setup, a 95 per cent yield will not be possible. This is elaborated in detail in a UNEP report on opportunities and limits of metal recycling (UNEP, 2013). However, relatively high metallurgical yields will be possible for these metals once it is assured that the relevant fractions in which they are contained are directed into state-of-the-art metallurgical processes.

The only way to develop a true circular economy for technology metals hence is to shift the open-loop systems of consumer products towards closed loops similar to those that prevail in industrial applications (Hislop and Hill, 2011). New business models, closer stakeholder cooperation and more transparency on material flows are a prerequisite. Deposit funding systems (e.g. for mobile phones or batteries) can incentivise consumers to hand back their old devices into an appropriate recycling chain. Leasing models or other innovative ways of product service systems are another option, especially for products with a longer lifetime. If a manufacturer or retailer retains ownership of his products and sells instead the access to its function he keeps track of the products throughout their entire time of use and eventually can take it back. As in the case example of PGM process catalysts he could then re-use certain parts internally and subcontract the remaining to recycling specialists, while still maintaining the ownership of the contained metals. By doing so he could partly secure his supply of critical metals for the next product cycle and largely rule out unethical sourcing (conflict metals). A manufacturer would also directly benefit from innovative design for recycling and disassembly in his products. He further could build up a strong form of continuous interaction with his customers, using their feedback for product improvement and service innovation which should have a positive impact on his economic performance. Hence, innovative business models have the potential to create win–win situations for improved resource security and business profit.

The above discussion has focused on the PGM, which, due to their high value and perfect technical recyclability, occupy a unique position among the critical metals. As shown in the previous section, for most other critical metals technical and economic challenges also need to be overcome. However, the PGM example underlines that creating closed-loop systems in combination with a carefully constructed infrastructure is the key to achieving a circular economy. Without this approach, research efforts to improve recyclability and economic viability will largely be wasted. As a contribution to the Raw Materials Initiative of the European Commission, Eurometaux and the Öko-Institute have proposed 10 measures to improve recycling and access to secondary raw materials (Eurometaux, 2010).

Conclusion and the way forward

As illustrated in Figure 3.11 consideration needs to be given both to the actual materials that are used and to how and why they are used in order to achieve high recycling rates of critical metals (see also UNEP, 2013). Figure 3.10 shows that losses occur at every stage of the product life-cycle, and the factors which influence the magnitude of these losses differ widely and must be addressed individually through appropriate measures ranging from product design and technical innovation in recycling, to new business models, training and education, and socio-political framework conditions. Substantially improving the recycling rates of metals in order to boost secondary supply requires innovation throughout the life cycle. However, the required quantum leap can only be achieved by a holistic system approach and adequate policy support which takes the interdependencies of life-cycle steps, impact factors and measures into account.

Innovation needs

In order to minimise the material losses, improved and systematic knowledge development and transfer is needed. In general, education on sustainability, legislative support and science-based innovation are needed. Education on sustainability will generate engaged professionals that are aware of the issues related to recycling and resource scarcity and that have the appropriate technical background to tackle these issues. Legislative support is needed to create incentives as recycling of some critical metals has macroeconomic benefits without being currently economically viable for every value-chain element. Support to transfer innovative laboratory-scale and pilot processes into industrially successful operations is required to reduce major investment risks. Legislation, knowledge development and cooperation will accelerate the

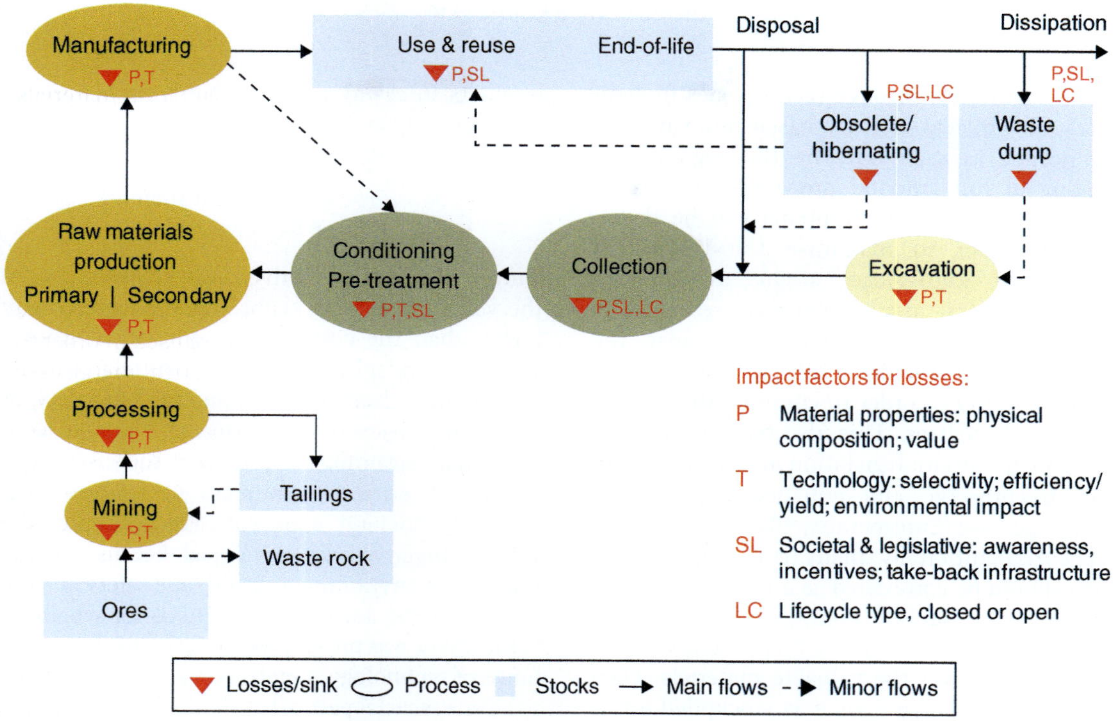

Figure 3.10 Life cycles of metals and products, and impact factors for losses at various stages. (After McLean et al., 2010.)

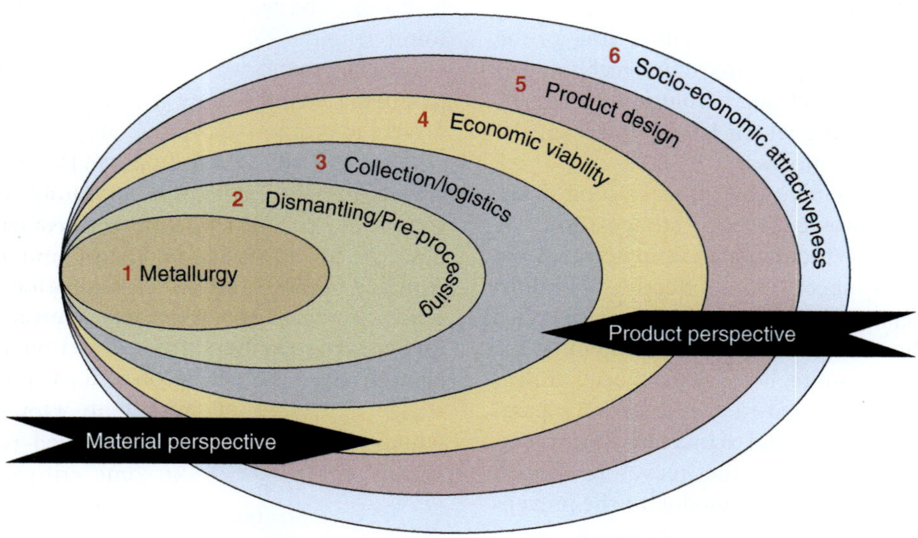

Figure 3.11 Recycling success factors – the material and product perspectives must be combined to ensure that the appropriate materials are used in the best way in products that meet the requirements of the consumer. The material perspective focuses on (technical) material properties and related impacts on recyclability, while the product perspective covers mainly socio-economic aspects such as consumer awareness, business models and design aspects.

realisation of breakthrough technologies in demonstration projects. Science-based innovation is also needed as it is only by thorough understanding of fundamental processes that sound solutions to technical problems can be developed. In order to maximise the impact of fundamental knowledge, alignment between competitive and pre-competitive research is essential. As the challenges faced are huge, the fundamental knowledge build-up needs to be tackled globally in order to combine the world's top research centres. This also requires support through appropriate legislation and regulation in order to preserve and enhance the existing knowledge base (Eurometaux, 2011).

The recognition by many governments that waste should be considered as a resource and not to consider disposal as an option is a major advance accomplished in recent years. As a result, waste is recycled to recover the valuable materials when technology is available, or it is stockpiled while waiting for suitable technology to be developed. For example, instead of sending indium-containing LCD screens to municipal waste incinerators it could be an option to store them in controlled intermediate depots in order to allow time for the development of appropriate recycling technologies and to build up an adequate volume of material suitable for recycling which later could form the base load for a new process.

In order to improve this process good documentation of waste stream compositions is needed in order to incorporate the material in the recycling flow sheet more effectively and to improve separation in temporary stockpiles (avoiding the formation of diluted complex waste streams). Furthermore, better and smarter collection is generally needed to ensure better separation in the early stages and to increase the fraction of materials that enter the recycling stream directly. This goes hand-in-hand with the prevention of dubious and illegal export of waste streams. Two major challenges ahead are to treat the complex waste streams available in landfills and to further improve collection and separation by more recycling-oriented design and tagging products in a way they can be optimally traced and recycled.

Resource security as a societal driver for recycling

Currently, recycling is driven either by value, when the value of recovered substances is significantly higher than the cost of separating them from a waste stream, or by societal concerns related to the environment, human health and human safety, or to volume aspects of waste streams. A new societal driver that is gaining importance is the sustainable supply of critical resources. Legislation best addresses societal concerns. For example, the WEEE Directive was developed because of the impacts on environment, health and safety and the volume of the devices and fractions for which no value driver was present. With regard to the importance of technology metals, resource aspects need to be an essential part of legislation in the future as well. The use of sub-standard recycling technologies, as well as the large amounts of electronic waste not entering the recycling chain or being lost along the way, are in urgent need of major improvement.

In the same way that environmental and health dimensions alone are not powerful enough drivers to promote recycling, market forces alone will not be sufficient to improve the recycling of critical metals. Favourable economic conditions cannot be relied upon, and the metals from EoL devices will inevitably be lost in the meantime. A policy framework for the sustainable supply of critical metals from secondary sources is therefore needed to address this societal concern. This should also include economic incentives to trigger recycling of important waste streams which contain critical metals.

Mining and recycling as complementary systems

As has been shown, some losses of metals are inevitable, even in an idealistic *static system* where demand remains at a constant level.

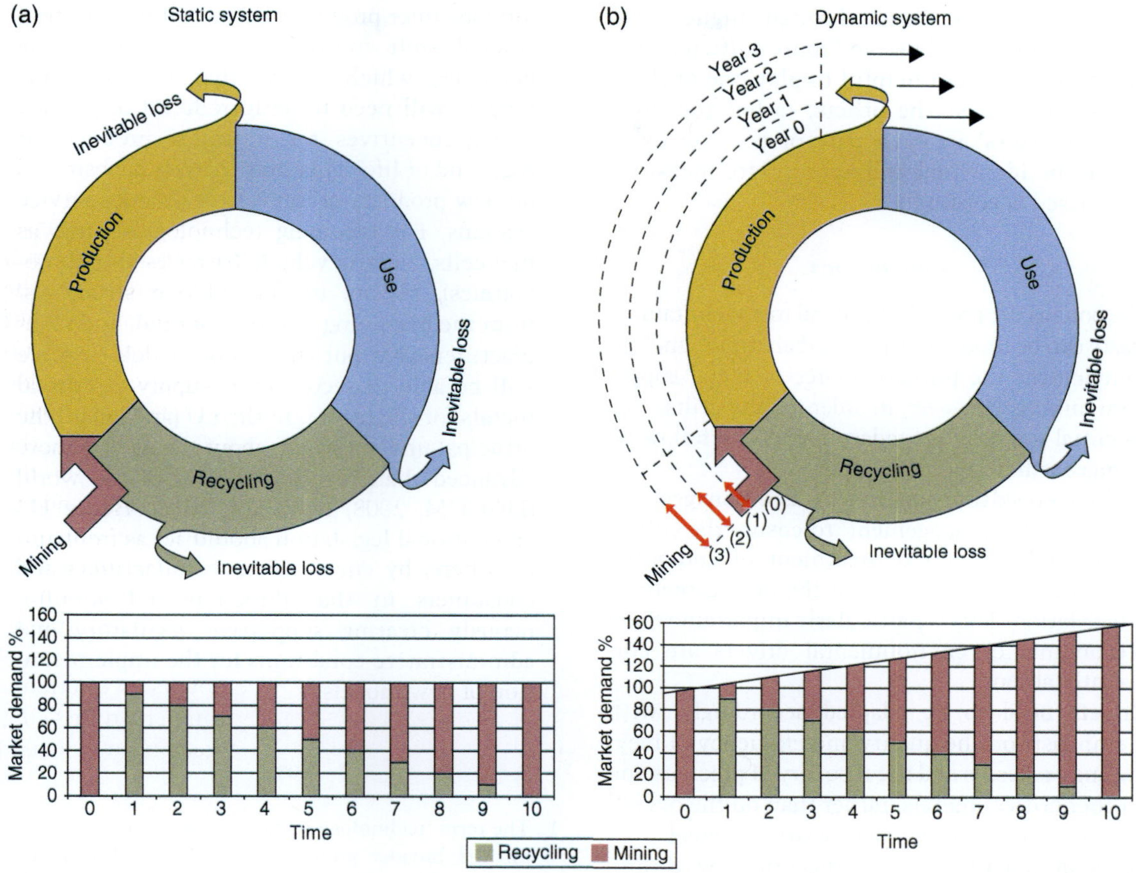

Figure 3.12 Mining and recycling are complementary systems to cover inevitable losses and market growth.

These losses would occur even with 100 per cent collection and using best available technology throughout the chain (Figure 3.12). In addition, economic constraints, which weigh raw material value against recycling costs, will increase losses further, usually significantly beyond the technical minimum. Such losses will therefore occur again in every new product/raw materials cycle and eventually the initial raw materials stock will be used up and will need to be complemented by supply from primary mined ores (Figure 3.12a).

In reality the system is *dynamic*, i.e. products have a lifetime that needs to be bridged before end-of-life recycling can take place and while market demand for raw materials may be growing. In this case even immediate recycling, if it were possible, would not be sufficient to meet demand (Figure 3.12b). Hence, although recycling is a cornerstone in securing supply of critical metals it cannot solve it alone. We need to use and combine the complete 'sustainability toolbox' in an optimal way, including mining, recycling, substitution and resource savings (both through higher efficiencies in material use and more sufficiency in life styles). It is not helpful to play one of these 'tools' off against the others, rather the 'low-hanging fruits' in each part of the life cycle should be identified and appropriate measures prioritised accordingly.

Conclusions

The potential for recycling critical metal-containing waste can be thought of as an 'urban mine', and can complement the primary source of metal supply from mining. However, in order to fully utilise this potential source of secondary supply, the following changes will need to take place:

Attitudes need to change from 'waste management' to 'resource management' to ensure the collection and appropriate treatment of end-of-life products and to encourage the enforcement of legislation. This is particularly important where economic or environmental drivers are currently absent.

Targets need to be adapted accordingly, with emphasis on the quality and efficiency of recycling processes and the recovery of precious and other critical metals, rather than on the overall mass of materials such as plastics or steel.

Recycling practice needs to reflect the new requirements. In place of the traditional structures of a scrap business, high-tech recycling can sit alongside clean-tech manufacturing and renewable energy generation in terms of company structures, appearance and stakeholder cooperation, with increased emphasis on transparency and business ethics. Reality in many cases is still far from this and the recycling industry as a whole has to undergo fundamental changes. This will probably leave many traditional players behind unless they adapt to the new requirements.

The manufacturers' vision needs to change as well. Rather than a burden imposed by legislation, recycling can be seen as an opportunity for manufacturers to sustainably increase access to the raw materials needed for their future production. To close the recycling loop for consumer products, there will need to be a gradual shift towards more industrial style practices, which means that new business models will need to be introduced to provide strong incentives for returning products at their end of life. This may include deposit fees on new products or innovative product service systems. For emerging technologies (such as fuel cells, electric vehicle batteries and photovoltaics), setting up 'closed-loop structures' from the beginning will be essential, and manufacturers that put successful models in place will be able to secure their supply of critical metals for the future. In the EU (and Japan) the principal understanding about the need is more advanced than in other regions of the world (EU-COM, 2008; EU-COM, 2012). Hence EU and national legislation should act as frontrunners here, by encouraging manufacturers and consumers in that direction and simultaneously creating supportive regulatory and administrative conditions for the implementation of new models.

Notes

1. The term 'technology metals' is used in this chapter for the broader group of speciality and precious metals which are – due to their specific physical-chemical properties – essential for the functionality of high-tech products.
2. At the end of 2011 the revision of the WEEE directive was agreed between the EU parliament, the council and the EU commission. It contains a number of improvements (e.g. own-collection category for small WEEE, increased collection targets, increased reporting requirements of input and output data of all actors involved in the recycling chain, stricter provisions to tackle illegal exports, development of harmonised standards for treatment operations) (EU Parliament News 2012; EU Legislative Observatory 2012)
3. For example, if one aims to achieve the required 65 per cent recycling rate for mobile phones, it will, with a narrow definition of recycling, be necessary to recover the plastic fraction. In a mechanical process this only can be achieved by intense shredding and sorting of the phones which will lead to high losses

of gold and other technology metals, which occur in traces only (Chancerel et al., 2009).

4. A comprehensive overview on the use and recycling challenges for REE is given in Schüler et al. (2011).

References

Achzet, B., Reller, A., Zepf, V., Rennie, C., Ashfield, M. and Simmons, J. (2011) Materials Critical to the Energy Industrie. An introduction. University of Augsburg.

Aurubis: www.aurubis.com/en/our-business/raw-materials/recycling/technology/

Basel Convention (2011) Where are WEEE in Africa? Findings from the Basel Convention e-Waste Africa Programme. Secretariat of the Basel Convention, Geneva/Switzerland (Dec. 2011)

Boliden: www.boliden.se

Böni, H. and Widmer, R. (2011) Disposal of Flat Panel Display Monitors in Switzerland, EMPA, March 2011.

Buchert, M., Hermann, A., Jenseit, W., Stahl, H., B., Osyguß, B. and Hagelüken, C. (2007) Optimisation of precious Metals Recycling: Analysis of exports of used vehicles and electrical and electronic devices at Hamburg port, Dessau: Federal Environmental Agency of Germany.

Chancerel, P. (2010) Substance flow analysis of the recycling of small waste electrical and electronic equipment – An assessment of the recovery of gold and palladium. Doctoral thesis, Technische Universitaet Berlin, Germany http://opus.kobv.de/tuberlin/frontdoor.php?source_opus=2590.

Chancerel, P., Meskers, C., Hagelüken, C. and Rotter, S. (2009) Assessment of Precious Metal Flows During Preprocessing of Waste Electrical and Electronic Equipment. Journal of Industrial Ecology 13 (5), 791–810.

Chi, X., Streicher-Porte, M., Wang, M.Y.L. and Reuter, M.A. (2011) Informal electronic waste recycling: A sector review with special focus on China. Waste Management 31, 731–742.

Dewulf, J., Van der Vorst, G., Denturck, K., Van Langenhove, H., Ghyoot, W., Tytgat, J. and Vandeputte, K. (2010) Recycling rechargeable lithium ion batteries: critical analysis of natural resource savings. Resources Conservation and Recycling 54, 229–234.

DOWA: www.dowa.co.jp/en/jigyo/metalmine_summary.html

EC (2010) Critical raw materials for the EU: Report of the Ad-hoc Working Group on defining critical raw materials, Enterprise and Industry Directorate General. (http://ec.europa.eu/enterprise/policies/raw-materials/documents/index_en.htm)

EU-ELV (2000) European Parliament and Council Directive 2000/53/EC of the European Parliament and the council of 18 September 2000 on end of life vehicles, Official. Journal of the European Communities, 21-10-2000, Brussels Belgium.

EU-WEEE (2003) European Parliament and Council Directive 2002/96/EC of the European parliament and of the council of 27 January 2003 on waste electrical and electronic equipment (WEEE). Official Journal OJ L 37 of 13.02.2003, Brussels Belgium.

EU-COM (2008) 699 Communication from the Commission to the European Parliament and the Council. The Raw Materials Initiative – Meeting our Critical Needs for Growth and Jobs in Europe, Brussels (2008).

EU-COM (2012) 82: Communication from the commission to the European Parliament, the Council, the European Economic and Social Committee and the Committee of the Regions. Making Raw Materials Available for Europe's Future Well Being - Proposal for a European Innovation Partnership on Raw Materials, Brussels 2012.

EU Parliament News (2012) Better management of e-waste www.europarl.europa.eu/news/en/headlines/content/20120106FCS34949/8/html/Better-management-of-e-waste.

EU Legislative Observatory (2012) European Parliament 2008/0241 (COD) – 19/01/2012 Text adopted by Parliament, 2nd reading, 19.1.2012.

Eurometaux (2010) Eurometaux's Proposals for the Raw Materials Initiative, Darmstadt / Brussels, 11th June 2010 (http://www.oeko.de/oekodoc/1069/2010-115-en.pdf and appendix).

Eurometaux (2011) Roadmap for securing a sustainable supply basis for Europe (draft), Brussels, Nov. 2011.

Gartner (2011) Gartner Says Worldwide Mobile Device Sales to End Users Reached 1.6 Billion Units in 2010; Smartphone Sales Grew 72 Percent in 2010 Gartner Newsroom, February 9, 2011. http://www.gartner.com/it/page.jsp?id=1543014 (July 15, 2011.)

GFMS, Umicore and Öko Institute (2005) Materials Flow of Platinum Group Metals; GFMS: London, 2005.

Graedel, T.E. (2011) On the Future Availability of the Energy Metals. Annual Review of Materials Research 41, 323–335.

Habashi, F. (1997) Handbook of Extractive Metallurgy. Wiley-VHC, Weinheim.

Hagelüken, C. (2006a) Improving metal returns and eco-efficiency in electronics recycling – a holistic approach for interface optimisation between pre-processing and integrated metals smelting and refining. In: Proceedings of the IEEE International Symposium on Electronics and the Environment, 8–11 May, San Francisco, 218–223.

Hagelüken, C. (2006b) Recycling of Electronic Scrap at Umicore's Integrated Metals Smelter and Refinery. World of Metallurgy – Erzmetall 59 (3), 152–161.

Hagelüken C., Buchert, M. and Ryan, P. (2009) Materials flow of platinum group metals in Germany. International Journal of Sustainable Manufacturing 1 (3), 330–346.

Hagelüken, C. and Meskers, C.E.M. (2008) Mining our computers – opportunities and challenges to recover scarce and valuable metals from end-of-life electronic devices. In: Reichl, H., Nissen, N., Müller, J. and Deubzer, O. (eds.) Electronics Goes Green 2008+. Fraunhofer IRB Verlag, Berlin, 623–628.

Hagelüken, C. and Meskers, C.E.M. (2010) Complex lifecycles of precious and special metals. In: Graedel, T. and Van der Voet, E. (eds.) Linkages of Sustainability; MIT Press: Boston.

Hagelüken, C. and Corti, C.W. (2010) Recycling of gold from electronics: Cost-effective use through 'Design for Recycling', Gold Bulletin 43 (3), 209–220.

Hagelüken, C. (2012) Recycling the Platinum Group Metals: A European Perspective. Platinum Metals Review 56 (1), 29–35.

Hislop, H. and Hill, J. (2011) Reinventing the wheel: A circular economy for resource security. Green Alliance, London 2011.

Johnson Matthey (2011) PGM Market review archive. www.platinum.matthey.com/publications/pgm-market-reviews/archive/platinum-2011/

Keller, M. (2006) Assessment of gold recovery processes in Bangalore, India and evaluation of an alternative recycling path for printed wiring boards. M.Sc. thesis, ETH Zürich, Zürich, Switzerland.

Kuper, J. and Hojsik, M. (2008) Poisoning the poor. Electronic waste in Ghana; Greenpeace International: Amsterdam.

McLean, H., Duchin, F., Hagelüken, C. et al. (2010) Mineral Resources: Stocks, flows, and prospects. In: Graedel, T., and van der Voet, E. (eds.) Linkages of Sustainability. Strüngmann Forum Report, Volume. 4. Cambridge, MA: MIT Press.

Meskers, C.E.M. (2008) Coated magnesium designed for sustainability? Ph.D. thesis, Delft University of Technology, Delft, the Netherlands, 2008.

Meskers, C.E.M., Hagelüken, C., Salhofer, S. and Spitzbard, M. (2009) Impact of pre-processing routes on precious metals recovery from PCs. Proceedings of the European Metallurgical Conference, 527–540.

Meskers, C.E.M., Vandenbroeck, K., Vliegen, J., De Ruijter, I., Dalle, T. and Rigby, P. (2010) Recycling technologies to close the loop for PV materials. In: Proceedings 25th European Photovoltaic Solar Energy Conference and Exhibition, 5th World Conference on Photovoltaic Energy Conversion, Valencia, 3683–3687.

Nakajima, K. Takeda, O. Miki, T. Matsubae, K. and Nagasaka, T. (2011) Thermodynamic analysis for the controllability of elements in recycling process of metals. Environmental Science and Technology 11, 4929–4936.

Nokia (2008) Global consumer survey reveals that majority of old mobile phones are lying in drawers at home and not being recycled. Nokia Press Room. July 8, 2008. http://press.nokia.com/2008/07/08/global-consumer-survey-reveals-that-majority-of-old-mobile-phones-are-lying-in-drawers-at-home-and-not-being-recycled/ (18th September 2011).

Oakdene Hollins (2010) Lanthanide Resources and Alternatives. A report for Dept. for Transport and Dept. for Business, Innovation and Skills, Aylesbury, UK, March 2010.

Prakash, S., Manhart, A., Amoyaw-Osei, Y. and Agyekum, O. (2010) Socio-economic assessment and feasibility study on sustainable e-waste management in Ghana, Öko-Institut Freiburg.

Puckett, J., Byster, L., Westervelt, S., Gutierrez, R., Davis, S., Hussain, A. and Dutta, M. (2002) Exporting Harm - The high-tech trashing of Asia; Basel Action Network and Silicon Valley Toxics Coalition: Seattle, February 25.

Puckett, J., Westervelt, S., Gutierrez, R. and Takamia, Y. (2005) The digital dump:-exporting re-use and abuse to Africa; Basel Action Network: Seattle, October 24, 2002.

Resnick Institute Report (2011) Critical Materials for Sustainable Energy Applications. Californian Institute of Technology, Pasadena, Sept. 2011.

Reuter, M.A., Boin, U.M.J., van Schaik, A., Verhoef, E., Heiskanen, K., Yang, Y. and Georgalli, G. (2005) The Metrics of Material and Metal Ecology. Elsevier, Amsterdam.

Rochat, D., Hagelüken, C., Keller, M. and Widmer, R. (2007) Optimal recycling for printed wiring boards. In: Hilty, L.M., Edelmann X., and Ruf, A. (eds.) India. R'07, World Congress – Recovery of materials and energy for resource efficiency.

Rochat, D., Rodrigues, W. and Gantenbein, A. (2008) India: Including the existing informal sector in a clean e-waste channel. Proceedings of the Waste Management Conference (WasteCon2008), Durban, South Africa.

Sander, K. and Schilling, S. (2010) Transboundary shipment of waste electrical/electronic equipment/electronic scrap – optimisation of material flows and control, Federal Environment Agency Report No. (UBA-FB) 1331, Dessau. Germany.

Schüler, D., Buchert, M., Liu, R., Dittrich, S. and Merz, C. (2011) Study on Rare Earths and their Recycling. Final Report for The Greens/EFA Group in the European Parliament. Öko-institut e.V. Darmstad, Germany.

Sepulveda, A., Schluep, M., Renaud, F., Streicher, M., Kuehr, R., Hagelüken, C. and Gerecke, A.C. (2010) A review of the environmental fate and effects of hazardous substances released from electrical and electronic equipment during recycling: examples from China and India. Environmental Impact Assessment Review 30 (1), 28–41.

Umicore/Rhodia (2011) Press release. CP-2011-18-R Umicore and Rhodia develop unique rare earth Recycling process for rechargeable batteries. 16th June 2011 www.rhodia.com

UNEP StEP Report (2009) Recycling – From E-waste to Resources. Berlin, Germany. (www.unep.org and www.step-initiative.org).

UNEP (2011) Recycling Rates of Metals – A status report, A report of the Working Group on the Global Metal Flows to the International Resource Panel. Graedel, T.E., Alwood, J., Birat, J.-P., Reck, B.K., Sibley, S.F., Sonnemann, G., Buchert, M. and Hagelüken, C. www.unep.org

UNEP (2013) Metal Recycling – Opportunities, Limits, Infrastructure. In: Reuter, M. et al. A report of the Working Group on the Global Metal Flows to the International Resource Panel.

US DOE (2010) Department of Energy. Critical Materials Strategy. Washington, Dec 2010.

US NRC (2008) National Research Council: Minerals, Critical Minerals, and the U.S. Economy. National Academic Press, Washington. (www.nap.edu/catalog/12034.html)

Van Schaik, A., and Reuter, M.A. (2010) Dynamic modelling of e-waste recycling system performance based on product design. Minerals Engineering 23, 192–210.

Van Schaik, A. and Reuter, M.A. (2012) Shredding, sorting and recovery of metals from WEEE: linking design to resource efficiency. In: Goodship, V. and Stevels, A. (eds.) Waste electrical and electronic equipment (WEEE) handbook, Woodhead Publishing Ltd, Cambridge, UK, 163–211.

Wang, F., Huisman, J., Meskers, C.E.M., Schluep, M., Stevels, A. and Hagelüken, C. (2012) The Best-of-2-Worlds philosophy: Developing local dismantling and global infrastructure network for sustainable e-waste treatment in emerging economies. Waste Management WM 05/2012; 32 (11), 2134–46.

Xstrata: http://www.xstrata.com/

4. Antimony

ULRICH SCHWARZ-SCHAMPERA

Federal Institute for Geosciences and Natural Resources (BGR), Stilleweg, Hannover, Germany

Introduction

Antimony metal, also referred to as regulus for impure forms of the element, and the natural sulfide of antimony, stibnite, have been known since 4000 B.C. The metal was used as a coating to harden copper between 2500 B.C. and 2200 B.C. The sulfide was used as eyebrow paint in pre-dynastic Egypt and early biblical times. Antimony is rarely found in nature as a native metal because of its strong affinity for sulfur and metals like copper, lead and silver. In fact, it is believed that the name 'antimony' is derived from the Greek words 'anti' and 'monos', which together mean 'not alone'.

Today, antimony has a range of industrial uses in batteries, chemicals, ceramics and glass but by far the most important is in fire retardants. The economically most important ore mineral and principal source for the production of antimony is stibnite.

Definitions and characteristics

Antimony (chemical symbol Sb) is a lustrous silvery white, brittle, crystalline semi-metal (or metalloid) that exhibits poor conductivity of electricity and heat. Antimony has a face-centred rhombohedral crystalline structure in the hexagonal system and forms elongated crystals with a perfect cleavage along (010). The melting point of antimony metal is very low (630.6 °C). Its common oxidation states are Sb^{5+}, Sb^{3+}, Sb^{0}, and Sb^{3-}, although other oxidation states have been recognised. Unlike typical metals, it is not malleable, but hard and brittle and can be crushed to a powder. Together with the other semi-metals (silicon, germanium, arsenic and tellurium), it is positioned in Period 5, Group 15 of the Periodic Table along the boundary between the metals and the non-metals. Selected key characteristics and physicochemical properties of antimony are listed in Table 4.1.

The geochemical properties of antimony are such that it tends to occur in nature with base metal Groups 11 (Cu, Ag), 14 (Sn, Pb), and with Groups 15 (Bi) and 16 (Se, Te) of the Periodic Table. In ore deposits it is closely associated with the precious metals, gold and silver, and in minerals, antimony usually occurs with sulfur in a trivalent state (Sb^{3+}). Natural antimony consists of a mixture of two stable isotopes that have the atomic weights 121 (57.25 per cent by weight of total) and 123 (42.75 per cent); in addition, 29 radioactive isotopes with mass numbers 104 to 136 are known.

Critical Metals Handbook, First Edition. Edited by Gus Gunn.

Table 4.1 Selected properties of antimony.

Property	Value	Units
Symbol	Sb	
Atomic number	51	
Atomic weight	121.75	
Density at 25 °C	6692	kg/m^3
Hardness (Mohs scale)	3.0	
Melting point	631	°C
Boiling point	1587	°C
Specific heat capacity at 25 °C	0.21	J/(g °C)
Electronegativity (Pauling scale)	2.05	
Electrical potential	0.21	V
Electrical resistivity at 25 °C	0.40	μΩ m
Thermal conductivity	25	W/(m °C)
Latent heat of fusion	163	kJ/kg
Latent heat of vaporisation	1602	kJ/kg

Abundance in the Earth

The abundance of antimony in the Earth's crust is estimated to be about 0.2 ppm for the continental crust, slightly less than arsenic and thallium. Concentrations in oceanic basalts vary from 0.02 to 0.8 ppm. Antimony is a moderately siderophile element which behaves like moderately incompatible and lithophile elements (such as the light rare earth elements) during magmatic processes. The volatile behaviour of antimony during subduction processes and crust formation is similar to that of lead. The meteoritic abundance of antimony is 0.142 ppm for chondrite (CI), while seawater contains 0.15 ppb. Geochemically, antimony is related to mercury and arsenic. In ore-forming processes, antimony is highly volatile and is chalcophile in character. An enrichment of about 150,000 times its crustal content is required to reach potentially economic concentrations of about three weight per cent, or 30 kg of antimony per tonne of antimony ore.

Mineralogy

There is a wide variety of antimony minerals in natural systems. More than 264 different antimony-bearing mineral phases have been defined so far.

Table 4.2 Antimony minerals with antimony concentrations greater than 70% Sb and additional antimony minerals (in bold) which typically occur in antimony-bearing ore deposits.

Mineral	Formula	% Sb
Antimony	Sb	100.00
Senarmontite	Sb_2O_3	88.39
Valentinite	Sb_2O_3	83.53
Nisbite	$NiSb_2$	80.58
Onoratoite	$Sb_8O_{11}Cl_2$	79.78
Cervantite	$Sb^{3+}Sb^{5+}O_4$	79.19
Stibiconite	$Sb^{3+}Sb^{5+}{}_2O_6(OH)$	76.37
Sarabauite	$CaSb_{10}O_{10}S_6$	75.62
Kermesite	Sb_2S_2O	75.24
Coquandite	$Sb_6O_8(SO_4)(H_2O)$	75.11
Stibnite	Sb_2S_3	71.68
Breithauptite	NiSb	67.47
Stibarsen	SbAs	61.91
Gudmundite	FeSbS	58.07
Ullmannite	NiSbS	57.29
Berthierite	$FeSb_2S_4$	56.94
Aurostibite	$AuSb_2$	55.28
Chalcostibite	$CuSbS_2$	48.81
Jamesonite	$Pb_4FeSb_6S_{14}$	35.39
Tetrahedrite	$(Cu,Fe)_{12}Sb_4S_{13}$	29.64
Famatinite	Cu_3SbS_4	27.63
Dyscrasite	Ag_3Sb	27.34
Boulangerite	$Pb_5Sb_4S_{11}$	26.44
Bournonite	$PbCuSbS_3$	24.91
Pyrargyrite	Ag_3SbS_3	22.48
Freibergite	$(Ag,Cu,Fe)_{12}(Sb,As)_4S_{13}$	18.93
Stephanite	Ag_5SbS_4	15.42
Vinciennite	$Cu_{10}Fe_4Sn(As,Sb)S_{16}$	3.83

Antimony substitutes for bismuth, lead, arsenic and sulfur in a variety of ore minerals and tends to concentrate in sulfide ores along with copper, lead and silver. Antimony rarely occurs in native form and the most important antimony minerals in economic deposits are the sulfide stibnite, various complex sulfosalts (berthierite, boulangerite, stephanite, jamesonite, bournonite, tetrahedrite, freibergite, gudmundite, ullmannite), antimonides (dyscrasite, breithauptite), and oxides (kermesite and valentinite/senarmontite) (Table 4.2).

Stibnite is the principal source of mined antimony and, although it is widely distributed, ore-grade concentrations are not common. In major antimony deposits like the Chinese Xikuangshan deposit in central Hunan, and the Antimony Line, Murchison belt, South Africa, stibnite is the dominant ore mineral (Figure 4.1). Complex sulfosalts occur as trace minerals in a wide variety of ore deposits. The mineralogy of antimony-bearing deposits is often characterised by complex intergrowths containing significant concentrations of metals like copper, lead, silver, bismuth, selenium, arsenic and mercury. The close mineralogical association with silver and the affinity for gold deposits makes antimony a useful indicator for the precious metal potential of some ore-deposit types. Where stibnite has been exposed at the surface, it is commonly weathered to an oxide phase such as bindheimite, kermesite, stibiconite or valentinite/senarmontite.

Major deposit classes

Antimony occurs in several different types of ore deposits of all ages, from the presently forming fumarole precipitates of active volcanoes, for example in New Zealand, to deposits in the Archaean volcanic strata of greenstone belts, for example in Canada and South Africa (Table 4.3; Figure 4.2). However, ore-grade antimony concentrations are not common and economically exploitable deposits of stibnite are generally small and discontinuous.

Most antimony deposits are of hydrothermal origin (Obolensky et al., 2007). Three main antimony deposit types can be distinguished, based on fluid generation and metal source: (i) low-temperature hydrothermal (epithermal) origin in shallow crustal environments associated with magmatic fluids; (ii) metamorphogenic hydrothermal origin in consolidated crustal environments derived from crustal fluids, triggered by, and with contributions from, magmatic heat and expelled fluids; and (iii) reduced intrusion-related gold systems. Antimony deposits may also be distinguished by their metal and mineralogical composition: (i) simple stibnite (plus gold) deposits; and (ii) complex polymetallic deposits with variable contents of elements of the 'epithermal suite' including gold, silver, tellurium, selenium, mercury, arsenic, antimony and thallium, and, locally, base metals (copper, lead, bismuth, zinc).

Antimony deposits are commonly associated with active continental plate margins and orogenic belts with steep geothermal gradients due to enhanced magmatic activity. Collision tectonics, mobilisation from the subducting slab, magmatic recycling, and metallogenetic processes within the continental crust or continental fragments are most important. The majority of antimony-bearing ore deposits is associated with the subduction-related western Pacific plate boundaries, especially in east and south-east Asia. Wu (1993) defined three so-called 'antimony belts' in China. A second antimony province can be delineated at the Nazca-South American plate boundary in Bolivia and Peru, which includes more than 500 antimony and antimony–gold deposits, and which continues along the western plate margins of North America. Other provinces are related to different metallogenic epochs in central Europe covering the Hercynian and Alpine belts.

The highest antimony concentrations commonly occur in low-temperature magmatic-hydrothermal systems in the epithermal environment. Antimony is typically enriched in the distal portions of these systems at shallow depths and close to the surface. It is closely associated with hydrothermal silica and carbon dioxide, and the common wall-rock is carbonate, either as sedimentary limestone or as a prominent hydrothermal alteration mineral phase.

Antimony-bearing deposits are generally associated with calc-alkaline to peralkaline, porphyritic felsic to intermediate volcanic and intrusive wall rocks, often in a volcanic cauldron setting. The original host rocks are variably affected by hydrothermal alteration processes such as silicification, carbonatisation, sericitisation, chloritisation and greisenisation. Volcanic host rocks are typically submarine, subaerial, and pyroclastic deposits of rhyodacite, dacite and andesite.

Figure 4.1 (a) Stibnite ore in association with carbonate-quartz alteration (white), Beta mine, Antimony line, South Africa; (b) Photomicrograph of stibnite (stb)- cinnabar (cn) intergrowth in quartz-carbonate matrix, Monarch mine, South Africa; (c) Same as B, crossed nicols; (d) Needles of stibnite, Antimony line, South Africa; (e) Sample of stibnite-gold ore, associated with coarse-grained quartz-carbonate alteration (white). Native gold is associated with the contact between stibnite ore and quartz-carbonate.

Table 4.3 Size and grade of antimony deposits (grades and tonnages are very variable between deposits and figures given are indicative only). (Source: BGR database.)

Deposit type	Deposit size range (tonnes)	Typical grade (Sb_2S_3%)	Estimated antimony metal content of known deposits (tonnes)	Per cent of total
Gold-antimony (epithermal) vein-type	10^4–10^6	0.1–3.5	580,000	20
Carbonate replacement	10^6–10^8	1.5–25	2,500,000	60
Reduced-magmatic	10^6–10^8	0.1–1.5	320,000	10
Polymetallic base metal vein	10^4–10^6	0.1–0.5	175,000	8
Hot springs	10^4–10^6	0.1–0.2	2500	2
			3,577,500	**100**

Subvolcanic to hypabyssal intrusive host rocks comprise granite–granodiorite–quartz–monzonite differentiates. Intrusion-related antimony deposits (skarns, replacement ores, vein-type) that form within the region of hydrothermal influence surrounding the magmatic source are related to plutons of diverse types with characteristics of I-, S-, and A-type granitoids (Hart, 2007).

The key characteristics and examples of the major types of antimony deposits are given in Table 4.4.

Gold–antimony (epithermal) deposits

Antimony belongs to a group of metals and metalloids, usually referred to as the 'epithermal' suite of elements, which includes gold, silver, tellurium, selenium, mercury, arsenic, antimony and thallium (Taylor, 2007). The geochemical association of these elements is a common and generic characteristic of the epithermal deposit group.

Epithermal deposits typically form within 1.5 km of the Earth's surface. They originate at active subduction zones and are closely associated with centres of magmatism, typically linked to the shallow crustal emplacement of magmatic porphyry copper systems. They comprise veins and disseminations in volcanic, sedimentary and metamorphic rocks. The deposits may be found in association with hot springs and frequently occur at the centres of young volcanism.

Three main sub-types of epithermal deposits are distinguished on the basis of alteration and ore mineral assemblages: the quartz-(kaolinite)-alunite or high-sulfidation sub-type; the intermediate-sulfidation sub-type; and the adularia-sericite or low-sulfidation sub-type (Hedenquist et al., 2000). Each sub-type is characterised by hydrothermal fluid of a particular oxidation state and acidity. The close association of antimony deposits with carbonate sequences and intense quartz-carbonate alteration indicate moderate acidity for the mineralising fluids. Temperatures of formation of epithermal deposits range from about 100 °C for hot spring or steam-heated deposits to about 350–400 °C for deeper vein and replacement deposits. Ore deposition is related to pronounced changes in the physical, thermal and chemical properties of the hydrothermal solutions which may occur over short distances. The principal commodities in epithermal deposits are precious metals (gold, silver), although co-enrichment in elements of the epithermal suite, as well as locally copper, lead, bismuth and zinc, gives rise to polymetallic deposits and provides potential for by-products, especially in deposits with high silver grades.

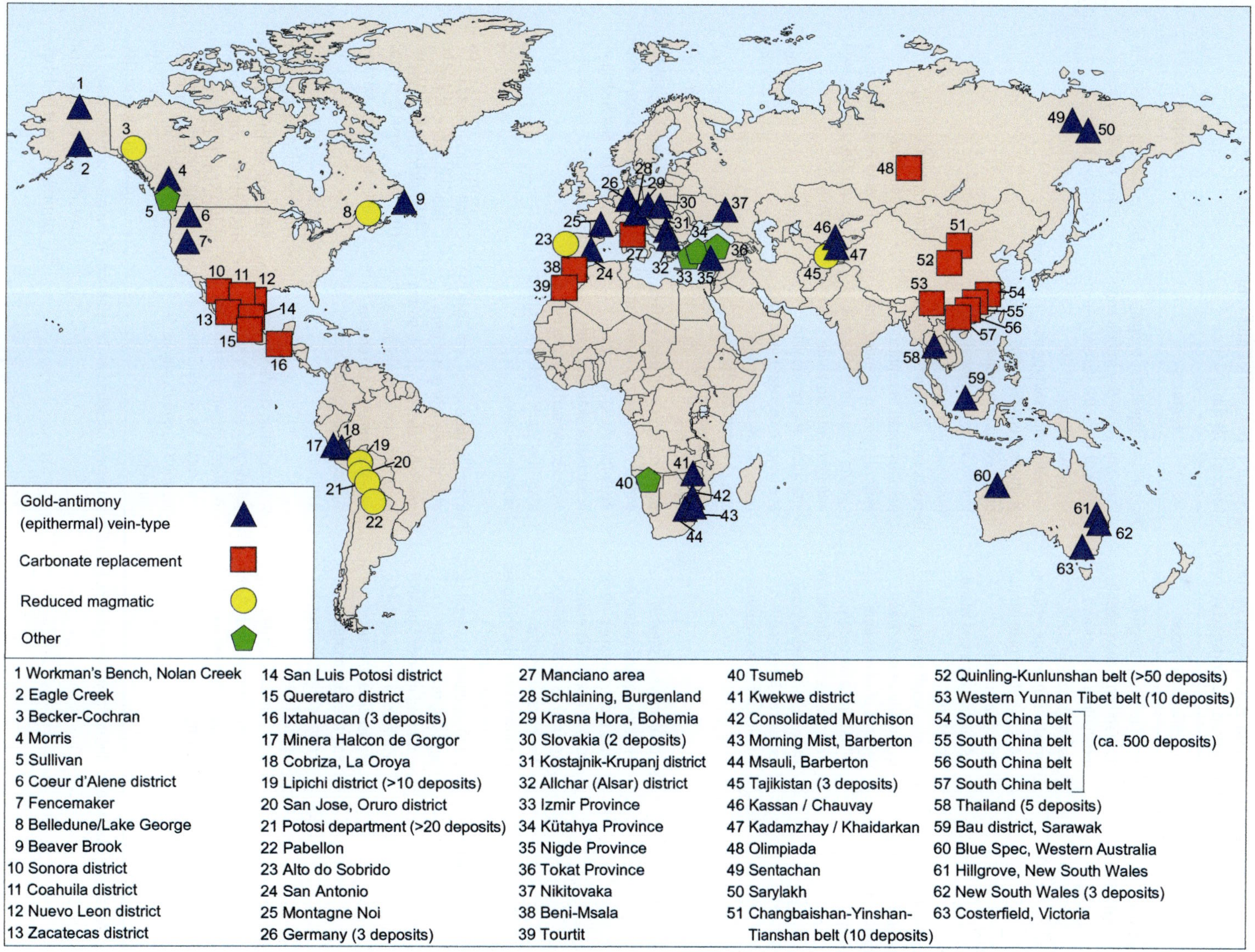

Figure 4.2 The global distribution of antimony mines, deposits and major occurrences. (Polymetallic base-metal vein deposits are not shown separately.) Some of the symbols on the map represent a single important deposit or resource, while others represent a cluster of deposits in one area or region. Antimony is also known to occur in Algeria, Bosnia and Herzegovina, Brazil, Burma (Myanmar), Ecuador, Greece, Honduras, Japan, Kazakhstan, New Zealand and Pakistan.

Table 4.4 Key characteristics and examples of antimony deposit types (grades and tonnages are very variable between deposits and figures given are indicative only). (Source: BGR database.)

Deposit type	Brief description	Features	Examples
Gold-antimony (epithermal) vein	Medium to large, low-grade stockwork-type quartz-stibnite ± tetrahedrite veinlets and disseminations in shale, limestone, quartzite, volcanic rocks, granite in greenstone belts of a potential subduction zone and island arc setting.	Deposits lack significant copper, lead, zinc and nickel sulfide and sulfosalt minerals; Sb-As-Hg-Au-Ag-Te assemblage of the low-sulfidation subtype in epithermal environments; veins of quartz cored by massive stibnite; transitions to mesozonal (orogenic) deposits.	Hemlo (Canada), Yanacocha (Peru), El Indio (Chile), Goldfield (USA).
Greenstone-hosted quartz-carbonate vein and carbonate replacement	Numerous small to large size, high-grade vein-stockwork deposits of almost pure stibnite in sedimentary, metasedimentary (limestone) and highly altered volcanic sequences. Syn- to post-collisional tectonic settings.	Deposits lack any significant base or precious metal assemblages; lenticular bodies of quartz and stibnite within limestone and intense quartz-carbonate alteration, at contacts with overlying shale, near high-angle faults; silicification extends tens of metres into the host metasedimentary or volcanic units.	Xikuangshan (China), Kadamdzhay (Russia), Antimony line (South Africa), Olympiada (Russia).
Reduced-magmatic	Regional arrays of sheeted auriferous quartz-carbonate veins around source plutons; small to intermediate size proximal As-Sb-Au veinlets. Weak post-collisional extension behind a thickened continental margin.	Deposits occur in a mineral system with an outward zonation of an Au-Bi-Te-W-As-Sb-Ag-Pb-Zn assemblage; skarn-like and replacement bodies and veins; associated with volatile-rich quartz monzonite melts.	Tintina gold province (USA, Canada), Timbarra, Kidston, (Australia), Niuxinshan (China).
Polymetallic base metal vein	Small to medium polymetallic deposits; structurally controlled in post-collisional vein breccia zones of clastic metasedimentary or magmatic-dominated terranes; mineralisation by basinal brines.	Polymetallic base metal and Ag-rich ores; densely intergrown ore minerals (telescoping); Sb hosted by stibnite, tetrahedrite and a variety of simple and complex sulfosalts; quartz-carbonate gangue.	Cobalt district (Canada), Bolivian antimony belt, Bau district (Malaysia).
Hot spring exhalative	Siliceous precipitates deposited by hydrothermal fluids, hot springs, fumaroles; volcanic activity.	Low temperature fluids forming sulfide and sulfosalt segregations in altered intermediate to felsic wall rocks; Sb co-enrichment with As and Hg, and locally Te, Se, Au, Ag.	Kudryavyi, Kuril islands (Russia), Merapi (Indonesia), Taupo volcanic zone (New Zealand).

Ag, silver; As, arsenic; Au, gold; Bi, bismuth; Hg, mercury; Pb, lead; Sb, antimony; Se, selenium Te, tellurium; W, tungsten; Zn, zinc .

Epithermal-type antimony deposits include stratabound mantos (e.g. Wadley, Sierra de Catorce antimony district, Mexico), vein deposits controlled by fault zones (e.g. Bolivian antimony belt), fault-controlled veins and limestone replacements (e.g. Bau district, Sarawak, Malaysia) and subordinate Carlin-type gold (arsenic, mercury, antimony) deposits (e.g. Getchell, Turqoise Ridge, Twin Creeks, Nevada; Zarshuran, Iran; Alšar, Macedonia). Hot spring deposits may have significant antimony enrichments (e.g. New Zealand, California, Nevada). Some deposits may be transitional to those related to Cordilleran granitoids. These include mesothermal lead–zinc–silver (antimony) veins (e.g. Coeur d'Alene district, Idaho; Bawdwin, North Shan State, Myanmar), hydrothermal antimony deposits in, and associated with, granitoids (e.g. Southern Bolivia, Sierra de Catorce, Mexico), and deposits in fault zones with scheelite-gold in silicate rocks (e.g. Yellow, Idaho).

Antimony deposits in Bolivia are typically epigenetic vein-type deposits crosscutting early Paleozoic fine-grained clastic rocks. The formation age is considered to be younger than early Cretaceous (Dill et al., 1995). They are dominated by stibnite, accompanied by pyrite, arsenopyrite and minor jamesonite, berthierite and base metal sulfides. Mexico hosts a number of epithermal-style mono- and polymetallic antimony deposits (Simmons et al., 2005). The vast majority occurs as vein-type and replacements in bedded carbonate rocks. The regional association with sub-volcanic rocks and co-enrichment with other metals and metalloids of the epithermal suite are indicative of a magmatic-hydrothermal origin. Stibnite predominates in stratabound mantos replacing limestone beds; gold may be a potential by-product.

Hot-spring deposits enriched in the epithermal element suite comprising silica sinters and siliceous muds are found at several locations, e.g. in the Coromandel and Taupo volcanic zones, New Zealand and at Monte Amiata, Italy. Hot-spring deposits also form gold-stibnite ores in veins and disseminations in sinters e.g. at Sulfur Bank and McLaughlin, California and Round Mountain, Nevada.

Greenstone-hosted quartz-carbonate vein and carbonate replacement deposits

Gold-quartz-carbonate vein deposits occur in deformed greenstone belts. Common lithological associations include tholeiitic basalts and ultramafic komatiite flows intruded by intermediate to felsic porphyry intrusions, accompanied by swarms of albitites in some areas (Dubé and Gosselin, 2007). They are distributed along major compressional to transtensional crustal-scale fault zones in deformed greenstone terranes, and mark the convergent margins between major lithological boundaries, such as volcano-plutonic and sedimentary domains. Large greenstone-hosted quartz-carbonate vein deposits are commonly spatially associated with fluvio-alluvial conglomerate distributed along major crustal fault zones. This association suggests an empirical time and space relationship between large-scale deposits and regional unconformities.

Although these deposits occur in Proterozoic and Palaeozoic terranes they are more abundant in Archaean terranes where their gold and antimony contents are greatest. They are epigenetic deposits characterised by networks of gold-bearing, laminated quartz-carbonate veins (Dubé and Gosselin, 2007). They are hosted by greenschist- to amphibolite-facies metamorphic rocks of dominantly mafic composition and formed at intermediate depth (5–10 km). The mineralisation is syn- to late-deformation and typically post- to syn-peak metamorphism. The deposits are typically associated with iron-magnesium-carbonate alteration. The gold is largely confined to the quartz-carbonate vein network but may also be present in significant amounts within iron-rich sulfidised wall-rock selvages or within silicified and arsenopyrite- and stibnite-rich replacement zones.

Examples of synorogenic hydrothermal antimony–gold deposits occur in Archaean greenstone terrains in Southern Africa (e.g. the Antimony Line, Murchison, South Africa; and the Kadoma and Kwekwe goldfields, Zimbabwe; Vearncombe et al., 1992; Buchholz et al., 2007). In the Murchison and the Barberton greenstone belts of South Africa the synorogenic mineralisation is sporadically developed in deformed thick quartz-carbonate alteration zones within highly deformed and modified komatiitic basalts. The antimony mineralisation consists of stibnite and minor berthierite. Gold is an important by-product of antimony mining in the so-called Antimony Line of the Murchison greenstone belt.

The largest antimony deposits in China are epigenetic carbonate-hosted stibnite deposits located in central Hunan, China (Wu, 1993). The replacement orebodies occur within Middle and Upper Devonian limestones which were subject to hydrothermal alteration in late Jurassic to early Cretaceous (Penga et al., 2003). The deposits are several metres thick, usually stratiform, but locally irregular in shape. Stibnite, the only ore mineral, is associated with intense silicification and carbonatisation.

Synorogenic antimony deposits may grade into more complex antimony–gold, antimony–arsenic–gold and antimony–tungsten deposits and their combinations (e.g. Xikuangshan, Hunan, China; Wadley, Sierra de Catorce, Mexico; Kadamzhai, Kyrgyzstan; Turhal, Turkey; Hillgrove, New South Wales, Australia). Mesothermal gold, transitional to synorogenic gold deposits, have moderate significance for the antimony production (e.g. Olimpiada, Russia). Hydrothermal deposits in intracratonic orogens may also be associated with granites and include multi-metal zoned tin, copper, lead, zinc, silver (antimony) skarn-replacement-vein systems (e.g. Dachang ore field, Guangxi, Guangdong). Synorogenic mercury–antimony deposits are known from the Khaidarkan, South Ferghana mercury–antimony belt in Kyrgyzstan and share characteristics with continental epithermal-type deposits (Obolensky et al., 2007).

Reduced magmatic gold systems

Reduced magmatic or intrusion-related gold–bismuth–tellurium–tungsten–antimony deposits are related to the brittle carapace at the top of small metaluminous, moderately reduced and fractionated post-collisional quartz monzonite plutons. The magmas have a reduced primary oxidation state that form ilmenite-series plutons and have characteristics of I-, S-, and A-type granitoids. The deposits typically form within the region of hydrothermal activity around the causative pluton. The systems include a wide variety of intrusion-related mineral deposit styles including skarns, replacements, veins and disseminations and they originate from a depth of five to seven km (Hart, 2007). The mineral systems show a pronounced zonation with proximal gold-tungsten-arsenic, intermediate arsenic–antimony–gold and distal silver–lead–zinc metal associations. Antimony-rich deposits typically occur as fault-filled veins. Stibnite predominates, but more complex sulfosalts, in association with bismuth, lead and tellurium, are usually present.

Type-examples of reduced intrusion-related gold systems are known in the Fairbanks area of central Alaska (e.g. Fort Knox deposit) and central Yukon (e.g. Tombstone gold belt; Hart, 2007). The deposits of the wider Tombstone gold belt across central Yukon and Alaska are hosted by, and formed from, reduced mid-Cretaceous plutons. Other examples are known from southern New Brunswick, the Bolivian polymetallic belt and the Yanshanian orogen of the North China craton (Thompson et al., 1999). Archaean examples of similar type have been suggested to occur in the southern Superior Province of Ontario and Quebec (Robert, 2001). Reduced intrusion-related gold–antimony systems share a number of characteristics with orogenic gold deposits as well as with epithermal-porphyry ore-forming systems. Critical features include the association with reduced intrusions, the lack of copper and the presence of tungsten.

Extraction methods and processing

Mining

Of the 18 countries that have produced primary antimony in the past decade, the most important have been, in order of decreasing production, China, Bolivia, South Africa, Russia, Tajikistan, Turkey and Australia. Antimony mining in those countries is dominated by one or a few stibnite-rich deposits, with possible by-products including gold, silver, tungsten and mercury. Underground extraction is used in one of the world's major antimony-producing mines, the Consolidated Murchison Mine in South Africa, which uses variations of shrinkage stoping. On the surface, the waste is removed from the high-grade ore, and the latter is trucked to a mill for the separation and concentration of gold and antimony. Where antimony is a minor constituent of a metalliferous ore the mining method, either open pit or underground, is designed to optimise the recovery of the principal metals, such as gold, lead or silver. In some countries (e.g. Bolivia, Peru, Zimbabwe) small mines work irregular and scattered antimony-bearing orebodies that cannot be readily exploited by large-scale mining methods. These mines are entered by a shallow shaft or short adit, developed by drifting in the vein, and stoped by simple overhand methods between raises driven on the footwall of the ore.

Ore processing, beneficiation and conversion to metal

Techniques for the processing of antimony-bearing ores encompass a wide range of techniques from traditional hand sorting, which is dependent on plentiful and cheap labour, to technologically advanced, capital-intensive mineral processing. The mineral processing stages generally include conventional crushing and grinding, followed by combined gravity concentration and flotation.

A typical stibnite deposit in China, such as Xikuangshan, contains stibnite, together with pyrite and associated gangue minerals such as quartz, calcite, barite, kaolinite and gypsum (Wu, 1993; Yang et al., 2006). The ore, which grades about 2.7% Sb, is treated by hand sorting, crushing, heavy medium separation and flotation. Hand sorting includes the recovery of antimony lump concentrates before and between the crushing stages. Crushing takes place in a two-stage closed circuit consisting of a jaw and cone crusher (Anderson, 2000). Gangue material is separated from the fine fraction using dense media and the heavy product is further prepared for flotation. A pure stibnite concentrate is produced following cleaning of the concentrate produced by flotation. Overall, 33 per cent of the ore is treated by hand sorting, seven per cent by heavy media separation and 60 per cent by flotation. Typical beneficiation routes, based upon the Xikuangshan South ore dressing plant and the Sunshine Mining and Refining Co. (closed in 2001) antimony process are shown in Figure 4.3.

At least six principal methods have been used to extract antimony from its ores. The chosen technique depends on the oxidation state (antimony sulfide, oxide or complex, mixed oxide–sulfide ore) and the ore grade. Stibnite ores can generally be processed more efficiently and at lower cost than oxide ores.

Roasting the ore to yield a volatile trioxide or the stable non-volatile trioxide is the only pyrometallurgical procedure suitable for low-grade ores (5 to 25 per cent antimony content). The control of volatilisation conditions produces a high-grade oxide that can be sold directly to consumers. Intermediate grade ores, containing 25 to 45 per cent antimony, are smelted at high temperatures in blast furnaces to produce crude antimony metal. Oxides, sulfides or mixed ores, residues, mattes, slags and briquetted fines or flue dusts can be used as blast-furnace charges. The method employs a high smelting column and comparatively low air pressure, and slag and metal are separated in a forehearth. Considerable quantities of slag are formed, which is desired because it tends to reduce volatilisation losses.

High-grade ores that contain 45 to 60 per cent antimony are melted in a crucible or reverberatory furnace with a reducing atmosphere to prevent oxidation and loss by volatilisation. On cooling, the solidified product is referred to as 'crude antimony', 'crudum', or 'needle antimony'. Alternatively, high-grade sulfide ores can be treated by a precipitation process in which scrap iron is used to reduce sulfur and yield impure antimony metal.

Oxide ores that contain about 30 per cent antimony are reduced in blast furnaces to crude

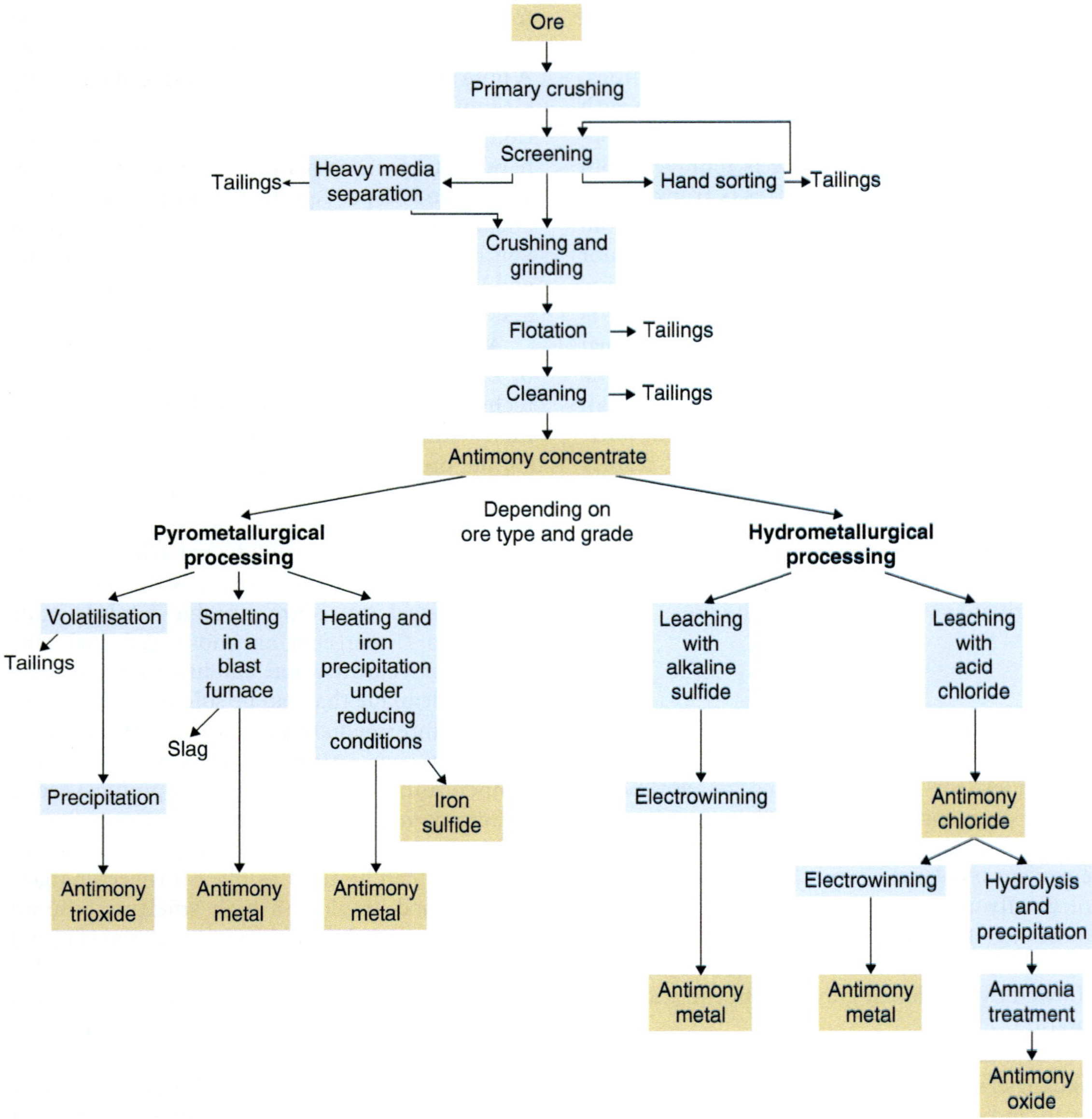

Figure 4.3 Generalised beneficiation flow diagrams, based on various ore dressing and beneficiation processes including the Xikuangshan South ore dressing plant, the Sunshine antimony process and the U.S. Antimony Corp. process. (Modified from Anderson (2000) and U.S. Antimony Corporation (2011).)

metal, while richer oxide ores, with about 50 per cent antimony, are reduced and refined to commercial-grade metal in reverberatory furnaces using coke and a suitable flux, such as soda ash.

Mixed sulfide and oxide ores are usually smelted in blast furnaces, although they are sometimes processed by selective leaching followed by electrolysis of the leachate to recover the metal (Butterman and Carlin, 2004). A typical process uses an alkali hydroxide or sulfide as the solvent. The filtered leach solution, which contains sodium thioantimonate, is electrolysed in a diaphragm cell using an iron or lead anode and an iron or mild-steel cathode. The cathode metal obtained ranges from 93 to 99 per cent antimony metal.

The metal produced by pyrometallurgical processes does not meet the quality requirement for commercial products and must be further refined. The major impurities are usually arsenic, copper, iron, lead and sulfur. The iron and copper concentrations may be lowered by treating the metal with stibnite or a mixture of sodium sulfate and charcoal to form an iron-bearing matte, which is skimmed from the surface of the molten metal. The metal is then treated with an oxidising flux that consists of caustic soda or sodium carbonate and sodium nitrate to remove the arsenic and sulfur. Lead cannot be readily removed from antimony, but material high in lead may be used in the production of antimony-bearing lead-base alloys.

Antimony metal is also recovered from slags and residues produced during the processing of other metals, such as lead and gold. In the cyanidation of gold ores stibnite is oxidised to various species which inhibit gold dissolution by consuming oxygen and cyanide.

For primary production, the antimony content of the ore has traditionally determined the pyrometallurgical recovery. The lowest grades of antimony sulfide ores containing 5–25 per cent antimony are volatilised to antimony trioxide, 25–40 per cent antimony ores are smelted in a blast furnace, and 45–60 per cent antimony ores are treated by liquation or iron precipitation (Anderson, 2000). The removal of antimony as the volatilized trioxide is the only pyrometallurgical method suitable for low grade ores. The combustion of the sulfide components of the ore supplies some of the energy and fuel requirements are minor. The volatilisation process includes sulfur combustion at about 1000 °C with coke or charcoal and the recovery of volatile antimony trioxide in flues, condensing pipes, a baghouse, a Cottrell precipitator or any combination of the above. Antimony sulfide may be inefficiently separated from the gangue of sulfide ore. An efficient method uses a reverberatory furnace and continuous liquation under reducing conditions. The oxide volatilisation process to recover additional antimony is used to treat residues containing 12–30 per cent antimony. The liquated product is sold for applications requiring antimony sulfide or is converted into metallic antimony. Antimony oxides are reduced to metal with charcoal in reverberatory furnaces at about 1200 °C. An alkaline flux consisting of soda, potash and sodium sulfate is commonly used to minimise volatilisation and dissolve residual sulfides and gangue. The loss of antimony from the charge by volatilisation is usually high (12–20 per cent) and lowered by the use of Cottrell precipitators or baghouses. Rich sulfide ore or liquated antimony sulfide (crude antimony) is reduced to metal by iron precipitation. This process includes the heating of molten antimony sulfide in crucibles with fine iron scrap. A light fusible matte with iron sulfide is formed to facilitate separation of the metal. A second fusion with liquated antimony sulfide follows for purification. Intermediate grades of antimony ores, liquation residues, mattes, slags and flue dusts are processed in blast furnaces at 1300 to 1400 °C in a process similar to that used for the recovery of lead. It is the favoured method of smelting of these ores to minimise volatilisation losses.

Hydrometallurgical methods are employed for simple as well as complex antimony ores. A two-stage process of leaching and subsequent electrodeposition is generally involved. Alkaline sulfide and the acidic chloride are used as solvent systems. The former predominates and is employed in the CIS, China and the USA (Anderson, 2000). The lixiviant is a mixture of sodium sulfide and sodium hydroxide which form a sodium thioantimonite (Na_3SbS_3) when applied to stibnite. The dissolution of elemental sulfur in sodium hydroxide is also used as a lixiviant for alkaline sulfide leaching. The electrodeposition of the antimony from the alkaline sulfide solution to cathode metal is carried out via electrowinning in diaphragm or non-diaphragm cells. The antimony metal product may attain a grade of 99.5%. The acidic chloride hydrometallurgy uses hydrochloric acid in conjunction with iron chloride ($FeCl_3$) to produce antimony chloride ($SbCl_3$). The iron chloride acts both as an oxidiser and as a chloridising agent. The dissolved antimony chloride can be electrowon from solution in diaphragm cells to produce cathode antimony metal. Alternatively, the antimony chloride solution can be treated by hydrolysis precipitation of the antimony from solution as a solid oxychloride (SbOCl, $Sb_4O_5Cl_2$). The precipitated solid is treated with ammonia to produce antimony oxide.

Specifications

The main antimony products in international trade are stibnite and subordinate stibnite-berthierite and tetrahedrite ores and concentrates, antimony metal, antimony trioxide and antimonial lead. Antimony ore concentrates contain 5 to 60% Sb. Chemical grade ores are sufficiently pure to be used directly in the production of antimony trioxide, antimony chloride or other compounds. The total impurity level must not exceed 0.25%, with arsenic and lead concentrations less than 0.1% (Anderson, 2000; Roskill, 2007). In the USA two grades of antimony metal are specified (Butterman and Carlin, 2004): grade A has a minimum content of 99.80% Sb, with maximum values of arsenic (0.05%), sulfur (0.10%) lead (0.15%) and other elements (0.05% each); and grade B metal must have a minimum content of 99.5% Sb, with maximum values of 0.1% arsenic, sulfur and other elements, and 0.2% lead.

Antimony metal is commonly traded in ingots and slabs weighing between 20 and 50 pounds and also in the form of granules, cast cake, powder, shot and single crystals. The Minor Metals Trade Association (MMTA) grade I specification for antimony metal used for the production of antimony trioxide is minimum 99.65% Sb, and maximum 0.15% As, 0.005% Se, 0.01% Bi, 0.02% Pb and 0.02% Fe (MMTA, 2012). High-purity antimony metal, used in thermoelectric devices and semiconductors, is produced in the form of ingots weighing 0.5 to 3.0 kg and grading 99.99% Sb (referred to as 4 N) to 99.99999% (7 N).

A number of antimony compounds are also traded on the international markets. These include antimony trioxide (Sb_2O_3; 83.5% Sb), antimony oxychloride (SbOCl; 70.3% Sb), antimony pentoxide (Sb_2O_5, 75.3% Sb), antimony trichloride ($SbCl_3$; 53.4% Sb), antimony trisulfide (stibnite Sb_2S_3; 71.1% Sb), sodium antimonite ($NaSbO_3$, 63.2% Sb) and stibine (SbH_3; 97.6% Sb).

Antimony trioxide, commonly referred to as ATO, is the most widely used antimony compound. Crude antimony trioxide grades below 98% Sb_2O_3, while commercial grades contain 99.2 to 99.9% Sb_2O_3. Several commercial specifications are available, each characterised by specific tinting strengths and/or the content of particular impurities such as arsenic, iron and lead (Amspec, 2011).

Uses

Various unique properties of antimony determine its use in a diverse range of products and applications. These properties include:

- low melting point enhancing workability at low temperatures;
- stability in air at room temperature and in water up to 250 °C;
- resistance to most cold acids;
- dissolution in some hot acids and in aqua regia;
- incompatibility with strong oxidising agents, chlorine, fluorine;
- reaction with materials that do not react with hydrochloric and nitric acids separately;
- metastability at rapid cooling providing exothermic reaction;
- high density (6692 kg/m^3), but weak bonding leading to low hardness and brittleness;
- two allotropic forms of antimony: stable metallic and amorphous grey;
- low electrical and thermal conductivity;
- expansion on freezing, like silicon, bismuth, gallium and germanium.

Table 4.5 Estimated global consumption of antimony by end-use in 2000 and 2011. (Data from Roskill, 2012.)

	2000		2011			Consolidated annual growth rate
Non-metallurgical uses	*Tonnes antimony*	*% of total*	*Tonnes antimony*	*% of total*	*% of non-metallurgical use*	%
Flame retardants	70,000	47.4	108,250	52.4	83.8	4.0
Plastic catalysts	6000	4.1	12,100	5.9	9.4	6.6
Heat stabiliser	1400	0.9	2700	1.3	2.1	6.1
Glass	16,000	10.8	1650	0.8	1.3	−18.6
Ceramics	1700	1.2	2550	1.2	2.0	3.8
Other	1500	1.0	1900	0.9	1.5	2.2
Sub-total	*96,600*	*65.4*	*129,150*	*62.5*	*100*	*2.7*
Metallurgical uses	*Tonnes antimony*	*% of total*	*Tonnes antimony*	*% of total*	*% of metallurgical use*	
Lead-acid batteries	40,000	27.1	53,600	25.9	69.2	2.7
Lead alloys	11,000	7.5	23,850	11.5	30.8	2.7
Sub-total	*51,000*	*34.6*	*77,450*	*37.5*	*100*	*3.9*
Total	**147,600**	**100**	**206,600**	**100**		**2.7**

Antimony is consumed in the following forms:

- antimony trioxide, used mainly in flame retardants and in PET (polyethylene terephthalate);
- sodium antimonite, used mainly in cathode ray tube glass;
- primary metal, used mainly in lead-acid batteries;
- antimonial lead, mainly recycled from and re-used in lead-acid batteries.

The early technical use for antimony was related to the development of cast metal printing types, mirrors, bell metal and pigments. The main antimony-producing countries in the 18th century were France, Germany and Italy. By the early 19th century the major uses of antimony were in pharmacology, agriculture, artillery, dyeing, pigments and paints for colouring glass, ceramics, cloth and paper, printing type, bearing and anti-friction alloy metal, vulcanising rubber, the manufacture of safety matches and in thermoelectric couples. Mine production increased sharply during World War I as shrapnel was hardened with 10 to 13 weight per cent antimony. In World War II antimony was used in the lead-acid batteries of military vehicles and in flame-retardant compounds for heavy textiles. Antimony trisulfide was used in artillery projectiles, in bomb fuses and for the generation of white clouds on detonation. Antimony alloyed with lead was used as long as the nineteenth century as bearing metal (Babbitt metal) and, alloyed with tin, to produce Britannia metal used in items such as eating utensils, teapots and candlesticks. Various antimony salts have long been used for medical and veterinary purposes. Today, antimony compounds are still used for the treatment of two parasitic diseases, schistosomiasis and leishmaniasis. The rapid growth in the use of plastics after 1950 led to today's dominant market for antimony-based flame retardants in a wide range of products. Pigments (glass and ceramic industries), lead-acid batteries and metal alloys together account for about 30 per cent of current antimony use.

The most important end uses of antimony in 2000 and 2011 are summarised in Table 4.5. In that period total antimony consumption grew by about 40 per cent from 147,600 tonnes to 206,600 tonnes. In fire retardants, the largest application, antimony use grew by 54 per cent in the same period, while use in catalysts for PET production more than doubled.

Table 4.6 Estimated world consumption of antimony by main product in 2001, 2006 and 2011. (Data from Roskill, 2012.)

	2001		2006		2011		2011
	Tonnes antimony	%	Tonnes antimony	%	Tonnes antimony	%	% of primary
Primary Product							
Antimony trioxide[1]	72,600	53.3	103,950	57.9	124,950	60.5	74.1
Sodium antimonate	15,600	11.5	12,150	6.8	4250	2.1	2.5
Antimony metal	16,000	11.7	23,000	12.8	39,450	19.1	23.4
Sub-total	*104,200*	*76.5*	*139,100*	*77.5*	*168,600*	*81.7*	*100.0*
Secondary Product							
Antimonial lead[2]	*32,000*	*23.5*	*40,500*	*22.5*	*38,000*	*18.3*	
Total	**136,200**	**100**	**179,600**	**100**	**206,600**	**100**	

[1]includes other antimony compounds such as tetroxide, pentoxide, oxychloride
[2]antimony content

Antimony trioxide

Antimony trioxide, ATO, is the most important antimony compound produced. Global consumption of antimony in ATO is estimated at almost 125,000 tonnes (equivalent to 150,000 tonnes ATO), which accounted for 74 per cent of 2011 primary antimony consumption (Table 4.6; Roskill, 2012). It is prepared by the volatilisation of antimony metal in an oxidising furnace. Of itself ATO has no flame-retardant property, but when it is combined with halogenated (brominated, chlorinated) flame-retardant compounds it provides the most effective and widely used flame-retardant system for plastics. The halogenated antimony compounds act as dehydrating agents and inhibit ignition and pyrolysis in the solid, liquid and gas phases. They also promote the formation of char on the substrate, which acts as a barrier and reduces oxygen availability and volatile-gas formation.

Flame retardants are by far the largest market for antimony today. The majority of flame retardants are used in plastics, with smaller amounts in rubber, textiles, paints, sealants and adhesives. The compounds account for 52 per cent of total antimony consumption and over 84 per cent of non-metallurgical antimony consumption (Table 4.6). The world demand for ATO in this market is estimated to have increased by a compound annual growth rate of four per cent, or 38,250 tonnes, between 2000 and 2011.

The other major markets for antimony in non-metallic uses are ATO as a heat stabiliser in PVC and as a catalyst in the production of polyethylene terephthalate (PET), which is used in the manufacture of synthetic textiles (polyester) and plastic containers, such as drinking bottles, and polyester film mostly used for packaging. ATO is also used in the glass and ceramics sector as a degassing agent and as an opacifier in porcelain enamels and pottery glazes.

Sodium antimonate

Sodium antimonate ($NaSb(OH)_6$) is mainly used as a fining and degassing agent in the production of high-quality clear glass. The antimonate decomposes in the molten glass generating large bubbles which rise to the surface scavenging much slower moving fine bubbles leading to the purification and homogenisation of the glass batch. Sodium antimonate is also a decolourant for glass as it removes traces of iron which can give rise to a greenish tint. It also has antisolarant properties which protect against colouring caused by sunlight or fluorescent lights during the lifetime of the glass.

Some sodium antimonate is also used in fire-retardant and smoke-suppressant compounds and formulations and as feed for metal and trioxide production. World consumption of sodium antimonate has steadily fallen from 15,600 tonnes in 2001, when it accounted for 11.5 per cent of total consumption, to 4250 tonnes in 2011, 2.1 per cent of total consumption (Table 4.6; Roskill, 2012).

Other non-metallurgical uses

Minor non-metallurgical markets include lubricants, ammunition primers, textiles, pharmaceuticals, pesticides, fluorescent lamps, fireworks and matches, zinc electrowinning and the refining of sour crude oil. Very high purity antimony (5 N to 7 N) is increasingly being used in the semiconductor industry as a dopant for ultra-high conductivity n-type silicon wafers which are utilised in diodes, infrared detectors and Hall-effect devices.

Antimony metal

The largest market for metallic antimony is in lead alloys adding hardness and smoothness of finish. The higher the proportion of antimony in the alloy, the harder and more brittle it will be. In lead-acid batteries, the addition of antimony improves the charging characteristics and reduces generation of unwanted hydrogen during charging. The property of some antimony alloys to expand on cooling is the basis of their use in some applications such as making typefaces for printing.

Other metallurgical applications of antimony metal are in solders, fusible alloys, type metals, lead weights, rolled and extruded antimony alloys, pewter, Britannia metal, shot and ammunition. It is also used to harden low-tin alloys, to reduce friction and wear in bearing alloys in machinery and to improve the durability of lead sheathing for cables.

Antimony metal is estimated to account for about 23 per cent of primary antimony consumption (Table 4.6). Primary consumption of antimony metal increased by 72 per cent between 2006 and 2011, from 23,000 tonnes to almost 40,000 tonnes (Roskill, 2012). Excluding its use in lead-acid batteries, alloys account for about 30 per cent of metallurgical consumption of antimony (Roskill, 2012).

Recycling

From most products, such as fire retardants, antimony compounds cannot be recycled because they are dissipated in use although recycling of PET containers makes an indirect contribution. However, antimony can be recovered from most applications where it is used as an additive in lead alloys. Secondary antimony is a significant source of supply in many countries and accounts for about 20 per cent of total antimony used and approximately 40,000 tonnes per year (Roskill, 2007).

Most secondary antimony metal is obtained from recycled lead-acid batteries which contain between 0.6 and 1.5% Sb. The recovery of antimony used in alloys for small-arms ammunition, semiconductors, bearings and solders cannot be recycled. The availability of secondary antimony depends almost entirely on the extent of secondary lead recovery and therefore on the market conditions for lead and lead battery scrap. It accounts for the entire US-sourced production and makes an important contribution in the EU, Canada, Japan, South Korea and Taiwan (Roskill, 2007). Despite stable lead recycling rates, the amount of secondary antimony recovery continuously decreased in the USA from 10,500 tonnes in 1995 to 3450 tonnes in 2011. The decline mainly reflects reductions in battery weight, lower antimony content and growth in antimony-free batteries. In industrialising countries, the recovery of secondary antimony has strongly increased due to the higher demand for lead-acid batteries and rising recycling rates. Secondary antimony recovery in highly industrialised countries, like the USA, Canada and the original EU-15, amounts to 5.6 kg to 8.7 kg per tonne of secondary refined lead production (Roskill, 2007). Southern European older generation battery recycling in Spain and Italy amounts to 15 kg per tonne, and Japan, South Korea and Taiwan to 10 kg per tonne. The use of antimony in lead-acid batteries may be much reduced in the future as alternative vehicle technologies are increasingly adopted worldwide.

Blast furnaces, reverberatory furnaces and rotary furnaces may be used for the recovery of antimony from secondary materials. Blast furnaces are used for a continuous and regular supply of coarse charge material (i.e. battery scrap) and produce a metal output of uniform composition. Reverberatory furnaces are suitable for finely divided feed material and produce an antimony-rich slag grading 5 to 9% Sb. Antimony oxide in the slag is subsequently reduced in a blast furnace to produce antimony metal. Rotary furnaces produce either a single grade of alloy or both, a high-antimony and a low-antimony alloy.

Substitution

There is a range of fire-retardant materials on the market. They include alumina trihydrate, magnesium hydroxide, calcium carbonate, zinc borate, zinc stannate, zinc hydroxyl stannate, melamine, and phosphorus-based, nitrogen-based, and phosphorus–nitrogen-based combinations. However, their performance is generally inferior to ATO-based fire retardants and accordingly the scope for substitution where high performance is required is restricted.

Several types of plastic catalysts and plastic stabilisers, which contain barium, cadmium, calcium, germanium, lead, tin, titanium, and zinc in various combinations, may compete with antimony but their use generally leads to increased production costs. Compounds of cadmium, chromium, tin, titanium, zinc and zirconium can substitute for antimony chemicals in paint, pigments and enamels.

A lead–calcium–tin alloy could be a possible alternative to lead-acid batteries, while lead-free solders based on tin–silver–copper alloys can substitute for antimony in solder. A wide range of materials, mostly metals, may substitute for antimony alloys in friction bearings, ammunition and cable sheathing (Butterman and Carlin, 2004).

Resources and reserves

Global antimony reserves compiled for this book have been estimated at about 3.4 million tonnes in 2011 (Figure 4.4). China, the largest antimony producer, has the largest reserves (1,225,000 tonnes, 36 per cent of total). Other producing countries with major antimony reserves include Thailand (525,000 tonnes), Russia (455,000 tonnes), Bolivia (420,000 tonnes) and Kyrgystan (385,000 tonnes). These five countries together account for approximately 87 per cent of world antimony reserves. Significant reserves also exist in Turkey, South Africa and Tajikistan. In the 1990s reserves were also reported in Guatemala, Canada, Peru, Vietnam, Laos and Mexico. A minor contribution to antimony supply originates as a by-product from gold and base metal refining, for example, from the Mississippi Valley-type lead deposits in the eastern United States, from base-metal–silver operations in New Brunswick, Canada, in La Oroya, Peru and in Coahuila and Nuevo Leon, Mexico and from gold deposits in Australia.

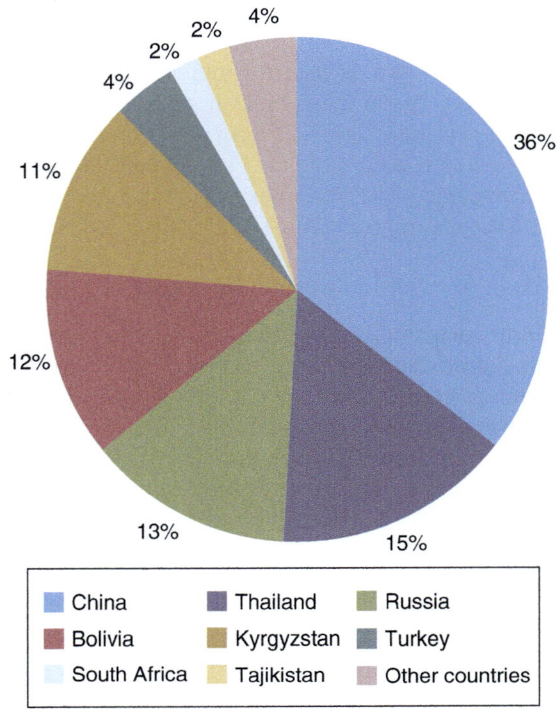

Figure 4.4 The distribution of world antimony reserves in 2011. (Data from: USGS, 2012; Roskill, 2011; Village Main Reef Ltd, 2012.)

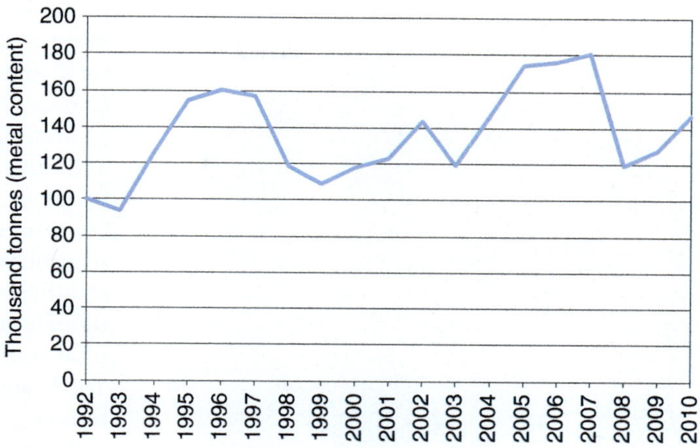

Figure 4.5 World antimony mine production between 1992 and 2010. (Data from British Geological Survey World Mineral Statistics database.)

It is significant to note that the USGS estimate of global antimony reserves in 2011 was 1.8 million tonnes, considerably less than the total presented here (USGS, 2012). This discrepancy highlights the difficulty of obtaining reliable and consistent global reserve estimates for any mineral commodity, as discussed in Chapter 1 of this book. Another factor which contributes to this uncertainty is that estimates of antimony resources are generally based on pure stibnite deposits and do not take account of antimony as a by-product of gold mining. Greenstone-hosted quartz-carbonate vein deposits are major sources of gold but often include significant quantities of stibnite, berthierite and tetrahedrite, which are not identified as potential antimony sources. The same is true for small-scale gold deposits widely present in Europe, Africa, Asia and the USA. Resource and reserve assessments for antimony should therefore be treated with caution as they likely underestimate the geological availability.

Production

Large mines with antimony-bearing ore are in production on all continents. Currently, the most important deposits are located in China, where the majority is produced at Xikuangshan in Hunan Province, and in Bolivia, Russia, South Africa, Tajikistan, Canada and Australia.

Since the nineteenth century antimony has been produced in more than 40 countries. In 1900 global antimony production was only about 7710 tonnes. However, in response to greatly increased demand for ammunition in the First World War, production rose sharply to 81,600 tonnes in 1916. This level was only reached again in 1994 when the annual production exceeded 100,000 tonnes for the first time. Since then, the worldwide production levels have been strongly affected by the global markets and have fluctuated between 103,000 tonnes in 1995 to 181,000 tonnes, an all-time high, in 2007 (BGS, 2012). The overall doubling of global production since the early 1980s is a consequence of growing demand for antimony in flame retardants and associated increases in production capacity, especially in China. Production levels fell back sharply in 2008 and 2009 in response to the global economic recession but have begun to recover since then (Figure 4.5).

China is by far the largest producer of primary antimony with production in 2010 of about 130,000 tonnes, equivalent to about 88 per cent of the world total (Figure 4.6). Bolivia is the next largest producer at about 5000 tonnes or three per cent of the total, followed by Tajikistan, Russia and South Africa, each contributing a further three per cent.

The refinery capacities for antimony are estimated to be in excess of 163,360 tonnes per annum

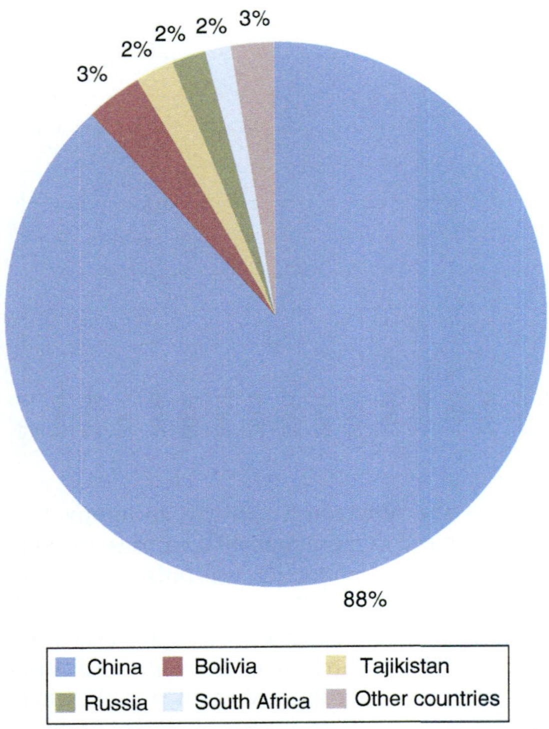

Figure 4.6 World antimony mine production in 2010, by country. (Data from British Geological Survey, 2012.)

(Table 4.7; Roskill, 2011). China has the largest share with 120,000 tonnes (nearly 74 per cent of total), followed by Russia, South Africa, Myanmar, Canada, Tajikistan and Bolivia, each contributing approximately four per cent of the total. Global antimony production exceeds the refinery capacity by about nine per cent, which is likely due to a contribution to supply from stock inventories. Antimony derived from recycling is also significant, while 'unofficial' production is thought to have accounted for about 16 per cent of total antimony supply in 2010 (Roskill, 2011). Total antimony supply in 2010 was 196,484 tonnes, a 10 per cent increase on 2009. The total global capacity for antimony production is estimated to be about 30 per cent higher than the current production level.

The Chinese government is pursuing a programme of closing antimony mines and smelters in order to improve the control of environmental and safety issues. In 2010 they also imposed an antimony production quota of 100,000 tonnes, comprising 62,520 tonnes for primary production and 30,480 tonnes from recycling. This compares with a quota of 90,180 tonnes in 2009. China used to operate about 50 industrial antimony mines and

Table 4.7 Global production capacity of refined antimony in 2010. (Source: Roskill, 2011.)

Company	Country	Capacity (tonnes/year antimony)
Hsikwangshan Twinkling Star	China	55,000
Liu Zhou China Tin Group Co Ltd	China	30,000
Hunan Chenzhou Mining Co. Ltd	China	20,000
Shenyang Huacheng Antimony	China	15,000
GeoProMining	Russia	6500
Village Main Reef (former Metorex)	South Africa	6000
Various (Shwe Zin Htut, Thu Ya Kan Chun)	Burma (Myanmar)	6000
Beaver Brook	Canada	6000
Comsup (former Anzob GOK)	Tajikistan	5500
Various (Emusa, COMISAL, Bernal Hermanos, CBPA)	Bolivia	5460
Mandalay Resources	Australia	2750
Cengiz & Özdemir Antimuan Madenleri	Turkey	2400
Kazzinc	Kazakhstan	1000
Various (Hong Xin, Siam)	Thailand	600
Kadamdzhai	Kyrgyzstan	500
SRS	Laos	500
US Antimony	Mexico	150
Total		**163,360**

many small production sites, with more than 300 antimony smelters of varying sizes and a large number of antimony mining and manufacturing companies. Now only six companies account for more than 90 per cent of China's reported production of antimony. Reported production of antimony in China fell in 2010 and is unlikely to increase in the future, despite the fact that the country is facing a serious shortage of antimony. The local government in Lengshuijiang, Hunan Province, which accounts for about 60 per cent of world antimony supply, closed almost all its mines and smelters in 2011 (USGS, 2012). No significant antimony deposits have been developed for about ten years and the remaining economic reserves are being rapidly depleted, with only five years of mining life remaining in Lengshuijiang (USGS, 2012).

The leading global producers of antimony ores and concentrates are listed in Table 4.8.

South Africa is among the largest antimony producers in the world. Since 1932, the Consolidated Murchison antimony deposits (Alpha/Gravelotte, Beta, Athens, and Monarch) have produced more than 550,000 tonnes of antimony in concentrate and more than 25 tonnes of gold. In March 2011, Consolidated Murchison was acquired by Village Main Reef Limited. At present, activities are focusing on ramping up production as more mining areas become available and existing shafts are deepened. Consolidated Murchison has a total antimony resource base of 200,000 tonnes of which more than 10 per cent is in the reserve category (Village Main Reef, 2012). The 2011 mineral resource and reserves for the underground operations are 3480 million tonnes of antimony ore at 1.87% Sb, which is equivalent to 65,130 tonnes of antimony. Inferred reserves comprise 6051 million tonnes of antimony ore at 2.34% Sb resulting in an additional 141,718 tonnes of antimony. Gold is the

Table 4.8 Leading global producers of antimony ores and concentrates in 2010. (Source: Roskill, 2011.)

Company	Country	Estimated production (tonnes antimony)
Hsikwangshan Shanxing Antimony Co. Ltd.	China	35,000
Liu Zhou China Tin Group Co Ltd	China	30,000
Chenzhou Mining Co. Ltd	China	15,000
GeoProMining	Russia	6500
Shwe Zin Htut, Thu Ya Kan Chun	Burma (Myanmar)	5897
Beaver Brook	Canada	5669
Comsup (former Anzob GOK)	Tajikistan	5370
Yunnan Muli Antimony Industry Co. Ltd	China	4500
Comite Boliviano de Productores de Antimonio	Bolivia	3800
Banxi antimony mine (Taojiang Jiutong Group)	China	3500
Village Main Reef (former Metorex)	South Africa	2257
Nandan Cheshan antimony mine	China	2000
Eti Holdings	Turkey	2000
Mandalay Resources	Australia	1106
Emusa and COMISAL	Bolivia	1000
Kazzinc	Kazakhstan	840
Hong Xin, Siam	Thailand	600
SRS	Laos	493
Kadamdzhai	Kyrgyzstan	480
Sontrarem	Morocco	280
Empresa Minera Bernal Hermanos	Bolivia	180
Doe Run Peru	Peru	120
Total		**126,592**

traditional by-product at Consolidated Murchison and significant gold reserves have been delineated. The estimated life of the mine is 11 years. The potential for reworking of the Consolidated Murchison Tailings Dump is also under investigation. Resources have been estimated at 6.18 million tonnes, with an average grade of 0.75% Sb and 0.46 g/t Au, containing a total of 46,461 tonnes of antimony and 2.84 tonnes of gold.

The Beaver Brook Mine in central Newfoundland, Canada, is North America's only producing antimony mine. It was re-opened in 2008 after it had been closed for 10 years and is now reported to be the largest antimony mine outside China and South Africa (Roskill, 2011). Resources are estimated at about 100,000 tonnes of antimony and the current production rate is 235 tonnes of concentrate per week. The stibnite concentrate is trucked to Halifax for shipping to China. In 2009, Hunan Nonferrous Metals Corporation (HNC), the largest antimony company in the world, acquired 100 per cent equity of Beaver Brook Antimony Mine Inc. At the end of 2009, China Minmetals Corporation, China's biggest metal trader, acquired 51% equity of Hunan Holdings Group (HNG), the state-owned parent of HNC. The state-owned Assets and Supervision Administration Commission of Hunan Province reserved the remaining 49 per cent of HNG. Small quantities of antimony concentrates and sodium antimonate were also produced as by-products from base-metal operations in Canada, for example, at Sullivan in British Columbia and Belledune in New Brunswick (Roskill, 2011).

Bolivia has a long history of antimony production from small, high-grade vein-type deposits which are typically associated with gold-, lead-, zinc- and tin-bearing ores. Three distinctive belts of orogenic gold and antimony deposits containing more than 500 known deposits and occurrences are recognised along the length of the Eastern Cordillera of the Andes. The gold ores, particularly those of the Caracota–Carma–Candelaria belt, may contain as much as 10 to 20% Sb and were originally mined for antimony. These deposits have been exploited on a small scale from pre-colonial days up to the present.

Projects under development

The growth in antimony consumption, together with the concentration of production and falling resource levels in China, have led to an increase in exploration activities for antimony-bearing deposits, mainly with antimony as a potential by-product. Most current projects are focusing on epithermal and orogenic hydrothermal targets.

Mandalay Resources Corp. operates two silver–gold–antimony mines at Cerro Bayo in Patagonia. It recently announced plans to expand its operations to four mines with future annual production of 1870 kg of gold and 1600 tonnes of antimony (Mandalay Resources Corporation, 2012; Roskill, 2007). The company is working epithermal vein-type mineralisation at Dagny, Fabiola, Delia NW and Yasna. Mandalay acquired the Cerro Bayo mining and concentrator complex from Coeur d'Alene Mines Corp. in August 2010.

In 2012 U.S. Antimony Corporation completed the construction of a flotation mill in the State of Guanajuato, Mexico, for processing the ores from its new mine at the Los Jaurez antimony–silver deposit in the Soyatal mining district and from other properties (U.S. Antimony Corporation, 2012). U.S. Antimony Corporation, through its wholly owned subsidiary USAMSA, also owns and operates a smelting facility at Estacion Madero in the Municipio of Parras de la Fuente, Coahuila, Mexico. Currently, crude antimony oxide and antimony metal are produced at this smelter for further processing at the company's smelter in Montana, USA. Concentrates and hand-sorted rock from Newfoundland, Peru, Honduras, Mexico, and other areas are processed at the Mexican facility.

There are several potential sources of antimony under investigation in Canada. These include the Lake George Mine in New Brunswick, owned by Apocan Inc. (an Amspec Chemical Corp. subsidiary), which is currently on care-and-maintenance (USGS, 2011). Exploration for antimony is also underway at other properties in British Columbia, Yukon and Newfoundland. Companies working on antimony as a by-product include Equity Silver Mines Ltd. which is assessing silver–gold–copper ores at a formerly important silver-producing site in central British Columbia. A flotation plant could produce 1700 tonnes antimony annually. Another is the Lead Smelting and Refining Factory in Trail set up by Cominco Ltd. It produced lead-antimony alloy as a by-product of silver-lead ores (USGS, 2011).

In Australia antimony resources are under investigation at a number of properties, some of which were former gold–antimony producers. These include the Hillgrove antimony–gold mine in New South Wales, which closed in 2009 and for which the present owner, Straits Resources Ltd, is currently seeking a buyer. The Costerfield gold–antimony mine in Victoria, owned by Mandalay Resources Corp., is an underground mining operation which produced 1577 tonnes of antimony in 2011 and is planning to exceed this in 2012 (Mandalay Resources Corporation, 2012). The company has identified new reserves in the Augusta deposit which is currently being mined and is also carrying out an economic analysis of high-grade antimony ores in the Cuffley lode which was discovered in 2011. Anchor Resources, owned by China Shandong Jinshunda Group Co Ltd., is exploring for antimony–gold mineralisation at the Bielsdown project in north-eastern New South Wales. Anchor has published new resource estimates for the Wild Cattle Creek antimony mine, located in the Bielsdown licence area, which was worked intermittently on a small scale from the late 1800s up to the 1970s (Anchor Resources Ltd, 2012). Northwest Resources Ltd is planning to develop a gold–antimony mine at its Blue Spec Shear project in Western Australia. A preliminary technical and economic assessment was completed in 2012 which concluded that the project has the potential to be a low-cost operation producing 1900 tonnes of antimony and 68,500 ounces of gold per annum over a mine life of five years (Northwest Resources Limited, 2012).

There are a number of active antimony-related ventures in Turkey. Production has historically been dominated by Ozdemir Antimony Mining Joint Stock Co from mines in the Turhal district of northern Turkey (Clarke, 2012). The Dutch corporation Advanced Metallurgical Group NV (AMG), which is Europe's largest ATO manufacturer, operates an antimony mine and adjacent smelter through its subsidiary Suda Maden AS. Tri-Star Resources plc has been investigating antimony resources at the historic small-scale workings at Göynűk in the Kutahya Province of north-west Turkey. The company has also been granted permission to construct a small-scale processing facility at Göynűk to treat stockpiled material (Tri-Star Resources plc, 2012a).

In Italy, Adroit Resources Inc. is investigating the possibility of reactivating antimony mining in the Manciano area of Tuscany. This area has several small and medium-sized deposits some of which were worked in the past (Adroit Resources, 2012). The company is aiming to develop a new mine producing 10,000 tonnes per annum antimony, together with gold and silver (Clarke, 2012).

In south-east Asia both Myanmar and Laos produce antimony ore which is exported to China. Tri-Star Resources is collaborating with RDP Singapore Ltd to undertake a technical assessment of the latter's antimony projects which include exploration rights around two producing deposits (Tri-Star Resources plc, 2012b).

World trade

Antimony is traded in the form of ores and concentrates, semi-refined unwrought antimony metal, antimony trioxide and antimonial lead. The international trade is traditionally characterised by the distinction between primary antimony production of ores, concentrates and unrefined metal and trioxide in industrialising countries and the refining and consumption of antimony in industrialised countries.

Over the last decade this pattern has changed due to the strong growth in the manufacturing sectors in the industrialising countries. In particular, China has pursued a value-added manufacturing policy and imposed export restrictions on crude antimony products. It has become the major producer and exporter of refined antimony metal, compounds and intermediate products

(Roskill, 2007). The combination of mine and smelter closures, production controls, export quotas, environmental restrictions and dwindling resources in China and the support provided by the government to Chinese companies operating overseas have, together, contributed to high levels of interest in antimony mining in Bolivia, Mexico, Australia and Canada and has led to increased trade in antimony concentrates. The main exporters of concentrate in 2010 were Myanmar, Russia, Canada, Tajikistan and Australia (Figure 4.7a); the overwhelming majority of shipments from these countries went to China, which accounted for about 90 per cent of the total (Roskill, 2007 and 2011). Recorded exports of antimony metal do not entirely reflect the trade volume as most of the exports from China are not included in the official export data. The main trade flows, once unofficial shipments are taken into account, are from China to Belgium, France, South Korea, Japan and the USA (Figure 4.7b; Roskill, 2011). Until 2005, South Africa was the main exporter of antimony oxide, but China now accounts for about 55 per cent of export shipments. Other leading exporters of antimony oxide are the companies converting antimony metal to oxide in Belgium and France.

Prices

Antimony is not traded on international metal exchanges and prices are agreed between producer or trader and consumer, depending on the quality and form of the product sold. Antimony metal produced by smelters, generally with purity between 99.5% and 99.6%, is traded in US dollars per pound. Further refining is carried out in smelters or by secondary processors and manufacturers to produce antimony trioxide and antimonial lead. The estimated value of primary antimony mine production in 2011, based on the annual average New York dealer price, was about US$ 1794 million.

Despite its geochemical scarcity, given its important industrial uses the commodity price of antimony is low when compared to more abundant minor metals such as molybdenum and tungsten. This may be attributed to the fact that mining costs for antimony are low because it occurs in highly concentrated, almost pure stibnite deposits, locally with gold as a by-product.

There has been significant price volatility in the antimony market over the past 40 years. In 1970 the rapidly rising production of plastics and the establishment of laws that regulated flammability of textiles and other materials, together with the concentration of mine production in a few countries, caused a considerable price spike (Figure 4.8). In 1974 sharply increased demand, especially for antimony trioxide, and supply disruptions from China resulted in another major price peak. Since the mid-1980s antimony price instability may be largely attributed to the concentration of production in China and to Chinese government policy. In early 1995 the imposition of a 20 per cent export tax on antimony ores and concentrates contributed to a steep price rise, which reached US$6000 per tonne. Other control measures, including the suspension of the issuance of new mining licences, the imposition of export quotas and a crackdown on illegal mining and smuggling, failed to rectify an oversupply situation and low prices persisted from 2001 to 2004. However, prices rose sharply between 2004 and 2008 as demand strengthened, stockpiles were depleted and Chinese control measures took effect. From early 2009, the antimony prices declined considerably because of the global economic crisis, with European prices at the end of the year falling to US$6050 per tonne for antimony trioxide.

In 2010 mine closures in China restricted global supplies and led to the doubling of antimony metal prices to more than US$12,000 per tonne at the end of the year. The upward trend continued in the first part of 2011 with prices climbing to more than US$16,000 per tonne by April. Thereafter, the price declined gradually to end 2011 at about US$13,000 per tonne. During the first three quarters of 2012 antimony metal has traded in the range US$12,000–14,000 per tonne.

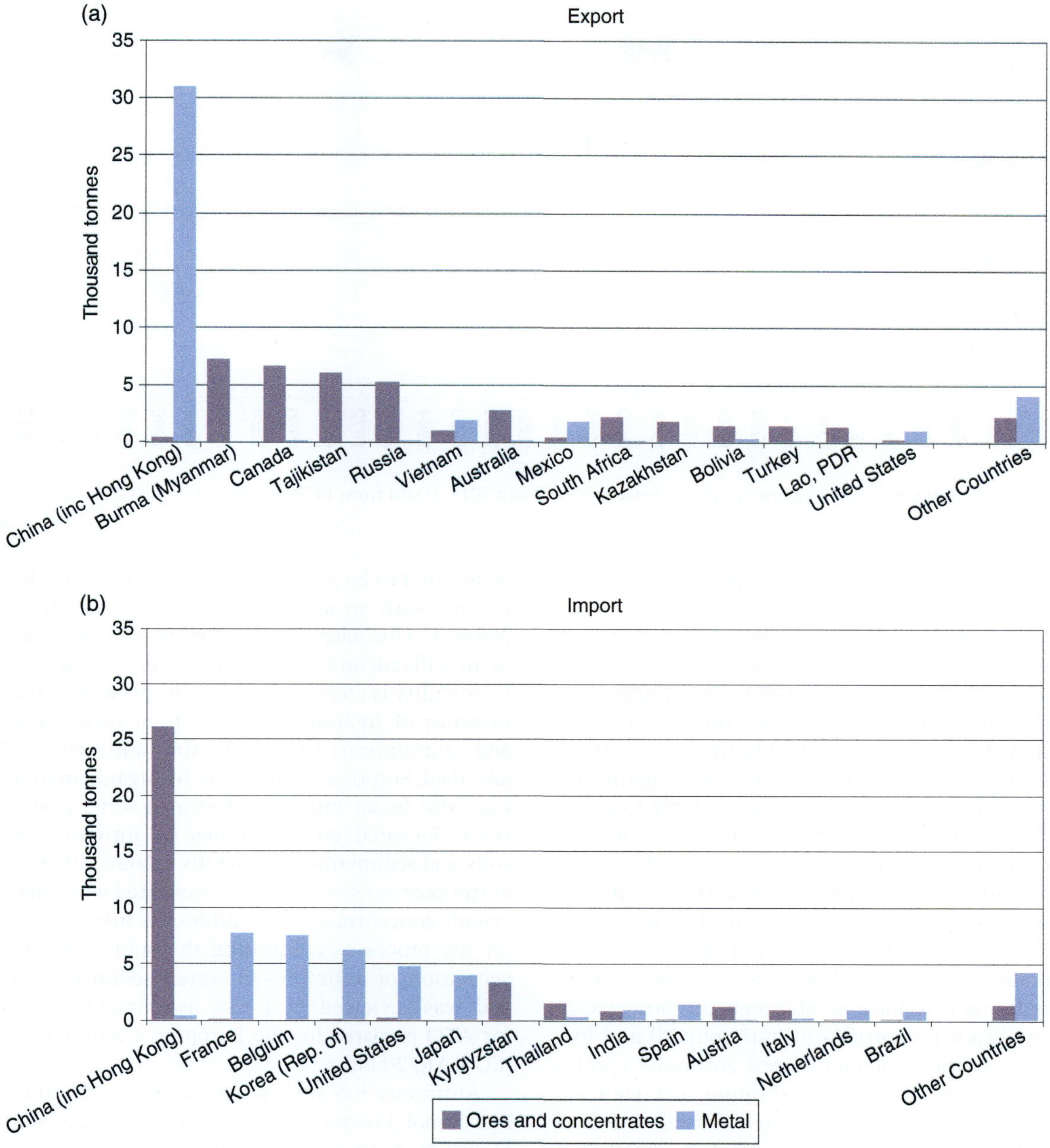

Figure 4.7 (a) Main antimony exporting countries, 2009. (Data from British Geological Survey World Mineral Statistics database and UN Comtrade, 2013) (b) Main antimony importing countries, 2009 (Data from British Geological Survey World Mineral Statistics database and UN Comtrade, 2013.)

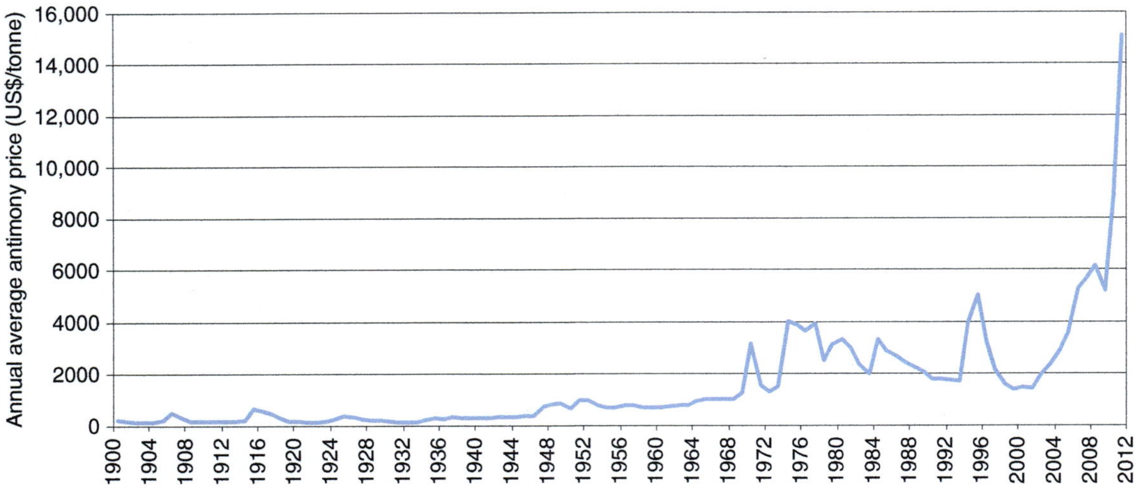

Figure 4.8 Annual average antimony prices between 1900 and 2011. (Data from BGR database.)

Environmental aspects

Antimony is widely dispersed in the environment as a result of natural processes such as volcanic eruptions and erosion of antimony-bearing rocks and minerals. Anthropogenic emissions of antimony from burning fossil fuels, mining, smelting, refining and waste incineration are also significant, but there is little modern data available to assess their magnitude (Belzile et al., 2011). Antimony is found in very low concentrations in soils, waters and air. It occurs mainly as antimony(III) and antimony(V) in environmental, biological and geochemical samples (Filella et al., 2002). Most studies report the dominance of antimony(V) under oxic conditions although the presence of significant proportions of antimony(III) and low concentrations of methylated antimony species are also documented (Nishimura and Umetsu, 2000). Typical concentrations of total dissolved antimony are less than 1.0 µg/l in unpolluted waters although the range is quite large, reflecting the wide range of conditions in natural systems and proximity to pollution sources (Filella et al., 2002). Much lower values have been reported in unpolluted groundwaters in southern Ontario, Canada (Shotyk et al., 2006). Antimony is not considered to be a highly reactive element in the oceans, with an average concentration of about 200 ng/l. Ultimately, antimony in the environment will end up either in soil or sediment, where its mobility is controlled mainly by pH and by the presence of hydrous oxides of iron, manganese and aluminium to which the antimony is adsorbed. Sorption of antimony by organic matter may also be an important factor in some conditions. Elevated concentrations of antimony in soils and sediments are generally related either to anthropogenic sources or are associated with high arsenic concentrations in sulfide ores. Information on the processes controlling the solubility and speciation of antimony in water, sediment and soil was reviewed in a risk assessment report for ATO prepared for the European Commission (EU RAR, 2008).

Antimony has no known biological function and is not known to be used by any organism. However, given its currently high level of use, there is concern about potential pollution and risks to human health. Long-term, low-level oral and dermal exposure from food, air and drinking water has negligible health effects (e.g. EU RAR, 2008). However, respiratory irritation, dermatitis, pneumoconiosis and gastrointestinal symptoms have been reported in workers in the antimony processing industry, but none of these symptoms have been directly linked with antimony exposure (Sundar and Chakravarty, 2010).

Antimony trioxide (ATO) is classified in the EU as 'suspected of causing cancer via inhalation' according to Regulation (EC) 1272/2008. The EU RAR (2008) concluded that there was no human health risk to consumers (through the use of products such as electrical and electronic equipment, PET bottles and flame-retarded items such as textiles and carpets) or to those exposed to ATO through environmental sources. It also stated that there was no need at that time for further information or testing and no need for risk-reduction measures beyond those being applied already. These conclusions have since been validated by international experts and are considered to remain valid today (e.g. OECD, 2008; Environment Canada, 2010).

Those most at risk from exposure to antimony are the workers in industries that process antimony ore and metal, and those that make antimony chemicals or are involved in the manufacture of products in which they are incorporated. The EU RAR assessed a wide range of possible health effects related to ATO exposure based on human exposure data and animal laboratory studies. Three areas of possible concern were identified: skin irritation for workers under conditions of perspiration; toxicity to the lung as a result of repeated long-term exposure through inhalation; and the development of tumours in the lungs of female rats as a result of inhalation exposure. It was concluded that additional data is only needed for repeated long-term inhalation exposure to better understand the risks to human health. Additional research is ongoing to investigate the mechanism and inhalation effects of ATO on lungs. In general, the utilisation of effective working practices and equipment, combined with strict adherence to guidelines, ensures that occupational exposure and environmental discharges are minimised and meet the official requirements. Long-term monitoring of workers in an ATO production plant has not identified any adverse health effects (International Antimony Association, 2012).

The US National Institute of Occupational Safety and Health (NIOSH) has set a recommended exposure limit for antimony of 0.5 mg/m^3 averaged over a 40-hour working week (NIOSH, 2010). In addition, various countries and the World Health Organisation (WHO) have established safe limits for the antimony content of drinking water at which no adverse health effects are likely to occur – for example, 5 μg/l in the European Union (European Commission, 1998), 6 μg/l in the USA (US Environmental Protection Agency, 2012) and 20 μg/l by WHO (WHO, 2003).

Given the increasing range of uses for antimony, the various routes through which it can get into the environment and the potential effects of a number of antimony species, it is clear that more research and monitoring is needed on its biogeochemical cycling and fate in the environment. Some studies have detected antimony in bottled waters and commercially available fruit juices although the antimony levels were all below the WHO guideline for drinking water and the source of the antimony is unknown (Shotyk et al., 2006; Hansen et al., 2010).

Outlook

The use of antimony in flame retardants is expected to remain its principal market in the future, although its application in the production of polyethylene terphthalate (PET) for plastic bottles and synthetic textiles and for the vulcanisation of rubber is likely to increase.

The demand for use in flame retardants has increased in recent years and is likely to continue to grow as fire regulations become more stringent and widely imposed worldwide, particularly in Asia, eastern Europe and Latin America. The trend of use of antimony in lead-acid batteries is difficult to assess. Alloys of antimony with lead and antimonial lead have shown a decrease in use as they are increasingly substituted on environmental grounds in some applications. However, the market for antimony in lead-acid batteries may expand significantly as the automotive sector, especially in Asia, continues to grow. The production rate of electric and hybrid vehicles with lead-acid batteries is also expected to increase. The demand for antimony in other metallic uses, such as bearings, tends to be stable. Health and environmental concerns may limit the use of antimony in some applications but the important role of antimony in catalysts for PET production is expected to continue to be important. On balance it is anticipated that worldwide demand for antimony will continue on an upward trend at least in the short term.

Antimony production in China is unlikely to increase in the near future and could even decrease as a consequence of mine and smelter closures. There is evidence of increased activity in several countries, such as Thailand and Myanmar, but assessing future production trends is difficult because of lack of information. No Asian country other than China produces more than a few thousand tonnes a year and there may be limited scope to expand output. Production in South Africa is likely to increase with the new ownership at Consolidated Murchison, while production in Bolivia will likely fluctuate in line with demand and will probably increase with the planned restart of the government-owned Vinto smelter. Australian production could increase by as much as 5500 tonnes per annum if the Hillgrove operation emerges from care and maintenance and if the new investors in Anchor Resources, China Shandong Jinshunda Group, bring the Bielsdown deposit on stream (Roskill, 2011). The new US Antimony Corporation operation in Mexico should be fully operational by 2013, but will only add a maximum of 1600 tonnes per annum. Production from Russia could be augmented by 2800 tonnes per year if RusAnt succeed in developing the Iliskoye deposit (Roskill, 2011). In total, new projects could add up to 11,200 tonnes per year antimony to world mine capacity within the next four to five years (Roskill, 2011).

Antimony prices are expected to remain at high levels as a result of reduced supply from China and increasing demand, especially as legislation for fire proofing becomes prevalent in industrialising countries.

References

Adroit Resources Inc. (2012) Valle Lupara. http://adroitresources.ca/valle-lupara/

Amspec (2011) Flame retardants: antimony trioxide. http://www.amspec.net/products/flame-retardants/antimony-trioxide/

Anchor Resources Ltd. (2012) Bielsdown Project, NSW. http://www.anchorresources.com/index.php/projects/nsw/bielsdown-nsw/

Anderson, C.G. (2000) A survey of primary antimony production. In: Young, C. (ed.) Minor Elements 2000: Processing and Environmental Aspects of As, Sb, Se, Te, and Bi. SME, Littleton, Colorado, 261–275.

Belzile, N., Chen, Y. and Filella, M. (2011) Human Exposure to Antimony: I. Sources and Intake. Critical Reviews in Environmental Science and Technology 41, 1309–1373.

British Geological Survey (2012) World mineral production 2006–10. (Keyworth, Nottingham: British Geological Survey).

Buchholz, P., Oberthür, T., Lüders, V and Wilkinson, J. (2007) Multistage Au-As-Sb Mineralization and Crustal-Scale Fluid Evolution in the Kwekwe District, Midlands Greenstone Belt, Zimbabwe: A Combined Geochemical, Mineralogical, Stable Isotope, and Fluid Inclusion Study. Economic Geology 102, 347–378.

Butterman, W.C. and Carlin Jr. J.F. (2004) Antimony. U.S. Geological Survey Open-File Report 03-019.

Clarke, G. (2012) Antimony (trioxide!) on the watch list. Industrial Minerals, September 2012.

Dill, H.G., Weiser, T., Bernhardt I.R., and Kilibarda, C.R. (1995) The composite gold-antimony vein deposit at Kharma (Bolivia). Economic Geology 90, 51–66.

Dubé, B. and Gosselin, P. (2007) Greenstone-hosted quartz-carbonate vein deposits. In Goodfellow, W.D. (ed.) Mineral deposits of Canada: A synthesis of major deposit types, district metallogeny, the evolution of geological priovinces, and exploration methods. Geological Association of Canada, Mineral Deposits Division, Special Publication No.5, 49–73.

European Commission (1998) Council Directive 98/83/EC of 3 November 1998 on the quality of water intended for human consumption. http://eur-lex.europa.eu/LexUriServ/LexUriServ.do?uri=CELEX:31998L0083:EN:NOT

Environment Canada (2010) Draft screening assessment for the challenge. Antimony oxide. Environment Canada Health Canada, March 2010. http://www.ec.gc.ca/substances/ese/eng/challenge/batch9/batch9_1309-64-4_en.pdf.

EU RAR (2008) European Union risk assessment report [draft]. Diantimony trioxide CAS No.: 1309-64-4. Luxembourg: Office for Official Publications of the European Communities. http://ecb.jrc.ec.europa.eu/DOCUMENTS/Existing-Chemicals/RISK_ASSESSMENT/REPORT/datreport415.pdf.

Filella, M., Belzileb, N., Chen, Y.-W. (2002) Antimony in the environment: a review focused on natural waters: I. Occurrence. Earth-Science Reviews 57, 125–176.

Hansen, C., Tsirigotaki, A., Bak, S.A., Pergantis, S.A., Stürup, S., Gammelgaard, B. and Hansen, H.R. (2010) Elevated antimony concentrations in commercial juices. Journal of Environmental Monitoring 12, 822–824.

Hart, C.J.R. (2007) Reduced intrusion-related gold systems. In: Goodfellow W.D. (ed.) Mineral deposits of Canada: A synthesis of major deposit types, district metallogeny, the evolution of geological priovinces, and exploration methods. Geological Association of Canada, Mineral Deposits Division, Special Publication No.5, 95–112.

Hedenquist, J.W., Arribas, A.R., and Gonzalez-Urien (2000) Exploration for epithermal gold deposits, Chapter 7. In: Hagemann, S.G. and Brown, P.E., (eds.) Gold in 2000. Reviews in Economic Geology 13, 245–277.

International Antimony Association (2012) http://www.antimony.be/en/detail_44.aspx.

Mandalay Resources Corporation (2012). Annual Report 2011. http://www.mandalayresources.com/index.cfm?pagepath=Investors/Annual_Reportsandid=20047

MMTA (2012) Minor Metals in the Periodic Table: Sb - Antimony http://www.mmta.co.uk/metals/Sb/1

NIOSH (2010) NIOSH Pocket Guide to Chemical Hazards: Antimony. http://www.cdc.gov/niosh/npg/npgd0036.html

Nishimura, T. and Umetsu, Y. (2000) Chemistry on elimination of arsenic, antimony, and selenium from aqueous solution with iron(III) species. In: Young, C. (ed.) Minor Elements 2000: Processing and Environmental Aspects of As, Sb, Se, Te, and Bi. SME, Littleton, Colorado, 105–112.

Northwest Resources Limited (2012) Positive scoping study for the Blue Spec Shear Gold-Antimony Project. http://www.nw-resources.com.au/images/north-west---iepho.pdf

Obolensky, A.A., Gushchina, L.V., Borisenko, A.S., Borovikov, A.A., Pavlova, G.G. (2007) Antimony in hydrothermal processes: solubility, conditions of transfer, and metal-bearing capacity of solutions. Russian Geology and Geophysics 48 (12), 992–1001.

OECD (2008) SIDS Initial Assessment Profile for diantimony trioxide. SIAM 27, 14-16 October 2008. http://webnet.oecd.org/hpv/UI/handler.axd?id=13e93c97-6605-4eac-961f-8af23cc6ad32.

Penga, J.-T., Hua, R.Z. and Burnardb, P.G. (2003) Samarium–neodymium isotope systematics of hydrothermal calcites from the Xikuangshan antimony deposit (Hunan, China): the potential of calcite as a geochronometer. Chemical Geology 200 (1–2), 129–136.

Robert, F. (2001) Syenite-associated disseminated gold deposits in the Abitibi greenstone belt, Canada. Mineralium Deposita 36, 503–516.

Roskill (2007) The Economics of Antimony. Roskill Information Services Ltd., 10th edition, London, pp. 231.

Roskill (2011) ANCOA Ltd. Study of the antimony market. 17 October 2011, London, pp. 16.

Roskill (2012) Antimony: Global Industry Markets and Outlook, 11th Edition 2012. Roskill Information Services Ltd..

Shotyk, W., Krachler, M. and Chen, B. (2006) Contamination of Canadian and European bottled waters with antimony from PET containers. Journal of Environmental Monitoring 8, 288–292.

Simmons, S.F., White N.C. and John, D.A. (2005) Geological characteristics of epithermal precious and base metal deposits. In: Hedenquist, J.W., Thompson, J.F.H., Goldfarb, R.J. and Richards, J.P. (eds.) Economic Geology 100th Anniversary Volume, 485–522.

Sundar, S. and Chakravarty, J. (2010) Antimony Toxicity. International Journal of Environmental Research and Public Health 7, 4267–4277.

Taylor, B.E. (2007). Epithermal gold deposits. In: Goodfellow W.D. (ed.) Mineral deposits of Canada: A synthesis of major deposit types, district metallogeny, the evolution of geological priovinces, and exploration methods. Geological Association of Canada, Mineral Deposits Division, Special Publication No.5, 113–139.

Thompson, J.F.H., Sillitoe, R.H., Baker, T., Lang, J.R. and Mortensen, J.K. (1999) Intrusion-related gold deposits associated with tungsten-tin provinces. Mineralium Deposita 34, 323–334.

Tri-Star Resources plc (2012a) Operational Update, June 20th. http://www.tri-starresources.com//news/detail/77

Tri-Star Resources plc (2012b) Antimony technical collaboration in Myanmar. http://www.tri-starresources.com//news/detail/82

UN Comtrade (2013) United Nations Commodity Trade Statistics Database, Department of Economic and Social Affairs/ Statistics Division, http://comtrade.un.org/db/

U.S. Antimony Corporation (2011) Processing of antimony ores: Metallurgy. http://www.usantimony.com/metallurgy.htm

U.S. Antimony Corporation (2012) U.S. Antimony reports start up of Mexican flotation mill. http://www.usantimony.com/USAC_news.htm

US Environmental Protection Agency (2012) Basic information about antimony in drinking water. http://water.epa.gov/drink/contaminants/basicinformation/antimony.cfm

USGS (2011) Minerals Yearbook 2010, Antimony. http://minerals.usgs.gov/minerals/pubs/commodity/antimony/myb1-2010-antim.pdf.

USGS (2012) Mineral commodity summaries 2012, Antimony. http://minerals.usgs.gov/minerals/pubs/commodity/antimony/mcs-2012-antim.pdf

Vearncombe, J. R., Barton, J. M., Cheshire, P. E., De Beer, J. H., Stettler, E. H., and Brandl, G. (1992) 1347 Geology, geophysics and mineralization of the Murchison Schist Belt, Rooiwater Complex and 1348 surrounding granitoids, Memoir of the Geological Survey of South Africa, 81, pp. 139.

Village Main Reef (2012) Annual report for 2011. www.villagemainreef.co.za/ir/f/Village-Main-Reef-annual-report-2011.pdf

WHO (2003) Antimony in drinking water. Background document for development of WHO Guidelines for Drinking-water Quality. Available from http://www.who.int/water_sanitation_health/dwq/chemicals/antimony.pdf

Wu, J. (1993) Antimony vein deposits of China. Ore Geology Reviews 8, 213–232

Yang, D.-s, Shimizu, M., Shimazaki, Li, X.-h. and Xie, Q.-I. (2006) Sulfur Isotope Geochemistry of the Supergiant Xikuangshan Sb Deposit, Central Hunan, China: Constraints on Sources of Ore Constituents. Resource Geology 56 (4), 385–396.

5. Beryllium

DAVID L. TRUEMAN[1] AND PHILLIP SABEY[2]

[1]*Consulting Geologist, Richmond, British Columbia, Canada*
[2]*Manager, Technology and Quality, Materion Natural Resources, Delta, Utah, USA*

Introduction

Beryllium, chemical symbol Be, is the fourth element in the Periodic Table. It is a constituent element in various gemstones, including emerald, chrysoberyl and aquamarine, which appear to have first been traded in Egypt about 100 BC (Harell, 2004).

Beryllium was discovered in 1797 by the French chemist Vauquelin, who suspected the existence of a common element in the minerals beryl and emerald. It was named glucina by Vauquelin in 1798 after its sweet-tasting salt. Klaproth (1801) proposed the alternative name beryllia to avoid potential confusion with the sweet salt of yttria. However, the name glucinium continued in use, largely in France, until the name beryllium was formally adopted by the International Union of Pure and Applied Chemistry (IUPAC) in 1949.

In 1828 elemental beryllium was isolated independently by two chemists, Wohler in Germany and Bussy in France, through the reduction of the salt, beryllium chloride. Its commercial use began in 1926 and, by 2006, the most recent year for which figures are available, grew to become a $700 million dollar a year business (Sabey, 2006).

Four countries currently have the resources and capability to mine beryllium ores, to process them and to manufacture beryllium products. These are the United States, Kazakhstan, China and India. At present, Materion Corporation[1] (formerly Brush Wellman Inc.) in the United States is the only fully integrated beryllium producer in the world, involved in the mining, ore processing, manufacture, sale and recycling of beryllium-bearing products. It is estimated that in 2010, Materion produced over 70 per cent of the world's mined beryllium ore (Merchant Research and Consulting, 2012). The remaining production comes from stockpiled ores in Kazakhstan and China, together with minor contributions from beryl sourced from South America, Africa and India.

Several countries, such as Japan and France, have the capability of processing intermediate compounds of beryllium, such as beryllium hydroxide, into final products.

Properties of beryllium

Beryllium is a silver-grey metal noted for its light weight with a density of 1846 kg/m^3, comparable to that of magnesium, and its remarkable stiffness. Beryllium has the unique property for a metal of being virtually transparent to X-rays. The fact that sound travels through beryllium

Critical Metals Handbook, First Edition. Edited by Gus Gunn.

Table 5.1 Selected properties of beryllium.

Property	Value	Units
Symbol	Be	
Atomic number	4	
Atomic weight	9.01	
Density at 25 °C	1846	kg/m^3
Melting point	1287	°C
Boiling point	2475	°C
Hardness (Mohs scale)	5.5	
Specific heat capacity at 25 °C	1.82	J/(g °C)
Latent heat of fusion	1350	J/g
Coefficient of linear thermal expansion	11.5×10^{-6}	/°C
Thermal conductivity	210	W/(m °C)
Electrical resistivity at 25 °C	0.38	μΩ m
Young's modulus	287	GPa
Mass magnetic susceptibility	1.26×10^{-8}	m^3/kg
Brinell hardness	600	MN/m^2

faster than any other metal is also important in certain applications.

When added to other metals, particularly copper, as an alloying element, beryllium provides controllable strengthening mechanisms, along with many other valuable attributes such as good electrical and thermal conductivity and very low friction against most bearing surfaces. It also has non-magnetic and non-sparking properties.

Beryllium is the first element of Group 2, the alkaline earth elements, in the Periodic Table. In natural systems its oxidation state is Be^{2+} and, with a high charge / ionic radius ratio, it tends to form strong covalent bonds. The beryllium nucleus also contains an extra neutron which can be easily dislodged by gamma radiation. This property is useful in nuclear applications and in the detection, evaluation and mining of beryllium deposits.

Beryllium has 12 known isotopes of which only one is stable, ^{9}Be. Two beryllium isotopes, ^{7}Be and ^{10}Be, are cosmogenic in nature and ^{10} Be can be used for age dating of very young and recent geological events (Dunai, 2010).

Physical properties germane to the end uses of beryllium are listed in Table 5.1.

Distribution and abundance in the Earth's crust

Beryllium has an average crustal abundance of about 2.1 ppm (Rudnick and Gao, 2004). Systematic studies of the beryllium contents of over 500 samples of various rock types by Beus (1966) gave mean beryllium values of less than 2 ppm in most sedimentary rocks and in mafic and ultramafic igneous rocks. Felsic igneous rocks returned higher values with 4 ppm in biotite granites and 9 ppm in two-mica granites. In various unusual rock types from the Khibiny and Lovozero alkaline complexes in the Kola Peninsula, north-west Russia, mean values of 4 ppm Be were reported in mela-nepheline syenites and 20 ppm Be in clinopyroxene-nepheline rocks.

Uses of beryllium

Many of the physical properties of beryllium, such as its rigidity, low weight, heat-absorbing capability and dimensional stability, are exploited in its applications as a metal, yet the greatest use is in its alloys with other metals and in ceramics. Figure 5.1 summarises the main end uses of beryllium, divided into four main sectors. It should be stressed that there is considerable overlap among these categories and hence an accurate breakdown of their relative importance is difficult to obtain. For example, a copper–beryllium alloy may find use in components of aerospace, military, electronic, automotive or other hardware. Similarly, electronic uses of copper–beryllium include connector terminals that may be used in applications as diverse as aircraft guidance systems, mobile device electromagnetic radiation shielding, automotive airbag impact sensors and medical diagnostic hardware.

The three primary forms of beryllium used commercially are:

- alloys containing small amounts of beryllium, especially copper–beryllium;
- pure beryllium metal;
- beryllia (BeO) ceramics.

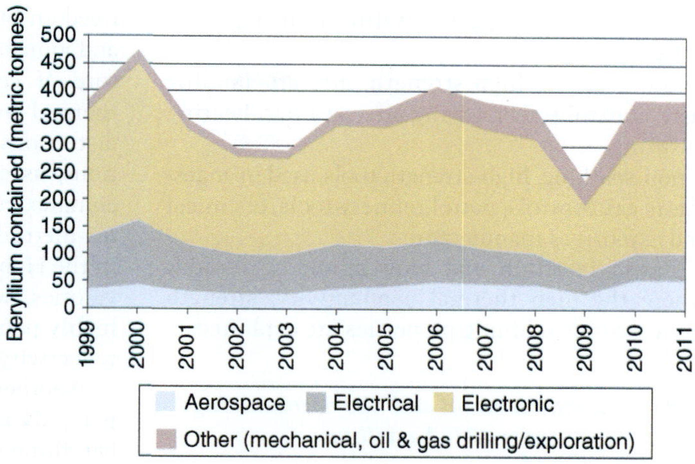

Figure 5.1 Consumption of beryllium by end-use sector, 1999–2011. (Data from BeST, 2012a.)

Alloys containing less than 2% beryllium, especially copper–beryllium

The largest end use of beryllium, comprising 75 per cent of the total world production, is in the form of alloys of copper containing less than two per cent beryllium. Smaller quantities of beryllium are also used to make alloys of aluminum and nickel. The beryllium content provides the alloys with valuable physical property enhancement as a result of the beryllium reacting with supplementary alloying elements, such as nickel and cobalt, to form local concentrations of intermetallic compounds known as berylides, which locate themselves within the crystalline structure of the metal matrix as precipitates. There, they produce local distortions or other irregularities in the crystal structure of the matrix that impede the smooth flow of dislocations caused by stresses applied to the alloy and thus offer enhanced strength, stiffness and hardness to the alloy, while retaining relatively good ductility, machinability, electrical and thermal conductivity. The highly predictable and precise attainment of a desired combination of those properties offers many options for the design of critical components.

The vast majority of copper–beryllium alloys are sold in thin strip between 0.1–0.4 mm and rod form in diameters from 0.5–7.5 mm, and they are utilised for the manufacture of components used in a myriad of applications, for example:

- connector terminals for high-reliability electrical and electronic connections between circuit boards, wires and components in electronics, telecommunications and appliance equipment, automobile systems such as airbag crash sensor and deployment mechanisms, anti-lock brake systems, dynamic suspensions, etc.;
- relays for controlling industrial, domestic and automobile electrical equipment;
- electromagnetic radiation shielding spring strips used to prevent 'leakage' from electronic wireless devices that can interfere with other equipment or become life threatening;
- diaphragms for pressure sensing in aircraft altimeters, medical stethoscopes and sphygmomanometers, aneroid barometers and automobile engine timing sensors;
- long service-life springs such as for fire sprinkler water-control valves.

Copper–beryllium alloys in thick plate, rod and tube form are used for the manufacture of:

- non-magnetic equipment components used in oil and gas exploration and production equipment such as directional drilling systems; coal and minerals mining equipment; mine detection and minesweeping;

- undersea cable signal amplification 'repeater' housings;
- low-friction, high-strength aircraft landing gear, control rod and wing aileron / flap-bearing bushings;
- non-sparking, high-strength tools used in anaesthetic gas controls; petrol refinery tools, chemical and explosives manufacture;
- plastic injection and blow-moulding moulds where the high thermal conductivity, strength and ease of machining properties are exploited.

Pure beryllium metal and alloys containing over 60% beryllium

The second largest consumption of beryllium, accounting for 20 per cent of the volume of beryllium produced, is in the form of beryllium metal of purity >99.5 per cent and alloys containing over 60 per cent beryllium. The principal properties of beryllium metal of commercial interest are its low density, high strength, high rigidity, structural stability at high temperatures, thermal conductivity, high transmission of sound and its transparency to X-rays. Beryllium is notable among metals in terms of its specific rigidity; i.e. the ratio of modulus to density, which is approximately fifty per cent greater than that of steel, while its density (1846 kg/m^3) is about 30 per cent less than that of aluminium.

Beryllium metal is typically used to produce components utilised in high-technology equipment. In situations where weight is a critical factor, such as for structures to be launched into space, it is imperative that such structures are rigid and not subject to distortions or resonant vibrations which might reduce the accuracy of their instrumentation. Beryllium metal is the optimal material for these purposes, principally because of its high specific rigidity. It also has attractive thermal properties that reduce thermal distortions, both at high temperatures experienced during launch and descent, and also at the extremely low temperatures of space.

Beryllium is an isotropic material, having uniform properties in all directions, which increases freedom of design. Its formability, machinability and joinability allow for relative ease of manufacture of complex structures. The isotropy and thermal properties of beryllium serve to minimise distortions in sophisticated dimensional applications, as exemplified by the ability to machine complex curved surfaces, such as the faces of the 6.3-metre diameter astronomical telescope mirrors in the Hubble and James Webb space-based observatories, which also rely upon its ability to be highly polished and accept coatings for enhanced reflectivity at operating wavelengths.

Another unique property of the metal is that it is highly transparent to X-rays. In thin foil form, beryllium is used as the window material for X-ray tubes and detectors to permit the X-rays through while maintaining vacuum. It is especially useful in security devices and high-resolution imaging technology, such as medical applications like mammography to detect breast cancer.

Beryllium is a very efficient moderator of neutrons, slowing and reflecting them, a property that finds application in materials test reactors and in fundamental particle-physics research. It is the material of choice for the wall-lining material relied upon to control the high-temperature gas plasma of fusion processes, such as that

Figure 5.2 Six of the 18 mirrors of the James Webb space telescope undergoing cryogenic testing at the Marshall Space Flight Center, Huntsville, Alabama, USA. (Photo: K. Hutchison, courtesy of Ball Aerospace and Technologies Corp.)

used in the current JET (Joint European Taurus) reactor and the larger-scale ITER (International Thermonuclear Experimental Reactor) fusion reactor project, now under construction in France.

The physical properties of beryllium can be modified by the addition of up to 62 per cent by weight of aluminium, thereby producing an alloy with enhanced ductility which may be easily machined by conventional metal-cutting techniques. This alloy finds many applications for lightweight high-strength components of aerospace and electronics systems such as structural components of aerospace and munitions guidance systems.

Beryllia (BeO) ceramics

Beryllium oxide, or beryllia (BeO), ceramic accounts for the remaining five per cent of beryllium consumption.

Beryllia ceramics have an exceptional combination of properties, the most important of which is the combination of high electrical insulation, coupled with a hardness only slightly lower than that of diamond, and thermal conductivity an order of magnitude greater than that of alumina. Its low dielectric constant and low loss index permit its use as an electronics circuit substrate with extremely good performance at high frequencies.

The coefficient of thermal expansion of beryllia is intermediate between that of silicon or gallium arsenide and that of typical metals, making it compatible with many adjacent structural components.

Beryllia is widely used as an electrical insulator, and in heat sinks for radio-frequency and radar equipment, automotive electrical systems and for laser bores and microwave waveguides. In all of those applications, the combination of heat extraction and electrical insulation is essential.

World production

Until 1962, beryllium mine production, mostly in the form of beryl from artisanal operations, had come from at least 14 countries. Artisanal mining of beryl, associated with gemstone mining, continues globally and provides a small proportion of the feedstock for USA operations, but a significant share for Chinese operations. Since 1962 a dedicated mine operated by Materion Corporation in the Spor Mountain region of Utah, USA, has provided an estimated 84 per cent of global production which is predominantly derived by the refining of the bertrandite ores extracted from that site, augmented with a small volume of imported beryl and recycled materials.

In 2010, three countries, the USA, Kazakhstan and China, produced most of the commercial beryllium and derived products. Production of beryllium in China (about 16 per cent of the world total) is from both local ores and imported beryl, while Kazakhstan derives its beryllium from historic inventories of Russian ore concentrates. A breakdown of global production by region for the main forms in which beryllium is used commercially is given in Table 5.2. The estimates presented include beryllium from all sources including both newly mined and stockpiled ore, government inventories and scrap.

Walsh and Vidal (2009) indicate that China has an annual capacity of about 1500 tonnes of beryl production, about half of which is located in Xinjiang Province and the remainder in Guangdong Province, with a fraction of this currently being worked. Processing of the ore from the Koktokay mine in Guangdong Province was carried out at the Fuyun facility, and was originally exported to the former USSR. Fuyun is reported to have an annual capacity of 100 tonnes/year of beryllium oxide and 1000 tonnes/year of copper–beryllium alloys. This area also hosts about 70 per cent of China's beryllium resources which are in the form of the mineral beryl.

In addition, China imports an estimated 800 tonnes of beryl annually, which, at an average BeO content of 11 per cent, would provide a total annual production of approximately 35 tonnes of beryllium at normal refinery efficiencies.

The British Geological Survey (BGS) compiles annual statistics for beryllium production from

Table 5.2 Estimated global production of beryllium from all sources by region in 2011 (Source: Beryllium Science and Technology Association, 2012a.)

	USA, Japan and Kazakhstan (a) kilograms	China (b) kilograms	Others (b) kilograms	Total kilograms
Be contained in metal >60% Be content	43,000	5000	1500	49,500
Be contained in alloys <60% Be content	283,000	59,000	5000	347,000
Be contained in BeO ceramics	2500	1000	0	3500
Total Be contained in all products produced in 2011	328,500	65,000	6500	400,000

(a) Total aggregated by an independent auditor from data provided by BeST members.
(b) Estimates.

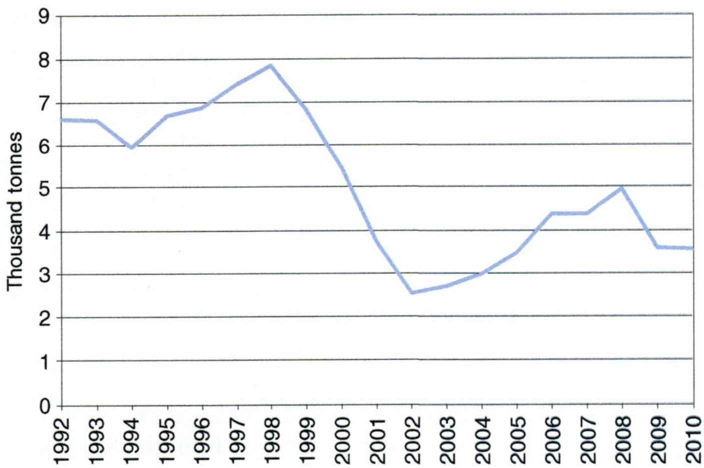

Figure 5.3 Global annual mine production of beryl for the period 1992–2010 (production from bertrandite ore is calculated as equivalent to beryl containing 11% beryllium oxide). (Data from British Geological Survey, 2012.)

currently operating mines (Figure 5.3). The data are shown as tonnes of beryl equivalent, based on a BeO content for beryl of 11 per cent.

The rapid decline in global production after 1998 is thought to be due in part to a slowing of the 'technology revolution' (Cunningham, 2004a), and ongoing miniaturisation of electronic devices (Figure 5.3). The drive for miniaturisation in electronics has significantly reduced the amount of copper alloys used, and hence that of beryllium, in this sector. This is illustrated graphically by the transformation of the mobile phone from near brick-size proportions of the early 1990s to its present very small size. However, it is also important to note that between 2000 and 2010 Materion Corporation temporarily stopped production of beryllium metal to allow construction of new facilities in Ohio for the refining of high-purity beryllium metal. During that period, Materion replaced production of metallic beryllium from newly mined ore with beryllium metal from the stockpile held by the US Defense Logistics Agency. Other beryllium users also took advantage of government surplus sales at that time, so the world requirement for newly mined beryl and beryllium ores was depressed temporarily. Materion's new plant, which commenced production in 2011, was built with US government financial assistance and is currently producing high-grade pure beryllium. Another contribution to the recent general trend of increasing production may be due to increases in demand for copper–beryllium alloy which may reflect the growing use of electronics, and consequently the number of copper–beryllium alloy connectors, in automotive applications (Jaskula, 2012a).

World trade

Global trade in beryllium in 2009 is summarised in Figures 5.4 and 5.5.

Chinese beryl and beryllium-containing products have been intermittently available in the global market place since at least the 1960s. The Ulba Metallurgical Plant JSC company in Kazakhstan has developed a growing beryllium-product export business, based upon a former Soviet facility at Ust Kamenogorsk. The Ulba operations rely on imported beryl concentrates, imported at that time from mines in Russia. However, the inventory is finite and, when depleted, Kazakhstan will need to develop an alternative source of beryl or beryllium-bearing material.

The large import of beryllium products into France and the USA reflects the intake of copper–beryllium intermediate shapes by the NGK Metals Corporation copper–beryllium alloy finishing facilities in Couëron, near Nantes, France and Sweetwater, Tennessee in the USA. These are sourced from the parent company's alloy production facility in Chita, Japan which imports refined beryllium oxide from Materion USA for reduction to copper–beryllium alloys (Jaskula, 2012b).

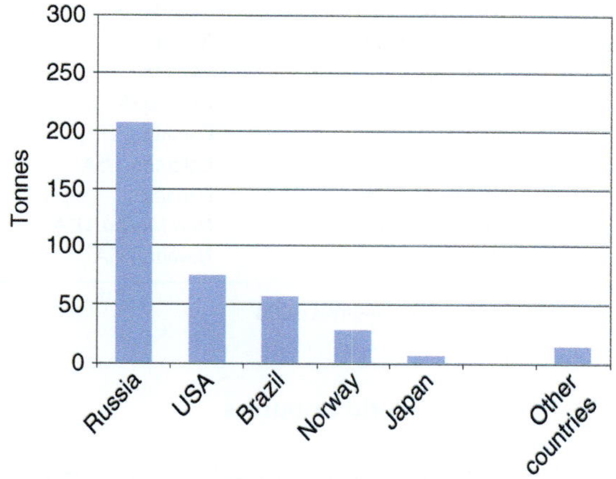

Figure 5.4 Global exports of beryllium by country in 2009. The data are shown in terms of contained beryllium metal in ores, concentrates and scrap. (Data from UN Comtrade, 2013.)

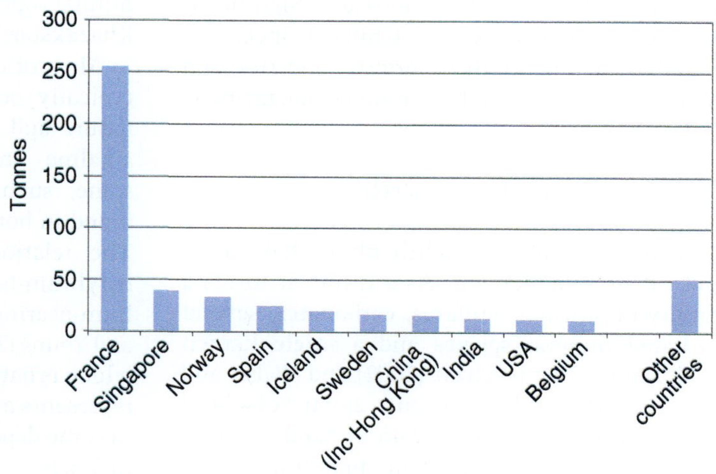

Figure 5.5 Global imports of beryllium by country in 2009. The data are shown in terms of contained beryllium metal in ores, concentrates and scrap. (Data from UN Comtrade, 2013.)

Table 5.3 Published beryllium resources (as beryllium oxide, BeO). (After Sabey, 2006.)

Deposit(s)	Location	Resource, tons BeO	Grade, % BeO
Various	North Carolina, USA	122,800	0.05
Spor Mountain	Utah, USA	72,315*	0.71
McCullough Butte	Nevada, USA	47,000	0.027
Various	Brazil	42,000	0.04
Strange Lake	Canada	42,000	0.08
Aqshatau	Kazakhstan	16,000	0.03–0.07
Thor Lake	Canada	13,300	0.76
Various, Black Hills	South Dakota, USA	13,300	na
Sierra Blanca	Texas, USA	11,300	>2.0
Lost River	Alaska, USA	>10,000	0.3–1.75
Yermakovskoye	Russia	>10,000	1.3
Seal Lake	Canada	6800	0.35–0.40
Gold Hill	Utah, USA	>5000	0.5
Tanco	Canada	1800	0.20
Boomer	Colorado, USA	<1000	2.0–11.2
Hellroaring Creek	Canada	<1000	0.10
Iron Mountain	New Mexico, USA	<1000	0.2–0.7
Mount Wheeler	Nevada, USA	<1000	0.75

*Remaining reserves 2004

World resources

Beryllium (BeO) resources in selected countries are shown in Table 5.3. The estimates for Spor Mountain, Utah, USA are proven reserves that meet international reporting standards such as NI 43-101 in Canada or the JORC code in Australia, but many of the other estimates are dated and would not qualify under present standards. Significant resources may also exist in China, France, various African and South American countries, and Afghanistan, but these have not been quantified.

Mineralogy of beryllium

Beryllium can behave as a lithophile, chalcophile and siderophile element. As a result, it forms a variety of silicates, sulfates, carbonates, several hydrated mineral species and a single known oxide. According to Grew (2002) and Walsh and Vidal (2009), beryllium occurs as an "essential ingredient in approximately 45 minerals" and as an "occasional constituent in approximately 50 other minerals". Selected beryllium-bearing minerals are listed in Table 5.4 according to their BeO content. The most important minerals from an economic standpoint are bertrandite and beryl. Other minerals which have been produced intermittently or on a small scale in the past include phenacite. Potential future production may come from the restart of former phenacite mining operations in the Kalesay deposit in Khazakstan.

Most of the other minerals listed in Table 5.4 typically occur in minor or trace amounts in nature and are most commonly found in sub-alkaline granitic pegmatite suites. However, some, such as bertrandite and phenacite, are found in both sub-alkaline and alkaline systems. The relationship between a large number of beryllium-bearing minerals and the main beryllium-bearing deposit types is provided by Barton and Young (2002). The high-grade mineral behoite, which is naturally occurring beryllium hydroxide, represents a potential work-environment concern in some deposits, for example in the Sierra Blanca in Texas.

Table 5.4 Selected beryllium minerals and their compositions. Those of current economic importance, bertrandite and beryl, are in bold.

Mineral	Formula	BeO content (wt %)
Bromellite	BeO	98.02
Behoite	$Be(OH)_2$	58.13
Hambergite	$Be_2(BO_3)(OH)$	53.5
Phenacite	Be_2SiO_4	45.5
Bertrandite	$\mathbf{Be_4Si_2O_7(OH)_2}$	**39.6–42.6**
Moraesite	$Be_2(PO_4)(OH)4H_2O$	25–28
Hurlbutite	$CaBe_2(PO_4)$	21.3
Berylonite	$NaBe(PO_4)$	19–20
Chrysoberyl	$BeAl_2O_4$	19.8
Euclase	$BeAl(SiO_4)(OH)$	17.0–21.8
Herderite	$CaBe(PO_4)(OH,F)$	15–16
Barylite	$BaBe_2(Si_2O_7)$	16
Beryl	$\mathbf{Be_3Al_2Si_6O_{18}}$	**11.0–14.3**
Helvite	$Mn_8(BeSiO_4)_6S_2$	11–14.2
Danylite	$Fe_8(BeSiO_4)S_2$	12.7–14.7
Genthelvite	$Zn_8(BeSiO_4)_6S_2$	12.6
Eudidymite	$Na(BeSi_3O_7)(OH)$	10.5–11.2
Gadolinite	$(Y,Ca)_2Fe(BeSiO_4)_2(O,OH)_2$	5.5–12.9

Beryllium deposits

Commercial production of beryllium has been derived from two principal types of deposit: granitic pegmatite deposits and hydrothermal-metasomatic deposits. Historically, beryllium has been extracted from pegmatitic deposits of all ages and distributed widely around the globe. However, since 1969 most of the world's beryllium production has been derived from one large low-grade hydrothermal deposit of Tertiary age located at Spor Mountain, Utah, USA. The locations of the most important beryllium deposits are shown in Figure 5.6, divided into those hosted in pegmatites and those of other types. It is important to note that pegmatites commonly occur in large numbers (a 'swarm') within a particular area (pegmatite 'field'). Within any one field mineral extraction may be focused on a few large deposits or, more commonly, small-scale mining is carried out at many sites.

Pegmatite deposits

Pegmatites are defined as very coarse-grained, crystalline rocks commonly of granitic composition. It is the large size of the individual crystals, which may attain metre or larger scale in pegmatites, that allows the processing of relatively coarse fragments, and/or hand sorting ('cobbing') of those minerals of economic interest.

Granitic pegmatites may be divided into two classes: those of the Lithium-Caesium-Tantalum (LCT) association and those of the Niobium-Yttrium-Fluorine (NYF) association (Černý, 1991a). For a detailed description of these classes, the reader is referred to Chapter 15. The LCT pegmatites generally occur in pegmatite 'swarms', whereas those of the NYF kindred tend to occur within restricted areas, commonly proximal to their parental rocks. The LCT pegmatites are the more important source of beryl.

Pegmatites have long been exploited as sources of mica, feldspar, quartz, gemstock (such as chrysoberyl, aquamarine, morganite, hiddenite, kunzite and polychrome tourmalines) and the elements tin, tantalum, niobium, lithium, rubidium, caesium and beryllium (London, 2008). In the LCT system micas, feldspars and quartz are ubiquitous but the rare elements in pegmatites display a zonal pattern of element enrichment and mineral variation resulting from increased fractionation with distance of the pegmatite from its parental source (Figure 5.7). No scale is given in Figure 5.7 as the field or swarm of pegmatites can be highly variable in extent, ranging from tens of metres to several kilometres (Černý, 1991a). Beryllium, usually in the form of beryl, makes an early appearance in the zoned sequence close to the source granite intrusion and continues to occur into the most highly fractionated and distal members of the pegmatite field. However, with increasing fractionation the elements sodium and caesium, for example, substitute for beryllium in the beryl lattice and lower the quality of the beryl as a feedstock.

The Harney Peak Granite of the Black Hills, South Dakota, is an excellent example of such a pegmatite field and was a primary source of beryl for the US industry until 1962 (Shearer et al., 1992).

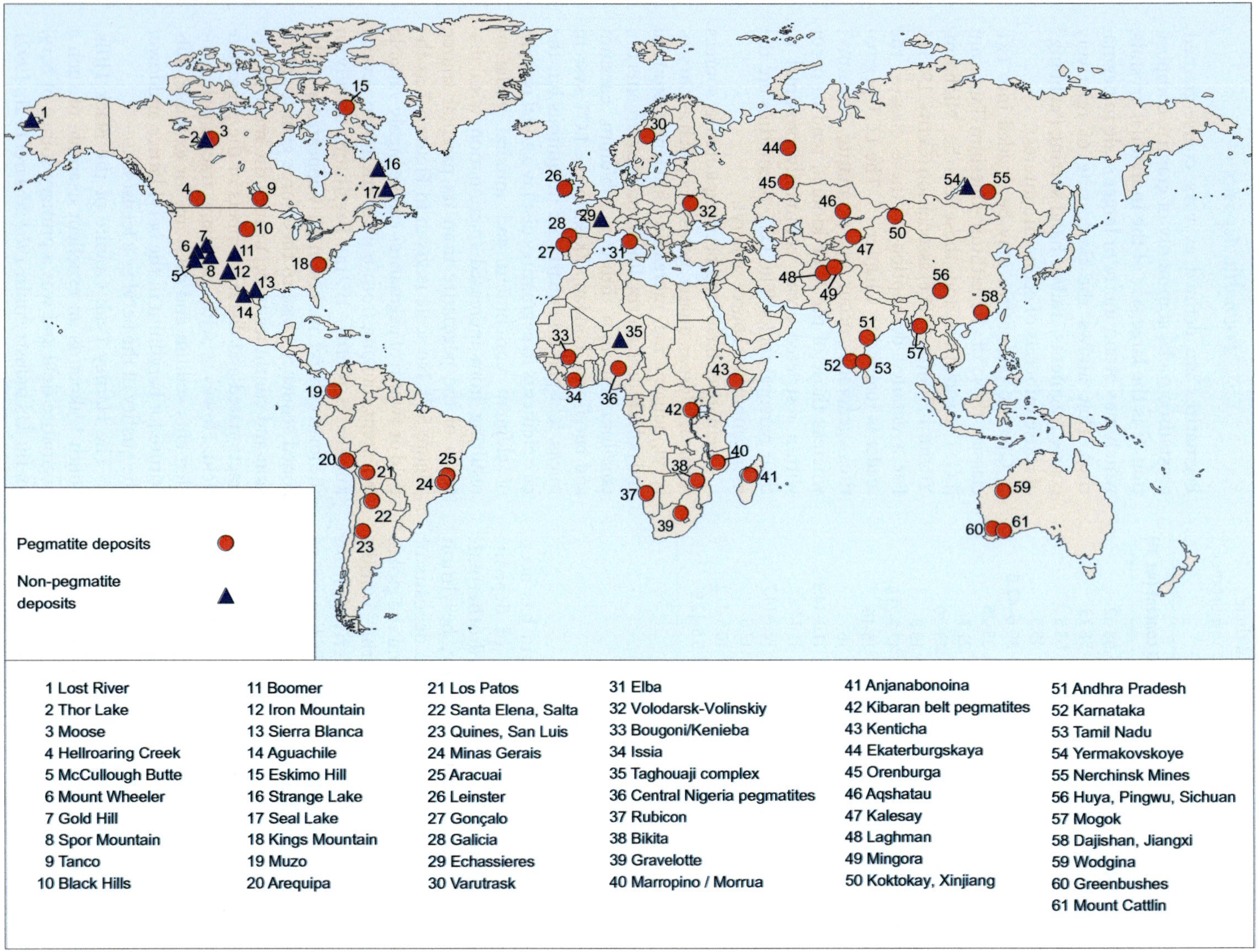

Figure 5.6 The global distribution of significant deposits of beryllium. Some of the symbols on the map represent a single important deposit or resource, while others represent a cluster of deposits in one area or region.

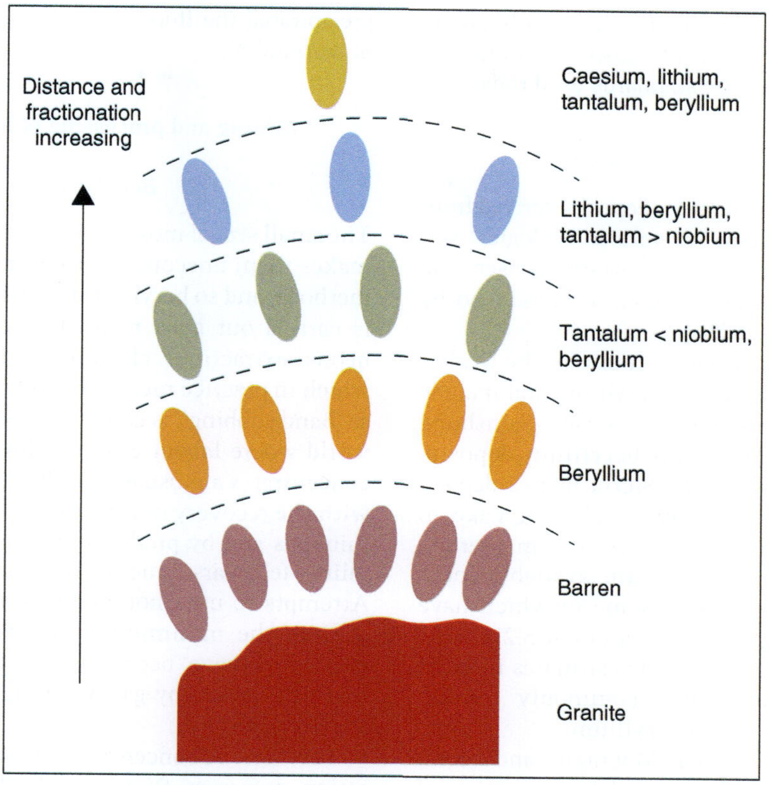

Figure 5.7 Rare-element zoning in pegmatite fields. (After Trueman and černý, 1982.)

The Arquehanna River pegmatites in the Aracuai area in Minas Gerais Province, Brazil, which have hosted intermittent beryl, tin, tantalum, lithium and gemstock production, display a convergence and overlap of three such suites of pegmatites.

Individual pegmatite deposits vary in size from metres to more than a kilometre in length and hundreds of metres in width. The largest, so-called 'giant', pegmatites may display internal zoning in a similar manner to the field or swarm in which they are located (Černý, 1991b). Examples include the Tanco pegmatite in Manitoba, Canada (Černý, 2005 and Linnen et al., this volume), the Greenbushes pegmatite in Western Australia (Fetherston, 2004) and the Bikita pegmatite in Zimbabwe (Černý et al., 2003).

No commercial beryllium production is known from NYF pegmatites but increased beryllium demand and the availability of suitable metallurgical processes could change this in the future. A novel and unique flotation process was developed by Bulatovic (1988) that recovered both phenacite and bertrandite from the Thor Lake deposits in Canada. However, these deposits remain undeveloped.

Pegmatite mineral deposits are difficult to evaluate for their contained beryllium resources. The minerals of particular interest occur in a wide range of sizes, from microscopic to metre scale, and are commonly widely dispersed, even within internally zoned pegmatites. Beus (1962) provides comprehensive procedures for evaluating pegmatitic beryllium deposits. These include stripping, trenching, drilling and underground development. Channel sampling may need to be conducted at intervals as closely spaced as two metres. This rigorous sampling process allowed Beus to categorise beryllium resources on the contemporary protocols of the USSR, but would probably also be compliant with resource-reporting standards used today.

Hydrothermal deposits

This broad category includes deposits derived from hydrothermal processes in various geological settings, including replacement, skarn, greisen and vein deposits, several of which are described by Barton and Young (2002).

Hydrothermal deposits account for the world's largest known resources of beryllium and include the Lost River Tin deposit in Alaska (Sainsbury, 1963), the Spor Mountain beryllium deposits in Utah (Davis, 1979), the Sierra Blanca deposit in Texas (Price et al., 1990) and Thor Lake in Canada's Northwest Territories (Trueman et al., 1988). Resource estimates are available for a number of these deposits, some of which have BeO grades in the per cent range (Table 5.2). These contrast markedly with grade estimates in beryl pegmatite deposits which commonly contain tens or hundreds of ppm beryllium.

The Lost River, Spor Mountain and Sierra Blanca deposits are related to felsic magmatism of latitic composition. The extrusive host rocks (termed 'ongonites' by Kovalenko et al., 1971) and their intrusive equivalents are enriched in fluorine, beryllium and uranium, while Lost River is also well known for its tin deposits (Sainsbury, 1963). The high fluorine content depresses the freezing point of these magmas and allows protracted enrichment of incompatible elements, such as beryllium, into the late hydrothermal phase. When the hydrothermal fluid encountered carbonate rocks fluorite was formed, while beryllium was deposited in the host volcanic rocks as bertrandite and phenacite and, in the Sierra Blanca deposit also as the rare mineral behoite (Sabey, 2006).

The Spor Mountain deposits formed from a low-temperature beryllium–uranium-fluorine-enriched hydrothermal fluid, circulating through Oligocene base surge rhyolite deposits (Lindsey et al., 1973, and Burt et al., 2006). The rhyolite deposits lie unconformably on Palaeozoic dolomites which are thought to have been the source of calcium that precipitated the fluorite and led to the deposition of bertrandite.

Mining and processing of beryllium

Beryl ores

The small size of most granitic pegmatite deposits makes them uneconomic to mine by mechanical methods, and so beryl extraction from pegmatites is carried out by a manual process of simultaneous extraction, cleaning and concentration which in practice means that this process, known as 'hand cobbing', is confined to those parts of the world where labour costs are low. The beryl ore concentrates are usually produced in conjunction with the recovery of gemstones or other valuable minerals and by-products, such as columbo-tantalite, feldspars, micas, and lithium minerals. Attempts to use photometric sorting of beryl to achieve the minimum 11.0% BeO required by processors have been unsuccessful due to the masking caused by gangue minerals adhering to the beryl surface.

Beryl can be concentrated by various novel flotation processes (Vidal et al., 2009). However, none of these are economically attractive for the extraction of beryllium on a commercial scale because of the relatively high cost of hydrofluoric and oleic acids, the principal reagents. The beryl flotation concentrates formerly produced and offered commercially from China were found to contain an undesirable mix of beryllium minerals that proved difficult to process because of excess remaining oleic acid.

Bertrandite ores

Materion has been mining bertrandite ores at Spor Mountain, Utah since 1969 (Figure 5.8), from eight open pits located along a complex fault-segmented ore body. An earlier attempt by Anaconda Minerals to conduct underground mining of the bertrandite ore was abandoned due to poor ground conditions.

Prior to mining, rigorous delineation of the bertrandite orebodies is undertaken and close

Figure 5.8 Materion Brush open-pit mining at Spor Mountain, Utah. (Courtesy of Materion Corp.)

control of grade, or beryllium content, continues through stockpiling of ore before delivery to the processing plant. Most assaying for grade control is carried out with beryllometers, which are hand-held, portable instruments that use a radioactive gamma radiation source to displace an excess loosely attached neutron in the beryllium nucleus. The neutron flux events are counted by a scintillometer that provides a quantitative analysis of the beryllium content at that location.

The shallow, westerly dipping orebodies at Spor Mountain are first drilled on 30-m centres and assayed at vertical intervals of 0.6 m. Overburden, hanging wall rhyolites and alteration clays are removed and a one-metre layer of clay is left on top of the ore. The overburden and waste are stockpiled for later restoration of the mined area. The orebody is next drilled and assayed on 7.5-m centres. The remaining clays are removed and the orebody is mined by loaders and scrapers guided by in-pit assaying utilising beryllometers.

Mined ore is gathered on a stockpile which is, in turn, drilled and assayed for grade control. There is no additional concentration process, and these 'direct shipping' ores, at a grade of about 0.265% Be are carried by truck 80 km to the Materion process plant, north of Delta, Utah.

Processing of beryl and bertrandite to beryllium hydroxide

There are essentially three processes for treating beryllium minerals to produce beryllium chemicals which are the starting points for beryllium-bearing commercial products:

- The Materion Corporation uses the Kjellgren–Sawyer sulfate extraction method to treat beryl ores in a mixed flowsheet, as described below.
- Walsh et al. (2009) reported that beryllium processing in the CIS and in China probably utilises the Kjellgren–Sawyer process as well. In the author's (PS) experience, however, the CIS and China have used the alkaline flux, or Degussa fusing process, in which beryl ores are mixed with quicklime, heated to 1500 °C and fritted. The frit or glass is then taken into solution in sulfuric acid and beryllium hydroxide precipitated with ammonia.
- The Copaux–Kawecki process, consists of heating beryl with sodium silicofluoride to produce sodium beryllium fluoride which is in turn aqueous leached to produce a beryllium solution (Walsh et al., 2009).

Different processes are required to treat the beryl and bertrandite which enter the Materion plant as hand-cobbed beryl concentrates and direct shipping ore, respectively (Figure 5.9).

- Beryl concentrates, after crushing, are heated to 1700 °C and quenched rapidly in water to form a frit or glass. The frit is heat treated at 1000 °C, ground to finer than 200 mesh (75 μm), and then leached with a concentrated sulfuric acid solution at 250 to 300 °C. This process extracts the beryllium, forming a beryllium sulfate solution.
- Bertrandite ore is crushed and wet-milled to yield a fine slurry, which is then leached with a sulfuric acid solution at about 95 °C to extract the beryllium, forming a beryllium sulfate solution.

After separating the solids from both streams of beryllium sulfate solutions, the two solutions are combined. Solvent extraction is used to remove additional elements that were extracted with the beryllium. During solvent extraction, the beryllium sulfate solution contacts an organic

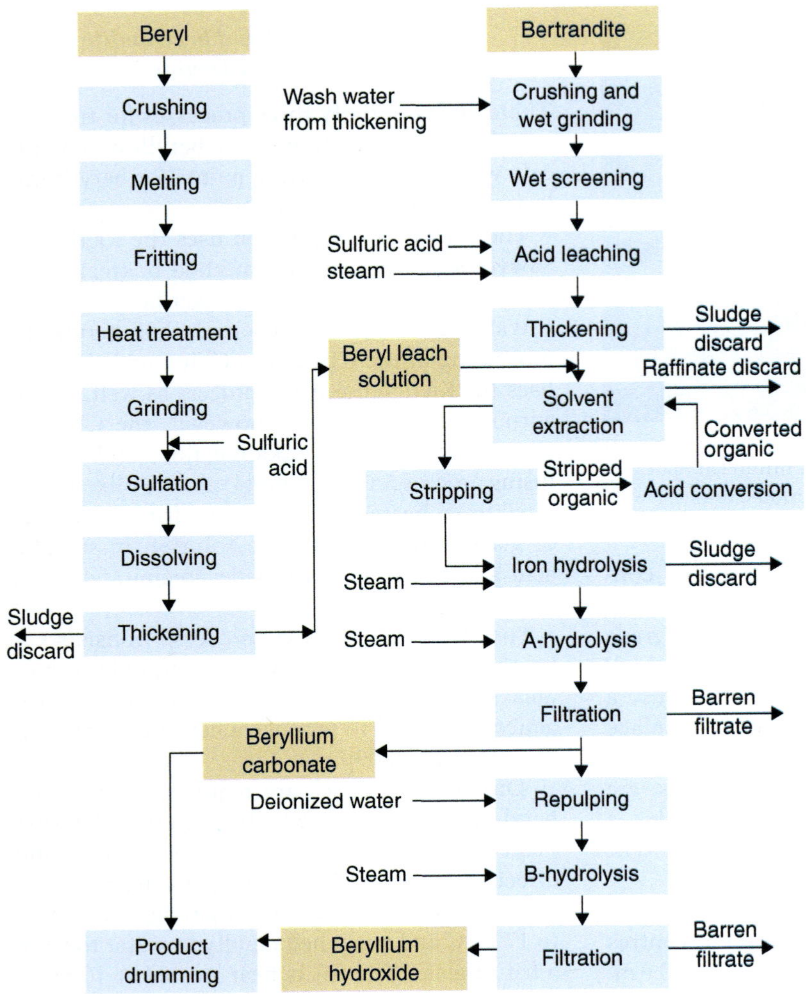

Figure 5.9 Beryllium hydroxide production from bertrandite and beryl. (Source: Materion Corp.)

solution of di-2-ethylhexylphosphoric acid in a kerosene-type organic solvent. Beryllium is selectively dissolved in the organic solution. The slow rate of extraction at room temperature is accelerated by warming the extractant and leach solution. The loaded organic phase is treated with an aqueous ammonium carbonate solution and beryllium is stripped from the organic phase, forming tetraammonium beryllium tricarbonate ($(NH_4)_4Be(CO_3)_3$).

In addition to beryllium, the solvent extraction step dissolves iron and small amounts of aluminum and fluoride present in the sulfate leach solution. Heating the strip solution to about 70 °C separates the iron and aluminum as hydroxide or carbonate precipitates. Two hydrolysis steps at 95 °C and 165 °C, respectively, remove the ammonia and carbon dioxide and result in the formation of beryllium hydroxide, which is filtered and then drummed for shipment to Materion's plant in Elmore, Ohio for further processing to finished products containing beryllium. The efficiency of this processing route is high, with approximately 80 per cent of the beryllium content of both beryl and bertrandite recovered (Walsh, 1979).

Production of metal and alloys from beryllium hydroxide

Beryllium hydroxide is the starting point for production of beryllium metal and a number of alloys and ceramics that are produced widely across the United States of America and also at other facilities in Europe and Asia. The majority of the beryllium hydroxide produced at Delta, Utah is processed at Materion's operations in Elmore, Ohio.

Beryllium metal is produced using the Schwenzfeir process (Figure 5.10). The hydroxide is reacted with ammonium fluoride to form ammonium fluoroberyllate ($(NH_4)_2BeF_4$) and then heated to produce amorphous beryllium fluoride (BeF_2). The fluoride is then reduced with magnesium at temperatures between 900 °C and 1300 °C to produce beryllium 'pebbles' and beryllium fluoride/magnesium fluoride slags. These are water leached and the remaining beryllium pebbles are melted under vacuum to remove entrained gas or trapped slags and then cast into ingots (Figure 5.11). Vacuum melting is also a start point for recycling of beryllium scrap (Stonehouse, 1985).

The ingots are subsequently machined into chips or cuttings that are ground and screened to produce a specific particle-size distribution of rounded particles. These are next compacted by cold, hot or isostatic pressing into 'blocks', that can be made to have specific combinations of properties designed for a variety of end uses (Figure 5.11).

Many beryllium alloys are produced commercially, the most important of which are those of copper, nickel and aluminum. These alloys, which vary in their beryllium content from 0.15% Be up to 62% Be, are produced in many forms including casting alloys, plate, rod, bar, tube, strip and wire products.

Commercial copper–beryllium alloys are manufactured by first making a master alloy containing 3.5–10% Be. The most widely used process utilises an electric arc furnace to produce a stoichiometric composition of 3.3–3.5% Be by adding calcined beryllium hydroxide to pure copper, together with a source of carbon, such as graphite, as a reducing agent. The master alloy is subsequently melted with additional copper and some critical alloying elements, such as nickel and cobalt, to produce a range of alloy compositions containing between 0.3–2.0% Be. By varying the composition, a controlled range of physical and mechanical properties can be obtained. Generally, two alloy families are used: those containing 1.8–2.0% Be, called 'gold alloys' due to their colour, which are used where strength is paramount; and those containing 0.10–0.25% Be called 'red alloys', which are used where electrical and thermal conductivity properties are critical.

Before use, alloys are heated to fully dissolve all of the elemental additions, and to homogenise and refine the size, shape and location of the beryllide precipitates, as well as soften the copper matrix. This is termed the solution-annealed state. The intermediate shapes are subsequently hot worked by such traditional processes as hot rolling, extrusion or forging to a semi-finished size and shape. Further cold forming is carried out by conventional metal working processes such as rolling to thinner gauge strip products ready for stamping to size/shape, or by cold forging or drawing to rod, wire or tube.

Solution-annealed material can be hardened by two complementary methods, work hardening and heat treatment, which alone, or in combination, provide a powerful control mechanism to obtain a desired combination of properties. The heat-treatment process can be carried out by the copper–beryllium manufacturer to deliver a product termed 'mill hardened' that is ready to be used by manufacturers to form the desired shape without further processing. In other cases, it can be more convenient for additional forming operations, such as machining and stamping, to be carried out by a component manufacturer starting from an alloy delivered in the more ductile and softer solution-annealed form.

Production of beryllium oxide from beryllium hydroxide

Beryllium oxide is produced by dissolving beryllium hydroxide in sulfuric acid to produce hydrated beryllium sulfate. The sulfate is

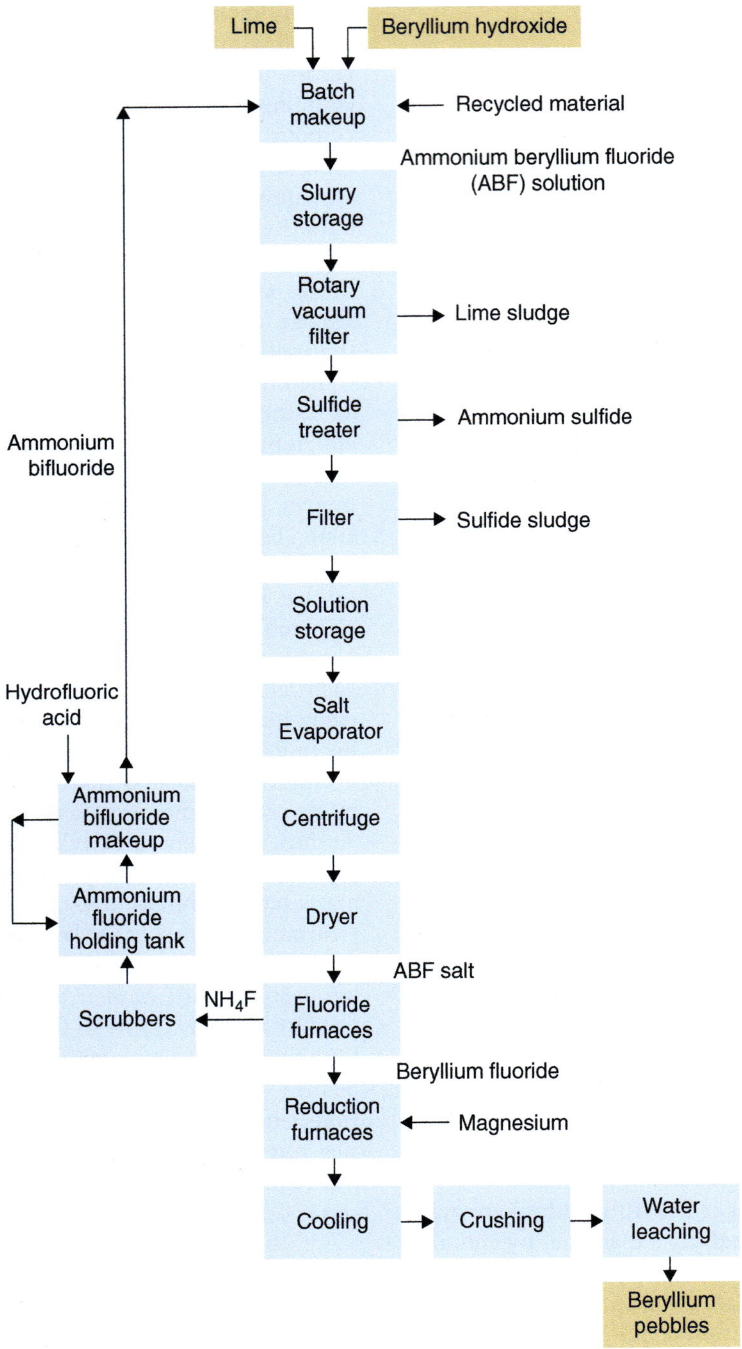

Figure 5.10 Production of beryllium metal (Source: Materion Corp.)

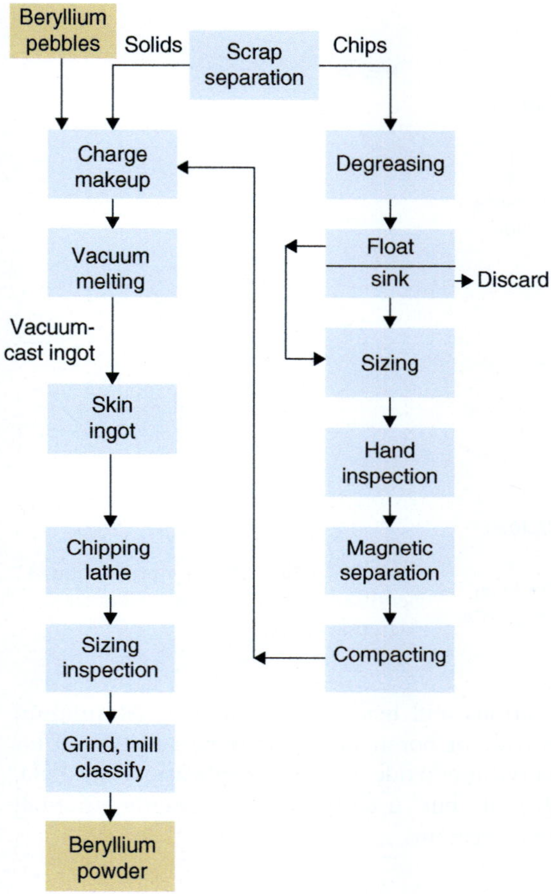

Figure 5.11 Final purification, grinding and sizing of beryllium powders. This is also a starting point for beryllium recycling. (Source: Materion Corp.)

precipitated, concentrated and calcined at 1430 °C to produce the oxide (Figure 5.12). The beryllium oxide, or beryllia, is supplied to the electronic and electrical sectors as fired ceramic components for use in many different applications.

Recycling

The pure beryllium metal components used in technological applications have extremely long lifetimes, and, therefore, return to the recycle stream very slowly. Some, because of applications in space or because of their sensitive military nature, do not return at all. When these components do finally return at end-of-life, they can be easily recycled. The aerospace industry is the largest source of pure beryllium new scrap, as many of the end uses only have short production runs. Because of its hardness, and the relative difficulty to cut it at high rotational surface speeds with conventional tooling, beryllium is considered a difficult metal to machine extensively. Accordingly, it is machined to final form from near net shapes, made by hot isostatically pressing or casting, resulting in significant amounts of chips (i.e. new scrap). A good example is the machining of the back reinforcing web structure of the James Webb telescope mirrors which resulted in 92 per cent of the original rough beryllium shapes being returned. When beryllium metal is recycled, the Beryllium Science and Technology Association (2012b) estimates a 70 per cent energy saving over the cost of producing newly won beryllium from ore.

Overall, 'new scrap' generated in the course of manufacturing beryllium-bearing products makes an important contribution to supply and may account for as much as 10 per cent of apparent consumption (Cunningham, 2004b). In Europe, a 45 per cent net recovery of new scrap copper–beryllium alloys from component manufacturers back to alloy producers has been reported (European Commission, 2010).

When it can be isolated and recovered, copper–beryllium or nickel–beryllium alloy end-of-life scrap is directly recycled to produce new alloys. A significant premium is paid by copper–beryllium manufacturers to stimulate the return of such scrap in order to take advantage of the energy conservation and sustainability advantages compared to extraction from ore.

It is generally not economic to recover beryllium metal from copper–beryllium alloys, such as those used in electronic components, because the beryllium content in each device is small and the beryllium content in the alloy is very low (<1.25%). Furthermore, the components are generally small and difficult to separate. The United Nations Environment Programme (UNEP, 2011) notes that most components of electronic and electrical

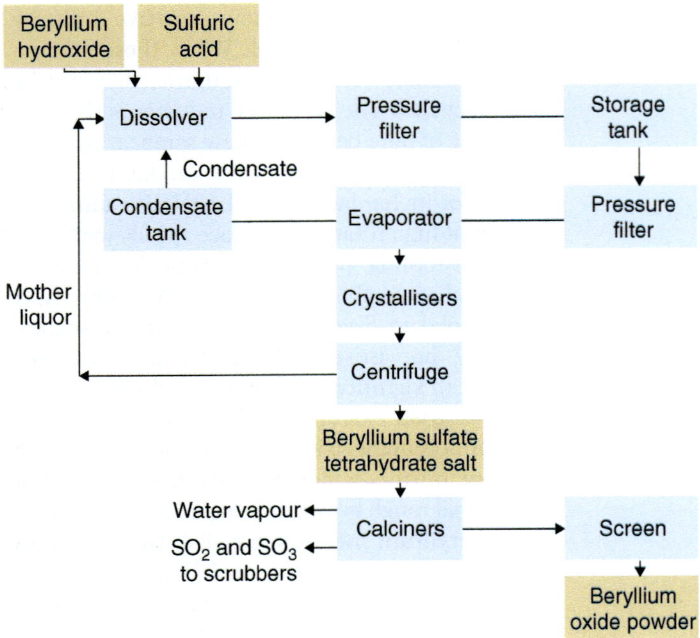

Figure 5.12 Flow sheet for production of beryllia or beryllium oxide. (Source: Materion Corp.)

devices are recycled for their copper value alone and end up in a copper smelter where the beryllium is captured in the slag and effectively lost. As a result, the recycled beryllium content of old scrap is low and the end-of-life recycling rate very low, at less than one per cent (UNEP, 2011).

Substitution

On account of its high cost beryllium is commonly used only where its specific properties are crucial and consequently substitution is difficult in many applications (Sabey, 2006). However, for some non-critical purposes certain composite materials, high-strength grades of aluminium, pyrolytic graphite, silicon carbide, steel, or titanium for example, may be substituted in place of beryllium metal or beryllium composites (Jaskula, 2012a). Many different copper alloys, such as phosphor bronzes, may be substituted for copper–beryllium alloys but these invariably result in some loss of performance, and energy loss due to the requirement to use thicker sections and hence higher weight. Aluminium nitride or boron nitride may be substituted for beryllium oxide in some applications (Jaskula, 2012a), but usually provide inferior thermal management.

Environmental aspects

A disease of the lungs, called chronic beryllium disease (CBD) or berylliosis, has been known in Europe and the USA since the 1930s and 1940s. Most reports were derived from the nuclear weapons industry and from the manufacture of fluorescent lamps containing beryllium-bearing phosphors. However, the health effects are now known to depend on numerous factors including the nature of the beryllium-bearing material (metal, alloy, ceramic, salt, etc.), the solubility and the level of exposure. According to the Beryllium Science & Technology Association, "Beryllium metal, copper–beryllium alloys (CuBe), aluminium–beryllium alloys (AlBe) and nickel–beryllium alloys (NiBe), in solid form and as contained in finished products, present no special health risks." Similarly, it has been shown that there is relatively little risk of CBD in the mining and processing of beryllium ores (Deubner et al., 2011).

In order to contract CBD, an individual must be exposed to airborne beryllium in the form of a dust, mist or fume and become sensitised to beryllium. This sensitisation is an immunological lymphocyte proliferation response in some individuals (Donovan et al., 2007). Only those persons who are genetically susceptible can become sensitised to beryllium (Wang et al., 1999 and 2001). Before the late 1980s, workers were diagnosed with CBD only when they exhibited clinical (observable) symptoms of CBD and changes in their chest X-ray or lung function test. During the late 1980s and early 1990s, the criteria by which CBD was diagnosed changed, and workers began to be diagnosed with CBD without clinical symptoms or measurable impairment. This diagnosis became possible as a result of the application of new technology in medical testing and evaluation. Workers diagnosed with CBD in the absence of X-ray or lung function changes or symptoms of disease are referred to as having sub-clinical CBD, meaning that they have no clinical symptoms or measurable impairment. Workers with subclinical CBD may never develop clinical CBD or may develop clinical CBD over time (Borak et al., 2006).

For situations where risk of inhalation exposure existed, an airborne standard was developed in the USA in the late 1940s that was generally adopted internationally and is still used today. This standard is two micrograms beryllium per cubic metre of air (2 $\mu g/m^3$), averaged over an eight-hour work period. Companies producing beryllium have more recently implemented a preventative model that incorporates an eight-hour exposure guideline of $0.2\,\mu g/m^3$. Scientific research and actual workplace experience in beryllium production facilities in the USA indicate that the use of engineering and work practice controls have been effective in preventing any health effects including sub-clinical CBD and clinical CBD.

There has been considerable debate as to whether beryllium should be regarded as carcinogenic to humans (Fulton and Goldberg, 2009). In the USA the National Toxicology Program (2002) lists beryllium and certain beryllium compounds as substances reasonably anticipated to be carcinogens, while the International Agency for Research on Cancer (2012) classified beryllium and beryllium compounds as carcinogenic to humans. The United States Environmental Protection Agency classifies inhaled beryllium as a probable human carcinogen (USEPA, 1998). However, the latest genotoxicity test results, together with expert reviews which highlighted inherent flaws in the epidemiological assessments used in previous studies, have indicated that beryllium metal is unlikely to be carcinogenic to humans (Hollins et al., 2009; Strupp, 2011a and 2011b; Rothman and Mosquin, 2011; and Boffetta et al., 2012).

Beryllium is typically present at concentrations of 0.5 to 2 ppm in soils and rocks throughout the world. Consequently it is commonly found at low levels in natural products like coal, wood, vegetables, foodstuffs and gemstones (ATSDR, 2002). Beryllium intake from the air and dust can be increased by 2–3 orders of magnitude in the vicinity of a point source, such as a coal-fired power plant. Beryllium is found in tobacco and therefore tobacco smoke is a potential source of exposure to beryllium in the general population. In plants and vegetables beryllium is found at low (ppb) levels that pose no risk to human health. Beryllium has been measured (fresh weight) in rice at $72\,\mu g/kg$, lettuce at $16\,\mu g/kg$, kidney beans at $2200\,\mu g/kg$, peas at $109\,\mu g/kg$ and potatoes at $0.59\,\mu g/kg$ (US Department of Health & Human Services, 2002).

It is estimated that within the United States about 45 per cent of airborne beryllium is due to anthropogenic releases (US Department of Health & Human Services, 2002). Natural sources, such as windblown dust and volcanic activity, account for the remainder of the beryllium released to the atmosphere. Electric utilities account for about 80 per cent of the anthropogenic emissions, with industry and metal mining contributing the rest.

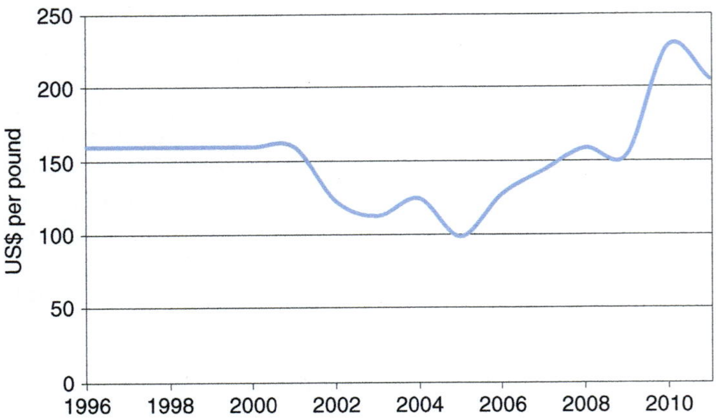

Figure 5.13 Annual average value of copper–beryllium alloy, per pound of contained beryllium, 1996–2011. (Compiled from USGS annual Mineral Commodity Summaries for beryllium from 1996–2012; and Cunningham, 1999.)

Beryllium released during the erosion of rocks and soils or derived originally from airborne emissions generally forms insoluble compounds and complexes: the mobility of these is further limited by adsorption onto organic matter, iron–manganese oxyhydroxides and clay minerals. Therefore, under most environmental conditions beryllium concentrations in groundwater and surface water are very low.

In air, beryllium compounds are present mostly as fine dust particles, which ultimately settle over the land and water surfaces of the globe. Fish do not accumulate beryllium from water into their bodies to any great extent. Concentrations of beryllium in drinking water range from 0.010 to 1.22 µg/l with an average of 0.19 µg/l (Kolanz, 2001). According to the World Health Organisation (2009), "Beryllium is rarely, if ever, found in drinking-water at concentrations of concern. Therefore, it is not considered necessary to set a formal guideline value. A health-based value for beryllium in drinking-water would be 12 µg/l."

Prices

As a result of its relatively small market, beryl is not traded on international commodity exchanges. Instead, prices are negotiated between buyers and sellers and the few posted prices for beryl containing an average of 11% BeO have remained unchanged for at least 35 years at US $1600 per metric tonne.

Long-term contracts for the supply of beryl concentrates normally specify BeO content and sieve specifications. A premium is often paid for BeO contents exceeding 11% as an incentive for suppliers to remove excess gangue minerals, usually quartz and feldspar, which consume excess acid during mineral-processing stages.

The USGS calculates the value of a pound of beryllium contained in copper–beryllium alloys, the most widely used beryllium product (Figure 5.13). The significant increase in prices over the period 2005–11 was driven by several key mining and refining cost factors. These include increases in the price of refining process chemicals, especially sulfuric acid and ammonia, the increase in the price of diesel fuel used in mining, depletion of resources that had low overburden removal requirements and the lower beryllium contents in the newly worked deposits.

Outlook

Since the early 1990s, much of the world's beryllium raw material supply has been derived from a single source in North America. China has developed substantial production capacity for the production of beryllium hydroxide and copper–beryllium hydroxide from imported beryl and domestic ores. A new, high-purity beryllium metal production facility was brought on stream in the USA in 2011 to replace the plant closed by Materion in 2000. In addition to providing new beryllium metal, this plant will also serve as an effective beryllium metal recycling facility. The mines at Yermakovskoye, in the Russian State of Buryatia, which were exploited during the Soviet era, are reported to be undergoing renovation with plans to re-open in 2017 (MBC Resources, 2011).

Copper–beryllium alloys are widely used to provide an unmatched physical property set to electronic and electrical components. This has led to an increasing demand for beryllium in both new consumer applications and in green technologies, such as wind power generation and hybrid or pure electric vehicles. The relevant physical property set for application in power generation is the unique combination of high strength, high conductivity and high resistance to loss of yield strength at elevated temperatures. In hybrid and pure electric vehicles, those properties provide a high current carrying capacity in applications such as battery-cell connectors and electric motor components. The same properties promote the use of the alloy in connector terminals used for solar power systems, where temperatures in roof-top or desert locations can exceed 75 °C for extended periods of time.

Ongoing research has demonstrated that by mixing uranium oxide with beryllium oxide, greater fuel rod utilisation efficiencies may be achieved in nuclear power generation as a result of the improved thermal conductivity of the fuel rods (McDeavitt et al., 2011). Nuclear power generation can be expected to grow and beryllium will continue to play a conventional role as cladding and moderators in these systems.

As a result of concerns over possible health issues related to the use of beryllium and the requirement for dust control in its handling, new applications for beryllium might have been expected to be taken up only slowly. In fact, the number of new uses for beryllium has increased, in parallel with a virtually equivalent reduction in the size of components due to miniaturisation, leading to a relatively flat demand for existing applications, but a steady increase in the number of new applications. It can be assumed that any substitution for beryllium-containing materials has already occurred, and future demand is based upon the real need for the properties of the materials that cannot be met by alternatives without loss of function. The major commercial producers of beryllium have seen steady growth, influenced by global economic cycles, but trending to a 3–5 per cent compounded annual growth rate.

Although one US company is the current market leader in the beryllium business, the resources of beryllium (Table 5.3), while not closely defined, are large and should be considered more than adequate to supply the small annual tonnages needed in the foreseeable future. It is anticipated that further installed production capabilities will emerge, notably in Russia, Kazakhstan and China, to respond to future market demands. Consumers of beryllium metal, alloys and ceramic would welcome alternative sources both for security of supply and for competitive market pricing.

Note

1. Materion Corporation includes: Materion Brush Beryllium & Composites, Materion Brush Performance Alloys, Materion Ceramics, Materion Natural Resources.

References

Agency for Toxic Substances and Disease Registry (2002) Toxicological Profile for Beryllium. ATSDR, Atlanta.

Barton, M.D. and Young, S. (2002) Non-pegmatitic Deposits of Beryllium: Mineralogy, Geology, Phase Equilibria and Origin. In: Grew, E.S. (ed.) Beryllium: Mineralogy, Petrology and Geochemistry, Reviews in Mineralogy and Geochemistry, v.50.

Beryllium Science and Technology Association (2012a) Production statistics. http://beryllium.eu/about-beryllium-and-beryllium-alloys/facts-and-figures/production-statistics/

Beryllium Science and Technology Association (2012b) Recycling of beryllium. http://beryllium.eu/about-beryllium-and-beryllium-alloys/facts-and-figures/recycling-of-beryllium/

Beus, A.A. (1962) Beryllium: Evaluation of Deposits During Prospecting and Exploratory Work. W.H. Freeman and Co., London.

Boffetta, P., Fryzek, J. and Mandel, J. (2012) Occupational exposure to beryllium and cancer risk: a review of the epidemiologic evidence. Critical Reviews in Toxicology 42 (2), 107–118.

Borak, J., Woof, S.H. and Fields, C.A. (2006) Use of Beryllium Lymphocyte Proliferation Testing for Screening of Asymptomatic Individuals: An Evidence-Based Assessment. Journal of Occupational and Environmental Medicine 48 (9), 937–947.

British Geological Survey (2012) World mineral production 2006–10. British Geological Survey Keyworth, Nottingham.

Bulatovic, S. (1988) United States Patent 4,735,710.

Burt, D.M., Wohletz, K.H. and Sheridan, M.F. (2006) "Water-laid Tuff" of the Utah Desert and Similar Surge Deposit Misinterpretations: A Possible Lesson for Mars. Lunar & Planetary Science Conference, XXXVII, No. 2295.

Černý, P. (1991a) Rare-element Granitic Pegmatites. Part II: Regional to Global Environments and Petrogenesis, Geoscience Canada, 18 (2).

Černý, P. (1991b): Rare-element Granitic Pegmatites. Part I: Anatomy and Internal Evolution of Pegmatite Deposits, Geoscience Canada, 18 (2).

Černý, P. (2005) The Tanco rare-element pegmatite deposit, Manitoba: Regional context, internal anatomy, and global comparisons. In: Linnen, R.L. and Samson, I.M. (eds.) Rare element geochemistry and mineral deposits. Geological Association of Canada Short Course Notes, 17, 127–158.

Černý, P., Anderson, A.J., Tomascak, P.B. and Chapman, R. (2003): Geochemical and morphological features of beryl from the Bikita granitic pegmatite, Zimbabwe. The Canadian Mineralogist 41, 1003–1011.

Cunningham, L.D. (1999) Metals prices in the United States through 1998: Beryllium. United States Geological Survey.

Cunningham, L.D. (2004a) Beryllium, United States Geological Survey Mineral Commodity Summaries.

Cunningham, L D. (2004b) Beryllium recycling in the United States in 2000. United States Geological Survey Circular 1196-P.

Davis, L.J. (1979) Beryllium and Uranium Mineralization in the Spor Mountain Area, Juab County, Utah, In: Basin and Range Symposium and Great Basin Field Conference, Rocky Mountain Association of Geologists and Utah Geological Association.

Deubner, D.C., Sabey, P., Huang, W. et al. (2011) Solubility and Chemistry of Materials Encountered by Beryllium Mine and Extraction Workers: Relation to Risk. Journal of Occupational and Environmental Medicine 09/2011.

Donovan E.P., Kolanz, M.E., Galbraith, D.A., Chapman, P.S. and Paustenbach, D.J. (2007) Performance of the beryllium blood lymphocyte proliferation test based on a long-term occupational surveillance program. International Archive of Occupational Environmental Health 81, 165–178.

Dunai, T.J. (2010) Cosmogenic Nuclides, Principles, Concepts and Applications in the Earth Surface Sciences, Cambridge University Press, Cambridge.

European Commission (2010) Critical Raw Materials for the EU. Annex V to the Report of the Ad-hoc Working Group on defining critical raw materials. http://ec.europa.eu/enterprise/policies/raw-materials/files/docs/report_en.pdf

Fetherston, J.M. (2004) Tantalum resources in Western Australia. Geological Survey of Western Australia, Mineral Resources Bulletin 22.

Fulton, G. and Goldberg, A. (2009) Medical Aspects of the Toxicity of Beryllium and Beryllium Alloys. In: Walsh, K.A. (ed.) Beryllium Chemistry and Processing, ASM International, Materials Park, Ohio.

Grew, E.S. (2002) Mineralogy, Petrology and Geochemistry of Beryllium: An introduction and list of beryllium minerals. In: Grew, E.S. (ed.) Beryllium: Mineralogy, Petrology and Geochemistry, Reviews in Mineralogy and Geochemistry. Volume 50.

Harrel, J.A. (2004) Archaeological Geology of the World's First Emerald Mine, Geoscience Canada, 31 (2).

Hollins, D. M., McKinley, M.A., Williams, C. et al. (2009) Beryllium and lung cancer: A weight of evidence evaluation of the toxicological and epidemiological literature. Critical Reviews in Toxicology 39 (S1), 1–32.

International Agency for Research on Cancer (2012). IARC Monographs on the Evaluation of Carcinogenic Risks to Humans. Volume 100C, A Review of Human Carcinogens: Arsenic, Metals, Fibres, and Dusts.

Jaskula, B.W. (2012a) Beryllium, USGS Mineral Commodity Summaries.

Jaskula, B.W. (2012b) USGS 2010 Minerals Yearbook: Beryllium.

Klaproth, M.H. (1801) Analytical Essays Towards Promoting The Chemical Knowledge Of Mineral Substances, G. Woodfall, London.

Kolanz, M.E. (2001) Introduction to beryllium: uses, regulatory history and disease. Applied Occupational Environmental Hygiene, 16, 559–567.

Kovalenko, V.I., Kuz'min, M.I., Antipin, V.S. and Petrov, L.L. (1971) Topaz-bearing quartz keratophyre (ongonite), a new variety of sub-volcanic igneous vein rock. Acad. Sci. USSR Doklady Earth Sci. Sec 199.

Lindsey, D.A., Ganow, H. and Mountjoy, W. (1973) Hydrothermal Alteration Associated with Beryllium Deposits at Spor Mountain, Utah, USGS Professional Paper 818-A.

Linnen, R.L., Trueman, D.L. and Burt, R. O. (2013) Niobium and tantalum. In: Gunn, Gus. (ed.) Critical Metals Handbook (Wiley-Blackwell).

London, D. (2008) Pegmatites. The Canadian Mineralogist, Special Publication 10.

MBC Resources (2011) (Division of Metropol Group, Moscow). News release 29 August 2011. Available from: http://www.steelguru.com/russian_news/Berillium_Hydrometallurgical_Plant_to_be_built_in_Buryatia/222584.html

McDeavitt, S.M., Ragusa, J., Revankar, S.T., Solomon, A.A. and Malone, J. (2011) A High-conductivity Oxide Fuel Concept, Nuclear Energy International, Globe Trade Media. http://www.neimagazine.com/story.asp?sectioncode=76andstoryCode=2060405

Merchant Research and Consulting Limited (2012) Beryllium Market Review. http://marketpublishers.com/report/industry/metallurgy/beryllium_market_review.html

National Toxicology Program (2002) 10th Annual Report on Carcinogens – December 2002.

Price, J.G., Rubin, J.N., Henry, C.D., Pinckston, T.L., Tweedy, S.W. and Koppenaal, D.W. (1990) Rare-metal enriched peraluminous rhyolites in a continental arc, Sierra Blanca area, Trans-Pecos Texas; Chemical modification by vapor-phase crystallization. GSA Special Paper 246.

Rothman, K.J. and Mosquin, P.L. (2011) Confounding after risk-set sampling in the beryllium study of Sanderson et al. Annals of Epidemiology 21 (10), 773–779.

Rudnick, R.L. and Gao, S. (2004) Composition of the Continental Crust. In: Treatise on Geochemistry, Volume 3 The Crust (Elsevier Pergamon), 1–64.

Sabey, P. (2006) Beryllium Minerals. In: Kogel, J.E., Trivedi, N.C., Barker, J.M. and Krukowski, S.T. (eds.) Industrial Minerals and Rocks, Society for Mining, Metallurgy and Exploration.

Sainsbury, C.L. (1963) Beryllium Deposits of the Western Seward Peninsula, Alaska, USGS Circular 479.

Shearer, C.K., Papike, J.J. and Jolliff, B.L. (1992) Petrogenetic links among granites and pegmatites in the Harney Peak rare-element granite-pegmatite system, Black Hills, South Dakota. Canadian Mineralogist 30, 785–809.

Stonehouse, A.J. (1985) Physics and chemistry of beryllium. (February, 2012). http://materion.com/~/media/Files/PDFs/Beryllium/BerylliumMaterials-MB/MB-003PhysicsandChemistryofBeryllium.pdf

Strupp, C. (2011a) Beryllium Metal I. Experimental Results on Acute Oral Toxicity, Local Skin and Eye Effects, and Genotoxicity. The Annals of Occupational Hygiene 55 (1), 30–42.

Strupp, C. (2011b) Beryllium Metal II. A Review of the Available Toxicity Data. The Annals of Occupational Hygiene 55 (1), 43–56.

Trueman, D.L. and Černý, P. (1982) Exploration for rare-element granitic pegmatites. In: Černý, P. (ed.) Granitic Pegmatites in Science and Industry: A short course. Volume 8. Mineralogical Association of Canada.

Trueman, D.L., Pedersen, J.C., de St. Jorre, L. and Smith, D.G.W. (1988) The Thor Lake rare-metal deposits, Northwest Territories. In: Recent Advances in the Geology of Granite-Related Mineral Deposits, Canadian Institute of Mining and Metallurgy Special Vol. 39.

UN Comtrade (2013) United Nations Commodity Trade Statistics Database, Department of Economic and Social Affairs/ Statistics Division, http://comtrade.un.org/db/

UNEP (2011) Recycling rates of metals – a status report. Report of the Working Group on Global Metal Flows to the International Resource Panel.

U.S. Department of Health and Human Services (2002) Agency for Toxic Substances and Disease Registry, Toxicological Profile for Beryllium.

U.S. Environmental Protection Agency (1998). Toxicological Review of Beryllium and Compounds.

Vidal, E.E., Foley, J. and Olson, D.L. (2009) Mineral Processing. In: Walsh, K.A. (ed.) Beryllium Chemistry and Processing, ASM International, Materials Park, Ohio.

Walsh, K.A. (1979) Extraction. In: Floyd D.R. and Lowe J.N. (eds.) Beryllium Science and Technology, Plenum Press, New York.

Walsh, K.A. and Vidal, E.E. (2009) Sources of Beryllium. In: Walsh, K.A. (ed.) Beryllium Chemistry and Processing, ASM International, Materials Park, Ohio.

Walsh, K.A., Vidal, E.E. and Mishra, B. (2009) Extractive Metallurgy. In: Walsh, K.A. (ed.) Beryllium Chemistry and Processing, ASM International, Materials Park, Ohio.

Wang Z., White, P.S., Petrovi, M. et al. (1999) Differential susceptibilities to chronic beryllium disease contributed by different Gly69 HLA DPB1 and -DPA1 alleles. The Journal of Immunology 163, 1647–1653.

Wang, Z., Farris, G.M., Newman, L.S.et al. (2001) Beryllium sensitivity is linked to HLA-DP genotype. Toxicology 165, 27–38.

World Health Organisation (2009) Beryllium in Drinking-water Background document for development of WHO Guidelines for Drinking-water Quality.

6. Cobalt

STEPHEN ROBERTS[1] AND GUS GUNN[2]

[1] School of Ocean and Earth Science, National Oceanography Centre, University of Southampton, UK
[2] British Geological Survey, Keyworth, Nottingham, UK

Introduction

Cobalt has been utilised by society since the Bronze Age, mainly to impart a rich blue colour to glass, glazes and ceramics. However, it was only isolated as a pure metal in 1735 by Swedish chemist Georg Brandt and demand for cobalt remained subdued until the turn of the 20th century and the development of cobalt–chromium alloys. In particular, the demand for cobalt increased considerably after the Second World War, driven by the use of high-purity cobalt in jet engines and gas turbines. Cobalt demand has further accelerated in the past 30 years, reflecting the increased use of cobalt as an essential constituent of materials used in high-technology industries including rechargeable batteries, superalloys and catalysts.

Physical and chemical properties

Cobalt (chemical symbol, Co) is a *d-block* transition metal, silver in colour, with an atomic number of 27, appearing in the first long period of the Periodic Table between iron and nickel. Cobalt has two main oxidation states (2^+ and 3^+) and one naturally occurring isotope (^{59}Co). Cobalt shows siderophile and chalcophile tendencies, has a high melting point of 1493 °C and is ferromagnetic (Table 6.1).

Distribution and abundance in the Earth

Estimates of the crustal abundance of cobalt vary between 15–30 ppm, not dissimilar to the other first-period transition metals such as scandium, copper, zinc and nickel. In particular, cobalt is most abundant in ultramafic rocks with an average concentration of about 110 ppm. Recent estimates of the upper continental crustal abundance of cobalt suggest a concentration of 15 ±1 ppm (Hu and Gao, 2008). The concentration of cobalt in sea water is very low (generally less than 10 ppt) which in part reflects its short residence time[1] of 340 years. In contrast, the residence times of other base metals are much longer, (nickel, 6000 years; copper, 5000 years; and zinc, 50,000 years) and their concentrations in sea water are much greater (Broecker and Peng, 1982).

Mineralogy

Pure cobalt is not found in nature, but, as a result of its chalcophile and siderophile properties, it preferentially bonds with iron, nickel, copper and sulfur rather than with oxygen into a number of sulfide and sulfarsenide phases. In particular, it forms cobalt sulfides and arsenides, such as cobaltite (Co,Fe)AsS, carrollite ($CuCo_2S_4$),

Critical Metals Handbook, First Edition. Edited by Gus Gunn.

Table 6.1 Selected properties of cobalt.

Property	Value	Units
Symbol	Co	
Atomic number	27	
Atomic weight	58.93	
Density at 25 °C	8800	kg/m³
Melting point	1493	°C
Boiling point	3100	°C
Electrical resistivity at 25 °C	5.81	µΩ m
Specific heat capacity at 25 °C	0.42	J/(g °C)
Latent heat of vaporisation	6490	J/g
Latent heat of fusion	263	J/g
Hardness (Mohs scale)	5.0	
Thermal conductivity	100	W/(m °C)

Figure 6.1 Skutterudite, a cobalt–nickel arsenide, from Bou Azer mine, Morocco. Maximum dimension of specimen is 7 cm. (BGS © NERC.)

Table 6.2 The most important cobalt-bearing minerals.

Mineral name	Chemical formula	Cobalt content (wt %)
Cobaltpentlandite	Co_9S_8	67.40
Heterogenite-2H	$Co^{3+}O(OH)$	64.10
Heterogenite-3R	$Co^{3+}O(OH)$	64.10
Linnaeite	$Co^{2+}Co^{3+}{}_2S_4$	57.95
Sphaerocobaltite	$CoCO_3$	49.55
Cattierite	CoS_2	47.89
Cobaltite	CoAsS	35.52
Erythrite	$Co_3(AsO_4)_2.8(H_2O)$	29.53
Carrollite	$Cu(Co,Ni)_2S_4$	28.56
Glaucodot	(Co,Fe)AsS	26.76
Safforite	$(Co,Fe)As_2$	21.25
Willyamite	(Co,Ni)SbS	20.78
Skutterudite	$(Co,Ni)As_{3-x}$	17.95
Kolwezite	$(Cu,Co)_2(CO_3)(OH)_2$	17.84
Siegenite	$(Ni,Co)_3S_4$	14.51

linnaeite $(Co,Ni)_3S_4$ and skutterudite $(Co,Fe,Ni)As_2$ (Figure 6.1), which are commonly associated with the iron sulfides, pyrite, arsenopyrite and pyrrhotite. However, it also occurs as a carbonate phase, sphaerocobaltite ($CoCO_3$), and a hydroxide, heterogenite (CoO(OH)) (Table 6.2).

Deposit types

Despite its low crustal abundance cobalt is concentrated by various geological processes to concentrations suitable for mining. However, mining of metallic ores chiefly for their cobalt content is rare, with the majority of cobalt production achieved through recovery as a by-product of copper and nickel mining in three principal geological settings: hydrothermal, magmatic and lateritic (Figure 6.2). A fourth, significant but as yet unexploited, cobalt resource resides in iron–manganese nodules and crusts developed on the ocean floor which may contain substantial quantities of base metals including cobalt. In the following sections the geology which accounts for the majority of the world's cobalt production is described, with a particular focus on the major cobalt-producing regions of the Democratic Republic of Congo (DRC) and Zambia, where cobalt is a by-product of copper mining, and in Canada, Russia and Australia where cobalt is produced mainly as a by-product of nickel mining.

Hydrothermal deposits

These are cobalt ore deposits derived from hydrothermal fluids which have interacted with a variety of mafic and/or ultramafic basement rocks or are substantially derived from within sedimentary basins. These include the deposits at Bou Azer in Morocco which are the only currently working mines that produce cobalt as a

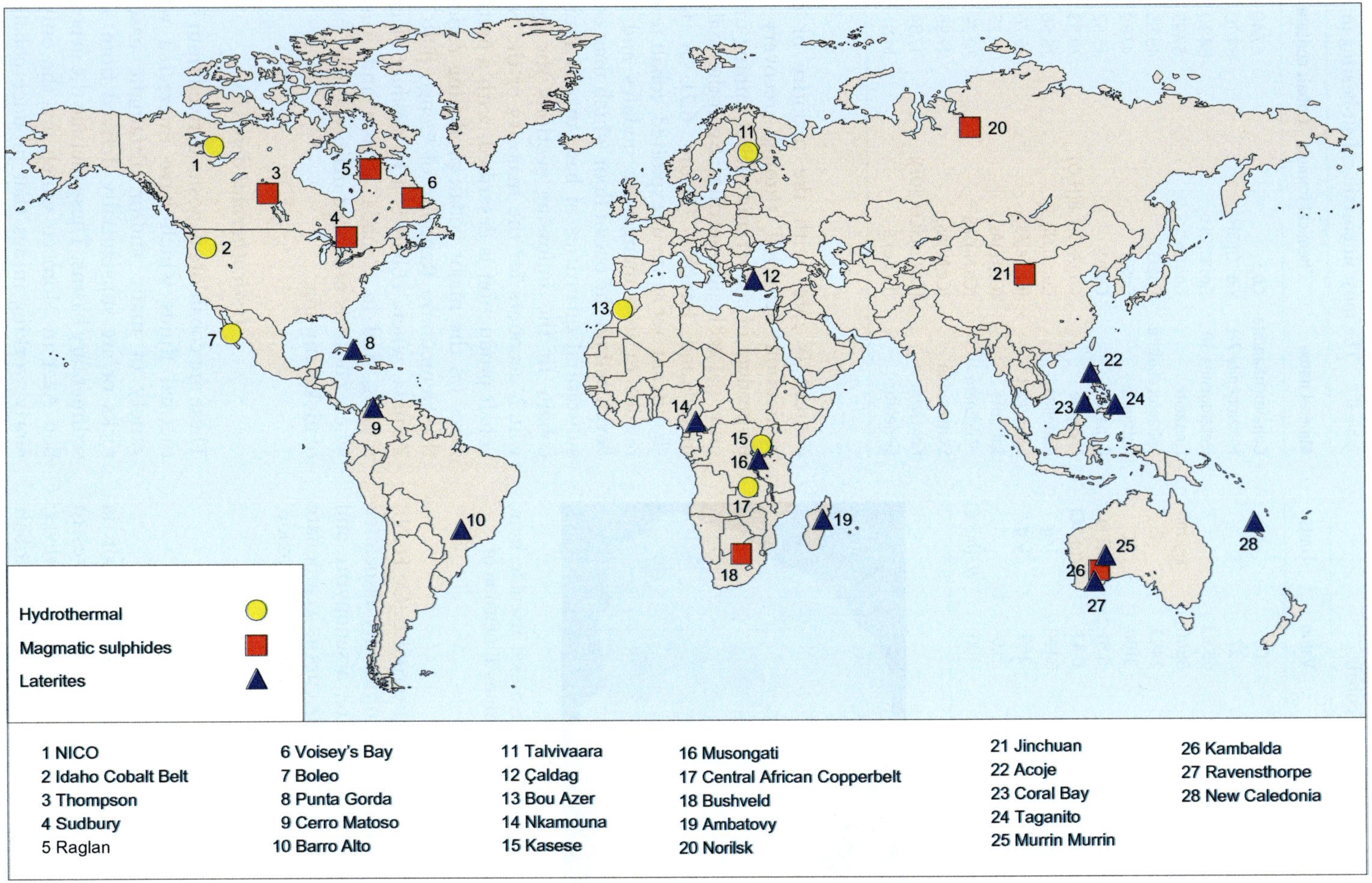

Figure 6.2 Major cobalt-producing mines and districts. The majority of new production is anticipated to be from laterite deposits.

primary product. The major producing deposits of the DRC and Zambia are also classified as hydrothermal in origin.

Bou Azer, Morocco

Located 320 km to the east of Agadir, more than 60 orebodies comprise the cobalt–nickel–arsenic–gold–silver mines of Bou Azer and adjacent areas. These mines produced about 1800 tonnes of cobalt in 2011 (Cobalt Development Institute, 2012a), with major by-products including nickel, gold and arsenic. These deposits are spatially associated with serpentinised ultramafic rocks of a Neoproterozoic ophiolite complex, which comprises a mantle sequence of serpentinised peridotites, ultrabasic and basic cumulates, stocks of quartz diorite, basic lavas, and a mixed volcanic and sedimentary sequence (Leblanc and Kroener, 1981).

The Bou Azer cobalt mineralisation is dominated by arsenides, sulfarsenides and sulfides in a quartz-carbonate gangue. Cobalt-bearing arsenide minerals include skutterudite ($CoAs_3$), safflorite ($CoAs_2$), loellingite ($FeAs_2$), nickeline (NiAs), rammelsbergite ($NiAs_2$), and sulfur-rich nickel–cobalt diarsenide, with accessory copper sulfides, molybdenite and gold. Cobalt-rich ores predominate in almost all the deposits, which contrasts with the high Ni/Co ratio of the underlying ultramafic rocks.

The cobalt mineralisation is located within 5 to 20 m thick and 50 to 600 m long veins, lenses and stockworks and is structurally controlled within shear zones. The shear zones are concentrated along irregular contacts between serpentinites and Precambrian volcanic rocks or quartz diorites (Leblanc and Billaud, 1982; Leblanc and Fischer, 1990). The timing of the Bou Azer mineralisation remains controversial. However, the present consensus is that the Bou Azer arsenide mineralisation postdates the obduction of the Neoproterozoic ophiolite fragments (Leblanc and Billaud, 1982; Oberthur et al., 2009).

The cobalt (nickel–iron–arsenic) mineralisation at Bou Azer developed through the leaching of serpentinites by magmatic fluids under moderately reducing conditions and at high fluid/rock ratios. Deposition of ore minerals occurred in response to increasing pH, by mixing between magmatic brines and meteoric water. The predominance of cobalt over nickel arsenide minerals in the Bou Azer mineralisation may be attributable to the different solubilities of nickel and cobalt in the hydrothermal system (Ahmed et al., 2009).

Idaho Cobalt Belt

The Idaho Cobalt Belt (ICB) in north-western United States is a north-west-trending zone of cobalt occurrences, approximately 64 km long and 10 km wide, centred on the former Blackbird mine. Past production from the Blackbird mine complex amounts to approximately 1.7 million tonnes of high-sulfide cobalt–copper ore, with the district as a whole potentially hosting 50 million tonnes of ore (Bookstrom et al., 2007). The ICB hosts cobalt–copper deposits in sequences of siliciclastic metasedimentary rocks of Mesoproterozoic age deposited within intracontinental extensional basins. Styles of mineralisation vary from stratabound, which is the most important, to locally discordant sulfide lenses, and discordant tourmaline breccias (Nash and Hahn, 1989; Bending and Scales, 2001; Bookstrom et al., 2007). Principal ore minerals are cobaltite and chalcopyrite which occur in a gangue comprising mostly quartz and biotite (Nash and Hahn, 1989). Other ore minerals include pyrite, pyrrhotite, arsenopyrite, glaucodot ($(Co_{0.5}Fe_{0.5})AsS$), safflorite ($(Co,Fe,Ni)As_2$), bismuthinite, native bismuth, bismuth tellurides and gold.

The deposits show a variety of deformational features and are often localised along faults, shear zones and fold axes. However, the origin of these deposits remains controversial with the stratabound cobalt–copper–gold sulfides considered to have formed either as synsedimentary deposits prior to deformation and regional metamorphism or during syntectonic metamorphism. It has also been suggested that they may be a variant of iron-oxide–copper–gold (IOCG) deposits (Slack, 2006).

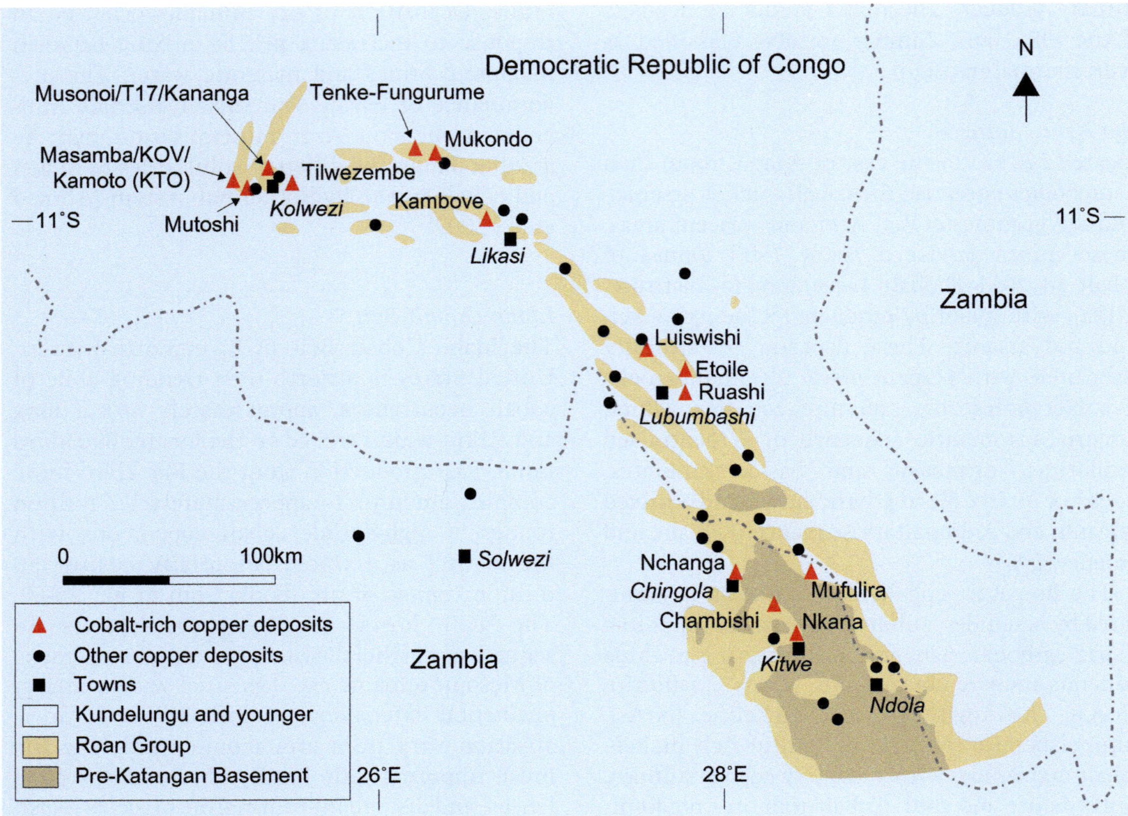

Figure 6.3 Simplified geology of the Central African Copperbelt in the Democratic Republic of Congo (DRC) and Zambia showing the location of selected cobalt–copper and copper mines and deposits. (Adapted from Trans Continent Exploration and Mining Company, 2011.)

Central African Copperbelt

The Neoproterozoic Katangan Copperbelt of Central Africa, located on both sides of the border between north-western Zambia and the Katanga Province of southern Democratic Republic of Congo (DRC), hosts one of the world's greatest concentrations of copper and cobalt (Figure 6.3). It currently produces about two thirds of the world's cobalt with a production of 70,000 tonnes in 2010 (BGS, 2012).

The copper–cobalt ores are hosted by siliclastic and carbonate sedimentary rocks and volcanic and plutonic mafic rocks of the Katangan supracrustal sedimentary succession, emplaced in a continental rift (Kampunzu et al., 2000; Cailteaux et al., 2005). Closure of the Katangan basin during the Lufilian Orogeny resulted in north-verging folds, thrusts and nappes of the 'Lufilian Arc'. The Zambian Basin appears to have had a protracted history with the onset of sedimentation at about 877 Ma[2], eclogite formation at about 600 Ma and final closure and uplift of the basin at about 530 Ma (John et al., 2003).

In Zambia the copper–cobalt deposits are hosted in para-autochthonous siliciclastic rocks close to basement terrains, whereas in the DRC the deposits and their host rocks define thrust sheets and nappes formed during the Lufilian Orogeny and the dominant lithological units are dolomites and dolomitic shales (Cailteaux et al., 2005). The majority of the copper–cobalt mineralisation was deposited prior to the Lufilian compressional tectonics both in the DRC and Zambia, evidenced by the folding and thrusting of orebodies.

The copper and cobalt ores comprise mainly disseminated sulfides, forming stratiform orebodies hosted in fine-grained siliciclastic or dolomitic sedimentary rocks. The copper is predominantly hosted in chalcopyrite, bornite, chalcocite and malachite. The cobalt occurs within cobaltite, carrollite, cattierite, cobalt pentlandite and siegenite, and as solid solution in pyrite (up to 20 per cent cobalt) (Annels and Simmonds, 1984). The primary sulfides are commonly overprinted by secondary supergene ore minerals, with heterogenite the most abundant oxidised cobalt mineral. The weathering process is economically significant because it strongly concentrates cobalt in the near-surface oxidised ore. Hydrothermal minerals associated with the mineralisation include potassium feldspar, phlogopite, sericite, muscovite, albite, carbonate, quartz, and rutile. These assemblages are indicative of calcium–magnesium, potassic, and sodic alteration (Selley et al,. 2005). Textural evidence indicates that alteration events occurred at multiple stages during the basin history, and can vary between and within deposits. These relationships result from a combination of: (1) the widespread and protracted nature of fluid flow; (2) variability in the composition of host strata; (3) variability in conditions at the sites of ore formation; and (4) the effects of subsequent regional metamorphism (Selley et al., 2005).

In Zambia, although the copper–cobalt deposits are predominantly located in the Lower Roan strata, significant examples are also found in the underlying basement (e.g. at Lumwana (Bernau et al., 2013) and in the Upper Roan (e.g. at Kansanshi, Broughton et al., 2002). Therefore, the long-lived Zambian Basin appears to preserve a protracted history of ore deposition. A role for tectonics in ore formation is recognised with copper–cobalt ores related to early sub-basin extensional faults (Annels, 1989; Selley et al., 2005), faults reactivated during basin inversion (McGowan et al., 2003, 2006) or the development of mineralized shear zones within the basement (Bernau et al., 2013). There is also evidence that structures developed during the Lufilian orogeny resulted in the modification and redistribution of high-grade ore horizons (Brems et al., 2009; Muchez et al., 2010).

The post-orogenic history is an important factor in the development of economic cobalt mineralisation in central Africa. During various episodes of weathering, uplift and erosion the sulfide ore deposits were partly oxidised, commonly down to a depth of about 100 m. This process was particularly common in the DRC and resulted in local cobalt enrichment in the upper part of the oxidised zone, often referred to as a 'cobalt cap', due to the downward leaching of copper and cobalt by meteoric fluids. These oxide ores can constitute a major part of the cobalt resource in a particular deposit; for example, they account for about half of the cobalt resource at Tenke Fungurume.

Other cobalt deposits in sedimentary basins

The Boléo district, located in the Santa Rosalía basin of Baja California, Mexico, consists of multiple, laterally extensive stratiform units of laminated claystone and claystone breccia that contain finely disseminated copper–cobalt–zinc sulfides and oxidised sulfides. The Boléo clastic sequence consists of a series of upward coarsening fan–delta cycles which include laterally extensive zones of conformable sulfide mineralisation with pervasive supergene oxide overprint in the basal claystone of each fan–delta cycle. Mineralisation extends for more than 90 km^2 with delineated reserves of 85 million tonnes grading 1.33% Cu, 0.08% Co, and 0.55% Zn, (Baja Mining Corp, 2010). An exhalative-infiltration model is proposed to explain the mineralisation with upflow and discharge onto the basin floor of saline metalliferous brines along basin growth faults. Mineralisation of the basal claystone is due to the downward infiltration of brine and replacement of diagenetic pyrite (Conly et al., 2006).

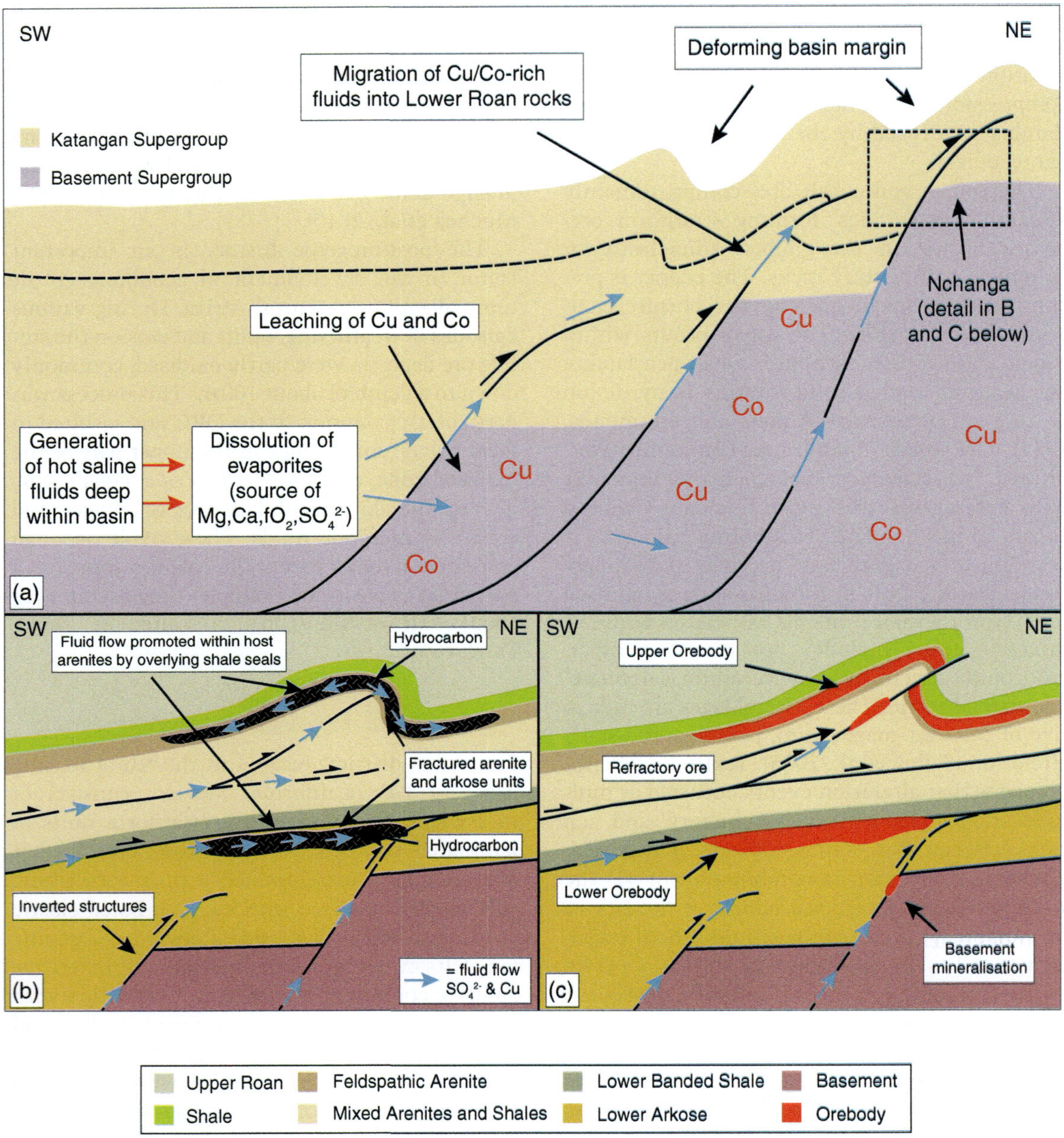

Figure 6.4 Schematic diagrams illustrating the development of the copper-cobalt deposit at Nchanga, Zambia. (a) Shows basinal brines migrating up reactivated faults at the margins of the Zambian Basin; (b) Shows migration of brines into gas-filled hydrocarbon traps, with overlying shales acting as seals; (c) Shows final distribution of orebodies at Nchanga. (Cu, copper; Co, cobalt; Mg, magnesium; Ca, calcium; SO_4^{2-}, sulfate; fO_2, oxygen fugacity.) (After McGowan et al., 2006.)

Iron-oxide–copper–gold deposits (IOCG)

IOCG deposits represent an enigmatic style of mineralisation recently defined as magmatic-hydrothermal deposits that contain economic copper and gold grades. They are structurally controlled, contain significant volumes of breccia and are commonly associated with pre-sulfide sodic or sodic-calcic alteration. IOCG deposits have a clear temporal, but not a close spatial, relationship to major magmatic intrusions (Groves et al., 2010). The most striking examples are the Olympic Dam deposit of South Australia, Candelaria in Chile and Salobo in Brazil. Cobalt in the form of carrollite and cobalt-rich pyrite (containing up to 3 wt% Co) occurs within the stratabound bornite–chalcopyrite–pyrite mineralisation at Olympic Dam with an estimated cobalt grade for the deposit of 0.02% (Williams, 1999). Significant cobalt reserves have also been reported in the NICO IOCG deposit located 160 km north-west of Yellowknife, in Canada. Fortune Minerals Ltd aims to produce 1800 tonnes of cobalt per annum from the NICO deposit (Fortune Minerals Limited, 2012).

Magmatic deposits

Concentrations of nickel and copper with recoverable by-product cobalt, typically between 0.04 and 0.08% Co, are present in mafic and ultramafic magmas, characterised by low total silica and alkalis and high magnesium contents. Sulfides such as chalcopyrite, pentlandite and pyrrhotite precipitate from these magmas either from lava flows or within vast magma chambers, particularly during early Earth history. In all the magmatic deposits the formation of an immiscible sulfide phase is an essential part of the ore-forming process. Copper, nickel and cobalt are efficiently scavenged into this sulfide phase on account of their chalcophile and siderophile properties.

Kambalda

The Kambalda Dome, located 700 km east of Perth, preserves a 3-km thick pile of tholeiitic and komatiitic basalts which host more than twenty nickel sulfide deposits (Lesher, 1989). These lenses of massive sulfide are composed of chalcopyrite, pentlandite and pyrrhotite, typically located toward the base of komatiitic lava flows within pronounced linear depressions with the sulfides containing between 1–4% Ni and 0.01–0.33% Co.

Sudbury Igneous Complex

The magmatic nickel–copper–platinum-group element (PGE) deposits of the Sudbury Igneous Complex (SIC) in Ontario, Canada were discovered in 1883 and by 2002 had produced in the region of 9.69 million tonnes of nickel, 9.59 million tonnes of copper and 69,600 tonnes of cobalt (Mudd, 2010). The SIC remains the world's largest nickel producer and also produced about 2209 tonnes of cobalt in 2011 (Xstrata, 2012); largely as a minor component of pyrrhotite and pentlandite which contain between 0.4–1.3% Co.

The Sudbury Igneous Complex is a layered intrusion emplaced more than 2600 Ma ago at the centre of the Sudbury structure, which comprises a series of rocks widely regarded to represent an eroded and tectonised remnant of an originally 200- to 250-km wide meteorite impact basin (Dietz, 1964; Therriault et al., 2002; Lightfoot and Doherty, 2001). Three major lithologies are generally recognised within the Sudbury Igneous Complex: traditionally termed norite, quartz gabbro and granophyre. At the base of the Sudbury Igneous Complex, the sublayer contains discontinuous kilometre-sized bodies characterised by abundant xenoliths and copper–nickel sulfide mineralisation. The sulfide mineralisation is mainly chalcopyrite and pyrrhotite–pentlandite, with more massive sulfides towards the base, becoming more disseminated upwards. Some of the xenoliths are identifiable as locally derived country rocks, others constitute a suite of ultramafic and mafic rocks of unknown source, which may be genetically linked to the Sudbury Complex. There is a clear genetic relationship between the Sudbury sulfide orebodies and the location of the contact sublayer and offset dikes, which also contain steeply plunging sulfide mineralisation (Lightfoot et al., 1997a, 1997b).

Voisey's Bay

The Voisey's Bay mine is located approximately 35 km south-west of Nain in northern Labrador, Canada. In 2012 cobalt production from Voisey's Bay was 1221 tonnes, derived from ores grading about 0.12% Co (Vale, 2012). The Voisey's Bay intrusion is a member of the Nain Plutonic Suite and comprises gabbroic-troctolitic rocks emplaced at 1.334 Ga within the boundary zone between the Churchill and Nain structural provinces (Amelin et al., 1999). It is widely regarded as a prime example of sulfide mineralisation associated with a magmatic conduit system.

The mineralisation associated with the Voisey's Bay intrusive system has been divided into four principal zones. From east to west, these are the Eastern Deeps, the Ovoid, Discovery Hill and Reid Brook. Disseminated to semi-massive sulfide mineralisation in the Eastern Deeps occurs within the Basal Breccia Sequence, an inclusion-rich unit characterized by gneissic and other rock fragments in a gabbronorite to ferrogabbronorite matrix; massive sulfide veins are also present. The Basal Breccia Sequence is enclosed by a variable-textured troctolite that locally contains xenoliths and up to 50 per cent interstitial sulfide. The Ovoid ore body is located to the west of the Eastern Deeps, and consists of a bowl-shaped accumulation of massive sulfide ($600 \times 350 \times 110$ m), separated from local gneiss by a thin zone of Basal Breccia. To the west, the Ovoid ore body narrows into disseminated, sulfide-bearing troctolitic rocks of the Discovery Hill zone dike. West of the Discovery Hill zone, sulfide mineralisation is principally in a steeply dipping dike in the Reid Brook zone and where abundant fragments of country rock are found in breccias around the mineralisation (Evans-Lambswood et al., 2000).

Petrological, geochemical and isotopic studies of the Voisey's Bay intrusion and associated country rocks have highlighted the importance of multi-stage magma contamination in the formation of the ores, with the Palaeoproterozoic Tasiuyak gneiss the probable source of certain key elements (Ryan et al., 1995; Li and Naldrett, 2000).

Norilsk

Igneous intrusions of Triassic age in the Norilsk region of northern Russia contain one of the largest known resources of nickel, copper and platinum-group element (PGE)-enriched sulfide mineralisation with significant cobalt production as a by-product. The sub-volcanic intrusions feed the lower Triassic members of a sequence of Permo-Triassic flood basalts. The sulfide ores occur in various forms: (1) disseminated within the intrusions themselves; (2) as concentrations within and close to the base of the intrusions; (3) disseminated within footwall rocks to the intrusions; and (4) in rich veins extending away from the contact into the underlying footwall rocks (Naldrett, 1989).

Geochemical and isotopic data suggest that the sulfide ore deposits of the Norilsk region were formed by injection of olivine-bearing melts containing immiscible sulfide with the sulfur derived from underlying Devonian evaporite country rocks.

Laterites

Nickel–cobalt laterite deposits contain about 70 per cent of world nickel resources and account for about 40 per cent of world nickel production, but they also contain appreciable concentrations of cobalt, between 0.025 and 0.18% Co (Berger et al., 2011). These deposits are the product of pervasive weathering of ultramafic rocks, which initially contain between 0.06 and 0.09% Co, exposed in tropical to sub-tropical environments. The formation of nickel–cobalt laterites is influenced by a number of geological variables including protolith composition, topography, structure and the duration of appropriate weathering processes.

Nickel–cobalt laterite deposits can be grouped in three main categories: (1) hydrous silicate

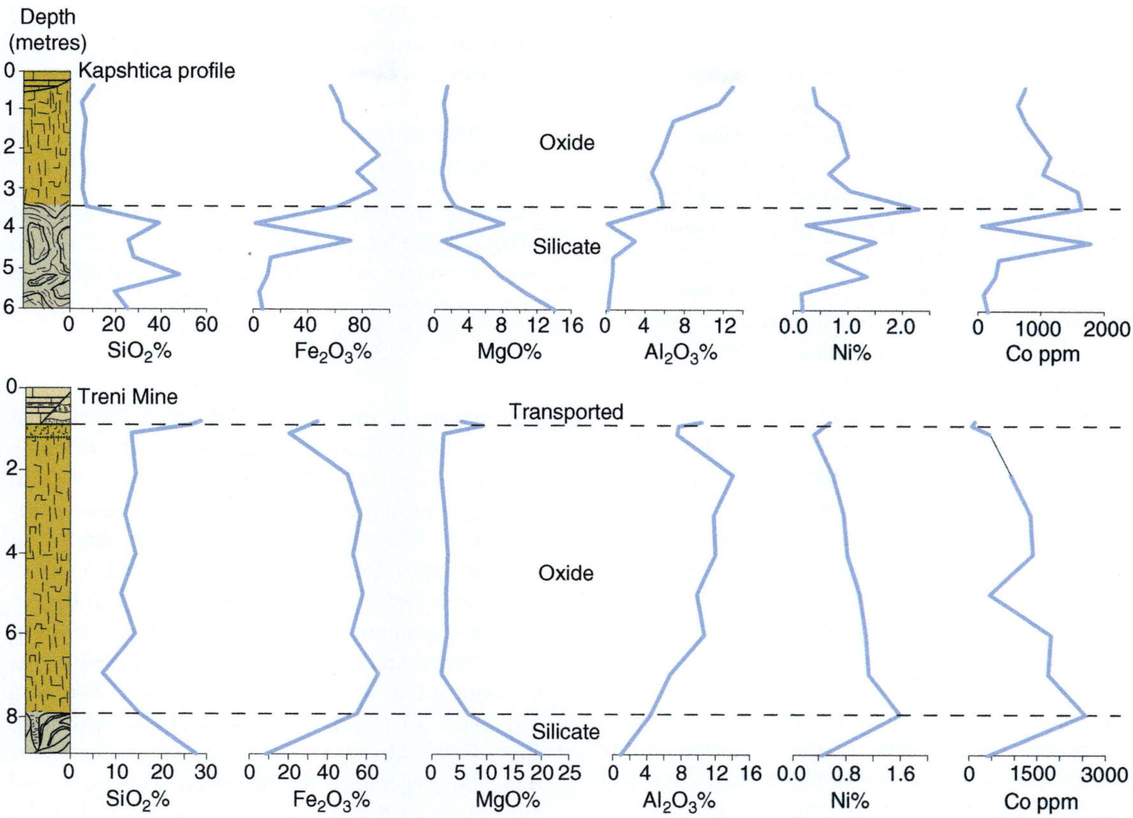

Figure 6.5 Geochemistry of a cobalt-bearing nickel laterite profiles. The profiles show the significant loss of SiO_2 and MgO at the silicate/oxide boundary and concomitant increases in Ni and Co. The profiles also indicate that the highest concentrations of Ni and Co within the profiles occur in close proximity to the silicate/oxide boundary. (SiO_2, silica; Fe_2O_3, iron oxide; MgO, magnesium oxide; Al_2O_3, aluminium oxide; Ni, nickel; Co, cobalt.) (After Thorne et al., 2012a.)

deposits, where hydrous magnesium–nickel silicates occur in the lower saprolite, overlain by oxide laterites; (2) clay silicate deposits, with largely smectitic clays developed in the mid or upper saprolite; (3) and oxide deposits, also known as limonite deposits, comprising largely iron-oxyhydroxides overlying altered bedrock, or, in some examples, hydrous silicate and smectitic clay deposits (Gleeson et al., 2003).

Climate exerts the major control on nickel laterite formation as relatively high temperatures and rainfall facilitate intense weathering of ultramafic protoliths, with the destruction of the primary nickel–cobalt-bearing minerals, olivine and serpentine, and the leaching of silica, magnesium and other mobile elements. Different types of nickel–cobalt laterites are developed in various climatic regimes. Deposits rich in clay silicates occur mainly in semi-arid regions with hydrous silicate-rich deposits developing in humid climates and oxide-dominant deposits formed in a range of climatic conditions (e.g. Freysinnet et al., 2005). Deposits which are currently located in cooler climatic regimes e.g. Bitincke, Albania (Thorne et al., 2012a) and Shevchenko in the Urals (Alexander et al., 2007), or in more arid climates e.g. Murrin Murrin, Western Australia (Gaudin et al., 2005), are considered to be palaeodeposits, originally formed during warmer or wetter climes in the past.

The primary control on nickel–cobalt laterite formation is the exposure of suitable protolith, which is typically an ultramafic rock, for example harzburgites and dunites of ophiolite complexes (e.g. Goro, New Caledonia), ultramafic igneous intrusions (e.g. Barro Alto, Brazil) or komatiitic extrusive rocks (e.g. Murrin Murrin, Western Australia). The majority of nickel–cobalt laterites form by the weathering of ophiolites, such that about 85 per cent of all nickel laterite deposits are found within accretionary terranes (Freyssinet et al., 2005.) The instability of the protolith constituent minerals, such as olivine, at surface temperatures and pressures, in conjunction with their elevated contents of nickel and cobalt, can result in the development of thick weathering profiles which possess important nickel–cobalt resources.

In addition to the importance of a suitable climate and protolith, the development of substantial nickel–cobalt laterite also requires stable geological conditions where the rate of chemical weathering is higher than the rate of physical erosion. As ultramafic rocks weather to form a laterite deposit, the weathering process and profile developed will have a finite duration and thickness depending upon geological variables such as uplift rates and the intensity of fracture development in the protolith. Consequently, where uplift is rapid thick laterite deposits are unlikely to be preserved.

Recent data suggest that the majority of peridotites presently weathering to form nickel–cobalt laterites experience distinct climatic conditions characterised by limited seasonality and annual precipitation of more than 1000 mm (Thorne et al., 2012b). The importance of minimum annual precipitation is well illustrated by the Oman ophiolite which experiences temperatures very similar to modern-day laterite deposits with a Cold Monthly Mean (CMM) temperature of 23 °C and Warm Monthly Mean temperature of 31 °C, but is presently subject to low annual precipitation rates (ca. 90 mm per annum) which are not conducive to nickel laterite deposit formation.

Manganese nodules and cobalt-rich ferromanganese crusts on the seafloor

Although not presently a commercial source of cobalt, nickel–cobalt–manganese nodules and crusts found on the seabed are a potential resource for the future.

Manganese nodules

Concretions of manganese and other transitional metal oxides, including cobalt, can accrete around a nucleus comprising a rock particle, a mineral grain or a fragment of pre-existing nodule in marine environments. Manganese–cobalt nodule formation involves the oxidation of a flux of dissolved or chemically absorbed manganese and iron through an oxidation gradient transforming manganese from Mn (II) to the oxidised tetravalent species. Two forms of manganese oxide, birnessite ($(Na_{0.3}Ca_{0.1}K_{0.1})(Mn^{4+},Mn^{3+})_2O_4 \cdot 1.5H_2O$) and todorokite ($((Mn,Mg,Ca,Ba,K,Na)_2Mn_3O_{12} \cdot 3H_2O)$), are the dominant manganese species although microcrystalline and amorphous iron and manganese oxides and oxide-hydroxides may constitute the bulk of the accreted material. In the Clarion–Clipperton Zone (CCZ) of the Pacific Ocean, nodules are abundant, with about 10 kg m^{-2}, and growth rates estimated at 2–8 mm/Ma, although elsewhere estimates vary between 1–24 mm/Ma (Cronan, 1992). Grade estimates for cobalt in the nodules are 0.74 wt% and the estimated resource for the CCZ is 78 million tonnes of cobalt (Cronan, 1992). In the Atlantic Ocean there appears to be a lower concentration of nodules, probably because of relatively high sedimentation rates.

Cobalt-rich ferromanganese crusts

Ferromanganese oxyhydroxide crusts, up to 250 mm thick, are found on hard-rock substrates throughout the ocean basins mainly on the flanks and summits of seamounts, ridges, plateaux and abyssal hills which are free of sediment

Sample 35D193	Layers	Thickness (mm)	Mineralogy	Element concentration (%)					
				Mn	Fe	Ni	Cu	Co	P_2O_5
	III	0–20	Fe-vernadite, Mn-ferroxyhyte, quartz, buserite, goethite, haematite, feldspars	20.0	18.9	0.33	0.10	0.57	1.4
	II	20–50	Fe-vernadite, Mn-ferroxyhyte, goethite, clayey materials, feldspars, apatite, quartz, calcite, haematite	16.9	16.2	0.38	0.18	0.38	2.0
	1-2	50–65	Fe-vernadite, Mn-ferroxyhyte, apatite	16.8	13.3	0.31	0.17	0.30	9.3
	1-1	65–105	Fe-vernadite, Mn-ferroxyhyte, goethite, apatite, asbolane, calcite, quartz, feldspar	14.6	11.9	0.33	0.09	0.25	8.2
	R	105–165	Asbolane, vernadite, todorokite, ferrihydrite, apatite, calcite, quartz	8.9	5.8	0.47	0.11	0.13	14.0

Figure 6.6 Thin section through a ferromanganese crust showing the mineralogical and geochemical variations within the sample. (Mn, manganese; Fe, iron; Ni, nickel; Cu, copper; Co, cobalt; P_2O_5, phosphorus pentoxide.) (After Glasby et al., 2007.)

accumulation. The cobalt content in most of the crusts from mid-Pacific seamounts and seamounts in the Exclusive Economic Zones of island nations ranges between 0.3 and 0.8%.

Iron–manganese crusts form by hydrogenetic precipitation from cold-ambient bottom waters or by a combination of hydrogenetic and hydrothermal precipitation in areas of hydrothermal venting, such as near oceanic spreading axes, volcanic arcs, and hotspot volcanoes. Iron–manganese crusts contain sub-equal amounts of iron and manganese, and are especially enriched in manganese, cobalt, lead, tellurium, bismuth and platinum relative to lithospheric crust and sea water. Cobalt is strongly enriched in hydrogenetic crusts because it is oxidised from Co^{2+} to the less-soluble Co^{3+} at the crust surface.

The element concentrations in the crusts generally reflect their abundance patterns in sea water. However, there are many complicating factors: for example, copper, nickel and zinc occur in comparable concentrations in sea water, but nickel is much more enriched in crusts than either copper or zinc.

Cobalt-rich manganese crusts from the far western Pacific Ocean are of particular interest because of their great thickness and high concentrations of cobalt and platinum, which make them of potential economic value. Crusts from the Magellan Seamount cluster, for example, are typically up to 15–20 cm thick and have cobalt contents up to 0.56% (Glasby et al., 2007).

Ferromanganese crusts from the Ioan Guyot in the Magellan Seamount cluster illustrate changes in mineralogy and composition during crust growth with manganese, iron and cobalt contents increasing in the crusts with time, whereas nickel and copper show no systematic changes. These results demonstrate that high cobalt contents occurred in the Magellan seamount crusts during the early Miocene. The age at which manganese crusts with high cobalt contents began to form in the oceans is difficult to establish. However, it is suggested that high cobalt contents did not appear in ferromanganese crusts from the far western Pacific until the early Miocene, following the formation of a well-developed Oxygen Minimum Zone (Glasby et al., 2007).

Extraction, processing and refining

The development of cobalt processing technologies and their use worldwide are reviewed in recent publications by Crundwell et al. (2011) and Fisher (2011). The production of cobalt from primary sources is most frequently linked with that of copper and nickel. About half of global cobalt production is either from the leaching of nickel-bearing laterites or the smelting of nickel sulfide ores, while much of the remainder of primary supply is derived from copper sulfide or oxide ores. The process flowsheet for cobalt recovery usually includes initial leaching of milled ore, a flotation concentrate or smelter matte, depending on the nature of the source material (nickel sulfide, copper–cobalt sulfide or laterite ore). The leaching is followed by a copper recovery stage and impurity removal before recovery of the cobalt and, finally, of the nickel, if any is present (Fisher, 2011).

Cobalt from nickel sulfide ores

Cobalt is nearly always present in nickel sulfide ores, occurring typically in pentlandite $(Fe,Ni,Co)_9S_8$, with the concentration of cobalt generally between 0.01 to 0.15%. The standard recovery of cobalt from sulfides involves production of a flotation concentrate, where nickel and cobalt behaviour is highly correlated, followed by smelting the concentrate to a cobalt–nickel–sulfur matte (Figure 6.7). Production of a matte suitable for refining is carried out either by roasting the concentrate followed by smelting or by flash smelting of the concentrate (Crundwell et al., 2011). About a quarter of nickel-sulfide smelting is carried out by the former method which involves production of a calcine in a fluidised-bed roaster followed by melting the calcine with a silica flux to produce a nickel–cobalt-rich matte and a metal-poor slag. The matte is further oxidised in a Peirce–Smith converter, while the slag is discarded. This approach yields high recoveries of nickel, cobalt, copper and precious metals, but it is expensive in terms of energy use.

Flash smelting accounts for about 75 per cent of nickel sulfide that is processed by pyrometallurgy. In this process roasting and smelting are carried out in a single furnace at 1300 °C by continuously blowing oxygen, air, sulfide concentrate and silica flux. The furnace matte produced, containing 20–40% Fe, is oxidised in Peirce–Smith converters to a low-iron matte with 0.5–4% Fe. Flash smelting is far more energy efficient than roasting/electric furnace melting, but its main disadvantage is that metal losses to the slag are much greater and an auxiliary settling furnace is generally required to recover metals from the slag.

In the refinery the first stage of cobalt recovery involves leaching of the matte using chlorine in hydrochloric acid, air in ammonia solutions, or oxygen in sulfuric acid. This is followed by solvent extraction to separate the cobalt and the nickel. Electrowinning or hydrogen reduction is then used for the recovery of the cobalt metal. Overall recovery of cobalt in nickel sulfide smelters varies between 30–80%, which is much less than the typical recovery rates for nickel (97%) and copper (95%).

Increasing environmental pressures in recent years have led to the development of alternative hydrometallurgical routes for the treatment of nickel-sulfide concentrates (Fisher, 2011). Several methods are in commercial operation including sulfate–chloride pressure leach at Voisey's Bay in Canada and bio-heap leach, under atmospheric conditions, at Talvivaara in Finland.

Cobalt from nickel laterite ores

Although cobalt is generally concentrated to some extent in nickel laterite ores due to its relatively high level in the precursor ultramafic rocks, most nickel laterite is smelted to ferronickel and the cobalt is not extracted. However,

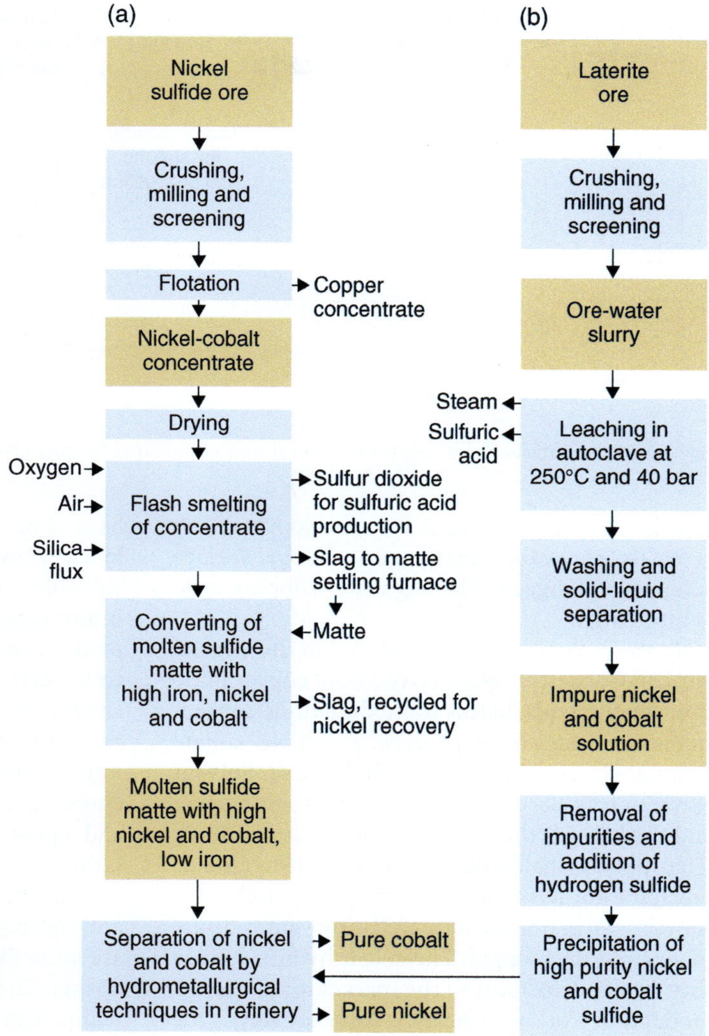

Figure 6.7 Generic flowsheet for the production of cobalt by (a) flash smelting and converting of nickel sulfide ore; and (b) high-pressure acid leaching of laterite ore.

in the 1950s in Cuba the introduction of hydrometallurgical techniques for the treatment of certain nickel laterite ores made the recovery of cobalt from these materials possible. Since that time high-pressure acid leach (HPAL) technology has been significantly improved and, with several operational plants in Australia, New Caledonia and Madagascar, it is now a major global source of nickel and cobalt. A simplified generic flowsheet for the production of cobalt by high-pressure leaching of goethite-bearing laterite is shown in Figure 6.7.

Cobalt from copper–cobalt ores in DRC and Zambia

In the major cobalt-producing region of the Central African Copperbelt, although the deposits formed primarily as sulfides, weathering has in many areas transformed the cobalt-bearing sulfide phases, such as carrollite (Co_2CuS_4), to heterogenite (CoOOH) and sphaerocobaltite ($CoCO_3$), which occur in association with malachite ($CuCO_3.Cu(OH)_2$) and chrysocolla ($Cu.SiO_2.H_2O$). The ore mined from these weathered zones

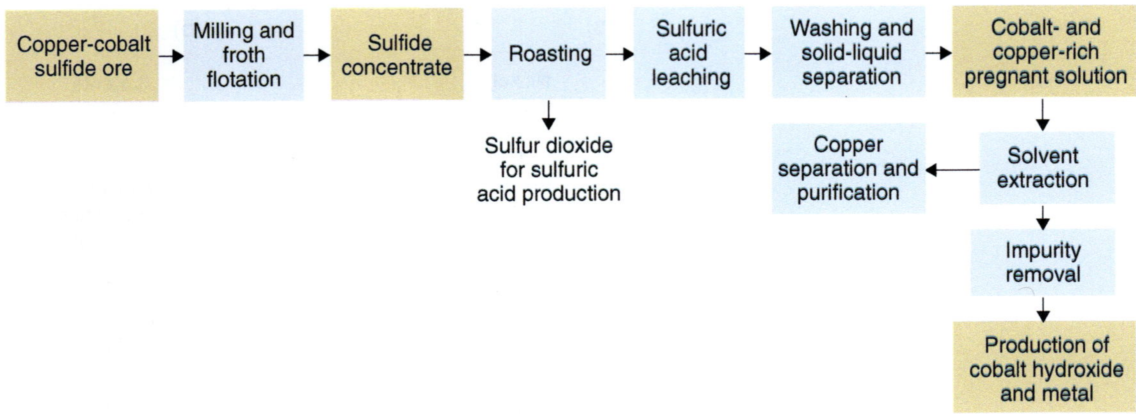

Figure 6.8 Schematic flowsheet for the extraction of cobalt from copper–cobalt sulfide ores.

typically contains about 0.3% Co and 3% Cu. It is easily excavated from open pits by surface scrapers and diggers, although the stripping ratio is often high.

In the past most cobalt produced in the Central African Copperbelt was derived from copper flotation concentrates. In this route the sulfide concentrates are roasted to produce a soluble sulfate calcine which is leached in sulfuric acid. Solvent extraction is used to separate cobalt and copper and the latter is then recovered by electrowinning (Figure 6.8). Cobalt is recovered by cobalt-hydroxide precipitation by addition of magnesia, following the removal of impurities, chiefly iron, aluminium, manganese and copper. The cobalt hydroxide may then be dried and sent to the market for use in the production of chemicals. Alternatively, the hydroxide is redissolved and cobalt metal recovered by electrowinning. However, the sulfide flotation process is inefficient for cobalt, with recoveries as low as 40 per cent for mixed oxide–sulfide ores. Consequently, direct whole ore leach (WOL), which provides much improved metal recoveries, has been increasingly used in recent years in the DRC for processing of dominantly oxide ores (Figure 6.9). In this process the ores undergo reductive leaching followed by solvent extraction to separate copper and cobalt. As in the 'traditional' method, cobalt hydroxide is precipitated by addition of magnesia. The hydroxide may be sold or redissolved and cobalt metal recovered by electrowinning. Although cobalt metal is more valuable than cobalt hydroxide or other intermediate cobalt salts which are traded, the production of large quantities of high-quality cobalt cathode is both technically challenging and capital intensive (Fisher, 2011). Since the global economic recession of 2008 and the consequent impact on metal prices, some companies have chosen a conservative approach for new projects and opted to produce an intermediate product, rather than cobalt metal.

The WOL method is dependent on abundant and cheap supplies of sulfuric acid, but these costs are more than offset by improved metal recoveries (Fisher, 2011). The successful utilisation of WOL technology has been an important factor in the revival of mining in the DRC in the last decade and this route is likely to become a major contributor to global cobalt supply in the near future.

Other sources of cobalt

As mentioned above, the flotation process routinely used for the processing of copper sulfide ores does not yield efficient recovery of cobalt. Consequently, tailings from these operations are potentially large resources of cobalt and have been considered for exploitation. The Kolwezi Tailings project in the Katanga Province of the

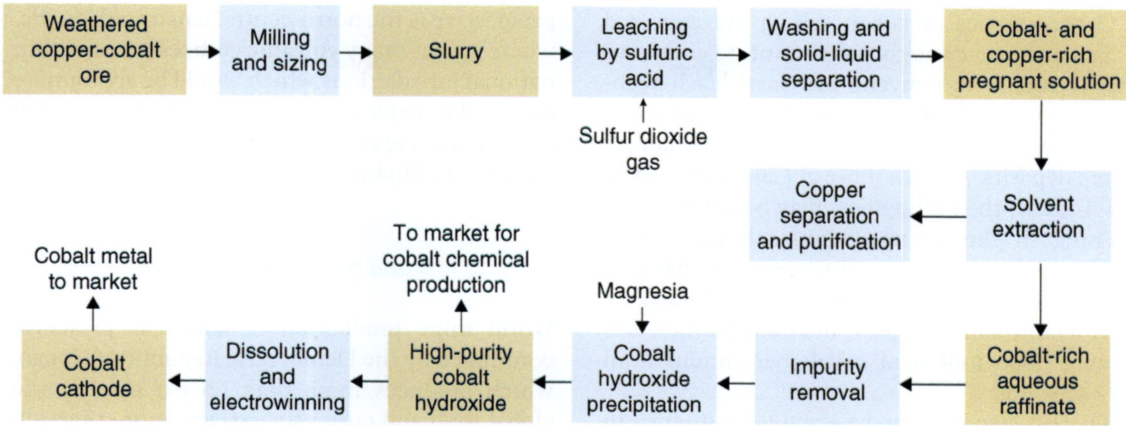

Figure 6.9 Schematic flowsheet for the extraction of cobalt from weathered copper–cobalt ore using whole ore leach (WOL).

DRC is the most advanced of this type with a reserve of 1.7 million tonnes of copper and 363,000 tonnes of cobalt in the tailings which grade 1.49% Cu and 0.32% Co. First Quantum Minerals (FQM) of Canada spent US$750 million to acquire and develop this project up to September 2009 when the government of the DRC revoked its exploration licence. After prolonged legal proceedings the Eurasian Natural Resources Corp (ENRC) agreed in early 2012 to purchase all the assets and claims of FQM in the Katanga Province including the Kolwezi Tailings project and its processing facility. ENRC is planning to begin production at Kolwezi in late 2013 with a targeted production capacity of 70,000 tonnes per annum of copper and 10,000 tonnes per annum of cobalt (ENRC, 2012).

Another source of cobalt which has long attracted interest are the slag stockpiles built up over many years from copper-smelting operations in both Zambia and the DRC. Cobalt follows iron during the smelting process and consequently over the years some operations have generated large slag dumps which contain significant amounts of cobalt. For example, the Nkana slag dump on the Zambian Copperbelt comprises about 20 million tons of slag grading between 0.3 and 2.6% Co (Jones et al., 2001). Mintek developed a process to recover the cobalt from these slags which was implemented at Chambishi in 1998. This involved carbothermic reduction in a DC arc furnace of the oxides of cobalt, nickel and copper leaving iron as oxide in the slag. The cobalt alloy is then processed by pressure leaching. The company Groupement du Terril de Lubumbashi (GTL) is processing part of a slag heap, known as Big Boy, derived from mining operations near Lubumbashi between 1924 and 1992, which comprises material from mines working both copper–cobalt and zinc ores. About one third of this heap, comprising about 4.5 million tonnes with a cobalt content of about 2.1 per cent, is being processed using this pyrometallurgical process. This operation has the capacity to produce about 5000 tonnes of cobalt per annum in an alloy which is transported to Finland for refining.

Since ENRC became owners of the Chambishi cobalt operation in 2010 there has been a move away from the expensive DC furnace processing of slag towards the hydrometallurgical processing for the recovery particularly of copper, but also cobalt, from concentrates supplied by ENRC's mining operations in the DRC. Cobalt production at Chambishi, the world's largest cobalt metal producer, is expected to reach 6000 tonnes per annum by the end of 2012 (Darton Commodities, 2012).

Other sources of cobalt which are expected to become increasingly important result from improvements to conventional heap-leach technology which is cheaper to set up than high-pressure acid leach or smelter operations. Previously uneconomic deposits, such as those at Çaldag in Turkey and Acoje in the Philippines, may be amenable to leaching in this manner. Bioleach technology under ambient pressure and temperature has also been successfully implemented at the Talvivaara polymetallic sulfide deposit in Finland which will produce 1800 tonnes of cobalt per annum from 2012 onwards.

There is also likely to be a revival of interest in cobalt arsenide ores as a result of the development of a high-temperature, pressure leach process for their treatment which delivers an environmentally acceptable stable arsenic residue. For example, Fortune Minerals Ltd is planning to develop an open-pit and underground mine to exploit the NICO cobalt–gold–bismuth deposit in the Northwest Territories of Canada.

Formation Metals Inc. is currently building a mine and concentrator at its Idaho Cobalt Project near Salmon in Idaho. This will be the only primary cobalt operation in the USA and is planning to produce 1500 tonnes per annum of high-purity cobalt metal over a minimum 10-year mine life. The ore reserves in this deposit are currently 2.64 million tons at 0.56% Co and 0.60% Cu (Formation Metals, 2012).

Mining considerations for iron–manganese–cobalt crusts and nodules

Although iron–manganese nodules and crusts are considered only as potential future sources of cobalt and nickel, speculation can be made regarding the geological, geochemical and oceanographic parameters that will ultimately determine the economic viability of an iron–manganese crust mine. Hein et al. (2009) outlined the assumptions which can be made that will likely characterise a sea-floor mine site including water depth, topography, crust thickness, and the extraction technique used. Overall, it appears that the most permissive area from a global perspective is the north-central equatorial Pacific, where a great many volcanic edifices occur within national jurisdictions which would be appropriate targets for exploration. Much smaller regional permissive areas exist in the South Pacific, Atlantic and Indian Oceans (Hein et al., 2009).

World production and trade

World mine production of cobalt is presently dominated by the Democratic Republic of Congo, which produces more than 65 per cent of the global total and currently extracts more than ten times as much as China, the second largest producer. Other significant producers, but each with less than five per cent of world production, include Zambia, Australia, Canada, Cuba, Russia, New Caledonia and Morocco. Global mine production of cobalt has grown dramatically during the past ten years (Figure 6.10), despite the global economic downturn, from about 47,000 tonnes/year in 2002 to 104,000 tonnes/year in 2010 (BGS, 2012). The increase in mine production has largely been supplied by the DRC, which has raised its output over the same period from 14,500 tonnes to 70,000 tonnes (Figure 6.11).

As with mine output, global production of cobalt metal has grown steadily over the past two decades (Figure 6.12). In contrast to mine production, refined metal production is dominated by China, which produces more than 40 per cent of the world total (Figure 6.13), with other substantive cobalt refining capacity in Finland (12 per cent) and Zambia (7 per cent). Notably, the DRC only refines 6 per cent of world cobalt metal production. About 36 per cent of refined cobalt production is based on imported material and processed by countries that have no indigenous cobalt mining production, e.g. Belgium.

Cobalt is traded as a variety of refined products, predominantly as speciality products and chemicals (chiefly cobalt hydroxide, carbonate, oxide and sulfate), but also as broken and cut cathodes, coarse powder, briquettes, ingots and rounds (Darton, 2012).

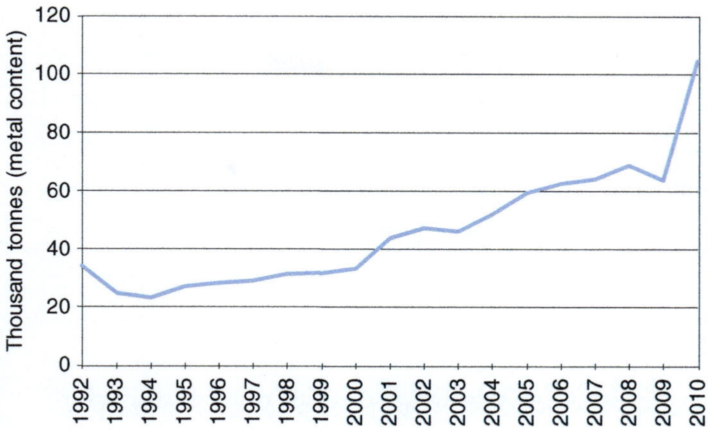

Figure 6.10 Global mine production of cobalt from 1992 to 2010. (Data from British Geological Survey World Mineral Statistics database.)

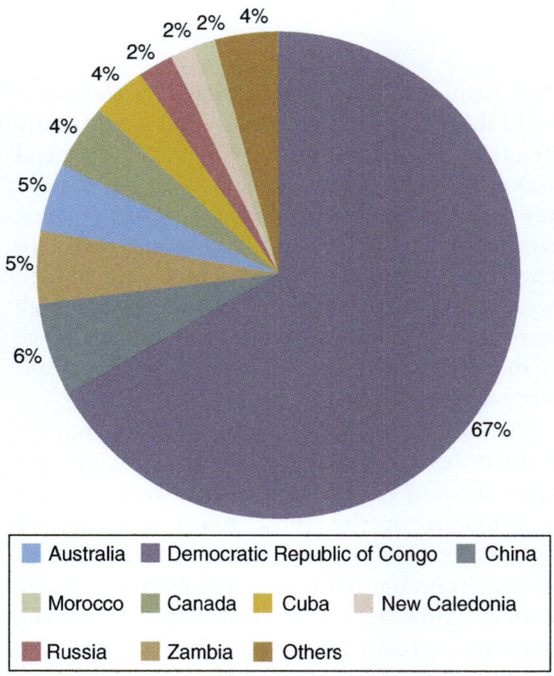

Figure 6.11 The distribution of global mine production of cobalt in 2010 by country. (Data from British Geological Survey, 2012.)

The DRC is the largest global exporter of both cobalt ores and concentrates, followed by Zambia (Figure 6.14). The DRC is also the largest exporter, by a significant margin, of refined metal, reflecting the increasing level of cobalt cathode metal production in recent years. China is the largest importer of both cobalt ores and concentrates and refined metal by significant margins (Figure 6.15). In 2009 China imported more than 283,000 tonnes of cobalt in ores and concentrates compared with Zambia, the second largest importer, with about 41,000 tonnes. Ores are shipped across the border from the Katanga province of southern DRC for processing in the Zambian Copperbelt, chiefly at Chambishi.

Resources and reserves

According to the US Geological Survey world reserves in 2011 are estimated at 7.5 million tonnes, dominated by the DRC with about 45 per cent of the current total, followed by Australia with about 19 per cent (USGS, 2012). Global cobalt resources are estimated to be in the region of 15 million tonnes, with the majority located within nickel-bearing laterite deposits. The remaining resources occur mainly in nickel–copper sulfide deposits hosted in mafic and ultramafic rocks in Australia (Kambalda district, Western Australia), Canada (Sudbury district, Ontario and Voisey's Bay, Labrador) and Russia (Norilsk-Talnakh district), and in the copper–cobalt deposits of the DRC and Zambia. In addition, as much as one billion tonnes of

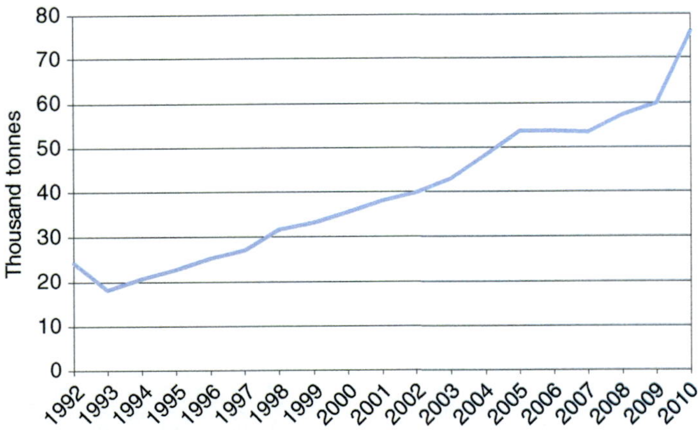

Figure 6.12 Global production of cobalt metal from 1992 to 2010. (Data from British Geological Survey World Mineral Statistics database.)

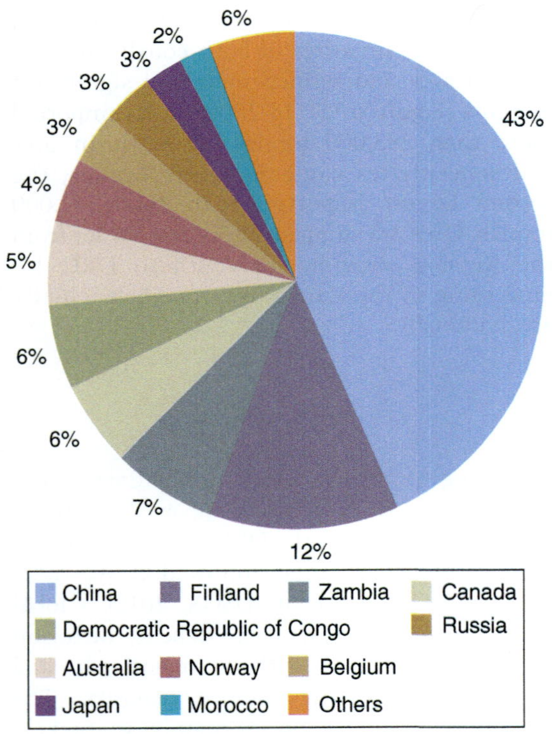

Figure 6.13 The distribution of global production of cobalt metal in 2010 by country. (Data from British Geological Survey, 2012.)

unidentified (hypothetical and speculative) cobalt resources may exist in manganese nodules and crusts on the ocean floor (USGS, 2012).

Uses

Although cobalt was first isolated in elemental form in 1730, use in pigments remained its only practical application until 1907 when the metal was first used in alloys. Cobalt continues to have few applications in its pure form and is most commonly used as an alloy constituent or chemical compound, where its chemical and wear resistance, magnetic properties and high temperature strength are used in a diverse range of materials with commercial, industrial and military applications.

At present the main uses of cobalt are in:

- Batteries: the largest use of cobalt (30 per cent of total) is in batteries. It is an important component in the three main rechargeable battery technologies: 1) lithium-ion batteries may contain up to 60 per cent cobalt as lithium cobalt oxide. However, this depends on the chemical construction of the lithium ion battery. Batteries composed of Li-Ni-Al-Co (NCA) may contain as little as 9 per cent cobalt; 2) nickel-metal hybrid batteries, used in current hybrid electric vehicles, contain up to 15 per cent cobalt; 3) cobalt oxide or hydroxide powder is used in nickel–cadmium batteries accounting for one to five per cent of the battery composition.
- Superalloys and magnet alloys: cobalt is widely used as an alloying metal in superalloys and magnets. Superalloys are primarily used in jet

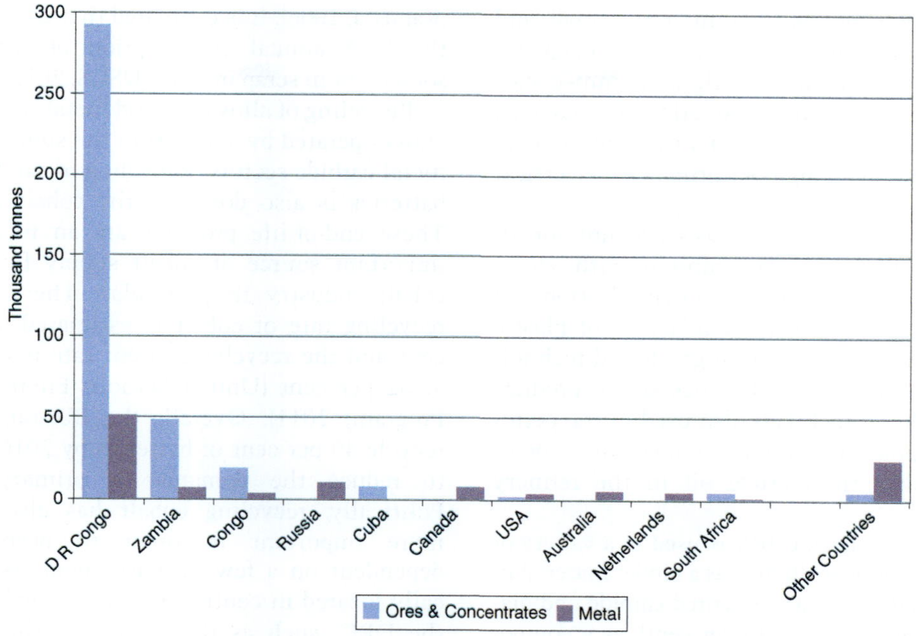

Figure 6.14 The main exporting countries of cobalt ores and concentrates and refined metal in 2010. (Data from British Geological Survey World Mineral Statistics database and UN Comtrade, 2013.)

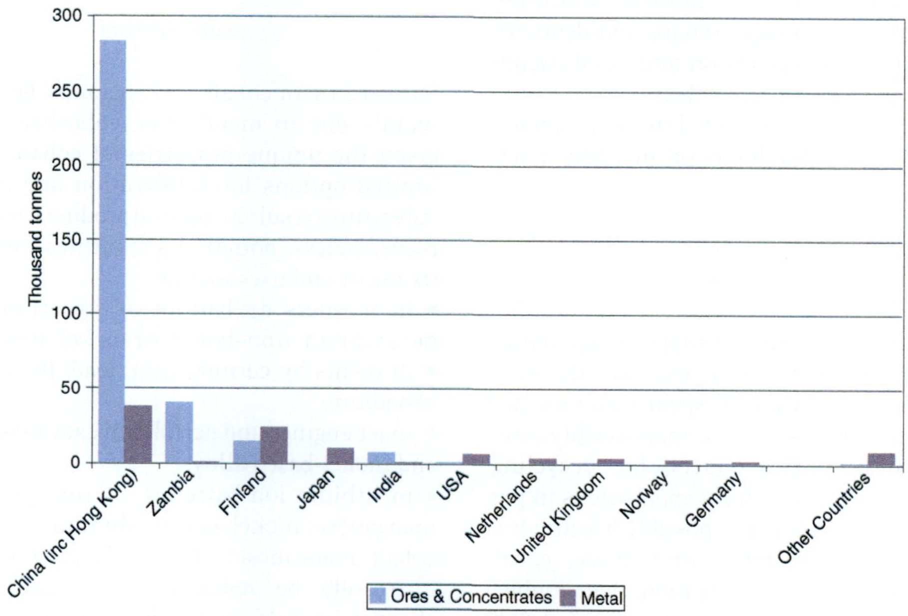

Figure 6.15 The main importing countries of cobalt ores and concentrates and refined metal in 2010. (Data from British Geological Survey World Mineral Statistics database and UN Comtrade, 2013.)

engines and turbines, and in automotive and chemical applications. Cobalt is used in magnetic alloys, either in high-strength samarium–cobalt magnets or lower powered AlNiCo magnets. These magnets are used in a variety of applications including high-performance electrical equipment.

- Catalysts: catalytic processes account for 10 per cent of all cobalt consumption, with one of the primary uses to increase polymerisation and oxidisation rates in the manufacture of plastic resins. Cobalt is also used in gas–liquid technologies where natural gas is processed to produce synthetic diesel fuel. It is also used in the petrochemical industry in the form of cobalt oxide to remove sulfur from crude oil in the refinery process.
- Other applications: cobalt is used in a variety of other applications including: as a binder material in hard materials, such as cemented carbide and diamond tool applications (13 per cent); as a component of high speed steels and other high-strength alloys (5 per cent); in pigments in glass, enamels, pottery and china (9 per cent); in medical applications as part of cancer treatments, as well as in the alloy vitallium (cobalt–chromium–molybdenum–carbon) used in prostheses systems and dentistry; and in electronic connectors on integrated circuits (containing up to 15 per cent cobalt).

Further information on cobalt uses is available from the Cobalt Development Institute (CDI, 2012a).

Recycling

Price volatility, geopolitics of supply and potential cost and environmental benefits drive the recycling of cobalt. Scrap metal, spent catalysts, and rechargeable batteries are the most readily amenable cobalt-bearing products for recycling, whereas cobalt recycling from applications in pigments, glass and paint is not possible because it is dissipated in use. In terms of tonnage, cobalt recovery from secondary feeds more than doubled in the period 1995 to 2005, from an estimated 4200 tonnes to more than 10,000 tonnes respectively (Kapusta, 2006). It is estimated that 24 per cent of the USA annual consumption of cobalt was sourced from scrap in 2011 (USGS, 2012).

Recycling of alloy and hard metal scrap is generally operated by and within the superalloy and metal carbide sectors. Recycling of catalysts and batteries is also done via the cobalt industry. These end-of-life products are an increasingly important source of cobalt supply for the EU cobalt industry in particular. The end-of-life recycling rate of cobalt is estimated at 68 per cent and the recycled content rate is estimated at 32 per cent (United Nations Environmental Program, 2011). Overall, the EU target is to recycle 40 per cent of batteries by 2016 in order to reduce the demand for primary cobalt. Politically, recycling cobalt has also become more important in order to become less dependent on a few primary suppliers, historically located in central Africa. Recent events in the DRC, such as the seizure of the Kolwezi Tailings Project by the government, have served to discourage foreign investment in the mining industry in that country.

Substitution

Substitutes for cobalt are constantly being sought mainly due to metal price volatility. However, given the unique properties of cobalt, there are limited options for substitution and almost all substitutes result in reduced product performance. Nevertheless, potential substitutes for cobalt in its major end uses include:

- in magnets by barium or strontium ferrites, neodymium–iron–boron, or nickel–iron alloys;
- in paints by cerium, iron, lead, manganese, or vanadium;
- in jet engines and petroleum catalysts by nickel and nickel-based alloys.
- in lithium ion batteries, by iron–phosphorus–manganese, nickel–cobalt–aluminium or nickel–cobalt–manganese. The cobalt content may ultimately be reduced or replaced by these cheaper metals and alloys, with technology development reducing the cobalt content from 60

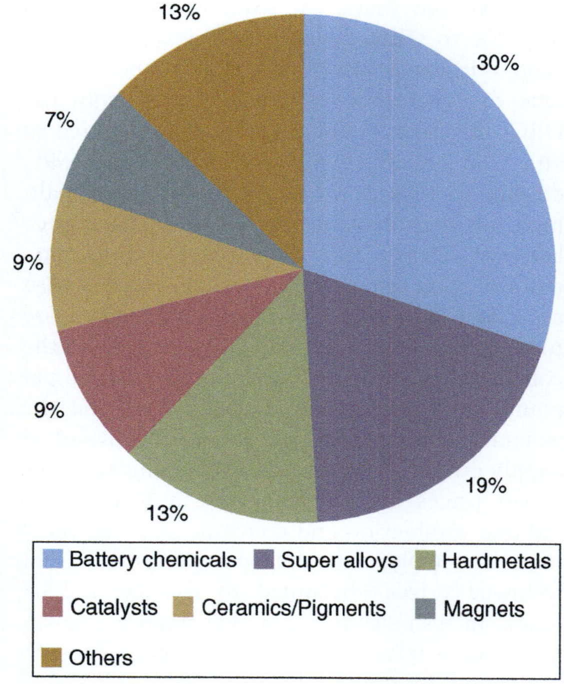

Figure 6.16 Estimated cobalt demand by end-use in 2011. (Data from Cobalt Development Institute, 2012a.)

per cent to less than 10 per cent. It is predicted that the use of existing cobalt cathode materials for Li-ion batteries will drop to approximately 45 per cent of current use by 2013 having previously accounted for almost all the market (European Pathway to Zero Waste, 2011).

Environmental issues

Cobalt in the environment is derived from both natural and anthropogenic sources. Natural sources of cobalt include erosion of cobalt-containing rocks, seawater, volcanic activity and biogenic emissions. Anthropogenic sources of cobalt include mining and processing of cobalt-containing ores, agricultural application of cobalt-containing fertilisers and deposition of atmospheric particulates from combustion of fossil fuels (Smith and Carson, 1981).

Measured atmospheric concentrations of cobalt in unpolluted areas are typically <1–2 ng/m^3 and surface and groundwater concentrations of cobalt are also low, typically less than 1–10 µg/l (Smith and Carson, 1981; Hamilton, 1994). The mean cobalt concentrations reported in surface waters throughout Europe is 0.333 ±1.01 µg/l [standard deviation] (Salminen et al., 2005). Cobalt concentrations in drinking water are generally less than 1–2 µg/l (Kim et al., 2006).

The average concentration of cobalt in soils throughout the world is about 8 ppm (Kim et al., 2006); European topsoil and subsoil concentrations were reported as 10.4 ±13.3 and 11.1 ±10.5 ppm, respectively (mean ± standard deviation) (Salminen et al., 2005). Elevated cobalt concentrations in soils around some mine sites in Ontario, Canada have been reported as high as 6450 ppm (Frank et al., 1976).

Cobalt metal is an essential component of vitamin B_{12}, cobalamin. Neither higher plants nor animals can synthesize vitamin B_{12}, but both require trace amounts. Non-ruminant animals are unable to synthesize vitamin B_{12} from inorganic cobalt, rather they require cobalt in the form of vitamin B_{12}. Ruminant animals (e.g. cows), on the other hand, have micro-organisms present in their stomach (rumin) that are able to synthesize vitamin B_{12} from elemental cobalt, and the vitamin B_{12} produced by the bacteria serves to meet the animals' cobalt requirement. Therefore, ruminant animals have a dietary requirement for elemental cobalt, while non-ruminant animals have a requirement for vitamin B_{12} (NRC, 2005). Vitamin B_{12} deficiency can have various effects on health including infertility, increased perinatal mortality, anaemia, fatty liver and decreased ability to fight disease. Characteristic signs of chronic cobalt overexposure for most species include reduced food consumption, decreased body weight, blood disorders, debility and increased disease susceptibility.

Estimates of human exposure to cobalt suggest that more than 99 per cent is through the ingestion of food, with an estimated daily intake of 5–40 µg/day, most of which is inorganic cobalt and almost all of which passes through the body unabsorbed. (Kim et al., 2006). The recommended Dietary Reference Intake (RDI) of Vitamin B_{12} for adults is 2 µg/day (NRC, 1998) and ingestion of cobalt within normal dietary ranges has not been associated with adverse health consequences. High repetitive oral doses of cobalt, on the other hand, have been associated with effects on red blood cells, thyroid and heart. Inhalation of high concentrations of cobalt is linked to lung disease such as asthma and pneumonia, but these effects appear restricted to workers exposed to high levels of cobalt in the air (U.S. EPA, 2000).

Additional information about the health, safety and environmental effects of cobalt is available from the Cobalt Development Institute (2012b).

Prices

Unlike other major industrial metals cobalt has only recently began trading on the London Metal Exchange. Prior to this, Western Mining Corporation began selling cobalt on its website (the Cobalt Open Sales System – COSS) in 1999 and in September 2000 it was joined by the OMG Group Inc. who also began selling its briquettes in this manner. Following these initiatives, other trading companies began to offer a buying and selling service through the internet. In 2008 BHP (now incorporating Western Mining Corp) suspended the COSS and the London Metal Exchange (LME) started the trading of cobalt (minimum Co content 99.3%) in February 2010. The LME offers a fully regulated market with which to trade spot and future cobalt contracts. The global contract is traded in 1-metric-tonne lots, minimum 99.3% cobalt metal, with delivery to warehouses in Asia, Europe, and the United States.

The price of cobalt is linked to supply, demand and the prevailing political environment of the key producer, the DRC (Figure 6.17). For example, in the early 1990s the price of cobalt peaked at around $33 per pound largely due to political and economic tensions in the DRC. However, during the next two years, the price of cobalt fell in response to weakened demand given the global economic downturn (Plunkert and Jones, 1999). Due to concerns over cobalt supply from the DRC, the price of cobalt more than doubled in late 1993 – early 1994. Between 1993 and 1995, despite increased world production of cobalt, high levels of demand supported a cobalt price between $20 and $30 per pound. However, forecasts of large increases in nickel demand with associated new sources of cobalt production, led to concerns about potential over-supply and the cobalt price fell to approximately $21.50 per pound by the end of 1995. From 1995 to 2002, the general trend in cobalt prices was downward as supply outpaced demand. However, during 2003, cobalt prices increased sharply in response to reduced production and concerns over tightness in global supply. With the commodity boom that followed prices continued to rise until 2008 when, in response to the global financial downturn, prices fell from near $50 per pound to $15 per pound. By mid-2011 these prices were stable at around $20 per pound with an early 2012 price of about $15 per pound. It seems probable that price volatility will continue, although this will depend on various factors including the timing of the opening of new mine capacity relative to mine closures and the pace of demand growth.

Outlook

There are many potential new sources of cobalt, both onshore, in Canada, Western Australia, the DRC, Zambia and Madagascar, and offshore in deep-sea nodules. A major recent development has been the increase in mining activity in the DRC where significant volumes of ore and concentrate are being mined for refining elsewhere, mainly China. It appears that future developments in the DRC will be an important factor in cobalt production. Given the importance of supply from the DRC and its relatively complicated and uncertain political environment it is perhaps not surprising that cobalt finds its way onto lists of critical metals,

Figure 6.17 The quarterly average price of cobalt, 1992–2011. (High Grade Metal Bulletin free market US$ per lb in warehouse.)

scoring high on economic importance, and with significant security of supply risks (European Commission, 2010).

Since 2002 cobalt consumption in Asia has increased significantly, while demand in the West has remained stable. The growth in demand is due chiefly to increased use in chemical applications, particularly rechargeable batteries and catalysts. The demand for rechargeable batteries for portable electronic devices and in automotive applications is likely to continue to grow, although the demand for cobalt will strongly depend on which battery technology is adopted by the car industry. Cobalt is used in both nickel–metal-hydride batteries, as currently utilised in hybrid electric vehicles, and in the more powerful lithium-ion batteries used in all electric and plug-in hybrid electric vehicles. The US Department of Energy analysed a number of scenarios for the future deployment of electric vehicles and the market share of various battery types and concluded that the availability of cobalt is more than adequate up to 2025, even without additional supply from the DRC (U.S. DOE, 2011).

Cobalt supply and demand forecasts suggest that the cobalt market will remain roughly in balance for the coming decade (CDI, 2012a). The Cobalt Development Institute (CDI) predicts a growth rate of 2.5 per cent per year, which lies in between the high and low growth rates modelled by Öko-Institut of 2.8 and 1.7 per cent (Öko-Institut, 2009). In addition to the growing demand for cobalt in batteries, its use in superalloys is expected to increase in response to continuing expansion of the global aerospace market. On the supply side there is a considerable degree of uncertainty over the opening of new mines over the next few years. The USGS lists over 80,000 tonnes of potential new mine capacity that is scheduled for completion by 2013. Some of this will augment or replace existing capacity, although some is unlikely to open or be fully utilised. Nonetheless, the existence of this potential mining capacity goes some way to alleviating scarcity concerns for cobalt even if demand growth is higher than predicted. Substitution of cobalt by cheaper metals in batteries and increased recycling of cobalt from metal alloys and batteries may also have significant impacts on the market in the future.

Acknowledgements

Gus Gunn publishes with the permission of the Executive Director of the British Geological Survey.

Notes

1. The amount of time in years required to replace a given ocean constituent from river supply.
2. Ma, million years before present.

References

Ahmed, A.H., Arai, S., and Ikenne, M. (2009) Mineralogy and paragenesis of the Co-Ni arsenide ores of Bou Azer, Anti-Atlas, Morocco. Economic Geology 104, 249–266.

Alexander, K., Herrington, R., Kirk, C. and Chlup, D. (2007) Geology and mineralogy of the Shevchenko Lateritic Ni-Co deposit, Kazakhstan. In: Andrew, C.J. et al. (ed) Proceedings of the 9th biennial meeting of the Society for Geology Applied to Mineral Deposits, Dublin, Ireland, 20th-23rd August 2007. Irish Association of Economic Geology, Dublin, 1393–1396.

Amelin, Y., Li, C., and Naldrett, A.J. (1999) Geochronology of the Voisey's Bay Intrusion, Labrador, Canada, by precise U-Pb dating of coexisting baddeleyite, zircon, and apatite. Lithos, An International Journal of Petrology, Mineralogy and Geochemistry 47, 33–51.

Annels, A.E. (1989) Ore genesis in the Zambian Copperbelt, with particular reference to the northern sector of the Chambishi Basin. Special Paper – Geological Association of Canada 36, 427–452.`

Annels, A.E. and Simmonds, J.R. (1984) Cobalt in the Zambian Copperbelt. Precambrian Research 25, 75–98.

Baja Mining Corp (2010) http://www.bajamining.com/static/downloads/Resource%20and%20Reserve.pdf

Bending, J.S. and Scales, W.G. (2001) New production in the Idaho cobalt belt; a unique metallogenic province. Transactions – Institution of Mining and Metallurgy. Section B: Applied Earth Science 110, 81–87.

Berger V.I., Singer D.A., Bliss, J.D. and Moring. B.C. (2011) Ni-Co Laterite Deposits of the World – Database and Grade Tonnage Models. US Geological Survey Open File Report 2011-1058, 26pp.

Bernau R., Roberts, S., Richards, M., Nesbitt, B., Boyce, A.J. and Nowecki, J. (2013) The Geology and Geochemistry of the Lumwana Cu (± Co ± U) deposits, NW Zambia. Mineralium Deposita 48, 137–153.

Bookstrom, A.A., Johnson, C.A., Landis, G.P. and Frost, T.P. (2007) Blackbird Fe-Cu-Co-Au-REE deposits. In: O'Neill J.M. (ed.) Metallogeny of Mesoproterozoic Sedimentary Rocks in Idaho and Montana - Studies by the Mineral Resources Program, U.S. Geological Survey, 2004–2007. U. S. Geological Survey Open-File Report 2007–1280, 11–20.

Brems, D., Muchez, P., Sikazwe, O. and Mukumba, W. (2009) Metallogenesis of the Nkana copper-cobalt South Orebody, Zambia. Journal of African Earth Sciences 55, 185–196.

British Geological Survey (2012) World Mineral Production 2006–2010 (Keyworth, Nottingham, British Geological Survey).

Broecker, W.S. and Peng, T.H. (1982) Tracers in the Sea. Lamont-Doherty Geological Observatory, Palisades, N.Y.

Broughton, D.W., Hitzman, M.W. and Stephens, A.J. (2002) Exploration history and geology of the Kansanshi Cu(-Au) deposit, Zambia. Special Publication Society of Economic Geologists (U. S.) 9, 141–153.

Cailteux, J.L.H., Kampunzu, A.B., Lerouge, C., Kaputo, A.K. and Milesi, J.P. (2005) Genesis of sediment-hosted stratiform copper-cobalt deposits, Central African Copperbelt. Journal of African Earth Sciences 42, 134–158.

Cobalt Development Institute (2012a). Cobalt Facts: Cobalt Supply and Demand 2011. http://www.thecdi.com/cdi/images/documents/facts/Cobalt%20Facts-Supply%20%20Demand-2011.pdf

Cobalt Development Institute (2012b). www.thecdi.com/hse

Conly, A.G. Beaudoin, G. and Scott, S.D. (2006) Isotopic constraints on fluid evolution and precipitation mechanisms for the Boléo Cu–Co–Zn district, Mexico. Mineralium Deposita 41, 127–151.

Cronan, D.S. (1992) Marine Minerals in Exclusive Economic Zones. Volume 5 of Topics in the Earth Sciences, Springer, pp. 212.

Crundwell, F.K., Moats, M.S., Ramachandran, V., Robinson, T.G. and Davenport, W.G. (2011) Extractive Metallurgy of Nickel, Cobalt and Platinum-Group Metals. Elsevier, Amsterdam, Netherlands, pp. 610.

Darton Commodities Ltd (2012) Cobalt Market Review 2011–2012. http://www.dartoncommodities.co.uk/

Dietz, R.S. (1964) Sudbury structure as an astrobleme: Journal of Geology 72, 412–434.

ENRC (2012) http://www.enrc.com/Media/press-releases

Evans-Lamswood, D.M., Butt, D.P., Jackson, R.S., Lee, D.V., Muggridge, M.G., Wheeler, R.I., and Wilton, D.H.C. (2000) Physical controls associated with the distribution of sulfides in the Voisey's Bay Ni-Cu-Co deposit, Labrador. Economic Geology 95, 749–770.

European Pathway to Zero Waste (2011) (http://www.environment-agency.gov.uk/static/documents/Business/EPOW-recovering-critical-raw-materials-T5v2.pdf.

European Commission (2010) Report of the Ad hoc Working Group on defining critical raw materials. http://ec.europa.eu/enterprise/policies/raw-materials/files/docs/report_en.pdf.

Fisher, K G. (2011) Cobalt processing developments. 6th Southern African Base Metals Conference 2011. The Southern African Institute of Mining and Metallurgy, 237–258.

Formation Metals Inc. (2012) Formation Metals company presentation, March 2012. http://www.formationmetals.com/s/Home.asp.

Fortune Minerals Limited. (2012) http://www.fortune-minerals.com/Projects/NICO/NICO-Gold-Cobalt-Bismuth-Deposit/default.aspx.

Frank, R., Ishida, K. and Suda, P. (1976) Metals in agricultural soils of Ontario. Canadian Journal of Soil Science 56, 181–196.

Freyssinet, Ph., Butt, C.R.M. and Morris, R.C. (2005) Ore-forming processes related to lateritic weathering. Economic Geology 100th Anniversary Volume, 681–722.

Gaudin, A., Decarreau, A., Noack, Y., and Grauby, O. (2005) Clay mineralogy of the nickel laterite ore developed from serpentinised peridotites at Murrin Murrin, Western Australia. Australian Journal of Earth Sciences 52, 231–241.

Glasby, G., Ren X., Shi X., and Pulyaeva, I. (2007) Co-rich Mn crusts from the Magellan Seamount cluster: the long journey through time. Geo-Marine Letters 27, 315–323.

Gleeson, S.A., Butt, C.R.M. and Elias, M. (2003) Nickel laterites; A review. Society of Economic Geologists (SEG) Newsletter, 54 (1), 12–18.

Groves, D.I., Bierlein, F.P., Meinert, L.D. and Hitzman, M.W. (2010) Iron Oxide Copper-Gold (IOCG) Deposits through Earth History: Implications for Origin, Lithospheric Setting, and Distinction from Other Epigenetic Iron Oxide Deposits. Economic Geology 105, 641–654.

Hein, J.R., Conrad T.A. and Dunham, R.E. (2009) Seamount Characteristics and Mine-Site Model Applied to Exploration- and Mining-Lease-Block Selection for Cobalt-Rich Ferromanganese Crusts. Marine Georesources and Geotechnology 27, 160–176.

Hu, Z. and Gao, S. (2008) Upper crustal abundances of trace elements; a revision and update. Chemical Geology 253, 205–221.

John, T., Schenk, V., Haase, K., Scherer, E., and Tembo, F. (2003) Evidence for a neoproterozoic ocean in south-central Africa from mid-oceanic-ridge-type geochemical signatures and pressure-temperature estimates of Zambian eclogites. Geology 31, 243–246.

Jones, R.T., Denton, G.M., Reynolds, Q.G., Parker, J.A.L. and van Tonder, G.J.J. (2001) Recovery of cobalt from slag in a DC arc furnace at Chambishi, Zambia. Presentation given at Copper Cobalt Nickel and Zinc Recovery conference, Victoria Falls, Zimbabwe, 16–18 July 2001.

Kapusta, J. (2006) Cobalt Production and Markets: A Brief Overview. Journal of the Minerals, Metals and Materials Society (JOM) 58, 33–36.

Kampunzu, A.B., Tembo, F., Matheis, G., Kapenda, D. and Huntsman-Mapila, P. (2000) Geochemistry and tectonic setting of mafic igneous units in the Neoproterozoic Katangan Basin, central Africa; implications for Rodinia break-up. Gondwana Research 3, 125–153.

Kim, J.H., Gibb, H.J. and Howe, P.D. (2006) Cobalt and inorganic cobalt compounds. Concise international chemical assessment document; Document 69. World Health Organization, Geneva, Switzerland.

Leblanc, M. and Kroener, A. (1981) The Late Proterozoic Opholites of Bou Azzer (Morocco); Evidence for Pan-African Plate Tectonics. In: Precambrian Plate Tectonics, Netherlands (NLD): Elsevier, Amsterdam, Netherlands, 431–451.

Leblanc, M. and Billaud, P. (1982) Cobalt arsenide orebodies related to an upper proterozoic ophiolite; Bou Azzer (Morocco). Economic Geology and the Bulletin of the Society of Economic Geologists 77 (1), 162–175.

Leblanc, M. and Fischer, W. (1990) Gold and platinum group elements in cobalt-arsenide ores; hydrothermal concentration from a serpentinite source-rock (Bou Azer, Morocco). Mineralogy and Petrology 42, 197–209.

Lesher, C.M. (1989) Komatiite hosted Ni-Sulphide deposits. Reviews in Economic Geology 4, 45–101.

Li, C. and Naldrett, A.J. (2000) Melting reactions of gneissic inclusions with enclosing magma at Voisey's Bay, Labrador, Canada; implications with respect to ore genesis. Economic Geology 95, 801–814.

Lightfoot, P.C., Keays, R.R., Morrison, G.G., Bite, A. and Farrell, K.P. (1997a) Geochemical relationships in the Sudbury igneous complex; origin of the main mass and offset dikes. Economic Geology 92, 289–307.

Lightfoot, P.C., Keays, R.R., Morrison, G.G., Bite, A. and Farrell, K.P. (1997b) Geologic and geochemical relationships between the contact sublayer, inclusions, and the main mass of the Sudbury Igneous Complex; a case study of the Whistle Mine Embayment. Economic Geology 92, 647–673.

Lightfoot, P.C. and Doherty W. (2001) Chemical Evolution and Origin of Nickel Sulfide Mineralisation in the Sudbury Igneous Complex, Ontario, Canada Economic Geology 96, 1855–1875.

McGowan, R.R., Roberts, S. and Boyce, A.J. (2006) Origin of the Nchanga copper-cobalt deposits of the Zambian Copperbelt. Mineralium Deposita 40, 617–638.

McGowan, R.R., Roberts, S., Foster, R.P., Boyce A.J. and Coller, D. (2003) Origin of the copper-cobalt deposits of the Zambian Copperbelt; an epigenetic view from Nchanga. Geology 31, 497–500.

Mudd, G.M. (2010) Global Trends and Environmental Issues in Nickel Mining: Sulfides Versus Laterites. Ore Geology Reviews 38 (1–2), 9–26.

Muchez, P., Brems, D., Clara, E. et al. (2010) Evolution of Cu-Co mineralizing fluids at Nkana Mine, Central African Copperbelt, Zambia. Journal of African Earth Sciences 58, 457–474.

Naldrett, A.J. (1989) Magmatic Sulphide Deposits. Oxford Monographs on Geology and Geophysics 14. vii +, New York, Oxford: Oxford University, pp.186.

Nash, J.T. and Hahn, G.A. (1989) Stratabound Co-Cu deposits and mafic volcaniclastic rocks in the Blackbird mining district, Lemhi County, Idaho. Special Paper – Geological Association of Canada 36, 339–356.

National Research Council (1998) Dietary Reference Intakes for Thiamin, Riboflavin, Niacin, Vitamin B6, Folate, Vitamin B12, Pantothenic Acid, Biotin, and Choline. Washington, DC: The National Academies Press.

National Research Council (2005) Mineral Tolerance of Animals, 2nd Revised Edition, National Research Council, National Academy of Science, The National Academies Press, Washington, D.C., USA.

Oberthur, T., Melcher, F., Henjes-Kunst, F. et al. (2009) Hercynian age of the cobalt-nickel-arsenide-(gold) ores, Bou Azer, Anti-Atlas, Morocco; Re-Os, Sm-Nd, and U-Pb age determinations. Economic Geology 104, 1065–1079.

Öko-Institut (2009) Critical metals for future sustainable technologies and their recycling potential. http://oeko.de/oekodoc/1070/2009-129-en.pdf.

Plunkert, P.A., Jones, T.S. (1999) Metals prices in the United States through 1998: Cobalt. http://minerals.usgs.gov/minerals/pubs/commodity/cobalt/210798.pdf.

Ryan, B., Wardle, R.J., Gower, C.F. and Nunn, G.A.G. (1995). Nickel-copper sulphide mineralisation in Labrador: The Voisey's Bay discovery and its exploration implications; Current Research, Report 95-1, Geological Survey, Department of Natural Resources, Government of Newfoundland and Labrador, 109–129.

Salminen R., Batista M.J., Bidovec M. et al. (2005) Geochemical Atlas of Europe. Part 1 – Background Information, Methodology and Maps. Espoo, Geological Survey of Finland.

Selley, D., Broughton, D., Scott, R. et al. (2005) A new look at the geology of the Zambian Copperbelt. In: Hedenquist, J.W., Thompson, John F.H.; Goldfarb, R.J. and Richards, J.P. (eds.) Economic Geology; One Hundredth Anniversary Volume 1905–2005, 965–1000.

Slack, J.F. (2006) High REE and Y concentrations in Co-Cu-Au ores of the Blackbird District, Idaho. Economic Geology 101, 275–280.

Smith I.C. and Carson B.L. (1981) Trace Metals in the Environment. Ann Arbor, Michigan, Ann Arbor Science Publishers.

Trans Continental Exploration and Mining Company (2011). http://www.tcemco.com/

Therriault, A.M., Fowler, A.D. and Grieve R.A.F. (2002) The Sudbury Igneous Complex: A Differentiated Impact Melt Sheet. Economic Geology 97, 1521–1540.

Thorne, R., Herrington, R.J. and Roberts, S. (2012a). The Formation and Evolution of the Bitincke Nickel laterite deposit, Albania. Mineralium Deposita 47 (8), 933–947.

Thorne, R., Roberts, S. and Herrington, R.J. (2012b) Climate change and the formation of Nickel laterites. Geology 40 (4) 331–334.

United Nations (UN) Comtrade (2013) Commodity Trade Statistics Database (http://comtrade.un.org/db/).

United Nations Environmental Program (UNEP) (2011) Recycling Rates of Metals – A status report, A report of the Working Group on the Global Metal Flows to the International Resource Panel. Graedel, T.E., Alwood, J., Birat, J.-P., Reck, B.K., Sibley, S.F., Sonnemann, G., Buchert, M., Hagelüken, C.

U.S. Department of Energy (2011) Critical materials strategy, December 2011.

U.S. Environmental Protection Agency (2000). Air Toxics Web Site – Cobalt: http://www.epa.gov/ttn/atw/hlthef/cobalt.html.

U.S. Geological Survey. (2012) Mineral Commodity Summaries: Cobalt. http://minerals.usgs.gov/minerals/pubs/commodity/cobalt/cobalmcs07.pdf.

Vale (2012) Annual Report 2012. http://www.vale.com/EN/investors/Quarterly-results-reports/20F/20FDocs/20F_2012_i.pdf.

Williams P.J. (1999) Fe-oxide-Cu-Au deposits of the Olympic Dam/Ernest Henry type. In: Hodgson, C.J., and Franklin, J.M., (eds.) New developments in the geological understanding of some major ore types and environments, with implications for exploration: Toronto, Prospectors and Developers Association of Canada, 1–43.

Xstrata (2012) http://www.xstratanickel.com/EN/Publications/Publications/xni%20brochure%202012_2%208_150%20dpi.pdf.

7. Gallium

THOMAS BUTCHER[1] AND TERESA BROWN[2]

[1]*Independent Consultant, New York, USA*
[2]*British Geological Survey, Keyworth, Nottingham, UK*

Introduction

The existence of gallium was predicted by Dmitri Mendeleev in 1871 as part of his development of the Periodic Table of elements, but it was actually discovered in 1875 by Paul Émile Lecoq de Boisbaudran. As a patriotic Frenchman, he named the new element 'gallia' after the Latin name for France.

Physical and chemical properties

As a pure metal, gallium is silvery-white in appearance. It is a relatively soft metal, with a low melting point but a comparatively high boiling point. Combined with a low vapour pressure, even at high temperatures, this gives it the longest liquid range of any metal. As a liquid it will wet glass and skin, and it will readily contaminate other metals by diffusing into their lattice structures, making it more difficult to handle than many other commodities.

When it solidifies from a liquid, gallium will expand and therefore should not be stored in a restricted container as this would rupture. As a solid, gallium will fracture conchoidally like glass. Gallium is magnetic and a good conductor of both electricity and heat. It exhibits a noticeable anisotropy in electrical resistivity dependent upon the orientation of crystals within its structure. Other selected properties of gallium are shown in Table 7.1.

Mineralogy and distribution

Gallium has an average crustal abundance of 19 ppm, which is greater than many better known metals, for example lead (ten ppm) or tin (two ppm). However, it does not occur in nature as a native metal but instead it substitutes for other elements in certain minerals, although usually at trace quantities. It is generally extracted as a by-product of aluminium or zinc production because it is rarely, if ever, found in sufficient quantities by itself to enable economic extraction.

There are a few minerals in which gallium forms a significant part. The main gallium-bearing mineral is gallite ($CuGaS_2$) which is similar to the copper sulfide mineral chalcopyrite ($CuFeS_2$), but with gallium substituting for iron in the crystal structure. Gallite is known to occur at Lubumbashi in the Democratic Republic of Congo and at Tsumeb in Namibia (Roskill, 2011), albeit in uneconomic quantities. Other gallium-bearing minerals have been identified, principally at Tsumeb in

Critical Metals Handbook, First Edition. Edited by Gus Gunn.

Namibia, as shown in Table 7.2, but all remain relatively rare (Geier and Ottemann, 1970; Scott, 1971 and Jambor et al., 1996). Gallium is more commonly found as a trace element in bauxite (the main ore of aluminium) and sphalerite (a zinc ore).

Sources of gallium

Currently there are no mines worked primarily for gallium. However, gallium is recovered as a by-product of processing bauxite or sphalerite and is found in uneconomic quantities elsewhere.

Bauxite

Deposits of bauxite are residual deposits, that is, they are formed by the weathering of pre-existing rocks in tropical or sub-tropical conditions. They are formed from a wide variety of source rocks and their composition varies considerably (Hill and Sehnke, 2006). The ratio of gallium to aluminium, and therefore the concentration of gallium, in bauxite increases with greater intensity of weathering. Gallium also appears to be more abundant where the bauxite was derived from alkali source rocks (Weeks, 1989).

It is generally accepted that the gallium in bauxite originates from minerals such as feldspar or nepheline. Weathering processes release both aluminium and gallium from these minerals and their similar geochemical properties result in the enrichment of both elements in bauxite (Dittrich et al., 2011). However, in certain environmental situations gallium also shows an affinity to iron and becomes decoupled from aluminium. Hieronymus et al. (2001) concluded that this happens where later remobilisation occurs as a result of differences in solubility between the two elements. In these circumstances gallium can become more concentrated in iron-rich horizons associated with bauxite.

The average gallium content in bauxite is reported to be approximately 50 ppm (Jaskula, 2011a) although it can vary from 10 to 160 ppm (Mordberg et al., 2001 and Bhatt, 2002). The production of alumina from bauxite ores using the Bayer process results in the concentration of gallium in the liquor and this process remains the primary source of the element (Roskill, 2011).

Sphalerite (ZnS)

Gallium concentrations in the zinc ore, sphalerite, are known to increase as the temperature of deposition decreases, although it can still be present in intermediate and higher-temperature deposit types (Stoiber, 1940 and Cook et al., 2009). Analyses from the 1940s revealed gallium contents in sphalerites from the Mississippi Valley province (a low-temperature deposit type)

Table 7.1 Selected properties of gallium.

Property	Value	Units
Symbol	Ga	
Atomic number	31	
Atomic weight	69.72	
Density at 25°C	5905	kg/m^3
Melting point	30	°C
Boiling point	2204	°C
Hardness (Mohs scale)	1.5	
Electrical resistivity at 25°C	approx. 14.00	μΩ m
Thermal conductivity	29	W/(m °C)

Table 7.2 Known minerals in which gallium forms a significant part.

Mineral name	Chemical formula	Ga content (%)	Locality
Gallite	$CuGaS_2$	35.3	Tsumeb, Namibia and Lubumbashi, DR Congo
Sohngeite	$Ga(OH)_3$	57.7	Tsumeb, Namibia
Gallobeudantite	$PbGa_3[(AsO_4),(SO_4)]_2(OH)_6$	14.6	Tsumeb, Namibia
Carnevallite	Cu_3GaS_4	20.9	Tsumeb, Namibia

which were greater than from most other types of deposits (typically 0.01–0.1 per cent), but gallium was absent from samples of European origin 'Mississippi Valley type' sphalerites. Stoiber (1940) concluded that although there is clear evidence for a correlation between gallium content and temperature, the composition of sphalerite from each metallogenic region is distinctive. Sphalerite from barytes veins in Central Kentucky was also found to contain up to 0.1% gallium (Stoiber, 1940).

Whilst Moskalyk (2003) reported that analyses of USA zinc deposits have indicated a typical content of 50 ppm gallium, earlier work by Hall and Heyl (1968) found gallium contents of sphalerites from the USA to be erratic, with their analysis ranging from 10 to 320 ppm. Analyses by Cook et al. (2009) found gallium contents of up to 366 ppm in a few samples of sphalerite from south-east European epithermal deposits and up to 273 ppm from a similar deposit in Japan, although they concluded that values in excess of 100 ppm are rare.

The processes used to extract zinc provide a secondary source of gallium although the actual recovery remains low (less than one per cent) (Roskill, 2011).

Other geological settings

Gallium also occurs in other aluminosilicate minerals, geothermal fields in volcanic zones (for example in Taupo, New Zealand (Christie and Brathwaite, 2002)), in association with phosphate ores and some coal deposits (Moskalyk, 2003). Rytuba et al. (2003), with reference to two particular deposits in the USA, noted that gallium can be enriched by hydrothermal processes in zones of advanced argillic alteration in high sulfidation epithermal precious metal deposits; although only one of the samples analysed contained >100 ppm gallium.

Gallium concentrations of 149–320 ppm have been reported in fly ash from the combustion of coal (Font et al., 2007). Whilst research has demonstrated that it is technically possible to extract gallium from fly ash, as yet it has not proved to be commercially viable (Roskill, 2011).

Apex Mine, St Georges, Utah, USA

Possibly the only attempt to open a mine primarily to extract gallium and germanium occurred in 1986/7 at the Apex Mine near the town of St Georges in Utah, USA. This historical mine had been worked many times in the past, principally for copper, although the deposit also contained small quantities of zinc and silver (Dutrizac et al., 1986). It was described by Bernstein (1986) as having quantities of gallium contained in jarosite (a sulfate mineral formed from the oxidation of iron sulfides) and limonite (a mixture of iron oxides and hydroxides). It did not prove to be economic and the operator went into liquidation towards the end of 1987 (Kramer, 1988). A subsequent attempt to re-open the mine also proved to be uneconomic.

Recovery methods and refining

Primary recovery

Most gallium today is produced as a minor by-product of alumina/aluminium production from bauxite ores. To a much lesser extent, the metal is produced as a by-product of the smelting and refining of zinc. In Russia, small amounts of gallium have also been produced in conjunction with the mining of both apatite-nepheline and nepheline-syenite.

Whilst a number of exploration companies have found potentially promising deposits of gallium minerals, so far none has been worked to produce gallium as a primary product. Despite the price of gallium having hit relatively elevated levels over the past couple of years, the metal has yet to be discovered in economically recoverable concentrations. One of the most promising such deposits, the Cordero Property in Humboldt County, Nevada in the USA (owned by Gold Canyon Resources Inc.) remains undeveloped. Research into other, potentially promising, sources of the metal, for example, fly ash (Font et al., 2007), waste from elemental phosphorous plants (Degerstrom, 2011), and claystone (Orbite Aluminae, 2011) has not yet led to any commercially attractive, or exploited, alternatives.

However, over the last couple of decades, companies have continued to improve gallium production either by upgrading existing methods of extracting the metal from Bayer liquors, or by developing their own, proprietary, or using others', technologies (Moss et al., 2011). Figure 7.1 outlines the processes involved in schematic form.

From alumina/aluminium processing

Over 95 per cent of all mined bauxite is converted into alumina using the Bayer process (Cardarelli, 2008), with an estimated 95 per cent of all the alumina so produced being smelted and refined to aluminium metal, mainly using the Hall–Héroult process. In the Bayer process, the reactions of aluminium trihydroxide and aluminium oxide hydroxide with aqueous caustic soda form the so-called 'Bayer liquor' (a sodium aluminate solution). This solution, with equilibrium concentrations of 100–125 ppm of gallium, is then recycled on a continuous basis.

To remove the gallium from Bayer liquor, the stream of liquid is tapped and the crude metal most usually extracted using proprietary techniques, employing ion-exchange resins (Srinivasa Rao et al., 2003), and then purified further. Both the types of resin used, and their longevity, have an important bearing on the cost of extracting the gallium from the Bayer liquor, as does the purity of the liquor itself. Extraneous and/or unexpected impurities in the liquor can drastically reduce the life of resins and drive up costs. Since neither all Bayer liquors, nor all bauxites, are equal, this is an important consideration (Moss, et al., 2011).

Non-proprietary extraction processes include:

The de la Bretèque process: since this process extracts the gallium (by direct electrolysis) using mercury as a cathode, it is not very popular.

The Bielfeld and Laspeyres process: similar to the de la Bretèque process, however, mercury loss is improved through the use of a sodium–mercury amalgam.

The Beja process: this process uses carbon dioxide to produce a gallium-bearing precipitate from which the gallium is then extracted.

The Frary and Pechiney processes: in both these processes, the gallium is recovered through electrodeposition.

From zinc processing

Once a gallium-bearing zinc ore has been leached with sulfuric acid to produce a zinc sulfate solution (akin to Bayer liquor), and before electrolysis to extract the metal itself, the impurities, which include gallium, are removed through the addition of antimony trioxide, zinc dust or proprietary reagents. The gallium is then extracted from the resulting separated solids or 'cement residues'.

Secondary recovery

Since one of the main uses of gallium is in compound semiconductor wafers, and their fabrication typically generates around 60 per cent scrap, recovered gallium from this waste is an extremely important source of the metal. The wastes from the manufacture of gallium arsenide (GaAs) and gallium nitride (GaN) wafers, because of their purity and availability, are the metal's most important secondary source. Whilst most of the scrap materials containing gallium are generated prior to the actual manufacture of the wafers themselves, both broken and damaged wafers are also recycled. At present, gallium is not recovered from post-consumer scrap, not least because, in items such as printed circuit boards, it is highly dispersed (Öko-Institut e.V. (2009).

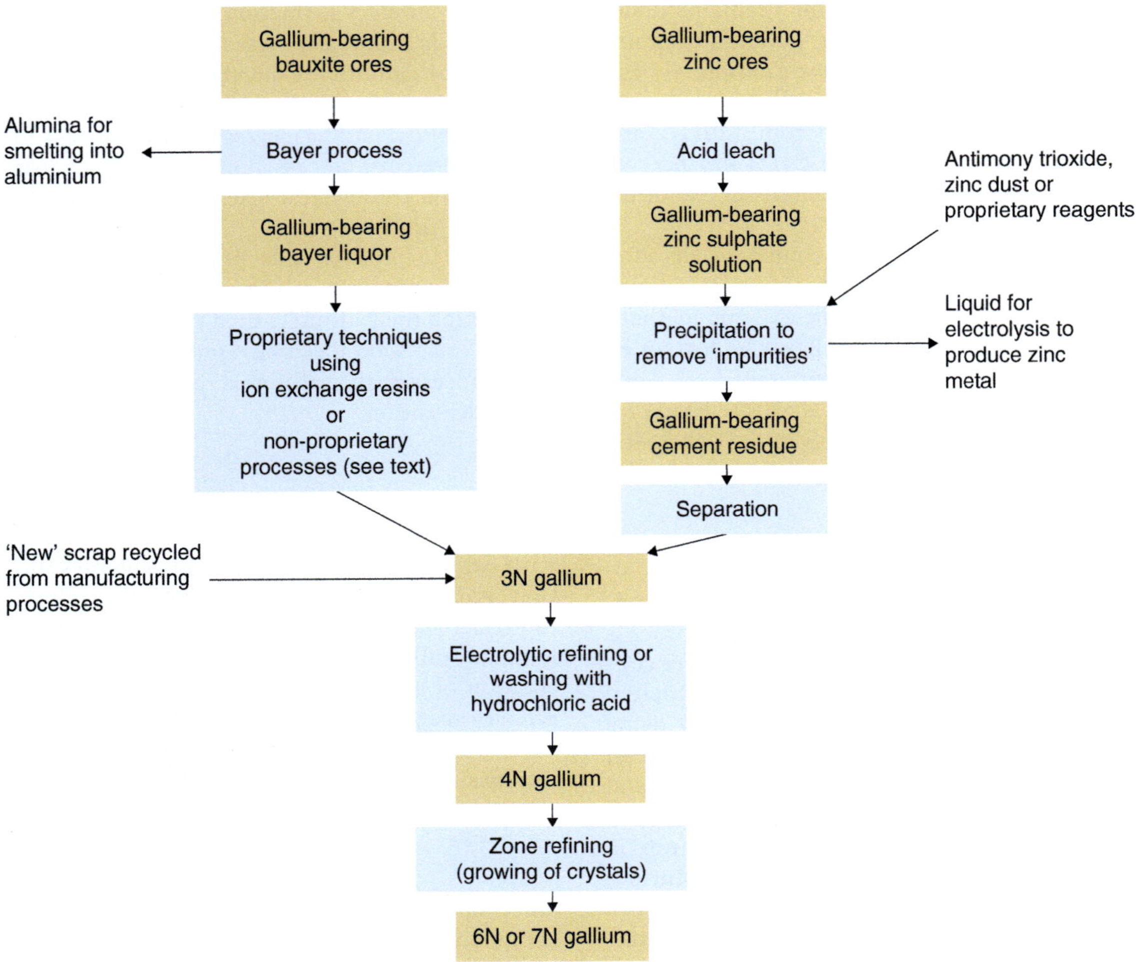

Figure 7.1 Schematic diagram of the processes involved in extracting gallium.

From compound semiconductor manufacture, the main sources of gallium scrap include (Burke, 2011):

- boule ends and chunks;
- broken wafers – from all stages of fabrication;
- cutting and polishing sludges;
- defective devices;
- dust and residues in wastewater filters;
- epitaxial growth wastes;
- kerf generated during wafer slicing;
- liquids following etching;
- residues in GaAs single crystal reactors;
- substandard GaAs ingots.

The gallium content in such scrap can be between one per cent (for example, from wastewater filters) and 99.99 per cent (frozen metal) (Kramer, 2002), depending at which stage in the fabrication process the scrap is generated.

Typically the recovery and recycling of gallium is economically viable from throughout the production process of gallium-containing compound semiconductors. However, many of the newer, and currently burgeoning, uses of gallium provide few opportunities for recycling, primarily because they do not generate as much waste. There is, for example, very little waste produced from the manufacture of:

- batteries and LEDs made using gallium trichloride;
- magnets containing gallium metal;
- phosphors containing gallium oxide.

For certain end-uses, such as batteries, phosphors and LEDs, there is currently little if any recycling, particularly end-of-life recycling. Although there is already some recycling of the gallium used in the solar industry, in CIGS (copper indium gallium selenium) technology in particular, it is likely to both increase in volume and improve in effectiveness over time. Since CIGS cells are manufactured with longevity in mind, it will, of course, be some time before a large volume of material becomes available for end-of-life recycling.

Refining and purification

Except for its use in chemicals, where it is usually of 4N–5N (99.99–99.999 per cent) purity, and in solar cells (5N), for most of its other end-uses, gallium must be of at least 6N (99.9999 per cent) through to 7N (99.99999 per cent) purity. The gallium (as does the arsenic) used in optoelectronic applications usually has to be of at least 6N purity, and in electronic applications the gallium needs to be of 7N purity.

Since the gallium from Bayer liquors and scrap will usually only have a purity of up to 3N (99.9 per cent), the metal needs to be further refined. Either electrolytic refining, or washing with hydrochloric acid will bring the metal up to 4N purity. Thereafter, in order to increase it to the purities of 6N or 7N, the gallium can, typically, be purified by fractional crystallisation. In this process melted gallium is seeded and increasingly pure crystals grown (since impurities tend to remain in the liquid phase they cannot contaminate the growing crystal).

Gallium in GaAs semiconductors

A brief description of how GaAs semiconductors are typically manufactured will serve as illustration of how, for one end-use (and, indeed, currently, one of the metal's largest single end-uses), gallium ends up in the finished product.

In the simplest terms, a semiconductor consists of a *substrate* on which one, or more, very thin surface layers (*epitaxial layers*) are deposited or 'grown'. Compound semiconductor substrates are, as their name suggests, composed of one or more different elements, in this case the metals gallium and arsenic. Ultrahigh-purity single crystals of GaAs are first grown and then sliced into wafers (the basis of electrical components) upon which the requisite epilayers are grown. Different end-uses require different qualities of substrate crystal, with integrated circuits (ICs) and microwave devices requiring the highest quality.

Substrate

Three of the most common manufacturing technologies for growing single crystals of GaAs substrate are:

- Vertical Gradient Freeze (VGF) - including vertical Bridgman (VB);
- Liquid Encapsulated Czochralski (LEC);
- Czochralski (CZ).

A fourth method, the horizontal (as opposed to vertical) Bridgman method (HB) is also used, for example, by Sumitomo Electric in Japan. These methods are shown diagrammatically in Figure 7.2. The LEC and VGF methods are currently the most commonly used, with the various VGF technologies dominating the market. All these methods produce cylindrical single-crystal ingots, or boules, which can be of varying diameters – most usually two inches, four inches, six inches and eight inches.

VGF: In the VGF method, a crucible containing chunks of GaAs, together with a seed crystal, is placed vertically in a furnace and, whilst remaining static, a temperature gradient is moved up the length of the crystal away from the seed. The single crystal propagates from the seed

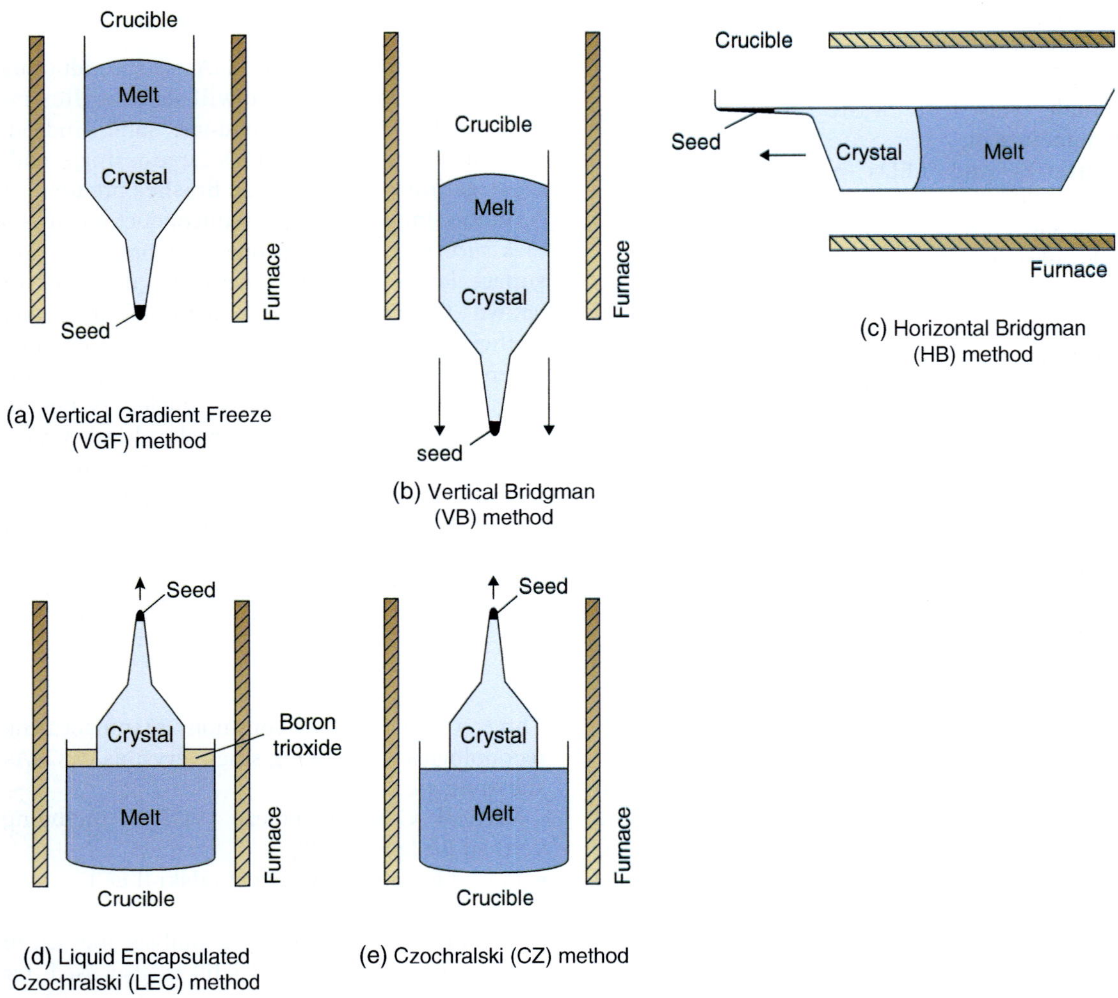

Figure 7.2 An illustration of most common manufacturing technologies for growing single crystals of GaAs substrate.

crystal in the bottom of the crucible. In contrast, in the Bridgman processes, the actual crucible is moved either vertically (VB) or horizontally (HB) relative to the furnace.

LEC: In the Liquid Encapsulated Czochralski method, named after the Polish chemist Jan Czochralski, either elemental gallium and arsenic, or polycrystalline GaAs chunks, are placed in a crucible together with a pellet of boron trioxide, set inside a high-pressure vessel and heated. A small crystal is then dipped in the melt, through the now molten layer of 'encapsulating' boron trioxide, and slowly withdrawn, bringing with it a layer of the melt, which, as it cools, assumes the crystalline structure of the seed.

CZ: In this method, since there is no encapsulation, the crystal is drawn directly from the melt.

Epitaxial layers

A number of different technologies, classified according to the phase (for example, liquid or vapour) of material used to form the epitaxial layer, can be used to lay down or grow the GaAs layers on the substrate. Some of the main methods of epitaxial growth are:

- MBE: molecular beam epitaxy;
- MOVPE: metal organic vapour phase epitaxy;
- LPE: liquid phase epitaxy.

In MBE, which takes place in a vacuum, extremely high purity gallium (together with equally pure arsenic) is heated and the gases of the two elements are 'beamed' at the substrate to condense (a physical deposition) on its surface, leaving a layer of gallium arsenide.

In MOVPE (also commonly referred to in the industry as MOCVD – metal organic chemical vapour deposition), the substrate is heated and a carrier gas 'transporting' precursors, for example trimethyl-gallium (a chemical compound of gallium) and arsine, reacts chemically with the surface of the substrate, leaving atoms of gallium and arsenic behind on its surface. As opposed to MBE, this is a chemical deposition of the two elements.

In LPE, the epitaxial layer forms after a saturated solution is placed in contact with the polished surface of the substrate.

Specifications and uses

Gallium is used in a number of different forms. The most common are:

- gallium metal;
- gallium antimonide;
- gallium arsenide;
- gallium chemicals;
- gallium nitride;
- gallium phosphide.

Gallium metal

As a pure metal, or alloyed with other metals, gallium is used in a number of applications:

- *Eutectic alloys*: alloyed with any, or all, of indium, selenium and zinc, gallium, with its low melting point, is used in devices ranging from fluid unions and heavy current switches, to rocker switches and sphygmomanometers (blood pressure meters).
- *Magnets*: the efficiency of neodymium–iron–boron (NeFeB) magnets is increased through the addition of a small amount of gallium. However, efficiency is not the primary reason for its use in NeFeB magnets. Gallium actually helps increase the fluidity of the alloy during the production process. It acts as a molten lubricant for the hot forming process.
- *Molecular Beam Epitaxy*: gallium metal, at purities of 7N and, sometimes, above, is used in the MBE process to grow epitaxial layers of the metal in the production of semiconductors.
- *Nuclear weapon pits*: alloyed with plutonium, gallium was used in the pits (the cores of implosion weapons) of the first nuclear weapons.
- *Thermometers*: alloyed with indium and tin in, for example, Galinstan, gallium provides a non-toxic substitute to mercury in thermometers.
- *Thin-film deposition*: metallic gallium of 4N purity is often used as the source material for thin film deposition. For example, by thermal evaporation in the manufacture of solar cells.

Gallium antimonide

Like gallium arsenide, gallium antimonide (GaSb) is a semiconducting compound (Dutta et al., 1997). It is used in both electronic and optoelectronic devices and, in particular, those functioning in the infrared. Some of the more important (and interesting) include:

- forward-looking infrared (FLIR) systems for night-time navigation;
- LEDs;
- missile homing guidance systems;
- thermal imaging;
- high-speed electronic circuits (Bennett et al., 2005).

Gallium arsenide

Currently, the greatest consumption of gallium is in gallium arsenide (GaAs) compound semiconductors. Compound semiconductors, in particular GaAs semiconductors, can provide a number of advantages over other semiconducting materials, for example, silicon. In ICs, for example, GaAs is significantly more

Table 7.3 Electron mobility in selected semiconductor materials, measured in centimetres squared per volt second. (Adapted from Tummala and Morris, 2001.)

Semiconductor	Mobility (cm^2/V-s)
Silicon (Si)	1500
Germanium (Ge)	3900
Gallium Arsenide (GaAs)	8500
Gallium Antimonide (GaSb)	4000
Indium Phosphide (InP)	4600
Indium Arsenide (InAs)	33,000
Indium Antimonide (InSb)	78,000

Table 7.4 Electron speed in selected semiconductor materials. (Adapted from Tummala and Morris, 2001.)

Semiconductor	Speed (kilometres per hour)
Indium Antimonide (InSb)	3000
Gallium Arsenide (GaAs)	300
Silicon Germanium (SiGe)	54–1400
Indium Phosphide (InP)	180
Germanium (Ge)	150
N-type Silicon	60
Gallium Phosphide (GaP)	10

efficient as a substrate than silicon, as shown in Tables 7.3 and 7.4. Not only is it faster than silicon (the electrons in GaAs travel faster than they do in silicon), but it can also operate over a much wider range of temperatures.

Pure GaAs substrates also have the great advantage of being semi-insulating, whilst silicon substrates are semiconducting. This is of particular importance as it allows the integration of a number of different devices on a single substrate. In addition, in contrast with silicon-based semiconductors, GaAs-based semiconductors operate at higher breakdown voltages and generate less noise at frequencies higher than 250 megahertz. Some of the most important uses of GaAs are in:

- cell phones;
- military applications;
- Infrared Emitting Diodes (IREDs), Laser Diodes (LDs) and Laser Emitting Diodes (LEDs);
- wireless communications.

Cell phones

Probably the most important ICs in a cell phone are the power amplifiers (PAs). The PAs in a cell phone are the vital components that amplify signals, both voice and data, to the appropriate power level for them to be transmitted back to the network base-station. The more advanced the generation used by the handset, the more PAs it needs. Whilst 2G handsets contain a single PA, 3G handsets can contain up to five PAs. Currently, the vast majority of PAs in cell phones are made using GaAs. It is also used in both the phone's switch and filter modules.

Military applications

GaAs is used in a number of different military applications, for example, in communications, night vision, radar and satellite.

In communications, GaAs is employed in a number of different contexts. In fibre optics, GaAs components, for example sensors, are used to facilitate the increasingly high speeds required in fibre-optic data communications. In military wireless communications, GaAs components are used, amongst other things, in point-to-point radios, wireless networks and cellular communications. Communications devices using GaAs will usually employ the semiconductor in one (or sometimes both) of two different ways. In the first, GaAs is used electronically in a device in which both the input and output are electrical, for example, in a field effect transistor (FET) or an analogue IC, also known as a MMIC (monolithic microwave integrated circuit), that functions as an amplifier, filter, phase shifter, frequency converter or mixer. Alternatively, GaAs can be used optoelectronically in a device in which the output or input is light rather than electrical, for example, an optoelectronic IC.

In modern night-vision technology, the semiconductor is used specifically in an optoelectronic context. In, for example, goggles using light amplification, Generation (Gen) 3 technology (as opposed to thermal imaging technology), GaAs is used to coat the photocathode (photomultiplier) that converts any available light into the required electrical energy to enable an image to be seen on a phosphor screen. In this instance, GaAs is converting photons (light) into electrons (electrical energy) in order for the wearer to see photons (light) emitted from the phosphor screen.

Military radars using GaAs MMICs vary in shape and type, and have many purposes, including (Fisher and Bahl, 1995):

- air defence;
- altitude measurement;
- guidance (Doppler systems);
- mapping (synthetic aperture);
- missile defence;
- surveillance: battlefield and long-range;
- weather monitoring.

As an example of a significant current defence use, Raytheon's enormous XBR radar, a phased array device some 85 metres (280 feet) tall, employs more than 45,000 of Raytheon's own GaAs transmit/receive modules (Defence Industry Daily, 2011).

As with radars, military satellites serve an array of purposes, not least surveillance, mapping, navigation and communications. All satellites need power and this can conveniently be provided by solar cells. Because of its ability to operate over a wider temperature range than silicon, and the fact that it has much higher radiation hardness, GaAs is ideal for use in the solar cells that power military satellites. GaAs-based solar cells are highly efficient (Savage, 2011). However, one of their main drawbacks has been their cost and hence their lack of general commercial exploitation, except in anything other than aerospace applications, especially satellites.

IREDs, LDs and LEDs

In each of IREDs, LEDs and LDs, GaAs is used in an optoelectronic capacity, because of its ability to convert electrical input into light output. Functioning in the infrared part of the electromagnetic spectrum (at wavelengths longer than those of visible light) IREDs emit infrared radiation. As an enabling technology for 'touchscreens' or multi-touch displays, IREDs are to be found in anything from PC and tablet PC displays, to self-service terminals, ATMs and smart phone devices.

LDs are an extremely efficient way of converting electrical signals into optical signals. In addition to their primary uses in optical storage and communications, LDs are used in bar scanners, materials processors and sensors. Whilst GaAs in LEDs is not used as a substrate nor as a dye to produce many colours, the ones it does emit are distinctive. As GaAs alone, or in combination with aluminium as gallium aluminium arsenide (GaAlAs), it can be used very effectively in the infrared (940–850 nanometres). Then, as wavelength decreases, at 660 nanometres it emits 'ultra-red'. Descending further down the wavelengths, gallium arsenide phosphide (GaAsP) emits 'High Efficiency Red' (635 nanometres), 'Orange' (605 nanometres), and 'Yellow' (585 nanometres).

Wireless communications

In addition to their use in military applications, and cell phones, GaAs MMICs (with such devices in each application operating in specific frequency ranges), are used in:

- broadband (high speed) satellite services;
- electronic toll collection systems;
- Global Positioning Systems (GPS);
- satellite TV;
- WiMAX;
- wireless LAN (WiFi, Bluetooth, etc.).

Finally, electronic devices employing GaAs are also found in commercial radars, ranging in size from those used in air traffic control and weather monitoring, to those needed in smart cruise controls and advanced collision warning systems for motor cars.

Gallium chemicals

The most important gallium chemicals currently commercially manufactured are: gallium nitrate, gallium trichloride, gallium trioxide, triethyl-gallium and trimethyl-gallium.

Gallium nitrate ($Ga(NO_3)_3$)

Produced using 4N purity gallium, this compound of gallium is used in two very different areas. As a pharmaceutical chemical, gallium nitrate is used in the treatment of navicular syndrome in horses, and arthritis and hypercalcemia (dangerously high levels of calcium in the blood) in humans. The chemical is also used in the manufacture of catalysts for the petrochemical industry and in the production of styrene.

Gallium trichloride ($GaCl_3$)

Typically produced with 4–5N purity gallium, gallium trichloride is used as a precursor in the production of organometallic gallium compounds, for example, trimethyl- and triethyl-gallium. However, as a compound itself, it reduces passivation (where a layer of oxide forms on a surface) and also improves ion transportation. Because of these two characteristics, gallium trichloride is used, in particular, as a cathode depolariser in lithium thionyl chloride batteries.

Gallium trioxide (Ga_2O_3)

Gallium trioxide is used as an intermediary, and in the manufacture of langasite (lanthanum gallium silicate) crystals for communications and piezoelectrics, and both LED and PDP phosphors.

Triethyl-gallium (TEGa, $Ga(C_2H_5)_3$) and Trimethyl-gallium (TMGa, $Ga(CH_3)_3$)

These two organometallic compounds of gallium are used as precursors for the MOVPE growth of epitaxial layers, although trimethyl-gallium is the most commonly used (SAFC, 2011). In particular, trimethyl-gallium is used in the deposition of the metal in thin-film contexts, especially in the production of LEDs.

Gallium nitride

As a semiconductor, gallium nitride (GaN) has, since the end of the last century, been widely used in both LEDs and LDs. However, it is also ideal for use in high-frequency and high-power electronic devices, and some of the most interesting advances in its use have been in these. It is increasingly being used in military radar, cable TV, aerospace applications, utility grids and wireless applications such as base stations. Amongst the advantages GaN offers over other semiconductors, some of the most important are:

- durability;
- greater electrostatic discharge resistance;
- greater power densities;
- greater power-added efficiency.

In addition, it offers:

- higher current capacity;
- higher operating voltages;
- inherently higher breakdown voltage;
- linearity;
- wide operating bandwidth.

Despite all these advantages, one characteristic of GaN poses significant problems. It is difficult to make. Whereas GaAs is composed of two metals, GaN is formed from a metal and a gas. Epitaxial layers of GaN have, for some time, been grown, using MBE and MOVPE, on a variety of substrates, for example, glass, sapphire (alumina), silicon carbide and silicon itself. However, combining the two elements to produce bulk GaN as a substrate has proved considerably more challenging. At present there are three main commercial technologies for growing bulk GaN crystal:

- *Ammonothermal*: using both high temperatures and pressures and an ammonia solution, this method uses liquid phase growth to 'create' crystals
- *Hydride Vapour Phase Epitaxy (HVPE)*: GaN crystal is grown using existing vapour phase technology
- *High Nitrogen Pressure Solution (HNPS)*: a method pioneered by Polish company TopGan.

The two main uses of GaN in optoelectronic devices are primarily in LEDs and, to a much lesser extent, in laser diodes.

Laser diodes

GaN is used in the production of both ultraviolet (UV) and blue laser diodes. Probably the most popular use of such lasers is in Blu-ray devices. In addition, GaN is a strong contender in the race to produce direct emission green laser diodes. The creation of such diodes poses a significant challenge, as they require the use of extremely pure semiconductor material. GaN may now provide a solution.

LEDs

In the last couple of years, the use of LEDs in general, and GaN-based LEDs in particular, has exploded. In 2010 alone, the GaN LED market grew by 60 per cent according to IMS Research (2011a). GaN-based LEDs have become the LEDs of choice. Not least because cheap, efficient, white-light LEDs can be made using LEDs emitting blue light (which is possible with GaN) combined with phosphors as illustrated in Figure 7.3. Whilst producing white light from a stack of red, green and blue LEDs is theoretically the most efficient, green LEDs remain inefficient and therefore white LEDs using this method are more expensive than using a blue LED with phosphors (USDOE, 2011).

However, apart from producing white light, commercially available GaN-based LEDs now cover the spectrum from green, through blue to ultraviolet. Although, as with green LDs, creating low-cost, high-performance green LEDs is proving a continuing challenge (Rensselaer, 2011).

In general, apart from using considerably less energy than conventional incandescent electric lamps and not containing any of the toxic heavy metals found in compact fluorescent light (CFL) bulbs, LEDs (in particular high brightness (HB) white LEDs) provide the following advantages:

- durable;
- fully dimmable;
- instantly on/off;
- colour saturation;
- longevity (up to 100,000 hours) (Focus Digital, 2011).

Although, currently, more than 80 per cent of LEDs are produced using a sapphire substrate (Öko-Institut e.V, (2009), there are various research efforts to find alternatives, two of which are glass and silicon.

Whilst LEDs are used in a very wide range of applications, their principal uses today are in: televisions and monitors, mobile appliances and solid-state lighting (SSL). Other significant uses are in signals (for example, traffic lights) and motor vehicles.

For Thin-Film Transistor (TFT) Liquid Crystal Display (LCD) televisions and monitors to work,

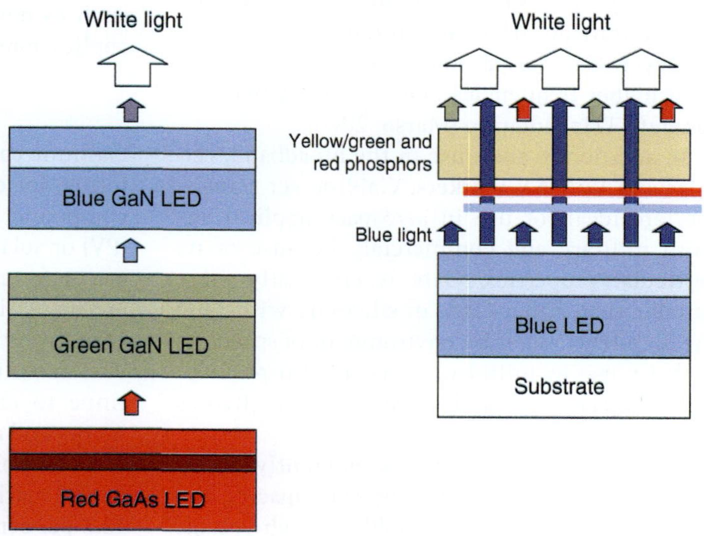

Figure 7.3 Two ways to produce white light from GaN LEDs (DenBaars, 2008).

there has to be a light source. GaN LEDs are now the light source of choice, replacing cold cathode fluorescent lamps (CCFLs). There can be anything from hundreds to thousands of individual LEDs in each backlit television. LED televisions and monitors are, literally, backlit (with the LEDs (coloured) directly behind the flat panel of the screen) or side-lit (with the LEDs (white) behind the plane of the screen, but ranged around its perimeter). LEDs are, similarly, also used in smaller, more portable, devices such as netbooks, notebooks, tablet computers, Personal Digital Assistants (PDAs), mobile gaming devices and some mobile phones.

As a major consumer of LEDs, the SSL market can be split into a number of segments. The most significant of these are: architectural, commercial, consumer portable (for example, torches), industrial, outdoor and residential.

Although the predominant use of GaN is in optoelectronics, its use in electronic devices is both widening and increasing. GaN is particularly suitable for use in military radars, not only because of its durability, but also because of its considerably greater power output than, GaAs. It is also capable of working across a very broad range of frequencies from 1–100 Gigahertz (Roosevelt, 2011). Successful tests have recently been completed for the U.S. Navy's Air and Missile Defence Radar (AMDR) using GaN modules (Strategic Defence Intelligence, 2011). However, GaN-based transistors can function at even higher frequencies and have been made to generate THz radiation (Marso, 2010).

In addition to their use in the broadband, cellular and WiMAX markets, GaN power transistors are ideal for use in aerospace applications (both military and commercial). Because of its particular properties, GaN is eminently more suitable than either GaAs or silicon to withstand the high-level radiation environment of space. As with its use in military radars, it also provides significantly improved power output (Barnes et al., 2011).

Finally, because of both its inherently higher breakdown voltage and its current capacity, the use of GaN is especially suitable in high-voltage environments, in the area of power conversion in electric and/or hybrid vehicles (Davis, 2009), and the power grid (Bradbury, 2011). In these contexts GaN can also reduce power loss (EE Times-Asia, 2011) and improve switching speeds (Briere, 2011).

Perhaps cognizant of the importance of such uses of GaN, in November 2011 the Berlin-based Ferdinand-Braun-Institut announced the launch of the EU project HiPoSwitch (Ferdinand-Braun-Institut, 2011), which will receive significant funding from the European Community. This project will focus on 'novel GaN-based transistors' as 'key switching devices' in these areas amongst others.

Gallium phosphide

Gallium phosphide (GaP) alone is used in a couple of visible spectrum LEDs: 'High Efficiency Green' (565 nanometres) and 'Pure Green' (555 nanometres). Compounded with GaAs it is also used in red, orange and yellow LEDs (see above). When compounded in the form of indium gallium aluminium phosphide (InGaAlP) and used in LEDs, a number of the wavelengths and, hence, colours between 'Super Red' (633 nanometer) and 'Pure Green' can also be achieved. Compounded with indium as indium gallium phosphide (InGaP), it is used in the manufacture of electronic devices such as power amplifiers for WiFi and WiMAX applications.

Photovoltaics

Gallium can be used in a number of different forms, for example as a metal or in a chemical compound, in the manufacture of photovoltaic (PV) or solar cells. Whilst GaAs-based solar cells are preferred in extraterrestrial applications and are increasingly being used in conjunction with concentrator photovoltaics (CPV) technology, the vast majority of solar cells for terrestrial use continue to employ crystalline silicon (c-Si, both mono- and multi-crystalline) technology. In 2010, this technology accounted for some 86.5 per cent of the market (Solarbuzz, 2011). The remaining 13.5 per cent of the market was accounted for by solar cells employing thin-film technology, or thin-film solar cells (TFSC).

The main TFSC technologies, apart from GaAs, are amorphous silicon, cadmium telluride (CdTe), copper indium selenide (CIS) and CIGS. All involve the deposition of a thin film, only a few micrometres deep, of semiconducting material on various different surfaces. Putting aside the merits of each of these different technologies, and the fact that CIGS technology is still developing, the quantities of gallium currently used in solar-cell production remain small. Whilst CIGS technology does continue to grow, until it becomes widely adopted, the production of solar cells requires considerably less gallium than is currently consumed in electronic devices or LEDs.

Substitution

As with many of the specialist metals, gallium has some unique properties which make it difficult to substitute in certain circumstances, or which results in poorer performance where substitution occurs. For example, silicon can be used instead of GaAs in photovoltaic cells but the conversion efficiency is reduced from 18–22 per cent to 8–15 per cent (Feneau, 2002). The use of gallium as a semiconductor typically results in increased speed, lower energy consumption and better resistance to radiation compared to silicon-based alternatives (Feneau, 2002).

Despite the advantages of using gallium, there are some specialist applications where other materials can be used. Germanium can be used as a substitute for GaAs in some electronics. Organic-based liquid-crystal displays can substitute for gallium in certain LED applications. Indium phosphide can be used for infrared laser diodes at a limited range of wavelengths and helium-neon lasers compete with GaAs in visible laser diode applications (Jaskula, 2011a).

However, there is no effective substitute for GaAs in integrated circuits for many applications (Jaskula, 2011a), although silicon- or silicon–germanium-based alternatives have been used in certain situations for many years.

GaN is the latest development in the field of semiconductors and is, if anything, likely to substitute for more silicon-based devices in future, particularly in high-power electronics because it is more durable, can operate at greater speeds and cope with higher temperatures (GaN Systems Inc., 2011).

Environmental aspects

Gallium does not occur in nature as a pure metal. As a compound it occurs at trace levels in many areas of the natural environment, including watercourses. It also occurs in very small quantities in the human body, where it has no proven benefit but is not harmful. Acute exposure (that is, the ingestion of large doses) to certain rare compounds, e.g. gallium (III) chloride, can result in throat irritation, breathing difficulties and chest pains (Lenntech, 2011). GaAs is potentially more toxic to humans, but this is caused by the arsenic not the gallium (Martin, 1993). Recent research has shown that GaN is non-toxic and potentially suitable for biomedical implant applications (Jewett et al., 2011).

As mentioned before, gallium is produced as a by-product from the Bayer liquors associated with the production of alumina from bauxite, from processing zinc ores and from recycling scrap. The main environmental issues associated with these processes result from the various chemicals used. Research has been undertaken, and is continuing, to improve the efficiency of gallium recovery processes. Frequently, these developments have associated health and environmental benefits, for example there has been a general shift away from electrolysis of Bayer liquors using mercury as a cathode.

An important aspect to consider is the environmental benefits which can be achieved through products which use gallium. For example, gallium is an essential component of LEDs, which are an alternative to traditional incandescent or compact fluorescent electric lamps. It has been estimated that replacing these traditional light bulbs with LED-based lighting can reduce electricity consumption by between 50 per cent and 75 per cent (Tsao, 2003 and Briere, 2010) and for this reason the LED market for lighting is likely to grow significantly in the future. The use of gallium in CIGS PVs also has the potential to improve further the efficiency of this technology (Masters, 2004 and Green et al., 2011).

World resources and production

Most commercial gallium today is produced as a by-product of aluminium processing, with a lesser amount being produced as a by-product of zinc processing. Assuming an average gallium content in bauxites, the minerals from which aluminium is most usually extracted, of around 50 ppm (Jaskula, 2011a), the USGS estimates that the global resources of gallium contained in bauxite alone exceed one million tonnes, and that global zinc reserves also contain significant amounts of gallium. Whilst these estimates refer to the total gallium contents of typical ores, currently very little gallium is actually recovered from either bauxite or zinc ores.

> *"In 2010, only about 10% of alumina producers extract gallium as a byproduct of alumina processing. The remainder of producers find it too expensive to extract the gallium and thus treat gallium as an impurity in the aluminum refining process."*
>
> *U.S. Department of Energy, Critical Materials Strategy, December 2010*

Based on world mine production in 2010 for bauxite of 211 million tonnes (Bray, 2011) and for zinc, of 12 million tonnes (Tolcin, 2011), together they could have potentially yielded (based upon a hypothetical 100 per cent recovery rate) some 11,170 tonnes of gallium or more (bauxite: 10,550 tonnes and zinc: 620 tonnes or more). However, with primary gallium production in 2010 estimated to be 106 tonnes (Jaskula, 2011a), 196 tonnes (Neo Material Technologies Inc., 2011) or 209 tonnes (Table 7.5), this indicates a recovery rate in the range of 1.00–1.98 per cent.

Table 7.5 Estimated virgin gallium capacity and production in 2010 (Data from Roskill Information Services Ltd, 2011 and company accounts.)

Country	Capacity (tonnes per year)	Production (tonnes)
China	141	118
Germany	32	32
Kazakhstan	25	18.7
Ukraine	15	15
South Korea	15-20	≥10
Russia	10	4-7
Hungary	8	4
Japan	10	4

Production in 2010

Obtaining figures for both primary and secondary gallium production capacity, and actual gallium production, is notoriously difficult. As with many of the critical metals, not only is the market relatively small, but its primary production and, to a lesser extent, its secondary production, are restricted to just a few companies. Because of both the nature and size of the market and their businesses, most of these companies prefer not to publish any data or only to publish information that is well out of date. With so little information available, many figures can only to be estimated or arrived at through discussion with market participants.

Primary production

China, Germany and Kazakhstan have both the largest capacity to produce primary gallium, and are the largest producers (Table 7.5).

China

China has, over the past ten years, established itself as, by far, the world's largest primary gallium producer. Production is currently controlled by four major concerns. In descending order of estimated production these are: the Aluminum Corporation of China Limited (Chalco); Zhuhai Fangyuan (owned by Golden Harvest); East Hope Mianchi Gallium Industry Co; and China Crystal Technologies Co Ltd.

Germany

The Ingal Stade primary gallium production plant in Stade, north-west of Hamburg, and on the same site as the Aluminium Oxid Stade (AOS) alumina refinery, is jointly owned 50:50 by 5N Plus (formerly by MCP) and Neo Material Technologies Inc. of Canada. In 2010, each of the co-owners received around 16 tonnes of 4N gallium.

Kazakhstan

Eurasian Natural Resources Corporation PLC (ENRC), quoted on the London Stock Exchange, operates the Pavlodar Alumina Refinery in northern Kazakhstan, which, in addition to being one of the world's largest producers of alumina, is one of the largest producers of virgin gallium. Production in 2010 was the same as 2009: 18.702 tonnes.

Other Countries

UKRAINE: RUSAL's Nikolaev Alumina Refinery on the Black Sea refines bauxite imported from Guyana and Kindia in Guinea.

SOUTH KOREA: Korea Zinc Company is one of the few producers (another is Dowa in Japan – see below) of gallium from the processing of zinc. The company publishes no figures for gallium production.

RUSSIA: Formerly owned by SUAL, and now owned by Mr Oleg Deripaska's Basel Cement, the Pikalevo alumina refinery, in the eponymous Russian industrial town, was in 2010 thought to be last and only significant producer of virgin gallium left in Russia.

HUNGARY: The Alumina Products Division of MAL Hungarian Aluminium Ltd extracts gallium from enriched Bayer liquor at its Ajka plant.

JAPAN: Whilst Japan may be the world's largest consumer of gallium, in 2010 it is thought that only Dowa Electronics Materials Co Ltd actually produced any virgin metal and is shifting its activities away from gallium and towards indium production.

Secondary production

There continues to be considerably more capacity for recycling and refining gallium than there is for primary production (Table 7.6). However, as with primary production, reliable figures for both capacity and production are very difficult to

Table 7.6 Estimated secondary gallium capacity in 2010. (Data from Roskill Information Services Ltd, 2011 and authors.)

Country	Number of companies	Refining (U) Recycling (R)	Total capacity (tonnes per year)
Japan	2	R	≤20
	4	R/U	≈150
	1	U	60
USA/Canada	1	R/U	≤60
		R	≤30
China	1	R/U	7/8N-15 and 6N-20
	2	U	60
Slovakia	1	R	8
		U	25
UK	1	R/U	20
Germany	1	R	≤10
		U	≤10
France	1	U	N/A

find. In addition, differentiating between those amounts produced from either recycling gallium metal or the waste, for example from wafer production, and those produced by upgrading 3/4N materials is extremely difficult.

Within the GaAs sector, the importance of recycling cannot be overemphasised. Typically, some 30–40 per cent of the gallium used by the sector comes from recycled materials. With advances in recycling technology, it is likely that there is a recovery rate of around 90 per cent from the gallium-bearing waste generated within the sector.

In the epitaxy sector, with recovery rates also around 90 per cent, it is estimated that, at any one time, in the region of 60 tonnes of gallium can be circulating in the liquid phase epitaxy (LPE) 'loop' in Japan.

Within the CIGS solar cell sector, and until yields improve significantly, recycling recovery rates remain around 50 per cent.

By far the largest recycler (as well as the world's largest consumer) of gallium is Dowa Electronics Materials Co Ltd. Other companies that, like Dowa, both upgrade and recycle gallium include: Nichia Corporation, Rasa Industries Ltd and Sumitomo Metal Mining. In 2010, two companies just refined gallium to higher purities: Furukawa Denshi Co Ltd and Nippon Rare Metal Inc. Sumitomo Chemical, whilst undertaking no recycling in 2010, remains one the world's largest producers of high-purity gallium at its Ehime plant. In the USA and Canada, Neo Performance Materials Ltd (formerly Recapture Metals Inc.) is the only recycler and refiner of gallium in North America.

Perhaps surprisingly, as the world's largest producer of primary gallium, China appears to have, comparatively, very little capacity to recycle the metal. In 2010, the largest participants in the Chinese secondary gallium market were: Nanjing Jin Mei Gallium Ltd (majority owned by the US company AXT Inc.), Sumika Electronic Materials (Shanghai) Co Ltd (owned by Sumitomo Chemical) and MCP Jin Shu (Shenzen) Co Ltd (now owned by 5N Plus and the country's largest producer of 6N purity gallium).

Elsewhere around the world, gallium was recycled and/or refined by: CMK in western Slovakia; Mining and Chemical Products Ltd (owned by 5N Plus) at their UK operation in Wellinborough; PPM Pure Metals GmbH (owned by France's Recylex SA) based in Germany; and Azelis Electronics in France.

Future supplies

In the foreseeable future, China will continue to be world's largest producer of primary gallium. At the end of 2010, it was thought that all the country's major producers had plans to expand both capacity and production, and during 2011 a number of these projects were initiated. In 2012, the projected annual capacity of primary gallium in China alone was approximately 300 tonnes.

Outside China, in the first such move by a major aluminium producer in many years, Rio Tinto Alcan has signed a memorandum of understanding with 5N Plus. Under this memorandum the two would discuss a project to extract gallium at Rio Tinto Alcan's Vaudreuil alumina facility in Quebec, Canada (5N Plus Inc., 2011).

This announcement was significant because, for most major alumina/aluminium producers, the risks of tapping into their Bayer liquor streams to extract gallium are not justified by the rewards from doing so. The concentrations of gallium may be quite small and the capital expenditure on an extraction unit may be significant, both of which reduce its economic viability. Furthermore, gallium remains only a by-product and any derived returns are generally small in contrast with those derived from processing the primary metal. As importantly, for those with Bayer liquor streams, the risks of disruption to the primary production processes cannot be dismissed. The same arguments can be made in relation to extracting gallium from the residues of processing zinc.

In 2007, there were at least ten companies in the Commonwealth of Independent States with some capacity to produce gallium. Today, there are only three: one each in Kazakhstan, Russia and Ukraine. It is not known whether the plant at the other seven facilities has either been dismantled or just mothballed. Some unused capacity, therefore, may remain elsewhere. However, gallium extraction may retain some attraction because at least one company in Russia has recently indicated it is both aware of the gallium contained within its apatite-nepheline ore resources and its potential value (PhosAgro, 2011).

Any increase in the supply of gallium from recycling will come from the thin-film solar sector, in particular CIGS. A great deal of capacity already exists to recycle the waste from both the manufacture of compound semiconductor wafers, and from the liquid phase epitaxy. Recovery rates for each are around 90 per cent. In contrast, recycling in the solar market is still growing and recovery rates, while still considerably less effective at around 50 per cent, are improving.

Amongst those who have recently become involved in recycling associated with CIGS, in Europe, Umicore has opened a new facility to recycle production waste from the manufacture of CIGS cells at its Hoboken facility, in Belgium. In the USA, Indium Corporation has a recycling facility in New York State and in Japan, a number of companies are already exploring end-of-life recycling of CIGS cells. Neo Materials also has significant capacity at its Canadian operations to treat all forms of CIG and CIGS scrap.

World trade

Gallium is traded in a variety of forms including unwrought metal, powders, waste and scrap, GaAs wafers (doped or undoped) and other articles containing the element. It is also traded at a variety of degrees of purity with some countries importing lower-grade materials for processing into higher-purity forms which are then exported.

Obtaining accurate data on the worldwide trade in gallium is extremely difficult, not least because the internationally recognised trade code systems record gallium trade combined with other elements such as hafnium, indium, niobium and/or rhenium. Individual countries identify some trade movements, but this gives only a partial picture of the trade that takes place.

For example, the United States Geological Survey reported that gallium (unwrought, waste, and scrap) imports for consumption in the USA amounted to 35.9 tonnes in 2009, decreasing from 41.1 tonnes in 2008. Imports to the USA of doped GaAs wafers amounted to 117 tonnes in 2009, decreasing from 165 tonnes in 2008 (Jaskula, 2011b). The countries supplying those 2009 imports are shown in Figure 7.4. However, figures for imports that are not consumed within the USA and statistics for exports from the USA are not reported.

Trade in 'unwrought gallium and powders thereof' to and from the United Kingdom in 2010 is shown in Figure 7.5. Imports amounted to 73.2 tonnes in 2010, which is an increase of nearly 350 per cent compared to the 16.4 tonnes imported in 2009. Exports of gallium in this form were 39.9 tonnes in 2010, an increase of 57 per cent from the 25.5 tonnes exported in 2009 (UK Trade Info, 2011). Details relating to the purity levels of this trade are not available. Information for gallium traded in other forms cannot be separated from other elements.

Prices

Prices for gallium will vary depending on the form and purity in which it is traded. Generally, higher-purity gallium metal will be approximately US$200 per kilogram higher in price than lower-purity grades (Jaskula, 2011b). Typical or average prices are quoted in trade publications, although usually this is for lower-purity forms of gallium metal rather than higher-purity metal or gallium compounds.

Historically, gallium prices have been very stable over long periods. It is believed that this general stability was a result of there being relatively few producers, supplying one main substantial market (GaAs wafers), and the

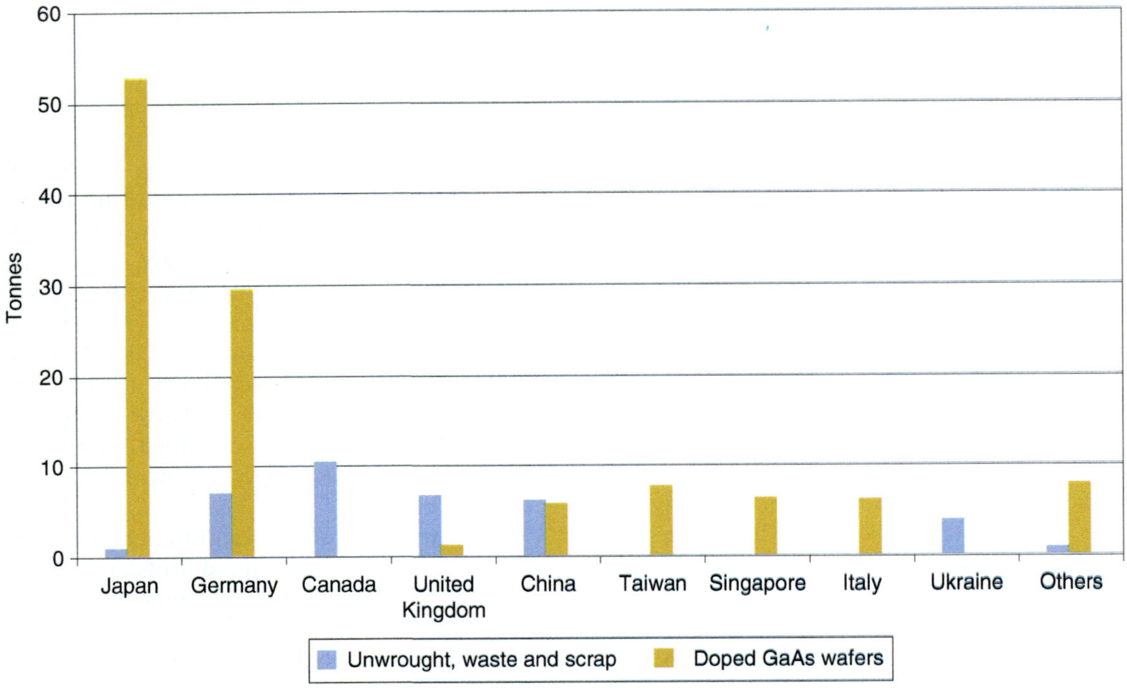

Figure 7.4 Sources of United States of America imports of gallium in 2009. (Data from Jaskula, 2011b.)

existence of significant spare production capacity (Mining Journal, 2010). The latter meant that output could readily be increased in response to higher prices, leading to an over-supply in the market and the consequent fall in prices. This general stability was interrupted in the final quarter of 2000 and first quarter of 2001 when prices for high-purity gallium metal increased to more than US$2000 per kilogram as a result of an apparent severe shortage in supply. However, this shortage was artificial and was driven by erroneously high future consumption projections, which led to hoarding and speculative inventories developing. These high prices were short lived and by July 2001, once forecasts had been re-evaluated, prices returned to more normal levels (Kramer, 2000 and 2001). Prices remained in the range US$300 to US$400 from 2001 to 2006.

Since 2007 there has been an unusual amount of movement in the price of gallium, as demonstrated by the typical price trend for 2006 to 2011 shown in Figure 7.6, which is based on quarterly averages. The price increase in early 2007 was attributed to Chinese producers claiming that there was a shortage in supply. As these fears subsided, and as a result of lower demand during the economic recession, the price reduced to the previous level of approximately US$400 per kilogram. During the first half of 2011, prices rose again to more than US$900 per kilogram before reducing to below US$700 per kilogram by November 2011. A price rise was seen in many of the minor metals, due in part to speculative buying, but for most of these prices decreased over the summer as the quantity of trade reduced (Metal Bulletin, 2011a). In the case of gallium, prices continued to fall as increases in capacity were expected (Metal Bulletin, 2011b) and demand remained weak (Metal Bulletin, 2011c).

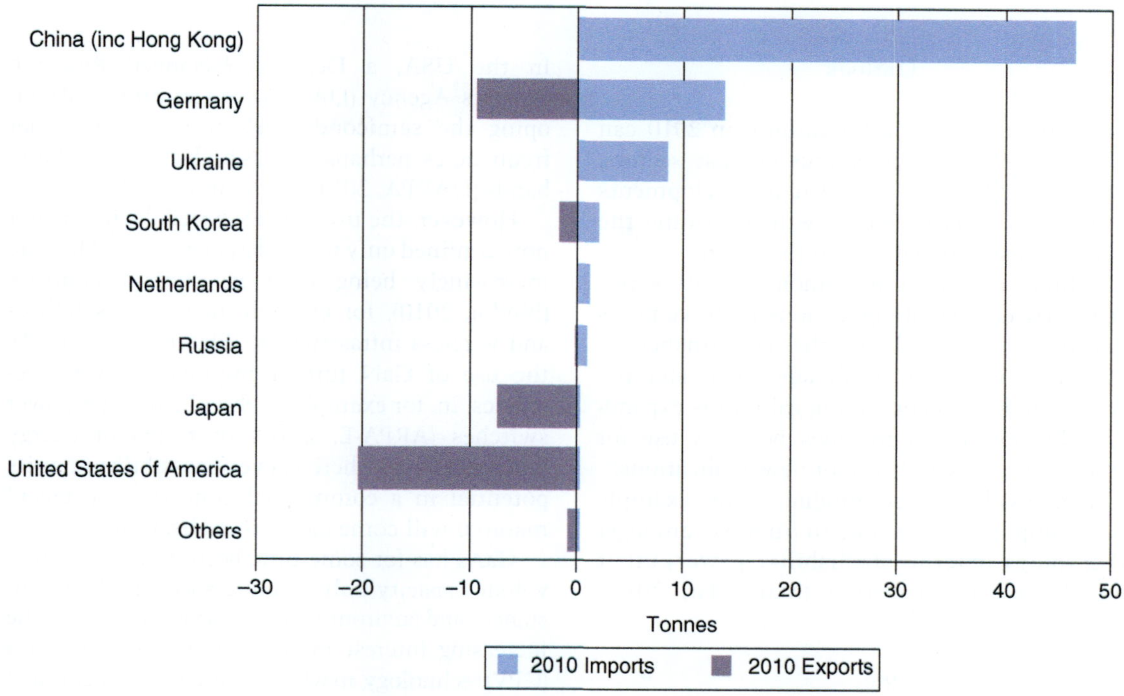

Figure 7.5 Imports and exports of unwrought gallium and gallium powders to and from the United Kingdom in 2010. (Data from UK Trade Info, 2011.)

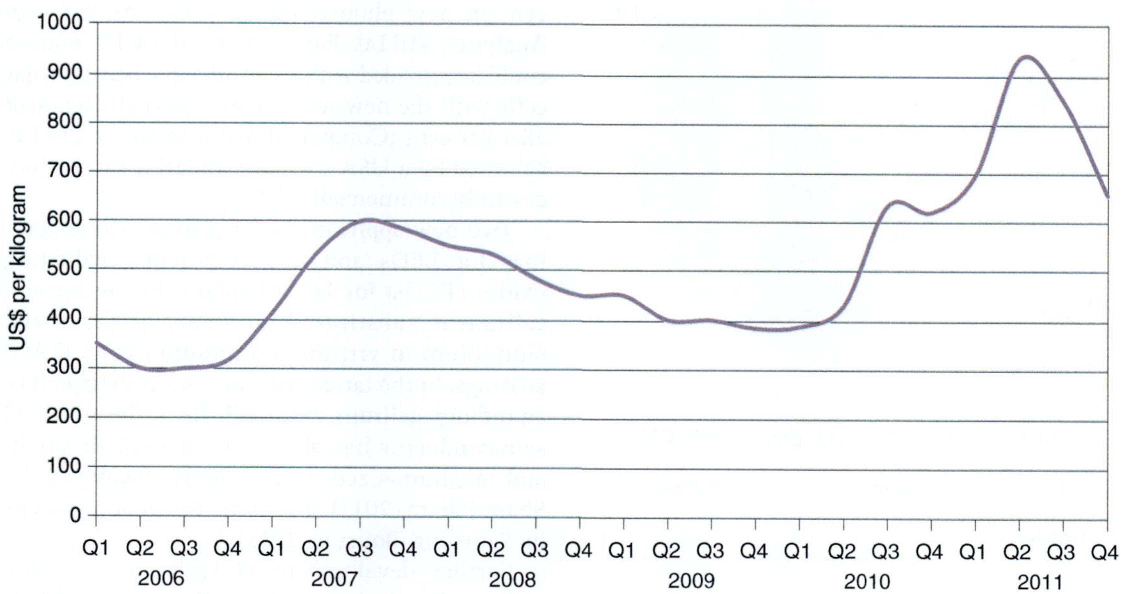

Figure 7.6 Quarterly average prices for gallium metal from January 2006 to November 2011. (Calculated from data quoted in Mining Journal, 2006–2011 and Metal Bulletin, 2011d.)

Outlook

The estimated demand for gallium in 2010 can be divided between its various end-use sectors, as shown in Figure 7.7, and it is developments within these sectors which will determine the overall demand for gallium in the future.

Gallium continues to demonstrate its versatility in a wide range of applications. Improvements continue to be made in the performance of devices in which it is already used, and the number of devices containing gallium is expanding. This expansion includes both its use for existing purposes and also for new applications.

GaN-based defence products, for example power amplifiers, continue to improve, amongst other things, in terms of reliability, power output and absence of distortion (Majumdar, 2011).

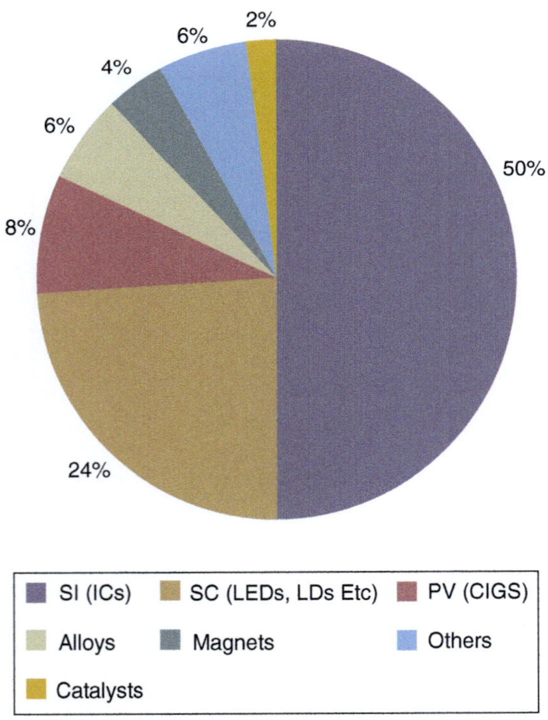

Figure 7.7 Estimated demand for gallium in 2010. Notes: SI, semi-insulating; SC, semiconducting; PV, photovoltaic.

In the USA, a Defence Advanced Research Projects Agency (DARPA) programme is developing the semiconductor's use at ever-higher frequencies, perhaps even as high as the Terahertz bands (DARPA, 2011 and Majumdar, 2011).

However, the use of GaN-based devices is not now confined only to the defence sector. They are increasingly being used in civilian contexts (Bindra, 2010), for example in CATV, satellites and wireless infrastructure (ElectroIQ, 2011). As the use of GaN further improves power electronics, in, for example, high-voltage, high-power switches (ARPA-E, 2010), in terms of energy efficiency alone there is every possibility that its potential in a commercial context as a 'green' material will come closer to realisation.

GaAs has for some time been used in a photovoltaic capacity only in quite specialised circumstances and environments, e.g. space. However, the increasing interest in concentrator photovoltaics (CPV) technology, in which sunlight is concentrated on solar cells, could indicate a promising area of growth for terrestrial applications. Indeed, the research firm Strategy Analytics believes that, by 2016, the CPV market will account for five per cent of new photovoltaic installations (Strategy Analytics, 2011a). Furthermore, the CPV market could be provided with a considerable boost if solar cells with the new record conversion efficiency of 28.4 per cent (Compound Semiconductor, 2011a), achieved by a USA company in 2011, can be successfully commercialised.

Two new applications for gallium are in coatings for LEDs and in transparent conducting oxides (TCOs) for LCD displays. In the former, gallium is substituted for some, or all, of the aluminium in yttrium aluminium garnet (YAG) coatings. In the latter, gallium, as a constituent of an indium, gallium, zinc oxide (InGaZnO (IGZO)) semiconductor has already been used in small- and medium-sized LCD panels developed by Sharp (Sharp, 2011) and in 3D television screens by Samsung (Bourzac, 2011).

Further development of technologies using gallium-doped zinc oxide (GZO), both in LEDs (Liu et al., 2011) and as a substitute electrode material (both as anode and cathode) for the more costly indium–tin oxide (ITO), could lead to considerable expansion of the metal's use in, for example, large area displays (Wang et al., 2010), solar cells (Ihn et al., 2010) and solid-state lighting.

As technological advances mean gallium is used in more applications, so the manufacturing processes that enable it to be used also advance. For example, in the field of GaN transistors, researchers announced in 2011 that they had, for the first time, succeeded in producing high-speed transistors containing GaN, but grown on a silicon wafer (Daimler, 2011), thereby offering both speed and functionality at high temperatures. In the field of GaN LEDs, Samsung of Korea announced in 2011 that "one of its research teams had figured out a way to grow crystalline gallium nitride (GaN) LEDs on regular glass" (Yirka, 2011).

Whilst these are some of the more interesting ways in which gallium may be used in future, demand for the metal is currently driven primarily by its use in cell-phone handsets, particularly in 'smart' phones, and in LEDs.

In mobile wireless communications, not only do the higher generations of the wireless standard require more power amps per handset, but as handsets become more sophisticated and the demand for data continues to grow, more GaAs components are required. It is estimated that, in 2014, 1.7 billion handsets will be shipped (Higham, 2011). In 2010, there was a 30 per cent growth in the demand for GaAs epitaxial wafers alone (Strategy Analytics, 2011b).

There continues to be significant demand for LEDs for backlighting in televisions, computer screens, etc., and the GaN LED market grew around 60 per cent in 2010 (IMS Research, 2011b; Compound Semiconductor, 2011b). The use of LEDs both in general and street solid-state lighting, because of their durability, energy savings and longevity, will become increasingly important in future (Figure 7.8).

After its use in wireless devices and LEDs, the third largest demand for gallium comes from its use in solar cells, in particular those using CIGS (as opposed to GaAs) technology. However, in 2010 its use in solar cells consumes comparatively little of the metal. Being both more environmentally friendly than CdTe (cadmium telluride) technology, the current market leader, and more efficient than silicon technology, CIGS technology is well placed to become a significant participant in the PV market. If it does, then demand for gallium from this sector would rise accordingly. However, it is difficult to make reliable forecasts for growth of demand for gallium from the CIGS sector of the solar-cell market as the industry is new and still evolving, both commercially and technologically.

Whilst announcements of involvement in CIGS cell manufacturing by organisations such as of Hyundai Heavy Industries and Saint Gobain, Samsung, and Taiwan Semiconductor Manufacturing Corp (TSMC) provide advocates with a degree of comfort (TSMC, 2011), announcements like that of Solyndra's bankruptcy (amongst others) in the USA (Church et al., 2011) illustrate some of the challenges the industry still faces.

In trying to reach any conclusions as to future demand for gallium (and, indeed, availability), there still remains a huge unknown – purchases by both traders and speculators. There is no way of discovering what these volumes are. However, these participants in the market should always be considered, especially as 'investors' become increasingly interested in a wider range of metals.

However, we can be confident that, with those resources of gallium that have hitherto been identified, there is more than sufficient to supply all the developing technologies for the foreseeable future. Moreover, since gallium is available from the alumina process and the plants undertaking this process are located in a diverse range of geographical locations, there continues to be potential to develop gallium production facilities in alternative areas.

Acknowledgements

Teresa Brown publishes with the permission of the Executive Director of the British Geological Survey.

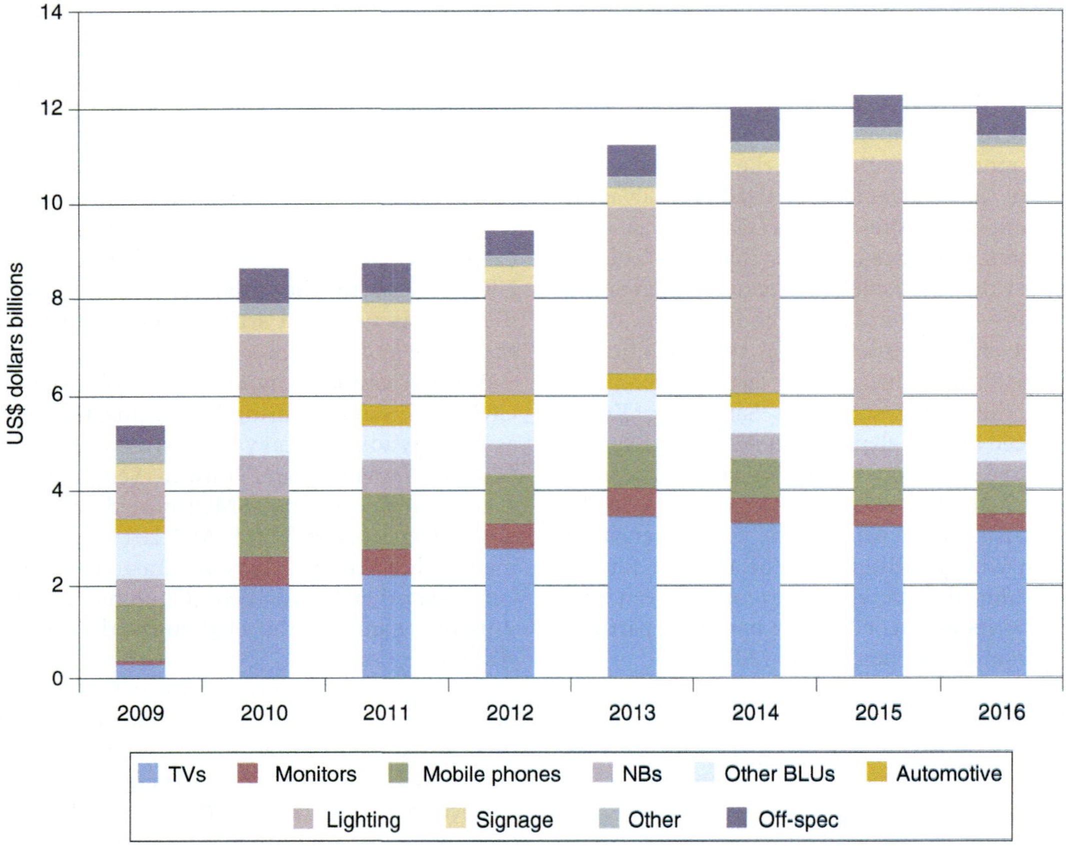

Figure 7.8 Forecast of packaged LEDs by application 2009–2016, revenue by segment. (Data from IMS Research, 2011c.) Notes: TVs, televisions; NBs, notebooks; BLUs, back-lit units.

References

5N Plus Inc. (2011) 5N Plus and Rio Tinto Alcan in Discussions Regarding Development of New Primary Gallium Source, http://www.5nplus.com/uploads/Eng_5NPlus_%20RioTintoAlcan.pdf

Advanced Research Projects Agency – ENERGY (ARPA-E). (2010) Transphorm Inc: High Performance GaN HEMT Modules for Agile Power Electronics. http://arpa-e.energy.gov/ProgramsProjects/ADEPT/HighPerformanceGaNHEMTModulesforAgilePower.aspx

Barnes, A., Vitobello, F., Daeubler, J., Hirche, K., Laetti, J. and Rostewitz, M. (2011) Europe improves reliability of GaN microwave devices for space applications. Compound Semiconductor Volume 17, No. 7, pp. 31.

Bennett, B.R., Magno, R., Boos, J.B., Kruppa, W. and Ancona, M.G. (2005) Antimonide-based compound semiconductors for electronic devices: A review. Elsevier, Solid-State Electronics 49, 1875–1895.

Bernstein, L.R. (1986) Geology and mineralogy of the Apex germanium-gallium mine, Washington County, Utah. United States Geological Survey, Bulletin 1577.

Bhatt, C.K. (2002) Estimation of gallium in a bauxite-ore deposit using an energy-dispersive X-ray fluorescence technique. Radiation Physics and Chemistry 65, 193–197.

Bindra, A. (2010) GaN Power Transistors Gain Ground In Military And Commercial Applications. MicroWaves&RF, http://mwrf.com/Articles/ArticleID/22528/22528.html

Bourzac, K. (2011) An Ultra-High-Definition 3-D TV. Technology Review. http://www.technologyreview.com/computing/37691/

Bradbury, D. (2011) Transphorm rolls out EZ-GaN switching transistor. Business Green. http://www.businessgreen.com/bg/news/2071714/transphorm-rolls-ez-gan-switching-transistor

Bray, E.L. (2011) Bauxite and Alumina in Mineral Commodity Summaries. United States Geological Survey.

Briere, M.A. (2010) GaN on Si based power devices: an opportunity to significantly impact global energy consumption. http://www.csmantech.org/Digests/2010/Papers/13.1.066.pdf

Briere, M. (2011) Commercial 600V GaN-Based Power Devices Coming of Age. Power Electronics Technology. http://powerelectronics.com/discrete-semis/gan-based-power-devices-coming-of-age-0711/

Burke, E. (2011) Waste Minimization, Pollution Prevention and Resource Recovery at a GaAs Manufacturer. Paper delivered at CS MANTECH Conference, 16–19 May 2011, Palm Springs, California, USA

Cardarelli, F. (2008) Materials Handbook: A Concise Desktop Reference. Springer-Verlag, London, UK, pp. 166.

Christie, T. and Brathwaite, B. (2002) Mineral commodity report 19 – beryllium, gallium, lithium, magnesium, uranium and zirconium. Institute of Geological and Nuclear Sciences Ltd, The New Zealand Ministry of Economic Development.

Church, S., Schneider, J. and Sandler, L. (2011) Solyndra, Solar-Panel Maker in California, Seeks Bankruptcy. Bloomberg. http://www.bloomberg.com/news/2011-09-06/solyndra-solar-panel-maker-in-california-files-for-bankruptcy-protection.html#

Compound Semiconductor (2011a) Gallium arsenide solar cells break SQ limits. http://compoundsemiconductor.net/csc/news-details.php?id=19734213

Compound Semiconductor (2011b) LED-TV explosion creates a boom in the LED Market. 2 November 2011 http://compoundsemiconductor.net/csc/news-details/id/19734184/name/LED-TV-explosion-creates-a-boom-in-the-LED-market.html

Cook, N.J., Ciobanu, C.L., Pring, A. Skinner, W., Shimizu, M., Danyushevsky, L., Saini-Eidukat, B. and Melcher, F. (2009) Trace and minor elements in sphalerite: a LA-ICPMS study. Geochimica et Cosmochimica Acta 73, 4761–4791.

Daimler Technology (2011) Gallium nitride transistors. http://www.daimler-technicity.de/en/gallium-nitride-transistors/

Davis, S. (2009) SiC and GaN Vie for Slice of the Electric Vehicle Pie. Power Electronics Technology. http://powerelectronics.com/passive_components_packaging_interconnects/resistors/sic-gan-vie-slice-electric-vehicle-pie-20091101/

Defence Advanced Research Projects Agency (DARPA). (2011) Nitride electronic next-generation technology (NEXT) program. http://www.darpa.mil/Our_Work/MTO/Programs/ Nitride_Electronic_NeXt-Generation_Technology_%28NEXT%29.aspx

Defence Industry Daily (2011) Floatin' Smokey: The USA's SBX Radar. http://www.defenseindustrydaily.com/Floatin-Smokey-The-USAs-SBX-Radar-04804/

Degerstrom, N.A., Inc. (2011) Metallurgical – Gallium. http://www.nadinc.com/metallurgical/gallium.html

DenBaars, S. (2008) What is LED lighting: Technology overview and introduction. Presentation at the DOE SSL Market Introduction Workshop 2008. Solid State Lighting and Energy Centre, University of California, Santa Barbara, USA.

Dittrich, T., Seifert, T. and Gutzmer, J. (2011) Gallium – can future demand be met by geological and technological availability? Department of Mineralogy, Technische Universität Bergakademie Freiberg, Germany. https://sga.conference-services.net/resources/1054/2590/pdf/SGA2011_0203.pdf

Dutrizac, J.E., Jambor, J.L. and Chen, T.T. (1986) Host minerals for the gallium-germanium ores of the Apex Mine, Utah. Scientific Communications in Economic Geology 81, 946–950.

Dutta, P. S., Bhat, H. L. and Kumar, V. (1997) The physics and technology of gallium antimonide: An emerging optoelectronic material. Journal of Applied Physics 81 (9) 1 May 1997, 5822.

EE Times-Asia (2011) GaN power transistors eliminate current collapse, cut power loss. http://www.eetasia.com/ART_8800639119_765245_NP_ ce147f95.HTM

ElectroIQ (2011) Gallium nitride extends beyond defence products. http://www.electroiq.com/articles/sst/2011/09/gallium-nitride-extends-beyond-defense-products.html

Feneau, C. (2002) From Ag to Zn. Umicore. http://www.umicore.com/en/aboutUs/fromAgToZn/main.html

Ferdinand-Braun-Institut (2011) More efficiency for power electronics. Ferdinand-Braun-Institut, Berlin, Germany. http://www.fbh-berlin.com/fileadmin/downloads/Press/PM_FBH_HiPoSwitch_Kickoff_2011_11_10_engl.pdf

Fisher, D. and Bahl, I.J. (1995) Gallium Arsenide IC Applications Handbook, Volume 1. Academic Press, Inc., San Diego, California, USA, pp. 81.

Focus Digital Displays, LLC (2011) Why a 100,000 Hour LED Half-Life Rating is Important To You. http://www.focusdigitaldisplays.com/100000-hour-led-halflife-rating-important

Font, O., Querol, X., Juan, R. , Casado, R., Ruiz, C.R., López-Soler, A., Coca, P. and García Peña, F (2007) Recovery of gallium and vanadium from gasification fly ash. Journal of Hazardous Materials 139 (3), 413–423.

GaN Systems Inc. (2011) Why Gallium Nitride? http://www.gansystems.com/technology/why-gallium-nitride

Geier, B.H. and Ottemann, J. (1970) New primary vanadium-, germanium-, gallium- and tin-minerals from the Pb-Zn-Cu-deposit Tsumeb, South West Africa. Mineralium Deposita 5, (1), 29–40.

Green, M.A., Emery, K., Hishikawa, Y. and Warta, W. (2011) Solar cell efficiency tables (version 37). Progress in Photovoltaics: Research and Applications. Issue 19, 84–92.

Hall, W.E. and Heyl, A.V. (1968) Distribution of minor elements in ore and host rock, Illinois-Kentucky Fluorite District and Upper Mississippi Valley Zinc-Lead District. Economic Geology 63, 655–670.

Hieronymus, B., Kotschoubey, B. and Boulèque, J. (2001) Gallium behaviour in some contrasting lateritic profiles from Cameroon and Brazil. Journal of Geochemical Exploration 72, 147–163.

Higham, E. (2011) Market trends in GaAs RF ICs. Semiconductor Today 6 (3), April/May, 94–95.

Hill, V.G. and Sehnke, E.D. (2006) Bauxite. In: Kogel, J.E., Trivedi, N.C., Barker, J.M. and Krukowski, S.T. (eds.) Industrial Minerals and Rocks, Society for Mining Metallurgy and Exploration Inc: Littleton, Colorado, USA.

Ihn, S-G., Shin, K-S., Jin, M-J. , Bulliard, X., Yun, S., Choi, Y.S., Kim, Y., Park, J-H., Sim, M., Kim, M., Cho, K., Kim, T.S., Choi, D., Choi, J-Y., Choi, W. & Kim, S-W (2011) ITO-free inverted polymer solar cells using a GZO cathode modified by ZnO. Solar Energy Materials and Solar Cells, January 2011.

IMS Research (2011a) IMS expects LED market to stall in 2011. Compound Semiconductor October 2011, 17 (7) pp. 6.

IMS Research (2011b) IMS Research Downgrades 2011 Packaged LED Market Growth to 1% Despite 29% Increase in Lighting Revenues. September 2011

IMS Research (2011c) Quarterly GaN LED Supply/Demand Report. August 2011.

Jambor, J.L., Owens, D.R., Grice, J.D. and Feinglos, M.N. (1996) Gallobeudantite, $PbGa_3[(AsO_4),(SO_4)]_2(OH)_6$, a new mineral species from Tsumeb, Namibia and associated new gallium analogues of the alunite-jarosite family. The Canadian Mineralogist 34, 1305–1315.

Jaskula, B.W. (2011a) Gallium in Mineral Commodity Summaries. United States Geological Survey.

Jaskula, B.W. (2011b) Gallium in 2009 Minerals Yearbook. United States Geological Survey.

Jewett, S.A., Makowski, M.S., Andrews, B., Manfra, M.J. and Ivanisevic, A. (2011) Gallium nitride is biocompatible and non-toxic before and after functionalization with peptides. Purdue University and North Carolina State University. Acta Biomaterialia.

Kramer, D.A. (1988) Gallium. United States Bureau of Mines Minerals Yearbook 1988, Volume 1, 407–411.

Kramer, D.A. (2000 and 2001) Gallium. In: United States Minerals Yearbook. United States Geological Survey.

Kramer, D.A. (2002) Recycling – Metals. In: United States Minerals Yearbook. United States Geological Survey.

Lenntech, B.V. (2011) Health effects of gallium. http://www.lenntech.com/periodic/elements/ga.htm

Liu, H.Y., Li, X., Liu, S. Ni, X., Wu, M., Avrutin, N., Izyumskaya, N., Özgür, Ü., Yankovich, A.B., Kvit, A.V., Voyles, P.M. and Morkoç, H (2011) GaN based light emitting diodes utilizing Ga doped ZnO as a highly transparent contact to p-GaN. Physica Status Solidi C 8 (5), 1548–1551. http://onlinelibrary.wiley.com/doi/10.1002/pssc.201000860/abstract

Majumdar, D. (2011) GaN Revolution. Defense News. http://www.defensenews.com/story.php?i=5819236

Marso, M. (2010) GaN for THz sources. 8th International Conference on Advanced Semiconductor Devices and Microsystems (ASDAM), 25–27 October 2010, pp. 147–154. ieeexplore.ieee.org/Xplore/login.jsp?url=http%3A%2F%2Fieeexplore.ieee.org%2Fiel5%2F5653062%2F5666309%2F05666326.pdf%3Farnumber%3D5666326andauthDecision=-203

Martin, R.B. (1993) The elements in the Environment. In: Downs, A.J. (ed.) Chemistry of Aluminium, Gallium, Indium and Thallium. Blackie Academic and Professional, Glasgow, UK.

Masters, G.M. (2004) Renewable and Efficient Electric Power Systems. John Wiley and Sons Inc., New Jersey, USA.

Metal Bulletin (2011a) Tellurium and gallium fall, other minors may follow. Metal Bulletin magazine, 8 June 2011.

Metal Bulletin (2011b) High-tech metals fall as China de-stocks, traders take profits. Metal Bulletin magazine, 16 June 2011.

Metal Bulletin (2011c) Indium, gallium prices fall again on weak electronics demand. Metal Bulletin magazine, 14 September 2011.

Metal Bulletin (2011d) Gallium Metal MB free market prices. http://www.metalbulletin.com

Mining Journal (2006–2011) Minor metals sections. Mining Journal magazine.

Mining Journal (2010) Minor metals: staying strong in summer. Mining Journal magazine, 9 July 2010.

Mordberg, L.E., Stanley, C.J. and Germann, K. (2001) Mineralogy and geochemistry of trace elements in bauxites: the Devonian Schugorsk deposit, Russia. Mineralogical Magazine 65 (1) 81–101.

Moskalyk, R.R. (2003) Gallium: the backbone of the electronics industry. Minerals Engineering 16, 921–929.

Moss, R.L., Tzimas, E., Kara, H., Willis, P. and Kooroshy, J. (2011) Critical Metals in Strategic Energy Technologies, European Commission Joint Research Centre Institute for Energy and Transport, pp. 66.

Neo Material Technologies Inc. (2011) Presentation given at MMTA's International Minor Metals Conference, 4–6 May 2011, Philadelphia, PA, USA.

Öko-Institut e.V. (2009) Critical Metals for Future Sustainable Technologies and their Recycling Potential. United Nations Environment Programme and United Nations University, pp. 38.

Orbite Aluminae Inc. (2011) Products and Markets. http://www.orbitealuminae.com/en/products/products-markets/

PhosAgro (2011) Presentation given for the non-deal roadshow. VTB Capital Conference "Russia calling!", Moscow, October 2011. http://www.phosagro.com/upload/1393/PhosAgro_Russia_Calling_2011_NDR.pdf

Rensselaer Polytechnic Institute (2011) Energy@Rensselaer: Zeroing in on the Elusive Green LED. News and Events, 25 April 2011. http://news.rpi.edu/update.do?artcenterkey=2860

Roosevelt, A. (2011) Raytheon's GaN Technology Advances Radar. Defense Daily 249, (31) 1.

Roskill (2011) Gallium: Global Industry Markets and Outlook, 8th Edition. Roskill Information Services Ltd., London, UK.

Rytuba, J.J., John, D.A., Foster, A., Ludington, S.D. and Kotlyar, B. (2003) Hydrothermal enrichment of gallium in zones of advanced argillic alteration – examples from the Paradise Peak and McDermitt ore deposits, Nevada. United States Geological Survey Bulletin 2209-C. http://pubs.usgs.gov/bul/b2209-c/

SAFC (2011) Triethylgallium – TEGa. http://www.safcglobal.com/safc-hitech/en-us/home/overview/products/gallium/triethylgallium.html.

Savage, N. (2011) Solar Cell Breaks Efficiency Record. IEEE spectrum. http://spectrum.ieee.org/green-tech/solar/solar-cell-breaks-efficiency-record.

Scott, J.D. (1971) Crystal structure of a new mineral, sohngeite. In Mineralogical Notes. The American Mineralogist 56, 355.

Sharp (2011) Sharp to Commercialize World's First Small- and Medium-Size LCD Panels using Oxide Semiconductor. http://sharp-world.com/corporate/news/110421.html

Solarbuzz (2011) Global PV Market. http://www.solarbuzz.com/facts-and-figures/market-facts/global-pv-market

Srinivasa Rao, K., Sarangi, D., Dash, P.K. and Roy Chaudhury, G. (2003) Preferential extraction of gallium from Bayer liquor using ion exchange chelating resin containing hydroxamic acid functional group. Journal of Chemical Technology and Biotechnology 78 (5), 555–561.

Strategy Analytics Inc. (2011a) Strategy Analytics: Accelerating CPV Penetration of Renewable Energy Markets. http://www.businesswire.com/news/home/20110912006520/en/Strategy-Analytics-Accelerating-CPV-Penetration-Renewable-Energy

Strategy Analytics Inc. (2011b) Strategy Analytics: Brisk 2010 GaAs Epitaxial Substrate Revenue Growth. http://www.businesswire.com/news/home/20111031006235/en/Strategy-Analytics-Brisk-2010-GaAs-Epitaxial-Substrate

Stoiber, R.E. (1940) Minor elements in sphalerite. Economic Geology 1 June 1940, 35 (4), 501–519.

Strategic Defence Intelligence (2011) US Navy's AMDR programme moves ahead. http://www.strategicdefenceintelligence.com/article/CPpUgH6rN66/2011/09/20/us_navys_amdr_programme_moves_ahead/

Tolcin, A.C. (2011) Zinc in Mineral Commodity Summaries. United States Geological Survey.

Tsao, J. (2003) Roadmap projects significant LED penetration of lighting market by 2010. Laser Focus World. http://lighting.sandia.gov/lightingdocs/TsaoJYLaserFocus200303.pdf

TSMC Solar Ltd. (2011) TSMC Solar readies CIGS PV fab for 1GW future. http://www.electroiq.com/articles/pvw/2011/10/tsmc-solar-readies-cigs-pv-fab-for-1gw-future.html

Tummala, R.R. and Morris, J.E. (2001) The Role of Packaging in Microelectronics in Fundamentals of Microsystems Packaging, McGraw-Hill, 44–79.

UK Trade Info. (2011) UK Trade Information. Revenue and Customs. Beyond 20/20 Web Data Server, Version 7.4, SP1. https://www.uktradeinfo.com/index.cfm HM

U.S. Department of Energy (USDOE) (2011) LED Basics. U.S. Department of Energy – Energy Efficiency and Renewable Energy. http://www1.eere.energy.gov/buildings/ssl/sslbasics_ledbasics.html

Wang, L., Matson, D.W., Polikarpov, E., Swensen, J.S., Bonham, C.C., Cosimbescu, L., Berry, J.J., Ginley, D.S., Gaspar, D.J. and Padmaperuma, A.B. (2010) Highly efficient blue organic light emitting device using indium-free transparent anode Ga:ZnO with scalability for large area coating. Journal of Applied Physics. http://jap.aip.org/resource/1/japiau/v107/i4/p043103_s1?isAuthorized=no

Weeks, R.A. (1989) Gallium, Germanium and Indium Resources. In: Carr, D.D. and Herz, N. (eds.) Concise Encyclopedia of Mineral Resources, Pergamon Press, Oxford, UK. pp. 133.

Yirka, B. (2011) Samsung researchers announce breakthrough in growing gallium nitride LEDs on glass. PhysOrg.com. http://www.physorg.com/news/2011-10-samsung-breakthrough-gallium-nitride-glass.html

8. Germanium

FRANK MELCHER[1] AND PETER BUCHHOLZ[2]

[1]*Federal Institute for Geosciences and Natural Resources (BGR), Stilleweg, Hannover, Germany, Present address: Chair of Geology and Ore Deposit Geology, University of Leoben, 8700 Leoben, Austria*
[2]*Mineral Resources Agency (DERA) at the Federal Institute for Geosciences and Natural Resources (BGR), Dienstbereich Berlin, Wilhelmstraße, Berlin-Spandau, Germany*

Introduction

Germanium (Ge, atomic number 32) is a chemical element in Group 14 of the Periodic Table.[1] The existence of an element (temporarily known as 'ekasilicon') with properties intermediate between the metal tin and the non-metal silicon had been predicted by the Russian chemist D.I. Mendeleev in 1871 due to the systematic nature of the Periodic Table of the elements. However, it was the German chemist Clemens Winkler who first detected it in 1886 as a component of the mineral argyrodite (Ag_8GeS_6) in silver ores from the Himmelsfürst mine near Freiberg, Germany. Typically, germanium is recovered as a by-product from zinc and copper ores and coal.

Physical and chemical properties

Germanium is a greyish-white, brittle semi-metal (metalloid) which has a bright lustre. Key properties are summarised in Table 8.1. Its electrical properties are those of a semiconductor, i.e. between a metal and an insulator, which makes germanium potentially useful for many technical applications. The oxidation states of germanium are +2 and +4, with a strong tendency towards quadrivalence.

Germanium has five naturally occurring isotopes, ^{70}Ge, ^{72}Ge, ^{73}Ge, ^{74}Ge and ^{76}Ge, the latter being slightly radioactive, decaying by double beta decay with a half-life of 1.78×10^{21} years. ^{74}Ge is the most common isotope, having a natural abundance of approximately 36 per cent. When bombarded with alpha particles, ^{72}Ge will generate stable ^{77}Se, releasing high-energy electrons in the process. Therefore, it is used in combination with radon for nuclear batteries.

Germanium is an essentially non-toxic element, except for a few compounds. Dissolved in drinking water, germanium in the ppm range may cause chronic diseases (Gerber and Leonhard, 1997).

Distribution and abundance in the Earth

Germanium is a rare element in rocks. However, it is present in trace quantities in most rock types because of its siderophile, lithophile, chalcophile and organophile characters. The crustal abundance is estimated at 1.5 ppm for oceanic crust and 1.6 ppm for continental crust (Taylor and McLennan, 1985). The average germanium content of the Earth is 13.8 ppm, with 37 ppm germanium in the core and 1.1 ppm in the primitive mantle (Dasch, 1996). The highest enrichment in common rock types is in deep-sea clays that average 2 ppm germanium.

Critical Metals Handbook, First Edition. Edited by Gus Gunn.

Coals and coal ashes carry on average 2.2 and 15 ppm germanium, respectively (Ketris and Yudovich, 2009). Considerable concentrations of germanium may be found in copper–zinc–iron–silver sulfides and sulfosalts, as well as in association with organic matter, e.g. in some coals found in the Russian Far East and in China. Germanium concentrations in waters range from 0.06 parts per billion (river water) to several 100 parts per billion in thermal waters.

Mineralogy

Germanium does not occur as a free metal in nature. About 30 germanium minerals are known to contain germanium ranging up to 70% in argutite, GeO_2. Many are sulfides, underlining the strong chalcophile character of germanium. There are four germanate minerals approved by the International Mineralogical Association (IMA) and, in addition, some oxides, hydroxides, sulfates and silicates of germanium are known. Most common are argyrodite, canfieldite, briartite, reniérite and germanite, even though they have been reported from only a limited number (<50) of occurrences (Table 8.2). Other phases are rare, or even unique to one deposit. Kipushi in the Democratic Republic of the Congo (DRC) is the type locality of reniérite (Figure 8.1b) and briartite, and Tsumeb in Namibia of 13 rare germanium-bearing species. The Polkovice mine in the Kupferschiefer of Poland is the type locality of morozevicite ($Pb_3Ge_{1-x}S_4$) and polkovicite ($(Fe,Pb)_3(Ge,Fe)_{1-x}S_4$), whereas barquillite ($Cu_2CdGeS_4$) was first described from the Fuentes Villanas Li-Sn mine in Spain.

Germanium is a substituting element in many sulfide structures, most notably in common minerals such as in zinc sulfides (up to 3000 ppm in sphalerite and wurtzite) and copper sulfides (e.g. enargite, tennantite, bornite and chalcopyrite (Bernstein, 1985, Höll et al., 2007 and Cook et al., 2009. Elevated concentrations have also been recorded from willemite ($ZnSiO_4$; up to 4000 ppm, Figure 8.1d; Saini-Eidukat et al., 2009), cassiterite (up to 3000 ppm), hematite (up to 7000 ppm) and goethite (up to 5310 ppm) (Bernstein, 1985).

Silicates may accommodate germanium due to the similarity of its ionic radius and valence to those of silicon; the maximum concentrations recorded are 700 ppm in topaz from pegmatite deposits. Quartz may carry up to a few ppm germanium.

Table 8.1 Selected properties of germanium.

Property	Value	Units
Symbol	Ge	
Atomic number	32	
Atomic weight	72.63	
Density at 25 °C	5323	kg/m^3
Melting point	938	°C
Boiling point	2820	°C
Crystal structure	cubic, diamond	
Hardness (Mohs scale)	6.0	
Specific heat capacity at 25 °C	0.32	J/(g °C)

Table 8.2 Properties of the most common germanium minerals.

Name	Formula	Typical Ge content (%)	Appearance	Crystal structure	Type locality
Argyrodite	Ag_8GeS_6	5–7	steel grey with red tint, tarnishes black	orthorhombic, pseudocubic	Freiberg, Germany
Canfieldite	$Ag_8(Sn,Ge)(S,Te)_6$	1–2	steel grey with reddish tint, tarnishes black	orthorhombic, pseudocubic	Colquechaca, Bolivia
Briartite	$Cu_2(Fe,Zn)GeS_4$	13–18	grey to grey-blue in reflected light	tetragonal	Kipushi, DR Congo
Reniérite	$(Cu,Zn)_{11}Fe_2(Ge,As)_2S_{16}$	4–8	orange-bronze, tarnishes reddish	tetragonal	Kipushi, DR Congo
Germanite	$Cu_{13}Fe_2Ge_2S_{16}$	5–9	reddish grey, tarnishes dark brown	cubic	Tsumeb, Namibia

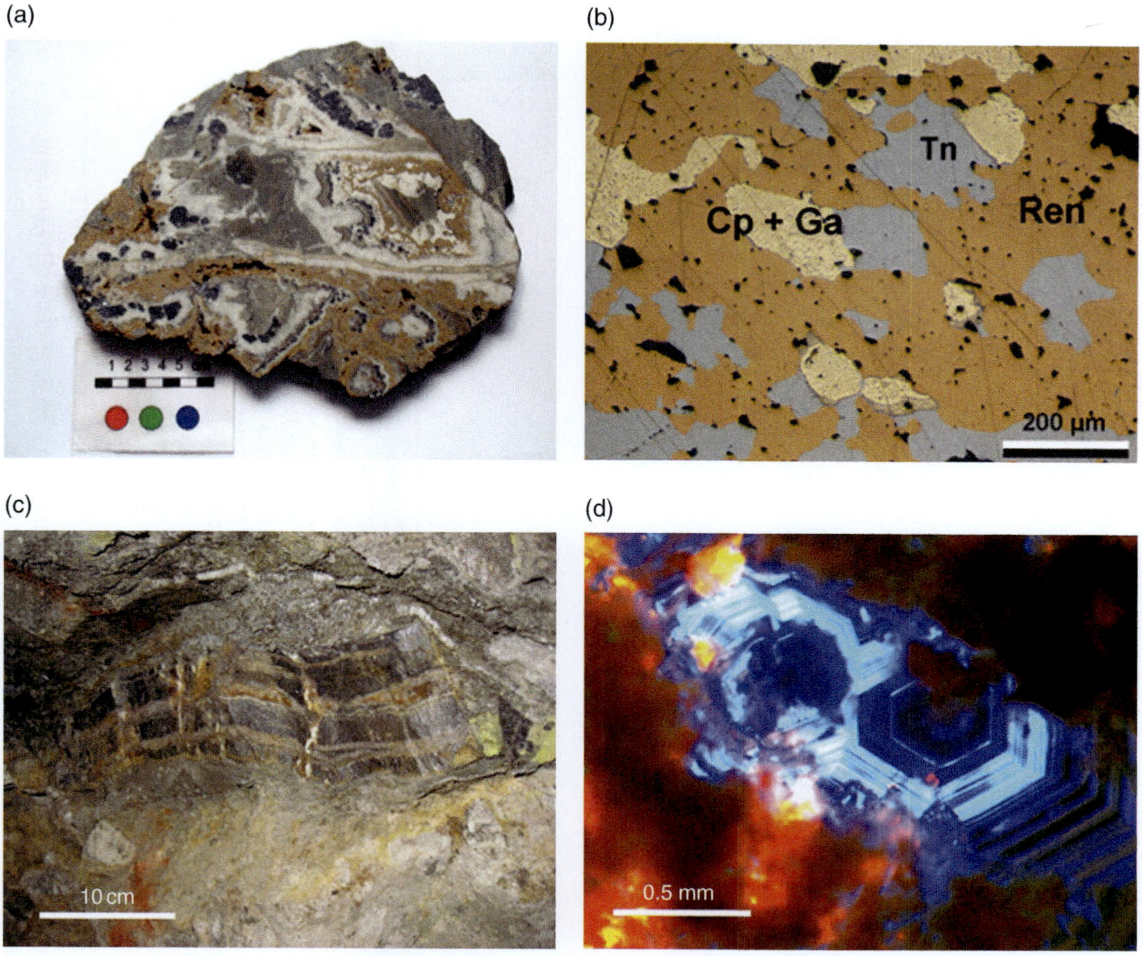

Figure 8.1 Photographs of germanium ores and minerals. (a) Solution collapse breccia cemented by different stages of sphalerite (different shades of brown) and galena (black) from the Bleiberg mine, Austria; grey colour is carbonate host rock, white colour is hydrothermal calcite. This represents a typical carbonate-hosted zinc-lead deposit of the 'Alpine type' grading 200–300 ppm Ge. (Courtesy of W. Prochaska, Leoben.) (b) Germanium ore from the Kipushi deposit, consisting of reniérite (Ren), tennantite (Tn), chalcopyrite (Cp) and gallite $CuGaS_2$ (Ga), forming myrmekitic intergrowth textures in chalcopyrite. Reflected light, oil immersion. (c) Banded Ge-rich sphalerite (grey to black with dark brown interlayers) at the Tres Marias mine, Mexico. Scale bar = 10 cm; photo taken in 2006. (d) Willemite ($ZnSiO_4$) carrying up to 1.5 wt% Ge from Tres Marias, Mexico, showing a bright blue luminescence colour in areas of low trace element concentrations, and weak colours in areas of high trace elements (Ge, Pb). Cathodoluminescence image. Scale bar = 0.5 mm.

Deposit types

Germanium is a trace metal in most types of oxidic and sulfidic metalliferous deposits, including banded iron formations (magnetite-hematite; average probably 20 ppm, in magnetite up to 78 ppm, Smirnov, 1977), manganese nodules (less than 10 ppm), shale-hosted sedimentary copper deposits (<20 ppm) and porphyry copper deposits (10 to 100 ppm). It is accumulated to economic concentrations in only a few deposit types (Figure 8.2). In most

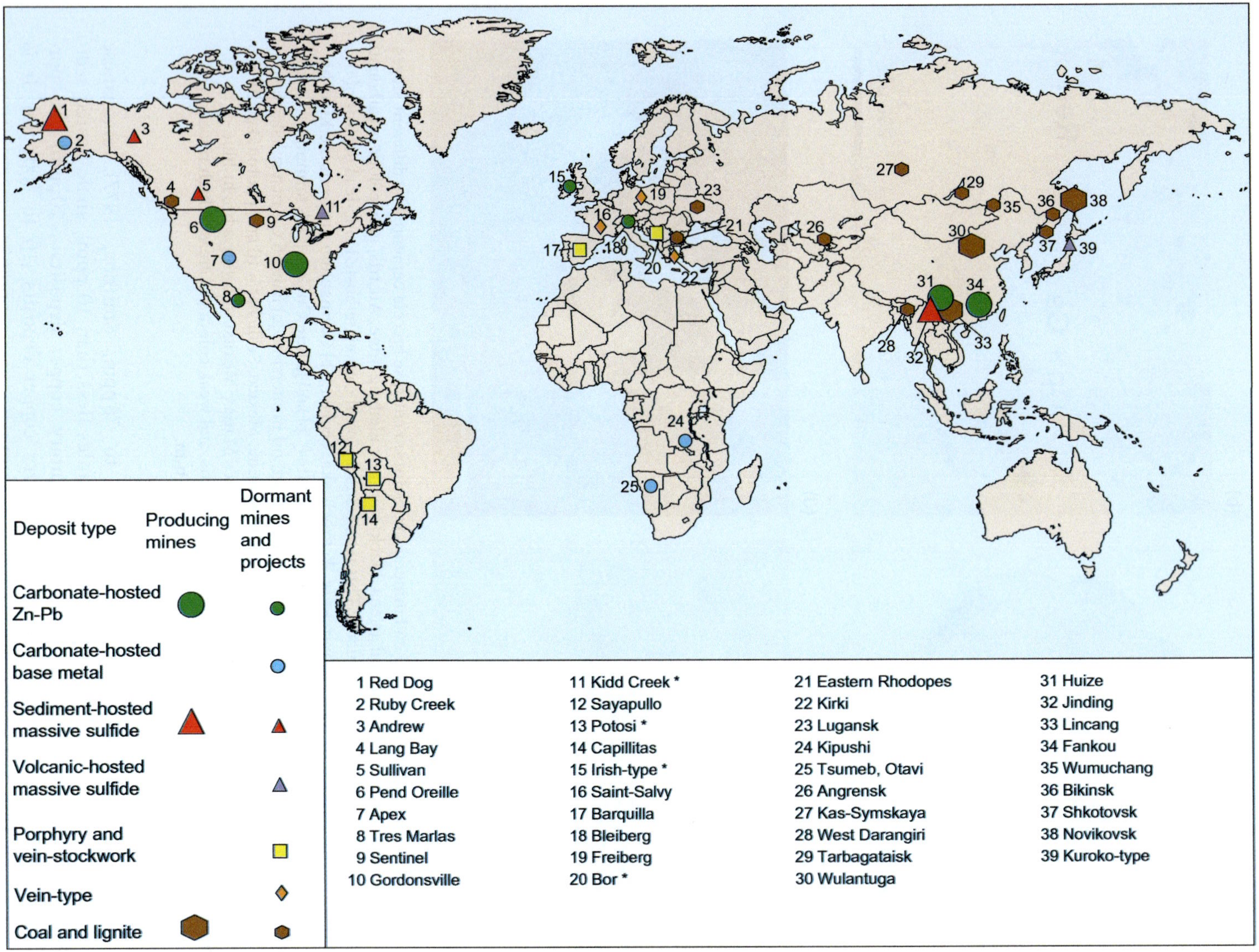

Figure 8.2 The global distribution of mines, deposits and major occurrences from which germanium may be produced. Locations marked * are producing mines where the deposit contains germanium, but which are not currently recovering the germanium.

Table 8.3 High-grade germanium concentrations and germanium potential in various deposit types. (Based on Höll et al., 2007.)

Class	Deposit type	Germanium-bearing species	Past production	Potential*	Typical ore grade (ppm Ge)
	Sulfide ores				
1	Volcanic-hosted Cu-Zn(–Pb) (–Ba) (Kuroko-type)	Sphalerite (bornite, Ge-sulfides)	low	medium	<<100 (–300)
2a	Porphyry and vein-stockwork Cu-Mo-Au	Cu-As sulfides, bornite, sphalerite	low	low	10–100
2b	Porphyry and vein-stockwork Sn-Ag	Argyrodite, sphalerite	low	medium	10–100
3	Vein-type (Ag-Pb-Zn)	Argyrodite, sphalerite	high (until 1993)	low	100–1000
4	Sediment-hosted Zn-Pb-Cu (–Ba)	Sphalerite, wurtzite	high	high	10–100
5a-c	Carbonate-hosted Zn-Pb (MVT, IRT, APT)	Sphalerite, wurtzite	high	high	100–1000
5d	Kipushi-type (KPT) polymetallic	Ge-sulfides (sphalerite)	high (until 2000)	medium	10–1000
6	Sediment-hosted stratiform Cu deposits	Sphalerite, pyrite, rare Ge sulfides	low	medium	1–20
	Oxide ores				
5e	Oxidation zones of KPT ores	Fe oxides/hydroxides, sulfates, arsenates	medium	low	100–1000
5f	Non-sulfide Zn-Pb	Fe hydroxides, willemite, hemimorphite	low	low	10–100
2c	Oxidation zones of Sn sulfides	Secondary Sn hydroxide, Sn oxide	low	low	10–100
7	Iron oxide ores	Fe oxides/hydroxides	none	low	10–50
	Coal and lignite deposits				
8a	Coal and lignite	Organic matter	medium	high	100–1000
8b	Coal and lignite ash		high	high	10–1000

*estimated future potential difficult to assess for many deposit types due to a lack of data on germanium grade in concentrates, recovery and reserves/resources.

Ag, silver; Au, gold; Ba, barium; Cu, copper; Fe, iron; Mo, molybdenum; Pb, lead; Sn, tin; Zn, zinc. APT, Alpine-type; IRT, Irish-type; KPT, Kipushi-type; MVT, Mississippi Valley-type.

of them, germanium may be recovered as a by-product from copper, zinc and lead production. In general, three modes of occurrence are distinguished (Table 8.3): sulfidic, oxidic and biogenic. Sulfidic concentrations of germanium are the most widespread and the most variable type, although the importance of germanium-rich coal deposits is likely to increase in the future.

Accumulation of germanium in sulfide deposits

In volcanogenic massive sulfide (VMS-) deposits, germanium concentrations are low to moderate (<100 ppm); however, due to their large tonnages, the refining of VMS zinc–copper–lead–gold–silver ores may yield significant cadmium, indium, gallium, tin, antimony, bismuth and also germanium (class 1, Tables 8.3 and 8.4).

Table 8.4 Key features of the main germanium-bearing deposit classes. (Modified from Höll et al., 2007.)

Class	Deposit type	Brief description	Features	Examples
1	Volcanic-hosted massive sulfide Cu-Zn(-Pb) (-Ba) (VMS)	Lenticular seafloor massive sulfide orebodies, often with footwall stringer zones	In extensional oceanic settings; associated with volcanic (basic to acid) rocks. Ge concentrations up to 100 ppm in recent ores, up to 370 ppm in Kuroko-type deposits	Kuroko-type ores (Japan); Neves Corvo Cu-Sn-Zn (–Ag-Se-In-Ge) (Portugal); Gorevskoe Pb-Zn (Russia); Ozernoe Zn-Pb (Russia)
2a	Porphyry and vein-stockwork Cu-Mo-Au	Medium to large, low-grade stockwork of quartz veinlets and disseminations in felsic intrusive rocks	In compressional tectonic settings; Ge-bearing sulfides occur in peripheral zones of porphyry systems, including late-stage epithermal veins	Capillitas (Argentina); Bor (Serbia); Chelopech (Bulgaria); Ladolam (Papua New Guinea)
2b	Porphyry and vein-stockwork Sn-Ag	Stockworks and arrays of ore veins in subvolcanic, felsic intrusions	Mineralization includes argyrodite and Ge-bearing Sn minerals	Potosí/Bolivian 'Ag-Sn belt'; Barquilla (Spain)
3	Vein-type Ag-Pb-Zn(-Cu)	Ore veins hosted by sedimentary and magmatic rocks	Heterogeneous group of vein deposits found in different tectonic settings	Freiberg (Germany); Noailhac-Saint Salvy (France); Kirki (Greece)
4	Sediment-hosted massive sulfides Zn-Pb-Cu(-Ba) (SMS)	Concordant lenses of stratiform massive to semi-massive sulfides and sulfates (baryte), often with footwall stringer zones and stockworks	Hosted by clastic marine sediments, including black shales, along continental margins or in intracratonic rift settings	Red Dog (Alaska, U.S.A.); Jinding/ Lanping (China)
5	Carbonate-hosted base metal deposits	Semi-massive to disseminated sulfide ores (Pb-Zn(-Cu)) in carbonate rocks	Stable carbonate platforms; low-temperature (commonly <300°C)	Worldwide from the early Proterozoic to Cenozoic
5a	Mississippi Valley-type (MVT) Zn-Pb-Fe(-Cu) (-Ba)(-F)	Hosted in carbonate successions since the Palaeoproterozoic; very common type	Signifcant age difference between host rock and mineralisation; sulfur derived from thermal sulfide reduction	Elmwood-Gordonsville district (Tennessee, U.S.A.); Pend Oreille (Washington, U.S.A.); Tres Marias (Mexico); Fankou, Huize (China)
5b	Irish-type (IRT)Zn-Pb(-Ag) (-Ba)	Hosted in Lower Carboniferous carbonate rocks	Bacteriogenic sulfur source	Navan, Tynagh, Lisheen (all Ireland)
5c	Alpine-type (APT) Zn-Pb	Hosted in Triassic limestones	Bacteriogenic and non-bacteriogenic sulfur sources	Bleiberg (Austria); Mežica (Slovenia); Cave de Predil (Italy)
5d	Kipushi-type (KPT) polymetallic Cu-Zn-Pb-Ag-As	Hosted in Neoproterozoic limestones (central and SW Africa)	High-temperature fluids (250-450°C); sulfur derived from thermal sulfate reduction. Highest Ge grades known	Tsumeb (Namibia); Kipushi (DR Congo); Kabwe (Zambia); Ruby Creek (Alaska, U.S.A.)

(Continued)

Table 8.4 Continued

Class	Deposit type	Brief description	Features	Examples
5e	Oxidation zones of KPT deposits	Extensive oxidation of primary sulfide ores	Ge in Fe-hydroxides, and forming secondary Ge minerals	Tsumeb (Namibia); Apex (Utah, U.S.A.)
6	Sediment-hosted stratiform Cu deposits	Stratiform disseminated sulfide mineralisation associated with carbonaceous sediments	Low-temperature diagenetic to epigenetic deposits formed from oxidized fluids of low salinity	Kupferschiefer (Poland, Germany); Central African Copperbelt (DRC, Zambia)
7	Iron oxide ores	Precambrian banded iron formations; Phanerozoic volcanogenic-sedimentary deposits	Ge in iron oxides from sedimentary deposits; low Ge in Fe skarn deposits	Hamersley Range (Australia), Krivoi Rog (Ukraine), Atasu type (Kazakhstan)
8	Coal and lignite	Extensive beds of coal and lignite close to bedrock contacts	Up to 3000 ppm Ge in fly ash from coal combustion	Lincang, Wulantuga (China); Tarbagataisk, Shotovsk (Russia); Lugansk (Ukraine); Angrensk (Uzbekistan)

Ag, silver; As, arsenic; Au, gold; Ba, barium; Cu, copper; F, fluorine; Fe, iron; In, indium; Mo, molybdenum; Pb, lead; Se, selenium; Sn, tin; Zn, zinc.

Porphyry copper and related vein-stockwork copper–(molybdenum–gold–silver–tin) deposits host large tonnages of ore, but commonly low grades of germanium (class 2a and 2b). Nevertheless, germanium minerals are reported from late-stage epithermal veins in Argentina, Peru and Bolivia, and from several porphyry copper deposits where germanium is hosted by bornite, chalcopyrite and pyrite, and also present as microphases. In tin–silver stockwork deposits of Bolivia and Peru (Potosi-type), germanium is an accessory component with ore shoots containing argyrodite associated with silver phases. However, no reliable information exists on average grades and expected tonnages in any of the above-mentioned deposits (Höll et al., 2007).

Polymetallic tin–silver and silver–lead–zinc vein deposits (class 3) contributed significantly to the global germanium production until the closure of the Noailhac-Saint Salvy deposit, French Massif Central, after 18 years of production, in 1993. The cumulative production was 500 tonnes germanium together with 0.35 million tonnes (Mt) zinc and 280 tonnes silver; zinc concentrates graded 700–800 ppm Ge, and sphalerite carried up to 2500 ppm Ge (Cassard et al., 1996). Other vein districts yielding scattered germanium values are known from the Freiberg district and Harz mountains in Germany, Kutna Hora (Czech Republic), Sardinia (Italy), Kirki (Greece) and South Korea, but their economic potential is limited (Höll et al., 2007).

A major share of today's global germanium production is from large-tonnage, but low-grade sediment-hosted zinc-lead deposits of the 'SEDEX class' (class 4). The sphalerite of the giant Red Dog deposit, Alaska (proven and probable reserves in 2006 were 85 million tonnes at 18.2% Zn, 5.6% Pb) averages about 100 ppm Ge (Kelley et al., 2004) which is partly recovered as a by-product in the Teck smelter in Trail, British Columbia, Canada. In 2007, the Trail facility produced about 40 tonnes of germanium from zinc concentrates sourced from Red Dog (600,000 tonnes per year zinc concentrate) and Pend Oreille, Washington (83,000 tonnes per year zinc concentrate) (Guberman, 2008). Since then, no production data have been released; however, Teck announced that 25 per cent of the zinc concentrates produced at the Red Dog mine are transported to the Trail smelter for further treatment.

Carbonate-hosted zinc-lead deposits, commonly referred to as 'Mississippi Valley-type'

(MVT, class 5), constitute a class of potentially economically interesting germanium deposits. There is discussion on alternative classifications or the application of MVT subsets (e.g. Leach et al., 2005); in addition to strictly epigenetic MVT sensu stricto (class 5a), an 'Irish-Type' (IRT, class 5b), an 'Alpine-type' (APT, class 5c; Figure 8.1a) and a 'Kipushi-type' (KPT, class 5d; Figure 8.1b) are distinguished by Höll et al. (2007) based on the relationships between ores and host rocks and geochemical characteristics of the ores and ore fluids (Table 8.4). In all carbonate-hosted zinc–lead sulfide deposits, germanium is invariably substituted into the sphalerite and wurtzite structures; with discrete germanium minerals usually absent. High germanium concentrations are reported from the Tri-State district (Tennessee, Missouri, Arkansas) on the Viburnum trend, USA (60–400 ppm in zinc ore concentrates), Bleiberg, Austria (on average 300 ppm; 126 tonnes germanium produced), Cave del Predil, Italy (250–450 ppm), Huize, China (up to 354 ppm), and Niujiaotang, China (up to 546 ppm) (Bernstein, 1985, Höll et al., 2007 and Ye et al., 2011). Only some of the deposits are actively mined at present, and only a few contribute to the global germanium production. Huize, for example, has expanded its annual production to 100,000 tonnes zinc and ten tonnes germanium. Some deposits have exceptional concentrations of germanium, e.g. on average 1000 ppm germanium in sphalerite at the small, high-grade Tres Marias deposit in northern Mexico (Figures 8.1c and d; Saini-Eidukat et al., 2009). The potential for germanium recovery from carbonate-hosted zinc-lead deposits is large, due to the widespread occurrence of these ores in carbonate platforms ranging from the Palaeoproterozoic to the Tertiary. However, local factors that are not well understood at present control the germanium distribution in these deposits.

The highest germanium concentrations are found in carbonate-hosted polymetallic hydrothermal deposits of the 'Kipushi-type' (KPT, class 5d). These enigmatic deposits are characterised by a complex elemental association copper–zinc (–lead–silver–arsenic–germanium–cadmium–vanadium–molybdenum–tungsten) reflected in a complex mineralogical composition, and by the presence of discrete germanium phases. At the type localities, Kipushi in Katanga, DRC (past production 1925–1993: 60 million tonnes at 6.8% Cu, 0.9% Pb, 11% Zn, on average 100–200 ppm Ge) and Tsumeb in the Otavi Mountainland, Namibia (past production 1947–1996: 28 Mt at 4% Cu, 12% Pb, 5% Zn, average 50–150 ppm Ge), the sulfide ores are composed of tennantite, sphalerite, galena, chalcopyrite, bornite, pyrite, and a large number of rare phases including the germanium minerals germanite, reniérite and briartite (Figure 8.1b; Melcher, 2003, Schneider et al., 2007 and Kampunzu et al., 2009). The ores are medium- to low-temperature (400–200 °C) hydrothermal in origin and frequently occupy breccia zones in carbonate successions. Similar deposits are known in the Neoproterozoic polymetallic ore districts of the Otavi Mountainland, Namibia, and the Katangan Copperbelt, Zambia and DRC, including the small, but high-grade Khusib Springs deposit in the vicinity of Tsumeb with average germanium concentrations of 100 ppm hosted by enargite, tennantite and germanium-bearing colusite (Melcher et al., 2006). None of the 'Kipushi-type' deposits are mined at present, and the high-grade zones in the known deposits that graded over 1% Ge have been mined out. However, for Kipushi, remaining resources down to the 1500-metre level are estimated at > 5 million tonnes zinc, >500,000 tonnes copper, and >100,000 tonnes lead from ores averaging 21.4% Zn, 2.1% Cu and 0.88% Pb with traces of germanium (68 ppm), cadmium, cobalt and silver (Kampunzu et al., 2009).

Although there is a potential for new KPT discoveries, base-metal exploration at present focuses on SMS, VMS and MVT-type zinc–lead ores, on porphyry copper ores, on sandstone-hosted copper ores, and on sediment-hosted copper ores of the Kupferschiefer type that are usually low in germanium (class 6). Kupferschiefer from Mansfeld in the Harz mountains of Germany contains 8–15 ppm Ge, which was partly recovered prior to mine closure in 1990.

Oxidised portions of germanium-rich sulfide deposits (class 5e) may carry exceptionally high germanium grades. However, as such deposits are rare and most of them mined out, their economic potential is limited. Germanium may be present as secondary germanium phases (Tsumeb, Namibia), or absorbed on iron hydroxides and oxides (up to 2.5% Ge at Tsumeb; 0.5% at Apex, USA). The little available data on germanium concentrations in supergene non-sulfide zinc–lead deposits (class 5f) suggest only minor significance for germanium recovery, although some of the deposits formed from germanium-bearing sphalerite. Some giant iron oxide deposits (class 7), e.g. Archaean and Palaeoproterozoic banded iron formations, carry appreciable germanium locked up in iron oxides and hydroxides (up to 100 ppm). However, the lack of appropriate technology and high costs will probably deter industry from recovering trace elements such as germanium. Iron ore wastes might constitute a future source of germanium; it was estimated that stockpiled wastes in the Ukraine contained at least 200 million tonnes of iron and 100,000 tonnes of germanium (Levine and Wallace, 2000); this, however, would imply on average 500 ppm Ge in the wastes, which seems unreasonable.

Appreciable quantities of germanium are contained in slag produced in the past from the smelters at Tsumeb (Namibia), Lubumbashi and Kolwezi (DRC). The Tsumeb smelter processed polymetallic ores from the Otavi Mountainland (Tsumeb, Kombat, Khusib Springs, Berg Aukas and others) and imported copper ores, whereas the Congolese smelters processed ores from the Congolese Copperbelt (sediment-hosted stratiform copper ores) in addition to germanium-rich KPT-type ores from Kipushi.

Enrichment of germanium in lignite and coal

Since V.M. Goldschmidt (Goldschmidt, 1930 and Goldschmidt and Peters, 1933) found high germanium concentrations in coal ashes from the Durham Coalfield, U.K. (up to 1.1% Ge), coal and lignite deposits have attracted both researchers and industry. In lignite, 97 per cent of the germanium is present in humus (Yang et al., 2003), mainly bound to humin (86–89 per cent) and humic acid (10–12 per cent). A process to recover germanium from coal ash was developed in the UK and used by Johnson Matthey between 1950 and 1974 utilising ashes containing between 0.5 and 1% Ge. The coals from Durham and Northumberland contained approximately 300 ppm Ge. Flue dusts from combustion were collected from coal gas works and power stations. However, when coal gas was replaced by natural gas and power stations were converted to oil power, germanium recovery ceased (Roskill, 1988). In the former USSR, 850 tonnes of germanium was mined from germanium-rich coals from the Novikovsk deposit (Sachalin) between the 1960s and the 1980s, and were burnt with less germanium-rich coals from the Tarbagataisk deposit at the Chita Heat Electropower Station (Kats et al., 1998). Germanium-rich ashes were subsequently transported to Angren (Uzbekistan) for upgrading and from there to Krasnoyarsk for further processing. Other germanium-rich coals from eastern Russia were processed in the Ukraine.

Today, 30 to 50 per cent of the primary germanium production is from lignite deposits in China, Russia and Uzbekistan (Seredin and Finkelman, 2008, Bleiwas, 2010 and Seredin, 2012). The first germanium-rich coal deposit in the former USSR was in the Angren valley (Uzbekistan) in the 1950s, followed by a number of deposits in various parts of Russia and China, with most of them in north-eastern China and the Russian Far East. According to Seredin and Finkelman (2008) and Seredin et al. (2013), the largest known germanium-rich coal deposit is in Wumuchang (Yumin / Yimin coal field, Inner Mongolia) with an estimated 4000 tonnes of germanium resources, followed by Bikinsk (eastern Siberia) with 2600 tonnes of germanium. Average germanium grades in ash range from 30 ppm (Wumuchang) to >1000 ppm (Shkotovsk, Russian Far East), and the thickness of individual germanium-bearing beds ranges up to 15–20 metres (Seredin and Finkelman, 2008). The Wulantuga deposit in the Shengli Coalfield, Inner Mongolia

(China) is the largest germanium producer at present (Du et al., 2009, Dai et al., 2012 and Seredin, 2012). The germanium-rich part of the deposit contains 12.3 million tonnes of lignite with >30 ppm Ge, and 1700 tonnes of germanium extractable from coal with Ge content >100 ppm (Du et al., 2009). The Lincang lignite deposit (1060 tonnes germanium resources, Hu et al., 2009) in Yunnan, southern China, is another significant germanium producer (16 tonnes per year of high-grade GeO_2). Evaluation of the Kas-Symskaya lignite deposit in western Siberia yielded resources of 11,000 tonnes germanium at an average grade of 205 ppm Ge in dry lignite (Evdokimov et al., 2004). Most germanium-rich lignite deposits are hosted by small fault-controlled coal basins in Mesozoic or Cenozoic sedimentary successions. Germanium was introduced by hydrothermal fluids from the basement, often connected with magmatic activity. Hydrothermal alteration is widespread, and metasomatic rocks associated with coal beds may carry significant germanium concentrations. In addition to germanium, the coals often carry elevated concentrations of tungsten, uranium, niobium, caesium, arsenic, antimony, mercury and thallium.

A review of trace-element data in coal provinces up to 1990 indicated that germanium concentrations mostly range from 0.5 to 50 ppm (Swaine, 1990). However, research in some coal provinces has identified higher germanium concentrations, e.g. in Bulgaria (Gouin et al., 2007). Germanium-bearing coal seams occur in a carbonaceous sandstone deposit with intercalated kaolinite at Lang Bay near the Pacific Coast, Canada (Queneau et al., 1986). A comprehensive study of coal fields in India identified elevated germanium concentrations only locally, e.g. in Assam (600 ppm Ge; Banerjee et al., 2000). Germany's lignites and coals seem to carry low germanium concentrations (e.g. 3–10 ppm Ge in ashes from Rhenish lignite; 1–2 ppm in hard coal from the Ruhr Basin, Feiser, 1966). However, elevated values of up to 50 ppm Ge in coal have been reported from some middle-German hard-coal basins (Leutwein and Rösler, 1956).

Extraction methods, processing and beneficiation

Extraction

Germanium is a by-product in some sulfide ores and in some coals; therefore, surface and underground mining methods that are commonly used to extract base metals (copper, zinc and lead) and coal are employed. Large sediment-hosted deposits, such as Red Dog, Alaska, and many lignite deposits operate in open-pits, whereas many MVT-type deposits and carbonate-hosted base-metal deposits have extensive underground workings, e.g. employing room and pillar techniques. The Apex mine in Utah was the only mine that produced primary germanium and gallium from an ancient underground copper mine with lead, zinc and silver as by-products (Taylor and Quinn, 1987); the valuable metals were concentrated in iron oxyhydroxides that were leached with sulfuric acid and sulfur dioxide, followed by the precipitation of cement copper and solvent extraction of the pregnant solution to separate the germanium and gallium. The mine and plant operated from 1986 to 1987, but closed due to decreasing germanium prices. A subsequent attempt to re-open the mine in 1990 also proved to be uneconomic.

Processing

After mining, the ores are processed to increase their base metal and germanium contents using conventional mechanical (grinding, sieving, magnetic separation) and flotation methods. Three methods of germanium extraction were used: (1) from germanite and reniérite ores, (2) from zinc ore concentrates, (3) from coal and fly ashes. Extraction from slag material will become important in the future (e.g., AUSMELT technology). A simplified flow diagram illustrating the consecutive steps of germanium processing is presented in Figure 8.3.

Originally, the Eagle-Pitcher smelter at Henrietta, Oklahoma, operated a Waelz kiln to fume zinc, cadmium, germanium and gallium from low-grade residues from the 1950s until 1968. GeO and GeS begin to sublime at 700 °C,

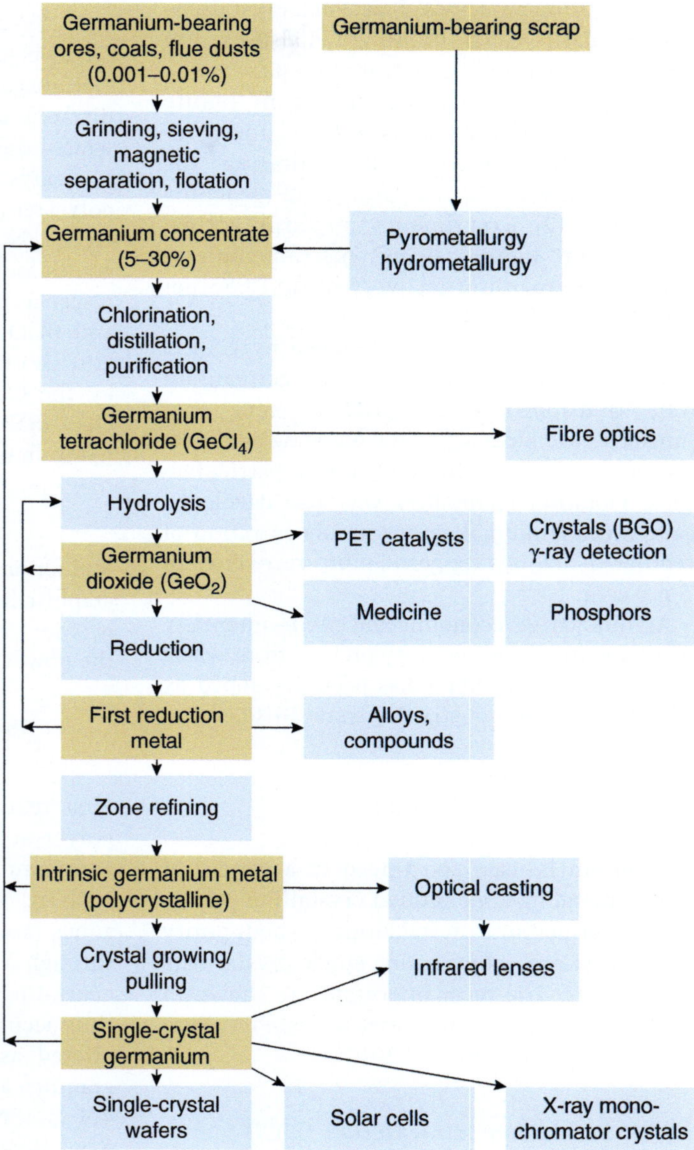

Figure 8.3 Simplified germanium processing flow diagram. (BGO, bismuth germanium oxides ($Bi_4Ge_3O_{12}$); PET, polyethylene teraphthalate.) (Based on Butterman and Jorgenson, 2005 and Naumov, 2007.)

whereas the boiling point of germanium tetrachloride ($GeCl_4$) is 85 °C. Today, germanium is mostly obtained from zinc smelter flue systems and dusts of smelters that process zinc ores. This process involves (1) refining zinc, (2) distillation under non-oxidising condition, (3) recovery of the distillation residue, (4) formation of germanium tetrachloride by leaching the distillation residue with hydrochloric acid, (5) hydrolysis of germanium tetrachloride into germanium dioxide (GeO_2) and (6) optional reduction of GeO_2 into metallic germanium with hydrogen at 760 °C (Global Industry Analysts, 2010). Metal powder is melted and moulded into metal bars, from which highly purified metal may be produced by zone refining polycrystalline processes. Fractional crystallisation of volatile $GeCl_4$ ('fuming') is also used to separate germanium from other metals.

The recovery of germanium from lignite includes burning the coal at 1200 °C and pyrometallurgical treatment after filtering the ashes, followed by sulfuric acid leaching (Seredin, 2012). Experiments to recover germanium from lignite using micro-organisms yielded a germanium recovery of up to 85 per cent (Yang et al., 2003).

Sirosmelt, a technique developed by CSIRO (Australia), was tested to extract germanium from sandstone-hosted lignite at Lang Bay, Canada (Queneau et al., 1986) due to its effectiveness for small-scale fuming of wet particulate feeds. Flotation technology was also developed using a cyanamid collector for concentration of weathered coal, giving germanium recoveries of 75 per cent.

An innovative, economic and environmentally friendly germanium recovery process from waste products of optical fibres has been developed by Bell Labs (Global Industry Analysts, 2010).

Specifications

Germanium is used and traded in a variety of forms, including zone-refined crystalline germanium, germanium tetrachloride, high-purity oxides, first-reduction ingots, single-crystal bars and castings. The most important are shown in Figure 8.3 (for further information see PPM Pure Metals, 2012 and Umicore, 2012).

Germanium tetrachloride, $GeCl_4$

$GeCl_4$ is recovered from germanium-bearing concentrates by dissolution in hydrochloric acid and purification through a series of fractional distillations. $GeCl_4$ is used to profile the refractive index of optical fibres for telecommunication. It is produced in two qualities, fibre-optic grade and ultrahigh-purity grade with less than five parts per billion metal impurities, used as a dopant for fibre optical products.

Germanium dioxide, GeO_2

$GeCl_4$ is hydrolysed with deionised water to precipitate germanium hydroxide. From this GeO_2 is filtrated and vacuum baked. GeO_2 is the most widely traded germanium product and is used as a catalyst for the production of high-quality polyester (PET). Both crystalline and amorphous GeO_2 (99.5%) can be used. Other crystalline grades are used for special applications, e.g. fluorescence grade for the production of phosphors and BGO grade (99.999%, 5N[2]) for the crystal growth of bismuth germanate (BGO, $Bi_4Ge_3O_{12}$) by the Czochralski technique. Electronic-grade GeO_2 (5N) is best suited for the production of germanium metal.

First reduction metal

Reduction of GeO_2 with hydrogen in ultra-clean graphite boats at 760 °C yields germanium metal powder. Metal bars are produced by melting of the powder at 1100 °C.

Production of zone-refined metal ('intrinsic' metal)

First reduction metal is purified to yield pure polycrystalline germanium known as 'intrinsic' germanium metal by a process known as zone refining, where a succession of small molten zones are created by passing the metal bar through a series of inductive coil heaters, concentrating impurities in the molten zone and thus accumulating at the end of the bar. Metal is offered as zone-refined polycrystalline ingots, granules and powder with purities ranging from 4N to 13N, the purity level needed for gamma X-ray detection.

Single crystals

Single crystals are required as optical components in infrared optical systems and electronics. Crystals up to 300 mm in diameter are grown by the Czochralski vertical pulling process, using a monocrystalline seed dropped in a bath of molten germanium. In the horizontal pulling process, a melting zone moves from a monocrystalline seed along a polycrystalline rod.

Uses

Germanium owes its usefulness to six properties:

- It is an intrinsic semiconductor, particularly effective at high frequencies and low voltages.
- It is transparent to infrared light.
- It is a glass-former, e.g. able to form three-dimensional networks of germanium-O tetrahedra.
- It has a high refractive index.
- It has low chromatic dispersion.
- It has an ability to catalyse the polymerisation in the production of plastic (polyethylene terephthalate (PET) mainly used for plastic bottles) without undesirable coloring.

Germanium was first used in industry after World War II when Karl Lark-Horovitz from Purdue University discovered its properties as a semiconductor (Haller, 2006). The point-contact diodes for radar pulse detection had already been used during the war. From 1948 until the 1970s, germanium transistors played a vital role in solid-state electronics, but then were replaced by high-purity silicon that has superior electrical properties. New applications arose in the late 1950s with the boom in nuclear physics and the need to develop nuclear radiation spectrometers with good energy resolution.

Global sales volumes of germanium at manufacturer level, which basically represent the germanium content in end-use sectors, were about 115 tonnes germanium in 2010. The major germanium users by global volume sales are the USA (37 per cent), Europe (21 per cent) and Asia-Pacific (14 per cent). Within Europe (including Russia), the largest volume sales of germanium in end-use applications are in France (33 per cent), Germany (24 per cent), and the UK (17 per cent); (Global Industry Analysts, 2010).

Over the past two decades, the major uses for germanium changed from dominantly infrared optics in 1990 to fibre optics in 2000 and new applications for PET (Table 8.5, Figure 8.4). Relative proportions of these three applications have continuously changed over the years without showing a major trend. Today, the three sectors, fibre optics (30 per cent), infrared optics (25 per cent) and catalysts for colorless PET (25 per cent) account for 80 per cent of the global end-use sectors (Guberman, 2013). For these three sectors, applications tend to vary with region, e.g. the PET sector is not important in the USA and Canada, but in Europe (29 per cent of global volume sales), Asia-Pacific (25 per cent), and Japan (21 per cent) (Global Industry Analysts, 2010). Fibre optics are mainly used in the USA (42 per cent of global volume sales) and the EU markets (21 per cent) followed by Asia-Pacific (12 per cent). The regional end-use distribution pattern of infrared optics is almost identical to the regional end-use pattern for fibre optics (Global Industry Analysts, 2010).

Compared to the dominant three end-use sectors, various applications in the electronic/solar industry consume about 15 per cent of the germanium produced, e.g. for silicon chips (SiGe chips), high-speed integrated circuits such as wireless communication devices or high-efficiency multi-junction photovoltaic cells (gallium arsenide on germanium substrate) for space applications. Although in small amounts only, other uses of germanium include gamma ray detectors, X-ray monochromators, thermo-photovoltaics, medical applications such as chemotherapy, and metallurgy (alloys with tin to increase hardness or silver to prevent tarnishing (Angerer et al., 2009).

Recycling, re-use and resource efficiency

Due to its high dispersion in most products and application in very low quantities, little germanium is recovered from post-consumer scrap ('old scrap'). Recycling from 'new scrap', however, is more widespread in the production of germanium-containing fibre-optic cables and infrared imaging devices. In the US economy, about 50 per cent of recycled material was from optical fibres (mainly new scrap). Most of the scrap is sent to germanium processors for recycling which are able to recycle material containing a minimum of

Table 8.5 Summary of the major end-use markets of germanium. (Data from Guberman, 2011.)

Market sector	End use (2010)*	Usage	Germanium product
Fibre optics	30%	Communication networks	GeO_2 dopant within the cores of optical glass fibres; Ge is deposited with silica on the inside of a pure quartz tube that is drawn into fine fibres
Infra-red optics	25%	Night vision systems; optical instruments such as camera lenses, IR spectroscopy; military, active car safety systems, satellite systems, fire alarms	Polycrystalline and single crystals; n-type Ge doped with P or Sb; Ge lenses and windows; GeO_2 is used as a component of glasses in wide-angle camera lenses and microscope objectives
Polymerisation catalysts	25%	Plastics (polyethylene terepthalate, PET) for water bottles	GeO_2 is used to make PET heat resistant (Asia)
Electronics and solar electrical applications	15%	Wireless devices, optical communication systems, hard disk drives, GPS	SiGe-based chips compete with GaAs and Si chips (since 1997)
Semiconductors		Transistors, rectifiers, lasers	Germanium doped with Sb, As, P (n-type semiconductor), Ge doped with Al, B, Ga, In (p-type semiconductor); amorphous semiconductors
Diodes		Light-emitting diodes (LEDs)	Ge substrate for high brightness LEDs
Transistors		Supercomputer	Si-Ge heterojunction bipolar transistors; silicon-on-insulator technology
Solar cells		Space-based photovoltaic solar cells	Polished Ge wafers for multilayer solar cells (improved efficiency compared to silicon)
Radiation detectors	5%	Detection of gamma radiation, e.g. air traffic control	Single crystals of ultra-pure Ge
Superconductors			Melting Ge and other metals to produce superconducting alloys, e.g. Nb_3Ge; vacuum sintering and annealing; thermal dissociation; electrolysis; chemical vapour deposition methods
Medicine		Chemotherapy, dietary supplement, antioxidant, super nutrient, diagnosis of diseases	Ge-organic compounds (Ge-132) found in medicinal plants; Bis-beta carboxyethyl Ge sesquioxide (Ge-Oxy 132); colloidal Ge

*estimated share of global annual sales volumes of germanium.

Al, aluminium; As, arsenic; B, boron; Ga, gallium; GeO_2, germanium dioxide; In, indium; Nb, niobium; P, phosphorus; Sb, antimony; Si, silicon.

2% Ge, including solutions, fibres, dust and filter mats from air-conditioning systems. Processing of scrap normally involves dissolution into germanium tetrachloride. The quality of products is not affected by recycled material (Kammer, 2009).

On a global scale, about 25–35 per cent of total germanium used is processed from recycled scrap, mainly new scrap (Butterman and Jorgenson, 2005). According to UNEP (2011) the global end-of-life recycling rate is less than one per cent (recycled germanium in old scrap as share of germanium metal content of end-of-life products).

Infrared optics: about 30 per cent of the germanium consumed for this end use is produced from recycled materials that accumulate during the manufacture of optical devices such as broken

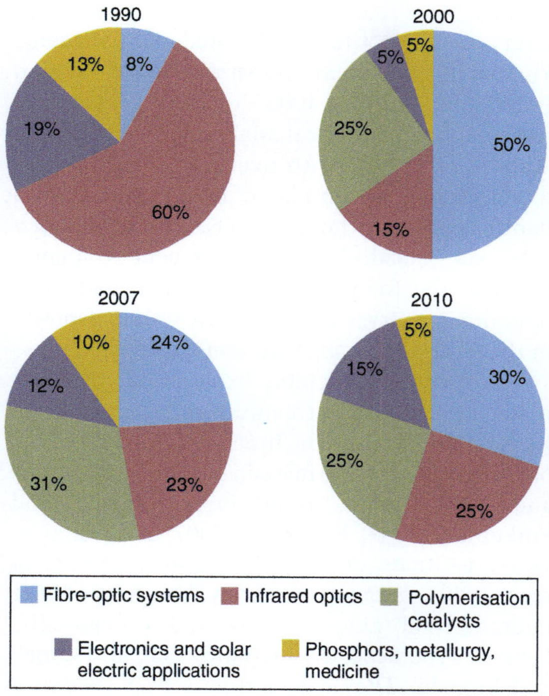

Figure 8.4 Comparison of global end-uses of germanium in 1990, 2000, 2007 and 2010. (Data from Guberman, 2011.)

lenses and glasses from night-vision devices. This compares favourably with the five per cent to ten per cent of broken lenses and glass from night-vision devices that were recycled as old scrap in the past (Jorgenson, 2006).

Fibre optics: fibre-optic cables for telecommunication contain up to 60 per cent of recycled material and the recovery rate of germanium from fibres is up to 80 per cent (Jorgenson, 2006). Optical silica fibres contain about 4% GeO_2, but only 0.03 per cent to 2 per cent of the cable is the optical core. For 2.5 km of optical fibre, about one gram of GeO_2 is used (Oakdene Hollins Ltd, 2011). Fibre-optic cables also have a long residence time in their applications, thus recycling volumes for old scrap are rather low. However, in the future, they may have considerable post-consumer recycling potential.

Electronics and solar electrical applications: during solar-cell production, about 50 per cent of waste accumulates as new scrap which is recycled. Old computers which are dismantled or shredded contain only minute amounts of germanium in their electronic parts. The major challenge is the effective collection of old devices and cost-effective separation of metals.

Polymerisation catalysts: estimated world polyester production in 2009 was about 50 million tonnes per year (Chemsystems, 2011). About a third of this production volume was used for PET bottle grade. In 2006, about 350,000 tonnes per year was produced by using germanium catalyst as a substitute for antimony catalyst, mainly in Japan. Applying germanium instead of antimony results in very low levels of antimony contamination in bottled drinking water. The average germanium metal consumption in polyester production is about 80 ppm per tonne PET (Thiele, 2006). The germanium remains in the product, mainly PET bottles, grading 10–70 ppm per tonne (Oakdene Hollins Ltd, 2011). PET bottles are normally re-used or shredded and sold for other applications. As PET post-consumer scrap, germanium is totally lost. Important producers of catalyst grade GeO_2 include Teck Metals Ltd. and Umicore.

Substitution

Since the beginning of the industrial use of germanium, attempts have been made to substitute this expensive and rare metal. In each of the applications listed in Table 8.5, substituting technologies have been developed. The use of germanium in fibre-optical systems is challenged by the invention of photonic crystal fibres consisting of an array of glass capillaries and solid rods stacked together.

Zirconium–indium-based fluoride glass for fibre-optic applications might also be a successful substitute, but research is at an early stage (Angerer et al., 2009).

Silicon is used as a substitute in certain electronic applications. Competing materials in infrared technologies are zinc selenide (ZnSe) and GASIR, an infrared-transmitting chalcogenide glass developed by Umicore. In the PET market with its low-cost products, replacement of expensive GeO_2 is possible using titanium-, antimony- and aluminium-based catalysts. Substitution for germanium substrates in space solar cells comes from Inverted Metamorphic Cells or advanced double junction cells grown on gallium arsenide (GaAs), and in LEDs by copper and silicon (blue LEDs) and gallium phosphide (GaP) and copper (red LEDs). Silicon–germanium (SiGe) chips may be substituted by gallium arsenide chips that are less expensive, but SiGe may achieve higher operational frequencies than its competitor products (Global Industry Analysts, 2010).

Environmental aspects of the life cycle of germanium and its products

Germanium has little or no effect upon the environment because it usually occurs as a trace element in rocks, ores and most products. Of the materials consumed or produced as a by-product during processing of germanium-bearing materials, arsenic and cadmium may present potential problems. However, these metals are separated out at the smelter stage and are readily controlled in the refineries. Acids and bases used in processing are neutralised and held in tailings ponds (Roskill, 1988). Germanium compounds also have a low order of toxicity, except for germanium tetrahydride, which is considered toxic (Roskill, 1988).

Resources and reserves

Global resource and reserve data for germanium are difficult to obtain, because details relating to trace-metal concentrations in many sulfide and coal deposits are not readily available, or are of poor quality.

The United States Geological Survey lists 450 tonnes germanium reserves for the United States (USGS, 2013). The combined germanium reserves and resources in China are estimated to be about 3782 tonnes (Xun, 2002), distributed in 33 germanium ore districts in 11 provinces or autonomous regions. More than 80 per cent of the germanium reserves and resources are located in Guangdong, Yunnan, Jilin, Sichuan and Shanxi. In the Russian Federation, about 4000 tonnes germanium reserves have been calculated for 21 deposits (Kats et al., 1998), with over 50 per cent of the total reserves accumulated in germanium-rich lignite deposits of the Russian Far East (Primorsky Krai and Sachalin), and 40 per cent in lignite fields in the Kemerovo and Chita regions. The remaining reserves are hosted in sulfide deposits in the Urals, Altai and Caucasus regions.

Following considerable exploration, the estimated reserves of germanium-rich coals in the Russian Far East and the Inner Mongolia Province of China are now estimated as 6000–7000 tonnes and 5600 tonnes, respectively (Seredin and Finkelman, 2008). Höll et al. (2007) give a conservative estimate of "a few thousand tonnes" of recoverable germanium from low-temperature hydrothermal zinc deposits and polymetallic Kipushi-type deposits, including slag at Tsumeb and Kipushi. They also estimate that at least a few thousand tonnes of germanium are present in the coal ash and flue dust produced annually at coal-fired power stations. However, extraction of germanium from these waste products is not generally economic under present conditions. Data on germanium resources in volcano-sedimentary magnetite-hematite deposits are not available.

A compilation of published and estimated reserves for active and semi-active mines and advanced exploration projects by Elsner et al. (2010) amounted to more than 27,000 tonnes germanium reserves and resources, including sulfide ores, coal and slag. With the addition of further projects to that database, e.g. the Kas-Symskaya coal field in Siberia with 11,000 tonnes germanium, the combined reserves and resources are estimated at 13,000 tonnes germanium from sulfide deposits and associated slags, and 25,000 tonnes from germanium-rich coals (Tables 8.6 and 8.7). The amount of germanium potentially recoverable from coal ash is unlimited, but the commercial recovery is currently not viable except for germanium-rich coals from Russia and China. The same holds for the extraction of

Table 8.6 Active and potential producers of by-product germanium from sulfide ores and selected projects under development. Annual germanium production capacity is the installed capacity, except where the value is in parentheses when it is potential capacity.

Deposit	Country	Mining Company[1]	Type	Ge resources (tonnes)	Ge grade (ppm)	Annual Ge production (maximum capacity) in tonnes per year	Status of operation
Huize	China	Yunnan Huize Lead and Zinc Mine[2]	Sulfide ore	500–600	40	10 (30)	production
Jinding (Lanping)	China	Sichuan Hongda Co., state[3]	Sulfide ore	3000	10–100	10	production
Fankou	China	Zhongjin Lingnan Nonfemet, Guangdong state[4]	Sulfide ore	600	100	15	production
Red Dog	AK, USA	Teck Resources[5]	Sulfide ore (SMS)	1200	15	40 (200)	production
Lubumbashi	DR Congo	GTL[6]	Slag	2250	100–250	2 (20)	production
Kipushi	DR Congo	Unclear[7]	Sulfide ore (KPT)	1500	68		care and maint. (1993)
Andrew and Darcy Zn deposits	Yukon, Canada	Overland Resources Ltd.[8]	Sulfide ore (SMS)	88	18		exploration
Tres Marias	Mexico	War Eagle Mining Company[9]	Sulfide and oxide ore (MVT)	150	150	(10)	exploration
Pend Oreille	WA, USA	Teck Resources[10]	Sulfide ore (MVT)	300	10–100	(50)	care and maint. (2009)
Gordonsville/ Elmwood	TN, USA	Nyrstar, Tennessee Valley Resources[11]	Sulfide ore (MVT)	800	20	(35)	care and maint.
Kolwezi	DR Congo	ENRC[12]	Slag		500 ?		development, suspended 09/2009
Tsumeb	Namibia	Emerging Metals Ltd.[13]	Slag	530	260	(10)	exploration
Current production						ca. 80	
Potential future production						ca. 380	

[1]Refineries and smelters not included

[2]Ge resource from Zhang (2003); Ge grade is taken as an average of seven analyses published by Han et al. (2007)

[3]Ore reserves + resources of 163.2 Mt grading 6.9% Zn (Metals Economics Group, 2011); Ge reserve calculated for a (conservative) estimate of 30 ppm in Zn ore concentrate

[4]Probably a major Ge source; sphalerite contains 30–170 ppm Ge (Höll et al., 2007). Ore reserves are 42 Mt at 9.97% Zn, production in 2008 was 117,200 tonnes Zn in concentrate (Metals Economics Group, 2011)

[5]Based on proven and probable reserves of 85 Mt grading 18.2% Zn and 4.6% Pb, and an estimated Ge content of 15 ppm.

[6]Groupement de Terril de Lubumbashi; 2009 production was 4590 tonnes Co, 3000 tonnes Cu, 20,000 tonnes Zn oxide

[7]Ge reserves calculated from 23.4 Mt remaining ore resources down to 1500 m level (Kampunzu et al., 2009).

[8]Mineral resource (May 2009) 8.95 Mt at 7.5% Zn equivalent; 132 ppm Ge in Zn concentrate (58% Zn). Indicated resources 4.1 Mt at 18.5 g/t Ge at 3% Zn cut-off (April 2008). http://www.overlandresources.com/pdfs/InvestorPresentationRoundUp2011Vancouver-28Jan11.pdf

[9]Ge reserve estimated for 1 Mt ore grading 150 ppm Ge in the Tres Marias mine; total reserves on the property may be significantly bigger. Previous production (1949–1992) was 62 tonnes Ge (Saini-Eidukat et al., 2009).

[10]Reserve estimate based on proven and probable reserves of 4.7 Mt grading 7% Zn and 1.2% Pb, and 100 ppm Ge in the Zn concentrate

[11]Reserve estimates for the mines are given as 39 Mt at 3.4% Zn and 20 ppm Ge. Returned residues from Zn production at the Clarksville, Tennessee, smelter were expected to contain up to 45 tonnes per year Ga and 35 tonnes per year Ge (Guberman, 2008).

[12]Eurasian Natural Resources Corporation (ENRC) purchased this deposit from First Quantum in 2012

[13]Purchased from Ongopolo in 2008

Table 8.7 Active and potential producers of by-product germanium from germanium-rich lignite and coal, and selected projects under development.

Deposit	Country	Company	Type	Ge resources (tonnes)	Ge grade in coal (ppm)	Ash content %	Status of operation
Lincang	China		Lignite	1060	850	5-20	production [1]
Wulantuga	China	Tongli Ge Refine Co., Ltd.	Lignite	1700 [2]	240 [2]	21	production [1]
Wumuchang	China		Lignite	4000 (?)	30-50		exploration
Novikovsk (Central and Southern)	Russia	SUEK [3]	Lignite	1665	700		closed; produced 850 tonnes Ge from 1960 to 1980s
Luchgorsky/Bikinsk	Russia	CJSC [4]	Lignite	2600	300	42	exploration
Tigninskiy/ Tarbagataisk	Russia	SUEK [3]	Lignite	340	53		closed
Pavlovsk (Spetzugli + Luzanovsk)	Russia	SUEK [3]	Lignite	1015	450	18-27	production
Smolianinovsky/ Shkotovsk	Russia	SUEK [3]	Lignite	880	1043	21	exploration
Rakovsk	Russia		Lignite	380	230		
Kas-Symskaya	Russia	Krasnoyarskaya mining co.	Lignite	11,000 [5]	205 [5]	17	exploration
Angren	Uzbekistan	JSC Uzbekugol	Hard coal	180	30	2-6	currently no Ge extraction
Church deposit, Sentinel project	ND, USA	Entrée Gold [6]	Lignite	165	40-70		exploration

[1]Resources and average Ge contents from Seredin & Finkelman (2008), unless otherwise stated
[2]Ge production from coal in China is about 30 tonnes per year, with 16 to 25 tonnes from the Lincang deposit (Mikolajzak, 2011) Seredin et al. (2013)
[3]SUEK, Siberian Coal and Energy Company
[4]CJSC, Luchegorsky Fuel Energy Company
[5]Evdokimov et al. (2004)
[6]Exploration project for uranium-molybdenum in lignite-bearing sedimentary strata

germanium from coking coal and iron ores, e.g. at the newly developed Yakovlevskoye deposit in the Belgorod Oblast, Russia (Kats et al., 1998), with a targetted capacity of 4.5 million tonnes iron ore per year (Mining Journal, 2006).

Production

As with data for reserves and resources, data for global germanium mine and refinery production are not readily available or are of poor quality.

Germanium is produced as a by-product from zinc mining. It is also extracted from coal ash in China, Russia, Ukraine and possibly Uzbekistan (see Tables 8.6 and 8.7). Estimated global germanium mine production in 2010 was probably more than double the reported refined germanium production (Mikolajczak, 2011; more than 300 tonnes in residues), which means that major amounts of germanium are not being extracted from residues.

Major zinc mines outside China, which contain significant amounts of germanium, are Red Dog (Alaska, USA), intermittently Pend Oreille (Washington, USA), and zinc–lead mines in Tennessee (USA). Minor production of about two tonnes germanium was from the DRC (period

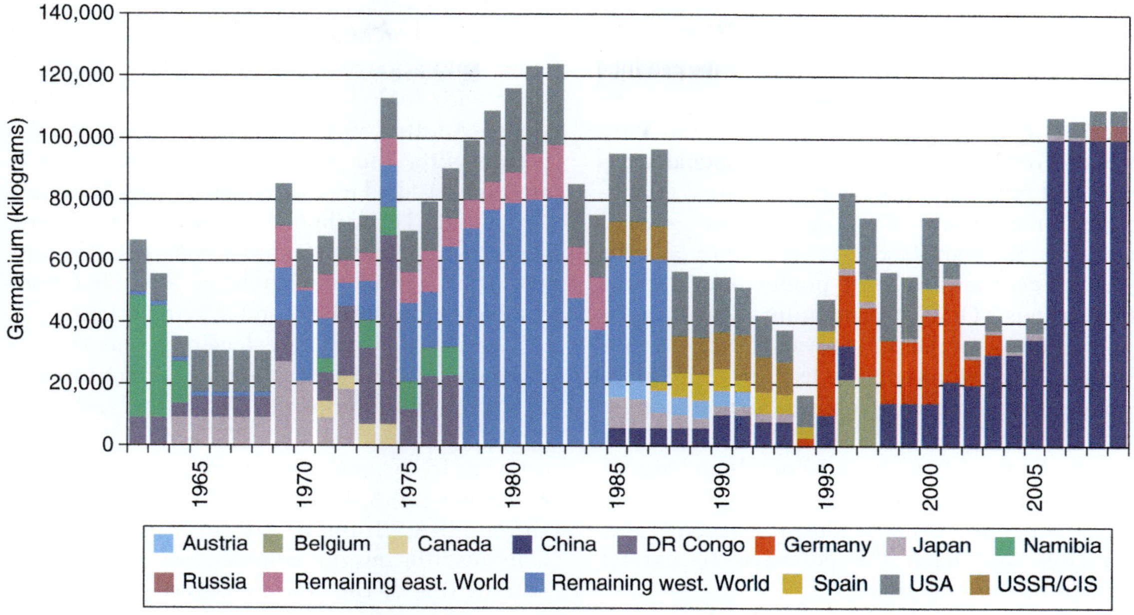

Figure 8.5 World germanium production, by country, 1962–2009. (Data from BGR database.)

2003 to 2009). This production was probably from the remelting of slags from zinc–copper mining in the Lubumbashi area (Yager, 2009).

World refinery production in 2012 is estimated at 128 tonnes germanium, produced in China (70 per cent), USA (ca. 2 per cent) Russia (ca. 4 per cent), and other countries including Canada, Spain, India, Finland and Australia (Guberman, 2011, 2013, and Mikolajczak, 2011). Production levels are highly volatile and the market is quite opaque. The currently installed global extraction and refining capacity is estimated at about 180 tonnes per year (Mikolajczak, 2011).

Before 1978, DRC, Namibia, Japan and USA dominated world refinery production of 60–80 tonnes per year. This was followed by a period of over-production (more than 100 tonnes per year) led by USA and other western producers. China started to enter the market around 1985. Since 2001, China has continuously taken the lead in global refinery production (Figure 8.5).

In China, five smelters/refineries account for the estimated 70 to 100 tonnes of germanium and germanium compounds produced (Guberman, 2009; Mikolajczak, 2011; Tse, 2010): Yunnan Lincang (ca. 25 tonnes germanium from coal ash, Yunnan); Nanjing Germanium Co. Ltd. (ca. 20 tonnes germanium, partly from coal ash, Jiangsu), Yunnan Chihong Zinc-Germanium Co. Ltd (ca. 18 tonnes germanium, Yunnan); Shaoquan Smelter (ca. 7 tonnes germanium, Guangdong); Nei Mongol Xilingol Tongtai Germanium Refine Co. Ltd. (Nei Mongol, Xilinhot); the latter may produce 100 tonnes germanium annually, if in full production (Seredin, 2012). The total Chinese production capacity is thus estimated to be 130 to 200 tonnes germanium. In Russia, only the Spetzugli lignite deposit in the Pavlovsk area is currently in production. According to Seredin et al. (2013), the designed annual capacity of the three high-germanium (240–850 ppm) coal deposits (Spetzugli, Russia; Lincang and Wulantuga, China; Table 8.7) is 150–170 tonnes of germanium.

The major germanium producer outside China is Teck Metals Ltd. with its metallurgical plant in Trail (Canada) with reported production of 40 tonnes germanium in 2007 (including germanium from local and imported zinc concentrates, e.g. from Red Dog, Alaska).

In addition to primary germanium production and refining, Umicore (Belgium) and Recylex (France) with its subsidiary PPM Pure Metals GmbH (Germany) are large global recyclers and refiners of zinc with combined germanium production. They also have special expertise in manufacturing and offer a wide range of germanium-based products for various applications. OJSC Germanium Krasnoyarsk, which operates the Krasnoyarsk Non-Ferrous Metals Plant, is among the leading germanium producers and the only reported major supplier from Russia (Global Industry Analysts, 2010). Xstrata, Vedanta Resources, Minerals and Metals Group – all among the five leading zinc producers outside China – operate several zinc refineries and have the potential to extract germanium as a by-product.

Special germanium-based technologies are produced by various companies. Prominent examples include: Applied Materials Inc. (the world's largest manufacturer of semiconductor production equipment, USA); Soitec SA (the world's leading manufacturer and supplier of Silicon-On-Insulator wafers for high-performance microprocessors, France); AXT Inc. (lighting display applications, wireless and fibre-optic communication, USA); and Germanium Corporation of America (USA).

Future supplies

There are no known high-grade deposits that would be economic for production of germanium as a major element (the only historic exception is the gallium–germanium Apex mine in Utah, USA, which was in production between 1985 and 1987). Future germanium supplies will, therefore, continue to depend upon the availability of germanium-bearing residues from the processing of zinc ores, from germanium-bearing coal ash and from recycling.

Exploration and mine expansions for zinc have been at a high level in recent years, mainly focussing on zinc sulfide ores from large SMS and VMS deposits and on non-sulfide zinc ores. About 100 zinc exploration projects are in the conceptual stage with little information about germanium grades. Another 50 exploration projects are in the prefeasibility and in the feasibility stage which are being studied in greater detail (Raw Materials Group, 2011). Although most of these zinc deposits have low germanium contents, it is possible that germanium could be extracted from residues of the smelting and refining stages.

Carbonate-hosted zinc–lead ores that are germanium enriched are currently of minor interest to exploration companies. With the restart of mining carbonate-hosted (MVT-type) zinc ores carrying 15–20 ppm germanium at Gordonsville, Tennessee, USA, by Strategic Resources Acquisition (SRA, Toronto, Canada) in 2007, the Gordonsville mining complex was expected to become one of the world's largest producers of germanium and gallium (Table 8.6). However, the mines were placed under care and maintenance in 2008 and SRA filed for bankruptcy in 2009. The mining complex produced a low-iron zinc concentrate, which was further smelted at Nyrstar's Clarksville smelter. The leachate contained 0.3–0.8% Ga and 0.2–0.5% Ge that was recovered through a series of process steps that include leaching, solution purification, concentration via precipitation, and sequential recovery. An SRA-owned germanium–gallium recovery plant in Tennessee was planned to reach completion sometime in early 2010. Another project of similar scale is the Tres Marias zinc mine in Mexico which grades 20% Zn and 150 ppm Ge (in some sections up to 245 ppm Ge). Other potential germanium-bearing deposits under exploration are Sierra Mojada (Mexico), Mehdiabad (Iran), and various MVT deposits in Ireland (Tatestown, Ballinalack, Keel; Navan Group).

The germanium-rich polymetallic sulfide ores of the Kipushi-type (KPT) are currently operating at low production levels; residues from smelting, especially slags, constitute important future resources of germanium in such deposits (Lubumbashi, Kolwezi smelters; Table 8.6).

Although there is potential to identify further deposits of this class, especially in the DRC and in Namibia, their remoteness and the political instability in the area are currently deterrents to investment. However, about 300 potential germanium-bearing sites have been reported for Africa alone, excluding coal (Deschamps et al., 2006).

Germanium may be recovered from copper ores in porphyry deposits and in sediment-hosted stratiform copper ores (Kupferschiefer/Germany, Central African Copper Belt of DRC/Zambia). For example, if appropriate production capacities were developed, germanium could be extracted from the residues of electrolytic zinc production.

Germanium-bearing coal and flue dusts constitute the most important germanium resource by volume (Seredin et al., 2013). The resources are huge (e.g. more than 25,000 tonnes germanium in germanium-rich lignite in Russia and China, Table 8.7). In addition, large germanium reserves are related to coking coals in the Kemerovo region, Siberia, that carry 1.6 to 3 g/t germanium (Kats et al., 1998). However, production from this source will depend on the development of germanium prices and technological improvements, especially in western countries. Within coal deposits, germanium is enriched in distinct zones of the coal seams. Selective mining, processing, transport and thermal use of coal from such zones would lead to increased germanium grade in ashes. To install this selective production route would be a major challenge and selective mining of germanium-rich zones would not be practical in most coal deposits. In addition, recycling of germanium will remain a major source of supply in future.

World trade

The major trade for germanium is related to the main zinc production and refining sites, notably within North America and China, and the overseas trade of zinc concentrates. For recycled material, the dominant markets are North America, Europe and Asia-Pacific. Trading of germanium-bearing slags, ashes and residues is constrained by the Chinese government by applying an export tax of ten per cent for these goods. The traded volumes and transport costs per unit are low, thus refined germanium and manufactured germanium compounds are easily shipped around the globe. Detailed trade routes are not available.

Between 2000 and 2006, the volumes of global annual sales did not vary significantly; they ranged between 88 tonnes (2003) germanium to 125 tonnes germanium (2006) (Global Industry Analysts, 2010). The share of sales volumes in these countries did not vary either. These data indicate that the market and trade have been relatively stable in the past.

World trade has also been influenced by sales and purchases of stocks from the Defense Logistics Agency, DLA Strategic Materials (formerly Defense National Stockpile Center, USA). However, the DLA suspended sales of germanium in 2011 (DLA, 2011). No sales of germanium metal were reported during fiscal year 2010 (USGS, 2011).

Prices

Since the 1950s, germanium prices were mainly influenced by its usage in high-tech applications. Before the 1970s, germanium was essential for the production of diodes, rectifiers and transistors, but subsequently substituted by electronic-grade silicon. This trend was offset by the massive demand of fibre optics in the telecommunications sector, in infrared night-vision devices and as a polymerisation catalyst. Due to these trends and tight supply, the first drastic price increase occurred at the beginning of 1979 when germanium prices tripled (Brown, 1998). Between 1982 and 1994, germanium prices stagnated despite massive purchases of germanium by the US National Defence Stockpile Programme. By mid-1995, germanium metal prices doubled, and those for germanium dioxide rose fourfold as a result of the strong increase in demand for fibre optics in IT applications, increased use for PET production

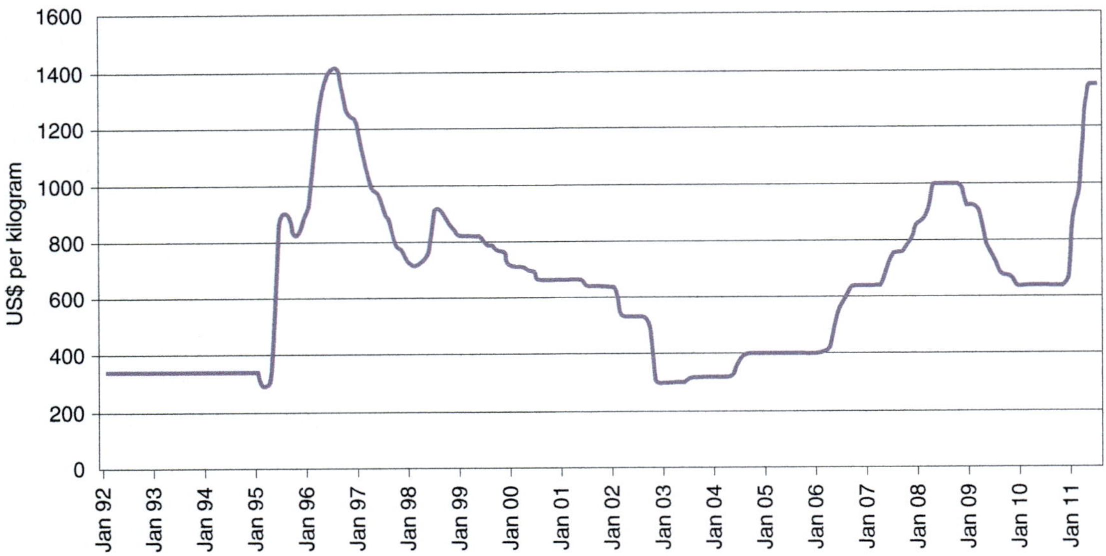

Figure 8.6 Germanium dioxide price (minimum 99.99%, monthly averages), 1992–2011. (Data from BGR database, 2012.)

and anticipated increase in satellite communication (Figure 8.6). Following strong downward price movement up to 2003, prices for germanium dioxide climbed and dropped due to the boom and bust movement of the global economy. The most recent price peak in 2011 is a function of global economic growth combined with a strong demand of germanium dioxide for fibre optics and supply shortages. The supply shortages are caused by the closure of the Chinese Shaoguan dioxide plant due to political clampdown on pollution- and power-intensive industries and strong demand from the Chinese defence sector. Fears of export restrictions by Chinese authorities supported higher prices. In 2012, germanium dioxide prices increased by 49 per cent from US$925 in mid-March to US$1375 per kilogram in late September. This can be explained by the interplay of several factors, namely Chinese export taxes, shutdown of three Chinese germanium dioxide plants and speculation of increased demand for germanium used in polymerisation catalyst following the 2011 earthquake and tsunami in Japan (Guberman, 2013).

The future price of germanium will depend on the Chinese market, breakthrough technologies (see below) and various substitution possibilities already developed.

Outlook

Supply challenges

Major germanium supply will be driven by primary zinc mine production and subsequent refining, and by extraction from coal ashes recovered from burning of germanium-rich coal. Processing costs versus germanium price will determine the production of germanium as a by-product. New demand could probably be met through the expansion of existing capacities and the installation of new processing capacities, e.g. at Red Dog, Pend Oreille, and Gordonsville/Elmwood (all USA). Further capacities could be built up in China, Russia, Mexico, the DRC and Namibia (Table 8.6). The total future production capacity up to 2020, based on new zinc exploration projects and announced production capacities of existing facilities, is estimated to be about 380 tonnes per year, which is three times the size of today's refined germanium production (Table 8.6).

Recycling of old scrap will not be a significant contribution to world supply in the near future. However, post-consumer scrap may increase slightly when solar fibre-optic cables or military devices are being replaced or if germanium-based solar cells are widely adopted and are recycled at the end of their lifetime after 15 to 25 years. Recycling of new scrap is already an important source of germanium and an individual recycling rate of 70–80 per cent was reached for optical-fibre production in the USA (Jorgenson, 2006).

Advances in recovering germanium from coal ash at coal-fired power plants, which is already an important source in China and Russia, could make a difference to future supply. The current annual production capacity of 150–170 tonnes germanium from three producing mines (Seredin et al., 2013) could be considerably enhanced if additional germanium-rich coal deposits enter into production (e.g. Wumuchang in China and Bikinsk in Russia). Commercial recovery of the metal appears not to be economically viable for many plants burning germanium-poor coal (Bleiwas, 2010), but it appears that few samples of ash from power plants and of coal deposits have been studied in detail for germanium. In 2008, War Eagle Mining Co. entered a research project with Sask Power International and Saskatchewan Power Corp. to sample fly ash from their plants. If successful, they have already arranged an offtake agreement. Similarly, War Eagle Mining Co. built a small-scale laboratory pilot plant for fly-ash germanium extraction from the Elcogas S.A. coal gasification plant in Puertollano (Spain) (fly-ash production of 12,000 tonnes per year; Mining Engineering, 2008). A weak acid solution recovered up to 83 per cent of the germanium from fly ash samples over a two-hour leach period. Other potential primary sources include germanium-rich magnetite from magnetite-hematite volcanic-sedimentary deposits in Ukraine, Kazakhstan and Russia (Smirnov, 1977, Kats et al., 1998), but information regarding extraction methods or the germanium content of pig iron slags using iron ores from such regions is not available.

Demand drivers

In fibre optics, germanium is used in Local Area Network (LAN) cables and for more sensitive video and audio transmissions. Programmes such as "Fibres-To-The-Home" (FTTH) in the USA may boost the demand of the fibre-optics technology (Buchert et al., 2009). Zirconium–indium-based fluoride glass is being studied as a substitute in fibre-optic applications and could impact on germanium demand, but the research is at an early stage (Angerer et al., 2009).

Future infrared optic applications, which could boost germanium demand, may include infrared night-vision systems or car-safety systems, such as adaptive cruise-control systems in the automotive industry, or fire rescue and detection systems.

Although germanium is an ideal substitute for antimony, cheaper catalysts have been developed (Thiele, 2006). According to Global Industry Analysts (2010) the use of germanium in PET production is expected to decrease in future because of other low-cost catalysts based on titanium as a substitute for germanium and antimony. However, new technical applications for germanium-based PET production may be of interest because of the high clarity and low crystallisation speed of germanium-based polyesters. This includes applications for TV screens or technical yarn.

A wide range of electrical applications use germanium, e.g. detectors and semiconductors which include transistors and diodes. As an example of a growing market, germanium-based gallium arsenide solar cells may replace silicon solar cells in future applications. If the technology becomes a commercial success in solar-cell production, demand for germanium could increase significantly. Germanium is also used in optoelectronics as a carrier substrate in certain high-end LEDs that are increasingly applied in displays, printers, High Definition Television (HDTV), traffic lights, etc.

To increase the number of transistors on a silicon chip, germanium is also used in small amounts to improve their performance. New generations of superchips may require more germanium for supercomputers based on silicon–germanium. They are also useful for chips used in wireless communication products where they are suitable for integrating mobile phone, e-mail, and internet access functions. Recently developed optical discs for video recording systems by Sony Corp., based on a germanium–antimony–tellurium (GeSbTe) alloy, have a storage capacity of eight gigabytes (Global Industry Analysts, 2010). Forecasts predict wide applications of germanium in further microelectronic applications (Depuydt et al., 2007).

Supply and demand scenario

The future demand for germanium is expected to be driven by the production of fibre-optic cables, which today account for about a third of total germanium demand. Between 2006 and 2030, the demand may increase eight-fold from 28 tonnes to 220 tonnes in 2030 or 8.6 per cent annually (Angerer et al., 2009). This additional demand could be met by the estimated 380 tonnes of new production capacities from zinc refining together with additional contributions from coal ash and from recycling. However, as technological breakthroughs of potential applications, which are currently in the research or development stages, cannot be foreseen, additional end-use sectors could further increase the demand for germanium up to 2030. The most promising applications are infrared night-vision systems, car-safety systems and fire rescue and detection systems. Other growth markets are likely to include germanium-based polyesters for TV screens and germanium-based gallium arsenide solar cells. In general, supply and demand was in balance in the past. Furthermore, the history of germanium production has demonstrated that in the past the existing production capacities were not used completely.

Acknowledgments

We kindly acknowledge the input of Kristof Dessein (Umicore), Vladimir V. Seredin (IGEM, Moscow), Sandro Schmidt, Doris Homberg-Heumann, Elke Westphale, Maren Liedtke and Maria Sitnikova (BGR). We also like to thank Prof. Dr.-Ing. Dr.h.c. mult. F.-W. Wellmer, Kristof Dessein (Umicore), and Teresa Brown and Gus Gunn (BGS) for their reviews that greatly improved the manuscript.

Notes

1. Much information, if not otherwise quoted, is taken from the reviews on germanium by Butterman and Jorgenson (2005) and from Höll et al. (2007) that resulted from an extensive literature study conducted by Kling et al. (2000) for the BGR.
2. Metal purity is often expressed by the 'N' notation, e.g. 3 N stands for 99.9% metal, and 5 N stands for 99.999% metal.

References

Angerer, G., Erdmann, L., Marscheider-Weidemann, F. et al.(2009) Rohstoffe für Zukunftstechnologien. Einfluss des branchenspezifischen Rohstoffbedarfs in rohstoffintensiven Zukunftstechnologien auf die künftige Rohstoffnachfrage. ISI-Schriftenreihe "Innovationspotenziale", Fraunhofer Institut für System- und Innovationsforschung. Fraunhofer IRB-Verlag.

Banerjee, N.N., Ghosh, B. and Das, A. (eds.) (2000) Trace Metals in Indian Coals. Allied Publishers Ltd., New Delhi. pp. 100.

Bernstein, L.R. (1985) Germanium geochemistry and mineralogy. Geochimica et Cosmochimica Acta 49, 2409–2422.

BGR database (2012) BGR database of commodity prices. Federal Institute for Geosciences and Natural Resources, Hannover, Germany.

Bleiwas, D.I. (2010) By-product mineral commodities used for the production of photovoltaic cells. USGS Circular 1365.

British Geological Survey (2012) World Mineral Production 2006–2010 (Keyworth, Nottingham: British Geological Survey.)

Brown, R.D. (1998) Germanium. In: Metal prices in the United States until 1998, United States Geological Survey.

Buchert, M., Schüler, D. and Bleher, D. (2009) Critical metals for future sustainable technologies and their recycling potential. United Nations Environment Programme and Öko-Institut.

Butterman, W.C. and Jorgenson, J.D. (2005) Germanium: Mineral Commodity Profiles. United States Geological Survey, Open-File Report 2004–1218.

Cassard, D., Chabod, J.C., Marcoux, E., Bourgine, B., Castaing, C., Gros, Y., Kosakevich, A., Moisy, M. and Villefond, L. (1996) Mise en place et origine des minéralisations du gisement à Zn, Ge, Ag, (Pb, Cd) de Noailhac-Saint Salvy (Tarn, France). Chronique de la Recherche Minière 514, 3–37.

Chemsystems (2011) Polyester shows rapid recovery. Abstract, Petrochemical Market Dynamics – Polyester & Intermediates, Nexant Limited. http://www.chemsystems.com/about/cs/news/items/PPE%20PCMD%20Aromatics%2020101.cfm

Cook, N.J., Ciobanu, C.L., Pring, A., Skinner, W., Shimizu, M., Danyushevsky, L., Saini-Eidukat, B. and Melcher, F. (2009) Trace and minor elements in sphalerite: A LA-ICPMS study. Geochimica et Cosmochimica Acta 73, 4761–4791.

Dai, S., Wang, X., Seredin, V.V. et al. (2012) Petrology, mineralogy, and geochemistry of the Ge-rich coal from the Wulantuga Ge ore deposit, Inner Mongolia, China: New data and genetic implications. International Journal of Coal Geology 90–91, 72–99.

Dasch, E.J. (ed.) (1996) Encyclopedia of Earth Sciences, Volume I. MacMillan Reference USA, New York.

Depuydt, B., De Jonghe, M., De Baets, W. et al. (2007) Germanium materials. In: Claeys, C. and Simoen, E. (eds.) Germanium Based Technologies, from Materials to Devices. Elsevier. 11–40.

Deschamps, Y., Milési, J.P., Hocquard, C. et al. (2006) Germanium potential in Africa. A predictive approach methodology. CAG21 – Colloquium of African Geology, 21st Maputo, Mozambique, 03-05/07/2006, pp. 283.

Defense Logistics Agency (DLA) (2011) Germanium, BOA sales. Strategic Materials. https://www.dnsc.dla.mil/pgm.asp?Commodity=Germanium

Du, G., Zhuang, X., Querol, X. et al. (2009) Ge distribution in the Wulantuga high-germanium coal deposit in the Shengli coalfield, Inner Mongolia, northeastern China. International Journal of Coal Geology 78, 16–26.

Elsner, H., Melcher, F., Schwarz-Schampera, U. and Buchholz, P. (2010) Elektronikmetalle – zukünftig steigender Bedarf bei unzureichender Versorgungslage? Commodity Top News, 33. http://www.bgr.bund.de/DE/Themen/Min_rohstoffe/Produkte/produkte_node.html#ctn

Evdokimov, A.P., Ozerskii, A.Y. and Ehanin, A.G. (2004) Germanium-bearing lignites of the southeastern rim of the West Siberian Platform. Razvedka i ohrana nedr 6, 26–29.

Feiser, J. (1966) Nebenmetalle. Die Metallischen Rohstoffe, ihre Lagerungsverhältnisse und ihre wirtschaftliche Bedeutung. 17, Band, Ferdinand Enke Verlag Stuttgart, 61–90.

Gerber, G.B. and Leonard, A. (1997) Mutagenicity, carcinogenicity and teratogenicity of germanium compounds. Mutation Research 387, 141–146.

Global Industry Analysts Inc. (2010) Germanium – a global strategic business report. MCP-2637.

Goldschmidt, V.M. (1930) Über das Vorkommen des Germaniums in Steinkohlen und Steinkohlenprodukten. Nachrichten von der Gesellschaft der Wissenschaften zu Göttingen. Mathematisch-Physikalische Klasse, 28, 398–401.

Goldschmidt, V.M. and Peters, C. (1933) Zur Geochemie der Germaniums. Nachrichten von der Gesellschaft der Wissenschaften zu Göttingen. Mathematisch-Physikalische Klasse 31, 141–166.

Gouin, J., Lerouge, C., Deschamps, Y. et al. (2007) Petrological and geochemical characterisation of germanium-bearing coals from the eastern Rhodopes, Bulgaria. Proceedings of the Ninth Biennial SGA Meeting, Dublin, 1521–1524.

Guberman, D.E. (2008) Germanium. USGS 2007 Minerals Yearbook, United States Geological Survey.

Guberman, D.E. (2009) Germanium. USGS 2008 Minerals Yearbook, United States Geological Survey.

Guberman, D.E. (2011) Germanium. USGS 2009 Minerals Yearbook, United States Geological Survey.

Guberman, D.E. (2013) Germanium: Mineral Commodity Summaries 2013, United States Geological Survey, 64–65.

Haller, E.E. (2006) Germanium: from its discovery to SiGe devices. Materials Science in Semiconductor Processing 9, 408–422.

Han, R.S., Liu, C.Q., Huang, Z.L., Chen, J., Ma, D.Y., Lei, L. and Ma, G.S. (2007) Geological features and origin of the Huize carbonate-hosted Zn-Pb-(Ag) District, Yunnan, South China. Ore Geology Reviews 31, 360–383.

Höll, R., Kling, M. and Schroll, E. (2007) Metallogenesis of germanium – A review. Ore Geology Reviews 30, 145–180.

Hu, R.Z., Qi, H-W., Zhou, M-F. et al. (2009) Geological and geochemical constraints on the origin of the giant Lincang coal seam-hosted germanium deposit, Yunnan, South West China: A review. Ore Geology Reviews 36, 221–234.

Jorgenson, J.D. (2006) Germanium recycling in the United States in 2000. USGS Circular 1196–V, Flow studies for recycling metal commodities in the United States.

Kammer, U. (2009) Recycling von seltenen Metallen und deren Verbindungen. In: Thomé-Kozmiensky, K.J. und Goldmann, D.: Recycling und Rohstoffe. Band 2, TK-Verlag, Neuruppin, 647–655.

Kampunzu, A.B., Cailteux, J.L.H., Kamona, A.F., Intiomale, M.M. and Melcher, F. (2009) Sediment-hosted Zn-Pb-Cu deposits in the Central African Copperbelt. Ore Geology Reviews 35, 263–297.

Kats, A.Ya., Kremenetsky, A.A., Podkopaev, O.I. (1998) The germanium mineral resource base of the Russian Federation. Mineral Resources of Russia, Economics and Management 1998 (3), 5–9.

Kelley, K.D., Leach, D.L., Johnson, C.A. et al. (2004) Textural, compositional, and sulfur isotope variations of sulfide minerals in the Red Dog Zn-Pb-Ag deposits, Brooks Range, Alaska: implications for ore formation. Economic Geology 99, 1509–1532.

Ketris, M.P, Yudovich, Ya.E. (2009) Estimations of Clarkes for Carbonaceous biolithes: World averages for trace element contents in black shales and coals. International Journal of Coal Geology 78, 135–148.

Kling, M., Höll, R. and Schroll, E. (2000) Literaturstudie über Germanium in Gesteinen, Mineralien, Erzen und Lagerstätten. Berichte zur Lagerstätten- und Rohstoffforschung, 39. Bundesanstalt für Geowissenschaften und Rohstoffe, Hannover.

Leach, D.L., Sangster, D.F., Kelley, K.D., et al. (2005) Sediment-hosted lead- zinc deposits: A global perspective. Economic Geology 100th Anniversary Volume 561–607.

Levine, R.M. and Wallace, G.J. (2000) The mineral industries of Armenia, Azerbaijan, Belarus, Georgia, Kazakhstan, Kyrgyzstan, Moldova, Russia, Tajikistan, Turkmenistan, Ukraine, and Uzbekistan. 2000 Minerals Yearbook, United States Geological Survey, 1–48.

Leutwein, F. and Rösler, H.J. (1956) Geochemische Untersuchungen an paläozoischen und mesozoischen Kohlen Mittel- und Ostdeutschlands. Freiberger Forschungshefte, C19, 1–196.

Melcher, F. (2003) The Otavi Mountain Land in Namibia: Tsumeb, germanium and Snowball Earth. Mitteilungen der Österreichischen Mineralogischen Gesellschaft, 148, 413–435.

Melcher, F., Oberthür, T. and Rammlmair, D. (2006) Geochemical and mineralogical distribution of germanium in the Khusib Springs Cu-Zn-Pb-Ag sulfide deposit, Otavi Mountain Land, Namibia. Ore Geology Reviews 28, 32–56.

Metals Economics Group (2011) Mine Search Database. Metals Economics Group, Canada.

Mikolajczak, C. (2011) The economic linkage between Zinc and Germanium. International Lead and Zinc Study Group, Lisbon, Portugal, September 29, 2011, presentation, pp. 24.

Mining Engineering (2008) War Eagle to sample fly ash for germanium. Volume. 60 (2), page 35.

Mining Journal (2006) Russia. http://www.mining-journal.com/reports/russia---2006-last-updated-september-2

Naumov, A.V. (2007) World market of germanium and its prospects. Russian Journal of Non-Ferrous Metals 48, 265–272.

PPM Pure Metals (2012) Germanium. http://www.ppm-puremetals.de/products/germanium/germanium.php?inhalt=inhalt_gr.htm

Oakdene Hollins Ltd. (2011) Study into the feasibility of protecting and recovering critical raw materials through infrastructure development in the south east of England. European Pathway to zero waste, Environment Agency, UK.

Queneau, P.B., Avotins, P.V. and Farris, L.F. (1986) Germanium recovery from Fargo Oil´s Lang Bay property. CIM Bulletin 79 (886), 92–97.

Raw Materials Group (2011) Raw materials data. Raw Materials Group, Sweden.

Roskill Information Services Ltd. (1988) The economics of germanium.

Saini-Eidukat, B., Melcher, F. and Lodziak, J. (2009) Zinc-germanium ores of the Tres Marias Mine, Chihuahua, Mexico. Mineralium Deposita 44, 363–370.

Schneider, J., Melcher, F. and Brauns, M. (2007) Concordant ages for the giant Kipushi base metal deposit from direct Rb-Sr and Re-Os dating of sulfides. Mineralium Deposita 42, 791–797.

Seredin, V.V. (2012) Proizvodstvo germaniya iz buryh uglei mestorozhdeniya Ulantuga. Energetik 2012, N3, 19–23 (in Russian)

Seredin, V.V. and Finkelman, R.B. (2008) Metalliferous coals: A review of the main genetic and geochemical types. International Journal of Coal Geology 76, 253–289.

Seredin, V.V., Dai, S., Sun, Y., Chekryzhov, I.Yu. (2013) Coal deposits as promising sources of rare metals for alternative power and energy-efficient technologies.

Applied Geochemistry (2013), 31, 1–11 doi: http://dx.doi.org/10.1016/j.apgeochem.2013.01.009

Smirnov, V. I. (1977) Deposits of Germanium. In: Smirnov, V.I. Ore Deposits of the USSR. Vol III. Pitman Publishing Pty. Ltd. 455–462.

Swaine, D.J. (1990) Trace Elements in Coal. Butterworths, London, pp. 278.

Taylor, S.R. and McLennan, S.M. (1985) The Continental Crust: its Composition and Evolution. Blackwell, Oxford.

Taylor, A. and Quinn, A.J. (1987) Musto Explorations – case study of financing of a small, high-technology mine. Transactions of the Institution of Mining and Metallurgy, Section A, Mining Industry 96, 23–29.

Thiele, U.K. (2006) Polyester Catalysts: A critical analysis of current technology and available alternatives. The European PET Conference, 11–12th October 2006, conference presentation, Barcelona. http://www.polyester-technology.com/Publication/BARCELONA-SCRIPT.pdf

Tse, P-K. (2010) The mineral industry of China. 2009 Minerals Yearbook, United States Geological Survey. 1–25.

Umicore. (2012) Electro-optic materials. http://eom.umicore.com/en/eom/

UNEP. (2011) Recycling rates of metals – a status report. A Report of the Working Group on the Global Metal Flows to the International Resource Panel.

U.S. Geological Survey, USGS. (2011) 2010 Minerals Yearbook: Germanium.

U.S. Geological Survey, USGS. (2013) Mineral Commodity Summaries: Germanium.

Xun, Z. (2002) Mineral Facts of China. Science Press, Beijing, China.

Yager, T.R. (2009) The Mineral Industry of Congo (Kinshasa). 2009 Minerals Yearbook, United States Geological Survey.

Yang, X., Yun, Z., Yuxia, G. and Banghui, G. (2003) A novel biotechnological process for germanium recovery from brown coal. Proceedings International Biohydrometallurgy Symposium IBS 2003, Athens.

Ye, L., Cook, N.J., Ciobanu, C.L. et al. (2011) Trace and minor elements in sphalerite from base metal deposits in South China: A LA-ICP-MS study. Ore Geology Reviews 39, 188–217.

Zhang, M. (2003) The geochemical feature and enrichment mechanism of Yunnan Huize Pb-Zn-germanium-Cd deposits. Masters Thesis. Chengdu University of Technology.

9. Indium

ULRICH SCHWARZ-SCHAMPERA

Federal Institute for Geosciences and Natural Resources (BGR), Stilleweg, Hannover, Germany

Introduction

Indium was discovered in 1863 by two German chemists, Ferdinand Reich and Hieronymous Theodor Richter, who were testing zinc ores from the mines around Freiberg, Saxony. They named it 'indium' from the indigo blue colour seen in its spectrum. In 1924, indium was found to have a valuable ability to stabilise non-ferrous metals, which was the first significant use for the element. However, the early applications of indium were few, the most important being in light-emitting diodes and in coating bearings in high-speed motors such as aircraft engines. Indium-containing semiconductors became important from the 1950s onwards, while the widespread use of indium-containing nuclear control rods increased demand during the 1970s. Since 1992 the major application of indium has been in the form of indium–tin oxide (ITO) in liquid-crystal displays. This use now dominates the market accounting for more than half of total indium consumption.

Physical and chemical properties

Indium is a soft, lustrous, silver-white metal, with a face-centred tetragonal crystalline structure. It is very malleable and ductile and retains these properties at very low temperatures approaching absolute zero, making it ideal for cryogenic and vacuum applications. Indium does not workharden, endures considerable deformation through compression, and it is easily cold-welded. Indium metal is not oxidised by air at ordinary temperatures, but burns to trioxide (In(III)) at high temperatures. On heating, indium reacts directly with metalloids (arsenic, antimony, selenium, tellurium) and with halogens, sulfur and phosphorus. It dissolves in mineral acids and amalgamates with mercury but is not affected by alkalis, boiling water and most organic acids. The chemistry of trivalent indium is characterised by covalent bonding. Indium is frequently used for glass coatings: as indium metal it forms a mirror surface with reflective properties equal to that of silver and with greater corrosion resistance; or in alloys to form transparent and conductive coatings. Selected key characteristics and physicochemical properties of indium are listed in Table 9.1.

Indium is a post-transition metal of Group 13 of the Periodic Table falling between gallium and thallium. The geochemical properties of indium are such that it tends to occur in nature with base-metal Groups 11 (Cu, Ag), 12 (Zn, Cd), 14 (Sn, Pb), and 15 (Bi) of the Periodic Table. Indium has two main oxidation states, +3 (III) and +1 (I). Naturally

Critical Metals Handbook, First Edition. Edited by Gus Gunn.

occurring indium consists of two isotopes, ^{113}In (4.3 per cent of total) and ^{115}In (95.7 per cent of total).

Abundance in the Earth's crust

The Earth's continental crust is estimated to contain about 0.05 ppm In, with 0.072 ppm In in the oceanic crust, about the same as silver. Indium is a highly volatile chalcophile element which behaves in a moderately to highly incompatible manner during mantle melting. The meteoritic abundance of indium is 0.08 ppm for chondrite (CI), while seawater contains 0.2–0.7 ppb In.

Table 9.1 Selected properties of indium.

Property	Value	Units
Symbol	In	
Atomic number	49	
Atomic weight	114.82	
Density at 25 °C	7290	kg/m^3
Hardness (Mohs scale)	1.5	
Melting point	157	°C
Boiling point	2072	°C
Specific heat capacity at 25 °C	0.23	J/(g °C)
Electrical resistivity at 25 °C	0.08	μΩ m
Thermal conductivity	82	W/(m °C)

Mineralogy

Indium minerals are rare in natural systems. Twelve indium mineral phases have been defined (Table 9.2). Roquesite is the most important indium mineral representing a trace component in the principal ore-forming minerals like bornite, chalcopyrite and sphalerite (Figure 9.1). More often, indium substitutes for elements with similar ionic radii, especially those having tetrahedral coordination with respect to the principal metal ion, in base-metal sulfides. High indium concentrations usually occur in sphalerite, chalcopyrite, stannite, tin-sulfosalts, tennantite and cassiterite. The incorporation of indium and the formation of indium-bearing sulfides may occur by (i) diadochic replacement and/or coupled substitution of iron, copper, tin and arsenic forming solid-solution series, (ii) the incorporation into the lattice of tetrahedral coordinated sulfides, and (iii) the formation of sub-microscopic inclusions of indium minerals (e.g. roquesite, indite) in sulfides or cassiterite.

Sphalerite is the most important indium-bearing mineral and the source of most indium currently mined. High indium concentrations in sphalerite are closely associated with elevated copper contents due to the formation of a dis-

Table 9.2 Compilation of defined indium minerals.

Mineral	Formula	Indium content (%)	Specific gravity	Appearance (colour, lustre)
Roquesite	$CuInS_2$	47.35	4.80	bluish-grey, metallic
Laforetite	$AgInS_2$	40.03	4.92	brown, metallic
Indite	$FeIn_2S_4$	55.50	4.59	iron-black, metallic
Sakuraiite	$(Cu,Zn,Fe,Ag)_3(In,Sn)S_4$	24.35	4.34	greenish-grey, metallic
Petrukite	$(Cu,Fe,Zn)_3(Sn,In)S_4$	6.05	4.07	grey, brown, metallic
Abramovite	$Pb_2SnInBiS_7$	11.41	9.0	silver-grey, metallic
Cadmoindite	$CdIn_2S_4$	49.58	4.85	black, dark brown, adamantine
Dzhalindite	$In(OH)_3$	69.23	4.37	yellow brown
Yanomamite	$InAsO_4 \cdot 2H_2O$	39.62	3.92	blue green, light yellow green, vitreous
Yixunite	Pt_3In	16.40	18.33	white, metallic
Damiaoite	$PtIn_2$	54.07	10.94	white, metallic
Native indium	In	100	7.29	yellow grey, metallic

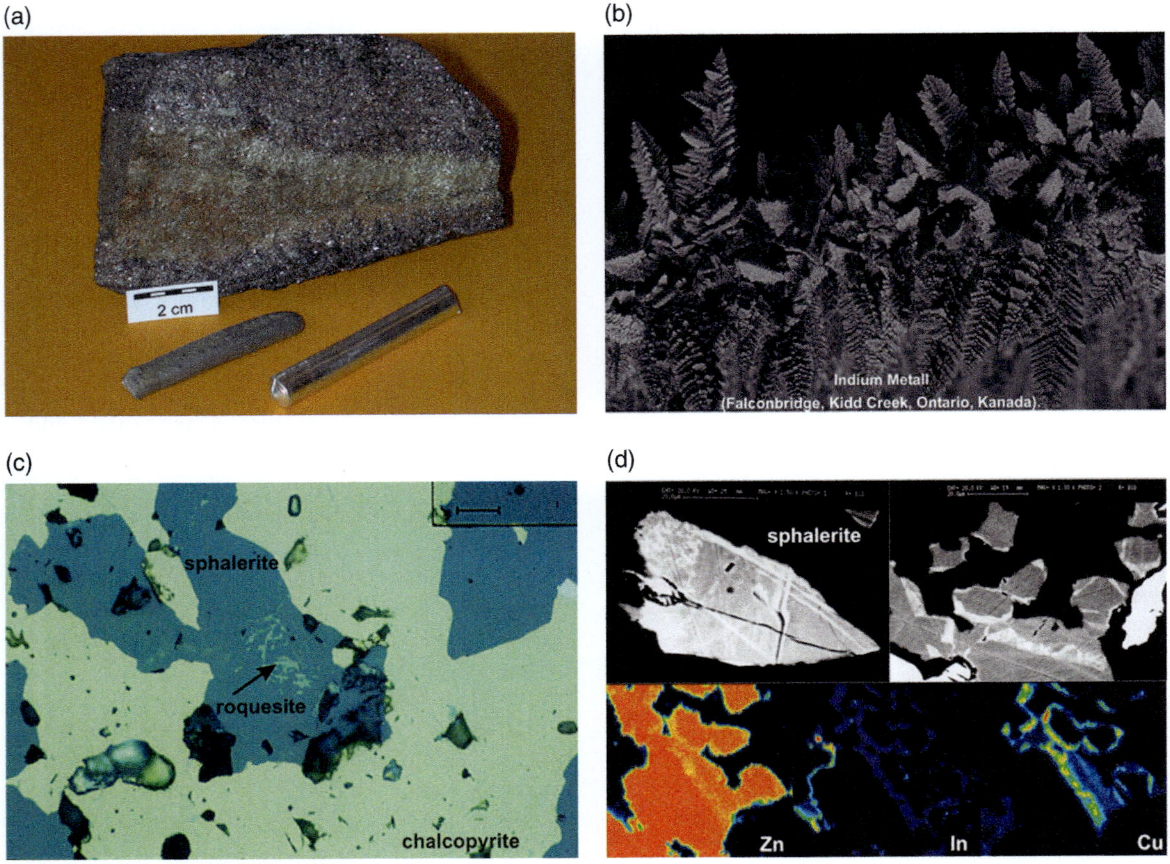

Figure 9.1 (a) Example of indium-rich ore (600 ppm In), Maranda J volcanic-hosted massive sulfide (VHMS) deposit, South Africa and indium rods (99.95% In); (b) Native indium metal from the former Kidd Creek processing plant, Canada (courtesy of Falconbridge Ltd); (c) Chalcopyrite-sphalerite ore with roquesite inclusions in sphalerite; Kidd Creek, Canada; (d) Sphalerite-roquesite solid solution series as light zonal enrichments in sphalerite grains from the Vai Lili hydrothermal vent field, southern Lau basin, south-west Pacific.

continuous binary solid solution of $CuInS_2$ (roquesite) forming solid-solution series members of $[Z_{n2-2x}Cu_xIn_xS_2]$ composition (Figure 9.1). Co-precipitation of indium with copper, zinc and iron in chalcopyrite accounts for the elevated concentrations in copper ores. The highest indium concentrations commonly occur in the chalcopyrite-rich high-temperature cores or at the base of the individual deposits. The mineralogy of indium-bearing deposits is characterised by complex intergrowths and replacement textures containing significant concentrations of metals such as tin, copper, zinc, lead, silver, bismuth, selenium and arsenic.

Major deposit classes

Indium occurs in different types of ore deposits of all ages, from the presently forming deposits at modern, actively spreading ridges and fumarole precipitates of active volcanoes to deposits in the Archaean volcanic strata of greenstone belts, e.g. in Canada and South Africa (Table 9.3). In decreasing order of

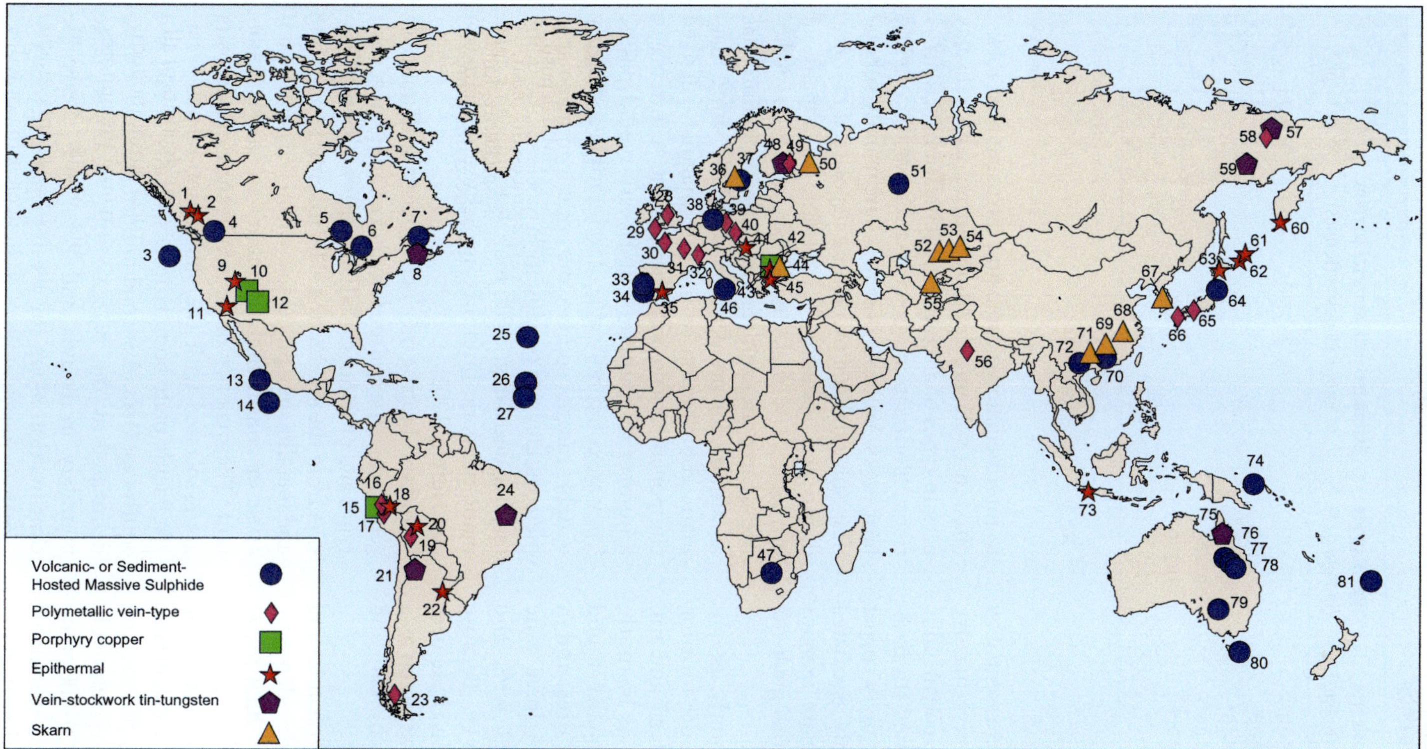

1 Equity Silver	16 Carahuacra	31 Vaulry	46 Palinuro	61 Kudryavy	76 Waterloo
2 Silver Queen	17 Sayapullo	32 Charrier (2 locations)	47 Murchison (2 locations)	62 Prasolov	77 Dry River South
3 Axial seamount	18 Morococha	33 Lagoa Salgada	48 Getmossmalmen (4 locations)	63 Toyoha (8 locations)	78 Balcooma
4 Sullivan	19 Bolivia (9 locations)	34 Neves-Corvo	49 Jungfrubergen	64 Kuroko	79 Broken Hill
5 Geco	20 Malku Khota	35 Cartagena	50 Kitel	65 Akenobe (6 locations)	80 Rosebery
6 Kidd Creek	21 Pirquitas	36 Gåsborn, Bergslagen	51 Southern Urals (6 locations)	66 Kyushu (5 locations)	81 Lau Basin
7 Brunswick Camp	22 San Roque, Rio Negro	37 Långban, Bergslagen	52 Kansay	67 Ulsan	
8 Mount Pleasant	23 Pinguino, Santa Cruz	38 Rammelsberg	53 Kyzyl Espe (4 locations)	68 Qibaoshan	
9 Crypto	24 Mangabeira	39 Freiberg (2 locations)	54 Sarykan (3 locations)	69 Dachang	
10 Bingham	25 Broken Spur	40 Bohemian Massif (7 locations)	55 Karamazar	70 Dachang	
11 McLaughlin	26 TAG, Mid Atlantic Ridge	41 Nagyborzsony	56 Tosham	71 Gejiu	
12 Central District	27 Snake pit	42 Elacite/Asarel, Sredna Gora	57 Deputaskoe	72 Laochang	
13 East Pacific Rise 21°N	28 West Shropshire	43 Sredna Gora (5 locations)	58 Lifudsin	73 Merapi	
14 East Pacific rise 13°N	29 Cornwall (2 locations)	44 Malko-Tarnovo (6 locations)	59 Pravourmiiskoe	74 Manus Basin	
15 Cerro de Pasco	30 Telhaie	45 Kirki	60 Mutnovsky	75 Baal Gammon	

Figure 9.2 The global distribution of indium mines, deposits and major occurrences.

Table 9.3 Size and grade of the major types of indium deposits (grades and tonnages are very variable between deposits and figures given are indicative only).

Deposit type	Deposit size range (tonnes)	Typical grade (indium g/t)	Estimated indium metal content of known deposits (tonnes)	% of total
Volcanic-hosted massive sulphide	10^6–10^8	20–200	13,750	43
Sediment-hosted massive sulphide	10^6–10^8	20–200	11,750	37
Epithermal	10^4–10^6	10–800	4500	14
Tin-tungsten porphyry	10^5–10^7	10–400	1250	4
Vein-type	10^4–10^6	10–350	520	2
Total			31,770	100

significance indium-rich deposits are represented by volcanic- and sediment-hosted exhalative massive sulfide deposits, epithermal deposits, polymetallic base metal vein deposits, granite-related tin–base-metal deposits, skarn deposits and porphyry copper deposits. These deposits are commonly associated with active oceanic or continental-plate margins and orogenic belts with steep geothermal gradients due to enhanced magmatic activity.

Indium deposits are most commonly associated with base-metal hydrothermal ore-forming systems enriched in zinc, copper, lead and tin, accompanied by trace metals such as bismuth, cadmium and silver. Indium is usually found primarily in zinc and copper sulfide ores, and in tin ores. Typical indium-rich ores have zinc concentrations between 10 and 22 wt% and copper concentrations at or above 2 wt%. Because of its relatively low concentration in these ores indium can only be economically extracted as a by-product under appropriate technical and processing conditions.

The most important deposits are volcanic- and sediment-hosted base-metal sulfide deposits, which are generally characterised by high metal abundance and large tonnages (Table 9.3). The concentration of indium in these ores is in the range 20–200 ppm. Even in a zinc concentrate, representing the most common commercial source, the indium concentration is relatively low at 70–200 ppm, but may reach 500–800 ppm.

Indium-bearing deposits are closely associated with calc-alkaline to peralkaline, porphyritic felsic to intermediate volcanic and intrusive host rocks, often in a volcanic cauldron setting, which display certain types of alteration (silicification, sericitisation, chloritisation, greisenisation) of the original host rocks. Volcanic host rocks are usually submarine, subaerial and pyroclastic deposits of rhyolite, dacite and andesite. Subvolcanic to hypabyssal intrusive host rocks comprise granite–granodiorite–quartz–monzonite differentiates. Indium has an affinity for ore deposits that have some magmatic contribution or indicate magmatic-derived components, as demonstrated by the occurrence of a number of epithermal and skarn-type deposits. They are commonly related to multiple and highly variable magma sources and combined magmatic-meteoric element precipitation processes including magma degassing, wall rock leaching and magmatic-meteoric fluid mixing (Schwarz-Schampera and Herzig, 2002).

Indium-bearing ores occur over a wide range of geological time and are found in a variety of deposit types and provinces. Collision tectonics, mobilisation from the subducting slab, magmatic recycling and metallogenic processes within the continental crust or continental fragments are most important. The majority of indium-bearing ore deposits are associated with the subduction-related western Pacific plate boundaries, especially in east and south-east Asia. A second indium province can be delineated at the Nazca–South American plate boundary in Bolivia and Peru with a possible continuation along the western plate margins of North America. Other provinces are related to different metallogenic epochs in central Europe covering the Hercynian and Alpine belts. Indium-rich deposits also occur in the Mesozoic strata of Patagonia, Argentina, in the Caledonian-Appalachian belt (New Brunswick, Canada) and in Archaean greenstone belts (Canada, South Africa).

Base-metal sulfide deposits

Base-metal sulfide deposits originate at oceanic spreading and rift zones, both at mid-ocean spreading axes and in back-arc rift zones. Volcanic-hosted massive sulfide (VHMS) deposits occur in submarine volcanic rocks of all ages, from the presently forming deposits in modern, actively spreading ridges to deposits in the pre-3400 Ma volcanic strata of the Pilbara Block in Australia (Galley et al., 2007 and Schwarz-Schampera et al., 2010). They are major sources of base and precious metals, and are the most important indium producers in the world. Polymetallic examples of the zinc–lead–copper group are found in Canada at Kidd Creek, Ontario and in deposits of the Iberian Pyrite Belt in Spain and Portugal. Deposits of the zinc–lead–copper group are characteristically dominated by felsic host rocks, with mafic volcanic rocks rare or absent. A distinct metallogenic zonation of the orebodies is characteristic of the majority of the ore deposits. An iron-rich sulfate- and quartz-rich exhalite top is typically underlain by a massive zinc–iron–lead–tin (sphalerite–pyrite–galena–cassiterite) orebody. A copper-rich keel and vein-type mineralization (chalcopyrite) underneath represent the high-temperature feeder zones of fluid upflow (referred to as the 'stockwork zone'). The deposits contain significant and recoverable amounts of trace elements like silver, tin, bismuth, cobalt and indium; bismuth, cobalt, and indium typically indicate elevated formation temperatures. Modern examples of this type have been discovered in the south-west Pacific (Lau basin, Manus basin).

Stratiform sediment-hosted base metal sulfide (SHMS) deposits are concordant, massive to semi-massive accumulations of sulfide and sulfate minerals that formed by exhalative processes on or immediately below the seafloor (Goodfellow and Lydon, 2007). The ore is syngenetic with the host siliciclastic sediments, with minor intercalations of carbonate layers. SHMS deposits occur in terranes dominated by sedimentary strata. Volcanic rocks (lava, tuff) may be a minor component of the associated strata, and synsedimentary intrusions (mafic sills, dikes) may be present regionally. Zinc, lead and silver are the primary metals recovered from these deposits, but tin and indium may be important by-products.

Active seafloor hydrothermal systems, which are modern analogues of ancient VHMS and SHMS deposits, may have elevated indium concentrations (Schwarz-Schampera and Herzig, 2002). The highest known values in active hydrothermal systems were detected in copper-rich sulfides associated with back-arc spreading centres propagating into island-arc crust (Lau Basin, Eastern Manus Basin). Back-arc spreading centres are characterised by bimodal volcanism and the different vent fields are closely associated with highly evolved magmas, derived from partial melting of an older oceanic crust. Assimilation of older crust may have been an important factor in developing highly fractionated arc-like magmas primarily enriched in incompatible elements such as indium.

Polymetallic vein-type deposits

Indium was first discovered in hydrothermal base metal vein deposits in the Erzgebirge region of south-east Germany (Schwarz-Schampera and Herzig, 2002; Seifert and Sandmann, 2006). Polymetallic vein-type deposits are structurally controlled and occur in faults, fault systems and vein breccia zones of clastic metasedimentary or magmatic-dominated terranes. The hydrothermal fluids were derived from basinal brines and may contain contributions of magmatic volatiles in magmatically active domains in syn- to post-collisional orogens. The deposits include lead-zinc-silver-, copper-zinc-lead-silver-tin-, and tin-tungsten-base-metal-dominated mineralisation. In the case of magmatic host rock lithologies, intermediate to felsic sub-volcanic intrusions clearly predominate. Polymetallic vein deposits may contain considerable and often recoverable indium concentrations and were among the most important indium producers worldwide until the 1980s, for example in Japan and Bolivia.

Base-metal-rich tin–tungsten and skarn deposits

Vein-stockwork deposits of tin and tungsten occur in a wide variety of structural styles that include individual veins, multiple vein systems, vein and fracture stockworks, skarns, breccias and replacement zones in altered wall rocks adjacent to veins. The deposits generally occur in or near granitic intrusions which have been emplaced at relatively shallow levels in the Earth's crust. The associated intrusions are highly fractionated and typically enriched in lithophile as well as highly volatile elements. Porphyry tin deposits usually formed in areas of great continental thickness related to collisional tectonic settings, although the deposits generally post-date the collision event. Vein stockwork tin–tungsten as well as porphyry tin deposits are important hosts for indium mineralisation (Sinclair et al., 2006).

Base-metal-rich epithermal deposits

Epithermal deposits originate at active subduction zones and are closely associated with volcanic processes above metalliferous intrusive rocks. The development of epithermal deposits is typically linked to the shallow crust emplacement of magmatic porphyry copper systems. Two principal subtypes of epithermal deposits are distinguished on the basis of alteration and ore mineral assemblages: the quartz-(kaolinite)-alunite or high-sulfidation sub-type and the adularia-sericite or low-sulfidation sub-type (Heald et al., 1987 and Hedenquist, 1987). Differing acidity and oxidation state largely determine the end-member hydrothermal environments of these two deposit types. Temperatures of formation for epithermal deposits range from about 100 °C for hot-spring or steam-heated deposits to about 350–400 °C for deeper vein and replacement deposits. Pronounced changes in the physical, thermal and chemical properties of the hydrothermal solutions may occur over short distances in these systems and promote ore deposition. Gold and silver are the principal commodities of epithermal deposits. Copper and other base metals (zinc, lead, tin) and rare metals (bismuth, indium) are important by-products, especially in deposits with high silver grades.

Indium deposition also occurs in active magmatic systems. Indium mineralisation is reported from fumaroles of high temperature fluids (500–940 °C) forming sulfide segregations in altered wall rocks (Kovalenker et al., 1993). Relative to the source magma, fumarole precipitates and gases are enriched by factors greater than 10^4 in indium and other trace metals.

The key characteristics and examples of the major types of indium deposits are given in Table 9.4.

Extraction methods and processing

Mining

The mining methods employed to extract indium are related to the recovery of other base metals in massive sulfide deposits. The exclusive extraction of indium is uneconomic and consequently primary indium is obtained solely as a by-product in the smelting of polymetallic ores containing zinc, copper and tin. The strict co-enrichment with zinc and copper makes copper-bearing zinc ores the most favourable ores for the indium recovery. The appropriate mining method typically depends on the physical and chemical properties of the minerals, tonnage and grade, the geometry and the depth of the orebody. The most common extraction technique for massive base-metal sulfide deposits is usually a combination of surface and underground mining, although all economically significant indium-bearing ores are mined by underground methods.

The mining of zinc-rich massive sulfide deposits usually starts as open pit mining, if the deposit is near surface. Overburden, altered volcanic or sedimentary wall-rock, and waste rock may be deposited and stored for later backfill. Massive sulfide deposits are usually a few hundred metres in diameter and comprise clearly defined orebodies. Open pit mining may reach considerable depths but rarely exceeds a few hundred metres before going underground.

If surface mining results in high production costs and geological conditions are complex, then the deeper parts of a deposit may be mined underground.

Table 9.4 Key characteristics and examples of the major types of indium deposits.

Deposit type	Brief description	Features	Examples
Volcanic-hosted massive sulfide	Concordant lenses of pyrite-sphalerite-chalcopyrite-galena in submarine volcanic sequences. Syngenetic volcanogenic exhalative origin.	Polymetallic examples of the Zn-Pb-Cu group dominated by felsic host rocks; distinct metallogenetic zonation; significant amounts of trace elements like Ag, Sn, Bi, Co, In.	Neves-Corvo (Portugal), Kidd Creek, Gai (Russia), Deerni, Hongtoushan (China).
Sediment-hosted massive sulfide	Concordant stratiform lenses of pyrite-sphalerite-galena-(chalcopyrite) in submarine sedimentary siliciclastic sequences. Syngenetic exhalative origin.	Massive polymetallic accumulations of sulfide and sulfate minerals; replacement processes immediately below the seafoor; minor intercalations of carbonate layers or volcanic rocks; synsedimentary intrusions may be present; In may be important by-product.	Red Dog (USA), Broken Hill, HYC (Australia), Dachang (Pb-Zn; China), Brunswick #12 (Canada).
Epithermal	Medium to large, low-grade stockwork-type quartz-sulfide veinlets and disseminations in felsic volcanic sequences in a subduction zone and island arc setting. Syngenetic exhalative origin.	Poylmetallic examples of the adularia-sericite or low-sulfidation subtype; wallrock equilibration of magmatic fluids; In is associated with Au-poor but Ag-rich Zn-Pb-Cu ore.	Toyoha (Japan), Equity Silver, Hemlo (Canada), Kirki (Greece), Lepanto (Phillippines), Cerro Vanguardia (Chile), McLaughlin (USA).
Porphyry	Medium to large, low-grade vein-stockwork deposits of Sn-W and Zn-Cu-Pb in subvolcanic felsic intrusive rocks and replacement zones in altered wall rocks adjacent to veins. Post-collisional tectonic settings.	Base metal-rich ores in concentrically zoned altered and mineralized granites; highly host rocks typically enriched in lithophile and highly volatile elements; In is associated with the polymetallic Zn-Cu-Pb-Sn zone.	Mount Pleasant (Canada), Altenberg (Germany), Dachang (Sn-W; China), Qibaoshan (China).
Polymetallic Vein	Small to medium polymetallic deposits; structurally controlled in post-collisional vein breccia zones of clastic metasedimentary or magmatic-dominated terranes; mineralisation by basinal brines.	Polymetallic Pb-Zn-Ag-, Cu-Zn-Pb-Ag-Sn-, Sn-W-base metal ores; densely intergrown ore minerals (telescoping); In hosted by sphalerite, partly replaced by chalcopyrite; quartz gangue more prospective than carbonate gangue.	Freiberg (Germany), Colquechaca, Pulacayo, Bolivar (Bolivia), Akenobe (Japan), West Shropshire (Great Britain).
Hot spring exhalative	Siliceous precipitates deposited by hydrothermal fluids, hot springs, fumaroles; volcanic activity.	High temperature fluids forming sulfide segregations in altered intermediate to felsic wall rocks; In co-enrichment with Zn and Cu (Cd, Te, Se).	Hydrothermal vent fields: Vai Lili, Whitechurch (Lau basin), Pacmanus (E. Manus basin); magmatic fumaroles at Kudryavyi, Kuril islands (Russia), Merapi (Indonesia).

Zn, zinc; Pb, lead; Cu, copper; Ag, silver; Sn, tin; Bi, bismuth; Co, cobalt; In, indium; W, tungsten; Te, tellurium; Se, selenium; Cd, cadmium.

Copper-rich massive sulfide deposits, which are the major hosts for indium, and copper-rich orebodies often occur in the deeper sections of an individual deposit. Block caving and stoping are underground mass mining methods appropriate to large-scale production from massive orebodies with large vertical and horizontal dimensions. Stoping is applicable when the deposit lies within stable hanging and footwall rocks, includes competent ores and host rocks, and has regular ore boundaries. Stoping techniques dominate the mining of most underground massive sulfide deposits.

Processing, beneficiation and conversion to metal

Indium is produced almost exclusively as a by-product of zinc mining and processing. It is extracted from the residues of concentrating and smelting zinc ores and from the recycling of dusts and gases produced during the smelting of zinc (Alfantazia and Moskalyk, 2003). A small proportion, less than 5 per cent of the total, is extracted from the residues of copper and tin treatment. Base-metal sulfides are generally separated by flotation techniques with indium typically enriched in zinc concentrates, comprising mainly sphalerite, with elevated copper concentrations at or above 2 wt%.

Techniques for indium recovery are capital intensive. In the past, when indium consumption was small, demand was satisfied by processing very large amounts of low-grade indium-bearing zinc ore. Low commodity prices and problems with separation from elements like copper, lead, tin, antimony and arsenic made the recovery of indium and other trace metals costly, inefficient and generally uneconomic. However, progress in the treatment of process slimes since the end of the 1990s has improved the beneficiation and reduced costs.

Impure zinc and copper concentrates, containing more than 80–90 per cent sphalerite and chalcopyrite respectively, are normally produced at or near the mine site using flotation techniques. The zinc concentrates are roasted, and the sulfur dioxide produced is used in the manufacture of sulfuric acid. The zinc calcine derived from the roasting is leached to provide a 'pregnant' zinc-rich solution from which cadmium is removed. The purified zinc-rich solution is sent to the electrolysis plant where zinc is plated onto aluminium cathodes. The jarosite process (formation of iron sulfate) for the zinc refining and iron fixation scavenges up to 60 per cent of the indium and germanium from solution. Remains from the jarosite process may contain up to 600 ppm In. As a consequence of the electrochemical series, the insoluble components (impurities), such as indium, settle as solids at the bottom of the electrolysis cells forming anode slime. In some ores this metal-rich slime may be an important potential source of indium and a number of other trace metals such as silver, cadmium and selenium. Where the appropriate technology is installed the pure metals are isolated from the anode slimes by further processing involving a combination of dissolution and precipitation methods.

Indium can be dissolved from anode slime or the processed intermediates from silver and cadmium recovery with hydrochloric acid or sulfuric acid. Anode slime residues may also be processed in lead smelters. The 'imperial smelting process' (ISP) enriches about 50 per cent of the indium with the lead fraction (bullion). The remaining indium stays with the waste blast-furnace slag and may be recovered by roasting in a rotary furnace and selective vaporisation (slag fuming). Indium is separated from the lead fraction during slagging and by electrothermal processes which also produce lead, antimony and tin. Electrolytic operations are employed to produce an indium anode and a slime containing 20–25% In. The slime is chemically treated to provide a crude indium metal of about 99% purity. Solvent liquid extraction leads to the electrolytic refinery of indium metal (Demarthe et al., 1990).

The Kidd Creek zinc smelter at Timmins, Ontario, represented a significant indium producer and developed a successful process for the recovery of indium from zinc ores of the nearby Kidd Creek VHMS deposit (Figure 9.3). The average indium concentration in the concentrate was approximately 270 ppm (Jorgensen and George, 2004). The indium flowsheet starts with the roasting of the

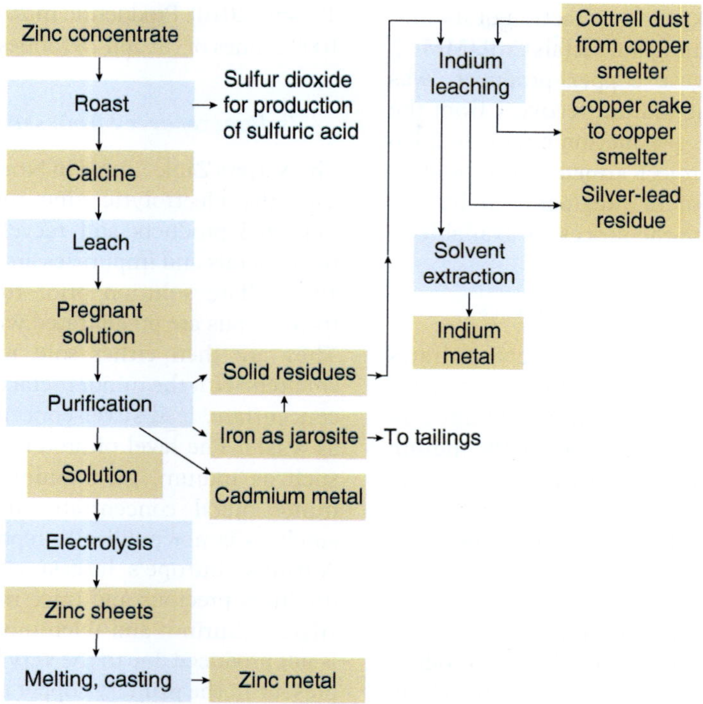

Figure 9.3 Generalised beneficiation flow diagram, based on the Kidd Creek operations. (Modified after http://www.mining-technology.com/projects/kidd_creek/kidd_creek7.html.)

sulfidic zinc concentrate which produces a zinc calcine. The oxidation process is followed by a two-stage dissolution process. The first stage selectively removes iron as jarosite from solution. The second stage introduces metal-rich Cottrell ash from the copper refinery to the solution and silver and lead are enriched in a residue. The indium-bearing solution is then treated by solvent extraction and complex precipitation–dissolution processes result in the recovery of pure indium.

At the Umicore facilities in Hoboken, Belgium, the 'Harris Process' is used to recover indium during the refining of lead. The process involves slag composition manipulation by selective oxidation using alkali-metal hydroxides and nitrites, and removal of impure compounds and elements found in lead bullion during pyrometallurgical processing. Indium was formerly produced at the La Oroya smelter in Peru from concentrates derived from Centromin SA, Peru's leading producer of silver, lead and zinc (Alfantzi and Moskalyk, 2003). The process used coke reduction of the lead bullion and the addition of lead and zinc chlorides to recover indium chloride ($InCl_3$) in the slag. The solution was purified by removal of tin and lead, and indium was recovered as a sponge by adding powdered zinc.

Indium is generally concentrated and recovered at a purity of greater than 99%. This low-grade indium standard is further refined to 4N (99.99%) and higher purity levels. Zone smelting and electrolytic refining are used for obtaining ultrapure indium and for the production of single crystals.

Indium production from copper ores

The production of indium from copper concentrates is poorly developed despite the fact that there are relatively high indium concentrations in the copper ores from many VHMS deposits. This is due chiefly to the very fine grain size of these ores which makes the extraction of indium uneconomic. Urals Mining and Metals Co (UMMC), Russia's second largest copper producer, was reported to be starting indium recovery from the residues of the copper production circuit in 2006 (USGS, 2010). The project aimed at the production of five tonnes of indium per year but no further information on the process is available.

Indium production from tin ores

Indium is recovered from some tin concentrates which contain about 100 ppm In. Bolivian tin ores are reported to contain elevated indium concentrations, although in 2010 none of the indium was being recovered domestically (Anderson, 2012). The indium-bearing concentrates were exported and processed in other countries, most likely in Japan and China.

At Capper Pass in the United Kingdom, which closed in 1991, the process of tin recovery involved the chlorination of electrorefined tin to produce a tin chloride slag containing less than 3% In (Alfantazi and Moskalyk, 2003). The majority of the tin was precipitated in two neutralisation stages in series, including recycling from the second stage. The recovery of indium involved cementation with zinc to produce a sponge material. The 95% purity sponge was then electrorefined to produce a 99.5% purity indium cathode that, in turn, was recast as the anode to yield 99.97% purity indium metal in a second stage of electrolysis.

An alternative vacuum refining technique at Novosibirsk Tin (NOKN), Russia`s single tin producer, produces a lead–tin alloy with indium concentrations up to 2000 ppm In. The indium-bearing alloy is melted in a 20-tonne vessel and the indium is recovered selectively from solution prior to further refining. This process is also applicable to other refinery residues with indium concentrations of at least 1000 ppm and may also be appropriate for other by-products such as lead, bismuth and gallium. The Novosibirsk Tin Plant is one of the largest producers of pure tin in Europe. The plant is treating tin concentrates from the Perevalnoye, Molodezhnoye und Pravourminskoye (Khabarovsk), Khinganskoye (Autonomous Jewish Region), Deputatskolovo (Yakutien) und Syryzhdatskoye (Kyrgyzia) deposits (Roskill, 2010). Production in 2011 was estimated at 1000 tonnes of tin and 25 tonnes of indium.

Indium recovery from secondary sources

The Xstrata Zink GmbH in Nordenham, Germany, uses the electrolytic zinc refinery process for imported products and recycled materials. The trace metals and impurities are separated from the zinc sulfate solution prior to electrolysis. Any trace metals are precipitated with the iron sulfate. They are then either sold to copper and lead smelters with the minor metal concentrate (copper concentrate, lead/silver concentrate) or discarded as waste. The level of production of trace metals such as indium, germanium and gallium from minor metal concentrates in copper and lead smelters is not publically reported. The German Aurubis, Europe's largest copper smelter, co-produces precious and trace metals such as gold, silver, tellurium and selenium. However, indium is not produced due to the very low concentrations present in the primary copper concentrates.

Specifications and uses

Indium is traded as indium metal in a wide variety of forms including pellets, powder, plates, sheets, wire and foils. The standard quality is 99.99% (termed '4 N'), but higher purities up to 5 N and 7 N (99.99999%) are also standard specifications.

Indium is valuable in a diverse range of products and applications because of its unique properties:

- low melting point;
- workability at very low temperatures down to absolute zero (−273 °C);
- ability to wet glass;
- suppleness, softer than lead;
- tight adhesion to other metals;
- dissolution in acids;
- amphoteric character;
- no reaction with water, boron, silicon or carbon;
- reaction with oxygen only at higher temperatures;
- oxidation by halogens or oxalic acid, to give indium(III) compounds.

The first large-scale application for indium was as a coating for bearings in high-performance aircraft engines during World War II. Subsequently production gradually increased as new uses were found in fusible alloys, solder and electronics. In the 1950s, tiny indium beads were used for the emitters and collectors of PNP alloy junction transistors. In the middle and late 1980s, the development of indium phosphide semiconductors and indium–tin oxide (ITO) thin films for liquid crystal displays (LCD) aroused much interest.

Indium is used for numerous purposes in a variety of applications. However, the end uses of pure indium metal are limited, with the majority used in some form of chemical compound. The major uses are in high-purity compounds of low-temperature alloys, (soft) solder and thin films (Figure 9.4). Other uses are in the manufacture of batteries, electrical components, semiconductors and in research.

The size of the indium market has increased rapidly in the past 20 years with most growth occurring in thin-film coatings such that this application accounted for 56 per cent of total consumption in 2010 (Mikolajczak and Harrower, 2012). Exceptional growth in demand has come from Japan, which accounted for about 60 per cent of indium consumption in 2011, followed by the Republic of Korea, China and Taiwan. More than two-thirds of the indium demand from these four largest consumers is used in applications for indium–tin oxide (ITO) and semiconductors.

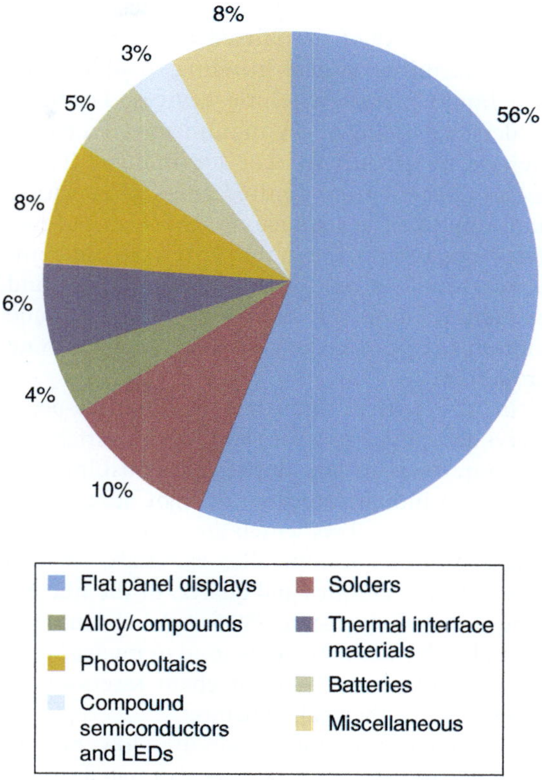

Figure 9.4 The main end-uses of indium in 2010. (Modified after Mikolajczak and Harrower, 2012, courtesy of Indium Corporation.)

Indium–tin oxide (ITO)

Indium oxide is a transparent conducting oxide which combines the three properties transparency, heat reflection and electrical conduction. Transparent materials which are also electrically conducting are quite rare because the mechanisms of conduction and the absorption of light energy are both determined by the free electron density of a material. Doping indium oxide with approximately 10 per cent tin oxide to form indium–tin oxide increases both the electrical conductivity and heat reflectivity without significantly affecting transparency. A film of ITO which is 5 μm thick will absorb less than 20 per cent of the visible light passing through. This unusual combination of properties makes ITO useful in a number of applications. Transparent heat-reflecting ITO films are used in glass coatings, solar collectors and streetlights. ITO films are also used for electrophoretic displays, electroluminescent displays, plasma display panels, electrochromic displays, field emission displays and as coatings on architectural glass, solar collectors, windscreen glass and cathode ray tubes. ITO is also used to improve the performance of low-pressure sodium vapour lamps.

Alloys and solders

Indium forms low melting point alloys with various metals including bismuth, tin, lead and cadmium. These alloys are used as solders and fusible alloys in the electronics industry and in a variety of specialised applications, e.g. metal forming, optical grinding, fusible links and liquid metal switching devices. Lead-free solders consisting of tin, indium and silver are a significant growth market for indium. These solders benefit from limited crack propagation and improved resistance to thermal fatigue. They also inhibit the mobilisation of gold components in electronic devices. Alloys of gold and palladium used in dentistry often include indium to improve their casting and machining properties.

In the optical industry, low-melting-point alloys are applied to lenses and act as a surface for machine tools to grip during the polishing process. The indium property of tight adhesion is also applied in alloys which are used as bonding agents between non-metallic materials, such as glass, glazed ceramics and quartz. In the nuclear industry an alloy of silver (80% Ag), indium (15% In) and cadmium (5% Cd) is used in nuclear control rods to control nuclear chain reactions by absorbing free thermal neutrons.

Global consumption of indium in low melting point alloys and solders was estimated to be about 200 tonnes in 2011, equivalent to approximately 14 per cent of total indium consumption (Mikolajczak and Harrower, 2012). This is a considerable increase from the consumption levels of the early 1980s when less than 10 tonnes per year was used for these purposes.

Semiconductors

Demand for very high purity indium is increasing rapidly in the photovoltaic industry and in the production of I-III-VI compound semiconductors (such as $CuInSe_2$, $CuInGaSe_2$ (CIGS), $CuInS_2$ (CIS), $CuInGaS_2$) for solar cells. Recent research indicates that there is potential for enhanced efficiency at decreasing production costs for thin films based on indium for the mass production of solar cells (ZSW, 2012).

Indium is also used in III-V compound semiconductors (InP, InSb, InAs, InN, $In_xGa_{1-x}N$) and III-VI compound semiconductors (InS, InSe, InTe) for the production of light-emitting diodes (LEDs) and laser diodes, photodetectors, fibre-optic telecommunications and optoelectronic integrated circuits. Indium antimonide, arsenide or phosphide compounds are used as a substrate. Several indium-bearing compounds are used as the epitaxial layer, such as indium gallium arsenide. Indium-based LEDs are used to optically transmit data in displays. Indium-based laser diodes are used in fibre-optic communications. Ultrapure metalorganics of indium include high-purity trimethylindium (TMI), which is used as the semiconductor dopant in III-VI compound semiconductors.

Others

Indium metal and indium alloys are used for a number of surface-coating applications. The most important of these is plating automobile and aircraft engine bearings. Other uses include decorative plating, corrosion inhibitors and plating on aluminium. Indium plates readily from either acid or alkaline solutions onto a wide variety of metallic substrates. The most commonly used plating solutions are indium sulfamate, indium cyanide, indium fluoroborate and indium sulfate. Indium sulfamate is an easily controlled, low-maintenance plating agent. Indium cyanide gives even bright deposits and exhibits excellent throwing power (measure of the efficiency of electrodeposition). Indium fluoroborate platings are usually used for applying heavy (25–75 μm), dense deposits of indium on items such as seal rings.

Indium is also used for electrodeless lamps, mercury alloy replacements and in alkaline batteries in which indium prevents the zinc from corroding and the build-up of hydrogen gas within sealed casings. Phosphors, like indium–boron oxide ($InBO_3$) and manganese–indium–based products, are used for monochrome cathode ray tubes. Small amounts of indium are also used in aluminium alloy sacrificial anodes (for salt-water applications) to prevent passivation of the aluminium. Indium wire is applied as a vacuum seal and a thermal conductor in cryogenics and ultra-high-vacuum applications. Indium is an ingredient in the gallium–indium–tin alloy Galinstan, which is liquid at room temperature but is not toxic like mercury.

Indium is also used as a thermal interface material in personal computers. Pre-shaped foil sheets are fitted between the heat-transfer surface of a microprocessor and its heat sink. The application of heat partially melts the foil and allows the indium metal to fill in any microscopic gaps and pits between the two surfaces, removing any insulating air pockets that would otherwise compromise heat-transfer efficiency.

Indium leukocyte scintigraphy, using the radioactive isotope ^{111}In, has a variety of medical applications (van Nostrand et al., 1988). These include early-phase drug development and the monitoring of activity of white blood cells in areas of infection.

Resources and reserves

Reliable estimates of global indium resources are not available. However, given that more than 95 per cent of indium is recovered as a by-product from zinc refineries, estimates of primary indium resources are generally derived from zinc resources worldwide. A minor proportion of global indium production is derived from the treatment of sulfur-rich copper and tin ores, with each accounting for less than five per cent of global supply. The close association of indium with copper-rich ore-forming systems and its co-enrichment with zinc and copper, however, make copper ores from massive sulfide deposits a potentially important indium resource. Slimes and other residues (e.g. anode slime, Cottrell ash) from the copper production process in indium-bearing ore deposits have been introduced into indium refineries and contribute to enhanced production of indium, e.g. at the Kidd Creek smelter until its closure in 2010.

Identified global zinc resources are about 1.9 billion metric tonnes, with zinc reserves estimated at 250 million tonnes (USGS, 2012a). Using a conservative estimate of 50 grams of indium per tonne of zinc ore, this is equivalent to a resource of 95,000 tonnes and a reserve of 12,500 tonnes of indium metal. Additional contributions to global indium resources and reserves may also be estimated from global copper resources. Table 9.5 provides estimates of global indium resources and reserves derived from the most recent USGS estimates for zinc and copper (USGS, 2012a and 2012b).

It is important to note that such estimates of indium resources and reserves are not well constrained because they treat all zinc and copper ores as if they had the same indium contents. In fact, the figures derived in this manner significantly underestimate global indium resources. If the average size and zinc grades of reserves in massive sulfide (VHMS and SHMS) deposits are examined it can be shown that known large VHMS deposits worldwide contain 13,750 tonnes of indium metal, based on an average indium content of 50 g/t. If the indium content is 20 g/t the indium reserves in these deposits is 5500 tonnes (Schwarz-Schampera and Herzig, 2002 and Alfantazi and Moskalyk, 2003). Similarly, it can be calculated that in the largest SHMS deposits there is an additional 11,750 tonnes of

Table 9.5 Estimated global resources and reserves of indium (tonnes) calculated from global zinc and copper resources and reserves reported by USGS (2012a and 2012b).

	tonnes
Resources in zinc ores[1]	95,000
Reserves in zinc ores[2]	12,500
Resources in copper ores[3]	30,000
Reserves in copper ores[4]	6300
Total indium resources	125,000
Total indium reserves	18,800

[1]calculated from global zinc resources with indium concentration of 50 g/t

[2]calculated from global zinc reserves with indium concentration of 50 g/t

[3]calculated from global copper resources with indium concentration of 10 g/t

[4]calculated from global copper reserves with indium concentration of 10 g/t

indium if the zinc ore contains 50 g/t In, or 4700 tonnes at 20 g/t In.

Among the large-tonnage massive sulfide deposits (i.e. those with more than 25 million tonnes of base-metal massive sulfide ore), six deposits in China account for some 317 million tonnes of ore with an average of about 3.2% zinc (VHMS) and about 188 million tonnes averaging about 5.6% zinc (SHMS), respectively. Given an estimated average indium grade of 50 grams per tonne this equates to 1100 tonnes of indium reserves in large-tonnage Chinese base metal sulfide deposits (VHMS, SHMS). The small size of these reserves is inconsistent with China being the largest world producer of indium, having continuously increased its production since 1998. In addition to supplies from domestic mines, other sources, such as stocks and residues from overseas, have to be inferred.

A study by the Indium Corporation of America (Mikolajczak, 2009) estimated primary indium reserves and resources of 26,000 tonnes of indium metal in western base-metal sulfide mines. China, Russia and other CIS member states account for another 23,000 tonnes of indium metal. An additional 15,000 tonnes of indium may come from tailings and residues (Mikolajczak, 2009).

The most recent USGS assessment of global indium reserves in 2008 was 11,000 tonnes (USGS, 2008). This was a major increase over the 2006 figure when reserves were estimated at 6000 tonnes of indium metal. China accounted for more than 75 per cent of global indium reserves in 2008. As a result of ongoing exploration programmes, Chinese reserves increased greatly in only one year, from 280 tonnes in 2006 to 8000 tonnes of indium in 2007. Peru, Canada, USA and Russia together had more than eight per cent of global indium reserves (USGS, 2008). Global indium resources in 2008 were 49,000 tonnes, less than half the amount estimated in 2012 (Table 9.5).

It is evident that a lack of reliable data has led to wide variation in global estimates of indium resources and reserves. However, it is clear that these estimates have increased in size over time, in tandem with growing worldwide demand.

Production

Global indium production was only about 40–50 tonnes in the early 1970s, but rapidly growing demand over the following three decades led to major increases in primary indium production. World production of primary indium increased by five per cent from 609 tonnes in 2010 to an estimated 640 tonnes in 2011, while between 2000 and 2011 mine production almost doubled (USGS, 2012c). The contributions to primary mine production of indium from individual countries is not available, although more than half of global indium production comes from China, where indium is produced mainly from tin–base-metal sulfide deposits (e.g. Dachang district), massive sulfide deposits (Yunnan) and from secondary sources (residues; Schwarz-Schampera and Herzig 2002).

The mining of indium alone as a primary commodity is uneconomic, and indium is a typical by-product of mining a major industrial metal, in this case zinc (Schwarz-Schampera and Herzig, 2002 and Niederschlag and Stelter, 2009). More than 95 per cent of global indium production comes from electrolytic zinc refining. In 2011 global zinc mine production grew by four per cent to 12.4 million tonnes, with China, Peru and Australia the largest producers (USGS, 2012a).

The largest producers of refined zinc were China, Canada and India. According to the USGS (2012a) China is the largest refinery producer of indium (340 tonnes, 53.1 per cent of total), followed by the Republic of Korea (100 tonnes), Japan (70 tonnes), Canada (65 tonnes) and Belgium (30 tonnes). Other producers accounted for an additional 35 tonnes. It is evident that the majority of refinery production comes from zinc smelters using electrolytic refining and the treatment of zinc-refinery slimes. However, not all zinc smelters process their own residues to recover indium and other trace commodities. In these cases the metal-rich residues are generally sold to secondary producers which possess the appropriate facilities. It should be noted that many of the large primary indium producer countries do not have significant indium-bearing ore deposits.

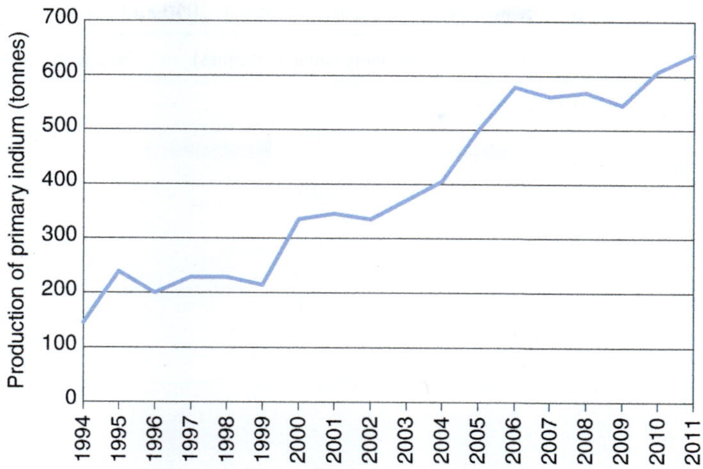

Figure 9.5 Global primary indium production, 1994–2011. (Data from USGS, 2012c.)

Table 9.6 Estimated indium mine production in 2009 from base metal zinc deposits. (Data from Roskill, 2010.)

	Zinc mine production	Mine production zinc concentrate	Estimated indium concentration and quantity in the concentrate		Share
Country	(million tonnes contained zinc)	(million tonnes sphalerite)	(ppm)	(tonnes)	%
Australia	1.3	1.9	15	29	6.2
Canada	0.7	1.1	37	40	8.6
China	2.8	4.2	50	210	45.0
Mexico	0.5	0.8	20	16	3.4
Peru	1.5	2.2	20	44	9.4
USA	0.7	1	20	21	4.5
Other countries	3.6	5.4	20	107	23.0
Total	11.1	16.6	29	467	100

Roskill (2010) concluded that about 470 tonnes of primary indium were recovered from zinc mining in 2009. It was estimated that worldwide zinc ore (sphalerite) concentrates contain an average of 15 to 50 g/t In. The Indium Corporation estimated global production of primary indium in 2010 to be about 500 tonnes (Indium Corp., 2010).

The levels of global indium production from primary sources between 1994 and 2011 are shown in Figure 9.5, while the estimated indium refinery production in 2009 from zinc-rich base-metal deposits is shown in Table 9.6.

During 2010 more than 90 producers worldwide delivered pure indium metal and indium compounds (Roskill, 2010). The majority of these are independent smelters and recycling companies which possess the technical knowledge and equipment for trace-metal recovery. There are only a few vertically integrated producers which have their own mining and smelting operations for indium recovery. These include Xstrata and Teck Resources producing in Canada, and state-owned consortia of mining companies and smelters in China. The number of indium producers and

Table 9.7 Important global indium producers. (Data from Roskill, 2010 and USGS, 2009.)

	Country	Refinery capacity (tonnes)	Secondary capacity (tonnes)
Nanjaing Germanium Factory	China	150	
Huludao Zinc	China	50	
Zhuzhou Smelter Group	China	?	
Dowa Metals & Mining Co	Japan	70	150
Asahi Pretec Corp	Japan		200
Mitsubishi Mat. Group	Japan		96
Korea Zinc	Rep. of Korea	100	
Umicore SA	Belgium	30	
Teck Resources Ltd	Canada	~75	
Xstrata plc	Canada	?	

specialised indium refiners (producing indium of purity greater than 4 N) has increased sharply since 2000, especially in industrialised countries like Japan, USA, Republic of Korea and Germany. In China too, the number of primary indium producers has grown rapidly with a total of 21 operating in 2010.

The Chinese raw material industry has the technical know-how and operating capacity for the recovery of various rare metals from different ore deposit types. Accordingly China currently dominates the international markets for most of these metals, including indium. In contrast, a number of refineries outside China have abandoned trace-metal production or now focus on the production of higher indium purities. Globally important companies with a significant share of indium refinery production are given in Table 9.7.

The annual refinery capacities for indium are estimated at 727 tonnes (Roskill, 2010). The shares are: China, 360 tonnes; Republic of Korea, 120 tonnes; with 50 tonnes each in Japan, Canada and Belgium and 40 tonnes in the USA. Russia, Ukraine, Kazakhstan, Germany and Peru have smaller installed capacities. The worldwide refinery capacity is estimated to be 17–25 per cent higher than the annual production of 540 to 600 tonnes indium (Roskill, 2010; Mikolajczak and Harrower, 2012).

A number of indium refineries have been expanded in recent years. For example, new capacity has been installed in South Korea, Japan, Canada, Bolivia, Brazil and Peru. Numerous Chinese companies improved their treatment lines to recover indium from residues containing less than 0.5% In, while other Chinese producers focused on high-purity indium metal. In 2006 the Canadian company Teck Resources Ltd completed the expansion of its indium production capacity to more than 75 tonnes per year at its Trail smelter in British Columbia (Teck Cominco Ltd., 2006). Teck accounted for all of Canada's indium production in 2011 (65 tonnes).

Kazzinc, the leading zinc and lead producer in Kazakhstan, also produces indium as a by-product (Kazzinc, 2012). The major zinc mines in Kazakhstan include Maleevsky, Shubinsky, Tishinsky and Shaimerden. The recovery and refining of trace metals is well developed at the Ust-Kamenogorsk smelter and Kazzinc also produces selenium, tellurium, thallium, mercury and antimony.

Production from residues and scrap

The recovery of indium from secondary sources like residues (new scrap) and end-of-life scrap (old scrap) exceeds the production from primary sources by at least 250 per cent (Roskill, 2010). The recycling of indium compounds and the recovery of secondary indium metal are particularly important in some producing countries such as Japan, China, Republic of Korea, UK, Canada, Philippines, Taiwan and Germany. In Japan the annual secondary indium production from treating ITO production waste is now more than 450 tonnes (Roskill, 2010). In Germany companies like PPM Pure Metals and Haines and Maassen specialise in the recycling of indium-bearing compounds. PPM produces and refines high-purity indium (up to 7 N).

Projects under development

The growing markets for LCD and photovoltaics have led to an increase in exploration activities for indium-bearing ore deposits. Current projects are focusing on magmatic base-metal-rich deposits, hydrothermal base-metal-vein deposits, and, most significantly, on massive base-metal sulfide deposits. Promising results with high trace metal and indium concentrations have been reported from the well-known Mount Pleasant deposit in New Brunswick, Canada and from a number of deposits in Argentina, Australia, Brazil, Bolivia, Finland, Portugal and the USA.

Argentex Mining is working on the definition of the mineral resources at Pingüino, Santa Cruz, Patagonia. Exploration revealed numerous base-metal veins with local indium concentrations in excess of 1000 g/t. Resource estimates total 7.32 million tonnes of ore grading 16.26 g/t indium. Inferred resources add another 35.4 million tonnes of ore with an average of 8.89 g/t indium (Argentex Mining Corporation, 2009). The Canadian company Marifil Mines Ltd. is exploring the lead–zinc–silver–gold–indium San Roque project in Argentina's Rio Negro province. The drill programme indicated 143 metres of ore containing 0.74 g/t gold, 27.2 g/t silver, 0.37% lead and 0.55% zinc and including 11.7 metres of 174 g/t indium (Marifil Mines, 2007). In 2011, Kagara Mining announced mining at the abandoned Baal Gammon mine in Queensland, Australia for copper, with potential by-products tin, silver and indium. The Baal Gammon deposit contains an indicated and inferred resource of 5.5 million tonnes grading 0.8% copper, 29 g/t silver, 29 g/t indium and 0.2% tin (Kagara Mining, 2012). The Australian company Crusader Holdings acquired exploration licences in the prospective Goiás tin district in Brazil, which includes the indium-rich Mangabeira deposit (Moura et al., 2007).

Exploration and resource assessment are continuing at the polymetallic Mount Pleasant deposit, New Brunswick. The companies Geodex Minerals and Adex Mining have confirmed significant potential for a wide variety of metals, including indium, and plan to start mining at Mount Pleasant in 2015 (Adex Mining Corp., 2012). The Canadian company Lithic Resources is exploring for zinc, copper, silver, indium and gold at its Crypto project in Utah. Total indicated and inferred indium resources are 283 tonnes and 516 tonnes, respectively (Lithic Resources, 2009). The South American Silver Corp. was developing the Malku Khota silver–indium–gallium project in Bolivia until August 2012 when it was nationalised by the Bolivian government. The deposit contains an indicated resource of 230.3 million ounces of silver and 1481 tonnes of indium, and an inferred resource of 140 million ounces silver and 935 tonnes indium (South American Silver Corp., 2011).

Several exploration targets in south-east Finland have been studied with respect to indium concentrations (Cook et al., 2011). Ores occur as massive indium-bearing magnetite–sphalerite veins (Getmossmalmen), as greisen-style veins (Jungfrubergen) and as polymetallic quartz veins (Korsvik-1 and -2, Sarvlaxviken area). The occurrences have bulk indium concentrations between 28 and 1160 ppm; roquesite is associated with copper-rich quartz veins. Indium is a significant trace metal in the ores of Lagoa Salgada, a 4 million tonne unexploited massive sulfide deposit in the Iberian Pyrite Belt, Portugal (de Oliveira et al., 2011). Indium concentrations reach up to 93 ppm in ores and are hosted by sphalerite, with a maximum of 0.8 wt% In.

Abandoned production

In 2010 the Kidd Creek smelter in Timmins, Ontario, which had an annual production capacity of 50 tonnes of indium metal, closed permanently. The smelter treated the copper and zinc ores from the nearby Kidd Creek VHMS deposit. The indium production came from the zinc concentrates, from residues from the copper production (Cottrell ash) and from imported concentrates. The smelter produced indium ingots with a 99.97% purity. The production of indium at Doe Run Peru's La Oroya metallurgical facility was suspended in June 2009 due to high sulfuric acid emissions and governmental restrictions. Some metal production at La Oroya restarted in 2012.

The abandoned lead–zinc–indium–silver Toyoha deposit, Japan, contained an average of 150 to 250 ppm In and produced about 500,000 tonnes of ore concentrates each year (Ohta, 2005). Annual production was about 50 to 55 tonnes of indium metal (USGS, 2002) and accounted for one third of the entire global production. Production from the silver-lead-zinc epithermal deposit, which is underlain by an active magmatic system, was finally suspended in 2006 when ore reserves were exhausted (Mining Journal, 2007). A number of other ore deposits in Hokkaido and Honshu, which are similar in style and mineralisation, are not economic. A recent resource evaluation of indium in Japanese ore deposits identified the Toyoha mine as the largest indium deposit in Japan with a total of 4700 tonnes of indium, followed by the Ashio mine with an estimated total of 1200 tonnes of indium (Ishihara et al., 2006).

The largest European copper mine, Neves-Corvo in Portugal, has a tonnage of 270 million tonnes of massive base-metal sulfides (Galley et al., 2007). This hybrid VHMS/SHMS deposit has very rich indium reserves which are associated with copper-, zinc- and tin-rich sulfide ores. Mining of the deposit commenced in 1989 when Somincor SA began to work copper and tin ores. In 2006 the present owner, the Lundin Mining Corp., started zinc production but this was suspended in 2008 due to the low zinc price. However, zinc production resumed in 2010 and a new zinc plant was commissioned in 2011 (Lundin Mining, 2012).

World trade

Most indium is traded in the form of semi-refined unwrought indium with 99.97% or 99.95% purity produced by electrolytic zinc refineries. Some zinc refineries ship the anode slime to other refineries which are equipped to extract and purify indium. Indium scrap and powders are also traded. High-purity metal (5N–7N) is produced in only few countries including Japan, USA, UK and Germany. The trade of unwrought indium is closely associated with zinc mining and refining capabilities, and is dominated by Canada, China, Bolivia, Russia, Kazakhstan, Ukraine and Brazil. It is difficult to obtain accurate data specific to indium trade because the trade codes record indium trade combined with other metals, such as gallium, germanium, hafnium, niobium, thallium and vanadium.

Since 2005 China has been the principal global source of indium. China's indium exports to the international markets between 2005 and 2011 are estimated at 1854 tonnes (Roskill, 2010; USGS, 2012c). This quantity represents more than 40 per cent of the estimated total indium which entered the trade market. However, since 2007 export restrictions, designed to improve the environmental performance of the indium sector, have reduced the quantity of indium available from China. From about 700 tonnes in 2005 it fell to only 88 tonnes in 2009, although the figure for 2010 and 2011 was 340 tonnes (USGS, 2012c). The number of companies that received export licences dropped to 18 in 2011 (USGS, 2012c). The other principal indium exporters include Canada, USA, Belgium, Brazil, Japan, Republic of Korea and Russia. Belgium, France and the Republic of Korea increased their exports in 2008 and 2009 (Roskill, 2010).

Japan, USA and UK have been the major importers of indium materials since 2005. In 2011 the USA's imports reached 150 tonnes, an increase of 43 per cent over 2009 (USGS, 2012c). The main suppliers to USA were China (31 per cent, 46.5 tonnes), Canada (25 per cent, 37.5 tonnes), Japan (16 per cent) and Belgium (9 per cent). Most of the imports were refined into 5N purity qualities and to indium compounds and alloys. Japan imports large quantities of indium from China, Canada, Republic of Korea and Taiwan. In 2011 Japanese indium imports reached a record high, increasing by 18 per cent to 494 tonnes from 418 tonnes in 2010. Indium is mainly imported into Japan in the form of residues, wastes, scrap and semi-refined metal.

Prices

Smelters usually produce indium of 99.95% purity which is further refined in the smelter or by secondary processors and tertiary manufacturers to higher purities. Indium metal (99.99%) is traded in US$ per kilogram. Unlike the major traded metals indium is sold by private contracts, often long term, between buyer and seller. The estimated value of primary indium metal consumed in 2011, based upon the annual average New York dealer price, was about US$ 82 million (USGS, 2012c).

The costs of indium exploration, mining and metallurgy are closely linked to those for the main commodities zinc, copper and tin, as well as additional trace metals (e.g. cadmium). The majority of the costs are included within the costs of the electrolytic zinc refining and indium production costs cannot readily be calculated separately. However, the recovery of indium from zinc-refinery residues requires significant additional investments which include technological facilities, chemicals, electricity and environmental compliance. The direct production costs for indium in these facilities are in the range of 50 to 100 US dollars per kilogram.

Since 1970 indium prices have followed an almost cyclical trend. The years 1975, 1980, 1988, 1995 and 2003 were characterised by rapid significant increases in global indium prices, followed by a slow gradual decrease (Figure 9.6). Between 1973 and 1980 demand increased for indium use in nuclear control rods. At the same time, supply from easily accessible indium deposits, such as those in Bolivia, gradually ceased. The indium price increased sharply from an annual average of 57 US$/kg in 1973 to 547 US$/kg in 1980. However, the installation of new extraction capacity, combined with decreasing global demand after the Three Mile Island accident and economic recession in the early 1980s, resulted in indium oversupply and decreasing prices. The average annual price in 1986 was only 84 US$/kg. Renewed interest in indium uses, largely from the Japanese electronics industry, started in 1988 when the annual average indium price reached 319 US$/kg. Since the mid-1980s, the development of indium phosphide semiconductors and thin-film ITO compounds for LCDs have dominated the indium markets. Low production costs in 1993 and 1994 resulted in a supply shortfall and subsequently led to a strong price increase in 1995 (annual average 388 US$/kg). Variable demand for LCDs, technological advances which reduced the amount of indium in LCD screens and the availability of supply from stockpiles in the Ukraine led to a price decline in the following years (Figure 9.6), with the annual average indium price falling to 97 US$/kg in 2002. Subsequently, a number of factors led to a marked rise in the indium price, which reached an annual average all-time high of more than 946 US$/kg in 2005. This increase was a response to improving global economic conditions, especially in emerging economies, strong demand for flat screens, technical innovations in thin-film photovoltaic cells, the closure of established indium producers, problems in global zinc production and speculation of a worldwide shortage of indium supply. As a consequence, production capacities were expanded and recycling began to make a significant contribution to global supply for the first time. Subsequently, increased production combined with weakening demand from Asia (Republic of Korea, Japan) and economic recession slowed down the markets and the indium price decreased gradually. In 2009, the average annual price was 500 US$/kg indium. From mid-2009 increasing ITO demand led to a general increase in prices. In 2011 the indium price rose approximately 43 per cent over 2010, to an annual average of 693 US$/kg.

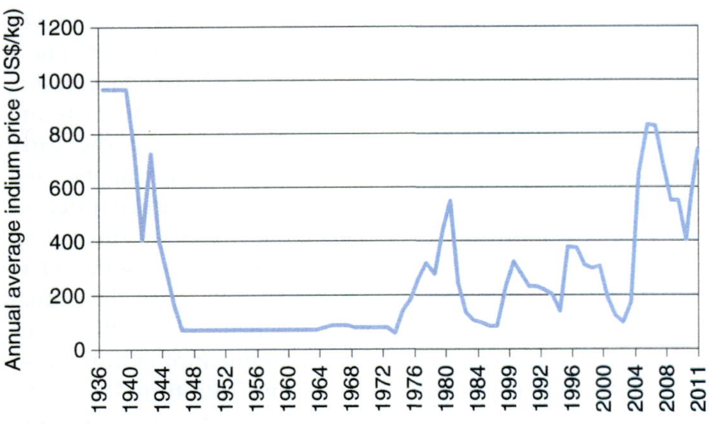

Figure 9.6 Annual average indium prices between 1936 and 2011. (Data from BGR database.)

Recycling and substitution

The recovery of indium from production waste ('new scrap') and, to a lesser extent, from the recycling of indium-bearing electronic scrap have become increasingly important in the last decade as the price of indium has risen. About two thirds (1000 tonnes) of global indium supply comes from recycling.

Most ITO producing countries have increased their recycling rates significantly in recent years. The recovery of indium from the indium–tin oxide (ITO) production and the sputter process has grown in parallel with the growth in production of flat screens and probably accounts for more than 50 per cent of the worldwide indium supply. The sputter process deposits ITO particles as a thin-film coating onto a glass substrate in a vacuum chamber. This process is highly inefficient and more than 70 per cent of the ITO target material is not deposited on the glass, with some of it remaining on the walls of the chamber. These large quantities of excess material, together with the grinding sludge, make its recovery economic. Indium recycling from electronic scrap is technically difficult because of its low indium content and only becomes economic at prices above US$ 400 per kilogram. A process to reclaim indium directly from scrap LCD panels has been developed (Sharp process; Yamasue et al., 2009). New technologies and increased capacities, together with improved collection of electronic scrap, may further reduce the recovery costs.

Price volatility and supply concerns have triggered research into the development of ITO substitutes (USGS, 2012c). However, given that the cost of ITO is a very small proportion of the total cost of a LCD panel and that there has been considerable recent investment in ITO sputtering technology and capacity, it is likely that ITO will remain the predominant transparent conductive coating for most LCD applications for the foreseeable future. Antimony tin oxide coatings, which are deposited by an ink-jetting process, are one possible alternative, while another is zinc oxide with aluminium doping (AZO) which has good transparency and electrical properties. Carbon nanotube coatings, applied by wet-processing techniques, have been developed as an alternative to ITO coatings in flexible displays, solar cells and touch screens (Glatkowski, 2005). Poly(3,4-ethylene dioxythiophene; PEDOT) has been studied as a substitute for ITO in flexible displays and organic light-emitting diodes (Galagan et al., 2011). Graphene quantum dots may replace ITO electrodes in solar cells and have also been explored as a replacement for ITO in LCDs (Wu et al., 2008). Gallium arsenide can substitute for indium phosphide in solar cells and in many semiconductor applications.

Environmental aspects

The environmental impacts of the mining and processing of sulfide-rich indium-bearing ores are dependent on a wide range of factors. Most VHMS deposits are small in size and have limited effects on the regional hydrological regime. However, soluble sulfate salt minerals derived from the weathering and oxidation of sulfide minerals in mine dumps and tailings piles are a potential source of metal contamination and generation of acidic drainage. The significance of this process and the related impacts depend on many factors including the size of the mineralised surface outcrop, the nature and extent of the mining operations and the disposal of waste products. The type of host rock, sedimentary or volcanic, does not seem to affect these characteristics. The most acidic and metal-rich mine water, draining from the base of well-consolidated tailings piles from VHMS deposits, has a pH of 2.6 to 2.7 and contains high concentrations of a range of metals, including some with well-known toxic effects, up to 21,000 µg/l Fe, 3600 µg/l Cu, 220 µg/l Pb, 3300 µg/l Zn, 30 µg/l Co, 10 µg/l Cd, and 311 mg/l sulfate (Taylor et al., 1995 and references therein).

Potential environmental concerns associated with the processing of base-metal ores are related chiefly to tailings ponds which may contain high abundances of lead, zinc, cadmium, bismuth, antimony and other reactants used in flotation and recovery circuits. Fine-grained and intergrown sulfide ores may require very fine grinding for effective beneficiation and can thus lead to the generation of particularly reactive fine-grained tailings. Today, tailings ponds are routinely lined with impermeable barriers, but in the past such measures were not common and there was potential for significant contamination of surface water and shallow groundwater. Today, some underground mines dispose of most of their tailings by backfilling and cementing mined stopes thereby virtually eliminating surface contamination.

Given the rapid increase in the use of indium in recent years and the anticipated future growth in consumption in electronic, photovoltaic and LED applications it is important to understand the natural and industrial cycling of indium and its toxicology. This will provide a basis for improving knowledge of its environmental behaviour and human health effects and thus contribute to its safe and sustainable use. Indium has no known physiological function in humans or animals and is not known to be used by any organism. Although the toxicity of indium metal is low and harmful effects are not reported, there is evidence suggesting that some indium compounds are toxic. There is no drinking water standard for indium in the United States but workplace exposure limits have been set for indium and its compounds in some jurisdictions (0.1 mg/m^3 time-weighted average; NIOSH, 2011; CAREX, 2012). There is little data available on environmental exposure to indium or its compounds. Very low concentrations of indium have been reported in air, seawater, rain water and food, although higher levels have been detected in seafood from contaminated waters near metal smelters (IARC, 2006; White and Hemond, 2012). The average daily intake of indium is estimated to be low, in the range 8–10 µg per day (Smith et al., 1978).

Indium phosphide, used to make semiconductors for laser diodes, solar cells, LEDs and integrated circuits, was the most commonly used indium compound until the last decade. Laboratory studies of rats and mice have provided clear evidence that indium phosphide is carcinogenic in animals, although it has not been adequately studied in humans (National Toxicology Program, 2001). On this basis indium phosphide was classified as 'probably carcinogenic' to humans (Group 2A) by the International Agency for Research on Cancer (IARC, 2006).

Given that indium–tin oxide (ITO) is the predominant form in which indium is currently used, there is considerable interest in establishing the risks associated with its use. Potential exposure to particles of ITO is greatest in the industrial workplace, particularly during the preparation of ITO targets, ITO film deposition and recycling. Most evidence for ITO toxicity has been derived from occupational inhalation exposure, but whether or not it is a potential health hazard to workers is not fully resolved (Tanaka et al., 2010). A number of clinical and epidemiological studies, mostly from Japan, but also from China and the United States, have reported adverse effects on workers exposed to ITO in indium processing and recycling facilities (National Toxicology Program, 2009; Omae et al., 2011). These investigations indicate that exposure to 'hardly soluble indium compounds', such as particulate ITO, can potentially cause a several types of lung damage. However, measures such as proper ventilation and the wearing of masks is sufficient to prevent health hazards in production. A number of studies to investigate the toxicity of indium in laboratory animals (mice, rats, hamsters, rabbits) have shown that indium compounds exhibit systemic toxicity in mammals with adverse effects on kidneys, liver, blood, lungs and the reproductive and developmental systems (Hoet at al., 2012 and references therein; White and Hemond, 2012; Tanaka et al., 2010; Nakajima et al., 2008).

Much remains unknown about indium's natural and anthropogenic cycling, but, given its rapidly increasing use, more research is required to investigate the extent of its toxic and ecotoxic effects and the factors that control them. The behaviour of inhaled indium in the human body and its potential effects on carcinogenesis and developmental and reproductive systems are poorly understood and should be given particular attention. Although releases of indium from the semiconductor and electronics industries are currently small, about twenty times less than that released by mining and smelting activities (White and Hemond, 2012), there is a need for long-term monitoring of the environment and the workplace to understand and quantify the risks associated with the increased use of indium. The quantity of indium released into the environment from mining, smelting and industrial activities already exceeds that from natural sources, such as volcanic eruptions and weathering, and it is therefore important that these are monitored to better understand indium's behaviour and transport and its toxic effects, especially those due to long-term low-level exposure (White and Hemond, 2012). Base-metal mines, smelters and refineries treating indium-bearing materials are employing environmental management principles, practices and equipment which allow them to meet or exceed relevant environmental requirements.

Outlook

The outlook for the indium market is healthy, with demand predicted to grow at a rate of 5–10 per cent per annum. Demand for LCD panels may increase more slowly, but it will continue as the main use of indium, especially for applications such as personal computers, tablets and mobile phones. The ITO sputtering process onto glass has reached maturity and plant has been installed widely and is performing well. New installations will probably use rotary ITO targets (instead of planar) for increased efficiency. Another market area with significant potential for growth is in thin-film photovoltaic cells where copper–indium–gallium–selenide alloy (CIGS) is one of a number of technologies that might become widely adopted as photovoltaic systems are increasingly used for electricity generation.

The fact that indium is a by-product metal has caused some concern because the availability of primary supply is dependent on the extraction, smelting and refining of base metals, chiefly zinc. Operators of a zinc smelter are very unlikely to increase their zinc production simply to deliver larger quantities of a by-product. It has been estimated that only about 25–30 per cent of the indium actually mined each year ends up as refined indium. The remainder is either lost in residues or goes to refineries that do not have the technology to recover indium. Nevertheless, worldwide capacity should be sufficient to accommodate the forecast steady growth in indium demand. Additional supply can be expected to come from increased zinc and copper mine production, particularly from ores containing high indium concentrations. Global mine production of base metals increased by 24 per cent over the five-year period 2006–2011 and, given the worldwide revival of exploration activities for base-metal sulfide deposits in known ore districts and the current interest in critical metals, it is likely that new indium reserves will be identified. Further contributions to supply will also come from the installation of new or expanded indium extraction plants at the base-metal smelters, from improved recovery efficiency at existing indium plants and from improvements in indium-recycling technology.

The possible influence of geopolitical factors on supply security cannot be discounted. Export restrictions have been imposed by some producer countries. For example, China has imposed export quotas and plans to increase domestic production of LCD screens. Stockpiling by governments and private institutions may also have some impact on global indium supply. Some governments, notably in Korea, Japan and China, have announced their intentions to stockpile indium although only in Korea is an official stockpile in place.

References

Adex Mining Corp. (2012) Mount Pleasant: property overview. http://www.adexmining.com/overview.php

Alfantazi, A.M. and Moskalyk, R.R. (2003) Processing of indium: a review. Minerals Engineering 16 (8), 687–694.

Anderson, S.T. (2012) The mineral industry of Bolivia. http://minerals.usgs.gov/minerals/pubs/country/2010/myb3-2010-bl.pdf

Argentex Mining Corporation (2009) Resource estimate for the Pinguino silver, zinc, lead, indiu, gold occurrence. (http://www.argentexmining.com/i/pdf/ATX_Pinguino_Mineral_Resource_Estimate_October_2011.pdf.)

CAREX (2012) Indium phosphide and other indium compounds. http://www.carexcanada.ca/en/indium_phosphide/

Cook N.J., Sundblad K., Valkama M., Nygård R., Ciobanu C.L. and Danyushevsky, L. (2011) Indium mineralisation in A-type granites in southeastern Finland: Insights into mineralogy and partitioning between coexisting minerals. Chemical Geology 284, 62–73.

Demarthe, J.M., Rousseau, A.M. and Fernandez, F.L. (1990) Recovery of Specialty Metals, Mainly Germanium and Indium, from Zinc Primary Smelting. Lead-Zinc '90, 151–160.

Galagan, Y, Rubingh, J.E.J.M., Andriessen, R., Fan, C.C., Blom, P.W.M., Veenstra, S.C., Kroon, J.M. (2011) ITO-free flexible organic solar cells with printed current collecting grids. Solar Energy Materials and Solar Cells 95, 1339–1343

Galley, A.G., Hannington, M.D. and Jonasson, I.R. (2007) Volcanogenic massive sulphide deposits. In: Goodfellow, W.D. (ed.) Mineral deposits of Canada: A synthesis of major deposit-types, district metallogeny, the evolution of geological provinces, and exploration methods. Geological Association of Canada, Mineral Deposits Division, Special Publication No. 5, 141–161.

Glatkowski, P.J. (2005) Carbon nanotube based transparent conductive coatings. www.walkingitaly.com/tuserg/tuserg_metallo/conductive_coatings.pdf

Goodfellow, W.D. and Lydon, J.W. (2007) Sedimentary exhalative (SEDEX) deposits. In: Goodfellow, W.D. (ed.) Mineral deposits of Canada: A synthesis of major deposit-types, district metallogeny, the evolution of geological provinces, and exploration methods. Geological Association of Canada, Mineral Deposits Division, Special Publication No. 5: 163–183.

Heald, P., Hayba, D.O. and Foley, N.K. (1987) Comparative anatomy of volcanic-hosted epithermal deposits: acid-sulfate and adularia-sericite types. Economic Geology 82, 1–26.

Hedenquist, J.W. (1987) Mineralization associated with volcanic-related hydrothermal systems in the circum-Pacific Basin. In: Horn, M.K. (ed.) Transactions of the Fourth Circum-Pacific Energy and Mineral Resources Conference, Singapore. American Association of Petroleum Geologists, 513–524.

Hoet, P., De Graef, E., Swennen, B. et al. (2012) Occupational exposure to indium; what does biomonitoring tell us? Toxicology Letters 213, 122–128.

IARC (2006). International Agency for Research on Cancer, IARC Monographs on the Evaluation of Carcinogenic Risks to Humans. Volume 86: Cobalt in Hard Metals and Cobalt Sulfate, Gallium Arsenide, Indium Phosphide and Vanadium Pentoxide.

Ishihara S., Hoshino K., Murakami H. and Endo Y. (2006) Resource evaluation and some genetic aspects of indium in the Japanese ore deposits. Resource Geology 56 (3), 347–364.

Jorgensen, J.D. and George, M.W. (2004) Indium. Open-File Report 2004–1300: 1–24 [http://pubs.usgs.gov/of/2004/1300/2004-1300.pdf].

Kagara Mining (2012) http://www.kagara.com.au/irm/ShowStaticCategory.aspx?CategoryID=251andHideTopLine=Trueandmasterpage=45.

Kazzinc (2012) http://kazzinc.vestnik.com.kz/#en/Operations/Products/Indium/.

Kovalenker, V.A., Laputina, I.P., Znamenskii, V.S. and Zotov, I.A. (1993) Indium mineralization of the Great Kuril Island Arc. Geology of Ore Deposits 35, 491– 495.

Lithic Resources (2009) Crypto project: resource estimate. http://lithicresources.com/properties/resources_estimate/.

Lundin Mining (2012) http://www.lundinmining.com/s/Neves-Corvo.asp.

Marifil Mines (2007)http://www.marifilmines.com/s/news.asp?daterange=2007/01/01...2007/12/31.

Mikolajczak, C. (2009) Availability of Indium and Gallium: 4 S: www.indium.com/_dynamo/download.php?docid=5529.

Mikolajczak C. and Harrower, M. (2012) Indium Sources and Applications. Minor Metals Conference, February 2012.

Mining Journal (2007) Report Japan 2005–6, updated October 2007. https://www.mining-journal.com/reports/japan—20056-last-updated-october-25

Moura, M.A., Botelho, N.F. and Carvalho de Mendonca F. (2007) The indium-rich sulphides and rare arsenates of the Sn-In-mineralized Mangabeira A-type granite, central Brazil. Canadian Mineralogist 45, 485–496.

Nakajima, M, Usami, M, Nakazawa, K, Arishima, K and Yamamoto, M. (2008). Developmental toxicity of indium: Embryotoxicity and teratogenicity in experimental animals. Congenital Anomalies 48, 145–150.

National Toxicology Program (2001) Abstract for TR-499 – Indium Phosphide 9CASRN 22398-80-7). Toxicology and Carcinogenesis Studies of Indium Phosphide in F344/N rats and $B6C3F_1$ Mice (Inhalation Studies).

National Toxicology Program (2009) Chemical Information Profile for Indium Tin Oxide [CAS No. 50926-11-9]. June 2009.

Niederschlag, E. and Stelter, M. (2009) 145 Jahre Indium - Ein Metall mit Zukunft? World of Metallurgy - Erzmetall 62, 17–23.

NIOSH (2011) NIOSH Pocket Guide to Chemical Hazards. Indium. http://www.cdc.gov/niosh/npg/npgd0341.html

de Oliveira D.P.S., Matos J.X., Rosa C.J.P. et al. (2011) The Lagoa Salgada Orebody, Iberian Pyrite Belt, Portugal. Economic Geology 106, 1111–1128.

Ohta, E. (2005) Toyoha Mine, the largest indium producer in the world. http://staff.aist.go.jp/ohta-e/toyoha.htm

Omae, K., Nakano, M., Tanaka, A., Hirata, M., Hamaguchi, T. and Chonan, T. (2011) Indium lung – case reports and epidemiology. International Archives of Occupational and Environmental Health 84, 471–477.

Roskill Information Services Ltd. (2010) Indium: Global industry markets and outlook (9th ed.): London, United Kingdom, Roskill Information Services Ltd.

Schwarz-Schampera, U. and Herzig, P.M. (2002) Indium: Geology, Mineralogy, and Economics. Springer-Verlag, Berlin.

Schwarz-Schampera, U., Terblanche, H. and Oberthür, T.H. (2010) Volcanic-hosted massive sulphide deposits in the Murchison greenstone belt, South Africa. Mineralium Deposita 45/2, 113–145.

Seifert, T.H. and Sandmann, D. (2006) Mineralogy and geochemistry of indium-bearing polymetallic vein-type deposits: Implications for host minerals from the Freiberg district, Eastern Erzgebirge, Germany. Ore Geology Reviews 28 (1), 1–31.

Sinclair, W.D., Kooiman, G.J.A., Martin, D.A. and Kjarsgaard, I.M. (2006) Geology, geochemistry and mineralogy of indium resources at Mount Pleasant, New Brunswick. Ore Geology Reviews 28, 123–145.

Smith, I.C., Carson, B.L. and Hoffmeister, F. (eds). (1978) Trace Metals in the Environment. Volume 5: Indium. Ann Arbor, Michigan; Ann Arbor Science.

South American Silver Corp. (2011) Economic assessment highlights for Malku Khota silver-indium project. http://www.soamsilver.com/may-16-2011-news-release.asp

Taylor, C.D., Zierenberg, R.A., Goldfarb, R.J., Kilburn, J.E., Seal II, R.R. and Kleinkopf, M.D. (1995) Volcanic-associated massive sulfide deposits. http://pubs.usgs.gov/of/1995/ofr-95-0831/CHAP16.pdf

Tanaka, A., Hirata, M., Homma, T. and Kiyohara, Y (2010) Chronic pulmonary toxicity study of indium-tin oxide and indium oxide following intratracheal installations into the lungs of hamsters. Journal of Occupational Health 52, 14–22.

Teck Cominco Ltd. (2006) Annual report 2005: Vancouver, British Columbia, Canada, Teck Cominco Ltd.

U.S. Geological Survey (2002) Minerals Yearbook 2002: Indium. http://minerals.usgs.gov/minerals/pubs/commodity/indium/indimyb02.pdf

U.S. Geological Survey (2008) Mineral commodity summaries: Indium. http://minerals.usgs.gov/minerals/pubs/commodity/indium/mcs-2008-indiu.pdf.

U. S. Geological Survey (2009): Minerals Yearbook 2008: Indium. http://minerals.usgs.gov/minerals/pubs/commodity/indium/myb1-2008-indiu.pdf

U.S. Geological Survey (2010) Minerals Yearbook 2010: Indium. http://minerals.usgs.gov/minerals/pubs/commodity/indium/myb1-2010-indiu.pdf

U.S. Geological Survey (2012a) Mineral commodity summaries: Zinc. http://minerals.usgs.gov/minerals/pubs/commodity/zinc/mcs-2012-zinc.pdf

U.S. Geological Survey (2012b) Mineral commodity summaries: Copper. http://minerals.usgs.gov/minerals/pubs/commodity/copper/mcs-2012-coppe.pdf.

U.S. Geological Survey (2012c) Mineral commodity summaries: Indium. http://minerals.usgs.gov/minerals/pubs/commodity/indium/mcs-2012-indiu.pdf

Van Nostrand, D., Abreu, S.H., Callaghan, J.J., Atkins, F.B., Stoops, H.C. and Savory, C.G. (1988) In-111-labeled white blood cell uptake in noninfected closed fracture in humans: prospective study. Radiology 167, 495–498.

White, S.J.O. and Hemond, H.F. (2012) The Anthrobiogeochemical Cycle of Indium: A Review of the Natural and Anthropogenic Cycling of Indium in the Environment. Critical Reviews in Environmental Science and Technology 42, 155–186.

Wu, J., Becerril, H.A., Bao Z., Liu Z., Chen Y. and Peumans P. (2008) Organic solar cells with solution-processed graphene transparent electrodes. Applied Physics Letters 92, 263–302.

Yamasue, E., Minamino, R., Numata, T. et al. (2009) Novel Evaluation Method of Elemental Recyclability from Urban Mine—Concept of Urban Ore TMR. Materials Transactions 50, 1536–1540.

ZSW–Zentrum für Sonnenenergie – und Wasserstoff-Forschung (2012) Results – Annual Report 2011, pp. 88.

10. Lithium

KEITH EVANS

Independent Consultant, San Diego, California, USA

Introduction

Lithium was discovered in 1817 by Johan Arfvedson when he analysed a sample of petalite and found that it contained "silica, alumina and an alkali". However, it was a year later when the pure metal was isolated independently by William Brande and Humphrey Davy. Its name is derived from the Greek word 'lithos', which means 'stone'.

Terminology

Concentration levels of lithium in pegmatites (and for the one known occurrence of jadarite) and ore concentrates are normally reported as percentage lithia (Li_2O). Brine grades are normally reported as parts per million (ppm), milligrams per litre (mg/lt) or weight per cent lithium (wt%Li). Global reserves are normally reported as tonnes Li.

Because a great range of chemicals are produced, production tonnages are often expressed as tonnes of lithium carbonate equivalents (LCEs). Lithium carbonate, much the largest volume chemical, contains 0.188% Li. Carbonate is, by definition, one LCE and other major products are metal (5.32 LCEs), bromide (0.425), hydroxide monohydrate (0.88), chloride (0.87) and butyllithium (0.576).

Properties and abundance in the Earth

Due to its high reactivity, lithium only occurs in nature in the form of compounds such as silicates in igneous rocks, in a number of clay minerals and generally as chloride in brines. The principal properties are shown in Table 10.1.

The average crustal abundance of lithium is 17 ppm, but it ranges from approximately 30 ppm in igneous rocks to an average of 60 ppm in sedimentary rocks. Seawater has an average concentration of 0.18 ppm.

Mineralogy and deposit types

Commercially viable concentrations of lithium are found in pegmatites, continental brines, geothermal brines, oilfield brines, the clay mineral hectorite and the newly discovered mineral, jadarite. Locations for the extraction operations and advanced stage projects mentioned are shown on the world map in Figure 10.1.

Critical Metals Handbook, First Edition. Edited by Gus Gunn.

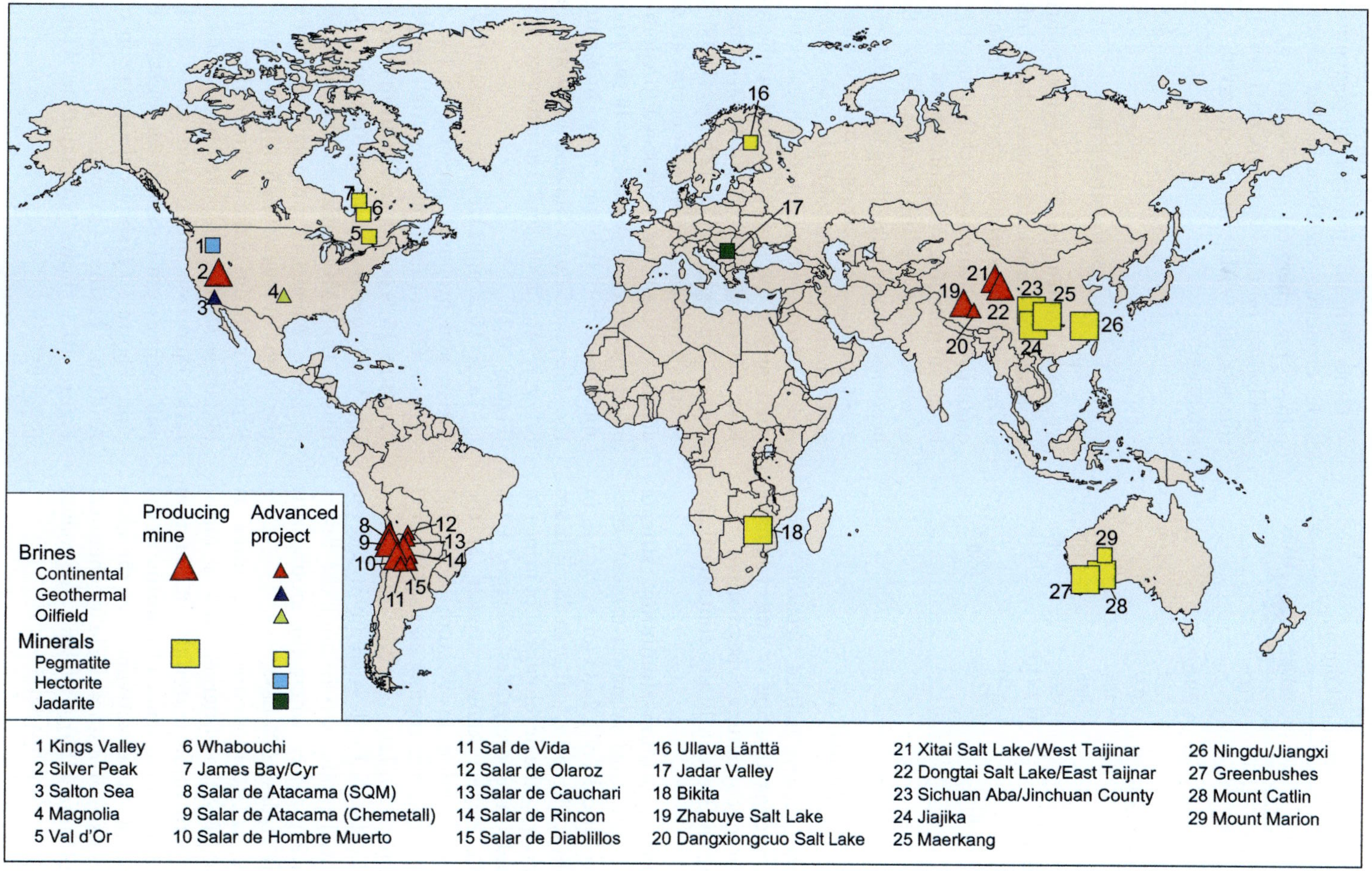

Figure 10.1 Location of producing lithium mines and selected advanced-stage lithium projects.

Table 10.1 Selected properties of lithium.

Property	Value	Units
Symbol	Li	
Atomic number	3	
Atomic weight	6.94	
Density at 25 °C	533	kg/m^3
Hardness (Mohs scale)	0.6	
Melting point	181	°C
Atomic radius	145	pm
Ionic radius of Li^+	90	pm
Electrical conductivity	11.7×10^6	S/m
Electric potential	3.04	V
Specific heat capacity at 25 °C	3.58	J/(g °C)

Pegmatites

Pegmatites are coarse-grained igneous rocks formed by the crystallisation of late magmatic fluids. Lithium-containing pegmatites are relatively rare and frequently also contain tin and tantalite.

Spodumene, a lithium–aluminium silicate ($LiAlSi_2O_6$), is the most common mineral containing a theoretical 7.9% Li_2O (or 3.7% Li). In North Carolina, the principal area for lithium production prior to the development of brines in Argentina and Chile, ore grades averaged 1.4% Li_2O and the richest known deposit, at the Greenbushes operation in Western Australia, has substantial reserves grading up to 3.9% Li_2O (Talison Lithium, 2011a).

Petalite ($LiAlSi_4O_{10}$) is less common and the only substantial production is from the Bikita operation in Zimbabwe where the ground petalite typically grades 4.2% Li_2O.

Lepidolite, a lithium-containing mica is equally rare. Bikita Minerals was a major producer of both lump material and flotation concentrate but the presence of fluorine in the product reduced its attractiveness to the glass industry. It is used on a small scale for chemical feedstock in China and this yields a by-product of caesium and rubidium salts.

Other lithium minerals in pegmatites have little economic significance but zinnwaldite, mined for chemical production by Metallgesellschaft on the German/Czech border before the second world war, is receiving renewed attention.

Continental brines

The term continental brines refers to brines in enclosed (endorheic) basins where inflowing surface and sub-surface waters contain modest quantities of lithium which has been released from surrounding volcanic rocks as a result of weathering (Ide and Kunasz, 1989). Lithium becomes concentrated along with other elements of economic interest, particularly potassium and boron, as a result of high evaporation rates. A description of the geology, climate and hydrochemistry of the Andean region can be found in Risacher et al. (2003). In China, some occurrences are open lakes (Zheng, 1989) but in South America the brines occur at shallow depths in aquifers composed predominantly of salt or of mixed sediments. In the Andes, the abundance of lithium may originate from the Altiplano-Puna Magma Body at depth (de Silva et al., 2006 and Houston et al., 2011) (Figure 10.2).

Table 10.2 indicates the variability of the active and proposed projects in the Andean region. Of note is the very high grade of the feed to SQM's potash and lithium pond system, the very low magnesium/lithium (Mg/Li) ratios at the FMC, Lithium One, Orocobre and Lithium America's proposed operations and the high and very high Mg/Li ratios at the Rincon Lithium (Salar de Rincon) and Comibol (Salar de Uyuni) proposed operations. A high ratio increases production costs as lithium and magnesium concentrate together in solution. Figure 10.2 shows the location of these Andean basins, known as salars, and Figure 10.3 shows the Salar de Atacama in more detail.

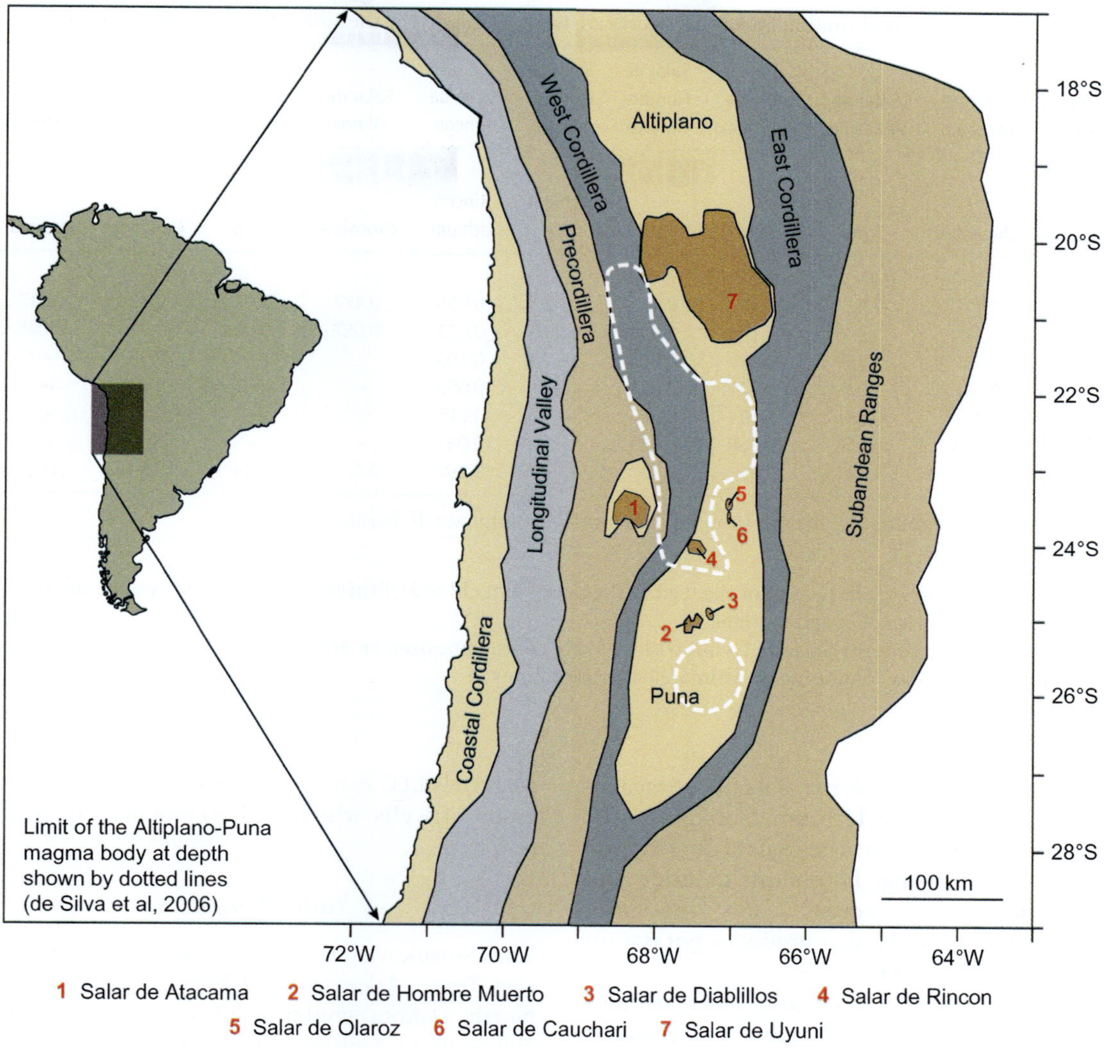

Figure 10.2 Location of salars in the Andean region where lithium extraction is active or proposed. (Compiled by John Houston.)

Foote Mineral Company (now Rockwood/Chemetall) commenced production of lithium from brines at its property in Clayton Valley, Nevada in 1966 and, although grades have declined significantly, it is still in operation. Lithium was also recovered at Searles Lake, California, where it was considered as a contaminant in the company's other products. As a result of a process change the recovery of lithium ceased.

In 1980, Foote negotiated an agreement with CORFO, a Chilean government agency, to evaluate and develop brines at the Salar de Atacama. The project came on stream in 1984 and the company's spodumene-based operation in North America ceased production. Two other brine operations commenced in the Andean region in the late 1990s. The first developed by Sociedad Quimica y Minera de Chile (SQM) located in

Table 10.2 Partial analysis of Andean brines. All figures are wt%. (Compiled by the author from published information.)

Salar name	Salar de Atacama	Salar de Atacama	Salar de Atacama	Salar de Hombre Muerto	Salar de Hombre Muerto	Salar de Rincon	Salar de Olaroz	Salar de Cauchari	Salar de Uyuni**	Salar de Diablillos
Company	Chemetall	SQM	SQM	FMC	Lithium One	Rincon Lithium	Orocobre	Lithium Americas Corp	Comibol	Rodinia
		(MOP)*	(SOP)*							
K	1.8	2.97	1.49	0.617	0.62	0.656	0.477	0.57	0.72	0.517
Li	0.147	0.305	0.11	0.062	0.057	0.033	0.057	0.064	0.045	0.046
Mg	0.96	1.53	0.82	0.085	–	0.303	–	–	0.65	–
Ca	0.031	0.04	0.02	0.053	–	0.059	–	–	0.046	–
SO_4	1.46	0.88	2.19	0.853	–	1.015	–	–	0.85	–
B	0.058	0.065	0.068	0.035	–	0.04	–	–	0.02	–
Mg/Li	6.53	5.02	7.45	1.37	2.19	9.29	2.4	2.34	18.6	3.68

K, potassium; Li, lithium; Mg, magnesium; Ca, calcium; SO_4, sulphate; B, boron.

Notes:
*MOP & SOP analyses are in respect of the feed to the potassium chloride/lithium chloride pond system and the potassium sulfate/boric acid pond systems
**Initial production from the Salar de Uyuni will probably commence from an area with a lithium grade of approximately 0.15% Li containing an estimated 400,000 tonnes Li

the northern portion of the Salar de Atacama and FMC's (formerly Lithium Corporation of America) operation at the Salar de Hombre Muerto in Argentina. Potassium chloride and boric acid are actual and potential co-products and SQM also produces potassium sulfate, a specialised fertiliser.

Brines in eastern China and Tibet are also found at high altitudes with high net evaporation rates but complex chemistries have delayed development compared with earlier projections.

Geothermal brines

Lithium-containing geothermal brines are known at Wairakei in New Zealand (13 ppm) at the Reykanes Field in Iceland (8 ppm) and El Tatio north of and draining into the Salar de Atacama in Chile (47 ppm). A very much higher concentration occurs in the Salton Sea Known Geothermal Resource Area (KGRA) in southern California (Duyvesterin, 1972), where Simbol Materials LLC is proposing production based on a group of wells which will co-produce electrical power.

Oilfield brines

The Smackover Formation extending through east Texas, Arkansas, Oklahoma, Wyoming and North Dakota contains brines grading up to 700 mg/lt Li (Collins, 1976). These brines are a major source of bromine recovered by two companies in Arkansas. Moderate grade brines also exist in the Paradox Basin in Utah and in west Alberta, Canada several companies are evaluating Devonian-age formation waters over an area of 4000 km^2.

Hectorite

Hectorite clay (0.53% Li), a magnesium–lithium smectite, is present in the sedimentary sequence in the Clayton Valley but large tonnages have been identified in the McDermott

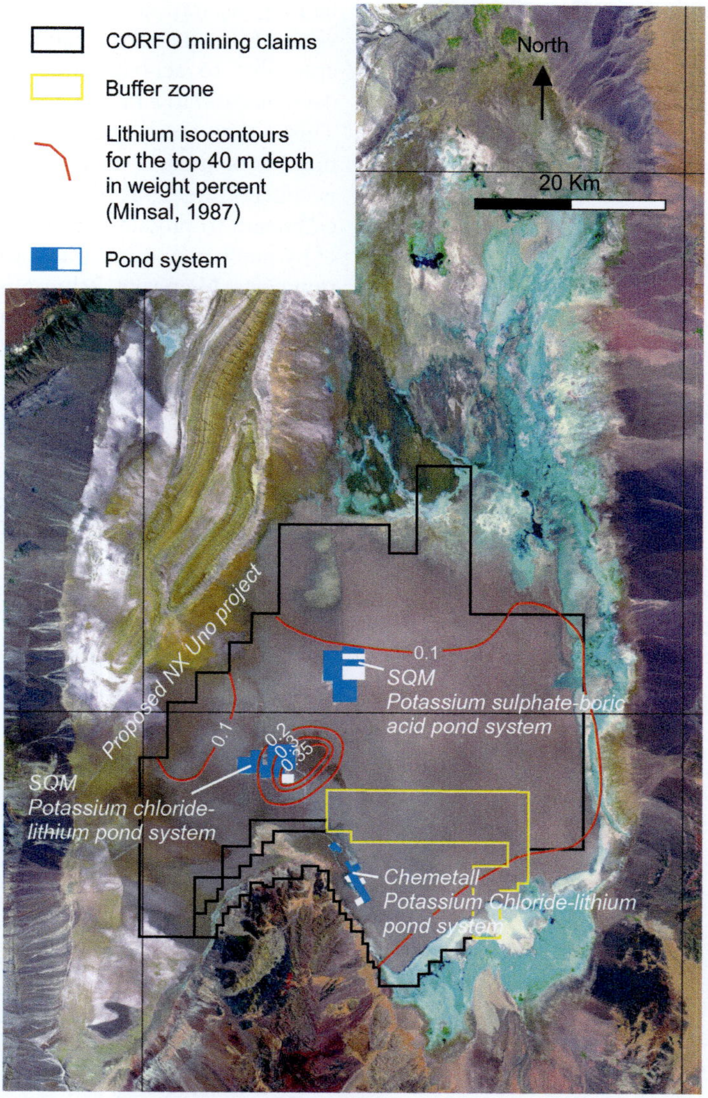

Figure 10.3 Landsat 7 ETM+image of the Salar de Atacama showing the outline of the mining claims held by SQM and Chemetall, together with the buffer zone between these claims and the lithium isocontours in red. Background is Landsat 7 ETM+, December 2001 (ETM+, Enhanced Thematic Mapper Plus). (Compiled by John Houston.)

caldera on the Nevada–California border. Similar clays have been reported in Argentina, Turkey and Mexico. Western Lithium USA Inc. is developing the northern Nevada hectorite deposit.

Jadarite

A newly discovered mineral, jadarite ($LiNaSiB_3O_7(OH)$), with a chemical composition of B_2O_3 47.2%, SiO_2 25.6%, Na_2O 15.0%, Li_2O 7.3% and H_2O 4.3%, was discovered in Serbia in 2004 by

Rio Sava Exploration, a subsidiary of Rio Tinto. The jadarite-containing beds are in three stacked layers.

Extraction methods and processing

Currently lithium is produced from pegmatite and continental brine sources. Photographs of these two widely varying extraction methods are shown in Figures 10.4 (Talison Ltd, Greenbushes mine, Australia) and 10.5 (SQM, Salar de Atacama, Chile).

It appears that most, if not all, spodumene-based pipeline projects will use an acid leach process to produce lithium carbonate – a process developed by Lithium Corporation of America for its former North Carolina operation. Following mining the ore is crushed, ground and spodumene and gangue are separated by flotation. Some projects where the spodumene is coarse can dispense with the flotation step and produce an acceptable product with density separation. The concentrate is then converted from its natural alpha form, which is not amenable to acid leaching, to its beta form by decrepitation in a kiln at a temperature of 1150 °C. This is then attacked by sulfuric acid to produce a number of sulfates including lithium sulfate. This is concentrated, purified and reacted with sodium carbonate to produce lithium carbonate. The sale of by-product sodium sulfate partly offsets production costs. Nemaska Lithium is proposing a different technology to take advantage of low electricity costs and high soda ash costs in Quebec and propose the recovery of both hydroxide and carbonate from lithium sulfate by electrolysis in a plant to be located near Montreal.

Details of the processes proposed for recovering lithium from geothermal brine (Simbol Materials LLC) and oilfield brines (Albermarle Corp) have not been published.

Western Lithium initially had a choice for recovering lithium from hectorite of two processes – one developed by Chevron and the

Figure 10.4 C3 Pit at the Greenbushes pegmatite in Western Australia, the world's leading source of spodumene. (Photograph courtesy of Talison Lithium Ltd.)

other by the U.S. Bureau of Mines. They chose the latter with some modifications and it involves the thermal decomposition of a mixture of ore, anhydrite and gypsum, water leaching and reacting the fluid with sodium carbonate. The process results in the co-production of a substantial tonnage of potassium sulfate.

Rio Tinto's preliminary proposal for recovering lithium carbonate from jadarite involves underground mining (room and pillar), multi-stage crushing and wet scrubbing, digestion in hot concentrated sulfuric acid followed by gangue removal by leaching, boric acid production from the liquor by crystallisation, magnesium and calcium removal from the boric acid plant liquor and precipitation of lithium carbonate by sodium carbonate addition. For each tonne of lithium carbonate produced, 4.5 tonnes of boric acid will be co-produced.

Methods of processing continental brines vary considerably depending on the overall chemistry. The basic aim is to concentrate the brines by solar evaporation to recover the economic products. This involves the precipitation of unwanted elements such as sodium chloride, excess calcium, sulfate and magnesium but avoiding the precipitation of chemicals of interest in an undesirable form such as potash in the form of carnallite rather than as sylvinite or the precipitation of lithium in the form of a complex salt. Excess magnesium in the concentrated brine that remains in solution increases lithium carbonate recovery costs by requiring larger quantities of sodium carbonate in converting the lithium chloride to lithium carbonate.

In a simple case, brine is pumped into a first set of ponds where the concentrating brine precipitates unwanted sodium chloride. At the appropriate level of concentration the brine is transferred to a second set of ponds in which sylvinite is precipitated. This is a mixture of sodium chloride and potassium chloride and when harvested the two components are separated in a flotation plant. The brine continues to evaporate

Figure 10.5 SQM's solar evaporation pond system at the Salar de Atacama covering an area equivalent to 7000 football pitches and in excess of 30 km^2. (Photograph courtesy of Sociedad Quimica y Minera de Chile S.A. (SQM).)

and, in the case of the Salar de Atacama, this reaches a concentration of six per cent lithium chloride and the material is transferred to chemical plants located near Antofagasta, where, following a series of purification steps, the lithium chloride is reacted with sodium carbonate to produce lithium carbonate. At the Chemetall operation the feed into the ponds is a mixture of two brine types.

At FMC's Hombre Muerto operation, solar evaporation plays a minor role and the process is basically an ion-exchange one with the selective extraction of lithium chloride. The process is being modified and will allow the recovery of potassium chloride.

At the Salar de Rincon a first set of solar ponds raises the lithium content to 2.5 grams per litre. It then passes to a reactor where both hydrated lime and sodium sulfate are added to precipitate magnesium hydroxide and calcium sulfate. Brine is then returned to a second set of solar ponds for further concentration to chemical plant feed grade. At Orocobre's operation at the Salar de Olaroz there is a similar treatment of brine at an early stage and SQM operates two pond systems – one for a high-sulfate brine and the other for a high-potash, high-lithium brine further south.

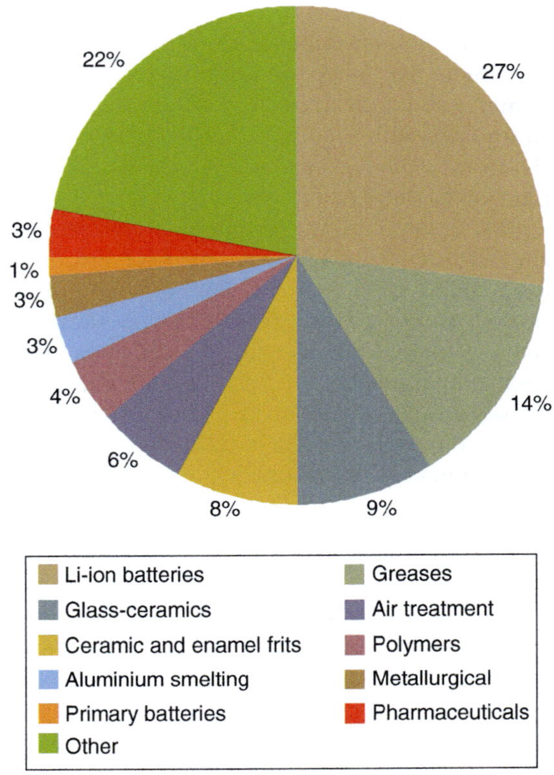

Figure 10.6 Lithium chemical uses, 2010. (Data from Roskill Information Services Ltd, personal communication.)

Specification and uses

The market for lithium is divisible into two major segments. The first is for mineral concentrates with spodumene concentrate the dominant product and lesser tonnages of petalite and lepidolite. The second segment is for a large range of lithium chemicals and metal (further detail is available from Harben, 2002).

Non-chemical demand – glasses, ceramics and glass ceramics account for a high percentage of demand where the principal value of lithia addition is to reduce melting temperatures and to provide thermal shock resistance. The concentrates contain silica and alumina – important in many glass and ceramic formulation. The other significant use is in continuous steel casting.

There are only two major sources – approximately 80,000 tpa of low iron content spodumene concentrate from Talison's Greenbushes pegmatite in Australia and approximately 50,000 tpa of petalite from Bikita in Zimbabwe.

Chemical demand – approximately 200 lithium-containing products are marketed. As is the case with non-chemical demand, Asian demand dominates, accounting for more than 50 per cent by value, with Europe and the Americas dividing most of the rest almost equally.

A breakdown by principal applications is shown in Figure 10.6 and the distribution by major products approximates to lithium carbonate 42 per cent, hydroxide 14 per cent, butyl-lithium five per cent, metal four per cent, chloride four per cent and others nine per cent. Carbonate is the precursor chemical for all other chemicals with the exception of metal and metal derivatives which are derived from lithium chloride.

Lithium carbonate – the main current use for carbonate is in glasses and ceramics, glass ceramics, enamels and glazes where quality and other factors rule out the use of mineral concentrates. The same factors apply as with mineral concentrates – the strong fluxing power of lithia based on its small ionic radius reducing the melting point, lowering the viscosity and thermal expansion and increasing density, workability, chemical durability and hardness. In the case of glass ceramics the thermal expansion is reduced to near zero and is used in cooktops, cookware and large telescopic lenses.

In aluminium electrolysis, carbonate is added to the cell converting to lithium fluoride to lower the liquefaction temperature, thus reducing the operating temperature and reducing emissions. Substantial tonnages of carbonate are used as an accelerator for quick-setting, high- and medium-alumina cements.

Carbonate is the principal chemical used for the production of cathodes for lithium-ion batteries although hydroxide is increasingly being used in this application.

Carbonate is sold in a variety of grades depending on purity, particle size, and particle-size distribution. Industrial grade is normally guaranteed at 99 per cent purity, with battery grade at 99.5 per cent purity. Several producers claim purity levels of 99.9 per cent.

Mobile phones require three grams of lithium carbonate for their batteries, while notebooks and power tools require 30–40 g. Motor vehicles require 0.6 kg per kilowatt hour, with a mild hybrid (one kilowatt hour) requiring two kg, a plug-in hybrid (16 kilowatt hour) requiring 15 kg and an all-electric vehicle (25 kilowatt hour) requiring 22 kg of lithium carbonate (or equivalent). These figures vary somewhat with differing battery chemistry, but, as examples, the Chevrolet Volt and its Opel equivalent (16 kilowatt hour) on this basis requires 9.6 kg and the Nissan Leaf (24 kilowatt hour) requires 14.4 kg. Currently, the main battery types have graphite anodes and variable lithium-containing cathodes of lithium cobalt, lithium manganese spinel, lithium iron phosphate and tri-element blends. Current lithium-ion batteries have storage capacities of 200 watt hour/kg with a theoretical potential of double this. The ultimate potential for lithium batteries is projected to be in a lithium–sulfur and lithium-oxygen battery with a lithium metal anode. Both are in the distant future. In the meantime the search goes on for greater energy storage (greater range) and faster charging time. Typically, the cost of carbonate in a lithium-ion battery is less than three per cent of the battery cost.

In addition to that for motor vehicles, demand for lithium-ion batteries could be considerable for the storage of intermittent sources of energy, such as solar and wind, as well as load-levelling in a wide range of applications from regional grids to individual factories and residences.

Lithium hydroxide monohydrate is used principally in multi-purpose lubricating greases. These greases are effective over a wide temperature range and have excellent water resistance. In its anhydrous form, hydroxide is most frequently used as a carbon-dioxide absorbent in closed systems such as submarines and spacecraft. Following the explosion of one of its oxygen tanks, a rigged system using the chemical saved the lives of the Apollo 13 crew.

Lithium chloride as a solid is used as a flux, particularly in welding aluminium, and as a solution is used for controlling humidity in food processing, pharmaceutical manufacturing and hospitals where it has a sanitising effect.

Lithium bromide has a high solubility in water, and brines with a 54 per cent concentration are used in large scale absorption-refrigeration systems.

Lithium metal is produced by the electrolysis of a mixture of potassium and lithium chloride, normally in a conventional sodium metal cell. Production is in the form of ingots, rods, granules, foils and powder and generally at several levels of purity based on sodium content. Lithium metal is highly reactive. It is used in primary (non-rechargeable) battery anodes having the highest electrochemical potential with the lowest mass and a very long shelf and operating life. A major use in metal is in the production of a wide range of organolithium compounds, the main one being butyllithium used in the manufacture of synthetic rubber. Increasing quantities of lithium–aluminium alloys are being used in aircraft, principally for weight saving.

Recycling

Potentially, after a period of probably eight to ten years following the large-scale introduction of large batteries, there will be a need to recover and recycle the valuable components. In many countries it will be mandatory as a result of legislation.

Electric vehicle (EV) batteries differ significantly from existing batteries in respect of weight, volume, energy content, modular structure, battery management system (BMS) and cell chemistry. Consequently, new recycling processes have to be developed or existing systems used for portable batteries need to be modified. In general there are two processes applicable for EV battery material recycling, each with fundamentally different approaches – pyrometallurgical and hydrometallurgical.

The focus with pyrometallurgical processes lies in the recycling of transition metals like valuable cobalt and nickel. After smelting the battery cells, an alloy of cobalt, nickel and copper is purged out for further hydrometallurgical refinement. Lithium will remain in the slag and is lost for high-value applications as extracting lithium from the slag is probably not economically viable. However, with hydrometallurgical processes it is possible to get high-purity lithium salts and transition metal salts back from the EV battery stream.

The recycling chain for EV batteries is fairly complex. Starting with dismounting the battery from the car, the battery must be discharged to a certain level for meeting transport and handling regulations. All the following process steps need to clearly separate the materials by battery chemistry. This separation is facilitated by labels with battery chemistry information, which should be harmonised worldwide and almost impossible to remove from the battery casing. The next step is the removal of the battery case. The BMS is brought into the existing electronics recycling scheme (for example, in Europe, under the Waste Electrical and Electronic Equipment (WEEE) Directive) and the cells can be removed from the battery modules. The cells can feed a smelter for the pyrometallurgical treatment (see above) or feed a shredder for the hydrometallurgical route.

For the hydrometallurgical process, after shredding, the resulting powders must be carefully separated. This is done by means of sophisticated separation steps, in which copper, aluminium as well as cathode material are separated.

The refined cathode material is the head feed for Chemetall's extraction process. A lithium-containing fraction as well as a transition metal fraction is obtained. The lithium solution is further refined and subsequently transformed by an electrochemical process to high-purity lithium hydroxide. This salt can be used directly for the production of new active cathode materials. The transition metal salts can also be used for producing the same product without further refining.

The hydrometallurgical process can be applied to all lithium-ion battery chemistries, including those without cobalt and nickel.

Substitution

Most of the applications of lithium have existed for decades and have grown at rates comparable with those of the world's economy. In glasses, glass ceramics and ceramics other fluxes can be used but none increases thermal shock resistance, for example, as much as with lithium carbonate. The market share held by lithium stearate in multi-purpose greases has not declined in the face of alternative formulations. In aluminium smelting the main competition has been the result of improvement in capturing emissions – particularly fluorine.

In primary batteries the competition is from other alkaline cells, while lead acid batteries are currently dominant in rechargeable batteries. In applications where weight and energy density are important lithium-ion is progressively replacing nickel cadmium and nickel metal hydride varieties and there is no current competition for lithium-ion in most electric vehicles. The competition here is the continued use of oil, natural gas, hydrogen and biofuels.

Environmental factors

Any new project and existing projects are faced with numerous environmental hurdles. For example, in the case of a new mining and processing project in west United States the company needs permits from the local authority, the State and the Federal governments. This requires intensive environmental baseline surveys including those for vegetation, wildlife, threatened or endangered species, cultural resources, air quality and water resources. This will result in a document in compliance with the National Environment Policy Act. Additionally, any company will require a host of permits regarding the mining and processing of the ore and air quality will be a major issue with processes requiring roasting or calcining ore.

Any pegmatite development will, in most jurisdictions, require similar permits as with any new mining operation and when onward processing from concentrate to carbonate is proposed the operation of a kiln will have an impact on air quality.

In conventional brine-based projects, most of the energy used to remove water and increase the concentration of elements of interest is solar in origin. The predominant waste product is sodium chloride, which can be safely deposited back on the salar's surface.

Used batteries containing lithium in cell phones and computers are, or should be, collected along with other electronic waste. The potential for recycling large lithium-ion batteries is discussed elsewhere. In glasses, glass ceramics and ceramics, the lithium is entrained in the products from which it cannot be leached after disposal.

Lithium products are used in the synthesis of many drugs, in small quantities in the treatment of manic depression and as a sanitiser in food-manufacturing facilities, laundries, swimming pools and hospitals. It presents few environmental problems in these applications.

World resources and production

Reserves and resources

In the mid-1970s the National Research Council in the United States created a committee to examine a range of issues including an assessment of world reserves and resources of lithium. The result was subsequently published (Evans, 1978) which estimated a total of 10.65 million tonnes of lithium in the western world, as little data were available concerning Russia and China.

Subsequently, there have been major discoveries, particularly of brines in western China and the Andes. However, the development of the potentially large-scale use of lithium-ion batteries in motor vehicles has caused lithium availability to become a major issue and concerns have been raised suggesting that resources are inadequate to support the large-scale electrification of vehicles.

The reserves and resources assessment has been updated several times (e.g. Kunasz, 2007; Evans, 2008; and Yaksic and Tilton, 2009) and the current version is shown in Table 10.3. With minor exceptions the tonnages are in situ. The only exceptions are in respect of the totals for North Carolina undeveloped and the extremely large pegmatites in the Democratic Republic of Congo where the tonnages used in the National Research Council report are quoted, which allowed for estimated mining and processing losses. The listing is of concentrations of lithium of economic grade with a

Table 10.3 Estimated world lithium resources compiled by the author based on previous work and published sources.

	Tonnes Li
Pegmatites	
North Carolina, USA *undeveloped	2,600,000
Manono, *D.R. Congo	2,300,000
Greenbushes, Australia (Talison)	1,500,000‡
Russia, numerous	1,000,000
China, numerous	750,000‡
Canada, others	430,000
North Carolina, USA (Former Operations)	230,000
Quebec Lithium, Canada (Canada Lithium)	230,000
Whabouchi, Canada (Nemaska)	187,000
Karalpa, Austria (E. Coast Minerals)	134,000
Lithium One, Canada (James Bay)	130,000
Mibra, Brazil (CIF Mineracao)	100,000
Mount Marion, Australia (Reed)	93,500
Mount Cattlin, Australia (Galaxy)	90,000‡
Brazil, other	85,000
Bikita, Zimbabwe	56,700‡
Lantiar, Finland (Keliber Oy)	14,000
Total Pegmatites	**9,930,000**‡
Continental Brines	
Salar de Uyuni, Bolivia (Comibol)	8,900,000
Salar de Atacama, Chile (SQM & Chemetall)	6,900,000‡
China & Tibet, numerous	2,600,000‡
Salar de Cauchari, Argentina (Lithium Americas)	1,520,000
Salar de Rincon, Argentina (Rincon Lithium)	1,400,000‡
Salar de Olarez, Argentina (Orocobre)	1,200,000
Sal de Vida, Argentina (Lithium One)	1,020,000
Salar de Hombre Muerto, Argentina (FMC)	850,000‡
Salar de Diablillos, Argentina (Rodinia)	530,000
Salar de Maricunga, Chile	200,000
Silver Peak, Nevada, USA (Chemetall)	40,000‡
Total Continental Brines	**25,160,000**‡
Others	
Hectorite	
Kings Valley, U.S.A (Western Lithium)	2,000,000
La Ventana, Sonora, Mexico (Bacanora)	180,000
Geothermal Brines	
Brawley, California (Simbol)	1,000,000
Jadarite	
Jadar, Serbia (Rio Tinto)	950,000
Oilfield Brines	
Smackover Formation, U.S.A (Albermarle)	850,000
Total Others	**4,980,000**
Total Overall	**40,070,000**

*Adjusted for mining losses
‡Includes reserves at producing operations (where published)

wfreasonable and realistic prospect of economic extraction.

The United States Geological Survey in 2011 estimated resources (a combination of reserves and resources) as 33 million tonnes Li (USGS, 2011), up from 25.5 million tonnes in 2010 (USGS, 2010) and 13.8 million tonnes in 2009 (USGS, 2009). Clarke and Harben (2009) in a 'Lithium Availability Wall Map' estimated a 'Broad Base Total Reserve' of 39.4 million tonnes Li (including both producers and developing projects). In a 2011 update of the Wall Map, Clarke (2011) quoted a reserve of 38.3 million tonnes and resources (all categories) of nearly 62 million tonnes Li. SQM estimated global reserves and resources at 18.8 and 56.4 million tonnes Li, respectively (Solminihac, 2009).

Unfortunately, many commentators have ignored the fact that reported figures are not always recoverable resources and, particularly because of the publication of scoping, pre-feasibility and feasibility studies, it is now possible to make estimates of recoveries through the various stages of production from most source types.

In the case of pegmatites, ore is generally clearly distinguishable from the host rock (as shown in Figure 10.7) and mining recovery should normally be high: Canada Lithium, for example, estimates 80–85 per cent. Recovery through flotation and chemical conversion is estimated at 67 per cent (Canada Lithium, 2011a). With a hectorite feed Western Lithium estimate a recovery of more than 87.2 per cent through the entire process (Western Lithium, 2011). Rio Tinto, with its proposed underground operation in Serbia, estimates a 40 per cent recovery from mining using a room-and-pillar system, 75 per cent recovery through beneficiation and 80 per cent recovery in the chemical plant (Kellie, 2009).

Significant losses can occur with a brine feed – a combination of losses in the pond system due to brine entrainment in salts precipitated in the solar ponds and in the chemical plant. Losses at

Figure 10.7 The Whabouchi spodumene pegmatite cutting through host rocks in the James Bay region, Quebec, Canada. (Photograph courtesy of Nemaska Lithium Inc.)

the Salar de Atacama are thought to be in the range of 30–35 per cent. Orocobre estimate such losses at the Salar de Olaroz at 26 per cent (Houston and Gunn, 2011). FMC's losses at the Salar de Hombre Muerto should be lower but there are no published data. Lithium One estimates process recoveries at its Sal de Vida project at 60 per cent.

The other major factor regarding most brine operations concerns the recovery from the host aquifer. Houston et al. (2011) suggest that a conservative estimate of one third should be assumed as there are no known examples of a salar-type aquifer of known initial volume being drained to the point where further abstraction ceases to become viable either technically or economically. This loss would not exist where the feed is from saline lakes. Recoveries from other potential lithium sources from geothermal and oilfield brines are not known.

Offsetting these reductions in recoverable lithium is the fact that exploration activity is at a high level. Numerous pegmatites in Namibia, Mozambique, Ireland and particularly Canada, where many pegmatites were partially explored during an earlier lithium exploration 'boom' in the 1960s, are currently estimated to contain 450,000 tonnes Li and most remain open along strike and at depth.

In Argentina numerous other salares including Rio Grande, Arizaro, Mariana, Antofalla, Ratones, Pozuelos, Pocitos, Huayatayoc and Salinas Grandes are known to be lithium containing. In Chile, the Geological Survey has identified numerous salares that could possibly be viable. These include Pedernales, La Isla and Quisquiro with lithium samples grading between 423 and 1080 mg/lt Li and Punta Negra, Aquas Calientes, Pajonales, Aquilar, Tara, Parinas and Pujsa with concentrations of between 220 and 620 mg/lt Li (Solminihac, 2010).

In early 2012 the Chilean government announced its intention to expand lithium production. As the mineral is classified as strategic under the 1973 Mining Code there were two options. The first was to change the law which required parliamentary approval and the second was to award Special Lithium Operation Contracts (CEOLs) similar in type to those held by Rockwood and SQM. The government decided on the second option and in September received three bids for 20-year concessions to produce up to 100,000 tonnes of lithium over 20 years, in return for an up-front payment and royalty payments of seven per cent on sales. SQM's bid of US$40 million appeared to be the clear winner but it soon emerged that the company was engaged in litigation with several Government agencies – grounds for exclusion from the process. Li3 Energy was the second highest bidder with an offer of US$17.5 million for a project based on a relatively small portion of the Salar de Maricunga with an estimated measured and inferred resource grading 1250 g/lt totaling 120,000 tonnes Li. Subsequently, the Government abandoned the concept of granting additional leases.

Companies other than Rio Tinto are exploring for additional jadarite deposits in Serbia and the second bromine producer from the Smackover Formation could follow Albermarle's lead. In Alberta, Canada, numerous companies are following up on a report by the Provincial Research Council and the Geological Survey that oil field brines of Devonian age contain a resource of approximately 500,000 tonnes Li (Bachu et al., 1955). Concentration levels, though, are low at about 140 ppm.

Despite the very low concentration of lithium in sea water, Korean companies and research organisations are funding a three-year project with a value of US$30 million to evaluate production economics from this source.

If the massive-scale electrification of motor vehicles actually occurs, the batteries are expected to have a life of eight to ten years. The lithium is not consumed in the battery and, if it is economically viable to recycle, the demand for virgin material will be greatly reduced.

Production

Worldwide lithium production (Li content) dropped in 2009 by 31 per cent compared to 2008 to less than 12,700 tonnes. However, output in 2010 recovered to 20,000 tonnes of Li content,

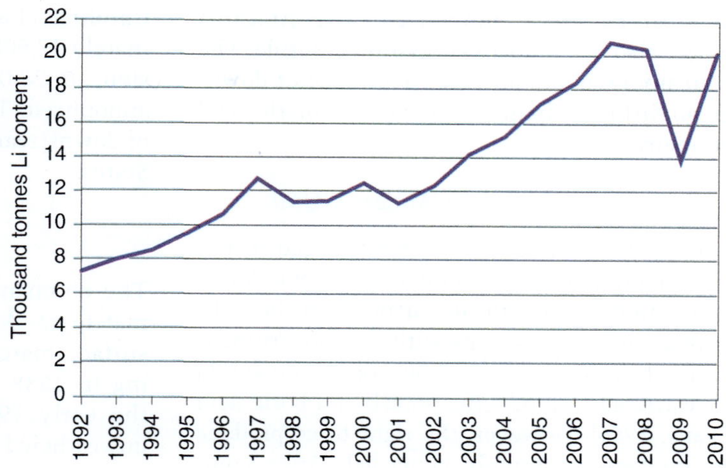

Figure 10.8 Worldwide annual production of lithium 1996–2010 (Li content). (Data from World Mineral Statistics Database, British Geological Survey.)

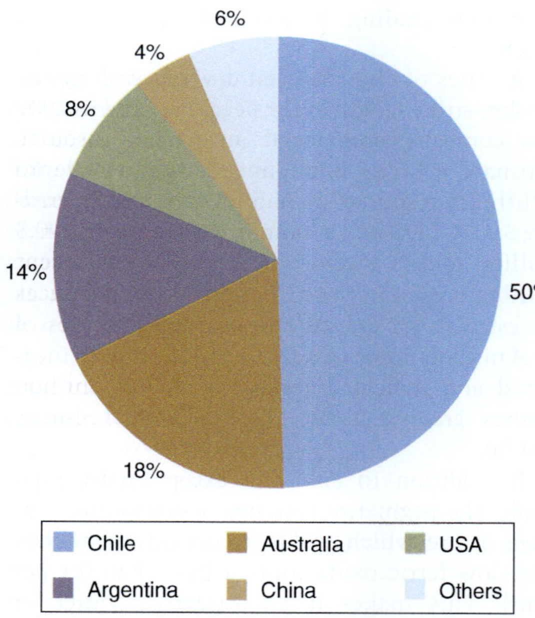

Figure 10.9 Lithium producing countries 2010 (based on tonnes Li content). (Data from British Geological Survey, 2012.)

which is just one per cent down compared to 2008. The trend over the previous decade has been for increasing production with worldwide output rising by 79 per cent between 1999 and 2008 (Figure 10.8). Chile was again the world's largest producing country in 2010, when measured by Li content, with output amounting to more than 9900 tonnes Li content, or 50 per cent of the world's total. Australia, the USA and Argentina are the next largest producers and together the top four represented 89 per cent of the world's total in 2010 (Figure 10.9).

Current producers

Rockwood Holdings/Chemetall

In the early 1980s Foote Mineral Co. reached an agreement with the Government of Chile which allowed it to evaluate and develop a block of claims at the southern end of the Salar de Atacama (Kunasz, 1983). Initially, the government through CORFO, a development agency, was a partner but they progressively sold their interest to Foote. Production commenced in 1984 with a capacity of 13,000 tonnes per annum of lithium carbonate and production from its spodumene-based operation in North Carolina ceased. Foote was later acquired by Chemetall which, in turn, was acquired by Rockwood Holdings.

The combined capacity of both brine operations in Nevada and Chile is currently 38,000 tonnes per annum of carbonate and hydroxide. A phased expansion to 65,000 tonnes per annum by 2020 has been announced. The Atacama project also co-produces potassium chloride. The company produces an extensive range of downstream lithium chemicals in North America and Germany.

Sociedad Quimica y Minera (SQM)

In 1983 the Chilean Government agreed to put the remaining CORFO claims up for bid. Amax Exploration and a Chilean partner won the right to enter into negotiations with the Government taking the position that the new operation would be restricted to producing potash and boric acid (Evans, 1986). However, the right to co-produce lithium was eventually conceded. After completing its feasibility study, Amax decided not to proceed and the project was acquired by SQM. Production commenced in 1996. SQM reduced the carbonate price by approximately 50 per cent because of the high concentration of lithium in the brine and their wish to make a rapid entry into the business

In late 2011, SQM announced that it was increasing carbonate capacity from 40,000 tonnes per annum by 50 per cent. Further expansions are possible at relatively low cost because with potash production at a very high level the volume of lithium in the pumped brine is so high that approximately 400,000 tonnes per annum of carbonate equivalent in a concentrated brine is returned to the aquifer through a series of injection wells. The company produces carbonate, hydroxide and chloride at a chemical plant located close to that of Chemetall near Antofagasta.

FMC Corporation

After an apparently successful negotiation with the Bolivian government regarding possible development of the Salar de Uyuni the agreement was strongly rejected by the local population. The company withdrew and eventually acquired the rights to develop the Salar de Hombre Muerto in Argentina. The project came on stream in 1997 and its pegmatite-based operation in North Carolina closed. Current capacity is approximately 17,500 tonnes per annum but an expansion to 23,000 tonnes per annum has been announced. The company produces a full range of downstream chemical products in the United States.

Talison Minerals

The company acquired the Greenbushes pegmatite in Australia in 2007. The weathered surface material was first mined for tin starting in 1888 and for tantalite commencing in the early 1940s. Lithium was discovered in unweathered pegmatite in 1949 and the operation has grown to be the world's largest producer of lithium which is contained in spodumene concentrates grading between 4.8% and 7.5% Li_2O.

A series of pits has been developed along the 3.5-km strike length of the pegmatite and in 2009 the company announced an overall resource estimate of 1.5 million tonnes Li subdivided into a 'lithium resource' of 35.5 million tonnes grading 3.31% Li_2O and a 'tantalite resource' of 190.8 million tonnes grading 1.1% Li_2O. More recent drilling results in the higher-grade area produces an estimate of proved and probable reserves of 31.4 million tonnes grading 3.1% Li_2O, with measured and indicated resources of 70.4 million tonnes grading 2.6% Li_2O (Talison Lithium, 2011a).

In addition to size and exceptionally high grade, the pegmatite contains a substantial tonnage of ore which yields a concentrate with a very low ferric oxide content (less than 0.1 per cent). This makes it an attractive source of lithium for the glass and ceramic industry and the only significant competition is the petalite produced by Bikita Minerals in Zimbabwe. However, the bulk of spodumene concentrate produced by the company, grading 6% Li_2O, is shipped to China for conversion into lithium chemicals by a number of companies. Talison is increasing its concentrate capacity to 740,000 tonnes per annum, most of which will be for chemical conversion and consideration is being given to developing the company's own chemical production capacity by 2015.

In early January 2012 Talison announced completion of its major expansion at Greenbushes and later in the month received a takeover offer from Rockwood at a 53 per cent premium over Talison's then current price. Two months later Chengdu Tianqui Industry Co offered a higher price of A$792 million and this was accepted. Tianqui is the major convertor of spodumene into chemicals and has been Talison's main customer. Conceivably, Tianqui will undertake some conversion to chemicals in Australia with a saving in freight costs; a possibility that Talison was considering. Earlier in the year the company had acquired a significant interest in Nemaska Lithium in Quebec.

Bikita minerals

The Bikita pegmatite in Zimbabwe is unique in many respects. The area developed initially was a classic zoned pegmatite with a massive lepidolite core which was the feedstock for a third of the USA stockpile of lithium hydroxide purchased by the Atomic Energy Commission commencing in 1953. Approximately half the material was subjected to an isotope separation process to breed tritium from the ^{6}Li for the hydrogen bomb programme. The entire stockpile of both virgin and depleted hydroxide was disposed of by the US Department of Energy in the mid-1990s.

Further north, the pegmatite becomes a mixture of large crystals of petalite together with spodumene and quartz. Initially, production was based on picking of coarsely crushed run-of-mine ore but current production is based on density separation. Petalite, the principal product, grades between 4.0 and 4.2% Li_2O and has an Fe_2O_3 content of 0.03% thus making it attractive to glass and ceramic customers. Production over the last three years (2010–2012) has totalled 47,000, 48,000 and 53,000 tonnes of a variety of grades of petalite.

Spain and Portugal

The British Geological Survey reports a substantial production of lepidolite in Spain and Portugal (5000 and 40,609 tonnes, respectively, in 2010; BGS, 2012). It is believed that these tonnages are of a quartz–feldspar–lepidolite mixture used in the local ceramics industry and not entering the international trade.

Brazil

Companhia Brasileira de Litio (CBL) in Brazil had the capacity to produce approximately 2000 tonnes per annum of LCE from locally sourced spodumene and Companhia Industrial Fluminense (CIF) is developing a pegmatite source at Mibra with a tantalite grade of 300 grams per tonne which also contains a lithium resource of 21.3 million tonnes grading 1.0% Li_2O.

China

Lithium-containing pegmatites have a wide distribution and include the deposits Maerkang, Daoxian, Jiajika, Kokalay, Jinjuan, and Ningdu and resources could total an estimated 750,000 tonnes lithium. Most appear to be high cost or have concentrates of poor quality, as there seems a distinct preference amongst the major chemical producers for Australian imports despite the high freight costs. Domestic spodumene production satisfies less than 25 per cent of internal demand.

China has a number of major brine lake sources. A consortium of Tibet Zhabuye Lithium Industry High Tech Group, Tibet Mineral Development, Yuxin Trading and BYD, the car manufacturer, have rights to the Zhabuye Salt Lake in Tibet. Qinghai Salt Lake Industry Group and Western Mining Group operate at the Dongtai Salt Lake, also known as East Taijinar. Qinghai National Security Co and CITIC have rights to West Taijinar also known as Xitai Salt Lake. The brines are generally complex with relatively low production levels in comparison with their reserves but one projection is that the two Lake Taijinar producers will increase capacity to 30,000 tonnes carbonate over the next five years. A new entrant will be Zong Chuan International Mining at the Dangxiongcuo (DXC) salt lake in Tibet.

Producers of lithium chemicals in China are numerous, many producing a full range together with metal. The major technical grade (generally 99.3 per cent purity) carbonate producers in China are Sichuan Tianqi Lithium, Qinghai Guoan, Qinghai Lithium Industry, and Tibet Minerals with a joint production of 8500 tonnes in 2010. Convertors to battery grade (generally 99.5 per cent purity) carbonate are Sichuan Tianqi, Sichuan Ni & Co, Ronghui and Xinyu Ganfeng. Production in 2010 approximated to 7000 tonnes. Convertors to lithium hydroxide were Sichuan Tianqi, Sichuan Ni & Co, Kinjian Non-ferrous, ABA Guansheng and Minfeng with a 2010 production of 10,800 tonnes. Sichuan Tianqi is the sole producer of lithium chloride at 1500 tonnes in 2010.

Production costs

Existing producers have not disclosed their production costs but a general assumption is that those incurred by the two producers at the Salar de Atacama are the lowest due to the high grades of the brine and exceptional climatic conditions.

The recent completion of feasibility studies for a number of projects based on pegmatites, continental brines and hectorite reveal cost estimates but no such data are available yet on geothermal brine, oilfield brine or jadarite.

Pegmatites

Canada Lithium, with a feed grade of 0.85% Li_2O, estimates a total capital cost of US$202 million for the mine, mill and chemical plant to produce 20,000 tonnes per annum of battery-grade carbonate. Average cash operating costs, after a major revision, are estimated at US$3,164 per tonne LCE (Canada Lithium, 2011a).

In October 2012, Nemaska Lithium announced the completion of a Preliminary Economic Assessment for the production of lithium hydroxide and carbonate from its Whabouchi pegmatite in Quebec. Figures released are in respect of an 18 years life of mine project with an initial capital cost of US$454 million. Production of 3.8 million tonnes of 6% concentrate will allow the production of 366,000 and 177,000 tonnes of battery grades of hydroxide and carbonate, respectively, and average costs of US$3400 per tonne and US$3500 per tonne. Annual production tonnages are estimated at approximately 20,700 and 10,000 for the two products but the plant will be designed to be flexible. Product will be recovered from lithium sulfate by electrolysis (Nemaska Lithium, 2012).

In Australia, Talison estimates the capital cost for its expansion from 315,000 tonnes per annum to 740,000 tonnes per annum of concentrates at between A$65 and A$70 million (Talison Lithium, 2011b).

Also in Australia, Reed Resources announced in October 2012 the results of a new Pre-feasibility Study giving more emphasis to hydroxide production rather than carbonate production from a chemical plant to be located in Malaysia. The project entails importing 147,000 tpa of spodumene concentrate grading 6% Li_2O, to produce 10,000 and 8800 tpa hydroxide and carbonate, respectively, from its Mount Marion pegmatite in Australia. The capital cost is estimated at US$83 million with production costs of US$3828 for hydroxide and US$4538 for carbonate. Spodumene concentrate delivered to the plant is priced at US$350/tonne (Reed Resources, 2012).

Altura Mining intends developing the Pilgangoora project also in Western Australia. A Scoping Study estimates a capital cost of A$96.3 million to produce 150,000 tpa of 6% concentrate. Cash operating costs are estimated at A$16 per tonne ore and A$90 per tonne concentrate ex-plant (Altura Mining Ltd, 2012).

Hectorite

Western Lithium USA estimates capital costs for the initial production of 13,000 tonnes per annum carbonate at US$248 million rising to

US$409 million when production is increased to 26,000 tonnes per annum. Cash costs for carbonate are estimated at US$3472/tonne but with co-product credits, principally from potassium sulfate, these are reduced to US$1,967/tonne (Western Lithium, 2011).

In January 2013, Bacanora Minerals announced the completion of a Preliminary Economic Assessment for its La Ventana hectorite deposit in Sonora, Mexico, with a resource of 600 million tonnes grading 3000 ppm Li. At a production rate of 35,000 tpa carbonate and initial capital costs of US$114 million, it estimates average operating costs of US$1958 per tonne (Verley and Vidal, 2013). The announcement did not disclose a processing method and the cost estimates differ greatly from those of Western Lithium.

Continental brines

In Argentina, studies have recently been published regarding anticipated costs at four somewhat similar brine-based projects.

At the Salar de Cauchari, Lithium Americas Corp, plans on producing 40,000 tonnes per annum carbonate with development in two phases. Capital for Phase 1 with a production of 20,000 tonnes per annum on stream in 2014 is estimated at US$217 million with an additional US$181 million required to double capacity three years later. Cash operating costs are estimated at US$1434/tonne (Lithium Americas, 2011).

At the adjacent Salar de Olaroz, Orocobre intends producing 16,400 tonnes per annum of carbonate at a capital cost of US$207 million and a cash operating cost of US$1512/tonne. Approximately three years into the project the co-production of 10,000 tonnes per annum of potash will commence. This will require an additional investment of US$14.5 million and the credit from this will reduce the carbonate production cost to US$1230/tonne (Orocobre, 2011).

A preliminary economic assessment of what was formerly Lithium One's Sal de Vida project (now acquired by Galaxy Resources) was released in October 2011. For a target production of 25,000 tonnes per annum of lithium carbonate and 107,000 tonnes per annum of potassium chloride, capital costs are estimated at US$356 million and production costs for carbonate, FOB Antofagasta at US$1537/tonne. The company points out the significant contribution that potash makes to project economies where sales revenues are sufficient to more than cover total operating expenses (Galaxy Resources, 2013a).

In late 2011, Rodinia Lithium announced the results of its Preliminary Economic Assessment for its proposed operation at the Salar de Diablillos. Two potential scenarios are considered. The first is for the production of 15,000 tonnes per annum lithium carbonate and 51,000 tonnes per annum potassium chloride with a capital cost of US$144 million and a cash cost of $1519/tonne for carbonate. The second option is to produce 25,000 tonnes per annum carbonate and 85,000 tonnes per annum potash. Capital for the larger case is estimated at $220 million with a cash production of $1486/tonne for carbonate. In both cases potash revenues are said to cover total operating costs (Rodinia Lithium, 2011).

No cost estimates are available for other projects.

Future supplies

The prospect of a major increase in demand for lithium, particularly battery-grade carbonate has resulted in a high level of exploration activity and several projects have reached an advanced stage.

Pegmatite-based projects

Canada Lithium Corp has acquired the Val d'Or property which was formerly operated as an underground operation between 1955 and 1965. The latest resource estimate for measured and indicated resources is 33.2 million tonnes grading 1.19% Li_2O (Canada Lithium, 2011b). Plant construction has commenced.

Nemaska Lithium, also in Quebec, is developing a moderately high-grade pegmatite with measured and indicated resources of 19.6 million tonnes grading 1.49% Li_2O. Initially the company planned to produce only spodumene concentrate, but the new target is to produce both, hydroxide and carbonate (Nemaska Lithium, 2012).

A third Quebec project near James Bay was initially developed by Lithium One with 11.75 Mt at 1.3% Li_2O of indicated resources and 10.47 Mt at 1.2% Li_2O inferred (Galaxy Resources, 2013b). The property was acquired by Galaxy Resources at the same time as they purchased the company's brine prospect in Argentina.

In Europe, Keliber Oy plans to produce 6000 tonnes per annum carbonate from a group of pegmatites but production, originally scheduled for 2010, has been delayed.

In Western Australia, Galaxy Resources is mining the pegmatite at Mt. Catlin with upgraded resources (in all categories) announced in March 2011 of 197,000 tonnes of contained Li_2O (Galaxy Resources, 2011). The target is to produce 127,000 tonnes per annum of concentrate grading six per cent and ship it to a company-owned chemical plant in an industrial zone near Shanghai. Carbonate (17,000 tonnes per annum) will be used in the production of e-bike batteries. At capacity, production will total 17,000 and 4000 tpa carbonate and hydroxide, respectively.

At Mount Marion in Western Australia, Reed Resources plans to produce 147,000 tpa of spodumene concentrate from a resource of 10.5 Mt grading 1.4% Li_2O and ship it to a plant proposed for Malaysia to produce both hydroxide and carbonate (Reed Resources, 2012).

Altura Mining, in a relatively recent discovery south of Port Headland in Western Australia, has announced a resource of 25.2 Mt grading 1.23% Li_2O. Current planning is to produce 150,000 tpa of 6% spodumene concentrate (Altura Mining Ltd, 2012).

Continental brines

A number of new projects are proposed in the Andes. In Chile, a joint venture between Minera Copiapo (a nitrate and iodine producer) and Korean interests proposed to develop the NX Uno project on the western margin of the Salar de Atacama with an initial target of 200,000 tpa of potash and 20,000 tpa lithium carbonate. Feed to the operation would have approximated to 0.15% Li. At the Salar de Maricunga Li3 Energy has announced an indicated resource of 120,000 tonnes Li grading 1250 mg/lt in claims covering 14 km^2. Neither project can now proceed due to the cancellation of the Government's proposal to award additional contracts for the production of lithium. Numerous other exploration programmes have, almost certainly, ceased in the country.

In Argentina, the evaluations of four salares have reached an advanced stage. Galaxy's (formerly owned by Lithium One) Sal de Vida project is adjacent to FMC's existing operation at the Salar de Hombre Muerto. The inferred resource comprises a clastic (sediment-filled) body containing 1.02 million tonnes lithium grading 695 mg/lt Li and 11.0 million tonnes of potash (Galaxy Resources, 2013a). A later discovery was a salt basin (similar to FMC's source) with significantly higher grades.

Rodinia Lithium at the Salar de Diablillos claims a recoverable inferred resource grading 556 mg/lt Li and 6206 mg/lt K containing 2.82 million tonnes of lithium carbonate and 11.27 million tonnes of potassium chloride (Rodinia Lithium, 2011). The company is considering two levels of production and potash sales are expected to cover all cash operating costs for the project.

At the Salar de Rincon (a salt-filled salar) Rincon Lithium, which is 100 per cent owned by the Sentient Group, is reportedly in the early stages of commercial production with a reputed target of 15,000 tonnes per annum carbonate. Orocobre, at the Salar de Olaroz and Lithium Americas Corp at the Salar de Cauchari (adjacent clastic salares) have somewhat similar deposits. Orocobre's resources are estimated to contain 1.2 million tonnes Li and 19.0 million tonnes of K grading 690 mg/lt Li and 5730 mg/lt K (Orocobre, 2011). Total resources, including inferred, at Cauchari are 1.52 million tonnes Li at a grade of 627 mg/lt Li and 13.3 million tonnes K at 5417 mg/lt K (Lithium Americas, 2011). As with all the Argentinian clastic salares the magnesium/lithium ratio in the brine is low, facilitating lithium recovery. Lithium Americas plans initial carbonate production at 20,000 tonnes per annum and doubling the tonnage within a few years. Orocobre plans initial production at 16,400 tonnes per annum with potash production starting two years after start up. Construction commenced in late November 2012 (Orocobre, 2013).

The Salar de Uyuni, in Bolivia, has attracted considerable attention in the world's press and is the world's most politicised potential source of lithium (Mares, 2010 and Wright, 2010). A strategic plan published by the government in October 2010 included a commitment of US$900 million to develop a project producing 30,000 tonnes per annum carbonate by 2014 with sales from a US$60 million pilot plant commencing in 2011. The overall sum included the cost of a lithium-ion battery plant (Achtenberg, 2010). Various countries have been courting the government for rights to participate in the development of the salar but, despite numerous memoranda of understanding, nothing concrete has emerged except that in July 2011 the government and Korean interests agreed to a joint venture to produce rechargeable battery parts. A later statement referred to the joint development of the country's lithium resources and Posco (Korea) is to be allowed to undertake testwork on a potential new process for lithium recovery at Uyuni. In January 2013 Bolivia commenced sales with a capacity of 40 tpm of lithium carbonate and 1000 tpm of potash.

The latest published estimate of resources at Uyuni is for 8.9 million tonnes grading 0.045% Li (Risacher, 1989) which includes an area of 240 km^2 bounded by the 1000 mg/lt isoconcentration line containing 430,000 tonnes Li at a grade of 0.15% Li. However, the aquifer is very thin and the area floods seasonally. Two deep holes drilled nearer the centre of the salar have indicated a salt and impermeable clay sequence with 12 salt horizons with a total salt thickness of 170 metres. It is claimed that the salt has high porosity throughout and that the lithium grades persist at depth. Comibol, the project manager, claims that this would indicate a resource of lithium in excess of 100 million tonnes.

Finally, as far as brine developments are concerned, Zong Chuan's DXC project in Western China is expected to produce 5000 tonnes per annum carbonate within a few years.

Geothermal brine

The brine in the Salton Sea Known Geothermal Resource Area (KGRA) in southern California has a uniquely high concentration of lithium and other potentially recoverable elements. Correspondence dated February 2008 states that one existing 228-megawatt facility with a lithium concentration of 200 ppm Li pumps 84,000 tonnes per annum of LCEs (W. Bourcier, personal communication). Simbol Materials, owned principally by private investors, plans initial production of 16,000 tpa carbonate plus other battery materials based on zinc and manganese with feed from a 50-megawatt power plant. Longer-term plans are to develop a total of four similar facilities.

Oilfield brine

Collins (1976) estimated a possible resource of 0.75 million tonnes Li in one tenth of the area underlain by the Smackover Formation which extends through North Dakota, Wyoming, Oklahoma, east Texas and Arkansas. Currently, the brine is exploited for its bromine content and one company, Albemarle Corporation, has announced its intention to recover lithium from its brine feed at Magnolia, Arkansas. Laboratory testing is complete and a pilot plant was under construction in mid-2011. The available brine feed ranges from 200 to 300 ppm Li and the current volume will be sufficient to produce 20,000 tonnes per annum carbonate. The company says its production will be competitive in terms of price and quality with the South American producers. Production of up to

3000 tonnes per annum could start within 12 months and reach capacity by 2015.

Hectorite

Western Lithium USA is developing the Kings Valley project located on the Nevada–Oregon border. The deposit comprises five hectorite-rich clay lenses at shallow depth within sedimentary and volcanic rocks in the moat of a caldera (Figure 10.10). When first drilled by Chevron the resource was estimated to contain 2.0 million tonnes of lithium grading between 0.31% and 0.37% Li. In a revised pre-feasibility study the company plans the initial production of 13,000 tonnes per annum carbonate commencing in 2015, increasing to 26,000 tonnes per annum in 2019 from a proved and probable reserve of 27.1 million tonnes grading 0.395% Li and 3.89% K in the southernmost lens (Western Lithium, 2011). In 2012, probably in recognition of potential oversupply in the short term and the project's relatively high costs, the company delayed its development plans for lithium production and is evaluating the production of low-grade lithium clays as a component in drilling muds.

In Sonora, Mexico, Bacanora Minerals is evaluating its La Ventana lithium deposit with an estimated 60 million tonnes of hectorite

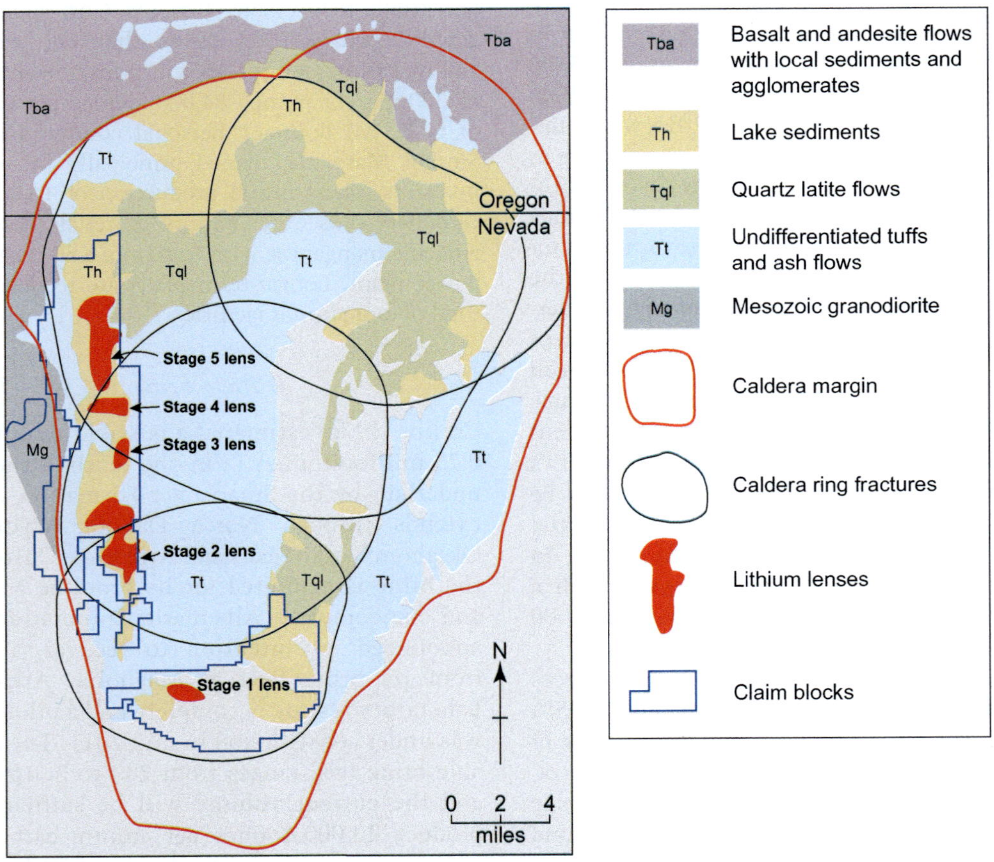

Figure 10.10 Geological map of Kings Valley showing the five hectorite-rich clay lenses. (Modified from Chmelauskas, 2010.)

clay grading approximately 3000 ppm Li – approximately 180,000 tonnes Li.

Jadarite

Rio Tinto owns the Jadar Valley project located 100 km south-west of Belgrade in Serbia. An exploration programme, which started in 1998, aimed at finding colemanite penetrated three vertically stacked zones containing a newly discovered mineral subsequently named jadarite. The lowest and thickest zone was estimated to contain, over an area of 4.5 km^2, 114.6 Mt grading 1.8% Li_2O and 13.1% B_2O_3. Test work indicates that, after mining and processing losses, this horizon at a depth between 300 and 600 metres and with a thickness of 9 to 20 metres will yield 6.4 million tonnes of boric acid and 1.4 million tonnes of lithium carbonate. Jardarite itself contains 7.3% Li_2O. Work on the project is continuing but, assuming ore production of one million tonnes per annum, the production would approximate to 27,000 tonnes per annum lithium carbonate and 133,000 tonnes per annum boric acid (Kellie, 2009).

World trade

Data for the trade in lithium-containing minerals are difficult to obtain because under most trade-code systems these minerals are grouped with other commodities. However, data for trade in lithium oxides and hydroxides can be obtained, as can data for lithium carbonate. The major importing and exporting countries for these are shown on Figures 10.11 and 10.12.

Talison Minerals in Western Australia currently dominates the lithium minerals concentrate market with Asia accounting for 50 per cent of demand with Europe being the second largest market and North America using less

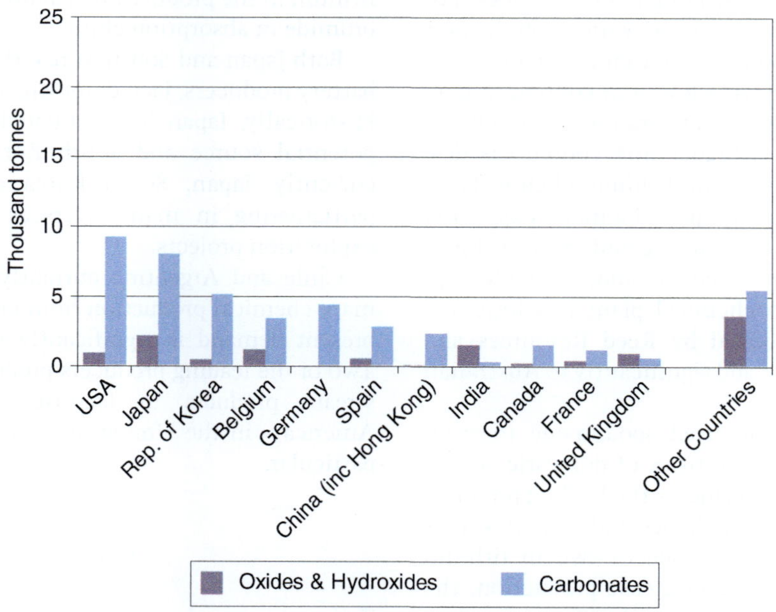

Figure 10.11 The main importing countries for lithium carbonate, oxides and hydroxides, 2009. (Data from British Geological Survey World Mineral Statistics database and UN Comtrade, 2013.)

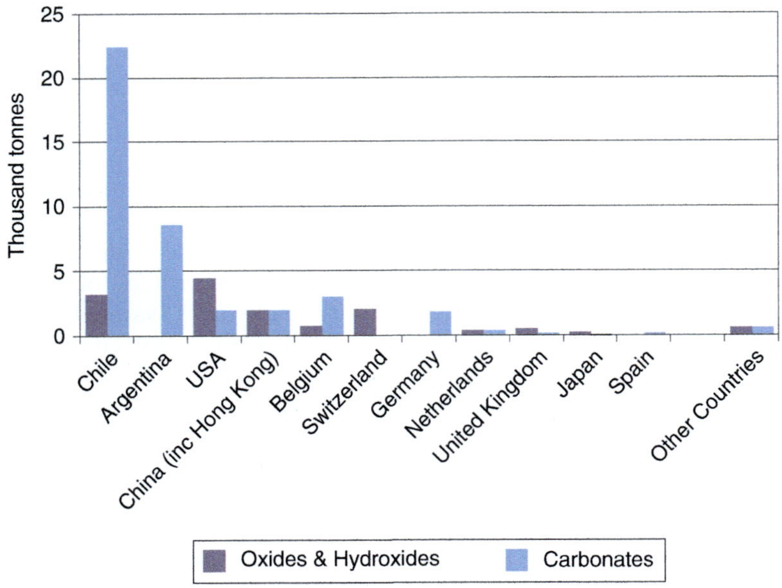

Figure 10.12 The main exporting countries for lithium carbonate, oxides and hydroxides, 2009. (Data from British Geological Survey World Mineral Statistics database and UN Comtrade, 2013.)

than 10 per cent. Although Talison does not disclose the breakdown of its various products it is possible that about 80,000 tonnes per annum of its concentrates have a low iron content, allowing its use in glasses and ceramics. The bulk of production (with a higher iron content) is destined for conversion to lithium chemicals in China. Talison has completed a major expansion to 740,000 tpa of concentrate and most of this is destined for chemical conversion. Galaxy's shipments to its own chemical plant in China and concentrates produced by Reed Resources and Altura Mining will add significantly to Australian exports.

China, using imported spodumene together with much lower quantities of domestic spodumene and domestic brine, is the leading producer of lithium-containing glasses and ceramics. It is also either the number one or two in lithium batteries, number one in grease production, the leading user of carbonate in aluminium production, the leading consumer of spodumene in cast-steel production, the leading user of butyllithium in the production of synthetic rubber and bromide in absorption chillers.

Both Japan and South Korea, the other leading battery producers, lack domestic lithium sources. Historically, Japan has evaluated seawater as a potential source and South Korea is doing so currently. Japan, South Korea and China are participating in many overseas advanced and exploration projects.

Chile and Argentina currently dominate primary chemical production from brine sources but present demand is significantly below capacity. Two of the leading producers produce their downstream products in facilities outside South America – in the United States and Germany in particular.

Prices

Talison reported an average sales price for all grades of spodumene concentrates for the last three months of 2012 of US$367 per tonne.

Elsewhere, price indications are published by Industrial Minerals (IM) magazine.

In early February 2013, prices for the highest grade spodumene, with a minimum of 7.5% Li_2O, ranged between US$720–770 per tonne CIF[1] Asia and between US$750–800 per tonne CIF Europe. IM also indicates US$300–400 per tonne for 5% Li_2O spodumene concentrates CIF Asia and US$440–490 per tonne CIF Europe. Finally, IM also shows a price of US$165–260 per tonne for 4.2% Li_2O petalite concentrates, FOB[2] Durban (Industrial Minerals, 2013).

Prices for lithium carbonate for large contracts delivered to the USA are indicated by IM to be in the range US$2.5–3.0 per pound. Lithium hydroxide prices are quoted by IM as US$6.5–7.5 per kilogram for 56.5–57.5% LiOH delivered under large contracts to Europe or the USA and US$6.0–6.6 per kilogram for Chinese lithium hydroxide of the same grade delivered under large contracts to Europe (Industrial Minerals, 2013).

A leading producer in China, in mid-2011, quoted prices in China, all in tonnes as follows – technical grade carbonate US$4700, battery-grade carbonate US$5300, hydroxide monohydrate US$5760, chloride US$4850 and metal at US$52,000.

A further indication of prices is in SQM's Annual Report for 2011 in which the company reports sales of 40,700 tonnes (a mixture of carbonate, hydroxide and possibly chloride) for revenues of US$183.4 million – an average of US$4506 per tonne.

Comments have been made by various observers regarding the 'dramatic' rise in lithium prices. The facts do not support this. Looking back in time to when Chilean brine-based carbonate entered the market in 1985 it was priced at US$1.45 lb (US$3200/tonne). United States inflation, as reported by the Bureau of Labor Statistics, from that time to the end of 2010 is calculated to be 71 per cent. Applying this to the 1985 price would give a current price of US$5500 tonne, which is in the same range as the figures quoted above.

A major reduction in price occurred when SQM entered the market, but they have gradually recovered and the same company cut prices in response to the major reduction in chemical demand in 2009 (70,000 tonnes) down from 91,500 tonnes in 2008. Again, prices are recovering.

Outlook

Estimating future lithium demand is complicated by the extreme difficulty in estimating future battery demand. Many estimates exist and in the case of vehicle demand, most estimators present ranges covering both 'ultra-green' and conservative scenarios. Difficulties stem from timing, the rate of market penetration of battery-powered vehicles, vehicle types (whether hybrids (HEVs), plug-in hybrids (PHEVs) or pure electric (EVs)) and battery chemistries.

Chemical demand in 2010 totalled between 102,000 and 105,000 tonnes of LCEs. Table 10.4, based principally on an SQM presentation in August 2010 concerning possible changes to the

Table 10.4 Lithium chemical production capacities in 2010. (Data from Solminihac, 2010.)

Company	Country	Source	Tonnes LCE
Continental Brines			
SQM	Chile	Atacama	40,000
Chemetall	Chile	Atacama	38,000
	Nevada	Silver Peak	
FMC	Argentina	Hombre Muerto	17,500
CITIC	China	Taijinaer L	5000
QLL	China	Taijinaer L	2000
Tibet	China	Zhabuye L	2500
Pegmatites			
ABA	China	Maerkang	2500
Jianxi	China	Ningdu	2000
Minfeng	China	Maerkang	2000
Ni & Co	China	Maerkang	5000
XLP	China	Kekeluhai /Talison	5500
Panasia	China	Talison	4000
Tianqi	China	Talison	9500
CBL	Brazil	Cachoeira	2300
Total			**137,800**

Table 10.5 Expanded production and new projects. (Compiled by the author from published sources.)

	Tonnes LCE
Continental Brines	
FMC, Hombre Muerto, Argentina	5500 Expansion
Chemetall, Atacama (Chile) & Silver Peak (Nevada)	12,000–27,000 Expansion
SQM, Atacama, Chile	20,000 Expansion
Chinese brine expansion	?
Zong Chuan, China	5000
Orocobre, Olaroz, Argentina	16,400
Sentient, Rincon, Argentina	15,000
Lithium Americas, Cauchari, Argentina	20,000–40,000
Lithium One, Sal de Vida, Argentina	25,000
Rodinia, Diablillos, Argentina	15,000–25,000
Comibol, Uyuni, Bolivia	20,000
Pegmatites	
Talison, Greenbushes, Australia	83,000*
Galaxy, Mount Catlin, Australia/China	21,000
Reed Resources, Mt. Marion, Australia	19,000**
Altura, Pilgangoora, Australia	19,000**
Canada Lithium, Quebec, Canada	20,000
New & Expansions, China	11,000
Keliber Oy, Finland	6500
Nemaska, Quebec, Canada	30,000**
Lithium One/Galaxy, Quebec, Canada	?
Geothermal Brine	
Simbol, Salton Sea, USA	16,000
Oilfield Brine	
Albermarle, Arkansas, USA	20,000
Hectorite	
Western Lithium, Kings Valley, USA	13,000–27,700
Bacanora, La Ventana, Mexico	35,000
Jadarite	
Rio Tinto, Jadar, Serbia	27,000***
TOTAL (excluding Lithium One/Galaxy and major Chinese brine expansions)	455,400–515,100

*Assumes 90% of estimate production is converted to carbonate at 85% recovery

**Assumes concentrate production is converted to carbonate at 85% recovery

***Not announced. Potential estimate if mined at 1.0 million tpa

Mining Law in Chile, estimates chemical production capacity in 2010 at 138,000 tonnes of LCEs indicating significant excess capacity of approximately 33,000 tonnes.

Table 10.5 lists expansions and targeted production levels announced by would-be producers announced through January 2013. Some of the expansions have been completed and some are under construction. All aim to be in production prior to 2020. If all were to proceed, total chemical capacity would increase to between 593,000 and 643,000 tpa of lithium carbonate equivalents.

Demand estimates for 2020 vary. Evans (2012) listed those of current producers and other responsible organisations at the fourth Lithium Supply & Markets Conference in Buenos Aires. They included those from FMC (dated 2011) of 278,000 tonnes, General Motors (dated 2010) of 215,000 to 243,000, SQM (dated 2011) of 200,000 to 270,000, Roskill Information Services (dated 2011) of 215,500 to 264,500 and Rockwood quoting an average of six estimates (dated 2011) of 220,000 to 320,000 tonnes.

As can be seen from Figure 10.13, demand is estimated to increase rapidly after 2020 with both high- and low-demand scenarios (Chemetall, 2010). Other estimates exist but this one is used for illustration as it is based on six market studies. Using this estimate, demand in 2030 should be between 400,000 and 750,000 tonnes per annum. The lower tonnage can be easily met by current and planned production, but to meet the higher scenario will require further expansions or additional sources.

Finally, it should be said that there is no shortage of sceptics on the issue of vehicle electrification with criticisms of costs, the lack of range and slow charging with many claiming that electric vehicles will only occupy a niche market. Most motor manufacturers do not appear to share these opinions.

Current sources, when the huge upside potential at the Salar de Atacama is taken into consideration, appear to be adequate to meet demand for very many decades. This though, has not reduced the enthusiasm of the Koreans in

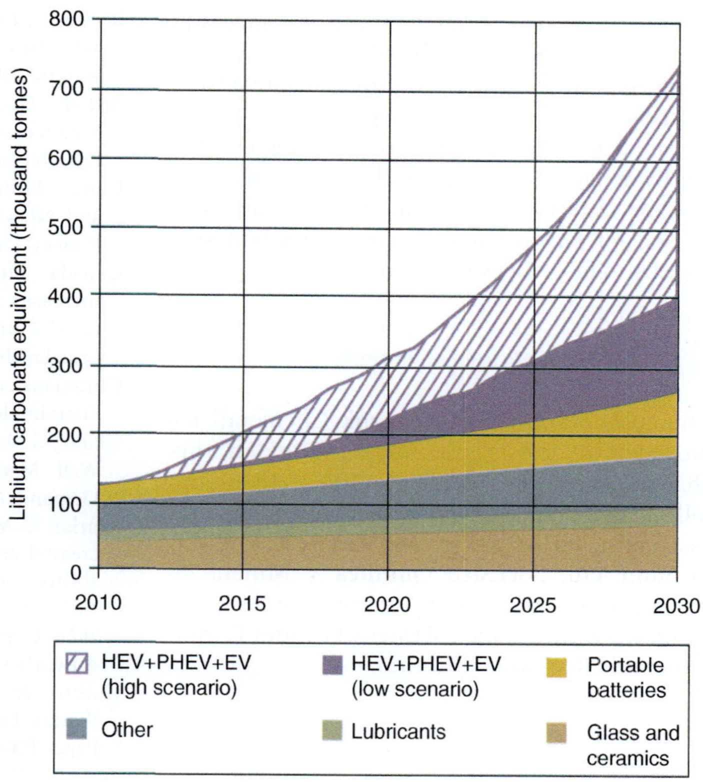

Figure 10.13 Forecast of the demand for lithium to 2030, with two different scenarios relating to the uptake of electric vehicles. (Courtesy of Rockwood Lithium.)

particular and, to a lesser extent the Chinese and Japanese, from acquiring interests in developing projects. These interests range from sales agency agreements for percentages of production to financing feasibility studies, to providing access to low-cost loans and to acquiring significant equity interests. Each of these countries, together with a French vehicle manufacturer, have heavily courted the Bolivian Government for a right to participate in the development of the Salar de Uyuni.

The new projects all claim, by including co-product credits in some cases, to have production costs that are competitive at current prices. With potential over-production until at least 2020, there seems little likelihood of significant price increases over and above inflationary costs.

All demand estimates have large batteries accounting for a high percentage of the total and, although large battery costs are projected to halve as volume increases, the initial cost of a battery-powered vehicle is still high in comparison with one powered by an internal combustion engine and government subsidies are necessary to promote sales. The biggest threat to vehicle electrification appears to be the successful development of biofuels, hydrogen and possibly natural gas.

The requirement for the production of fusion energy is not included in any demand projections although potential demand from this application was the main reason for establishing a National Research Council committee in the United States in the mid-1970s. Any generating system will almost certainly employ a deuterium-tritium (DT) reaction, with the tritium obtained from lithium's ^{6}Li isotope.

Two major research projects are underway: the International Thermonuclear Experimental

Reactor (ITER), located in the south of France; and the National Ignition Facility (NIF) at Livermore, California. The former aims to generate fusion reactions in a doughnut-shaped vessel call a tokomak and the other proposes igniting DT pellets by means of 192 lasers. Lithium requirements will be determined by the ultimate design(s) chosen for reactors and commercialisation still appears to be decades in the future.

Acknowledgements

The author would like to thank Chemetall for provision of the section on recycling in this chapter. The author would also like to thank the following for the provision of photographs and assistance with figures: John Houston, Talison Lithium Ltd, Sociedad Quimica y Minera de Chile S.A. (SQM), Roskill Information Services, Nemaska Lithium Inc., Western Lithium Corporation and Rockwood Lithium.

Notes

1. CIF, Cost, Insurance, Freight. The seller's price includes the cost of the goods, the insurance of the goods to their destination port, and the cost of freight.
2. FOB, Free on Board. The seller is responsible for the cost of delivering goods to the ship. The buyer is responsible for transportation and insurance costs from that point.

References

Achtenberg, E. (2010) Bolivia bets on state-run lithium industry. North American Congress on Latin America. https://nacla.org/node/6799

Altura Mining Ltd. (2012) Positive results from Pilgangoora lithium project scoping study. ASX announcement, 19 November 2012. http://alturamining.com/files/announcements/2012/2012%2011%2019%20-%20Pilgangoora%20Lithium%20Project%20Scoping%20Study%20Positive%20Results.pdf

Bachu, S., Brulotte, M. and Yuan, L.P. (1955) Resource estimates of industrial minerals in Alberta formation waters. Alberta Research Council.

British Geological Survey (2012) World Mineral Production, 2006–10. (Keyworth, Nottingham, British Geological Survey.)

Canada Lithium Corp. (2011a) Feasibility Study Update and NI 43–101 Technical Report, Quebec Lithium Project, La Corne Township, Quebec.

Canada Lithium Corp. (2011b) Canada Lithium increases measured and indicated mineral resource to 33.2 million tonnes grading 1.19%. http://www.canadalithium.com/s/Home.asp or www.sedar.com

Chmelauskas, J. (2010) Kings Valley Lithium Project, Shareholder Presentation.

Clarke, G.M. and Harben, P.W. (2009) Lithium Availability Wall Map – Lithium Availability a Question of Demand. AABC, Long Beach, California, USA.

Clarke, G.M. (2011) Lithium Availability Wall Map created and published as a part of "Lithium availability: prospective demand being answered." Advanced Automotive Battery Conference, Mainz, 2011. Copies available on application to: gmclarke1@ntlworld.com

Collins, A.G. (1976) Lithium Abundance in Oilfield Waters. United States Geological Survey Professional Paper 1005.

de Silva, S.L., Zandt, G., Trumball, R., Viramonte. J.G. Sales. G. and Jiminez, N. (2006) Large ignimbrite eruptions and volcano-tectonia depressions in the Central Andes: a thermomechanical perspective. In: Troise, C., De Natale, G. and Kiburn, C.R.J. (eds.) Mechanisms of Activity and Unrest at Large Calderas. Geologist Society, London. Special Publication No. 269, 47–63.

Duyvesterin, W.P.C. (1972) Recovery of Base Metals from Geothermal Brines. Geothermics Journal of Geothermal Research and its Applications 21 (5/6).

Evans, R.K. (1986) Further Developments at the Salar de Atacama. Industrial Minerals, International Congress.

Evans, R. K. (1978) Lithium Reserves and Resources Energy 3 (3).

Evans, R. K. (2008) Know Limits. (An Abundance of Lithium) Industrial Minerals.

Evans, R.K. (2012) An overabundance of lithium? Lithium supply and markets conference, Buenos Aires, January 2012.

Galaxy Resources Ltd. (2011) Galaxy extends Mt Cattlin mine life after resource upgrade. ASX announcement /

Media release, 22 March 2011. http://www.galaxyresources.com.au/investor_asx.shtml

Galaxy Resources Ltd. (2013a) Sal de Vida, Argentina. http://www.galaxyresources.com.au/projects_sal_de_vida.shtml

Galaxy Resources Ltd. (2013b) James Bay, Quebec, Canada. http://www.galaxyresources.com.au/projects_james_bay2.shtml

Harben, P.W. (2002) Lithium Minerals and Chemicals. In: The Industrial Minerals HandyBook, 4th Edition. Metal Bulletin plc, London, 184–192.

Houston, J., Butcher, A., Ehrens, P., Evans, K. and Godfrey, L. (2011) The evaluation of brine prospects and the requirement for modifications to filing standards. Economic Geology, 106, 1225–1239.

Houston, J. and Gunn, M. (2011) Technical Report for the Salar de Olaroz lithium-potash project, prepared for Orocobre Ltd. www.sedar.com

Ide, Y.F. and Kunasz, I.A. (1989) Origin of Lithium in the Salar de Atacama, Chile. In: Ericken, G.E., Cañas Pinochet, M.T. and Reinemund, J.A. (eds.) Geology of the Andes and its relation to hydrocarbon and mineral resources. Circum-Pacific Council for Energy and Mineral Resources. Earth Science Series 11, 165–172.

Industrial Minerals. (2013) Industrial Minerals Prices. http://www.indmin.com/Prices/Prices.aspx

Kellie, R. (2009) Jadar Lithium Project, Serbia. Proceedings of Lithium Supply and Markets Conference, Santiago, Chile. 28 January 2009. Chief Geologist, Rio Tinto Exploration.

Kunasz, I.A. and Pavlovic, P.Z. (1983) The Salar de Atacama - a new centre for lithium production. In: Cope, B.M. and Clarke, G.M. (eds.) Proceedings of the 5th Industrial Mineral International Congress. Metal Bulletin plc, London

Kunasz, I. (2007) Lithium Resources. In: Kogel, J. (ed.) Industrial Minerals and Rocks 7th Edition, Society for Mining, Metallurgy and Exploration Inc. (SME), Englewood, Colorado, USA, 599–613.

Lithium Americas. (2011) NI 43–101 Technical Report, Preliminary Assessment and Economic Evaluation of the Cauchari-Olaroz Lithium Project. In: WorleyParsons, A.R.A (ed.) www.sedar.com

Mares, D.R. (2010) Lithium in Bolivia – can resource nationalism deliver for Bolivians and the world? In: James A. (ed.) Baker III Institute for Public Policy.

Nemaska Lithium Inc. (2012) Nemaska Lithium announces positive preliminary economic assessment (PEA) for Whabouchi mine and lithium hydroxide/carbonate plant. Press release 2 October 2012. http://nemaskalithium.mwnewsroom.com/press-releases/nemaska-lithium-announces-positive-preliminary-eco-tsx-venture-nmx-201210020823456001

Orocobre Ltd. (2011) Orocobre, the next low cost lithium producer. Investor Presentation 14 December 2011. http://www.orocobre.com.au/PDF/16Dec2011_Investor%20Presentation.pdf

Orocobre Ltd. (2013). Salar de Olaroz. http://www.orocobre.com.au/Projects_Olaroz.htm

Reed Resources Ltd. (2012) Shareholder agreement and PFS results Mt Marion lithium. 17 October 2012. http://www.reedresources.com/announce-blog.php?id=656andprojpg=mtmar

Risacher, F. (1989) Economic Study of the Uyuni Salar UMSA-ORSTOM-CIRESU PACT.

Risacher, F., Alonso, H. and Salazar, C. (2003) The Origin of Brines and Salts in Chilean Salars-a Hydrochemical review. Earth Science Reviews 63, 249–293.

Rodinia Lithium Inc. (2011) Rodinia Lithium files favourable preliminary economic assessment technical report. http://www.rodinialithium.com/ and www.sedar.com

Solminihac, P. (2009) SQM Lithium Resources and View of the Lithium Industry. Presentation to Lithium Supply Markets conference, Santiago, Chile, January 2009.

Solminihac, P. (2010) Resources de Lithio en el Mundo y Chile. Presentation by SQM concerning possible changes in the Mining Law in Chile.

Talison Lithium Ltd. (2011a) Annual Information Form for year ended 30 June 2011. http://www.talisonlithium.com/home and www.sedar.com

Talison Lithium Ltd. (2011b) Talison Lithium commences expansion to double capacity at the Greenbushes lithium operations. News Release.

UN Comtrade (2013) United Nations Commodity Trade Statistics Database, Department of Economic and Social Affairs/ Statistics Division, http://comtrade.un.org/db/

U.S. Geological Survey (2009). Mineral commodity summaries 2009, Lithium. http://minerals.usgs.gov/minerals/pubs/commodity/lithium/mcs-2009-lithi.pdf

U.S. Geological Survey (2010). Mineral commodity summaries 2010, Lithium. http://minerals.usgs.gov/minerals/pubs/commodity/lithium/mcs-2010-lithi.pdf

U.S. Geological Survey (2011). Mineral commodity summaries 2011, Lithium. http://minerals.usgs.gov/

minerals/pubs/commodity/lithium/mcs-2011-lithi.pdf.
Verley, C.G. and Vidal, M.F. (2013) Preliminary Economic Assessment for the La Ventana lithium deposit, Sonora, Mexico for Bacanora Minerals. 24 January 2013. http://www.bacanoraminerals.com/reports/pdf/pea_ventana.pdf
Western Lithium Corp. (2011) Western Lithium Pre-feasibility Study indicates $550 million NPV and supports low cost lithium production in Nevada. News Release.
Wright, L. (2010) Lithium Dreams. The New Yorker.
Yaksic, A. and Tilton, J. (2009) Using the Cumulative Availability Curve to Assess the Threat of Mineral Depletion: The Case for Lithium Policy 34.
Zheng, M. (1989) Saline Lakes on the Quinghai-Xizang (Tibet) Plateau. Beijing Scientific and Technical Publishing House, Beijing.

11. Magnesium

NEALE R. NEELAMEGGHAM[1] AND BOB BROWN[2]

[1] 'Guru', Ind LLC, 9859 Dream Circle, South Jordan, Utah, USA
[2] Publisher, Magnesium Monthly Review, Prattville, Alabama, USA

Introduction

Magnesium has a variety of special properties which make it useful in broad range of applications. On account of its high strength and light weight it is the most commonly used structural metal after steel and aluminium. The fact that the density of magnesium (1738 kg/m^3, Table 11.1) is only 63 per cent of that of aluminium and 23 per cent of that of steel has promoted its use in many transportation applications where whole life-cycle energy considerations are important and may be a critical factor for energy conservation in the 21st century. About half of the magnesium used today is in the production of alloys, most commonly with aluminium. These have important applications in the food industry (for example, in drinks cans) and in the aerospace and military sectors. Magnesium metal is also unique in that it can be used as the reducing agent in the production of several special metals, such as beryllium, zirconium and uranium, which are strategically important for defence and nuclear reactor applications, and titanium, which is used in the chemical and aerospace industries.

Magnesium is the only structural metal that can be obtained economically from both the lithosphere and hydrosphere. Magnesium can be produced from oxide resources or from chloride raw materials originating from saline water. This has resulted in the development of two main-stream production technologies, unlike many other metals which use minor variations of a single process technology.

Concern over the security of supply of magnesium has arisen since the late 1990s, not due to constraints on the availability of raw materials or technology, but rather due to changes in the pattern of the global supply chain. Over the past 20 years or so the production of magnesium metal has increased in China, whilst production in many other countries has declined or ceased. During the twentieth century western world production increased to about 400,000 tonnes per year, while Chinese production was reported to be about 3000 tonnes per year in 1991. However, by 2008, the primary magnesium metal production capacity of the western world, including Russia, had decreased to about 120,000 tonnes per year while in China capacity increased to more than 600,000 tonnes per year.

Physical and chemical properties

Magnesium (chemical symbol Mg) has an atomic number 12, an atomic weight of 24.32 and it exhibits a valency of two. It occurs in three stable isotopes, ^{24}Mg, ^{25}Mg and ^{26}Mg, of

Critical Metals Handbook, First Edition. Edited by Gus Gunn.

Table 11.1 Selected properties of magnesium.

Property	Value	Units
Symbol	Mg	
Atomic number	12	
Atomic weight	24.31	
Density solid Mg at 25 °C	1738	kg/m^3
Density liquid Mg at 700 °C	1584	kg/m^3
Melting point	650	°C
Boiling point	1090	°C
Hardness (Mohs scale)	2.5	
Specific heat capacity at 25 °C	1.05	J/(g °C)
Latent heat of fusion	368	kJ/kg
Latent heat of vaporisation	5251 ± 251	kJ/kg
Thermal conductivity	156	W/(m °C)
Electrical resistivity at 25 °C	44.8	nΩ m
Young's modulus (at 20 °C, 99.8% pure Mg)	44.7	GPa

which the former accounts for about 79 per cent of the total. It makes strongly ionic compounds with common halide, sulfate or phosphate anions as well as making weakly ionic but covalent compounds of magnesium oxide and nitride.

Magnesium, in its structural applications, is typically used as an alloy casting or as a wrought alloy. The mechanical properties of these materials can vary quite widely, with tensile strength in the range 150 to 400 megapascals (MPa). The cast alloys have lower tensile strength and elasticity than wrought alloys or shapes made by powder metallurgy using nanotechnology. However, the properties of die cast magnesium shapes are adequate for many automotive components and for portable electronic devices where 'light-weighting' is important.

Magnesium can be used as a reducing agent because of its high position in the electrochemical series. It is one of the most electronegative elements, only being surpassed by the alkali metals, higher alkaline earth metals and several rare earth metals. It forms more stable compounds (oxides, sulfates, carbonates, nitrates, halides, etc.) than most other common metals.

Distribution and abundance in the Earth

Magnesium is the fifth most abundant element in the hydrosphere and the most abundant structural metal ion in the ocean. It is therefore unique among the structural metals in that it can be extracted from either the hydrosphere or the lithosphere. In contrast, aluminium is only sparingly soluble in the oceans, and is extracted from lithospheric resources only. In the upper crust magnesium is the eighth most abundant element (2.5% MgO by weight (Rudnick and Gao, 2004)). It is an important constituent of major rock-forming silicate minerals, such as pyroxene and olivine. Consequently, rocks rich in these minerals, such as gabbro and basalt, have higher magnesium contents, commonly exceeding 5% MgO.

Irrespective of the source of the raw material, an energy input is required to effect the conversion of the mineral into metal. The nature and cost of the energy source is a particularly important consideration in the choice of the process route for the production of primary magnesium.

Mineralogy

The solid mineral phases from which magnesium is extracted from the lithosphere are: dolomite ($CaCO_3.MgCO_3$), magnesite ($MgCO_3$), periclase (magnesium oxide, MgO), hydromagnesite ($3MgCO_3.Mg(OH)_2.3H_2O$), brucite ($MgO.H_2O$), and various silicates of magnesium, such as olivine ($(Mg,Fe)_2SiO_4$), serpentine ($3MgO. 2SiO_2. 2H_2O$) and biotite ($K(Mg,Fe^{2+})_3AlSi_3O_{10}(OH,F)_2$).

The solid mineral phases of hydrospheric origin, found in evaporite deposits, from which magnesium is extracted are: magnesium sulfate (epsomite, $MgSO_4.7H_2O$), kieserite ($MgSO_4.H_2O$), langbeinite ($K_2SO_4.2MgSO_4$), kainite ($KCl. MgSO_4.3H_2O$) and carnallite ($KCl. MgCl_2.6H_2O$) and bischofite ($MgCl_2.6H_2O$). Many of these were initially identified in Stassfurt, Germany in the mid-nineteenth century. It should be noted that magnesium is the second most abundant cation in solution in the oceans and terminal lakes.

Sodium is present in larger amounts usually providing the ionic balance for the chloride ions, with the ionic balance for magnesium provided by sulfate and chloride ions.

Deposit types

Of the potential lithospheric sources dolomite (calcium–magnesium carbonate) is the most commonly used. Thus, calcined dolomite, MgO. CaO, with a minimum magnesium content of 8 per cent, is one of the main sources of magnesium. Alternative sources are magnesite and magnesium silicates, which have been used as raw material for magnesium production during the past two decades. The magnesium silicate can be in the form of olivine or from the tailings of asbestos extraction where it is present as serpentine minerals.

In saline lake water or brines magnesium is present at lower concentrations, typically less than 0.7 per cent by weight, while in sea water the magnesium content is 0.13 per cent. Two major undesirable components, water and sodium chloride, must be removed from these resources before attempting to extract magnesium. Other major contaminants are potassium and sulfate.

Extraction methods, processing and beneficiation

A historical review is given to provide background to the development of production processes that are currently in use. Brown (2000, 2003a) provides an in-depth history of magnesium, including a review of production methods, up to 2003. Neelameggham (2013) provides detailed information on the fundamental chemistry and production technology needed to produce magnesium from different raw materials.

The processes used in the production of magnesium have gone through several major changes which almost mirror changes to the global supply chain over the same period – whether it be during the two World Wars, the Cold War through the 1980s or the emergence of the global economy since the 1990s, and through the recent commodity rise and fall between 2006 and 2009. In 1935, John A. Gann, Chief Metallurgist of The Dow Chemical Company, noted the following – ".. our light metals, occur only in the form of compounds so stable that their discovery, isolation, commercial production, and use were forced to await some of the modern advances in chemistry and engineering. Under such conditions, the evolution of a new industry is often a romance in which scientific and industrial difficulties and near failures adds to the thrill of success." (Gann, 1935). This statement has been verified time and again in the development of magnesium production processes and also in the development of new alloys and uses of magnesium.

There are two main routes for making magnesium metal:

1. The electrolytic method – magnesium chloride is the feed material to this process. The magnesium chloride is derived from magnesium oxide sources (by chlorination of magnesia or magnesite) or from chloride sources (by dehydration of brines or hydrous carnallite) (Figure 11.1).
2. Metallothermic reduction – the dolomite feed material is calcined and then reduced, either by silicon or one of its alloys, or by other metals such as aluminium. One of the most important processes of this type, which uses ferrosilicon as the reductant, is known as the Pidgeon process (Figure 11.2).

All magnesium metal production utilises the following general steps, although the specific physical and chemical features of the source raw materials lead to considerable variation in the process technology employed:

- Raw material upgrading: in the electrolytic process this involves removal of sulfates, and other impurities from the magnesium chloride. For the metallothermic process the proportion of dolomite in the feed should be as high as possible with the least amount of non-reacting components, such as silicates.
- Removal of impurities: for example, in the magnesium chloride feed preparation all non-chloride anions (such as sulfates, borates, etc.) are

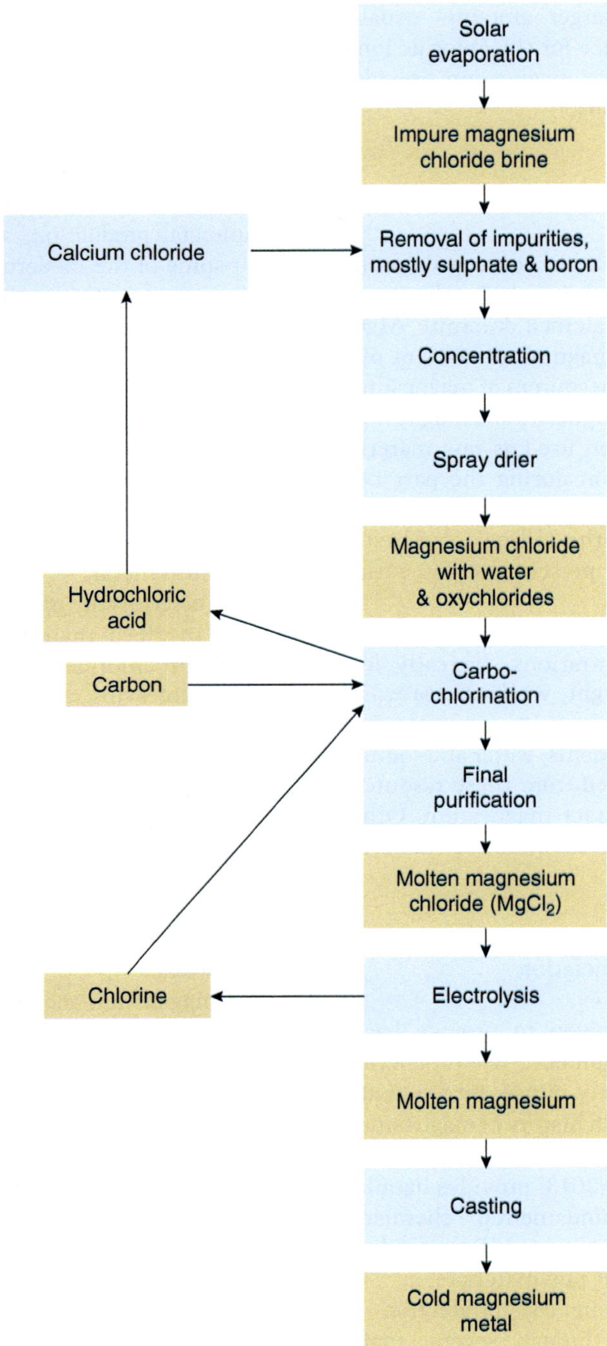

Figure 11.1 Schematic flowsheet for U.S. Magnesium's electrolytic process in Utah. (Modified from Holywell, 2005.)

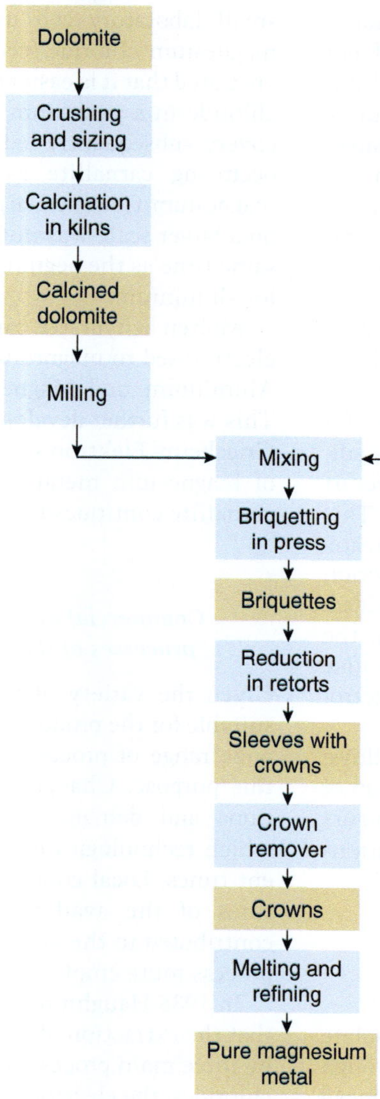

Figure 11.2 Schematic flowsheet for the production of magnesium by the Pidgeon process.

removed and also most of the water by preparing either close to a monohydrate of magnesium chloride ($MgCl_2.1.3H_2O$) or essentially anhydrous magnesium chloride, depending on the electrolytic process chosen. In the case of the calcined dolomite reduction process, the contents of calcium oxide and magnesium oxide are maximised by reducing the content of other components which do not contribute to the process. This is accomplished by using high-grade dolomite as the raw material feed, thereby reducing the proportion of other components which do not take part in the reaction and reduce retort production capacity.

- Removal of impurities undesirable in the finished metal: for example, one of the most undesirable impurities in the finished metal is nickel, which should be kept below 0.001 per cent to avoid the product magnesium from becoming corrosive. In practice, impurity control is achieved mainly through measures aimed at avoiding the incorporation of impurities into the process. The unalloyed magnesium normally marketed has several grades, depending on the customer requirement, normally in the range from 99.8 per cent pure to 99.95 per cent. About 30 to 40 per cent of the magnesium is marketed as alloys containing 91 to 97 per cent magnesium.

- Separation from other process products: for example, in the Pidgeon process a calcium silicate residue containing iron remains in the retort after removal of the magnesium product. This by-product, which is mostly calcium silicate, may be used as a substitute cement in construction either by itself or with other ingredients added. Different by-products may be derived from the electrolytic process. Typically, these include chlorine gas and excess mixed chloride electrolyte from the electrolysis.
- Melting, refining and casting metal and alloys: the typical product is magnesium in the form of ingots of the pure metal, although some plants supplying magnesium for chemical applications supply magnesium as a powder.

Nineteenth-century magnesium production processes

Humphrey Davy was the first person to isolate magnesium. In 1808 he passed a current through moistened magnesium sulfate and produced magnesium at a mercury cathode. He also converted red hot magnesium oxide with potassium vapours collecting the magnesium into mercury. In 1828, Bussy reduced magnesium chloride with potassium metal in a glass tube and, when the potassium chloride was washed out, small globules of magnesium were present.

In 1833 Faraday electrolysed impure magnesium chloride in a molten state to get magnesium metal. However, it took two more decades before Robert Bunsen made commercial quantities in a small laboratory cell using molten anhydrous magnesium chloride feed. In 1852 Bunsen demonstrated that it is easier to dehydrate magnesium chloride in a potassium chloride bath. This discovery subsequently led to the use of naturally occurring carnallite as a source for making magnesium. Commercial production of magnesium on a larger scale was initiated in 1886, about the same time as the beginnings of Hall–Heroult cell for aluminium.

Molten dehydrated carnallite ($KCl.MgCl_2$) was electrolysed to magnesium metal in 1886 by the Aluminium und Magnesium Fabrik, Germany. This was further developed by Chemische Fabrik Griesheim–Elektron starting in 1896. Production of magnesium metal by electrolysis of molten carnallite continues to the present day.

Commercial magnesium production processes of the twentieth century

Given the variety of raw materials potentially suitable for the production of magnesium metal a wide range of processes has been developed for this purpose. Changing global economic conditions and demand patterns have determined which technologies have been adopted at different times. Local conditions and the specific features of the available source materials have contributed to the development of the particular process route employed in any one area.

In 1938 Haughton and Prytherch (1938) noted that the extraction of magnesium was carried out by three main processes: the electrolysis of fused chlorides; the electrolysis of the oxide in solution of molten fluorides; and the carbo-thermal process which involves direct reduction of the oxide by carbon in an arc furnace with a hydrogen atmosphere followed by re-distillation in an inert atmosphere. During World War II, the silico-thermic reduction of oxide technology was developed which gave rise to the important Pidgeon process, which was named after its inventor, Lloyd Pidgeon (Pidgeon and Alexander, 1944).

The Pidgeon process (Figure 11.1) uses a briquetted charge of finely ground calcined dolomite and ferrosilicon that is placed in a horizontal alloy retort with a diameter of ten inches. The retort is externally heated to 1200 °C in a furnace and a vacuum is applied to it. The silicon of the ferrosilicon reduces the magnesium oxide in the charge and the resulting magnesium vapours flow to a condenser in the portion that extends from the furnace. This is a cyclic process with the retorts being charged and discharged several times per day. The Chinese were introduced to the Pidgeon process in the late 1980s. Since then they have built hundreds of silico-thermic plants based on this process and China is now the world's largest magnesium producer.

Between 1977 and 2002 Alcoa produced magnesium by Magnetherm technology, a variation of silico-thermic reduction (Jarrett, 1981 and Faure and Marchal., 1964). This process was used at Addy in Washington, USA, and in Marignac in France, where its use was discontinued in 2002. The Rima Group in Brazil, the only producer of magnesium metal in the southern hemisphere, uses another variation of silico-thermic reduction, known as the Bolzano process, which was used in Bolzano, Italy, up to 1992. In this method briquettes of dolime (calcined dolomite) and ferrosilicon are processed in a larger retort which is internally heated.

Electrolytic magnesium technology, chiefly developed in Germany, has been described in detail by Strelets (1977). This technology was subsequently improved in the Soviet Union by VAMI (the Soviet Aluminium and Magnesium Institute) and at the Soviet magnesium plants in Solikamsk, Russia, Zaporozhye, Ukraine, and Kazakhstan (Muzhzhavlev et al., 1965 and 1977). Improved versions of these electrolytic methods are still in use – at the Dead Sea Magnesium Plant in Israel since 1995 and at Solikamsk in Russia. Both use carnallite as the starting material, which is derived from mining evaporite deposits in Russia and from solar evaporation in the Dead Sea.

An economic comparison of the carbo-thermal, silico-thermal and electrolytic processes was made by the U.S. Bureau of Mines (Dean, 1965). Information on these three processes, which were in operation in the US between 1930–1960, is presented in detail in the U.S. Bureau of Mines Reports of Investigations (Elkins et al., 1968). Some of the technical aspects of these analyses are still valid 40 years later.

The carbo-thermal process developed by Hansgirg was in commercial production only between 1930–1950. Efforts to revive it in an economically viable form have continued to the present day, mostly in the laboratory, but sometimes on a pilot-plant scale (Hansgirg, 1932 and Brooks et al., 2006). In the 1930s, magnesium oxide was converted to magnesium using calcium carbide as a reductant, by the Murex process (Beck, 1939 and Emley, 1966). Although it is not used commercially today, this process may be applicable whenever the calcium carbide required for reduction is cheaper than ferrosilicon.

The demand for magnesium fell sharply after World War II, giving rise to competition between the available production technologies. Carbothermal reduction and fluoride-fluxed magnesium oxide melt electrolytic processes were no longer viable on account of the high costs associated with specific technical issues. The silico-thermic process continued to take a backseat to the electrolytic conversion of magnesium chlorides until the mid-1990s. However, broader Chinese engagement in the global economy from the early 1990s has had a major impact on the supply chain for magnesium metal. The availability of very low-cost ferrosilicon in China led to a revival of the silico-thermic method, making it the dominant process of the first decade of the twenty-first century.

Specifications and uses

For structural applications, magnesium is alloyed with other metals to give suitable properties for castings and for wrought or formed material. The alloy elements used are identified by the use of prefixes (A for aluminium, M for manganese, Z for zinc, etc.). The tensile properties of cast magnesium are much lower than wrought magnesium. Typical casting alloys, such as AM50, AM60 and AZ91, contain 5 to 9% aluminium, 0.5 to 0.7% zinc, and 0.2 to 0.5% manganese. Typical wrought alloys contain less aluminium, such as AZ31 and AM30, as well as the M1 alloy (99.3% Mg and 0.5–0.7% Mn).

In terms of volume the largest use of magnesium has for many years been as an additive to impart stiffness to aluminium alloys. For example, aluminium foil, which is made from essentially pure aluminium, can be rolled into very thin, flimsy sheets. However, with the addition of magnesium or other alloying additives, aluminium becomes a structural metal. Adding as little as one per cent magnesium provides aluminium alloys with enough stiffness and good forming properties making it suitable for the body of beverage cans. The beverage can top is stiffer with three per cent magnesium, and the closures are stiffer still with 4.5 per cent magnesium. Magnesium is present in most of the aluminium alloys used in structural extrusions and sheet. Since the global production of aluminium alloys, from primary and secondary sources, is about 50 million tonnes per year it is easy to see how this use consumes the largest proportion of magnesium metal production.

Design engineers select materials to perform specific functions in large volume applications such as in the automotive and the 3-C (computers – cameras – communications) industries based on the following criteria, as described by Avedesian (1999):

(a) High strength to weight ratio to minimise life-cycle energy consumption.
(b) Good damping capacity.
(c) Suitability for high-speed production processes.
(d) Long life for the tooling and machinery parts used in the fabrication process, e.g. the wear of steel moulds by molten magnesium is low compared with molten aluminium.
(e) Electromagnetic shielding properties.
(f) Long lifetime of the finished magnesium parts as shown by low corrosion rates in each application. This depends on the composition of the magnesium alloys as well as effective design, for example, by isolating joints between dissimilar metals.

Magnesium metal has a long history of use in the automotive and aerospace sectors dating back to the 1920s. Each of the 24 million Volkswagen Beetles manufactured had a magnesium transmission and engine block. All the engines in German aircraft during the 1930s and throughout World War II were joined to the fuselage by the use of forged magnesium engine bearers.

Many new uses of magnesium and magnesium alloys were triggered by the need to reduce the weight of the automotive to improve fuel economy. The special properties of some wrought alloys have resulted in the development of specific niche applications. For example, the AZ31 alloy has been used to make printing plates for over sixty years making use of masked chemical etching techniques. However, the complexity of the conventional multi-step process for making wrought alloys kept the growth of the market for wrought magnesium low for a long time. However, the twin roll casting technique, which was originally developed in the late 1970s, reappeared in a modified form in the early 2000s, initially in Australia, later in Taiwan and recently in South Korea. This method with fewer processing steps in making magnesium sheet is now facilitating more applications for wrought magnesium. As a result, there is now potential for growth in this market in automobiles with a magnesium frame and body.

Interest in using magnesium alloys for military applications has recently been revived, for example, in plates in military armour, and in lightweight applications, including signal system electronic and radio machinery (Jones and DeLormo, 2008; Jones and Kondoh, 2009). In addition, magnesium powder has found a new application in the 'magnesium ready to eat' (MRE) food packages used by military personnel in field situations, where the reaction between metallic magnesium and water provides the energy needed for warming the food.

A significant market has developed over the past 30 years for using magnesium powder to reduce the sulfur content of molten iron to the low levels required for making high-quality steel sheet. The high electronegativity of magnesium provides the potential to reduce oxides which cannot be reduced by carbon or aluminium. Magnesium is also essential to the production of some strategic metals used in specialist defence and nuclear technologies, such as beryllium, uranium and zirconium. Magnesium is also the preferred reductant in the production of titanium from titanium chloride. In the organic chemicals industry magnesium has been used in making Grignard reagent, an important organic synthesis intermediary, for over a century.

Magnesium metal is the preferred reducing agent in the preparation of metals such as titanium, zirconium, uranium and beryllium from the respective chlorides or fluorides. In one such application, the Kroll Process, magnesium is used as the reducing agent in the production of titanium metal from titanium tetrachloride. The system is designed so that the magnesium chloride generated in the reduction process is recovered and is utilised to produce magnesium, making this process essentially a 'closed-loop' system in respect to magnesium. Global production of titanium metal is estimated to be about 120,000 tonnes per year of which more than 90 per cent is made by the Kroll process. Since one tonne of magnesium is required to produce one tonne of titanium in the Kroll process it is estimated that about 120,000 tonnes per year of magnesium is produced and re-used in this process. Based on the volumes of magnesium they buy, it appears that titanium producers purchase about one to five per cent of their magnesium requirement to replace process losses.

A review of the world usage divided among different categories with emphasis on North America was presented by Slade (2011). This author suggested that the outlook for magnesium was brighter in 2011 than in 2010. In terms of usage by region, China is projected to continue to be the largest user with over one third of the total. North American consumption in 2011was estimated to be 135,000 tonnes with a similar level of demand in Europe. Other important consumers are in Asia (Japan, Korea, etc.), Russia, and Brazil. The main market segments are aluminium alloying, die casting, permanent mould casting, sand castings, wrought products, cathodic protection, desulfurisation of iron, nodular iron production, as a reducing agent for titanium, and other metal production.

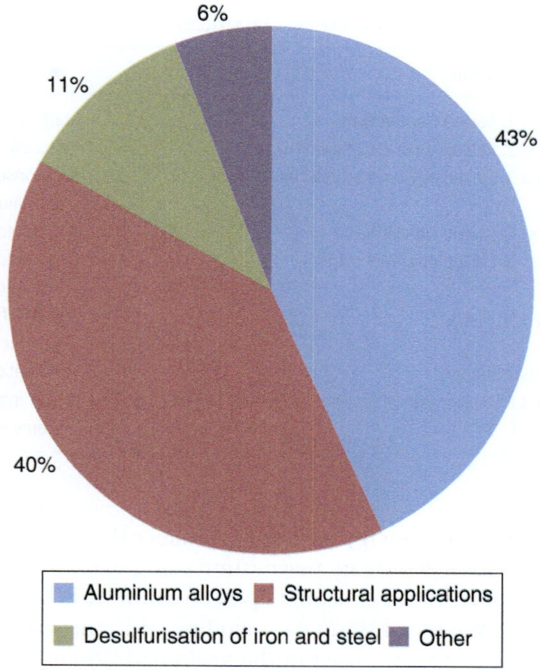

Figure 11.3 The main sectors of magnesium use in the USA in 2011. (Data from United States Geological Survey, 2012.)

In 2011 in the United States the main use of magnesium metal (43 per cent of total) was in aluminium-based alloys used for packaging, transportation, and other applications (USGS, 2012). Structural uses of magnesium (castings and wrought products) accounted for 40 per cent of primary metal consumption, with desulfurisation of iron at 11 per cent and other uses at six per cent (Figure 11.3).

Recycling, re-use and resource efficiency

It is very important to utilise natural resources, particularly metals, in a sustainable manner. For magnesium, it is particularly important to optimise

Table 11.2 Magnesium scrap classification. (After Fechner et al., 2009.)

Type of Scrap	Description
Class 1: Sorted clean returns	Start up shots, gates, runners, sprues, biscuits, trimmings, rejected parts
Class 2: Sorted clean returns with inserts	Rejected castings containing metal inserts
Class 3: Sorted oily painted returns	Post-consumer parts or parts rejected after painting/coating operations, may contain non metallic inserts
Class 4: Sorted clean chips	Generated during dry machining of magnesium products
Class 5: Sorted oily/wet chips	Generated during machining of magnesium products using oil or oil/water emulsions
Class 6: Dross	Salt free furnace cleaning products, mostly oxides from the melt surface, also residues from the bottom of the crucible
Class 7: Sludge	Salt containing cleaning residues mostly from the bottom of the crucible
Class 8: Mixed and off-grade returns	Mixed magnesium–including post consumer scrap–off grade magnesium and items not included above

the use of the energy invested to produce the metal. Production of primary magnesium using an electrolytic process can consume up to 35 kilowatt-hours per kilogram (kWh/kg), including the thermal energy equivalent used in the feed preparation. By comparison, re-melting magnesium metal scrap uses less than ten per cent of the energy used to make primary metal.

The low density of magnesium makes it attractive as a material of construction, particularly for transport applications. However, the low weight of magnesium can also be a penalty according to some methods of calculation that are based on weight rather than volume. Albright and Haagensen (1997) noted that the energy consumption (at the primary plant site only) to produce magnesium, aluminium and steel are estimated at 35, 30 and 11 kWh/kg, respectively. However, when these data are considered on a unit volume basis, which is the common design concern (that is, components must be designed or packaged into a given space or volume), the life-cycle advantage of magnesium is clearer. On this basis, the above numbers correspond to 63, 81, and 87 kWh/litre for magnesium, aluminium and steel, respectively. For recycled materials, the relative energy used is even more favourable for magnesium, with recycled magnesium energy consumption about one kWh/litre.

Magnesium is a metal that is easily recycled in many of its primary forms. It is also capable of being recycled from secondary sources when it appears as an alloy with aluminium. The recycling of aluminium cans (Used Beverage Containers, (UBC)) has become a highly technical and controlled process. The average magnesium content of the UBC scrap is about two per cent. By careful handling, most of the magnesium is recovered in the recycled aluminium. In the author's estimate, in 2010, approximately 740,000 tonnes of cans were recycled, saving nearly 15,000 tonnes of magnesium in the recovered alloy.

Magnesium scrap classification follows a system devised by the magnesium alloy producers which was standardised by the International Organisation for Standardisation (ISO) in 2005 (Table 11.2). This classification was developed in the 1980s to ensure the purity of magnesium metal and alloy derived from secondary sources and thereby to reduce problems associated with corrosion.

Class 1 scrap is the most frequently and most easily recycled. This material consists of gates and runners from die casting, together with scrap castings and used parts that have been isolated from other scrap. Class 1 magnesium scrap recovery is mostly carried out by special smelters and refiners who take the scrap from the die casters and return clean ingot with closely specified alloy chemistry. This is called 'toll melting' since the customer is only charged a conversion fee.

It is estimated by the author that the total magnesium recycling capacity in the world is about 200,000 tonnes per year with actual recycling of about half that amount in 2010. North America has a Class 1 recycling capacity of 75,000 tonnes per year. In addition, it has been estimated that 30,000 tonnes per year could be recycled internally by the die casting companies. Class 1 scrap recycling capacity in Europe has been estimated to be 75,000 tonnes per year (Fechner et al., 2009). Asia, particularly China, has many primary plants and many of these utilise scrap returned by their clients.

There are two basic methods of recycling magnesium: flux-free recycling and recycling with flux. The fluxes used are chloride-based salts which have been used for many years to separate oxide and other solid inclusions from the molten magnesium. In recent years, specially designed furnaces have been developed to enable melting to be done without flux. Instead they use inert atmospheres to prevent the molten magnesium from oxidising during melting and refining.

The quality and characteristics of scrap may vary widely, even from one producer of magnesium die castings to another. Even Class 1 scrap can vary and, consequently, recycling contracts are typically based on a preliminary evaluation of representative material. This is followed by a detailed quality plan produced in collaboration with the generator of the scrap to ensure continuing conformance to the agreed specification.

The value and cost to the die caster are highly dependent on the quality of the metal recovered. Major components of the recycling costs are:

1. value of input material – related chiefly to its price and to transportation costs;
2. processing – related to the costs involved in pre-treatment, melting and casting; and
3. residue disposal – costs associated with the digestion and disposal of waste residues.

The industry recycling programmes for magnesium are applied to clean and compact scrap of known composition. This material is remelted, refined and re-used for manufacturing magnesium components. The metal from the other scrap classes can be recovered, but it is mainly used for the production of iron desulfurisation reagents or as an alloy element in an aluminium alloy production process.

The processing of scrap with a low metal content and of sludges from the melting and recycling process still poses major problems. As the utilisation of magnesium alloys in the automobile industry grows, and more attention is directed to the management of scrap from end-of-life vehicles, more technical developments will be required to cover the treatment of all magnesium scrap classes. Several large European, North American and Asian research organisations are working on these problems.

Substitution

Magnesium alloys are used in many applications including the aerospace, automotive and the 3-C industries. The properties of magnesium, such as its strength-to-weight ratio, its low fabrication costs and other special properties, such as electromagnetic shielding, thermal and electrical conductivity, and damping characteristics, combine to make it difficult to find a direct substitute for magnesium.

From time to time, some plastics or composites have been proposed as potential replacements for magnesium especially for applications with short product lives. However, where longer lives and higher durability are required magnesium parts generally deliver superior performance – for example, in notebook computer cases and automotive instrument panels.

Some potential problems when magnesium components are being joined can be addressed by choosing fasteners with similar thermal expansion coefficients, such as aluminium or magnesium fasteners. Neelameggham (2013) notes that the use of fasteners made of aluminium or magnesium would also overcome the problems caused by the inappropriate use of steel fasteners which are the root cause of many galvanic corrosion issues with magnesium, as well as often-publicised creep problems. When a steel fastener is used, its lower thermal expansion against the increased expansion of magnesium causes stresses when the temperature of the part increases, and when the part cools the bolt tension gets reduced, which is termed creep. When a steel fastener is used for holding magnesium alloy base and top part, whenever an aqueous solution gets between the steel and magnesium a galvanic cell is created causing magnesium to corrode more rapidly However, if an aluminium or magnesium fastener is used these problems are avoided and the longevity of the parts increased.

Environmental aspects

As the lightest available structural metal, the use of magnesium provides the benefit of reducing the weight of automobiles and thereby reduces gasoline consumption. This results in substantial reduction in the carbon dioxide emissions from an automobile during its lifetime. Carbon dioxide (CO_2) from vehicle engines is one of the major sources of anthropogenic emissions of 'Greenhouse Gases '(GHG).

Notwithstanding the benefits from the use of magnesium in reducing carbon dioxide emissions, the emissions of GHG during the production and processing of magnesium alloys must also be considered through the use of tools such as Life-Cycle Analysis (LCA). The climate control aspects of GHG release to the atmosphere from metal production has already prompted producers to use sulfur dioxide and other reagents instead of sulfur hexafluoride to provide the cover gas used in the production of magnesium alloys. The cover gas provides an inhibitive and tenacious film of fluoride and/or sulfur compound to prevent reaction of the molten alloy with oxygen.

The issue of GHG emissions has dwarfed other production-related pollution control issues relating to air, water and solid-waste emissions over the past two decades. Even basic pollution controls are not regulated in a consistent manner worldwide. The development of regulations and controls started in the US and Europe in 1970, and producers in other countries are now being asked to perform to similar standards by automotive industry consumers.

Non-greenhouse-gas regulations – electrolytic magnesium production

Many of the major magnesium production plants that were based on electrolytic technologies have been shut down. These include Dow Chemical in the US, Hydro Magnesium in Norway and Canada, and Noranda Magnesium, also in Canada. Some of the original LCA studies were done on the Hydro Magnesium plants in Porsgrunn, Norway and Becancour, Québec, Canada (Albright and Haagensen, 1997). The Porsgrunn plant in Norway used sea water and dolomite as raw material which was converted to magnesium oxide and then to magnesium chloride which was used to feed the electrolytic cells. At times, the Porsgrunn plant also utilised magnesium chloride brine from Germany as a source of magnesium. In Becancour the magnesium chloride was produced by treating magnesite with hydrochloric acid.

The environmental regulations of the 1980s were devoted to issues which controlled chlorinated hydrocarbons in water, especially polychloro biphenyl compounds (PCBs) and dioxin. This required Hydro Magnesium to modify the electrolytic process in Porsgrunn to develop the use of hydrogen to convert the anode gas chlorine into hydrogen chloride and hydrochloric acid. This process was selected for the plant built in Becancour which operated from 1991 until 2006.

The published estimates for the production capacities of each of these plants were about 40,000 metric tonnes per annum (Burstow, 1999). Chlorinated hydrocarbon (CHC) emissions were actually reduced by half at both plants. Dioxin emissions also decreased substantially. Albright and Haagensen (1997) also discussed some of the global warming issues which came to prominence

Table 11.3 Life-cycle inventory for magnesium production. (After Albright and Haagensen, 1997.)

Item	Pure Mg	Mg Alloy AZ91
Total energy (MJ/kg metal)	144	151
Global warming effect (kg CO_2 eq/kg)	19	19
Acidification (kg/kg metal)	0.02	0.025
Winter smog (kg/kg metal)	0.015	0.017
Solid waste (kg/kg metal)	0.5	0.5
Dioxins to air (μg/kg metal)	0.24	0.21
Chlorinated hydrocarbon to air (mg/kg metal)	13.7	12.4

in the early 1990s following the Kyoto Protocol. Albright and Haagensen (1997) produced a LCI (Life Cycle Inventory) for Hydro Magnesium plants. Separate calculations were made for both pure magnesium production and for magnesium alloy production (Table 11.3). In regard to the emissions and wastes at Hydro Magnesium plants, it was reported that "The solid wastes from Hydro's plants consists mostly of inorganic salts and minerals from the raw materials. This waste is disposed of at authorized sites and does not cause any negative environmental impact. The treatment of gas and waste water on site, along with incineration of residues containing chlorinated hydrocarbons, form a portion of the (LCI) analysis as well."

Designers of Australian Magnesium's process also spent considerable effort in meeting the environmental regulations in Australia. This process, using magnesite feed material, was tested in a 1500 tonne per year pilot operation, but the commercial plant was not constructed.

In 1997 when Noranda was permitting its Magnola magnesium project, utilising an asbestos waste (serpentine) as raw material, the company was required to meet strict controls on emissions of chlorinated hydrocarbons, dioxins and chlorine. Similar concerns must be addressed in the design and operation of electrolytic magnesium reduction plants today (Brown, 2003b). Improved scrubbing technology, together with various process modifications, has made great progress in reducing emissions. Although the construction of electrolytic plants continues to be slow, the research and development of the new equipment and methods should be encouraged on a global basis.

Non-greenhouse-gas regulations – thermal magnesium

Thermal magnesium plants had to address common environmental regulations on airborne particulate emissions from calciners and ferrosilicon furnaces. In addition, they had to address disposal of solid wastes comprising mostly reduction furnace residues and casting sludges containing fluxes. Recent thermal processes tend to make by-products for sale from the solid residue from retorts in an effort to provide added revenues.

Greenhouse-gas emission studies

Casting cover-gas issues

For most of the twentieth century magnesium production, both electrolytic and thermal, used either sulfur or sulfur dioxide as a cover gas while casting magnesium to prevent oxidation. When this was done in an uncontrolled fashion sulfur dioxide was released into the work area. The early work by Reimers (1934) was further developed by Hanawalt, and others from Dow, leading to the use of non-toxic sulfur hexafluoride (SF_6) as cover gas and this became an acceptable alternative. Accordingly from the 1970s onwards magnesium companies worldwide started switching to using sulfur hexafluoride.

Following Kyoto Protocol studies, the advantage of SF_6 as a cover was questioned because of its global warming potential. SF_6 was found to have a global warming potential 23,900 times that of carbon dioxide (Gjestland et al., 1997). This problem was addressed by a partnership between the U.S. Environmental Protection Agency (US EPA) and the magnesium community of producers and users (Bartos, 2001 and Bartos, 2002). Several alternative fluorine-containing cover gases were developed as a result, but most of the companies started reverting back to using sulfur or sulfur dioxide in a controlled fashion.

Life-cycle analysis

When comparing the use of various structural metals in automobiles, it has become customary to normalise the advantages and disadvantages by carrying out Life-Cycle Analyses of the material usage with respect to reduction in fuel usage. There have been several life-cycle analyses of magnesium production along with downstream fabrication of parts for use in automobiles.

Most of the major GHG-related environmental problem areas in the production of the metal were studied showing the effects of the electrolytic magnesium and the thermal magnesium production methods. Further sub-division assessed the preparation of the magnesium-containing feed materials and the actual processing of this material to produce magnesium metal. All of these areas are impacted by secondary and upstream processes which impact on the magnesium production. For example, the method of onsite electricity generation for the production of the input materials may be responsible for significant emissions, e.g. if coal is used for the electricity required to make 75 per cent ferrosilicon used in the Pidgeon process.

Environmental concerns related to magnesium production and magnesium processing first appeared in the early 1990s when Life-Cycle Analysis articles started to be published. Subsequently various studies have compared the environmental impacts of the two main magnesium production processes. As the measurement technology improves and the information is better understood so the environmental challenges can be more clearly identified. The pressures from the climate control aspects might impact on the status of the differing processes used for magnesium production. Academic studies on the global warming impacts of present-day magnesium production processes have been published (Ramakrishnan and Koltun, 2004a, 2004b and Cherubini et al., 2008).

The earliest LCA papers were prompted by the environmental work that was done to get construction permits for new electrolytic magnesium plants planned for Canada and Australia in the late 1980s. The work focused on the GHG impact, including the effects of SF_6 and CO_2 emissions from furnaces. Some of the work was conducted as part of investigations into the advantage of lighter weight magnesium cars in regard to Life-Cycle emissions.

The initial calculations were part of a study to assess the environmental impact of a magnesium engine block supply chain. This study was broadened to include an engine block made from magnesium alloy ingots produced at Becancour and secondary ingots produced in the USA. The study concluded that "the use stage of a passenger car contributes significantly to the total GHG impact of magnesium components over their entire life cycle. A significant reduction in the GHG impact during the use stage, and hence over the entire life cycle, may be achieved from a significant mass reduction of a car by using magnesium components" (Tharumarajah and Koltun, 2005).

Casting operations in China mostly used very small amounts of SF_6 in the ingot casting operations, and this is an advantage for these processes. The majority of the plants built in China during the 1990s used very simple melting and casting operations with sulfur powder directly dusted on the cast ingots to prevent oxidation during solidification. When the plants increased in size, most of the operations retained the sulfur dusting for convenience and simplicity.

Most of the small Chinese Pidgeon process plants that used coal as the main energy source and large quantities of hand labour are now disappearing. Larger, more efficient plants, with advanced technology and equipment, are being built. The Chinese are approaching the problem of addressing the global warming potential by improving the production plants and working hard in downstream processing to produce lightweight magnesium products for use in bicycles, motorcycles, cars, trains and airplanes. China has been carrying out industrial restructuring for the purpose of optimising its magnesium industry and upgrading its competitiveness. Some of these measures have been announced in the Chinese 5-year plans, as noted in China Magnesium Industry and Market Bulletin (Sunlight Metal Consulting (Beijing) Co. Ltd., 2011).

The Chinese government has announced new regulations requiring existing smelters to consume a maximum of 5.5–6.0 tonnes of coal for one tonne of magnesium smelting, and for new smelters, a maximum of five metric tonnes of coal. Chinese magnesium smelters, over the past few decades, have been consuming as much as 11–18 metric tonnes of coal for each tonne of magnesium smelting, although some smelters in the last two years were able to cut the usage to around 5–6 metric tonnes coal, following the state's advocacy of a low-carbon economy, according to Chinese Magnesium Association (CMA) (Shukun et al., 2010).

Much of the fuel-based pollution in China is currently being addressed and, wherever it is available, natural gas is being designed into many of the newer plants. It should be noted that many of the Pidgeon plants in the USA and Canada used natural gas as a fuel for their retorts between 1940 and 1970.

Cherubini et al. (2008) concluded that world magnesium production contributes to the global warming potential with an emission of about 25.5 million tonnes of CO_2 equivalent per year. This will be lowered with improved energy efficiency and as SF_6 is increasingly substituted. In the EU car manufacturers are addressing this issue as a matter of urgency. They are cutting carbon dioxide emissions faster than expected and will reach the European Union targets ahead of time (D'Errico et al., 2010).

Magnesium is produced by a highly modified vertical furnace process by Rima Metallurgical in Brazil. This is a silico-thermic process which uses an electrical resistance furnace for reducing the briquettes of calcined dolomite with ferrosilicon, instead of the fuel-fired retorts of the Pidgeon process. Not only does Rima produce its own ferrosilicon, but the company also states that its process is very 'green', noting that calcination of dolomite and ferrosilicon production are carried out using charcoal instead of coal, and using sulfur dioxide instead of SF_6 in their casting process (Fernando 2011). For these reasons, in combination with their use of hydroelectric power, Rima's process probably has the lowest carbon footprint for magnesium production in the world.

Dubreuil et al. (2010) made a cradle-to-grave comparison of magnesium, aluminium and steel usage in automobiles. He concluded that magnesium is the preferred metal in terms of its overall carbon dioxide emissions.

It is also important to note that the raw materials used in making magnesium, such as dolomite and magnesium chloride, are not toxic. All plants have magnesium in the centre of the chlorophyll molecule essential for photosynthesis through the green pigment in the leaves. Magnesium is therefore a common additive to fertilisers. Magnesium is also important to all living cells where it plays a major role in the functioning of numerous enzymes.

World resources and production

Global resources of magnesium, on land and in the sea, are huge compared to current production and usage. Figure 11.4 shows the world production of magnesium since 1992 and Figure 11.5 shows the distribution of production by country in 2010.

There have been a number of closures of formerly important magnesium producers in the West in the last two decades. These include Pechiney in France, which stopped producing magnesium in 2001 from its thermal magnesium plant. In 1997 Dow Magnesium ceased production of magnesium in the USA after a period of 83 years. Dow initially used brine in Michigan as the source of magnesium and latterly sea water in Freeport, Texas. At its peak Dow produced about 100,000 tonnes per year, utilising partly hydrated magnesium chloride cell feed in its electrolytic process. For more than 25 years up to 2001, Alcoa operated a thermal magnesium production plant in Addy, Washington, USA, with a production

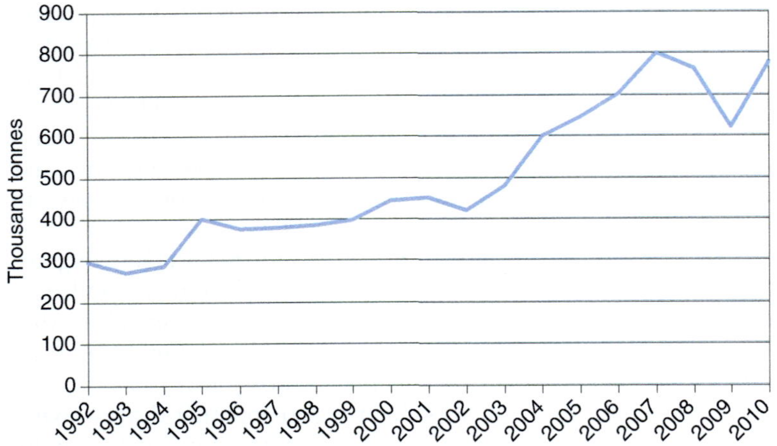

Figure 11.4 World production of magnesium metal, 1992–2010. (Data from British Geological Survey World Mineral Statistics database.)

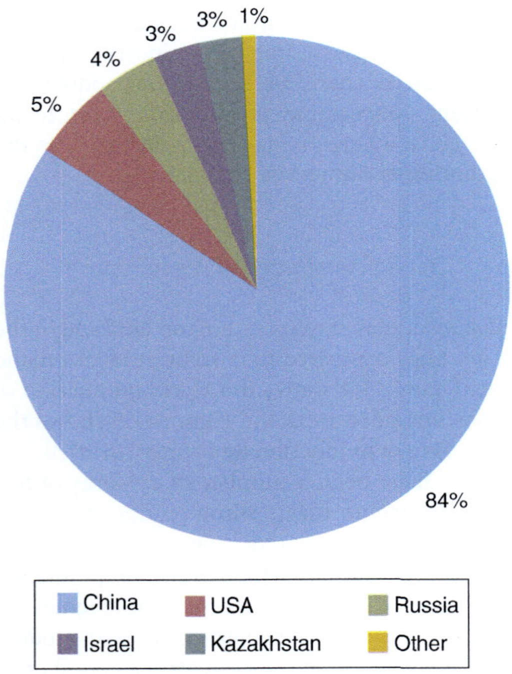

Figure 11.5 World production of magnesium metal by country in 2010. (Data from British Geological Survey, 2012.)

capacity of 27,000–36,000 tonnes. In Japan, Ube Industries stopped its thermal magnesium production at the end of 1994.

It is of historical interest to note that magnesium was produced at Clifton Junction, UK during and prior to World War II, and also that the Magnesium Elektron company, with its headquarters in the UK, has operated in downstream wrought magnesium applications for over 70 years. Other countries which formerly produced magnesium include Canada, Italy, Norway and Yugoslavia (Serbia). The magnesium plant in Becancour, Quebec, Canada, which operated between 1991 and 2006, produced 40,000 to 50,000 tonnes per year magnesium. The Magnola Metallurgie, Noranda magnesium plant, also in Canada, operated for a short time between 1999 and 2001. Both were electrolytic magnesium producers.

The only continuing magnesium producers in the West are:

1. US Magnesium, which has a present capacity of about 50,000 tonnes per year, producing from the Great Salt Lake brines since 1972.
2. Rima Metallurgical, Brazil, which has been in production since the late 1970s and currently produces about 20,000 tonnes per year of thermal magnesium.
3. Dead Sea Magnesium, Israel, which has been producing about 30,000 tonnes per year by electrolysis of carnallite since 1995.
4. Solikamsk, Russia, which has been in operation since 1936. Solikamsk's cumulative magnesium production passed the one million

tonne mark in 2009. Its annual production is typically between 15,000 and 20,000 tonnes.

Total world magnesium production was estimated at 775,000 tonnes in 2010, up from 619,000 tonnes in 2009. 84 per cent of the world's magnesium in 2010 was produced by China (Figure 11.5). The demand for magnesium grew at an estimated rate of 16 per cent worldwide, including estimates for the magnesium used within China and CIS. About 40 to 45 per cent of usage continues to be in aluminium alloying. The global economic downturn has led to a worldwide decrease in the die-casting market, which now accounts for between 28 and 35 per cent of total usage. The principal market for die castings continues to be the automotive industry.

Future supplies

A new thermal reduction plant using the Pidgeon process was commissioned in Perak, Malaysia in 2010 (CVM Minerals, 2010). It has an initial design capacity of 15,000 tonnes per year and was developed with the assistance of a Chinese company.

South Korea has recently started primary magnesium production from dolomite by the thermal method. Posco, the South Korean steel producer, will be the main producer, with a projected output of 10,000 tonnes per year and a plant start-up in 2012. Posco plans two further expansions to bring the magnesium plant's capacity to 100,000 tonnes a year by 2018 (Eun-Seong Min, 2011). They have demonstrated the production of a 16-tonne coil of wrought magnesium alloy sheet using the twin roll casting machinery originally developed by Fata Hunter. Posco plans to utilise the sheet magnesium in the automotive industry.

Since 2006 the Ural Asbestos Company has been studying the construction of a magnesium plant using asbestos waste converted to magnesium chloride for an electrolytic process. The parent company is interested in this project and has been involved in raising capital for the project since 2009 but there were no major announcements in 2011.

Molycor Gold Corp., with its high-grade dolomite property at Tami-Mosi in Nevada, completed a resource analysis and a preliminary economic assessment for a 30,000 tonnes per year magnesium plant for the North American market, and is in the process of raising funds (Molycor Gold Corp., 2011). Molycor, which was renamed Nevada Clean Magnesium Inc. in 2012, plans to use a silicothermic reduction method via an updated and automated Bolzano process.

World trade

Global import and export data for magnesium are illustrated in Figures 11.6 and 11.7, respectively. The status of the Netherlands as the largest importer of magnesium is a function of the way in which theses data are collected and can be explained because Rotterdam is a free port through which many of the transactions pass. USA, Japan and Germany are the leading importers, which reflects the importance of magnesium use in the automotive and aluminium sectors in these countries.

The export statistics indicate that China exports about 180,000 tonnes per year of their annual production of 700,000 tonnes per year, indicating the high level of domestic consumption in China of over 500,000 tonnes per year. The high level of exports from Austria is due mainly to recycled magnesium – as Austria is not a producer of primary magnesium or its alloys.

Prices

It should be noted that magnesium is not traded on any metal exchanges. As a result, prices are the product of negotiated contracts which reflect variations in the supply–demand balance and local conditions in different parts of the world.

An internal export tax of ten per cent is levied on magnesium and magnesium alloys that are shipped out of China. Furthermore, there is steady upward pressure on production costs in China due to increases in the price of

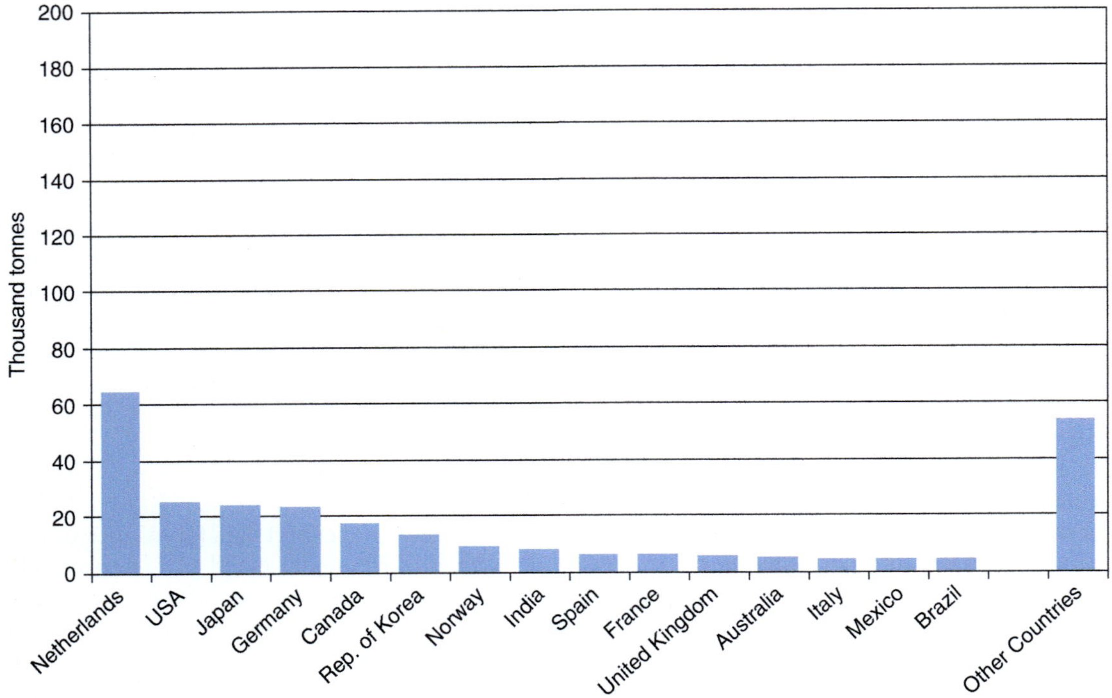

Figure 11.6 The top importing countries of magnesium metal in 2009. (Data from British Geological Survey World Mineral Statistics database and UN Comtrade, 2013.)

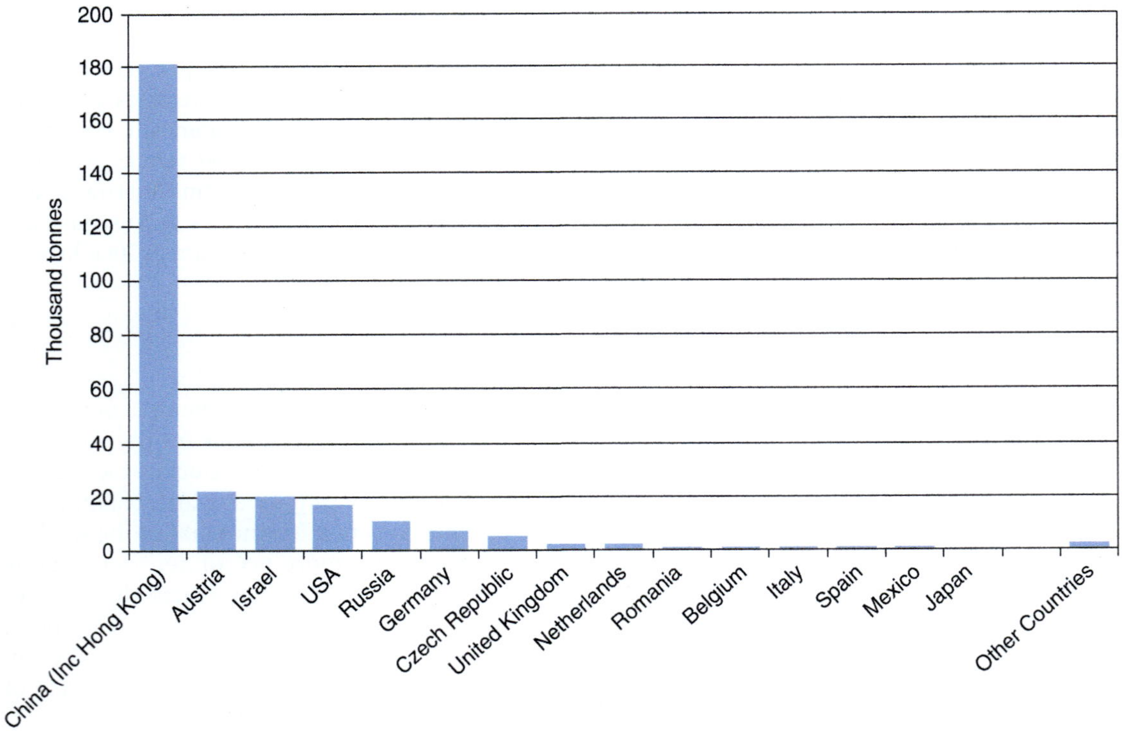

Figure 11.7 The top exporting countries of magnesium metal in 2009. (Data from British Geological Survey World Mineral Statistics database and UN Comtrade, 2013.)

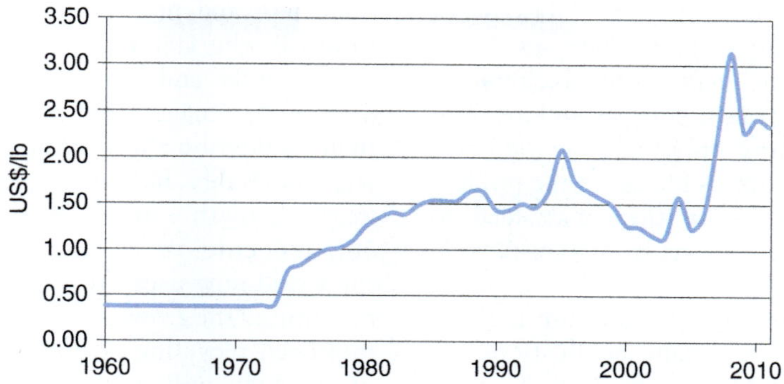

Figure 11.8 Year-end prices of magnesium in USA, 1960 – 2011. (Prices are U.S. spot Western price for 99.8%-pure magnesium ingot, compiled from United States Geological Survey (1999) for 1960–1998, and from United States Geological Survey Annual Mineral Commodity Summaries for 1999 to 2011.)

electricity and labour and the need to comply with environmental regulations. Since the mid-1990s prices in the USA have been higher than in the rest of the world mainly because of anti-dumping duties levied against metal produced in China, Russia and Canada. According to the World Trade Organisation (WTO), when any imported product is proven in a court of the International Trade Council to be sold at less than the cost of production, the importing countries can levy an anti-dumping duty on the item being imported.

The long-term price trend for magnesium metal in the USA is shown in Figure 11.8. Following a long period of abundant cheap supplies magnesium prices rose steadily from the early 1970s after the oil crisis. In the following decade the price of magnesium more or less trebled in response to escalating energy costs and increased demand. New suppliers in North America and Russia entered the market from early 1990 causing the price to fall back, stabilising at just below US$1.50 per lb until about 1994. However, increasing demand in the USA, together with the effects of the anti-dumping duties, led to rapid price escalation during 1995 (USGS, 1999). In the following year imports from Russia and Israel resumed resulting in lower prices in the USA. Prices began to rise sharply again in 2006 in the face of increasing demand and restrictions on imported supplies. However, demand slumped in 2008 as a result of the global economic recession, leading to a fall in the price to about US$2.35 per lb in the USA at the end of 2011.

Outlook

The 21st century may see changes in magnesium production processes not because of lack of raw materials or lack of process know-how but rather from the continued pressure to balance market needs and environmental costs of process components, as it did in the nineteenth and twentieth centuries. There is continuing international research and development into reducing the GHG emissions associated with the production of magnesium. In China, while the older Pidgeon process produced emissions of about 36 kg CO_2 per kg of magnesium, this is being reduced to about 20 kg CO_2 per kg of magnesium by better energy management, as noted above. In contrast, the electrolytic magnesium plant in Canada, which was operated with hydroelectric power and used magnesite as the starting material, produced 6.9 kg CO_2 per kg magnesium (Das, 2010). The other western electrolytic process in the US and elsewhere will have higher emissions due to the use of calcium carbonate for removing sulfate impurity from the brine, and for evaporation and dehydration of the purified brine. The use of power derived from fossil fuels also contributes to higher emissions using this technology. Alternative raw materials, such as sulfate minerals (gypsum, epsomite, etc.), are currently being investigated by the authors for use in the production of magnesium where they may lead to further significant reductions in carbon dioxide emissions.

The unique properties of magnesium make it essential for some important applications. In particular, the low density of magnesium alloys means that they are especially useful when 'lightweighting' is an important factor, such as in the transport sector. The high reactivity of magnesium is important in some applications which together currently account for less than 20 per cent of total magnesium production. These include magnesium use in the desulfurisation of iron and steel, in the production of titanium and other reactive metals from their compounds and in the production of specialised flares for military applications. Most of the magnesium used in the production of titanium is recycled and is not counted in the total world production. The total volume of magnesium utilised each year is growing and is likely to continue to do so, especially as new applications are developed. The main growth areas are likely to be in the structural uses of magnesium, either in die-cast metal components or wrought-alloy sheet components.

It is important to note that the quantity of magnesium present in seawater, brines, dolomite and other minerals is very high in relation to present and anticipated future demand for magnesium. However, due to a variety of other factors the number of locations where primary magnesium is produced has decreased since the 1980s. China now dominates global output using a technology that may not have the lowest environmental footprint, although China has made major improvements in its environmental controls and energy reduction in its process industries in recent years.

The US Government currently imposes an anti-dumping duty on magnesium metal and alloys imported from China with the result that the price for magnesium in the US is higher than elsewhere and primary metal production continues in the US, whilst producers in Canada, France, Italy and Norway have shut down. If demand for magnesium in China grows faster than production the volume of Chinese exports will inevitably fall, which could limit global market growth – unless other producers expand output or enter the market. The fact that most of the world now uses lower-priced Chinese metal has limited the growth of the magnesium market. Improved recycling of magnesium from end-of-life products will contribute to improved supply security and to the reduction of the environmental impacts associated with magnesium use.

A significant portion of U.S. demand for magnesium depends on aluminium alloys. Aluminium production in the United States is increasing as aluminium companies are opening new capacity and upgrading rolling mills to make car body sheet to help make lighter cars. In mid-2010, only about 50 per cent of the US primary aluminium production capacity was in use. A recent report on the aluminium sheet business by Novelis, a large USA aluminium sheet maker, stated that the can sheet market is mature and usage remains level in the USA (Martens, 2011). In contrast, in Europe and Japan the use of aluminium cans is increasing. It was also noted that the automotive industry will see an acceleration of aluminium use, and that, in the USA there will be a significant increase in body structures made up of some percentage of aluminium by the end of this decade.

The level of use of magnesium die casting varies around the world. In the past 25 years diecast magnesium has been used mainly in the construction industry for lightweight hand tools and in the automotive industry. Magnesium die casting usage in China has grown markedly in recent years compared to other countries, with the largest growth in the 3-C industries and in the automotive sector. Over the past six years, the Chinese automotive industry has seen an increase of more than 500 per cent in the number of cars produced.

In 2011, it is forecast that auto production demand in the United States for die castings will be up ten per cent, partially due to the increase in the US Fuel Economy standards from 27 to 35 miles per gallon. The use of new magnesium part designs that incorporate crash energy management and improved fit and finish over assembly of steel stampings provides some examples of the move towards increased magnesium use. The inner lift gate for the Lincoln automobile is a 7.7 kg die casting which replaces six steel stampings weighing 13.6 kg. General Motors has developed a 10-kg magnesium die cast battery tray for the Chevrolet Volt. This is a single piece casting that replaces a steel assembly which was twice as heavy.

Governments and industry worldwide are currently funding major research programmes into developing new uses for magnesium. There is particular interest in the automotive sector where manufacturers already use magnesium for individual components, but where new alloys and processing methods are needed before it can become economically and technologically feasible as a major automotive structural material.

An international collaboration between Canada, China and the United States, called the Magnesium Front End Research and Development (MFERD) programme, was initiated in 2007. The goal of this programme is to advance the manufacture of magnesium-intensive vehicles, by developing technologies and knowledge that will lead to a vehicle front-end body structure that is 50 to 60 per cent lighter, equally affordable, more recyclable, and of equal or better quality than today's vehicles. Toward this end, databases are being developed that will enable further alloy and manufacturing process optimisation, and a life-cycle analysis of the magnesium alloys is being conducted.

In Europe there are several active research groups including the Magnesium Innovation Centre (MagIC) in Geesthacht, Germany and BCAST (the Brunel Centre for Advanced Solidification Technology) in London. These groups focus on twin roll casting, although MagIC is also one of the lead institutes in the relatively new field of developing magnesium alloys for medical implants. Another important research centre is the Magnesium Research Institute (MRI) in Beer-Sheva, southern Israel, which was established in the mid-1990s. The MRI is concentrating on developing new products, processes and markets for magnesium use and applications.

References

Albright, D. and Haagensen, J. (1997) Life Cycle Inventory of Magnesium, Proceedings of the 54th Annual World Magnesium Conference (IMA), Toronto, Canada, June 1997, 32–37.

Avedesian, M. (1999) Magnesium and Magnesium Alloys. In: Avedesian, M. and Baker, H. (eds.) ASM Specialty Handbook, ASM International, Materials Park, Ohio, USA, 1–4.

Bartos, S.C. (2001) United States Environmental Protection Agency SF6 emission reduction partnership for the magnesium industry: an update on early success. In: Hryn, J. Magnesium Technology, TMS Annual Meeting, 43–48.

Bartos, S.C. (2002) Building a Bridge for Climate Protection: United States Environmental Protection Agency and the Magnesium Industry, Proceedings of the 59th Annual World Magnesium Conference (IMA), Montreal, Canada. May 2002, 22–24.

Beck, A. (1939) The Technology of Magnesium and Its alloys. F.A. Hughes and Co. Ltd, London, 1–19.

Brooks, G., Trang, S., Witt, P., Khan, M.H.N. and Nagle, M. (2006) The carbo thermic route to magnesium. JOM The Member Journal of TMS (The Minerals, Metals and Materials Society) 58 (05), 51–55.

Brown, R.E. (2000) Magnesium Industry Growth in the 1990 Period. In: Kaplan, H.I., Hryn, J. and Clow, B. (eds.) Magnesium Technology 2000, Nashville, TN, TMS Annual Meeting, 3–12.

Brown, R.E. (2003a) A History of Magnesium Production. http://www.magnesium.com/w3/databank/index.php?mgw=196.

Brown, R.E. (2003b) M-Cell Modernization Improves US Magnesium Process and Environmental Performance. Light Metal Age, June 2003, 6–12.

Burstow, C. (1999) A Strategic Analysis of the Global Magnesium Industry, Metal Bulletin Research, London.

Cherubini, F., Raugei, M. and Ulgiati, S. (2008) LCA of Magnesium Production: Technological Overview and Worldwide Estimation of Environmental Burdens. Resources, Conservation and Recycling 52, 1093–1100.

CVM Minerals Ltd. (2010) Company profile, Message and Vision. http://www.cvmminerals.com/ceo.html

Das, S. (2010) Life cycle energy and CO_2 Analysis of Solid Oxide Membrane Primary Magnesium Production Technology, U.S. DOE Report No. DE-AC5-00OR22725.

Dean, K.C. (1965) Economic and Technical Evaluation of Magnesium Production Methods - Part 1. Metallo Thermic. USBM R.I. 6656.

Dubreuil, A., Bushi, L., Das, S., Tharumarajajah, A. and Xianzheng, G. (2010) A Comparative Life Cycle Assessment of Magnesium Front End Autoparts. http://aluminumintransportation.org/downloads/MFRED-LCA-Study.pdf

Elkins, A., Dean, K.C. and Rosenbaum, J.B. (1968) Economic Aspects of Magnesium Production. In: Henrie, T.S. and Baker Jr, D.H. (eds.) Electrometallurgy, Proceedings of the Extractive Metallurgy Division Symposium on Electrometallurgy, AIME, Cleveland, OH, 1968 – Dec, 173–184.

Emley, E.F. (1966) Principles of Magnesium Technology, London, Pergamon.

D'Errico, F., Fare, S. and Graces, G. (2010) The Next Generation of Magnesium Based Material to sustain the IPCC Policy. Unpublished paper from Italy/Spain.

Eun-Seong Min. (2011) Vision of POSCO Magnesium – from Raw Materials to Products. International Magnesium Association Annual Meeting, Prague, May 2011.

Faure, C. and Marchal, J. (1964) Magnesium by the Magnetherm Process. Journal of Metals (1964 September), 721–723.

Fechner, D., Hort, N., Blawert, C., Dietzel, W. and Kainer, K.U. (2009) Magnesium Recycling: State-of-the-Art Developments, Part I. Light Metal Age (4), 46–50.

Fernando, F. (2011) Rima's Process: Green Magnesium From a Fully Integrated Plan. Proceedings of the 68th Annual World Magnesium Conference (IMA), Prague, Czech, May 2011, 80–84.

Gann, J.A. (1935) Magnesium in Modern Uses of Non-ferrous Metals. In: Mathewson, C.H. (ed.) Proceedings of American Institute of Mining and Metallurgy.

Gjestland, H., Westengen, H. and Phahte, S. (1997) Use of SF_6 in the Magnesium Industry: An Environmental Challenge. Proceedings of the Third International Magnesium Conference, The Institute of Materials, London, 33–41.

Hansgirg (1932) Carbothermal production of magnesium. U.S. Patent, 1884993, 1932 - October.

Haughton, J.L. and Prytherch, W.E. (1938) Magnesium and Its Alloys. Chemical Publishing Co, pp. 170.

Holywell, G.C. (2005) Magnesium: the first quarter millennium. JOM The Member Journal of TMS (The Minerals, Metals and Materials Society) 57 (07), 25–33.

Jarrett, N. (1981) Advances in the Smelting of Magnesium. American Institute of Mining and Metallurgy, Englewood, Colorado, USA.

Jones, T. and DeLorme, R.D. (2008) Development of a Ballistic Specification for Magnesium Alloy AZ31B. Technical Report / ARL-TR-4664, Army Research Laboratory.

Jones, T. and Kondoh, K.(2009) Initial Evaluation of Advanced Powder Metallurgy Magnesium Alloys for Armor Development, Technical Report/ARL-TR-4828, Army Research Laboratory.

Martens, P. (2011) http://www.novelis.com/en-us/Pages/Video-Gallery.aspx.

Molycor Gold Corp. (2011) Positive preliminary economic assessment study completed for Tami-Mosi magnesium project. Molycor Gold Corp. http://www.molycor.com/positive-preliminary-economic-assessment-study-completed-for-tami-mosi-magnesium-project/

Muzhzhavlev. K.D., Lebedev, O.A., Frantas'ev, N.A., Olyunin., G.V., Sheka, T.S. and Dolgikh, T.K. (1965) Results of tests on individual elements in the design of magnesium electrolyzers. Tsvetn. Metal 38 (5), 57–60.

Muzhzhavlev, K.D., Kostrev, S.P., Schegolev, V.I., Ivanov, A.B., Romanenko, O.N., Yazev, V.D. and Vasiliev, A.V. (1977) Diaphragmless electrolyzer for producing magnesium and chlorine. USA Patent 4058448.

Neelameggham, N.R. (2013) Magnesium metal production technology. In: Pekguleryuz, M., Kainer, K. and Kaya, A. (eds.) Fundamentals of Magnesium Alloy Metallurgy, Woodhead Publishing, Cambridge, UK.

Pidgeon, L.M. and Alexander, W.A. (1944) Thermal production of magnesium – pilot plant studied on the retort ferrosilicon process, Transactions American Institute of Mining and Materials Engineers 159, 315– 352.

Ramakrishnan, S. and Koltun, P. (2004a) A comparison of the greenhouse impacts of magnesium produced by electrolytic and Pidgeon process. In: Luo, A. (ed.) Magnesium Technology 2004, Charlotte, NC, TMS Annual Meeting, 173–178.

Ramakrishnan, S. and Koltun, P. (2004b) Global Warming Impact of the Magnesium Produced in China Using the Pidgeon Process. Resources, Conservation and Recycling 42 (1), 49–64.

Reimers, H.A. (1934) Method for inhibiting the oxidation of readily oxidizable. USA Patent 1,972,317.

Rudnick, R.L. and Gao, S. (2004) Composition of the Continental Crust. In: Holland, H.D. and Turekian, K.K. (eds.) Treatise on Geochemistry, Volume 3, The Crust. Elsevier, Pergamon.

Shukun, M., Xiuming, W. and Jinxiang, X. (2010) China Magnesium Development Report in 2009. IMA 67th Annual world magnesium Conference, Hong Kong, 2010-May, 3–10.

Slade, S. (2011) Presentation at Magnesium Technology 2011. TMS Annual Meeting, San Diego, CA.

Strelets Kh, L. (1977) Electrolytic Production of Magnesium. (Translated from Russian by J. Schmorak.) Israel Program for scientific Translations, NTIS, Springfield, VA.

Sunlight Metal Consulting (Beijing) Co. Ltd. (2011) China magnesium bulletin industry and market. http://www.lightmetal.com.cn/en/content.asp?id=4.

Tharumarajah, A. and Koltun, P. (2005) Lifecycle Assessment of Magnesium Component Supply Chain. Proceedings of the 62nd World Magnesium Conference (IMA), Berlin, Germany May 2005, 67–73.

UN Comtrade (2013) United Nations Commodity Trade Statistics Database, Department of Economic and Social Affairs/ Statistics Division, http://comtrade.un.org/db/

United States Geological Survey (1999) Metal Prices in the United States through 1998. In: Plunkett, P.A. and Jones, T.S. (eds.) Magnesium, 79–81.

United States Geological Survey (2012) Mineral Commodity Summaries: Magnesium Metal.

12. Platinum-group metals

GUS GUNN

British Geological Survey, Keyworth, Nottingham, UK

Introduction

The six chemical elements normally referred to as the platinum-group elements (PGE) are, in order of increasing atomic number: ruthenium (Ru), rhodium (Rh), palladium (Pd), osmium (Os), iridium (Ir) and platinum (Pt). Metallurgists and engineers more commonly refer to them as platinum-group metals or PGM. Minerals in which one or more of the PGM comprise an essential part are referred to as platinum-group minerals, also often abbreviated to PGM. In this chapter PGM is used to refer to platinum-group metals.

The PGM are rare precious metals, although unlike gold, they are used in a diverse range of industrial applications as well as in jewellery. Today, platinum and palladium are the most commercially important of the PGM with their largest application being in the automotive industry where they are used to reduce harmful emissions from vehicle exhaust systems. Rhodium, which is also used in autocatalysts, is the third most important PGM, albeit with consumption an order of magnitude less than that of platinum and palladium. Ruthenium and iridium are used in smaller amounts although they are becoming increasingly important in a variety of new technologies such as data storage, medical implants and renewable energy.

The name platinum is derived from the Spanish 'platina' meaning little silver. Archaeological evidence indicates that there is a very long history of platinum working in South America (Scott and Bray, 1980). The platinum, which was used for making ornaments, was derived from sands and gravels in rivers draining the Pacific lowlands of Ecuador and Colombia. Its discovery is often credited to the Spanish military leader Don Antonio de Ulloa who brought it to Europe from South America around 1750. Palladium was discovered in 1803 by the English chemist William Hyde Wollaston. It was named after the asteroid Pallas, identified in 1801, and which itself was named after the Greek goddess of wisdom.

Properties and abundance in the Earth

The PGM are transition metals, which together with iron, cobalt and nickel, are found in Groups 8, 9 and 10 of the Periodic Table. Osmium, iridium and ruthenium are sometimes referred to as the IPGM or iridium sub-group, and platinum, palladium and rhodium as the PPGM or palladium sub-group, on account of their

Critical Metals Handbook, First Edition. Edited by Gus Gunn.

Table 12.1 Selected properties of the six platinum-group metals (PGM) compared with gold (Au).

Property				Value				Units
Name	Platinum	Palladium	Rhodium	Iridium	Ruthenium	Osmium	Gold	
Symbol	Pt	Pd	Rh	Ir	Ru	Os	Au	
Atomic number	78	46	45	77	44	76	79	
Atomic weight	195.08	106.42	102.91	192.22	101.07	190.23	196.97	
Density at 25 °C	21450	11995	12420	22550	12360	22580	19281	kg/m^3
Melting point	1769	1554	1960	2443	2310	3050	1064	°C
Electrical resistivity at 25 °C	106	105	45	51	71	92	22	nΩ m
Hardness (Mohs scale)	4–4.5	4.75	5.5	6.5	6.5	7	2.5–3	

similar geochemical behaviour under magmatic conditions. Together with gold and silver, the PGM have long been referred to as noble or precious metals.

All six PGM are chemically similar with strong siderophile and chalcophile tendencies, preferentially bonding with iron, nickel, copper and sulfur rather than with oxygen. The distribution of PGM in the Earth is thus controlled by the presence of metallic phases, being strongly concentrated in the Earth's core after planetary accretion. They do not dissolve in strong acids, but they do react with oxygen at high temperatures to form volatile oxides. The physical properties of the PGM vary considerably (Table 12.1) but all display properties typical of metals, including the ability to form alloys, to conduct heat and electricity, and some degree of malleability and ductility. Platinum, iridium and osmium are the densest known metals, being significantly denser than gold. Platinum and palladium are highly resistant to heat and to corrosion, and are soft and ductile. Rhodium and iridium are more difficult to work, while ruthenium and osmium are hard, brittle and almost unworkable. All PGM, commonly alloyed with one another or with other metals, can act as catalysts which are exploited in wide range of industrial applications. Platinum and palladium are of major commercial significance, while, of the other PGM, rhodium is the next most important.

The PGM are very rare in the Earth's crust with platinum and palladium present at similar concentrations of approximately five parts per billion (ppb), by weight. Rhodium, iridium and ruthenium are even scarcer at about one ppb. Relative to other rock types the PGM are enriched in ultramafic lithologies, such as peridotite, in which platinum and palladium concentrations are commonly 10–20 ppb.

Mineralogy

In nature, the PGM are chiefly held either in base-metal sulfide minerals, such as pyrrhotite, chalcopyrite and pentlandite, or in PGM-bearing accessory minerals. Cabri (2002) provides mineralogical and chemical data for more than 100 different platinum-group minerals, although many other poorly characterised forms are also known. Some are very rare and are known only from one deposit.

The PGM rarely occur as native metal but commonly form a wide variety of alloys with one another or with other metals, notably with iron, and less commonly with tin, copper, lead, mercury and silver. In other platinum-group minerals the PGM are bonded to sulfur, arsenic, antimony, tellurium, bismuth and selenium (Table 12.2).

Major deposit classes

Enrichment of PGM occurs in deposits of several types developed in a limited range of geological settings, commonly associated with nickel and

Table 12.2 A selection of the more common platinum-group minerals.

Group	Name	Formula	Group	Name	Formula
Alloys	Isoferroplatinum	Pt_3Fe	Arsenides	Sperrylite	$PtAs_2$
	Osmiridium	OsIr		Stillwaterite	Pd_8As_3
	Rustenburgite	$(Pt,Pd)_3Sn$	Antimonides	Genkinite	$(Pt,Pd,Rh)_4Sb_3$
	Tulameenite	Pt_2FeCu		Isomertieite	$Pd_{11}As_2Sb_2$
Sulfides	Braggite	(Pt,Pd)S		Stibiopalladinite	Pd_5Sb_2
	Cooperite	PtS		Geversite	$Pt(Sb,Bi)_2$
	Hollingworthite	RhAsS	Tellurides	Kotulskite	PdTe
	Laurite	$(Ru,Os)S_2$		Merenskyite	$(Pd,Pt)(Te,Bi)_2$
Bismuthides	Froodite	$PdBL_2$		Moncheite	$(Pt,Pd)(Te,Bi)_2$

copper. A map showing the distribution of major deposits, mines and districts, and other selected PGM occurrences worldwide is shown in Figure 12.1. Magmatic PGM deposits, found in mafic and ultramafic igneous rocks, are of two principal types (Table 12.3): PGM-dominant deposits which are associated with sparse, dispersed sulfide mineralisation (Maier, 2005); and nickel–copper sulfide deposits in which the PGM occur in association with sulfide-rich ores (Naldrett, 2010). A variety of other, less economically important deposit types is also known (Table 12.4).

In simple terms PGM-sulfide deposits are derived from magmatic processes of crystallisation, differentiation and concentration. The magma becomes saturated with sulfur and an immiscible sulfide liquid separates from the magma as disseminated droplets. On account of their chalcophile behaviour the PGM are concentrated strongly in the sulfide liquid and are, therefore, scavenged from the silicate liquid. An ore deposit may form if this PGM-enriched sulfide liquid accumulates in sufficient quantities at a particular location.

PGM-dominant deposits

In these deposits the PGM are the main economic products with minor nickel and copper derived from sparsely disseminated (up to five volume per cent) sulfides, chiefly pyrrhotite, pentlandite and chalcopyrite. Four classes of PGM-dominant deposits are recognised:

- Merensky Reef type;
- chromitite reef type
- contact type;
- dunite pipes.

All of these are typically developed in the Bushveld Complex in South Africa, which is the largest layered igneous complex in the world.

Bushveld Complex

The Bushveld Complex (ca 2054 Ma[1]) hosts about 90 per cent of all known platinum resources in the world and accounts for more than 75 per cent of global platinum production. It extends about 450 km east–west and 350 km north–south, and is about 9 km thick (Figure 12.2). The Bushveld Complex is divided into three main sectors: the Western Limb, the Eastern Limb and the Northern Limb (Cawthorn et al., 2002). It is saucer-shaped with the layered mafic and ultramafic rocks dipping into the centre of the intrusion where they are covered by felsic lithologies. The platinum-bearing horizons are located within part of the complex, known as the Critical Zone, which comprises packages, or cyclic units, of repetitively layered chromitite, pyroxenite, norite and anorthosite on a scale of centimetres to tens of metres.

Platinum is currently mined from three horizons in the Bushveld Complex: the Merensky Reef, the UG2 Chromitite Reef, and the Platreef. The Merensky Reef has been exploited for many years with PGM production

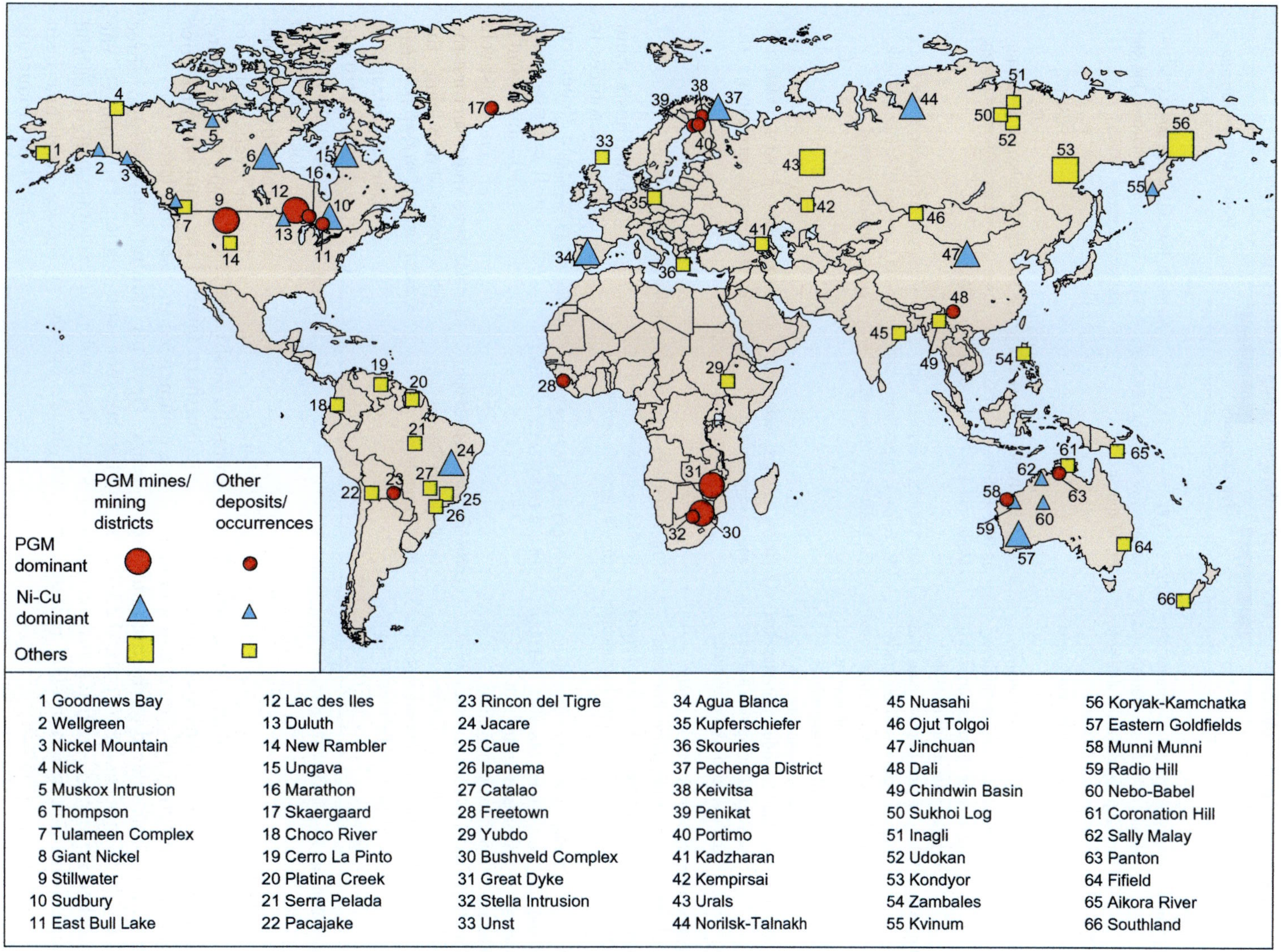

Figure 12.1 The distribution of the main platinum-group minerals mining districts, mines and deposits. Selected occurrences of other types are also shown (some nickel mines which produce by-product platinum-group minerals are omitted).

Table 12.3 Key characteristics and examples of the major platinum-group minerals deposit types.

Deposit type	Description	Typical grades	Major examples
PGM-dominant			
Merensky type	Extensive, laterally continuous, thin layers, or 'reefs', of ultramafic rocks in large layered mafic-ultramafic intrusions; with minor disseminated nickel and copper sulfides.	5 – 7 g/t Pt + Pd; Pt/Pd = 3 in Merensky Reef. Major source of Pt, Pd and Rh.	Merensky Reef, Bushveld Complex, South Africa; Great Dyke, Zimbabwe; J-M Reef, Stillwater Complex, USA.
Chromitite type	Similar morphology to Merensky type but comprising thin layers of massive chromite with sparse base metal sulfide minerals.	4 – 8 g/t Pt + Pd; Rh 0.3 – 0.6 g/t in UG2. Pt/Pd = 2.5 in UG2. Major source of Pt, Pd and Rh.	UG2, Bushveld Complex, South Africa; Lower Chromitites, Stillwater Complex, USA.
Contact type	Extensive (km) discontinuous PGM mineralisation with low-grade nickel and copper in basal contact zones of layered intrusions. Commonly heterogeneous, brecciated, xenolithic.	1-4 g/t Pt + Pd in Platreef, with Pt/Pd ca. 1; by-product Ni and Cu. Major PGM resources.	Platreef, Bushveld Complex, South Africa; Duluth Complex, USA; East Bull Lake, Canada; Portimo, Finland.
Dunite pipes	High grade platinum mineralisation in discordant pipe-like bodies of dunite, up to 1 km in diameter.	3-2000 g/t. Largely worked out and no longer mined.	Onverwacht, Driekop and Mooihoek, Bushveld Complex, South Africa.
Nickel-copper-dominant (with by-product PGM)			
Associated with meteorite impact	Nickel-copper sulfide deposits in impact melt rocks and underlying radial and concentric fracture and breccia zones.	1-10 g/t Pt + Pd, Pt/Pd < 1. Numerous deposits worked for Ni, with lesser Cu and PGM.	Sudbury, Canada.
Related to rift- and continental-flood basalts	Nickel-copper sulfide deposits in sub-volcanic sills which feed flood basalts associated with intracontinental rifting.	2-100 g/t PGM. Average 7.31 g/t Pd and 1.84 g/t Pt in Noril'sk.	Norilsk, Russia; Jinchuan, China.
Komatiite related	Nickel sulfide deposits related to komatiitic (magnesium-rich) volcanic and intrusive rocks of Archaean and Palaeoproterozoic age.	Commonly a few hundred ppb, locally greater than 1 g/t. Pt/Pd generally <1.	Kambalda, Western Australia; Pechenga district, Russia; Thompson Belt, Canada; Ungava Belt, Quebec, Canada.

Cu, copper; Ni, nickel; Pd, palladium; Pt, platinum; Rh, rhodium.

starting in 1928 in the Rustenburg area. The UG2 Chromitite Reef was not mined on a commercial scale until 1985 on account of metallurgical problems related to the high chrome and low sulfide contents. However, the UG2 is now worked at several localities and accounts for a significant proportion of PGM production from the Bushveld. The Platreef is restricted to the Northern Limb of the complex where mining commenced in 1926 but lasted only a few years. Open-pit mining restarted in 1992 at Mogolakwena and higher PGM prices have spurred considerable exploration on the Platreef in the last decade.

Table 12.4 Key characteristics and examples of other platinum-group minerals deposit types.

Deposit type	Description	Typical grades	Examples
Alaskan or Alaskan-Ural type	PGM enrichment in small zoned ultramafic complexes.	Generally low grade, locally very high.	Nizhny Tagil, Urals; Tulameen Complex, Canada.
Placer	An accumulation of dense platinum-group minerals, normally in sand and gravel deposits in a river system.	Generally low grades with sporadic rich pockets.	Urals, Russia; Choco River, Colombia; Goodnews Bay, Alaska; Southland, New Zealand.
Ophiolites	PGM enrichment chiefly with podiform chromite deposits in lower sections of ophiolite complexes. Hydrothermal processes may enrich PGM.	Values in g/t range reported from Unst.	Unst, UK; Zambales, Phillippines; Veria, Greece; Kempirsai, Russia; Al Ays, Saudi Arabia.
Laterites	Residual enrichment of PGM caused by intense tropical weathering of ultramafic rocks.	ca 0.5 g/t Pt.	Yubdo, Ethiopia.
Hydrothermal	Polymetallic veins and breccias and/or hydrothermal remobilisation of nickel, copper and PGM in mafic-ultramafic rocks.	Up to tens of g/t Pt + Pd. Hundreds of g/t at Waterberg.	New Rambler, USA; Waterberg, South Africa; Rathbun Lake, Canada; Goodsprings, Nevada.
Unconformity-related	Hydrothermal PGM enrichment with gold ± uranium in metasedimentary rocks at, or close to, an unconformity with younger, oxidised sedimentary rocks. Commonly focused along faults.	0.19 g/t Pt, 0.65 g/t Pd, 4.31 g/t Au at Coronation Hill. Tens/hundreds g/t Pt and Pd at Serra Pelada.	Coronation Hill, Australia; Serra Pelada, Brazil.
Porphyry deposits	Local PGM enrichment with gold in porphyry style copper deposits hosted by igneous intrusions.	Generally less than 1 g/t PGM.	Elatsite, Bulgaria; Skouries, Greece; Santo Tomas II, Phillippines.
Shales and other sedimentary rocks	PGM enrichments in metalliferous black shales. Generally thin but may be laterally extensive.	Generally less than 1 g/t, but local very high values in the Kupferschiefer.	Kupferschiefer, Poland; Yukon, Canada; Sukhoi Log, Siberia.
Carbonatite and alkaline complexes	PGM enrichment in small carbonatite-phoscorite igneous complexes. Disseminated copper-palladium-rich sulfide mineralisation in gabbro at Coldwell.	Few hundred ppb in Palabora; few g/t in Brazilian examples. 0.76 g/t Pd, 0.23 g/t Pt and 0.27% Cu in Marathon deposit.	Palabora, South Africa; Catalão and Ipanema, Brazil; Kovdor, Russia. Coldwell Complex, Canada; Marathon, Canada.

Au, gold; Cu, copper; Pd, palladium; Pt, platinum.

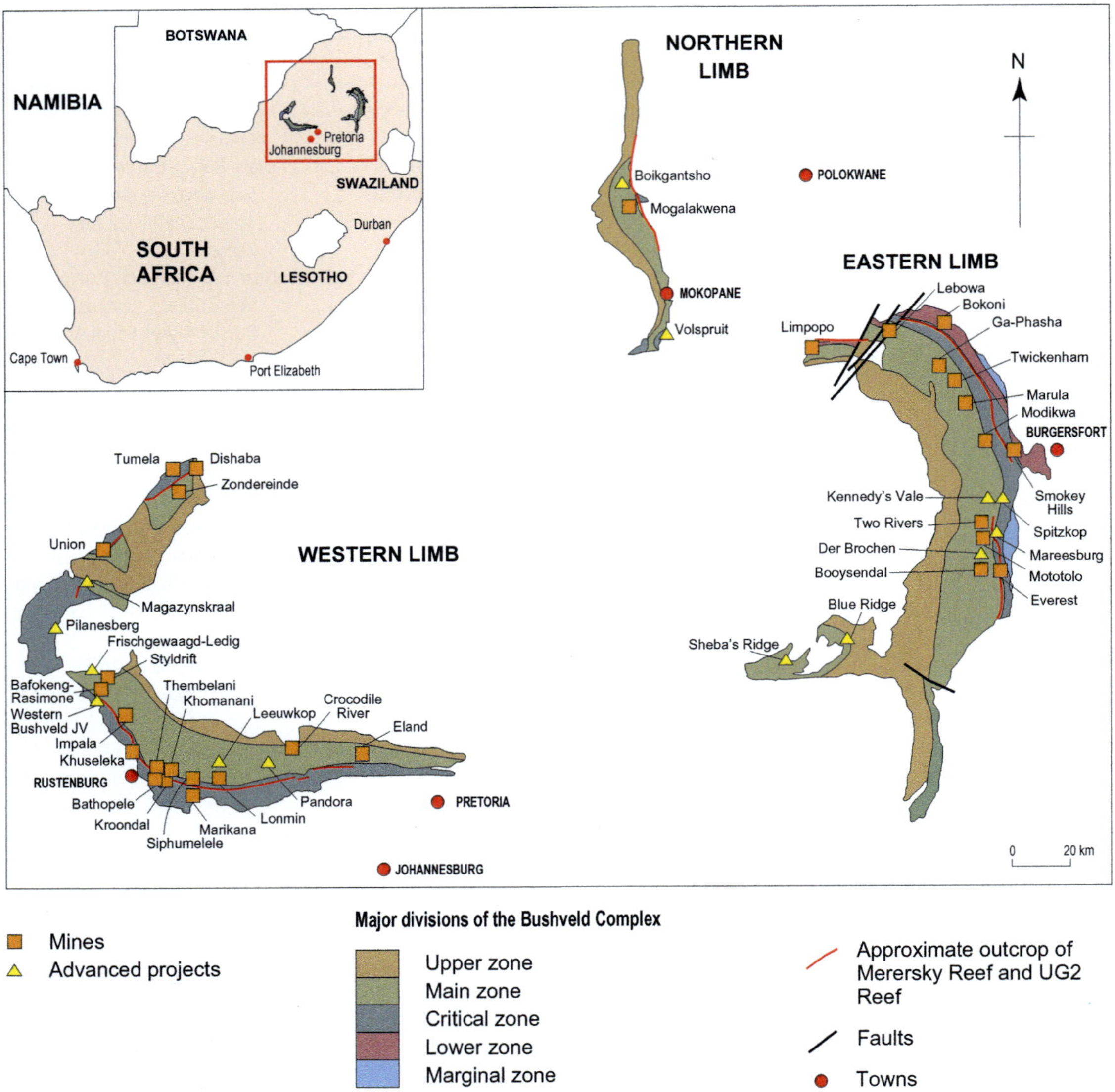

Figure 12.2 The geology of the Bushveld Complex showing the location of platinum mines and advanced projects.

Merensky Reef type

The Merensky Reef is the type example of a PGM-dominant deposit associated with a minor amount of sulfide mineralisation. Other important examples are found in the Great Dyke, Zimbabwe and the Stillwater Complex, USA. This deposit class is a major source of global production of platinum and palladium.

These deposits occur as thin, laterally extensive, stratiform zones of sparsely disseminated sulfides within major layered mafic-ultramafic igneous intrusions. In the Bushveld Complex these deposits are typically located about 2 km above the base of the intrusion. There is no consensus on the processes responsible for their genesis but the most widely accepted model involves mixing of a residual magma after partial crystallisation with a new pulse of magma. This leads to sulfide saturation and separation of immiscible sulfide droplets which scavenge PGM from the silicate magma (Lee, 1996 and Maier, 2005).

The Merensky Reef comprises a pegmatoidal pyroxenite layer about 1 m thick, bounded on both sides by thin chromitite layers and containing up to three per cent disseminated sulfides. It can be traced laterally for hundreds of kilometres and is found on both the Eastern and Western limbs of the Bushveld Complex. The grade of the Merensky Reef averages about 5–7 g/t Pt + Pd (Maier, 2005), although the mineralogy is quite variable and includes a wide range of alloys, sulfides, tellurides and arsenides.

Chromitite Reef type

The UG2 Reef, one of many thin near-massive chromite layers in the Critical Zone of the Bushveld Complex, is generally regarded as the largest PGM resource in the world. The ruthenium and rhodium contents of the UG2 are also significantly higher than the Merensky Reef, and hence it is an important source of these metals. The UG2 comprises a chromitite layer between 0.7 and 1.3 m in thickness with sparse disseminated sulfide mineralisation. The UG2 has a similar lateral extent to the Merensky Reef, occurring below it at depths from 30 m up to about 400 m. The average grade of the UG2 varies from approximately 4–8 g/t Pt + Pd. The platinum-group minerals in the UG2 comprise mainly sulfides and alloys, including laurite, cooperite, braggite and Pt–Fe alloy.

Contact type

This type of deposit is exemplified by the Platreef in the Bushveld Complex. The Platreef is a layer of ultramafic rocks, chiefly pyroxenite, up to 300 m thick, that can be traced about 35 km along the base of the Northern Limb of the Bushveld Complex (Kinnaird and MacDonald, 2005). It has generally higher base-metal contents than Merensky-type deposits, particularly nickel and copper, up to about three volume per cent. The Platreef is a major PGM resource with mineralisation in both the pyroxenite and the underlying rocks. The mineralisation is variable in thickness and is best developed along a 25-km section of strike length. It varies considerably within this interval, in part according to the nature of the footwall rocks which include extensive sections of granite-gneiss basement, dolomite, banded iron formation, chert, quartzite and shale. The mineralisation in the Platreef is less laterally continuous than the reef styles and the PGM contents are also generally lower, typically in the range 1–4 g/t. The PGM mineralogy of the Platreef is dominated by tellurides, bismuthides and arsenides, but this varies significantly both laterally and vertically.

The PGM-dominant mineralisation in the Platreef may be classified as a contact-type PGM deposit, which comprise basal accumulations of magmatic sulfides associated with abundant xenoliths and brecciation, and coarse pegmatitic textures. Other examples of this class are recognised in the USA, Finland, the Kola Peninsula of Russia and in Ontario, Canada (Table 12.3), although these deposits are not mined for PGM at present.

Dunite pipes

These are discordant pipe-like bodies, up to 1 km in diameter, of varying composition, which are found widely within the Bushveld Complex (Scoon and Mitchell, 2004). However, in only four of them, located in a 20-km belt on the Eastern Limb to the north-west of Burgersfort, has significant PGM mineralisation been discovered. High-grade platinum ore, locally up to 2000 g/t Pt, occurs in discrete zones of the pipes within unmineralised dunite. Discovered in 1924 and largely mined out by 1930, there is little modern information on these deposits.

Other PGM-dominant deposits

The Great Dyke (2575 Ma) is a layered mafic-ultramafic intrusion up to 11 km wide that can be traced for 550 km in a north-north-easterly direction across central Zimbabwe (Oberthür, 2002). The Great Dyke is a major PGM resource and also contains important deposits of chromite. The platinum mineralisation is associated with two layers: a 30–60-m zone of disseminated mineralisation in the Lower Sulfide Zone (LSZ); and a 2–8-m PGM-rich interval in the Main Sulfide Zone (MSZ). Currently only the MSZ is economic to exploit. The Hartley Complex, one of five zones within the Great Dyke, is by far the largest and economically most significant and contains approximately 80 per cent of its PGM resources.

The Stillwater Complex (2700 Ma) in Montana, USA, is another of the world's major layered intrusions that hosts important PGM mineralisation (Zientek et al., 2002). The PGM-bearing J-M Reef can be traced for 45 km along strike and approximately 1.6 km down-dip. The reef consists of a 1–3-m thick pegmatitic peridotite and troctolite containing sparsely disseminated sulfides (0.5–1 volume per cent) and platinum-group minerals. In contrast to deposits in the Bushveld Complex and the Great Dyke, the J-M Reef is relatively enriched in palladium with a Pt/Pd ratio of about 0.3. The reef contains an average of 20–25 g/t Pd + Pt over a thickness of about two metres. Most of the palladium is contained within pentlandite, while the platinum occurs mainly in moncheite, braggite, cooperite and Pt–Fe alloy.

The Lac des Iles Intrusive Complex (2700 Ma) comprises a suite of mafic-ultramafic rocks situated approximately 85 km north of Thunder Bay, Ontario, Canada. The deposit, mined chiefly for palladium, is similar in some respects to contact-type PGM mineralisation, but the processes responsible for its formation remain unclear (Maier, 2005). Open-pit mining commenced at Lac des Iles in 1993, with underground operations following in 2006.

Nickel–copper-dominant deposits

Magmatic nickel–copper deposits are the most important source of nickel worldwide (Naldrett, 2010). Copper, cobalt and the PGM, mainly palladium, are important by-products. Gold, silver, chromium, sulfur, selenium, tellurium and lead are also recovered from some deposits.

The dominant ore minerals are sulfides, pyrrhotite, pentlandite and chalcopyrite, which generally constitute more than ten per cent by volume of the host rock. Nickel grades typically range from 0.5–3.0% Ni, with attendant copper in the range 0.2–2.0%. PGM contents vary widely from a few ppb up to, exceptionally, 10 g/t. The size of the deposits ranges from a few hundred thousand tonnes of ore up to a few tens of million tonnes. Globally two nickel–copper districts are predominant, Sudbury in Ontario, Canada and Norilsk-Talnakh in the polar region of Russia, with each containing more than ten million tonnes of nickel metal.

Magmatic sulfide deposits occur in diverse geological settings in rocks ranging in age from Archaean to Permo-Triassic. Eckstrand and Hulbert (2007) recognised four principal classes:

1. A meteorite-impact mafic melt with basal sulfide ores. Sudbury is the only known example.
2. Rift- and continental-flood basalt, with associated dykes and sills. Important examples include the Norilsk-Talnakh district in Russia and Jinchuan in China.
3. Komatiitic (magnesium-rich) volcanic flows and related intrusions. Important examples include Kambalda in Australia, Thompson and Raglan in Canada and Pechenga in north-west Russia.
4. Other mafic-ultramafic intrusions. Although deposits of this class contain significant resources of nickel and copper (e.g. Voisey's Bay, Labrador, Canada), their PGM contents are generally very low.

The origin of these deposits involved derivation of the metals from primitive magmas which ascended to high levels in the crust along major faults where they were contaminated by sulfur-bearing sedimentary rocks, such as evaporites or carbonaceous shales. This led to segregation of an immiscible sulfide liquid that settled to the lower part of the magma chamber scavenging chalcophile elements such as nickel, copper and PGM. Where these sulfide droplets became sufficiently concentrated they formed a magmatic sulfide deposit, either in a basal or marginal setting (Naldrett, 2010).

Deposits related to meteorite impact

The origin of the nickel–copper–PGM deposits of the Sudbury Igneous Complex (SIC) is related to a meteorite impact at 1850 Ma (Lightfoot et al., 2001). The impact crater was originally about 200 km in diameter with the mineralisation located at the base and in the footwall of the impact melt. Following later deformation the SIC now occupies an elongate basin 65 km long and 27 km wide. The nickel deposits, which are present on both the north and south flanks of the basin, are hosted by a basal mafic noritic unit containing abundant fragmental material, referred to as the 'sublayer'. Radiating and concentric fracture and breccias zones, known as 'offsets', in the underlying footwall locally host sulfide-rich orebodies enriched in copper and PGM at depths up to 400 m below the sublayer. The sulfide ores have typical magmatic sulfide mineralogy and textures. A wide range of platinum-group minerals is present of which the most abundant are tellurides (michenerite and moncheite) and arsenides (sperrylite).

Deposits related to rift- and continental-flood basalts

The Norilsk–Talnakh district in the Taimyr Peninsula of Russia is the world's largest producer of nickel. The ores here are also very rich in PGM and consequently this area is the source of the vast majority of Russia's PGM and is also the world's largest palladium producer and the second largest platinum producer (Johnson Matthey, 2012). The resource in the region has been estimated at more than 1.3 billion tonnes of ore averaging 1.77% Ni, 3.57% Cu, 0.06% Co, 1.84 g/t Pt and 7.31 g/t Pd (Naldrett, 2004). The Norilsk–Talnakh deposits are associated with the huge Permian flood basalt suite known as the Siberian Traps, which developed in major intracontinental rift zones. The nickel–copper–PGM ores occur as flat-lying sheets at the base of gently dipping gabbroic sills, in the underlying sedimentary rocks and in proximity to the upper contacts of the sills.

Another major deposit of this type is Jinchuan in China which contains more than 500 million tonnes of ore with an average grade of 1.1 wt% Ni and 0.7 wt% Cu (Su et al., 2008). It has been mined for nickel and copper since 1964, although in recent years PGM production has become increasingly important and Jinchuan currently accounts for about 95 per cent of China's PGM production.

Deposits related to komatiitic rocks

Nickel–copper deposits with low PGM contents associated with komatiitic volcanic flows and sills are widespread in Archaean and Palaeoproterozoic terranes in Australia, Canada, Brazil, Finland and Zimbabwe (Lesher and Keays, 2002). Both small high-grade deposits (1.5–4.0% Ni) and much larger low-grade deposits (ca. 0.6% Ni) have been mined at many localities in the Eastern Goldfields of the Archaean Yilgarn Craton of Western Australia (Hoatson et al., 2006). Other important komatiite-associated deposits which are mined for nickel and by-product PGM include the Thompson Nickel Belt of Manitoba, Canada and the Pechenga district of the Kola Peninsula in Russia.

Other deposit types

High PGM concentrations are known in several other geological environments. Although very high grades may occur locally, most deposits are small and mining of PGM is not currently economic. Alluvial placer deposits of PGM, mainly related to Alaskan-type intrusions, were worked for many decades in Colombia, Alaska, British Columbia, the Urals and central New South Wales (Tolstykh et al., 2005). They were the world's principal source of platinum until the discoveries at Sudbury and in the Bushveld Complex, but now significant PGM production from alluvial deposits is restricted to the Russian Far East (Johnson Matthey, 2012).

Extraction and processing

Extraction methods

The method used to mine PGM-bearing deposits is dependent on their size, grade, morphology and the value of any co-products. Underground extraction uses a variety of standard mining methods depending on the characteristics of the orebody. The deepest currently operating platinum mine is Zondereinde, owned by Northam Platinum Ltd, on the western limb of the Bushveld Complex where mining takes place at a maximum depth of 2.2 km. Due to the high temperatures at this depth, 'hydropower' is now widely used in the deep-level mines on the Bushveld. This technology, pioneered by Northam Platinum, uses water, cooled on the surface to 5 °C, to power its drilling machines and cool working places. Typically, these underground operations use labour-intensive drilling and blasting techniques, though increased mechanisation is being introduced into the workplace. Ore is blasted using explosives then transported to the shaft using an underground rail system. Old workings are backfilled with waste material to improve ventilation by forcing air to travel through only those areas that are being worked, as well as providing more roof support. Underground mining is also employed in the Norilsk-Talnakh, Sudbury and Stillwater deposits.

Surface mines are generally cheaper and safer to operate than underground mines. Open-pit mining is used where the ore is near-surface (typically less than 100 m) and for lower-grade ore bodies, which would not be economic to mine underground. This method typically involves removing the overburden, digging the ore or blasting with explosives, then removing the ore by truck or conveyor belt for further processing.

Processing

After mining, the ores are processed to increase their PGM content. Concentration is normally carried out at, or close to, the mine site and involves crushing the ore and separating platinum-bearing and gangue minerals, using a range of physical and chemical processes. The concentrate may be transported further afield for smelting and refining to produce pure metals. Different procedures are used for processing sulfide-poor ores (e.g. Merensky and UG2) and sulfide-dominant ores (e.g. Norilsk) due to their contrasting chemical, mineralogical and physical properties (Cole and Ferron, 2002).

The processing flowsheets for PGM-dominant and nickel–copper-dominant ores are shown schematically in Figure 12.3.

PGM-dominant ore

Comminution and concentration

Crushing, followed by fine grinding (milling), is used to facilitate separation of the ore minerals from the gangue (Merkle and McKenzie, 2002). Some mines employ additional processing stages after comminution to optimise PGM recovery. Where the PGM occur within the crystal structure of pyrrhotite, then pyrrhotite may be magnetically separated, collected and treated separately. Dense media separation may be used to remove lighter silicate minerals from the denser chromite and platinum-group minerals after initial crushing and prior to milling. This significantly reduces the volume of ore to be milled and further treated without reducing PGM recoveries.

Froth flotation normally follows milling. In this process, water is added to the powdered ore to produce a suspension and air is blown upwards through the tanks. Chemicals are added to the mix, making some minerals water-repellent and

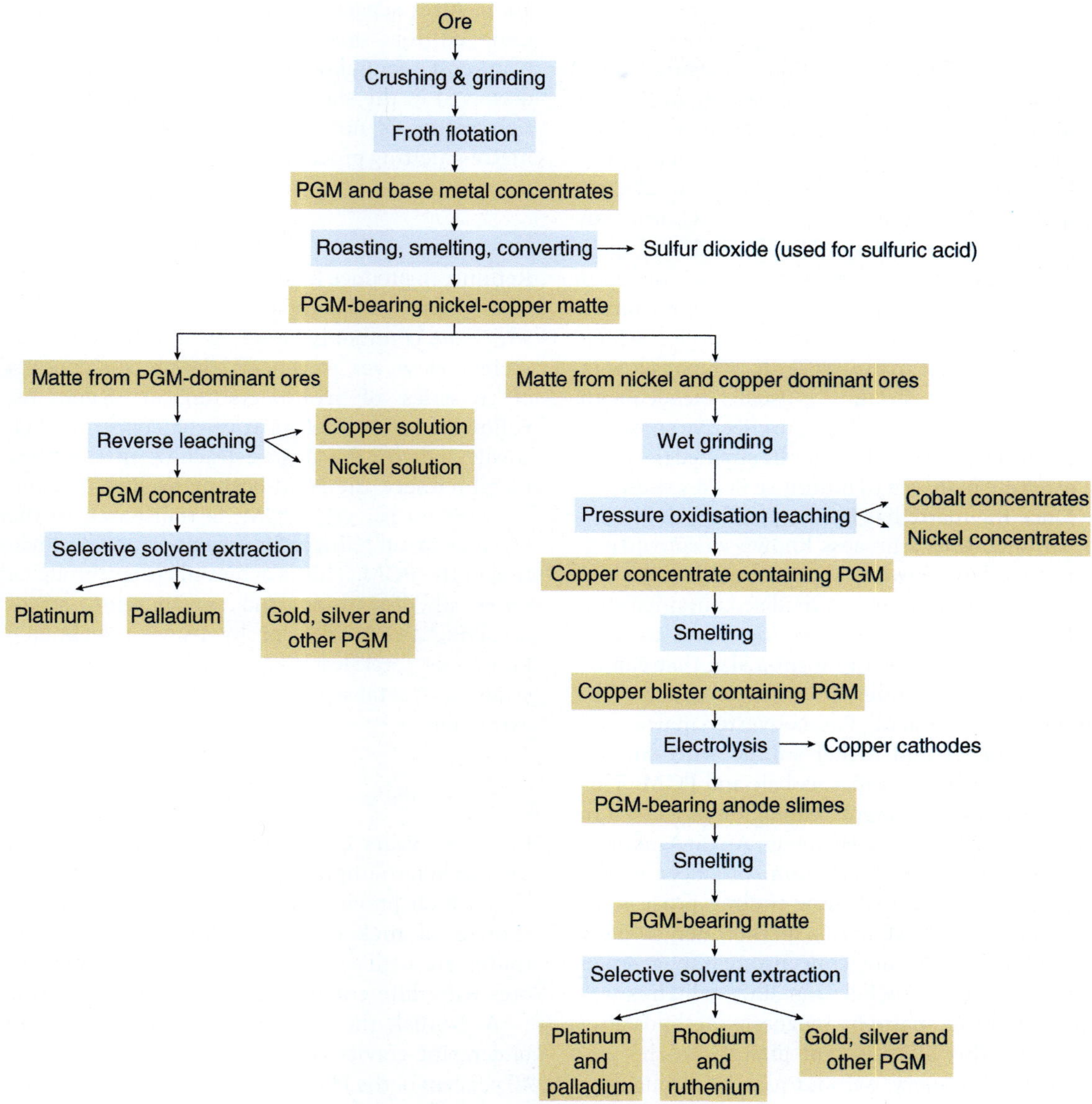

Figure 12.3 Schematic flowsheet for the processing of PGM-dominant and nickel–copper-dominant ores.

causing air bubbles to stick to their surfaces. These minerals collect in the surface froth and are removed as a metal concentrate.

UG2 and Merensky ores are normally treated separately on account of their contrasting and variable mineralogy. Merensky ore is the more sulfide rich (up to about three volume per cent base-metal sulfides), while UG2 ore generally comprises 60–90 per cent chromite. Many operations employ two stages of milling and floating conducted in series, known as mill-float-mill-float (MF2) operations, in order to maximise PGM recovery.

Smelting

Matte-smelting of the flotation concentrate produces a silicate melt (slag) from which an immiscible sulfide melt (matte) containing the PGM separates on account of its greater density (Jones, 2005). Matte smelting takes places in electric furnaces at temperatures of about 1350 °C, although higher temperatures may be needed for UG2 concentrates. Fluxes such as limestone are added to the smelter to reduce the melting temperature of the platinum and base-metal sulfides which accumulate in the matte. UG2 ore, which is low in sulfur and high in chromite, is usually blended with Merensky concentrate to ensure effective smelting. The percentage of chromite in the smelted ore must be carefully managed to minimise PGM losses.

Once the matte has been tapped off, the liquid metal undergoes a process known as converting. This involves blowing air, or oxygen, into the matte for several hours to oxidise contained iron and sulfur. Silica is added to the matte to react with the oxidised iron to form a slag that can be easily removed, while the sulfur is collected to produce sulfuric acid. The converter matte consists of copper and nickel sulfide with smaller quantities of iron sulfides, cobalt and PGM. This is usually cast into ingots which are processed in the base metal refinery. However, Anglo American Platinum, the largest platinum producer in the world, uses a different slow-cooling process for the recovery of PGM. In this method, about 95 per cent of the PGM are concentrated in a small volume of copper–nickel–iron alloy, which is separated from the matte by magnetic methods after crushing. This alloy is sent directly to the precious metal refinery, without passing through the base-metals refinery, thereby reducing processing times considerably.

A potentially important new PGM smelting process, known as ConRoast, has been developed by Mintek (Jones, 2005). In this process the sulfur is removed prior to smelting by roasting the sulfide concentrates in a fluidised bed. The derived sulfur dioxide is efficiently collected and the sulfur-free product is then smelted in a DC arc furnace in which the PGM are collected in an iron alloy. After separation from the slag the iron is removed from this alloy prior to PGM refining. ConRoast technology ensures that chromium is dissolved in the slag, thereby allowing chromite-rich materials not amenable to the traditional matte-smelting process to be treated (Jones, 2009).

Refining

Refining to produce high-purity platinum products is a lengthy and complex process the details of which are commonly closely guarded commercial secrets. However, in general, the process consists of a series of hydrometallurgical operations. Following removal of most of the copper, nickel, cobalt and iron by reverse leaching in the base-metal refinery, the PGM-bearing residue, containing over 60 per cent PGM, is transferred to the precious metal refinery for separation and purification of the PGM. This is achieved either through a series of dissolution and precipitation stages involving PGM salts, or by using a technique known as total leaching, which is followed by sequential metal separation with the aid of solvent extraction.

Nickel–copper-dominant ore

In recent years considerable research has been undertaken to improve PGM recovery from these ores which previously were regarded chiefly as sources of nickel. Several different processing routes are utilised to optimise the treatment of ores with different sulfide contents and textures.

At Norilsk, the ore is crushed and milled before undergoing gravity concentration to recover up to 40 per cent of the PGM. Froth flotation is then used to produce a metal concentrate which is sent to the smelter where it is roasted, smelted and converted to produce a copper–nickel–PGM matte, together with slag as a waste product. The matte is granulated, wet-ground and then treated in the base-metal refinery where pressure oxidation leaching is used to produce concentrates of nickel, copper and cobalt. The copper concentrate, which includes almost all the PGM and gold, is then smelted in furnaces to produce copper blister (98.5% to 99.5%

Cu) with PGM and gold. The copper is then further refined using electrowinning to produce pure cathode copper with the PGM remaining in the anode slimes. The anode slimes are then combined with other slimes produced from the production of nickel, and with the gravity concentrate produced after the initial crushing and milling. This mixture is smelted again to produce a PGM-bearing matte. The matte is then pressure-leached to produce a silver concentrate and two PGM concentrates, one containing platinum and palladium and the second the other PGM.

Treatment of the matte produced by smelting to extract the PGM is carried out in various ways. At Vale's Sudbury operations, the matte is treated by flotation and magnetic separation to produce separate nickel, copper and precious-metal concentrates. The precious-metal concentrates are treated in a two-stage pressure leaching process to dissolve first the nickel and cobalt, and then the copper, selenium and tellurium. This concentrate, containing 60–80% PGM+gold, together with small quantities of silver and base metals, is then refined to produce high-purity individual elements.

The copper concentrates derived from many copper ores may also contain small quantities of PGM. In processing these ores the PGM will be present, together with gold and silver, in the anode slimes at the copper refinery. Most large copper refineries therefore operate a precious-metals recovery plant to treat these slimes.

Specifications and uses

PGM are sold in many forms including pure metal and a wide variety of compounds, solutions and fabricated products. These are produced by a small number of specialist companies including Johnson Matthey, BASF, Umicore, Heraeus and Tanaka Kikinzoku.

Demand for PGM has grown rapidly in recent years, with annual mine production doubling since 1980 and increasing by an order of magnitude since 1960. The catalytic properties of the PGM are the basis of their most important applications in emission control systems and in industrial process catalysts. Other important properties that make PGM useful for many additional purposes include their high strength, high melting point and resistance to corrosion.

Because of the wide range of applications and the differences in the contributions of recycling to PGM supply between sectors it is important to distinguish between net and gross demand for PGM. Net demand is defined as gross demand less the amount of PGM recovered from recycling. This means that in applications where recycling rates are very low then net demand is approximately equal to gross demand. Conversely, where recycling rates are high then net demand is small. The main sectors in which PGM are used and the net demand in each are shown in Table 12.5 and Figure 12.4.

Uses of platinum, palladium and rhodium

The largest net use of platinum, palladium and rhodium is in autocatalysts, which are used to convert noxious emissions (carbon monoxide, oxides of nitrogen and hydrocarbons) from car exhaust systems to harmless non-toxic products. Demand for PGM in autocatalysts has grown markedly from the 1970s onwards as many countries followed the lead of USA and Japan and introduced legislation setting standards for emissions from motor vehicles.

The quantities and proportions of platinum and palladium used in autocatalysts have varied considerably over time in response to technological changes and price variations. Currently palladium is the major constituent of catalysts used in gasoline-powered vehicles while platinum is still the chief active component in catalysts and particulate filters fitted to diesel-powered vehicles. Consequently, this is the main contribution to platinum demand in Europe where more than 50 per cent of all new cars have a diesel engine.

Although global demand for rhodium in 2011 (906,000 troy ounces gross or 28.2 tonnes) was little more than ten per cent of that for palladium (Johnson Matthey, 2012), rhodium is a critical component of autocatalysts which account for

Table 12.5 The main uses of platinum and palladium, and global consumption by sector in 2011. (Data from Johnson Matthey, 2012.)

Sector	Platinum: Main uses	Platinum: Consumption (thousand troy ounces)	Palladium: Main uses	Palladium: Consumption (thousand troy ounces)
Autocatalyst	Catalysts for vehicle exhaust emission control.	3105	Catalysts for vehicle exhaust emission control.	6030
Investment	Exchange traded funds, ingots, bars, coins.	460	Exchange traded funds, coins.	(565)*
Jewellery	Fabrication of of platinum jewellery.	2480	Fabrication of palladium jewellery, white gold.	505
Chemical	Catalysts for production of nitric acid and other bulk and speciality chemicals.	470	Catalysts for production of nitric acid and other bulk and speciality chemicals.	445
Electrical	Hard disk coatings, thermocouples.	230	Multi-layer ceramic chip capacitors, hybrid integrated circuits, plating.	1380
Petroleum	Catalysts for petroleum refining and production of petrochemicals.	210	–	–
Glass	Vessels for glass manufacture, bushings for fibre glass production.	555	–	–
Dental, medical and biomedical	Anti-cancer drugs, implants such as heart pace-makers.	230	Alloying agent, mainly with minor platinum, gold, silver.	550
Other	Spark plug tips, oxygen sensors for engine management, fuel cells.	355	Emission control catalysts in stationary applications.	105
	Total global consumption	**8095**	**Total global consumption**	**8450**

1000 troy ounces is approximately 31.1 kilograms.
*in 2011 palladium investment was negative and 565,000 ounces were supplied back to the market.

about 80 per cent of its use. Other areas where rhodium is important include glass manufacture and the chemicals sector.

Platinum has long been widely used in jewellery, especially in Japan, and in recent years it has become increasingly popular in Europe and China. In 2011 jewellery was the second most important use of platinum accounting for about 31 per cent of total net consumption.

Like gold and silver, there is a demand for platinum as a safe haven for storing wealth. Various platinum coins, ingots and bars have been sold over the past thirty years, but, today, investors are increasingly purchasing platinum and palladium through exchange-traded funds (ETFs) which were launched in Europe in 2007 and in USA in 2010. Total net holdings of platinum and palladium in ETFs at the end of 2011 were estimated to be 1.32 million troy ounces of platinum and 1.7 million troy ounces of palladium (CPM Group, 2012).

Platinum and palladium are widely used in catalysts in the production of chemicals and pharmaceuticals. However, net PGM demand for these purposes is relatively small because the catalysts are recycled with a very high degree of efficiency and new supply is needed only to replace catalysts at the end of their useful lifetime or when additional

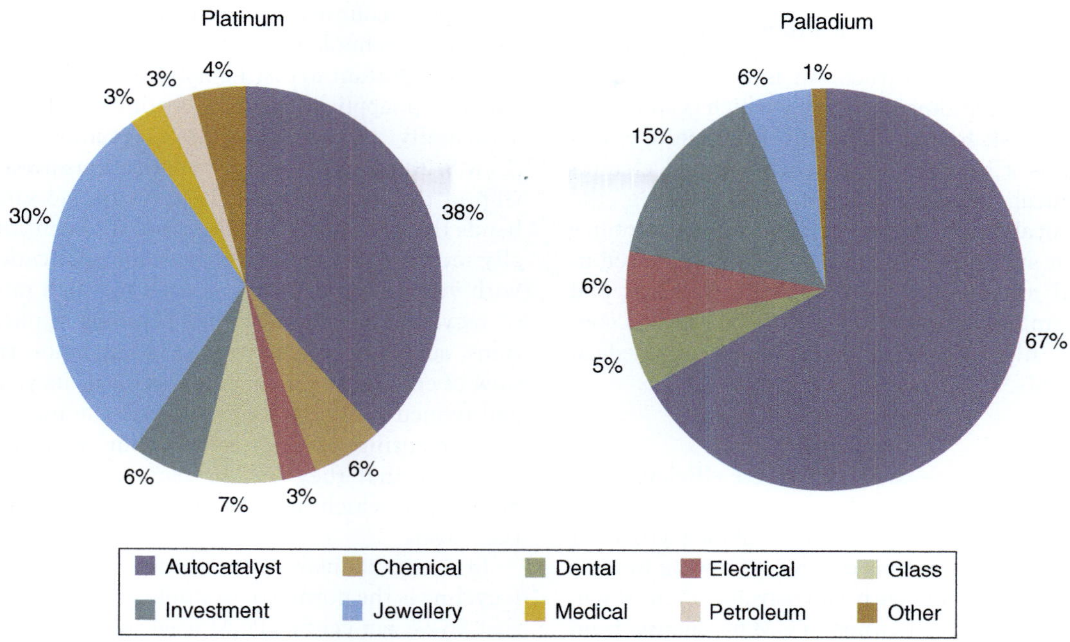

Figure 12.4 The main applications of platinum and palladium in 2011. (Data from Johnson Matthey, 2012.)

capacity is installed. Most important is the use of platinum in the conversion of ammonia to nitric acid for the manufacture of fertilisers and explosives. Platinum-based catalysts are also indispensable to the refining of petroleum and the production of petrochemical feedstocks used in the manufacture of plastics, synthetic rubber and polyester fibres. Some palladium is also used in petrochemical catalysts but the main use is in fine chemistry (i.e. the manufacture of vitamins, dyes, antibiotics, etc.) and in pharmaceuticals (Hagelüken, 2008).

The electronics industry is the second largest consumer of palladium where it enjoys a significant cost advantage over rivals such as gold. A large proportion of this palladium is used in multi-layer ceramic capacitors (MLCC), while other important applications are in hybrid integrated circuits, used mainly in the automotive sector, and in plating of connectors inside computers. Platinum is also used as a component of the magnetic coating on computer hard disks for which demand has escalated in recent years.

Platinum and platinum–rhodium alloys are used in equipment that hold, channel and form molten glass. These materials are able to withstand the abrasive action of molten glass and do not react with it at high temperatures. Palladium and, to a much lesser extent, platinum are also important components of dental alloys used in inlays, bridges and crowns. On account of its chemical inertness, high electrical conductivity and its ability to be fabricated into very small complex components, platinum is an important component of various biomedical devices that are implanted in the body. Platinum electrodes are used in heart pacemakers and in implantable cardioverter defibrillators (ICD) used in the treatment of irregular heartbeat.

The ability of platinum, in certain chemical forms, to inhibit cell division is the basis of its use in a range of anti-cancer drugs. The first of these, cisplatin, was introduced in 1977 and is widely used in the treatment of testicular, ovarian, bladder and lung cancers. Several other platinum-based drugs are now in use or undergoing testing.

Uses of ruthenium, iridium and osmium

The main use of ruthenium is in perpendicular magnetic recording technology which is employed in hard disk drive manufacture. Ruthenium is also used as a catalyst in the chemical and electrochemical industries. Iridium is used in the chemical industry, for crucibles in the electronics sector and in spark plug tips. Osmium is used in small amounts in an alloy with platinum and iridium, which is used in pen tips, electrical contacts, filaments in light bulbs and in medical implants.

Recycling, re-use and resource efficiency

The durability of PGM in use, combined with their recent high prices, has ensured a growing interest in using them efficiently and recycling them whenever possible. Furthermore, PGM production from secondary sources, such as scrap and end-of-life products, reduces the need to mine primary ores thereby contributing to improved supply security. Recycling also produces less carbon dioxide than mining and processing primary ores.

In 2010 the proportions of PGM supply from recycling increased as economies began to recover from the global economic recession and PGM prices rose. For platinum, approximately 23 per cent (1.84 million troy ounces) was derived from recycling, while the corresponding figures for palladium and rhodium were 19 per cent (1.85 million troy ounces) and 27 per cent (236,000 troy ounces), respectively, (Johnson Matthey, 2011). The largest contributions came from autocatalysts, stimulated by national car-scrappage schemes, from platinum jewellery and from the recovery of palladium from electronic scrap. In the EU the Waste Electrical and Electronic Equipment (WEEE) Directive (2003), which is intended to promote the recovery of electrical scrap, has had a positive impact on PGM recycling. However, only a small proportion of the recycling potential of PGM from this source is currently achieved in Europe because of low collection rates, high exports of old equipment to developing countries and use of inappropriate treatment technology (UNEP, 2011).

It is important to distinguish 'closed-loop' and 'open-loop' applications as their differences fundamentally impact on PGM recycling rates (GFMS, 2005). In closed-loop applications ownership of the metal effectively stays in the same hands throughout its life cycle and it is continually recycled and reused in the same application with very transparent flows, ensuring high rates of recycling. In contrast, in open-loop applications, such as autocatalysts and electronics, the flow of end-of-life products is less clear: recycled and refined PGM are sold back into the market, supplementing primary supply but with less assurance that they will be recycled when the products in which they are used reach the end of their lives.

In some industrial applications closed-loop recycling is the norm. For example, it is estimated that 95–98 per cent of PGM used in glass manufacture are recycled and re-used in glass making (Johnson Matthey, 2011). High recycling rates are also achieved in the manufacture of chemicals and in oil refining where the spent catalysts are returned directly to a recycling company.

For many consumer applications PGM recycling rates are much lower due mainly to inefficient collection and losses during product dismantling. Although the technology required to handle these materials is both technically efficient (more than 95 per cent recovery) and environmentally friendly, it is highly specialised and not widely available. UNEP (2011) reports that end-of-life recycling rates for platinum and palladium are 80–90 per cent in industrial applications, but, where the products do not enter an appropriate closed-loop recycling chain, they are much lower at about 50–55 per cent in vehicle applications and less than ten per cent in electronic goods.

In some uses, such as medical and biomedical, spark plugs, sensors and hard disks PGM are effectively dissipated in use and not available for recycling. In contrast, recycling rates for PGM in jewellery and investment items are very high.

Overall, if gross demand and recycling rates in closed-loop applications (chemicals, oil refining and the glass industry), which are not normally included in published statistics, are considered recycling contributes more than 50 per cent of total PGM supply.

Substitution

In many applications one PGM can substitute for another. This has happened in the past in response to price differentials between palladium and platinum. For example, in recent years palladium has been increasingly used in three-way catalysts for petrol cars with manufacturers also moving to increase the proportion of palladium to platinum in diesel catalysts which formerly used platinum as the sole catalytic metal.

In response to the price spikes in 2001 there was also significant substitution of palladium in MLCC electronics applications by nickel and copper, albeit with some reduction in performance level. Consequently, in many high-tech applications palladium remains the material of choice.

The prevailing high prices have stimulated research in thrifting of the PGM, i.e. using less material in a particular application with little or no reduction in performance. For example, as technology has advanced autocatalysts have become more efficient and smaller quantities of PGM are required to achieve the same performance. However, at the same time, emission standards around the world have become increasingly stringent and so the amounts used have remained nearly constant. There is also considerable interest in reducing the amount of PGM used in fuel cells in order to reduce unit costs and stimulate uptake of this technology (Johnson Matthey, 2011).

Environmental issues

In metallic form the PGM are generally regarded as inert, non-toxic and non-allergenic. However, some of their compounds, particularly certain PGM-chlorinated salts, are both highly toxic and allergenic, and DNA damage due to PGM exposure has been reported (Gagnon et al., 2006).

The greatly increased use of PGM in recent years has raised the level of PGM emissions into the environment. The largest contribution has been from vehicle emissions following the introduction of PGM-bearing autocatalysts from the 1970s onwards. High PGM concentrations have been recorded in dust, silt, soils and waters close to major highways and in urban areas (Ravindra et al., 2004). Furthermore, complexation of PGM derived from autocatalysts has been shown to give rise to mobile species with possible associated increase in bioavailability (Colombo et al., 2008). Several studies suggest a variation between individual PGM in environmental mobility and uptake rate in plants and animals, with palladium greater than rhodium, which is, in turn, greater than platinum (Ek et al., 2004). Further research is required to provide a better understanding of PGM emissions from autocatalysts, their dispersion and bioavailability and their effects on animals and plants. However, it is important to note that continual improvements to the design of catalytic converters and vehicle exhaust systems have led to significant reduction of specific PGM emissions, provided fuel of appropriate quality is used and vehicles are well maintained.

In general, the levels of PGM in the environment are so low that health effects are limited. However, the allergic effects of platinum salts, termed platinosis, to occupationally exposed workers, such as those employed in PGM refineries, are well known. The symptoms include irritation of the nose and upper respiratory tract, sometimes accompanied by asthmatic symptoms. Some cases of contact dermatitis due to palladium allergy have been reported, while sensitisation to palladium is also a concern to miners, dental technicians and chemical workers (Kielhorn et al., 2002). Further research is required into the occupational health effects on platinum miners and workers involved in the manufacture and recycling of autocatalysts.

The main environmental issues associated with the mining and processing of sulfide-bearing ores of any type, including PGM ores, include: the generation of acid mine drainage from mine workings, ore dumps, treatment plant and tailings; release of associated trace elements which may be harmful to human health and the environment (e.g. arsenic, lead, antimony); discharge of chemicals used in the mining and processing of the ores and concentrates; and gaseous emissions (chiefly sulfur dioxide) associated with the smelting and refining processes. The South African government has begun to implement a programme of increasing environmental regulation aimed at delivering improved environmental performance, including more stringent controls on air and water quality and emissions of greenhouse gases. The platinum mining industry in South Africa has been proactive in its environmental management and reporting of sustainability issues including water and energy utilisation and greenhouse gas emissions (Mudd, 2012). Environmental management systems operated by the major companies, certified against international standards, share common goals. These are aimed at minimising pollution of all types, promoting efficient use of energy, reducing waste, and minimising the footprint on the land, including effective remediation and rehabilitation at closure.

World resources and production

Resources and reserves

The Merensky, UG2 and Platreef in South Africa dominate world platinum resources. It is difficult to estimate the total quantity of platinum in the Bushveld Complex but a recent study suggested that the proven and probable reserves of PGM (platinum, palladium and gold) exceed 250 million troy ounces (Cawthorn, 2010). This amount alone is adequate to meet global demand at current levels for several decades with existing technology and current prices. However, the true availability is almost certainly very much greater because companies only report reserves that are required to underpin their short-and medium-term plans. Cawthorn (2010) suggests that there is about 800 million troy ounces in the upper two kilometres of the Bushveld Complex. In the longer term there is significant potential to mine at greater depths and there are several other geological units that could be exploited for PGM if market conditions were favourable. There are also considerable PGM resources in other countries, notably in the Great Dyke in Zimbabwe, the Stillwater Complex in the USA, and in nickel–copper deposits, for example in the Norilsk-Talnakh district of Russia and the Sudbury area of Canada.

The rise in the price of platinum over the last 20 years has led to increased levels of exploration, both on new targets and on several deposits previously thought to be uneconomic. The latter category includes the Skaergaard Complex in Greenland, the Duluth Complex in the USA, and the Penikat and Keivitsa complexes in Finland. At the Marathon copper–PGM deposit in western Ontario the construction of a new open-pit mine and processing operation is planned, which, subject to permitting, could produce up to 200,000 troy ounces of PGM annually (Marathon PGM Corporation, 2011).

Proven and probable ore reserves in the Norilsk-Talnakh area are about 55 million troy ounces of palladium and 13 million troy ounces of platinum (Norilsk Nickel, 2010). In the Stillwater Complex in Montana, USA, proven and probable reserve of approximately 19.9 million troy ounces combined palladium and platinum have been reported (Stillwater Mining Company, 2011).

Production

In 2010 there was a modest increase in global mine production of platinum (total in 2010, 193,000 kg or 6.8 million troy ounces) (Figure 12.5). In South Africa there had been six years of steady growth from 2001 in response to increased demand for autocatalysts until it fell back in 2007 and 2008 with onset of the global economic recession. In 2010 South African platinum production was at the same level as in 2009.

South Africa is the largest platinum producer, accounting for about 76 per cent of world production in 2010 (Figure 12.6). Production is dominated by two companies, Anglo American Platinum and Impala Platinum, who together control about two thirds of global production of

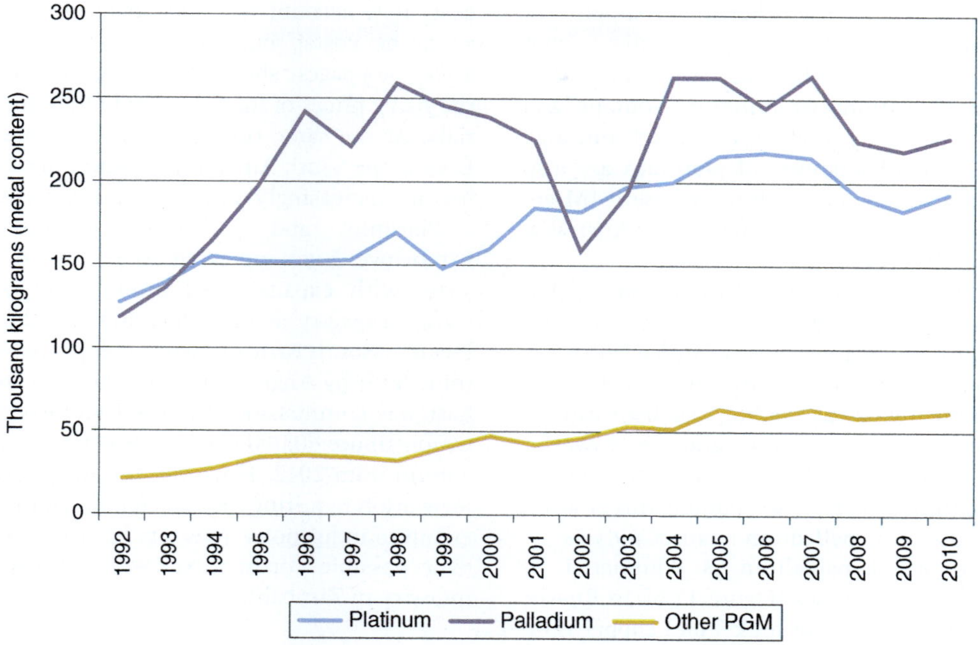

Figure 12.5 World mine production of platinum, palladium and other PGM, 1992–2010. (Data from British Geological Survey World Mineral Statistics database.) (1 kilogram equals 32.151 troy ounces.)

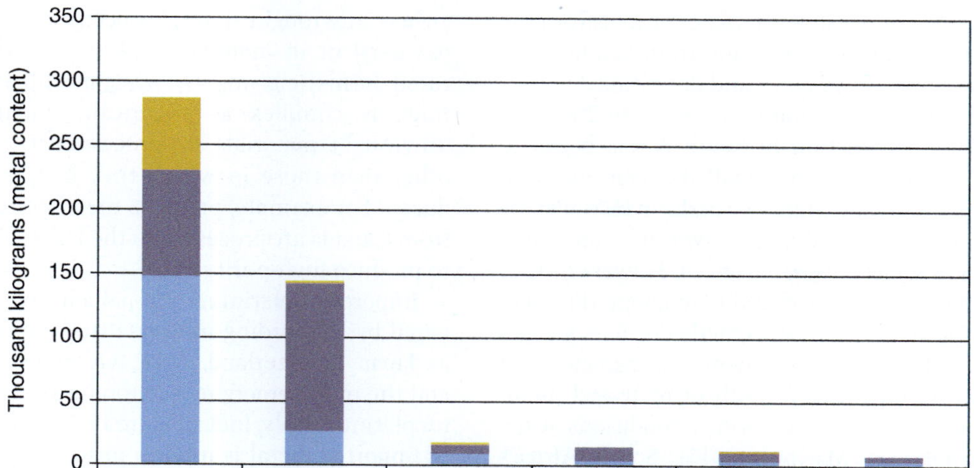

	South Africa	Russia	Zimbabwe	USA	Canada	Other
Other PGM	57,292	2200	1550	0	400	0
Palladium	82,222	115,700	6968	11,600	6200	4205
Platinum	147,790	26,500	8703	3450	3600	3076

Figure 12.6 World mine production of platinum, palladium and other PGM in 2010 by country. (Data from British Geological Survey, 2012.) (1 kilogram equals 32.151 troy ounces.)

platinum. Russia is the other major supplier of platinum, with minor production from Zimbabwe, Canada (Lac des Iles mine and Sudbury), USA (from two mines in Montana) and Australia (as a by-product of nickel mining). Relatively small amounts of platinum are also produced at Jinchuan in China, where PGM are also extracted from ores mined in Australia, Spain and Zambia.

Over the last decade mine production of palladium has declined (Figure 12.5) and levels remain significantly lower than those recorded before the recession. There was a sharp fall in 2002 as a result of the high price, leading to substitution of palladium by platinum in autocatalysts. However, as the price differential was reversed and vehicle sales continued to rise, so manufacturers again began to use more palladium in autocatalysts.

Production of palladium is dominated by Russia and South Africa (Figure 12.6). In Russia, PGM are by-products of nickel and copper mining, with Norilsk Nickel the largest global producer accounting for more than 40 per cent of mine supply. World palladium mine production increased in 2010 to 227,000 kg (approximately 8 million troy ounces), with higher output from the main producing countries and from smaller producers such as Zimbabwe and Botswana.

Following the economic recession in 2008 and 2009 PGM production in South Africa began to recover in 2010 with mothballed operations coming back on stream, improved performance at some operations and several major new mines and mine-expansion projects well advanced. However, mine production decreased in 2011 by about three per cent (120,000 troy ounces), chiefly due to stoppages imposed by the government's Department of Mineral Resources as a result of increased monitoring and enforcement of safety conditions at the mines (Johnson Matthey, 2012). South Africa's platinum mining industry continues to face these and other major challenges to ensure its future profitability and growth. In recent years operating costs have escalated, increasing from under US$300 per troy ounce in 2003 to just above US$800 per troy ounce in 2010 (CPM Group, 2011). This is due to a range of factors including rising wages (labour costs now account for more than 50 per cent of operating costs), currency fluctuations, strikes, safety stoppages, shortages of skilled labour and the rising prices of fuel, steel and other raw materials. At the same time electricity supplies from Eskom, the South African state power utility, have become increasingly unreliable and expensive.

Platinum and palladium supplies from Zimbabwe have increased markedly in recent years with expansion of mining and concentrating capacity at three locations on the Great Dyke by South African-owned companies. A new mine built by Anglo American Platinum at Unki East was commissioned in 2011 and is expected to contribute 60,000 troy ounces of platinum per annum from 2012. However, recent government legislation requiring the transfer of majority ownership to indigenous investors has cast a shadow over possible further expansion of the PGM industry in Zimbabwe.

World trade

Platinum is traded in many forms, generally in either unwrought (ores, concentrates, sponge, powders) or in manufactured or semi-manufactured forms (e.g. ingots, wire, mesh). Platinum trade is complex as significant quantities of unwrought materials are processed in countries other than those in which they have been produced. For example, precious-metal concentrates from Canada are processed in the UK and Norway to produce high-purity PGM.

Imports of platinum and palladium are dominated by the leading industrialised nations, such as Japan, Switzerland, USA, Germany and UK, and the major emerging economies of China and, for platinum only, India (Figure 12.7). The majority of imported metal is used to make catalytic converters for vehicles, where the market has grown rapidly in recent years, especially in China. However, significant proportions are also used in electronics, jewellery and for investment.

Exports of platinum are dominated by South Africa, which is also the world's leading producer (Figure 12.8). South Africa exported about

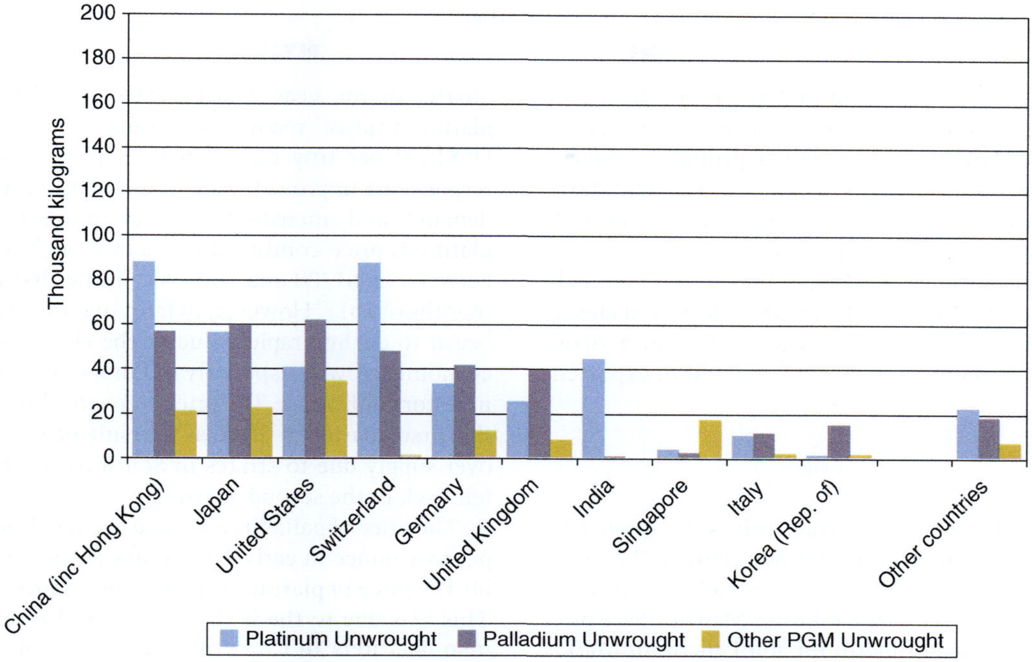

Figure 12.7 Major importers of platinum metal, concentrates and intermediate products, 2009. (Data from British Geological Survey World Mineral Statistics database and UN Comtrade, 2013.) (1 kilogram equals 32.151 troy ounces.)

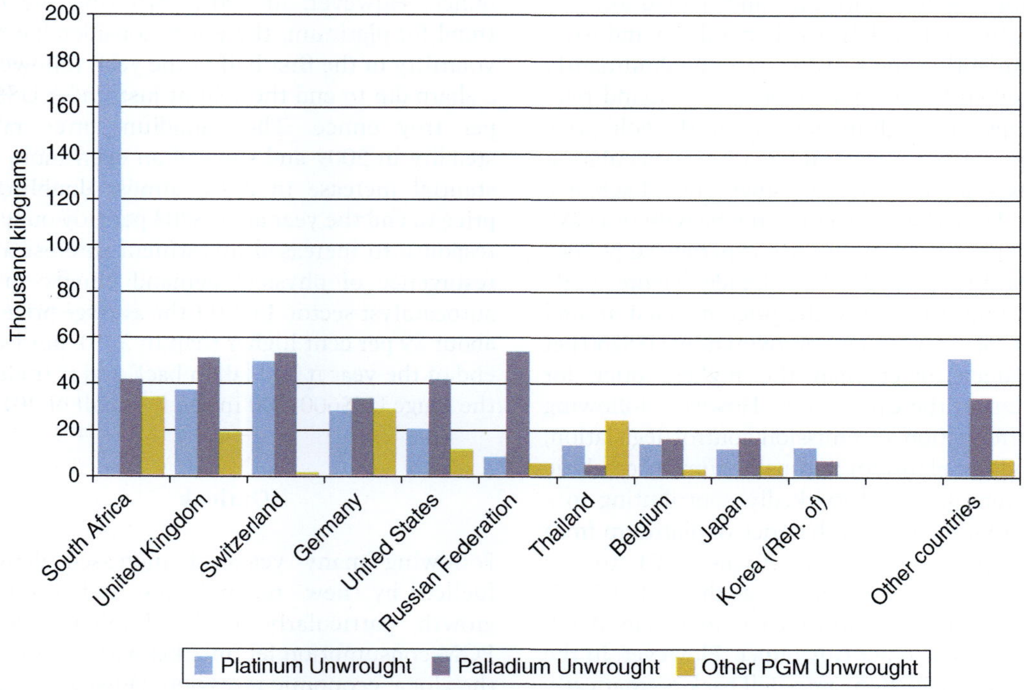

Figure 12.8 Major exporters of platinum metal, concentrates and intermediate products, 2009. (Data from British Geological Survey World Mineral Statistics database and UN Comtrade, 2013.) (1 kilogram equals 32.151 troy ounces.)

180,000 kg of platinum in 2009, more than three times the quantity from the next biggest exporter Switzerland, which is not a primary producer. Other leading exporters include the UK, USA, and Germany. However, of these countries, only the USA is a primary producer.

Russia is the leading exporter of unwrought palladium (ca 54,000 kg in 2009), followed closely by Switzerland and the UK. USA, South Africa and Germany are also major palladium exporters.

Prices

A number of daily market prices are quoted for the pure (minimum 99.9 per cent) PGM in US dollars per troy ounce. These include the European free market price, the Johnson Matthey base price and the London Platinum and Palladium Market prices. In the spot market platinum and palladium are sold for cash and immediate delivery in sponge, plate or ingot form.

Unlike gold, platinum and palladium are important and widely used metals in industry. Consequently, their prices are predominantly determined by normal supply and demand relationships, although movements in the gold price do have some effect and periodic speculation increases price volatility. Since 2007 Exchange Traded Funds have also invested heavily in PGM.

The price of platinum has experienced periods of volatility over the last decade (Figure 12.9). Throughout the 1990s, the price hovered around US$400 per troy ounce, approximately US$250 per troy ounce lower than the highest price for platinum in the early 1980s. However, following the introduction of emission-control legislation, demand for platinum and palladium in catalytic converters increased markedly contributing to a substantial increase in the price of platinum from US$420 per troy ounce in August 2001, to just over US$1500 per troy ounce at the end of 2007. The price reached an all-time record high in March 2008 of US$2276 per troy ounce. However, in the second half of 2008 the price fell back dramatically with the onset of the global financial crisis and associated massively reduced demand, falling to levels below US$800 per troy ounce during the last quarter of the year. During 2009 and 2010 the platinum price recovered strongly, peaking at US$1500 per troy ounce in December 2010, in response to improved industrial and autocatalyst demand and increased investor interest. The platinum price continued to trade at high levels, between US$1700 and US$1900 in the first eight months of 2011. However, in late August the price began to decline rapidly due to the slowdown in economic growth, especially in Europe, and falling investor confidence. The price recovered briefly in the first quarter of 2012 as a result of concerns over supply due to strikes in South Africa, but it fell back in the second quarter.

The price of palladium peaked at over US$1000 per troy ounce in early 2001 and was almost double the price of platinum at the time (Figure 12.9). This was due to the high level of use by vehicle manufacturers in catalytic converters combined with restrictions on exports from Russia. Since then palladium has traded mostly between US$200 per troy ounce and US$400 per troy ounce. However, in 2008, mirroring the price trend for platinum, there was considerable price volatility in the first half of the year, followed by a sharp dip to end the year at just under US$200 per troy ounce. The palladium price rallied steadily in 2009 and showed an even more substantial increase in 2010, almost doubling its price to end the year at US$803 per troy ounce, in response to increased investment interest and a resurgence of physical demand chiefly in the autocatalyst sector. In 2011 the average price was about 39 per cent higher than in 2010, but by the end of the year it had fallen back and it traded in the range US$600–700 in the first half of 2012.

Outlook

Following many years of increased demand, fuelled by new technologies and economic growth, particularly in the developing world, PGM consumption fell in 2008 with the onset of the global economic recession. However, in 2010 industrial demand was revitalised, especially in the automotive sector as vehicle production exceeded pre-recession levels. In 2011 and the

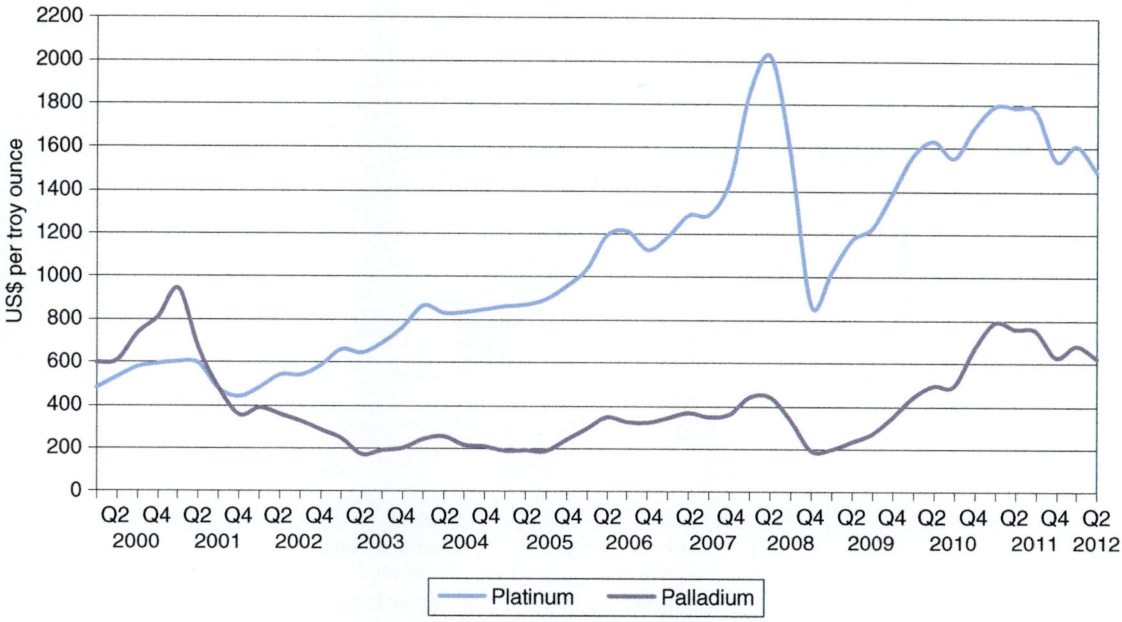

Figure 12.9 The prices of platinum and palladium between 2000 and end June 2012. (Johnson Matthey base price (unfabricated) US$ per troy ounce, quarterly average prices from Metal Bulletin.)

first half of 2012 global economic growth slowed again and the high levels of industrial demand for PGM have not been sustained.

China uses large and increasing quantities of PGM to support its rapidly growing economy, the modernisation of its industries and the spread of prosperity among its population. China is the largest vehicle market in the world and consequently a major consumer of palladium in its predominantly gasoline-powered vehicle fleet. China is also the largest market for platinum jewellery and has major glass and electronics manufacturing sectors. However, China meets only a small proportion of its PGM demand from domestic sources (Figure 12.10). In 2009 domestic production accounted for less than six per cent of total platinum demand, with approximately equal proportions from mining and recycling. For palladium, a larger proportion (almost 32 per cent) was domestically sourced although almost 90 per cent of this was from recycling.

Per capita vehicle usage in China and India remains very low compared with the USA. As vehicle ownership increases in these and other developing countries so demand for PGM in autocatalysts will increase significantly. At the same time increasingly stringent environmental legislation is being introduced in many countries to control vehicle emissions. This will require increased PGM loadings in autocatalysts and hence further raise demand. PGM market growth is also expected in emission control systems for stationary applications, such as diesel generators, gas turbines and manufacturing processes.

Given the continuing global economic uncertainties, the increased demand for platinum and palladium from the investment sector seen in 2010 may well continue in the short and medium term. The increased availability of exchange-traded funds (ETFs) makes investment in platinum and palladium simpler and easier than in the past.

Considerable growth in demand for platinum is anticipated in the fuel-cell sector as the requirement for clean energy assumes ever-growing global importance. Today fuel-cell manufacturing capacity is small and the units are relatively expensive. The largest markets are in the provision of auxiliary or back-up power in camper vans and boats, in

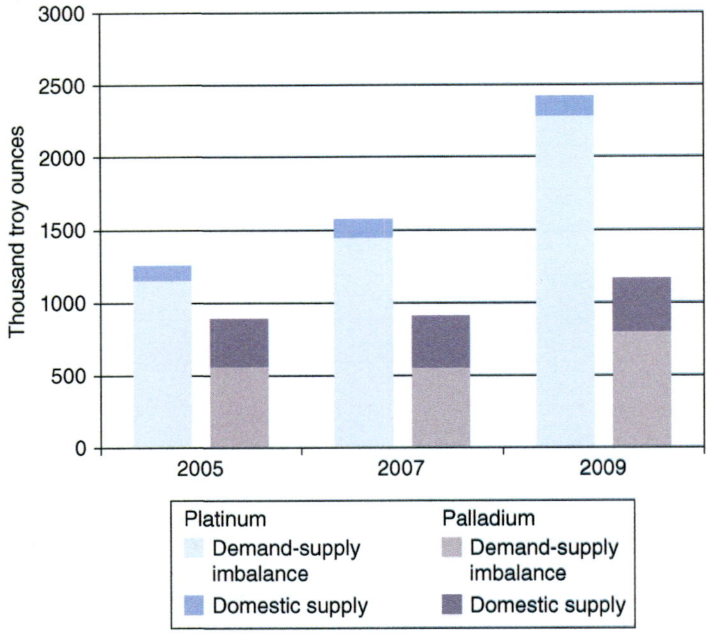

Figure 12.10 The consumption of platinum and palladium in China. (Data from CPM Group, 2011.)

combined heat and power supply for houses and businesses, and in materials-handling vehicles, such as forklift trucks. The largest markets for future use of fuel cells are likely to be light-duty vehicles, although development of the supporting hydrogen infrastructure may constrain growth to some extent. The potential fuel cell market in consumer electronics, such as portable computers, tablet devices and mobile phones, is huge.

Demand for platinum and rhodium in glass manufacturing has been growing rapidly since 2000, driven largely by new technologies. Further growth is anticipated in the production of fibre glass, glass for solar photovoltaic applications and for LCD displays, especially in portable electronic devices. Other growth areas include the use of PGM in medical and biomedical applications and in specialist alloys. For example, ruthenium, together with rhenium, is increasingly being used in nickel-based superalloys in turbine blades for jet engines.

The high level of concentration of PGM production in South Africa and Russia has in the past given rise to concerns about supply security (House of Lords, 1982). However, these concerns subsided following the break-up of the Soviet Union in 1991 and the end of the apartheid era in 1994. More recently, the continued concentration of PGM production in the same two countries and the burgeoning demand has once again given rise to worries about the security of future supplies. Of particular significance for the long-term future of the South African mining industry is black empowerment. The Youth League of the African National Congress has been highly vociferous in calling for nationalisation of the mining industry. Although it is generally accepted that, in principle, exploitation of South African mineral wealth should benefit its citizens, nevertheless it is widely considered that any move towards nationalisation would damage South Africa's reputation as a safe destination for international investment and would threaten the future of the industry. The recent and ongoing labour disputes affecting South Africa's mining industry are very likely to have short-term impacts on global platinum supply as well as longer-term effects on investor confidence.

The physical availability of PGM resources is not likely to be a constraint on economic growth in any sector in the long term. New supplies from both primary (mining) and secondary (recycling) sources are likely to be available with existing technology. In addition to the possibility of mining to greater depths on the Merensky and UG2 Reefs, alternative cheaper sources are expected to make increasing contributions to PGM supply from South Africa. These include the development of new open-pit mines, including some exploiting chromite seams at higher levels in the Bushveld Complex, and the recovery of PGM from chromite mine tailings. Full commercialisation of the ConRoast smelting technology will also increase the range of materials that can be smelted, while also providing important environmental benefits through reduced pollution.

Mine production of PGM from Russia has been relatively stable in recent years despite falling grades. Supplies have been regularly supplemented by major sales, of up to a million troy ounces of palladium per annum, from state stocks. Market analysts have recently predicted that supply from Russian state sales is likely to decline significantly in the near future, contributing to possible future shortfalls in palladium supply.

In contrast, if nickel prices remain high, PGM supplies from nickel mines in Canada, Australia, Botswana and elsewhere might be expected to increase. In addition, new supply could be derived from mining PGM-dominant ores in several countries including Canada, Greenland, Australia and Brazil, although bringing new PGM production capacity on stream is a lengthy and expensive process. As with other critical metals there may also be increased vertical integration in the PGM supply chain with industry users investing in mining and processing capacity to ensure security of supply.

An increasing proportion of PGM supply is expected to be derived from secondary sources, stimulated by increased regulation and fiscal incentives from governments to promote recycling and new initiatives to encourage closed-loop recycling in sectors where hitherto open-loop practices have prevailed. The economic and environmental benefits of efficiently recovering the PGM resource tied up in vehicles still on the road will become increasingly important as the global market continues to expand. Hagelüken and Grehl (2012) have suggested that globally up to 2010 there may be as much as 3000 tonnes of PGM in the autocatalysts of vehicles being used, equivalent to more than six years of the combined global mine production of platinum, palladium and rhodium in 2010. However, it is important to stress that open-loop systems are highly complex with many points at which PGM losses may occur. Consequently, improvements in recovery rates cannot be expected to be easily achieved and such systems are not likely to be adopted widely or quickly without legislative intervention.

As demand increases there will be technical challenges related to mining and processing new ore types and lower grade ores, to mining at greater depths, and to achieving more efficient recycling. Given timely and adequate investment in education and research, especially in South Africa, technological innovation will, in the long term, continue to ensure sustainable PGM supplies with improved environmental performance throughout the supply chain. However, given the major problems currently faced by the mining industry in South Africa, it is almost inevitable that there will be some impact on global platinum supply, even if only in the short term. PGM resources in other countries will become more attractive to mining companies if supplies from South Africa are constrained and if demand from industry, especially the automobile sector in developing countries, continues to grow.

Acknowledgements

Gus Gunn publishes with the permission of the Executive Director of the British Geological Survey.

Note

1. Ma, million years before present

References

British Geological Survey (2012) World Mineral Production 2006–2010 (Keyworth, Nottingham: British Geological Survey.)

Cabri, L.J. (2002) The Platinum-Group Minerals. In: Cabri, L.J. (ed.) The geology, geochemistry, mineralogy and mineral beneficiation of platinum-group elements, Canadian Institute of Mining, Metallurgy and Petroleum, Special Volume 54.

Cawthorn, R.G., Merkle, R.K.W. and Viloen, M.J. (2002) Platinum-group element deposits in the Bushveld Complex, South Africa. In: Cabri, L.J. (ed.) The geology, geochemistry, mineralogy and mineral beneficiation of platinum-group elements, Canadian Institute of Mining, Metallurgy and Petroleum, Special Volume 54.

Cawthorn, R.G. (2010) The Platinum Group Element Deposits of the Bushveld Complex in South Africa. Platinum Metals Review 54, 205–215.

Cole, S. and Ferron, C.J. (2002) A review of the beneficiation and extractive metallurgy of the platinum-group elements, highlighting recent process innovations. In: Cabri, L.J. (ed.) The geology, geochemistry, mineralogy and mineral beneficiation of platinum-group elements, Canadian Institute of Mining, Metallurgy and Petroleum, Special Volume 54.

Colombo, C., Oates, C.J., Monhemius, J. and Plant, J.A. (2008) Complexation of platinum, palladium, and rhodium with inorganic ligands in the environment. Geochemistry: Exploration, Environment, Analysis 8, 91–101.

CPM Group (2011) The CPM Platinum Group Metals Yearbook 2011.

CPM Group (2012) The CPM Platinum Group Metals Yearbook 2012.

Eckstrand, O.R. and Hulbert, L. (2007) Magmatic nickel-copper-platinum group element deposits. In: Goodfellow, W.D. (ed.) Mineral Deposits of Canada: A Synthesis of Major Deposit-Types, District Metallogeny, the Evolution of Geological Provinces, and Exploration Methods. Geological Association of Canada, Mineral Deposits Division, Special Publication No. 5, 205–222.

Ek, K.H., Morrison, G.M. and Rauch, S. (2004) Environmental routes for platinum group elements to biological materials - a review. Science of the Total Environment 334–335, 21–38.

Gagnon, Z.E., Newkirk, C. and Hicks, S. (2006) Impact of platinum-group metals on the environment: a toxicological, geotoxic and analytical chemistry study. Journal of Environmental Science and Health, Part A, 41, 397–414.

GFMS. (2005) Materials Flow of Platinum Group Metals. Umicore, Öko-Institut, GFMS, London.

Hagelüken, C.J. (2008) Recycling of spent catalysts. In: Ertl, G., Knötzinger, H., Schüth, F. and Weitkamp, J. (eds.) Handbook of Heterogeneous Catalysis, 2nd edition, Wiley-VCH, Weinheim, Germany, 1846–1863.

Hagelüken, M. and Grehl, M. (2012) Recycling and loop concept for a sustainable usage, In: Sehrt, U. and Grehl, C. (eds.) Precious Materials Handbook, Umicore AG and Co KG, Hanau-Wolfgang, Germany. 38–87.

Hoatson, D.M., Jaireth, S.A. and Jacques, A.L. (2006) Nickel sulfide deposits in Australia: Characteristics, resources and potential. Ore Geology Reviews 29, 177–241.

House of Lords (1982) Strategic Minerals. 20th report of the House of Lords Select Committee on the European Communities, Session 1981–82, HMSO.

Johnson Matthey (2011) Platinum 2011.

Johnson Matthey (2012) Platinum 2012.

Jones, R.T. (2005) An overview of South African PGM smelting. Nickel and Cobalt 2005: challenges in extraction and production. 44th Annual Conference of Metallurgists, Calgary, Canada. http://www.mintek.co.za/Pyromet/Files/2005JonesPGMsmelting.pdf

Jones, R.T. (2009) Towards commercialisation of Mintek's ConRoast process for platinum smelting. Nickel and Cobalt 2009: Advances in the processing of nickel, cobalt and PGMs using pyrometallurgical techniques. 48th Annual Conference of Metallurgists, Calgary, Canada. http://www.mintek.co.za/Pyromet/Files/2009Jones-ConRoast.pdf.

Kielhorn, J., Melber, C., Keller, D. and Mangelsdorf, I. (2002) Palladium – a review of exposure and effects to human health. International Journal of Hygiene and Environmental Health 205, 417–432.

Kinnaird, J.A. and MacDonald, I. (2005) An introduction to the mineralisation in the northern limb of the Bushveld Complex. Applied Earth Science (Transactions of the Institution of Mining and Metallurgy) 114, B194–198.

Lee, C.A. (1996) A review of mineralization in the Bushveld Complex and some other layered mafic intrusions. In: Cawthorn, R.G. (ed.) Layered Intrusions, Elsevier Science, Amsterdam, 103–146.

Lesher, C.M. and Keays, R.R. (2002) Komatiite-associated Ni-Cu-PGM deposits: geology, mineralogy and geochemistry. In: Cabri, L.J. (ed.) The geology, geochemistry, mineralogy and mineral beneficiation of platinum-group elements. Canadian Institute of Mining, Metallurgy and Petroleum, Special Volume 54.

Lightfoot, P.C., Keays, R.R. and Doherty, W. (2001) Chemical evolution and origin of nickel sulfide mineralization in the Sudbury Igneous Complex, Ontario, Canada. Economic Geology 96 1855–1875.

Maier, W.D. (2005) Platinum-group element (PGM) deposits and occurrences: Mineralisation styles, genetic concepts, and exploration criteria. Journal of African Earth Sciences 41, 165–191.

Marathon PGM Corporation (2011) Project Overview. http://www.marathonpgmproject.com/Project-Overview.html

Merkle, R.K.W. and McKenzie, A.D. (2002) The geology, geochemistry, mineralogy and mineral beneficiation of PGE. In: Cabri, L.J. (ed.) The geology, geochemistry, mineralogy and mineral beneficiation of platinum-group elements. Canadian Institute of Mining, Metallurgy and Petroleum, Special Volume 54.

Mudd, G.M. (2012) Sustainability reporting and platinum group metals: a global mining industry leader. Platinum Metals Review 56, 2–19.

Naldrett, A.J. (2004) Magmatic Sulphide Deposits: Geology, Geochemistry and Exploration. Springer, Berlin.

Naldrett, A.J. (2010) From the mantle to the bank: the life of a Ni-Cu-(PGE) sulphide deposit. South African Journal of Geology 113, 1–32.

Norilsk Nickel (2010) Mineral Reserves and Resources Statement. http://www.nornik.ru/en/our_products/MineralReservesResourcesStatement/

Oberthür, T. (2002) Platinum-group element mineralization in the Great Dyke, Zimbabwe. In: Cabri, L.J. (ed.) The geology, geochemistry, mineralogy and mineral beneficiation of platinum-group elements. Canadian Institute of Mining, Metallurgy and Petroleum, Special Volume 54.

Ravindra, K., Bencs, L. and Van Grieken, R. (2004) Platinum group elements in the environment and their health risk. The Science of the Total Environment 318, 1–43.

Scoon, R.N. and Mitchell, A.A. (2004) The platiniferous dunite pipes in the eastern limb of the Bushveld Complex: Review and comparison with unmineralised discordant ultramafic bodies. South African Journal of Geology 107, 505–520.

Scott, D.A. and Bray, W. (1980) Ancient platinum technology in South America. Platinum Metals Review 24, 147–157.

Stillwater Mining Company (2011) Annual report for 2010.

Su, S., Li, C., Zhou, M-F., Ripley, E.M. and Qi, L. (2008) Controls on variations of platinum-group element concentrations in the sulphide ores of the Jinchuan Ni-Cu deposit, western China. Mineralium Deposita 43, 609–622.

Tolstykh, N.D., Sodorov, E.G. and Krivenko, A.P. (2005) Platinum-group element placers associated with Ural-Alaska type complexes. In: Mungall, J.E. (ed.) Exploration for platinum-group element deposits. Mineralogical Association of Canada, Short Course Series Volume 35, 113–143.

UNEP. (2011) Recycling rates of metals – a status report. A Report of the Working Group on the Global Metal Flows to the International Resource Panel.

United Nations (UN) Comtrade (2013) Commodity Trade Statistics Database (http://comtrade.un.org/db/).

Zientek, M.L., Cooper, R.W., Corson, S.R. and Geraghty, E.P. (2002) Platinum-group element mineralization in the Stillwater Complex, Montana. In: Cabri, L.J. (ed.) The geology, geochemistry, mineralogy and mineral beneficiation of platinum-group elements. Canadian Institute of Mining, Metallurgy and Petroleum, Special Volume 54.

13. Rare earth elements

FRANCES WALL

Head of Camborne School of Mines and Associate Professor of Applied Mineralogy, Camborne School of Mines, University of Exeter, Penryn, UK

Introduction

The rare earth elements (REE) are defined according to the International Union of Pure and Applied Chemistry (IUPAC) as the 15 lanthanides (also called lanthanoids) together with yttrium and scandium. In practice, the element scandium behaves rather differently in nature to the other rare earths and its inclusion in the official definition is not particularly helpful. Moreover, one of the lanthanides, promethium has a short half-life and is thus exceptionally rare. This leaves a group of 15 elements comprising 14 lanthanides and yttrium (Y), which do form a coherent group in nature as well as in technological uses. The IUPAC definition divides the REE into the light REE (lanthanum (La) – samarium (Sm)) and heavy REE (europium (Eu) - lutetium (Lu)). Yttrium behaves as a heavy REE. The terms 'mid', 'middle' and 'medium REE' are also sometimes used to denote Sm – dysprosium (Dy). Together with the actinides, the lanthanides are members of the 'f elements' or 'f block' in the Periodic Table. Some authors prefer to use the abbreviation REY instead of REE to denote Y and the lanthanides.

The name rare earths is older than the term lanthanides and was used to describe the difficulty experienced by nineteenth-century chemists in separating the elements from each other (rare) and to denote that the REE are usually stable as oxides (earths) rather than metals. The first rare earth, a mixture of REE called 'yttria', was isolated from a mineral found at Ytterby, Sweden by Finnish chemist, J. Gadolin in 1794; the others followed over the next 153 years, until promethium was found in 1947 (Gupta and Krishnamurthy, 2005, and Kaltosoyannis and Scott, 1999). Separation of REE from each other is still a significant challenge in the processing and refining of rare earth ores. The term didymium, used originally for an 'element' that proved to be a mixture of praseodymium (Pr) and neodymium (Nd), is still used today, particularly in the USA, to refer to the REE, mainly La, Nd and Pr, remaining after extraction of cerium (Ce). The United States Geological Survey (USGS) uses didymium to refer to 75 per cent neodymium, 25 per cent praseodymium.

Physical and chemical properties

The REE are soft, silver-coloured metals that tarnish quickly in air and have high melting points (Table 13.1). The lanthanide series results from filling of the 'f' electron shell in the atoms, with configurations ranging from $5d6s^2$ for La and $4f^15d^16s^2$ for Ce to $4f^{14}5d\,6s^2$ for Lu at the end of

Critical Metals Handbook, First Edition. Edited by Gus Gunn.

Table 13.1 Selected properties of the rare earth elements.

Element name	Chemical symbol	Atomic number	Atomic weight	Density at 25 °C (kg/m^3)	Melting point (°C)	Cation radius (pm)	Colour of +3 cation in solution
scandium	Sc	21	44.96	2992	1541	87.0	Colourless
yttrium	Y	39	88.91	4475	1522	107.5	Colourless
lanthanum	La	57	138.91	6174	918	121.6	Colourless
cerium	Ce	58	140.12	6711	798	119.6	Colourless
praseodymium	Pr	59	140.91	6779	931	117.9	Yellow-green
neodymium	Nd	60	144.24	7000	1021	116.3	Rose
promethium	Pm	61	144.91	7220	1042	-	Pink
samarium	Sm	62	150.36	7536	1074	113.2	Yellow
europium	Eu	63	151.96	5248	822	112.0	Colourless
gadolinium	Gd	64	157.25	7870	1313	110.7	Colourless
terbium	Tb	65	158.93	8267	1356	109.5	Pale pink
dysprosium	Dy	66	162.50	8531	1412	108.3	Pale yellow-green
holmium	Ho	67	164.93	8797	1474	107.2	Yellow
erbium	Er	68	167.26	9044	1529	106.2	Pink
thulium	Tm	69	168.93	9325	1545	105.2	Greenish tint
ytterbium	Yb	70	173.04	6966	819	104.2	Colourless
lutetium	Lu	71	174.97	9842	1663	103.2	Colourless

Compiled from Gupta and Krishnamurthy (2005); pm, picometres.

the series. This gives rise to the magnetic and spectroscopic properties that make REE so useful in many applications. These inner electrons are shielded and so maintain the distinct elemental properties in various bonding situations. The electronic properties give the REE distinct and sharp absorption and emission spectra, including distinct colours in solution (Table 13.1). Most REE are strongly paramagnetic and the strong magnetism that results from their combination with transition metals such as iron and cobalt is one of their most important features.

Although the physical and chemical properties of the REE are similar, many of their properties change systematically and smoothly through the lanthanide series. One of the most important properties in determining chemical behaviour is cation size and, conversely to what might be expected, the size of lanthanide cations (and atoms) *decreases* smoothly with increasing atomic number (Figure 13.1). This phenomenon is called the lanthanide contraction and arises because the f electrons do not screen the other electrons from the positive pull of the nucleus and, therefore, as the atomic number (and thus number of protons in the nucleus) increases, the electrons are pulled close to the nucleus. The Y^{3+} cation is most similar in size to Ho (Figure 13.1), whereas Sc^{3+} is much smaller at just 0.87Å.

In nature, REE almost always occur in the +3 valency state (i.e. they form the oxide REE_2O_3) but there are two important exceptions in natural environments. In oxidising environments, such as weathered deposits and seawater deposits, Ce forms Ce^{4+} (CeO_2), a much smaller cation, and in reducing environments, Eu forms the larger Eu^{2+} cation (Figure 13.1). In the laboratory, +2 valencies are also known for Sm, Tm, Yb; Pr_6O_{11} is a mixed synthetic oxide ($4PrO_2{\cdot}Pr_2O_3$).

Distribution and abundance in the Earth's crust

The rare earth elements range in abundance in the crust from about the same level as copper and lead down to the same levels as bismuth. They

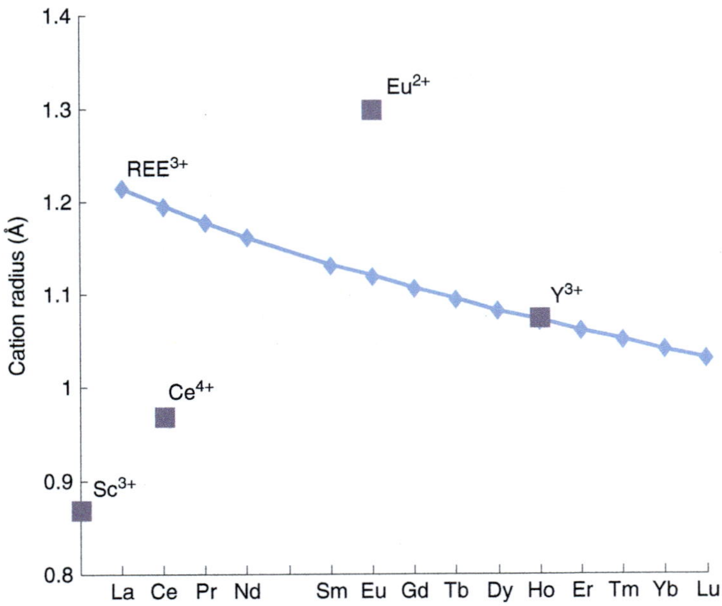

Figure 13.1 The lanthanide contraction demonstrated by plotting the radii of 3+ cations in nine coordination. The radii of the cations of the other most common lanthanide oxidation states are also shown, as are the radii of Y^{3+} and Sc^{3+}. (Modified after Gupta and Krishnamurthy, 2005.)

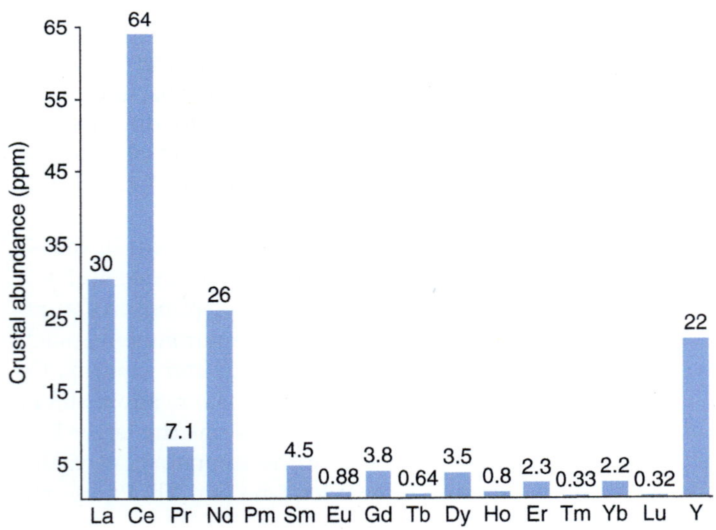

Figure 13.2 Crustal abundances of rare earth elements (Data from Taylor and McClennan, 1985). The light rare earths, lanthanum (La), cerium (Ce) and neodymium (Nd), are the most abundant rare earth elements in the Earth's crust.

are considerably more abundant than precious metals such as gold and the platinum group elements. Light REE are up to 200 times more abundant than heavy REE and odd atomic number REE are more abundant than the even number elements (Figure 13.2). This 'zig-zag' odd–even pattern is called the Oddo Harkins effect and it is usual practice to remove it from graphical presentations of REE concentrations in rocks and minerals by chondrite normalising the data, which involves dividing the absolute REE concentrations by the accepted REE values in chondrite meteorites (e.g. McDonough and Sun, 1995). Crustal abundances determine which REE are most abundant in ore deposits and which are only ever found in small quantities subsidiary to the other REE. For example, Tm is heavy and odd atomic number and thus has a low abundance (Figure 13.2); Ce is light and even atomic number and is the most abundant of the REE, usually predominant in minerals. Heavy to light REE ratios are often used and are best calculated using La to represent the light REE and either Y or Yb to represent the heavy REE.

Mineralogy

There are over 200 rare earth minerals (Miyawaki and Nakai, 1996) approved by the International Mineralogical Association (2012). There are also many others that can contain significant substitutions of REE in place of their main constituents. However, many REE minerals are rare and the list of minerals likely to be encountered as major constituents of an ore deposit is rather less extensive (Table 13.2).

Rare earth minerals are named with a suffix that indicates the predominant REE in the REE site (Bayliss and Levinson, 1988). Monazite-(Ce), for example, is the most common form of monazite. If La, Nd, or Sm are dominant in the REE site instead of Ce, they form different mineral species called monazite-(La), monazite-(Nd) or monazite-(Sm). In practice, almost all light REE minerals will have the suffix-(Ce), because Ce has the highest crustal abundance, but they will also contain significant La, Pr and Nd. Likewise, Y is almost always the predominant REE in heavy REE minerals because of its higher crustal abundance compared with the other heavy REE. Only in rare cases have other heavy REE species been described, e.g. keivyite-(Yb). It is possible to find more than one mineral species within a single mineral grain if the mineral is zoned. Most minerals are either light or heavy REE-enriched rather than equally rich in all REE because of the variation in cation size. Pr, Eu, Gd, Tb, Dy, Ho, Er, Tm and Lu are almost never dominant in any mineral because their absolute abundance is too low compared with neighbouring members of the series and there are no geochemical processes that can separate and concentrate these individual REE sufficiently for them to form their own minerals.

Monazite-(Ce) is one of the most common and widespread REE minerals. It occurs as an accessory mineral in granite, gneiss and other igneous and metamorphic rocks. It is highly resistant to weathering, robust during transport and thus survives to become incorporated into sedimentary rocks and concentrated in mineral sand placer deposits. Monazite also occurs in hydrothermal deposits and may form during weathering. It is almost always strongly enriched in the light REE (Figure 13.3). Examples of monazite composition in a variety of rock types are given by Förster (1998), Smith et al. (2000), Wall and Mariano (1996), Wall and Zaitsev (2004). The other principal constituents of monazite are thorium, calcium and silicon. Thorium substitution into monazite must be coupled with either calcium or silicon in order to main charge balance. Coupled substitution of thorium and silicon for REE moves the composition towards the end member huttonite ($ThSiO_2$), whilst coupled substitution of calcium and thorium changes the composition towards cheralite ($CaTh(PO_4)_2$). Monazite derived from granite (e.g. monazite in placer deposits) tends to have higher thorium and uranium than monazite from carbonatite. For example, van Emden et al. (1997) found an average of 8.79 and standard deviation of 0.08 wt% ThO_2 in 500 monazite grains from Western Australian mineral-sand deposits.

The other most common light REE minerals that are important in REE ore deposits are the fluorcarbonate group of minerals including bastnäsite-(Ce), synchysite-(Ce) and parisite-(Ce) (Figure 13.3). These minerals have layered structures and are commonly syntaxially intergrown with the each other; they also frequently occur as sheaves of fine needle-like crystals.

The most common heavy REE mineral is xenotime-(Y), which occurs as an accessory mineral in granite, gneiss, and other igneous and metamorphic rocks. It is also resistant to weathering and robust enough to be incorporated into sedimentary

Table 13.2 The most common rare earth minerals, including an indication of the rare earth oxide (REO), thorium (Th) and uranium (U) contents and beneficiation process.

Mineral	Formula	Wt % REO	Th, U	Other REE variants	Beneficiation
CARBONATES AND FLUORCARBONATES					
ancylite-(Ce)	$SrCe(CO_3)_2(OH)\cdot H_2O$	43	-	La	HCl acid dissolution (exp)
bastnäsite-(Ce)	$CeCO_3F$	75	-	La, Nd, Y	F
huanghoite-(Ce)	$BaCe(CO_3)_2F$	40	-		
parisite-(Ce)	$CaCe_2(CO_3)_3F_2$	50	-	Nd	F
synchysite-(Ce)	$CaCe(CO_3)_2F$	51	-	Nd, Y	F
PHOSPHATES					
cheralite	$Ca,Th(PO_4)_2$	variable	M		-
churchite-(Y)	$YPO_4.2H_2O$	51	V	Nd	-
florencite-(Ce)	$(Ce)Al_3(PO_4)_2(OH)_6$	32	-	Sm	-
monazite-(Ce)	$CePO_4$	70	V	La, Nd, Sm	F or GME
xenotime-(Y)	YPO_4	61	V	Yb	GME
OXIDES					
aeschynite-(Ce)	$(Ce,Ca,Fe,Th)(Ti,Nb)_2(O,OH)_4$	32	V	Nd,Y	-
cerianite-(Ce)	CeO_2	100	V		-
loparite-(Ce)	$(Ce,La,Nd,Ca,Sr)(Ti,Nb)O_3$	30	-		Chlorination
yttropyrochlore-(Y)	$(Y,Na,Ca,U)_{1\text{-}2}Nb_2(O,OH)_7$	e.g. 17	V		-
SILICATES					
allanite-(Ce)	$CaNdAl_2Fe^{2+}(SiO_4)(Si_2O_7)O(OH)$	38	V	La, Nd, Y	-
britholite-(Ce)	$(Ce,Ca,Sr)_2(Ce,Ca)_3(SiO_4,PO_4)_3\ (O,OH,F)$	e.g. 23	V	Y	-
eudialyte	$Na_{15}Ca_6Fe_3Zr_3Si(Si_{25}O_{73})(O,OH,H_2O)_3(Cl,OH)_2$	e.g. 9	-		Have leached REE^{exp}
fergusonite-(Ce)	$CaNdAl_2Fe^{2+}(SiO_4)(Si_2O_7)O(OH)$	53	-?	Nd, Y, β-fergusonite-(Ce), Nd,Y	Hot caustic digestion and acid dissolution^{exp}
gadolinite-(Ce)	$Ce_2Fe^{2+}Be_2O_2(SiO_4)_2$	60	V	Y	treated with acid (small scale)
gerenite-(Y)	$CaNdAl_2Fe^{2+}(SiO_4)(Si_2O_7)O(OH)$	44	-		-
kainosite-(Y)	$Ca_2Y_2(SiO_3)_4(CO_3).H_2O$	38	-		
keiviite-(Y)	$Y_2Si_2O_7$	69	-	Yb	-
steenstrupine-(Ce)	$Na_{14}Ce_6(Mn^{2+})_2(Fe^{3+})_2Zr(PO_4)_7Si_{12}O_{36}(OH)_2.3H_2O$	31	V		F and leach^{exp}
FLUORIDES					
fluocerite-(Ce)	CeF_3	83	-	La	-

Minerals currently mined commercially are in bold.
F = flotation; GME = combination of gravity, magnetic and electrostatic processes; exp = technique not yet applied on a working mine.
U, Th contents: M = usually has major (wt%) Th; V = Th and U are variable from <1 wt% to major (e.g. 10 wt%) amounts. Other minerals usually have <1 wt% Th and U but almost all will contain trace quantities of these elements.

rocks and concentrated in placer deposits such as mineral sands. Compositions are always enriched in the heavy REE (Figure 13.3).

Cerium, when oxidised to its +4 state, can form separate Ce minerals, such as cerianite (Table 13.2), where Ce is dominant and concentration of the other light REE are very low. In reducing environments, Eu^{2+} substitutes for Ca in feldspar and fractionates from the other REE. This gives rise to Eu anomalies with respect to

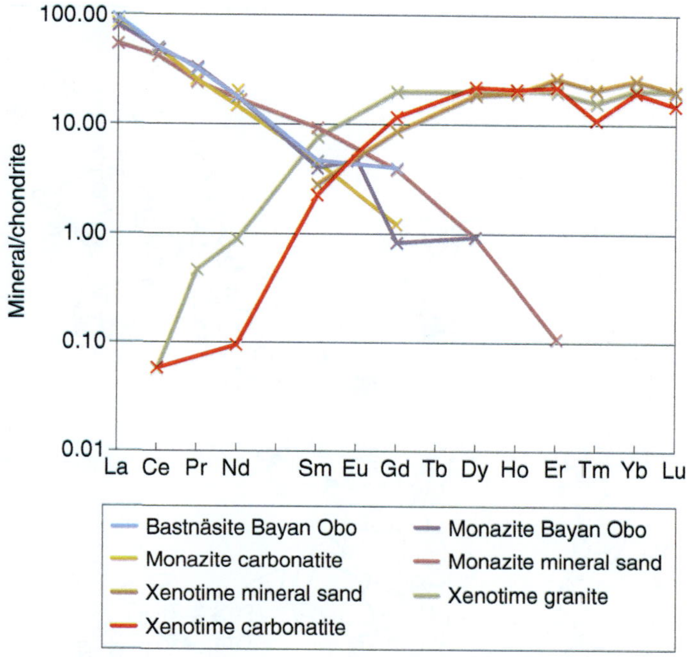

Figure 13.3 Compositions of the most common light REE minerals, bastnäsite-(Ce) and monazite-(Ce), and the most common heavy REE mineral, xenotime-(Y). Following usual precedent, values are divided by values of a chondrite meteorite (McDonough and Sun, 1995) in order to remove the Oddo-Harkins 'zig-zag' effect. (Data from Förster (1998), Smith et al. (2000), van Emden et al. (1997), Wall and Mariano (1996), Wall et al. (2008).)

the concentrations of neighbouring REE in minerals but does not result in discrete Eu minerals.

The most abundant non-REE mineral with significant REE substitution is the calcium phosphate, apatite (fluorapatite is the official mineral name for the most common fluorine-bearing variety), which occurs in many different rock types. In most, it is present in minor quantities but in carbonatites and alkaline igneous rocks apatite can be one of the main rock-forming minerals. It is also the main constituent of sedimentary phosphorites. The REE can substitute into the two Ca sites in apatite, often in a coupled substitution with silicon or sodium, and can reach levels of tens of weight per cent (wt%), although REE contents of about 1 wt% are more normal in alkaline rocks and carbonatites, with lower levels in other rocks, such as phosphorites.

Zircon and fluorite are also important hosts for REE. Zircon favours the heavy REE but only contains trace or minor quantities. Fluorite usually only contains trace quantities of REE, but it can occasionally take up either light or heavy REE and has yttrofluorite and cerfluorite species.

Deposit types

There are many potential REE ore deposits in a wide variety of rocks. Orris and Grauch (2002) listed 822 occurrences of REE divided into 14 different deposit types and Figure 13.4 illustrates 78 deposits divided into nine different categories. For the purposes of the discussion here, the REE deposits are divided into carbonatite-associated deposits, including weathered carbonatite; alkaline igneous rocks, including alkaline granites; other hydrothermal deposits; ion adsorption deposits; placer deposits and seafloor deposits. REE are also produced as by-products of other minerals and can be recovered from waste. The various types of deposits have particular characteristics of size and grade (Figure 13.5). Carbonatites tend to be medium to large tonnage and high grade, whereas alkaline rock deposits are generally larger tonnage but lower grade, tending to have higher proportions of heavy REE. Mineral sands are low grade but REE minerals are by-products; ion adsorption deposits are small and low grade but relatively rich in heavy REE.

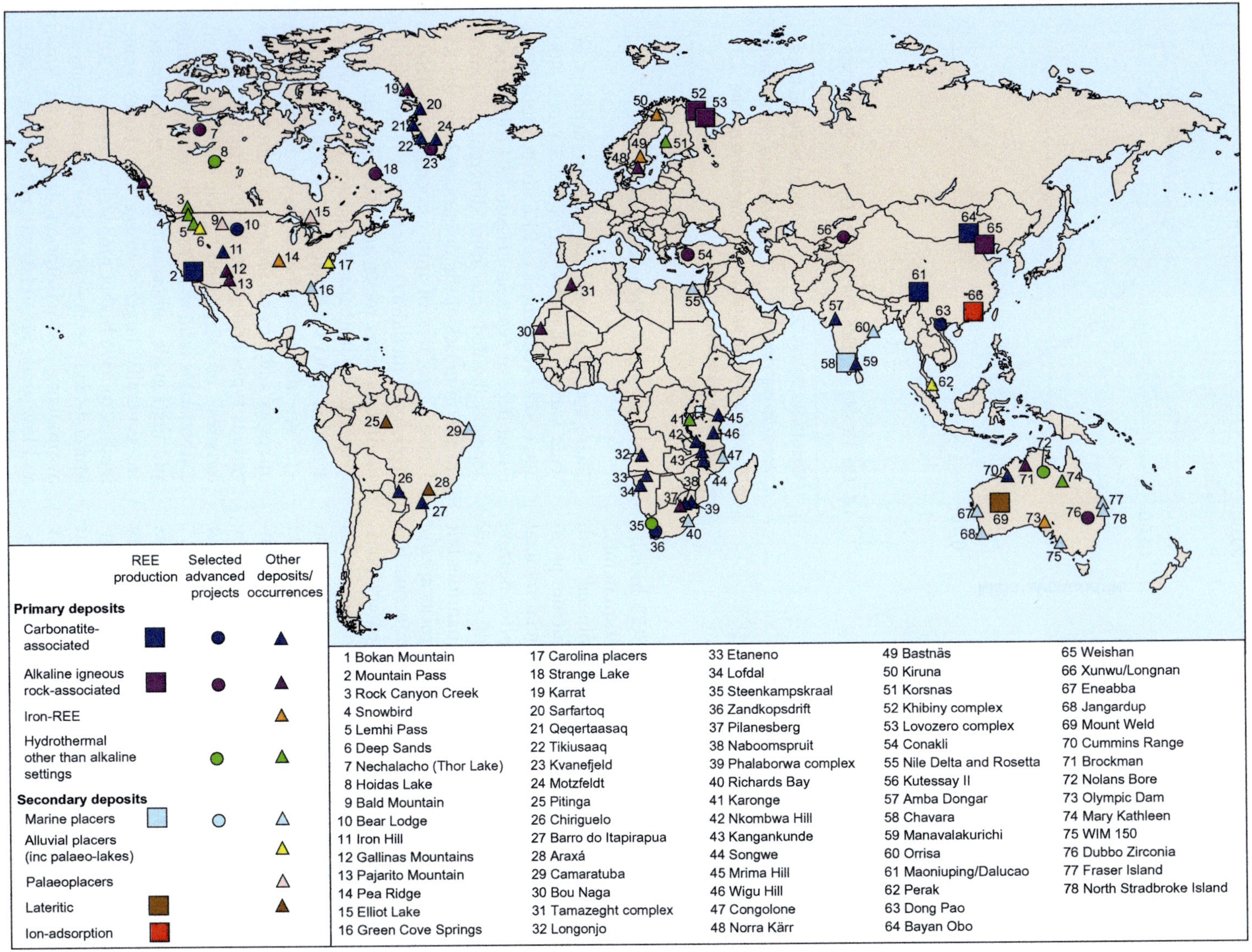

Figure 13.4 Global distribution of rare earth deposits (Source: British Geological Survey, 2011).

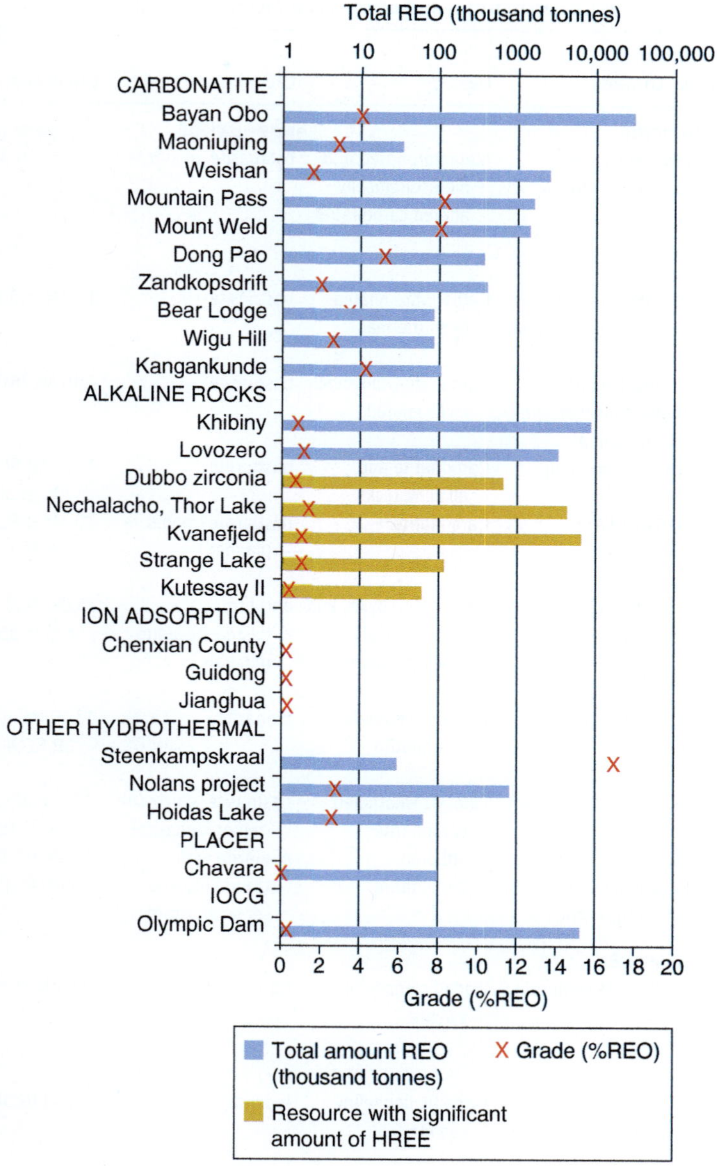

Figure 13.5 Comparison of the size and grade of rare earth mines and advanced development projects. (Data from Table 13.3.)

Hydrothermal deposits in an advanced stage of development are high grade and medium size.

Carbonatite-related REE deposits

Carbonatites, which are igneous rocks composed of more than 50 per cent carbonate, are the main economic source of REE (Figure 13.4, Table 13.3). Although they are igneous rocks, carbonatites are not typical of mid-ocean ridges and plate collision zones where most volcanoes form, but instead are found in the middle of continents, often associated with extensional plate tectonic activity, such as rift valleys or continental breakup. Intrusive

Table 13.3 Rare earth mines and advanced projects

Name, Location	Geology	Ore minerals*	Grade and size	Notes on company, mining, processing
Carbonatite				
BAYAN OBO, Inner Mongolia, China	Metamorphosed and metasomatically altered carbonatite	bastnäsite, monazite	750 Mt at 4.1% REO	Baotou Steel Rare Earth (Group) Hi Tech Co Ltd. Open pit. Flotation, sulphuric acid extraction. REE by-product of iron ore.
MAONIUPING, Mianning County, Sichuan, China	Carbonatite related. hydrothermal	bastnäsite	1.2 Mt at 2.89% REO	Jiangxi Copper Sichuan Rare Earth Company. Open pit. Flotation.
DALUCAO, Liangshan Autonomous Prefecture, Sichuan, China	Carbonatite/alkaline rocks related	bastnäsite	1.86 Mt ?REO	Dechang Houdi Rare Earth Mining Co., Ltd. Open pit.
WEISHAN, Shandong Province, China	Carbonatite and alkaline rocks	bastnäsite	12.75 Mt REO, grade >1.6% (Wu et al. 1996)	Shandong Weishan Lake Rare Earth Co. Ltd. Open pit.
MOUNTAIN PASS, California, USA	Carbonatite	bastnäsite (monazite planned)	20 Mt at 8.24% REO, 5% cut-off	Molycorp Inc. Open pit. Flotation. Processed to REE products on site.
Bear Lodge, Crook County, Wyoming, USA	Carbonatite dykes and veins in alkaline igneous complex	ancylite, bastnäsite-group minerals	15.88 Mt at 3.454% REO, 1.5% cut-off. 0.5484 Mt REO	Rare Element Resources (RER). Open pit. Crushing, attrition, screening and hydrochloric acid leaching.
MT WELD, Western Australia	Lateritic weathered carbonatite	monazite	17.49 Mt at 8.1% REO, 1.416 Mt REO#	Lynas Corp. Open pit. Flotation. Concentrate sent to Kuantan, Malaysia.
Dong Pao, North Vietnam	Lateritic weathered carbonatite	bastnäsite, synchysite, (minor monazite)	7.4 Mt at 5.22% REO, 3% cut-off, for No. 3 orebody	Toyota Corp, Sojito, Vinacomin joint venture.
Zandkopsdrift, Namaqualand, northern Cape, South Africa	Weathered carbonatite	monazite with crandallite and some churchite	21 Mt at 1.99% REO at 1% cut-off, 415,000 t REO	Frontier Minerals. Open pit. Flotation. REE extraction at Saldanha Bay.
Alkaline rocks				
KHIBINY, Kola Peninsula, Russia	Agpaitic nepheline syenite	fluorapatite	9 Mt REO [O&G]	Apatit GOK Open Pit and underground. By-product of fertiliser manufacture if nitric acid used.
LOVOZERO, Kola Peninsula, Russia	Agpaitic nepheline syenite	loparite	3.4 Mt REO#, 0.8–1.5% REO[O&G]	Lovozerskiy Mining Company. Underground mine. REE by-product of Nb.
Dubbo Zirconia, Toongi, central west NSW, Australia	Alkaline trachyte dyke	bastnäsite, ancylite	73.2 Mt at 0.75% REO. 0.549 Mt total REO.	Alkane Resources Ltd. Sulphuric acid leaching followed by solvent extraction. Also Zr, Nb, Ta
Nechalacho, Thor Lake, North West Territories, Canada	Peralkaline layered nepheline syenite with hydrothermal upgrade of REE	bastnäsite, allanite, parisite, monazite, synchysite. HREE in fergusonite and zircon.	107.59 Mt at 1.26–1.48% REO, 4.3 Mt REO	Avalon Rare Metals Inc. Undergound mine. Also Nb.

Name, Location	Geology	Ore minerals*	Grade and size	Notes on company, mining, processing
Kvanefjeld, Ilimaussaq Gardar Province, Greenland	Agpaitic nepheline syenite	steenstrupine, (eudialyte)	619 Mt at 10,585 ppm REO, 6.55 Mt REO	Greenland Minerals and Energy Ltd and Westrip Holdings. Open pit. Carbonate pressure leaching to recover U, precipitation of REE.
Strange Lake, NE Quebec/NW Labrador border, Canada	Peralkaline granite (hypersolvus, subsolvus and pegmatite) with additional hydrothermal concentration of REE	yttropyrochlore, kainosite, allanite, gadolinite, gerenite, keiviite: (fluorocarbonates & monazite)	114.8 Mt at 0.999 % REO, 43% HREE	Quest Rare Metals.
Zeus, Kipawa alkaline complex, Témiscamingue, Quebec, Canada	Alkaline compex with syenite and granite	eudialyte, Y-bearing titanite, britholite	67,200 t REO, significant HREE	Matamec Explorations. Open pit. Mild crushing and leaching. Also Zr.
Kutessay II, Aktyuz Ore field, Chu Oblast, Kyrgyz Republic	3 mineralised zones, (I, II, III). Massive pockets of REE carbonates	monazite, xenotime-(Y), bastnäsite-(Y), parisite- (Y)?, synchysite-(Y), yttrofluorite, fluorcerite	16.27 Mt at 0.264% REO[BGS] 50:50 LREE:HREE	Stans Energy Corp. Former REE mine. Produced all 15 REE.
Ion adsorption				
Ion adsorption deposits, 214 deposits in Jiangxi, Hunan, Guangdong, Guangxi,and Fujian, China	Clay-rich weathered granites and other silicate rocks	REE are adsorbed onto kaolinite and other clays	most deposits <10,000 t, 0.03–0.35% REO[BGS], HREE rich	Multiple companies. In-situ leaching or small scale mining and leaching.
Hydrothermal				
Steenkampskraal, Western Cape Province, South Africa	Hydrothermal monazite in metamorphosed gneisses	monazite	249,500 t at 17% REO, 29,400 t REO)#	Rareco (Great Western Minerals Group). Underground mine. Gravity separation and flotation.
Nolans project, Northern Territories, Australia	Sub-vertical dykes, veins and stockworks of fluorapatite in gneiss.	REE-bearing fluorapatite (especially Nd and Eu), cheralite	30.3 Mt at 2.8% REO, 848 kt REO	Arafura Resources. Open pit. Heavy media. REE carbonate recovery from acid leach.
Hoidas Lake, Saskatchewan, Canada	REE in apatite and allanite veins	apatite, allanite	2.8 Mt at 2.139–2.568% REO	Great Western Minerals Group.
Placer				
Manavalakurichi, Tamil Nadu, India	Marine placer?	monazite	total reserves in India amount to 10Mt monazite.#	Indian Rare Earths Ltd (IREL), Federal Department of Atomic Energy.

(*continued*)

Table 13.3 Continued.

Name, Location	Geology	Ore minerals*	Grade and size	Notes on company, mining, processing
Chavara, Kerala, India	Marine/alluvial placers	monazite		IREL
Orissa, Orissa, India	Alluvial placers	monazite		IREL
Chhatapur, Ganjam district, Orissa, India	Alluvial placers on coast	monazite		Toyotsu Rare Earths (IREL and Toyota Tsusho).
Other				
SARECO, Kasatomprom's nuclear facilities, Ust-Kamenogorsk, Kazakhstan**	Y-rich U ore tailings, U ores, in situ leaching solutions and REE deposits	?	?	Sumitomo, Kazatomprom. Processing at Kazatomprom facility.
Pitinga, Amazon Region, Brazil	Processing tailings at a tin mine	xenotime-(Y)	8.5% REO in tin tailings	Neo Material Technologies, Mitsubishi.

OPERATING MINES in capitals
source Roskill, company websites
**also a second source in Kazakhstan but too few details known to list
*all minerals are -(Ce) varieties unless specified
O&G Orris & Grauch, 2002.
Nb, niobium; Ta, tantalum; U, uranium; Zr, zirconium.

carbonatites occur as plugs, ring dykes, dykes and veins; they are frequently surrounded by an alkali metasomatic aureole called fenite and are commonly (but not necessarily) associated with alkaline or ultrabasic igneous silicate rocks. A typical igneous carbonatite assemblage of calcite, apatite, magnetite with accessory pyrochlore is unlikely to constitute an economic REE deposit although it may contain REE minerals as late-stage minor components. The weight per cent levels of light REE required for an economic deposit commonly occur in magnesium- and iron-rich carbonatite dykes and veins emplaced in pegmatoid-type, fluid-rich environments late in the development of the igneous complex. The original minerals have commonly been replaced although their shapes may have been maintained, forming 'pseudomorphs'. Such deposits can be high grade but frequently consist of narrow dykes and veins. Minerals include bastnäsite, monazite, and ancylite. Kangankunde, Nkombwa Hill, Wigu Hill, Songwe, Lofdal, Bear Lodge, Qeqertaasaq and Tikiusaaq are examples (Figures 13.4 and 13.5, Table 13.3). The two most famous REE-rich carbonatites at Mountain Pass, USA and Bayan Obo, China (Table 13.3) are different and atypical in this respect.

The world's largest REE deposit at Bayan Obo, Inner Mongolia, China has produced REE since the late 1980s and currently accounts for the majority of global REE supply. The ore minerals are bastnäsite-(Ce) and monazite-(Ce) set in an iron oxide-rich metamorphosed stratiform dolomite together with a variety of other REE minerals, notably aeschynite, fluorite and aegirine. There are three ore bodies (main, east and west) along strike of the H8 dolomite horizon. The exact nature of Bayan Obo has been controversial, and, although most researchers now agree on a carbonatite-related origin, there is no doubt that the deposit has been highly altered and reworked (Smith and Wu, 2000 and Wu, 2008).

The other best known REE deposit is the Mountain Pass carbonatite, USA (Table 13.3) (Castor, 2008). This is a large dyke-like igneous intrusion of carbonatite, which is associated with an igneous silicate rock called shonkinite. The

Figure 13.6 (a) Backscattered electron image of monazite in weathered carbonatite at Mount Weld, Australia. There is a narrow band of later monazite on the main monazite grain. The associated minerals are iron oxides (Feox) and clays. The grains are set in resin. (Sample courtesy of Lynas Corporation). (b) Steenstrupine crystals from a sodalite- and feldspar-rich late vein in nepheline syenite. Taseq Slope, Ilimaussaq alkaline complex, Greenland. (Sample and photo courtesy Henrik Friis.)

carbonatite contains bastnäsite, and other fluorcarbonate minerals (e.g. synchysite and parisite) together with some monazite. It has been proposed that much of this deposit consists of igneous minerals (all light REE enriched) that precipitated directly from the carbonatite magma, rather than being subject to the more common late and post-magmatic processes in carbonatites.

Weathering or alteration by hydrothermal fluids dissolves the soluble carbonates and thus concentrates the less-soluble REE. The monazite deposit at Mount Weld, Western Australia is produced by weathering of carbonatite, as is the Zandkopsdrif deposit in South Africa. These deposits can have fine-grained and complex mineralogy (Figure 13.6) but tend to have low Th. They are of similar size and grade to other carbonatite deposits (Figure 13.5) but richer in REE than the original unweathered carbonatite.

Alkaline igneous rocks

Alkaline igneous rocks are also typical of intracontinental tectonic settings. Variants termed 'peralkaline' and 'agpaitic' have particularly high sodium and potassium contents compared with silicon and aluminium and contain characteristic mineral suites, including complex sodium, titanium, zirconium silicates. Some of these minerals can host significant quantities of REE, although few of them have been beneficiated on a commercial scale. These minerals may form during the original crystallisation of the magma or during later hydrothermal alteration that upgrades the original magmatic REE concentrations. The rocks may contain polymetallic deposits with zirconium, niobium and tantalum as well as REE (Table 13.3). Examples include the agpaitic nepheline syenite complex at Illimaussaq (Kvanefjeld), Greenland (Table 13.3) which contains steenstrupine (Figure 13.6), The perovskite mineral, loparite, is mined primarily as a source of niobium from the layered agpaitic nepheline syenite at Lovozero, Kola Peninsula, Russia but contains about 1.04–1.25% REO, which can be extracted (Petrov, 2004) (Table 13.3). Eudialyte in a separate intrusion at Lovozero has also been considered as a potential source of REE. The agpaitic nepheline syenite at Khibiny, Kola Peninsula, Russia contains the world's largest igneous apatite mines. The apatite contains 0.66 wt% REE, which can be extracted during fertiliser production if a nitric acid process is used rather than the more usual sulfuric acid production route (Petrov, 2004). The Kutessay II deposit in Kyrgyzstan is notable for its Y-variety minerals (Table 13.3) and high proportion of heavy REE. Its geology is poorly described but it appears to be highly altered and associated with granitic intrusions, although the REE minerals are mainly carbonates (Stans Energy Corp., 2012).

Other hydrothermal veins

A number of deposits have apparently formed by precipitation of REE-bearing minerals from hot-water-based solutions without any direct relationship to the emplacement of igneous rocks. It is notable that most of these contain apatite, in addition to REE minerals. They can be as large as some of the carbonatite deposits and are of variable grade (Figure 13.5). For example, the monazite-apatite vein at Steenkampskraal, South Africa (Andreoli et al., 1994) is related to local geological structures but has no apparent igneous source.

Iron-oxide–apatite deposits, including iron-oxide–copper–gold (IOCG) deposits

The Olympic Dam iron-oxide–copper–gold deposit, South Australia, which is mined currently for copper, uranium, gold and silver, also contains a large, low-grade deposit of REE in its steeply dipping, dyke-like bodies of hematite breccias five km long and one km deep within fractured granite. Hydrothermal bastnäsite, florencite, monazite, xenotime and britholite host an average of 5000 ppm REE with variable light to heavy REE ratios (Oreskes and Einaudi, 1990), giving a large but low-grade deposit (Figure 13.5). Bayan Obo has also been proposed as an IOCG-type deposit but has higher grades of REE (Wu, 2008). Other iron-ore deposits that contain apatite are also potential deposits of REE. For example, apatite in apatite–magnetite ores at Kiruna, Sweden contains 2000 to 7000 ppm REE with 'moderate' light to heavy REE-enrichment but negative Eu anomalies. The REE content of apatite in other iron ores is about 1000 ppm (Frietsch and Perdahl, 1995).

Placer deposits (mineral sands)

Monazite and xenotime are among a set of resistate minerals that survive weathering, erosion and transport and can be redeposited in river, estuarine or shallow marine environments. Sedimentation processes tend to concentrate the heavier minerals together and thus deposits of quartz with elevated concentrations of zircon, ilmenite, rutile, monazite and xenotime can be formed. There are well-known deposits along the coast of southern India in Kerala, and Orissa; in Western Australia north and south of Perth; and in South Africa at Richards Bay.

Elliot Lake in Ontario, Canada is a palaeoplacer uranium and REE deposit in conglomerate.

Ion adsorption deposits

Ion adsorption deposits are formed on weathered granites, and, to a lesser extent on pyroclastic rocks and lamprophyres, in the sub-tropical climates of Jiangxi, Hunan, Guangdong, Guangxi, and Fujian provinces, south of 28°N in southern China (Bao and Zhao, 2008). There are 214 deposits recorded over an area of 90,000 km^2. Individually, each deposit tends to be small, with less than 10,000 tonnes of ore in each deposit, and low grade, varying from 300 to 3500 ppm REO (British Geological Survey, 2011). The weathering profiles are typically 15 to 35 m thick (Bao and Zhao, 2008). The deposits contain REE adsorbed onto the surface of clays as well as REE minerals. A key controlling factor in their formation is the presence of REE in the fresh granite in minerals that are easily weathered such as parisite (Table 13.2) (Bao and Zhao, 2008) rather than resistant minerals such as monazite. The radioactivity of these deposits is low. Furthermore, the deposits, being clay, are easy to mine and in some cases the adsorbed REE can be released from the clays in situ by ion exchange.

Seafloor deposits

Deep-ocean manganese nodules, iron-manganese crusts and deep sea muds are potential marine sources of REE, of similar size to deposits such as Mountain Pass. The Clarion–Clipperton Manganese Nodule Zone (CCZ) in the north-east Pacific contains a resource of 211 million tonnes at 0.1% REO or 21 million tonnes REO. The Prime Iron–Manganese crust zone (PCZ) in the central Pacific contains 7500 million tonnes at 0.3% REO or 23 million tonnes REO. These deposits have higher proportions of heavy REE (e.g. 6.5 to 10 per cent) than carbonatite-related deposits and have very low thorium contents of 11–14 ppm (Hein et al., 2011). Deep-sea sediments in the Pacific Ocean that have REE held in iron oxyhydroxides and phillipsite have also been proposed as large, low-grade, low-thorium REE deposits (Kato et al., 2011).

By-products, co-products and waste products

An important point when considering future deposits, is that there is potential for production of REE as by-products and co-products of other commodities. In the deposits described above, REE at Bayan Obo are a by-product of iron ore, monazite from mineral sands is a co-product with ilmenite, zircon and rutile, and xenotime is produced from tin concentrate. REE can be extracted from loparite during processing for niobium and from apatite during fertiliser production (e.g. at Lovzero and Khibiny – see Alkaline igneous rocks, above). There is more potential in other phosphate deposits used for fertiliser production. Simandl et al. (2011) calculated that, assuming an average REE content of 460 ppm in phosphorite, the world's annual phosphate production of 170 million tonnes represents over 70,000 tonnes of contained REE. At the high REE prices of mid-2011, the REE could have a higher market value than the phosphate in some deposits (Simandl et al., 2011). Bauxites and waste from their processing also contain REE. The potential for extracting REE from the red mud wastes has been considered by various researchers (Red Mud Project, 2006).

Extraction methods, processing and beneficiation

Mining

Most REE mines, such as Bayan Obo, Mountain Pass and Mount Weld, are open-cast operations, involving conventional blast, load and haul techniques. No underground mines have ever been designed for the exclusive production of REE but there is, or has been, production of REE from a few underground mines. For example, the loparite (niobium) mines at Lovozero, Kola Peninsula, Russia, the former thorium mine at Steenkampskraal, South Africa, now being re-opened primarily for REE, and the former uranium mines at Elliot Lake, Ontario, Canada.

Different mining techniques are used for the beach sand placer deposits because they are generally much less consolidated than carbonatites, alkaline rocks or hydrothermal deposits. They are also often under water. Mining techniques include dredging and excavation by bucket wheel or by excavator. Some crushing may be required (e.g. beach placers, now inland in Western Australia). There are few details available of mining techniques for ion adsorption deposits in China but many are small-scale operations, with much of the mining done by manual labour. The clay deposits are excavated and leached to extract REE or are leached in situ.

Beneficiation

In most mining operations, it is more economic to concentrate the ore mineral and remove as many of the waste (gangue) minerals as possible before attempting to extract the elements of value. This rule applies in most REE mining operations, where both bastnäsite and monazite are liberated from accompanying phases by crushing and grinding and then concentrated by flotation. The flotation process demands a very fine grain size of about 50 µm and is thus energy intensive. At Mountain Pass the ore is crushed and ground and then processed in long series of flotation stages. The reagents used are hot, which is unusual in flotation processes (Figure 13.7). At Mount Weld, the ore is crushed in a ball mill and then sent to the flotation circuit. At Bayan Obo,

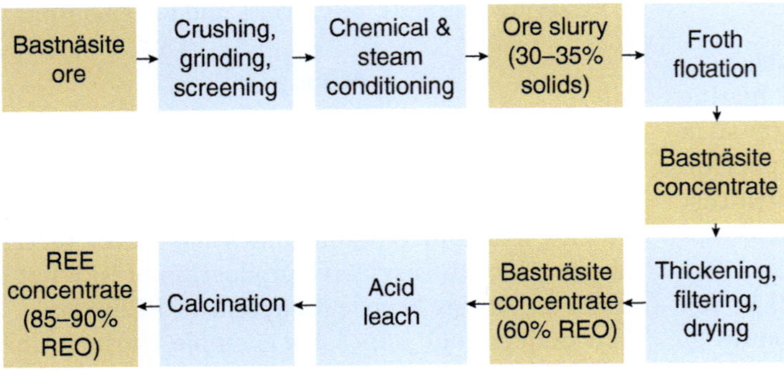

Figure 13.7 Schematic summary of the beneficiation of bastnäsite ores.

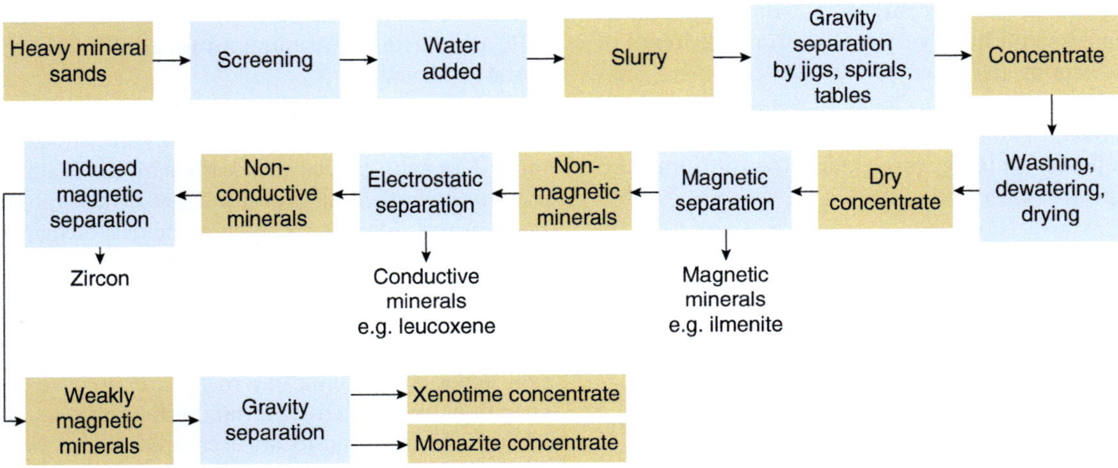

Figure 13.8 Schematic summary for the extraction of monazite and xenotime from heavy mineral sands.

the ore is more complex and the beneficiation circuit also produces magnetite, fluorite, hematite and niobium oxide as by-products (Gupta and Krishnamurthy, 2005). However, the rare earth minerals, bastnäsite and monazite, are again recovered in a series of flotation stages.

There is an alternative beneficiation route for REE minerals produced from mineral sands deposits. Since the grains are unconsolidated (or much less consolidated), these deposits are amenable to gravity, magnetic and electrostatic (high tension) separation methods (Figure 13.8). Many, especially contemporary beach deposits, require no crushing or grinding. Monazite and xenotime are separated from quartz, which is the main constituent of the sands; ilmenite, leucoxene, rutile and zircon are also usually produced as co-products. The combinations and flowsheets vary in detail according to the minerals present and products desired.

Processing of ion adsorption clays is distinct from other REE deposits in that no beneficiation is required and the REE are leached directly from the ore, in situ or in ponds, in neutral or slightly acid solutions with an ion-exchange agent such as ammonium sulfate or EDTA (ethylenediaminetetraacetic acid) (Bao and Zhao, 2008). In situ leaching was developed in Ganzhou, Jiangxi in an attempt to avoid the despoliation caused by pond leaching. However, it is more suited to the softer ionic clays in the areas around Dingnan and Longnan. Pond leaching is still used for the harder granite clay deposits in Guangdong.

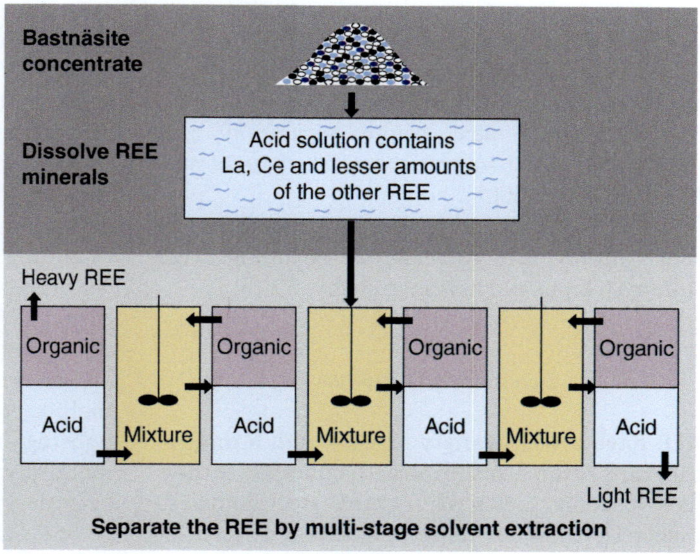

Figure 13.9 Schematic flowsheet for the extraction of rare earths from bastnäsite at Mountain Pass. (After Molycorp Inc., 2011b.)

The exploitation of new types of deposits will require different methods of concentration. There are no commercial-scale processes demonstrated for any of the REE minerals apart from monazite, xenotime, bastnäsite and loparite. Other minerals have been mined on a much smaller scale or separated in laboratory tests (Table 13.2). During development of new deposits, consideration of geometallurgy and potential extraction methods is critical. For example, flotation tests on ancylite from the Bear Lodge deposit, USA were unsuccessful. A flowsheet involving crushing, scrubbing (attrition), hydrochloric acid leaching to dissolve the ancylite (and presumably other carbonates), then a separate stage to remove uranium and thorium, and final production of REE oxalate product has been devised (Richardson et al., 2010).

Extraction and separation of the REE

Separation of the individual REE is the most difficult part of the extraction process. There are many routes for dissolving the various REE minerals and then separating the REE from each other (Gupta and Krishnamurthy, 2005). Most involve an initial acid attack to create soluble REE salts such as sulfates. Multiple stages are then used to separate the REE. Selective oxidation can be used for Ce, and also for Pr and Tb, which have a potential +4 oxidation state, whilst selective reduction works for Eu, Sm and Yb because they have a potential +2 valency. Otherwise, the processes exploit the property of the decreasing cation size across the REE series (i.e. the lanthanide contraction) during processes of fractional crystallisation, fractional precipitation, ion exchange and solvent extraction.

At the Molycorp operation at Mountain Pass, USA, bastnäsite concentrate is roasted in air at 620 °C to drive off carbon dioxide and oxidise Ce^{3+} to Ce^{4+}. This calcined product is then treated with hydrochloric acid to dissolve the non-cerium REE. The cerium concentrate can be sold. Further processing is then used to separate Eu (originally for cathode ray tubes) and other REE (Gupta and Krishnamurthy, 2005). The processing and extraction operation is now being updated to make it more water, resource and energy efficient (Molycorp Inc., 2011a). In the new process, the bastnäsite will be dissolved in acid and then subjected to counter-current solvent extraction using immiscible organic and acid solvents (Figure 13.9).

In China, bastnäsite concentrate is processed with 98 per cent sulfuric acid at 500 °C in a rotary kiln. This converts REE to sulfates. They are then converted to hydroxides and dissolved in hydrochloric acid before separation and purification using solvent extraction (Gupta and Krishnamurthy, 2005).

Xenotime is treated with concentrated (93 per cent) sulfuric acid to make water-soluble REE sulfates (Gupta and Krishnamurthy, 2005).

Specifications and uses

REE have a wide variety of uses and, although they are often used in small quantities, they have become pervasive in many technologies, especially those for improving energy efficiency and in digital technologies (Table 13.4). Cerium oxide polishing powders, still used today for virtually all polished glass products, and the Ce metal alloys in lighter flints were some of the early applications of REE in the 1950s. Europium gave the red colour to colour televisions and was one of the main products from the Mountain Pass mine during the 1970s and 1980s. The small headphones first used in the 'Walkman', and in all portable music devices since, are thanks to the small high-strength REE-bearing permanent magnets, which are now essential components in all computer disk drives and many other technologies. The uses of REE can be grouped into eight categories (Table 13.4). Catalysts form one of the largest market sectors by volume (Table 13.4). For example, Ce is a component of catalytic converters in cars because of its ability to form non-stoichiometric CeO_2, which helps oxidise unburned hydrocarbons. Lanthanum and cerium halides help to stabilise the zeolite structures used to crack petroleum.

The use of REE in permanent magnets is the highest value market sector and is forecast to increase in size, not least because of the use of Nd-bearing magnets in large wind turbines. Neodymium–iron–boron ($Nd_2Fe_{14}B$) is the strongest permanent magnetic material known and samarium–cobalt ($SmCo_5$) is also an important magnetic material. Small quantities of Dy and Tb are substituted into $Nd_2Fe_{14}B$ magnets in order to improve their properties and performance at higher temperatures.

REE make good phosphors, and this is another high-value sector. In addition to Eu, Y is used in various phosphor applications, including televisions and computer monitors. Yttrium is also important in yttrium aluminium garnet (YAG) lasers in high-temperature superconductors, and in specialist alloys. Other applications of REE in 'green' technologies include their use in low-energy fluorescent lighting, their use as the 'metal', e.g. La, in metal hydride batteries used in rechargeable batteries, including in hybrid vehicles, and use in fuel cells. Magnetic refrigeration may be an important future application.

Recycling, re-use and resource efficiency

Low prices have been the main contributing factor to a low rate of recycling or re-use of REE products. Less than one per cent of REE, mainly magnets, were thought to have been recycled from end-of-life products in 2010 (Kara et al., 2010). The recent rising prices and supply problems have encouraged new research and development in this area. There is also research taking place to reduce the amounts of REE used in various processes. Recycling is often difficult because of the way that REE are incorporated as small components in complex items or are part of complex materials. The processes required are energy intensive and complex (Schüler et al., 2011). Some REE are incorporated in components that typically have short lifetimes such as mobile phones, computer disk drives and low-energy light bulbs but others are in products with lifespans of 10–20 years or more, such as wind turbines and vehicles.

The advantages of recycling are that the REE are already separate from radioactive elements (and thus one of the main problems in mining raw materials is avoided), the energy

Table 13.4 Uses of rare earth elements, estimated global rare earth oxide (REO) demand in 2012 and forecast REO demand in 2016. (Data from Kingsnorth, 2013.)

Category	Application	Principal elements used	Demand 2012 (tonnes REO)	Demand 2012 (market share %)	Forecast demand 2016 (tonnes REO)	Forecast demand 2016 (market share %)
Magnets	Motors Disc drives Power generation Actuators Microphones and speakers MRI Automotive parts Communication systems Electric drive Frictionless bearings Magnetic refrigeration	Nd, Pr, Tb, Dy, Sm	22,500	20	33,000	21
Metal alloys	Hydrogen storage (NiMH batteries, fuel cells) Steel Aluminium/magnesium Cast iron Superalloy	Ce, La, Pr, Nd, Sm, Sc, Y	22,000	19	30,000	19
Catalysts	Catalytic converter Chemical processing Diesel additives Petroleum refining	Ce, La, Pr, Nd, Y	22,000	19	26,000	15
Polishing	Polishing compounds	Ce, La, Nd	19,000	16	25,000	15
Glass	Optical glass UV resistant glass Thermal control mirrors Colourisors	La, Ce, Pr, Nd, Eu, Gd, Dy, Ho, Er, Y	7,500	7	9,000	6
Phosphors and pigments	Display phosphors Medical imaging Lasers Fibre optics Fluorescent lighting Optical sensors Pigments LEDs	Eu, Tb, Y, La, Dy, Ce, Pr, Gd, Nd, Ce, Er, Eu	8,500	7	9,000	7
Ceramics	Capacitors Sensors Colourants Scintillators	Y, Ce, La, Pr, Nd	6,500	6	8,000	5
Other	Water treatment Fertiliser Medical tracers Coatings Nuclear reactors	Gd	7,000	6	20,000	12
Total			**115,000**	**100**	**160,000**	**100**

and resources requirement can be lower for recycling than mining, and the dependency on imports of REE is reduced (Schüler et al., 2011), particularly important for Europe, Japan and USA. The disadvantage is that recycled materials, such as magnets, may not perform so well as new materials (British Geological Survey, 2011). Re-use is also potentially possible, for example, of rare earth magnets. The problem is that product design changes rapidly and new sizes and styles of magnets are usually required, so this kind of re-use would require a 'cradle-cradle' design philosophy from the outset.

Substitution

For many applications of the REE, there are either no known substitutes or the substitutes do not perform as well (Haxel et al., 2002). There is also the problem that some substitutes may be as expensive as the REE. An example of substitution is the use of less strong samarium–cobalt magnets or, where size is not so important, iron magnets in some applications of neodymium–iron–boron. The hardest applications for which to find substitutes are the uses that require the optical and chemical properties of REE, such as colouring and decolourising agents, catalysts and phosphors. However, reduction of the amount of REE used is proving possible in certain applications. For example, the amount of REE in some catalysts used to crack petroleum was reduced by up to a half in 2011 as prices escalated in response to concerns about the security of supplies from China.

Environmental aspects

In general, the toxicity of the rare earths themselves is thought to be low. However, there are some health risks documented (U.S. Environmental Protection Agency, 2012). The main reported concern is the use of cerium polishing powders in lenses and other glass objects and jewellery. Once in the body, Ce tends to accumulate in the bone, liver, heart, and lung (U.S. Environmental Protection Agency, 2009) and prolonged exposure has been associated with lung problems such as interstitial lung disease or pneumoconiosis. Cerium poisoning causing endomyocardial fibrosis has been reported in Kerala, India, where the local population eat root vegetables and thus ingest soil containing high levels of Ce. However, a study of an area in Uganda with elevated Ce in soil in Uganda did not find a similar link (Smith, 1998). Cerium is common in other products but the exposure is much less, e.g. as catalysts for auto-emissions control, decolourisers for glass. There is now some evidence of other REE from anthropogenic sources in the natural environment. Gadolinium anomalies have been found in river waters following use of Gd in medical imaging (Kulaksız and Bau, 2007). These levels of Gd have not been correlated with any adverse health effects but the Gd-based contrast agents administered during medical imaging can cause nephrogenic systemic fibrosis in patients with impaired kidney function (U.S. Food and Drug Administration, 2010). Limits for levels of REE permitted in discharges are generally lacking and there is little regulation of factories that use REE. In China, REE have been used in fertilisers for many years and also fed to livestock as a growth promoter, with the conclusion that there is no harmful effect to humans or animals (Baotou National Rare-earth Hi-tech Industrial Development Zone, 2011). REE-bearing fertilisers have been distributed on 67,000 km^2 of land in China thus providing an opportunity for a future case study of REE distribution in the environment.

The main environmental concerns regarding REE mines and processing facilities are not usually the REE themselves but the presence of the radioactive elements, thorium and uranium, and the chemicals used to extract and separate the REE from the ore minerals. Thorium, and to a lesser extent uranium, may be present as discrete mineral phases that can be separated from the REE minerals at the beneficiation stage. However, there will almost always be some thorium, and lower amounts of uranium, incorporated into the REE mineral lattice and hence present in any final concentrate and in the factory where extraction of REE takes place. Processing of monazite from mineral sands, which can contain about 10 wt% ThO_2, has been banned in some countries, such as Australia. The ore at Mountain Pass contains 0.02 wt% Th and 0.002 wt% U; the tailings at Mount Weld will contain 0.05 wt% ThO_2 and 0.003 wt% UO_2 (Schüler et al., 2011). Areas around Chinese rare earth mines have been polluted and the Chinese government is now trying to regulate its industry and encourage larger companies to take over the small REE leaching operations on ion adsorption deposits, which have low thorium levels but have caused chemical pollution. China's Ministry of Environment Protection brought new rules into effect in October 2011 to limit pollution from rare earth mines (China Daily, Europe, China Daily Europe, 2011). The new rules govern chemicals used in the processing and production of rare earths, such as an emission cap on ammonia nitrogen in water of 25 mg/l from 1 January 2012, reduced from previous limits of 300–5000 mg/1. In addition to these considerations, processing uses considerable energy to grind ores fine enough for flotation. Processes that rely on acid dissolution may also have high carbon footprints as a result of the energy used to produce the acid and require appropriate safety measures. All of the processing routes require water.

World resources and production

The best estimates of world reserves of REE are about 114 million tonnes of REO (Cordier, 2011) (Table 13.5). China has the largest proportion (48 per cent) of reserves, mainly in the Bayan Obo deposit, but also in other carbonatite and

Table 13.5 Production of rare earth oxide (REO) concentrates in 2010 and estimates of reserves. (Production data from Roskill Information Services Ltd; reserve data from Cordier, 2011.)

Country	Comment	REO production (tonnes)	REO reserves (tonnes)
China	Mainly from bastnäsite from Baotou and ion adsorption clay from southern provinces	120,000	55,000,000
	stockpiled ore in Sichuan	ca.10,000	
	'unoffcial' sources	ca.15,000	
Russia (reserve: CIS)	In chemical concentrates from mine output in 2009	1,898	19,000,000
USA	From stockpiled ore. Mining restarted in December 2010	1,883	13,000,000
India	Chemical concentrates from minerals sands tailings	25-50	3,100,000
Brazil	Monazite stockpiled by-product from rutile production	550*	48,000
Australia	Some mining has taken place at the Lynas operation at Mount Weld – the ore was stockpiled until the concentrator was completed.	?	1,600,000
Others	Small amounts of monazite and xenotime from south-east Asia, probably shipped to China for processing	?	22,030,000
Total			113,778,000

alkaline rock deposits. It is significant to note that a large proportion of the world's heavy REE reserves is in ion adsorption deposits (Tables 13.3 and 13.5) and also some mineral sands in China. The USA has reserves at Mountain Pass and in various projects under development such as Bear Lodge, Wyoming (Tables 13.3 and 13.5). CIS, including Russia, has significant reserves in its large alkaline complexes at Khibiny and Lovozero and at Kutessay II in Kazakhstan (Tables 13.3 and 13.5, Figure 13.4). Additional resources include the weathered carbonatite deposit at Tomtor in Siberia. Other reserves and resources are in the large alkaline complexes in Canada and Greenland. There are many carbonatite complexes in Africa that contain REE (Figure 13.4), although they tend to be smaller than the carbonatites currently mined. The reserve figure will vary with time, as new exploration projects are completed and as economics and new technologies permit the consideration of additional deposit types in the reserve category. There are also potentially large amounts of REE in lower-grade deposits such as Olympic Dam, for example (Table 13.3, Figures 13.4 and 13.5) and in alkaline complexes as well as carbonatites (Table 13.3, Figures 13.4 and 13.5). The reserve figure does not include novel sources of REE such as ocean-floor deposits.

An estimate of the long-range extractable geological resource is difficult to obtain. One hypothetical approach is to consider apatite, which is much more abundant than the common rare earth minerals and therefore is probably host to a higher proportion of crustal REE. In alkaline igneous rocks and carbonatites, apatite commonly contains 1 wt% total REE. Beneficiation and extraction technologies are well established for apatite. Assuming there are about 2500 alkaline complexes at the Earth's surface, averaging five km^2 in area and containing 3 wt% apatite, with an REO content of 1 wt%, the total REO content of these complexes would be 10,000 million tonnes of REO. This is equivalent to 50,000 years supply of REO at a projected annual consumption rate of 200,000 tonnes per year.

Production of REE has increased consistently since the 1950s but the producers have changed markedly during this time. Early production of REE came from mineral sands in Western Australia and India; then the Mountain Pass deposit in USA was discovered and became the world's main supplier of REE during the 1960s and 1970s, set up particularly to produce Eu for televisions. In the 1980s Bayan Obo in China became the main supplier, joined by other Chinese producers, to reach the current position where China supplies some 97 per cent of the world's REE (Figure 13.10). As recently as 1992 though, there was a wider variety of producers, with USA, Australia and Russia significant, and China producing just 38 per cent of the world's rare earths (Figure 13.10). By 2010, only Russian loparite and Indian mineral sands were still contributing any appreciable REE to the total.

Estimates of current production (Table 13.5) vary slightly because of factors such as unofficial (illegal) production from China and inconsistency in recording production that is stockpiled rather than sold (e.g. monazite in Brazil is a waste product of the benefication of rutile and is stockpiled). Other potential producers such as Australia do not record their stockpiles as 'production' and so figures are harder to compare. Some mineral sand producers return monazite to the ground, diluted with waste material rather than stockpile radioactive material.

As noted above there are many REE resources outside China. If demand is sufficient and prices high enough, the other deposit types could be used, including apatite as described above, and, in the longer term, seafloor deposits. By-product REE from other ores and extraction from waste are also potential future sources of supply.

Future supplies

Exploration activity for REE has increased greatly in recent years. Many deposits are now under active exploration, including most of those shown in Figure 13.4, and many more. A new trend may be that non-carbonatite sources become mines, especially targeting heavy REE. With the higher prices, production of REE as a

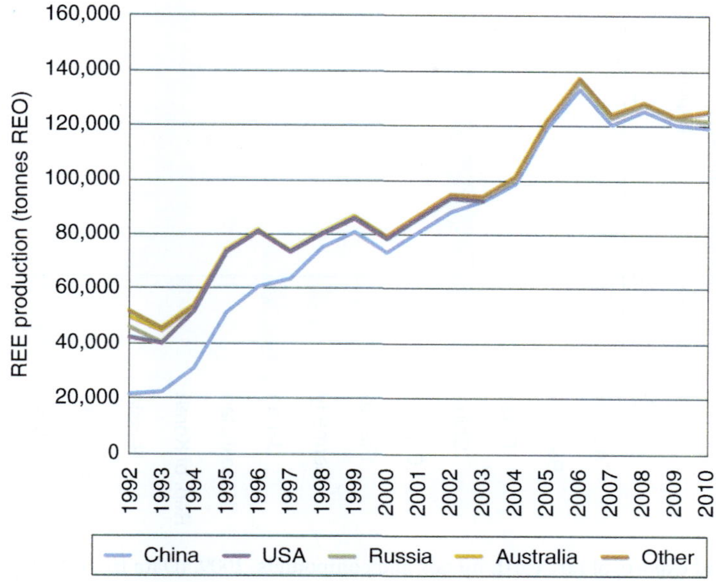

Figure 13.10 Rare earth production between 1992–2010. (Data from British Geological Survey World Mineral Statistics database.)

by-product may also become economic in more cases. Recycling is also increasingly economic and, together with Government and company funding to reduce reliance on China, it is likely that more recycling will take place.

Two new projects have recently started production. The re-opening of the Mountain Pass mine is the first major development. The second project, at Mount Weld in Western Australia, commenced mining in 2011 and the new REE facility in Malaysia, which processes the concentrates from Mount Weld, began operation in early 2013. Market analysts list up to 20 projects on a variety of deposit types that are at an advanced stage of exploration and development (Table 13.3). It is noteworthy that although the first two mines opening (Mountain Pass and Mount Weld) are in carbonatite, there is a wider range of deposits in the next phase of projects likely to reach fruition. REE have been produced from the niobium ore, loparite, mined at Lovozero, Kola Peninsula, Russia, with REE production taking place at the Silmet factory in Estonia. It is not clear if there is still any REE production from this source now that the Silmet factory has been sold to Molycorp.

World trade

China is by far the main exporter of REE metals and compounds (Figure 13.11), reflecting its dominance in the mining of REE. China has pursued a strategy of building up its own REE industry so that it manufactures the high-value REE intermediate products and attracts foreign companies to operate in China rather than exporting raw materials. The USA, Germany, France and Austria are the largest importers. None of these countries has mined REE in recent years and so they are totally reliant on Chinese exports. This situation will change in USA as Mountain Pass re-opens but Europe will still need to import practically all of its REE. The imports to Estonia are related to the Silmet factory. This factory has now been bought

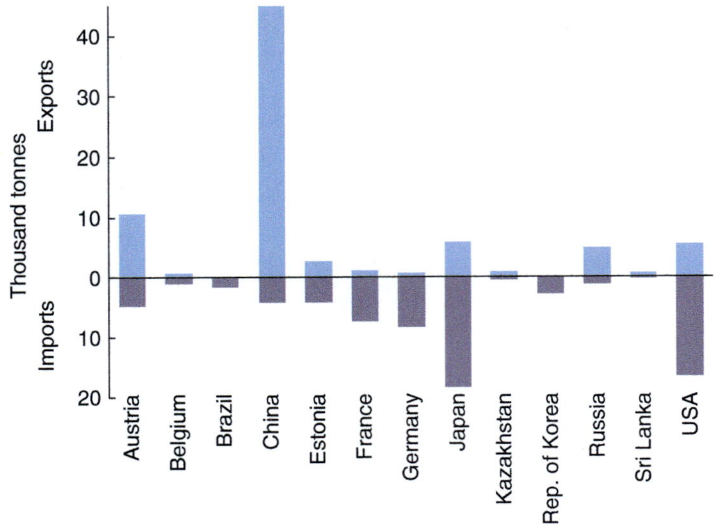

Figure 13.11 Imports and exports of rare earth metals and compounds, 2009. (Data from British Geological Survey World Mineral Statistics Database and UN Comtrade, 2013.)

by Molycorp, the operator of Mountain Pass, so that it now has three sites that can process REE: Mountain Pass (California, USA), Tolleson (Arizona, USA) and Sillamae (Estonia, previously AS Silmet).

There is no international metals exchange for REE and many mines and deposits are now vertically integrated into the supply chain, with large companies such as Toyota and Siemens undertaking joint ventures to source raw materials and thus gain security of supply. Mines such as Mount Weld have their own processing factories so that they will sell individual REE oxides and metals rather than intermediate products. REE are similar to industrial minerals in this respect. As a result of this, there is little trade in REE mineral concentrates or leach products (carbonates or oxalates).

Demand for REE is predicted to continue to rise. The U.S. Department of Energy (2011) highlights five REE (neodymium, europium, dysprosium, terbium and yttrium) as critical in the near- (up to 2015) and medium-term (2015 to 2025) for use in energy technologies. It is notable that none of the most critical REE are the main constituents of any common REE minerals and in order to produce these elements it will be necessary to also produce more La and more Ce or Y.

Prices

Prices for REE vary considerably (Figure 13.12). The main controlling factors are the nature and purity of the product, the abundance of the particular REE, and demand from current uses. The cheapest products are mixtures of the REE, such as mischmetal (La 35%, Ce 65%) or mixed REE carbonates (Figure 13.12). Generally, the oxides are easier and cheaper to produce than the metals. The cheapest pure REE are the most abundant elements, La and Ce, which are used in many applications. Although Nd is abundant it is more expensive because of the demand for $Nd_2Fe_{14}B$ magnets. The most expensive REE are Eu, Dy and Tb because of their low abundance in the majority of REE ores and high demand for these elements in phosphors and magnets.

There is no international metal exchange for REE. Prices are determined by the actual trades that take place and are available on various commercial and company web sites, quoted in Chinese currency and US dollars FOB[1] China.

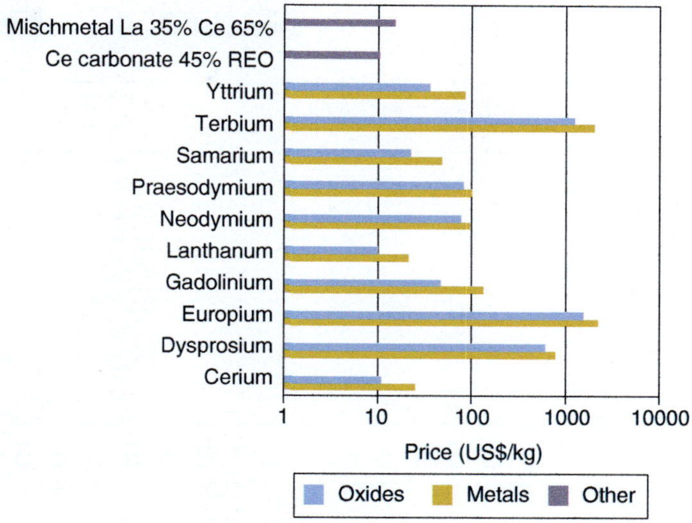

Figure 13.12 Prices of rare earth metals, oxides, mischmetal and mixed light REE carbonate in January 2013. Prices are FOB China, 99% metal min., except Y 99.9% min.; oxide: FOB China, 99% min., except Eu 99.9%, Y, 99.999%. (Data from Metal Pages, www.metal-pages.com.)

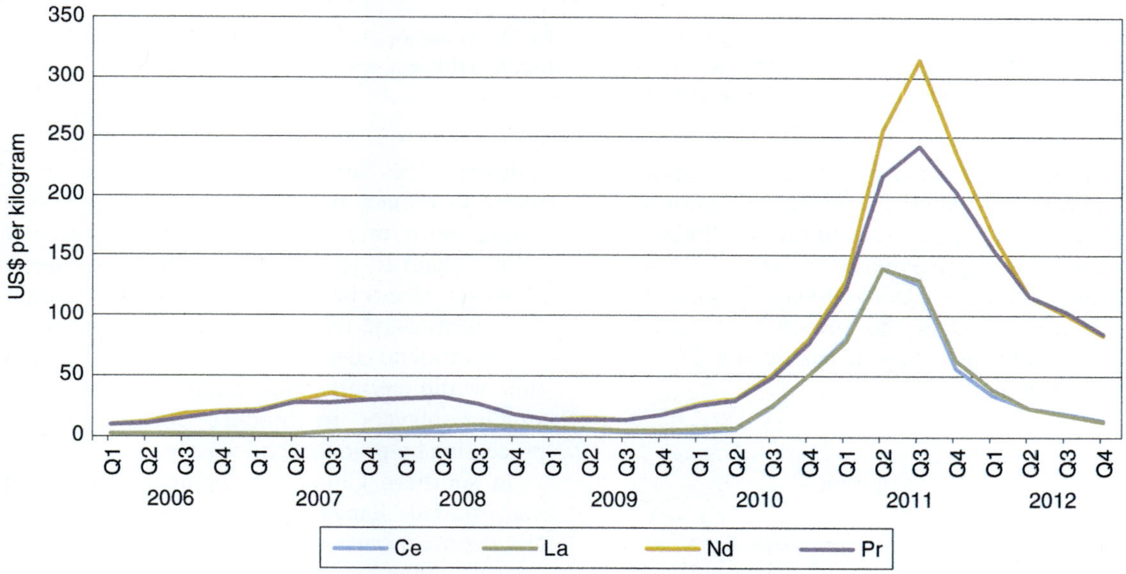

Figure 13.13 Quarterly average prices for selected LREE. Prices are for oxide FOB China 99% min. (Data from Metal Pages, www.metal-pages.com). Ce, cerium; La, lanthanum; Nd, neodymium; Pr, praseodymium.

These are export prices but between 60–70 per cent of all rare earth consumption is now in China and here the REE may be traded at different internal prices, which may impact on the price of part-processed or finished goods for export.

Prices of all of the REE increased very rapidly during 2010 and 2011 owing to supply problems as China restricted its export quotas (Figures 13.13 and 13.14). Demand reduced and prices fell at the end of 2011 and continued to decline throughout 2012. Nevertheless, prices

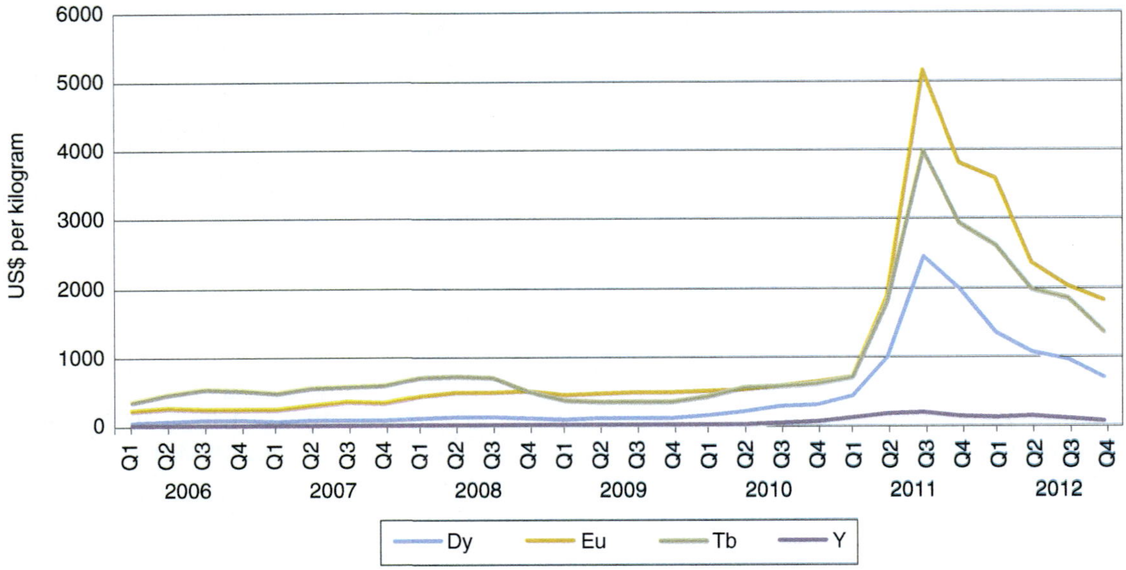

Figure 13.14 Quarterly average prices for selected HREE and Y. Prices are for oxide FOB China 99% min. for Dy and Tb, 99.9% min. for Eu, and 99.999% min. for Y. (Data from Metal Pages, www.metal-pages.com). Dy, dysprosium; Eu, europium; Tb, terbium; Y, yttrium.

for most REE at the end of 2012 remained significantly higher than the levels in 2008 prior to the onset of the global economic recession. For example, at the beginning of 2008 the price of dysprosium oxide was about US$93/kg compared with US$615/kg at the end of 2012. Over the same period the prices of La, Ce and Tb oxides more than doubled.

Outlook

There is a wide variety of rare earth deposits, containing plenty of REE for the foreseeable future. The challenge is to diversify supply quickly enough to overcome the current market problems and allow the current and new uses of REE to continue and expand. China is building zones around the lesser known Maoniuping and Weishan mines for 'deep processing' of rare earths similar to that at Baotou near the Bayan Obo mine. It aims to make a vertically integrated industry. This kind of vertical integration is likely to remain important in the market as a whole, with more end users buying into the supply chain to secure the resources they need. However, it will be interesting to see if a market for intermediate mixed carbonate leached products or mineral concentrate does develop because this would permit more flexibility and allow smaller players to produce REE, including processing by-products and wastes.

In southern China developments are taking place outside Jiangxi and Guangdong, traditionally the main centres for HREE production. State-owned companies and research institutes (e.g. Grirem) are involved in developing new sites for mining, separating and downstream processing in Hunan, Fujian and Guangxi. In Guangxi new mining licences have been granted that could replace exhausted HREE resources in other parts of the south.

The desire of end users to prevent another supply crisis may help to increase the number of REE producers, including the encouragement of projects in Europe. Considerable research is required to establish production from new types of deposits and to conduct more REE separation and purification outside China. The most attractive deposits to find will be those with high proportions of the 'critical REE', i.e. the most highly sought after heavy REE and Nd (i.e. Nd, Eu, Dy, Tb, Y).

China is bringing in new environmental regulations in order to reduce pollution from rare earth mines but has long-standing problems to overcome. With a diverse range of deposit types available, once there are alternatives to China, responsible sourcing from the most environmentally friendly mines may also become an issue. Recycling rates will increase if the price for REE remains sufficiently high.

Note

1. FOB, Free on Board. The seller is responsible for the cost of delivering goods to the ship. The buyer is responsible for transportation and insurance costs from that point.

References

Andreoli, M.A.G., Smith, C.B., Watkeys, M., Moore, J.M., Ashwal, L.D. and Hart, R.J. (1994) The geology of the Steenkampskraal monazite deposit, South Africa; implications for REE-Th-Cu mineralisation in charnockite-granulite terranes. Economic Geology 89, 994–1016.

Bao, Z. and Zhao, Z. (2008) Geochemistry of mineralisation with exchangeable REY in the weathering crusts of granitic rocks in South China. Ore Geology Reviews 33, 519–535.

Baotou National Rare-earth Hi-tech Industrial Development Zone (2011) Rare Earth: an introduction. http://www.rev.cn/en/int.htm Downloaded 11.12.11.

Bayliss, P. and Levinson, A.A. (1988) A system for nomenclature of rare-earth mineral species: Revision and extension. American Mineralogist 73, 422–423.

British Geological Survey (2011) Rare Earth Elements Mineral Profile, British Geological Survey, Nottingham, United Kingdom. http://www.bgs.ac.uk/mineralsuk/search/home.html

Castor, S.B. (2008) The Mountain Pass rare-earth carbonatite and associated ultrapotassic rocks, California. The Canadian Mineralogist 46, 779–806.

Cordier, D.J. (2011) Rare Earths, Mineral Commodity Summaries. United States Geological Survey, 128–126

China Daily Europe (2011) China's rare earth industry faces reshuffle: experts. http://europe.chinadaily.com.cn/china/2011-03/06/content_12126297.htm

Förster, H-J. (1998) The chemical composition of REE Y-U-rich accessory minerals in peraluminous granites of the Erzgebirge-Fichtelgebirge region, Germany. Part II: Xenotime. American Mineralogist 83, 302–1315.

Frietsch, R. and Perdahl, J-A. (1995) Rare earth elements in apatite and magnetite in Kiruna-type iron ores and some other iron ore types. Ore Geology Reviews 9, 489–510.

Gupta, C.K. and Krishnamurthy, N. (2005) Extractive Metallurgy of Rare Earths. CRC Press, Boca Raton, USA.

Haxel, G.B., Hedirk, J. B. and Orris, G.J. (2002) Rare Earth Elements – Critical Resources for High Technology, USGS Fact Sheet 087-02. US Department of the Interior, United States Geological Survey.

Hein, J.R., Conrad, T. and Koschinsky, A. (2011) Comparison of land-based REE ore deposits with REE-rich marine Fe-Mn crusts and nodules. Mineralogical Magazine 75, 1000.

International Mineralogical Association (2012). http://www.ima-mineralogy.org/Minlist.htm

Kaltosoyannis, N. and Scott, P. (1999) The f Elements. Oxford Chemistry Series, Oxford University Press, Oxford, United Kingdom.

Kara, H., Chapman, A., Crichton, T., Willis, P. and Morley, N. (2010) Lanthanide Resources and Alternatives. A report for Department for Transport and Department of Business, Innovation and Skills, Oakdene Hollins Research and Consulting.

Kato, Y., Fujinaga, K., Nakamura, K. et al. (2011) Deep-sea mud in the Pacific Ocean as a potential resource for rare-earth elements. Nature Geoscience, Advance Online Publication, DOI: 10.1038/NGEO1185

Kulaksız, S. and Bau, M. (2007) Contrasting behaviour of anthropogenic gadolinium and natural rare earth elements in estuaries and the gadolinium input into the North Sea. Earth and Planetary Science Letters 260, 361–371.

Kingsnorth, D.J. (2013) Rare earths forecast update. Presentation to AusIMM 2013 Critical Minerals Conference, Rare Earths – Is Supply Still Critical, Perth, Western Australia, 4–5 June 2013.

McDonough, W.F. and Sun, S. s- (1995) The composition of the Earth. Chemical Geology 120, 223–253.

Miyawaki, R. and Nakai, I. (1996) Crystal chemical aspects of rare earth minerals. Rare Earth Minerals: Chemistry, origin and ore deposits. In: Jones, A.P., Wall, F. and Williams, C.T. (eds.) Mineralogical Society Series 7, Chapman and Hall, London, 21–40.

Molycorp Inc. (2011a) The Many Uses of Rare Earths in Advanced Technologies. http://www.molycorp.com/GreenElements/RareEarthsManyUses.aspx.

Molycorp Inc. (2011b) Separating rare earths at Mountain Pass http://www.molycorp.com/resources/science-center

Oreskes, N. and Einaudi, M.T. (1990) Origin of rare earth element-enriched hematite breccias at the Olympic Dam Cu-U-Au-Ag deposit, Roxby Downs, South Australia. Economic Geology 85, 1–28.

Orris, G.J. and Grauch, R.I. (2002) Rare Earth Element Mines, Deposits, and Occurrences. United States Geological Survey Open-File Report, 02–189.

Petrov. S.V. (2004) Economic deposits associated with the alkaline and ultrabasic complexes of the Kola Peninsula. Phoscorites and Carbonatites from Mantle to Mine: the Key Example of the Kola Alkaline Province. In: Wall, F. and Zaitsev, A.N. (eds.) Mineralogical Society Series 10, Mineralogical Society, London, 469–490.

Red Mud Project (2006) The Red Mud Project. http://www.redmud.org/Bibliography.html#rareelements.

Richardson, M.P., Noble, A.C., Roman, R. and Clark, J.G. (2010) Technical report: preliminary economic assessment (scoping study) of the Bear Lodge Rare-earths project - a national instrument 43-101 report. Crook County, Wyoming. Prepared for Rare Element Resources Ltd.

Schüler, D., Buchert, M., Liu, R., Dittrich, S. and Merz, C. (2011) Study on Rare Earths and their Recycling. Öko-Institute. V., Final Report for The Greens / EFA Group in the European Parliament.

Simandl, L., Simandl, G.J. and Fajber, R. (2011) Rare Earth Elements (REE) recovery as a by-product of fertiliser production from sedimentary Phosphate deposits – Conceptual evaluation. Mineralogical Magazine 75, 1878.

Smith, B. (1998) Cerium and Endomycardial fibrosis in tropical terrain. Project summary report. Technical report WC/98/26, British Geological Survey.

Smith, M. and Wu, C. (2000) The Geology and Genesis of the Bayan Obo Fe-REE-Nb Deposit: A Review. In: Porter, T. M. (ed.) Hydrothermal Iron Oxide Copper-Gold and Related Deposits: A Global Perspective, Volume 1, PGC Publishing, Linden Park, South Australia, Australia, 271–281.

Smith, M.P., Henderson, P., Campbell, L.S. (2000) Fractionation of the REE during hydrothermal processes: Constraints from the Bayan Obo Fe-REE-Nb deposit, Inner Mongolia, China. Geochimica et Cosmochimica Acta 64, 3141–3160.

Stans Energy Corp. (2012) Kutessay II Geology, Historical Geological information of the Kutessay II REE-Th-Nb-Ta deposit. http://www.stansenergy.com/projects/?page_id=258

Taylor, S.R. and McClennan, S.M. (1985) The Continental Crust: Its Composition and Evolution. Blackwell Scientific Publications, Oxford.

United Nations (UN) Comtrade (2013) Commodity Trade Statistics Database (http://comtrade.un.org/db/).

U.S. Department of Energy (2011) Critical materials strategy, December 2011, DOE PI-0009.

U.S. Environmental Protection Agency (2009) Toxicological review of cerium oxide and cerium compounds (CAS No. 1306-38-3) In: Support of Summary Information on the Integrated Risk Information System (IRIS) 2009, EPA/635/R-08/002F, Washington, DC.

U.S. Environmental Protection Agency (2012) Rare Earth Elements: A Review of Production, Processing, Recycling, and Associated Environmental Issues. EPA/600/R-12/572, Cincinatti, OH.

U.S. Food and Drug Administration (2010) FDA Drug Safety Communication: New warnings for using gadolinium-based contrast agents in patients with kidney dysfunction. http://www.fda.gov/Drugs/DrugSafety/ucm223966.htm#sa

van Emden, B., Thornber, M.R., Graham, J. and Lincoln, F.J. (1997) The incorporation of actinides in monazite and xenotime from placer deposits in Western Australia. The Canadian Mineralogist 35, 95–104.

Wall, F. and Mariano, A.N. (1996) Rare earth minerals in carbonatites: a discussion centred on the Kangan—kunde carbonatite, Malawi. Rare Earth Minerals: Chemistry, origin and ore deposits. In: Jones, A.P., Wall, F. and Williams, C.T. (eds.) Mineralogical Society Series 7, Chapman and Hall, London, 193–225.

Wall, F. and Zaitsev, A.N. (2004) Rare earth minerals in Kola carbonatites. In: Wall, F. and Zaitsev, A.N. (eds.) Phoscorites and Carbonatites from Mantle to Mine: the Key Example of the Kola Alkaline Province. Mineralogical Society, 43–72.

Wall, F. Niku-Paavola, V.N., Storey, C., Mueller, A. and Jeffries, T. (2008) Xenotime-(Y) from carbonatite dykes at Lofdal, Namibia: unusually low light to heavy REE ratio in carbonatite, and the first dating of xenotime overgrowths on zircon. Canadian Mineralogist 46, 1251–1267.

Wu, C. (2008) Bayan Obo Controversy: Carbonatites versus Iron Oxide-Cu-Au-(REE–U). Resource Geology 58 (4), 348–354.

14. Rhenium

TOM A. MILLENSIFER[1], DAVE SINCLAIR[2], IAN JONASSON[2] AND ANTHONY LIPMANN[3]

[1]*Executive Vice President and Technical Director of Powmet, Inc., Rockford, Illinois, USA*
[2]*Formerly research scientist at Geological Survey of Canada, Ottawa, Ontario, Canada*
[3]*Managing Director, Lipmann Walton & Co Ltd, Walton on Thames, Surrey, UK*

Introduction

Rhenium is reported by many sources to be the last natural element to be discovered (Habashi, 1996). Its discovery in 1925 is credited to Ida Tacke, Walter Noddack and Professor Otto Berg. They also claimed the discovery of masurium, Ma, the neighbour of rhenium, which today is known as technetium, Tc. However, this was disputed and the discovery of technetium was not confirmed until 1936.

Much of the early work developing rhenium processing was done in the USA by Kennecott and many patents related to rhenium recovery from molybdenite processing were granted to this company. Shattuck Chemical of Denver Colorado subsequently licensed the Kennecott processes and began rhenium recovery in 1960 from the roasting of molybdenite concentrates derived from mines operating in western USA. Since then rhenium production has taken place in several other countries with Chile now the dominant supplier. In recent years production of rhenium from primary sources has grown substantially to current levels of about 44.8 tonnes (MMTA, 2012a). Today the main use of rhenium is in superalloys for gas turbines in aircraft and in land-based applications, while its use in petroleum-reforming catalysts accounts for about 10 per cent of total consumption.

Physical and chemical properties

Naturally occurring rhenium, element number 75 in the Periodic Table, consists of two isotopes: ^{187}Re, which accounts for 62.6 per cent of the total, and ^{185}Re, which makes up the balance of 37.4 per cent. ^{187}Re is radioactive with a half-life of approximately 4.3×10^{10} years. However, the beta radiation emitted is very weak, 0.3 MeV, and cannot penetrate human skin. The chemical properties of rhenium resemble the metals in the manganese group (Group 7) of the Periodic Table. The physical properties, however, are much more similar to those of the refractory metals of Groups 5 and 6, particularly molybdenum and tungsten. It was for this reason that the Noddacks used molybdenum concentrates as their source for larger quantities of rhenium (Habashi, 1996). Rhenium is considered a refractory metal because of its high melting point of about 3200 °C, with only tungsten having a higher melting point. However, in contrast to other refractory metals, rhenium does not form carbides. Platinum and osmium have greater specific gravities. Selected properties of rhenium are listed in Table 14.1.

Rhenium exhibits several valences from –1 to +7, with the most common being +7, +6, +5 and +4. It easily changes from one valence to another, a property which makes it ideal for use as a

Critical Metals Handbook, First Edition. Edited by Gus Gunn.

Table 14.1 Selected properties of rhenium. (Modified after Treichel, 2000.)

Property	Value	Units
Symbol	Re	
Atomic number	75	
Atomic weight	186.21	
Density at 25 °C	21023	kg/m³
Melting point	3180	°C
Boiling point	5926	°C
Hardness (Mohs scale)	7.0	
Specific heat capacity at 25 °C	0.14	J/(g °C)
Electrical resistivity at 25 °C	18.40	nΩ m
Thermal conductivity	48	W/(m °C)
Young's modulus	463	GPa

catalyst. In the 1960s it was observed that when rhenium is alloyed with molybdenum or tungsten, the resultant alloy exhibits or retains the 'best' properties of both metals individually and none of the 'poor' properties of either (J. Port, Cleveland Refractory Metals, personal communication, 1965). For example, heating tungsten above its recrystallisation temperature, 1200 °C, and then cooling it to room temperature causes it to become brittle. The addition of rhenium to tungsten and molybdenum significantly reduces the brittle characteristics, increasing the recrystallisation temperature, the ductility and the ultimate tensile strength of the alloys (Leonhardt, 2009).

Rhenium does not have a ductile-to-brittle transition temperature. It retains its ductility from low sub-zero temperatures to very high temperatures, making it ideal for use in space propulsion applications. Rocket-thruster nozzles have been reported to withstand more than 100,000 thermal fatigue cycles from room temperature to above 2225 °C without evidence of failure (Wooten and Lanshaw, 1989).

Distribution and abundance

Rhenium is one of the most dispersed of the naturally occurring elements in the Earth's crust, with estimates of average crustal abundance ranging from 0.2 to 2 ppb Re. The rhenium content of common igneous rocks, from basalts to granites, is typically in the range of 0.2 to 1 ppb Re (e.g. Morris and Short, 1969). Comparable concentrations of rhenium occur in most sedimentary rocks, except in reduced, organic-rich rocks, which can have rhenium contents several orders of magnitude greater than the crustal average (e.g. Lipinski et al., 2003).

Table 14.2 Rhenium-bearing minerals. Most rhenium is produced from molybdenite.

Name	Formula	Rhenium content
Rheniite	ReS_2	74%
Tarkianite	$(Cu,Fe)(Re,Mo)_4S_8$	49 to 56%
Dzhezkazganite	$ReMoCu_2PbS_6$	22%
Molybdenite	MoS_2	<10 ppm to 11.5%
Castaingite	$CuMo_2S_5$	up to 1%
Uraninite	UO_2	up to 2700 ppm
Gadolinite	$Y_2Fe^{2+}Be_2Si_2O_{10}$	up to 1 ppm

Mineralogy

Because rhenium is such an extremely rare element, rhenium minerals are also rare and relatively few minerals concentrate rhenium to a significant degree (Table 14.2). It should be stressed that minerals rich in rhenium are generally found in very small quantities and are not commercially viable sources of rhenium at the present time. For example, rhenium occurs at elevated levels in some minerals deposited around high-temperature volcanic fumaroles and, in at least two locations, it is present as the sulfide mineral rheniite. One of these locations is Usu volcano, Hokkaido, Japan (Bernard and Dumortier, 1986); the other is Kudriavy volcano on Iturup Island in the Kurile volcanic arc, western Pacific ocean (Korzhinsky et al., 1994) (Figure 14.1). At the Kudriavy volcano, rheniite is being actively deposited as a condensate from volcanic gases at a rate of up to 20 tonnes rhenium per year (Naumov, 2007). However, any commercial recovery of this rhenium in the near future is unlikely due to the remoteness of the area and the extreme

Figure 14.1 Crystals of rheniite, ReS_2, about 1 mm in size, from a fumarole on Kudriavy volcano, Kurile Islands, western Pacific. (Photo courtesy of R. Lavinsky, iRocks.com.)

conditions under which the rheniite is being deposited (up to 900 °C).

Rheniite has also been identified in other types of occurrences, including the Pagoni Rachi porphyry-type molybdenum–copper–tellurium–silver–gold prospect in northern Greece (Voudouris et al., 2009) and the Phoenix nickel–copper–platinum-group element (PGE) deposit in Botswana (Maier et al., 2008). Other rhenium sulfide minerals include dzhezkazganite, recognised in the copper ores of the Dzhezkazgan deposit, Kazakhstan (Genkin et al., 1994) and tarkianite, identified in the Hitura nickel–copper–PGE deposit, Finland and other nickel–copper–PGE deposits (Kojonen et al., 2004).

Although rhenium has an affinity for sulfide phases, its concentration in most sulfide minerals is relatively low. However, the ionic radius of Re^{4+} is very close to that of Mo^{4+}, which allows for a limited substitution of rhenium for molybdenum in molybdenite and other molybdenum minerals such as castaingite. Furthermore, rhenium is similar geochemically to molybdenum, which it commonly accompanies through magmatic and related hydrothermal processes, and is commonly concentrated in molybdenite associated with various types of granite-related deposits, particularly porphyry deposits. The rhenium content of most molybdenites is generally within a range of a few ppm to several thousand ppm Re (Table 14.3), although contents as high as 4.7 wt% Re occur in molybdenite at the Pagoni Rachi prospect (Voudouris et al., 2009) and up to 11.5 wt% Re has been recorded in molybdenite deposited in high-temperature fumaroles (Bernard et al., 1990). However, molybdenites, from which rhenium is currently recovered, commercially contain in the order of 200 to 1000 ppm Re. Other minerals in which rhenium may be concentrated in significant levels (i.e. >1 ppm Re) include uraninite, which can contain up to 2700 ppm Re (Min et al., 2005), and gadolinite (Noddack and Noddack, 1931).

Deposit types

Porphyry deposits

Porphyry deposits are large, low-grade deposits in which ore minerals occur in extensive zones of fracturing and brecciation associated with porphyritic granitic intrusions. They are the world's most important source of copper and molybdenum, and are major sources of gold and silver. They also account for roughly 85 to 90 per cent of primary rhenium production, which is recovered as a

Table 14.3 Rhenium content of molybdenite in various deposit types worldwide.

Country	Deposit	Re (ppm) Average	Range	References
Porphyry copper deposits				
Canada	Lornex-Valley	330	294–350	Sinclair et al. (in prep.)
Canada	Gibraltar	443	238–750	Sinclair et al. (2009)
Canada	Island Copper	1784	1704–1863	Sinclair et al. (2009)
Chile	Chuquicamata	220	194–250	Giles and Schilling (1972); Mathur et al. (2001)
Chile	Collahuasi	410	368–448	Mathur et al. (2001)
Chile	El Salvador	630		Giles and Schilling (1972)
Chile	El Teniente	390	182–1154	Mathur et al. (2001)
Chile	Escondida	1355		Mathur et al. (2001)
Kazakhstan	Kounrad	664	24–1930	Ivanov et al. (1969)
Mongolia	Erdenet	535		Gerel and Munkhtsengel (2005)
Peru	Cerro Verde	3280	3061–3497	Mathur et al. (2001)
United States	Bagdad	460	330–642	Barra et al. (2003)
Porphyry copper-gold deposits				
Armenia	Kadzharan	340	80–1610	Ivanov et al. (1972)
Armenia	Agarak	820	57–6310	Magakian et al. (1984)
Canada	Snowfield	3579		Pretium (2011)
Iran	Sar Cheshmeh	1000	900–1160	Shariat and Hassani (1998)
Mongolia	Oyu Tolgoi	1500		H. Stein (pers. comm., 2007)
United States	Bingham	360	120–2000	Giles and Schilling (1972)
United States	Pebble	1130	900–2260	Rebagliati et al. (2009)
Uzbekistan	Kal'makyr-Dalnee	1200	150–2100	Ivanov et al. (1969); Golovanov et al. (2005)
Porphyry copper-molybdenum deposits				
Canada	Schaft Creek	590		Bender et al. (2007)
Peru	Toquepala	790	387–1496	Giles and Schilling (1972); Mathur et al. (2001)
United States	Sierrita-Esperanza	600	90–1800	Giles and Schilling (1972)
Porphyry molybdenum deposits				
Canada	Endako	35	15–67	Sinclair et al. (2009)
United States	Climax	35	11–80	Giles and Schilling (1972)
United States	Henderson	14	7–18	Giles and Schilling (1972)
United States	Thompson Creek	120		Sinclair et al. (in prep.)
Porphyry tungsten-molybdenum deposits				
Canada	Northern Dancer (Logtung)	22		Sinclair et al. (2009)
Canada	Sisson	9	6–12	Sinclair et al. (in prep.)
Kazakhstan	Verkhnee Qairaqty	57	47–66	Ivanov et al. (1972)
Vein molybdenum deposits				
Australia	Merlin	1062	300–2098	Horton (2010)
Canada	Playter	402	185–1047	Kilpatrick and Grieco (2010)
Japan	Daito	132	116–188	Ishihara (1988)
Norway	Knaben	14	1–28	Fleischer (1959); Giles and Schilling (1972)

by-product from the processing of molybdenite concentrates.

The average rhenium content of molybdenite from porphyry deposits varies widely, from less than 10 ppm to nearly 4000 ppm Re (Table 14.3), and is inversely related to the molybdenum content (Figure 14.2). Porphyry copper and copper–gold deposits have the lowest molybdenum grades, in the order of 0.02% Mo or less, but molybdenites from these deposits have the highest average rhenium contents, ranging as high as 3858 ppm Re at the Kemess South porphyry copper–gold deposit in British Columbia and 3280 ppm at the Cerro Verde porphyry copper deposit in Peru. Molybdenite from the now-closed Island Copper porphyry copper deposit on Vancouver Island, British Columbia, averaged nearly 1800 ppm Re. In comparison, average rhenium-in-molybdenite contents from porphyry copper–molybdenum deposits range from about 90 ppm Re to 790 ppm Re. For example, molybdenite from the Sierrita-Esperanza deposit in Arizona averages about 600 ppm Re and in the Collahuasi deposit in Chile it contains 410 ppm Re.

Porphyry molybdenum deposits have the highest molybdenum grades, ranging from 0.07% Mo (Endako, British Columbia) to as high as 0.24% Mo (Climax, Colorado), but have the lowest rhenium contents in molybdenites, typically in the range of 10 to 100 ppm Re. For example, molybdenite from both Endako and Climax contain about 35 ppm Re. Molybdenites from porphyry molybdenum deposits with rhenium contents in excess of 100 ppm Re, such as at Mount Tolman in Washington (182 ppm Re)

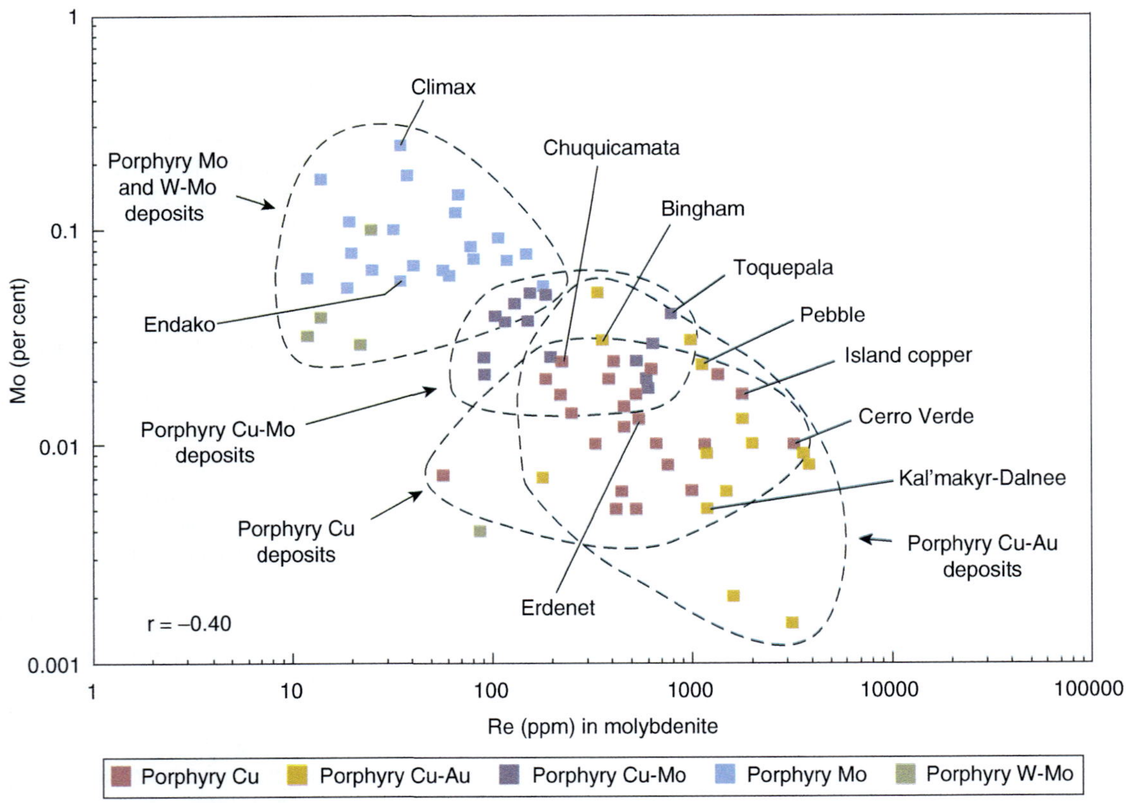

Figure 14.2 Rhenium content of molybdenite versus molybdenum grade in porphyry deposits. (Revised after Sinclair et al., 2009.) (Cu, copper; Mo, molybdenum; Au, gold; W, tungsten)

and at Thompson Creek in Idaho (120 ppm Re) are exceptional. Rhenium content of molybdenite from tungsten-dominant porphyry deposits is also relatively low.

Vein deposits

Vein deposits occur in various structural settings such as faults, fault systems and breccia zones and, in some cases, include replacement zones in associated host rocks. Like porphyry deposits, they are commonly associated with granitic intrusions. Unlike porphyry deposits, they are typically small, but are highly varied in size and metal contents. In the past, they have been an important source of molybdenum in some countries, including Canada, Norway and Japan, but up to the present time there has been no recorded production of rhenium from a vein deposit. However, this will change if the Merlin deposit in Australia can be brought into production.

In many vein deposits, molybdenite has rhenium content comparable to molybdenite from porphyry molybdenum deposits, typically less than 100 ppm Re. For example, the rhenium content of molybdenite from vein molybdenum deposits in the Preissac-Lacorne area in Canada averages 25 to 30 ppm Re and molybdenite from the Knaben deposit in Norway contains an average of 14 ppm Re. However, rhenium-in-molybdenite contents in vein deposits are highly varied, ranging as high as 188 ppm Re in Japanese vein molybdenum deposits. At the Playter deposit in Ontario, rhenium content of molybdenite averages about 400 ppm Re, although it ranges as high as 1047 ppm. Molybdenite from the Merlin deposit averages 1062 ppm Re.

The recently discovered Merlin deposit in Australia is an unusual vein-type deposit. The molybdenum mineralisation is broadly stratabound within metamorphosed Proterozoic sedimentary rocks, but the molybdenite mineralisation post-dates the metamorphism. The host rocks are brecciated and molybdenite occurs as fine-grained, semi-massive to massive concentrations filling open spaces and replacing host-rock fragments, and as disseminated grains in the metasedimentary host rocks. Also, whereas most vein molybdenum deposits, such as Knaben and Playter, consist of molybdenite disseminated in quartz veins, the Merlin deposit is distinguished by the absence of quartz. Also, in contrast to molybdenum deposits in quartz veins, which typically contain an average of 0.1 to 0.3% Mo, the grade of the identified mineral resource at the Merlin deposit averages 1.3% Mo (Horton, 2010). Such a high content of molybdenum, combined with the high rhenium content of the molybdenite, makes Merlin an exceptionally rich molybdenum–rhenium deposit. This, along with the geology, particularly the absence of quartz, makes it unique among vein-type deposits.

Sediment-hosted copper deposits

Sediment-hosted copper deposits consist of disseminated to veinlet copper sulfides that occur in zones more or less concordant with the stratification of their sedimentary host rocks such as the Kupferschiefer deposits at Lubin in Poland and the redbed-type deposits at Dzhezkazgan in Kazakhstan. They account for nearly one-quarter of the world's production and reserves of copper and are important sources of silver and cobalt. They also account for 10 to 15 per cent of world rhenium production.

The nature of the occurrence of rhenium in sediment-hosted copper deposits is not well understood. Rhenium is recovered from the processing of copper concentrates and some rhenium may be present in the main copper minerals, such as chalcopyrite, bornite and chalcocite. However, rhenium is also present in other minerals. For example, at Dzhezkazgan rhenium is present in the sulfide mineral dzhezkazganite, which is closely associated with the copper ores (Genkin et al., 1994) and, in the Lubin deposit, rhenium-bearing minerals include castaingite and molybdenite (Kucha, 1990).

The rhenium content of the ore at Dzhezkazgan averages 1.4 ppm Re (Hitzman et al., 2005). In the Lubin deposit it averages approximately 1 ppm Re (H. Kucha, personal communication, 2007).

Uranium deposits

For a short time, beginning in about 1969, significant amounts of rhenium were recovered as a by-product from the mining of uranium deposits in sedimentary rocks in the Falls City area south of San Antonio in Texas (Millensifer, 1997). At the Palangana deposit, also in southern Texas, recovery of rhenium was investigated (Goddard, 1984), although no production was reported. Currently, some rhenium is recovered from in situ solution mining of sediment-hosted uranium deposits in the Central Kyzylkum region of Uzbekistan (Chekmarev et al., 2004), although this probably accounts for no more than about one per cent of current world production.

The nature of the occurrence of rhenium in uranium deposits is not well understood. At Palangana and other sediment-hosted uranium deposits, rhenium is closely associated with molybdenum and is likely concentrated in jordisite, an amorphous molybdenum mineral with the same composition as molybdenite. However, at the Sun Valley uranium mine in Arizona, rhenium is also associated in part with uranium (Petersen et al., 1959) and some rhenium may also be present in uranium minerals such as uraninite. Rhenium can also be sequestered in carbonaceous material in sediment-hosted deposits.

The rhenium content of uranium deposits varies from <1 to 5 ppm Re. In the uranium deposits in the Central Kyzylkum, the rhenium content is generally in the range of 0.5 to 2 ppm Re (Venatovskij, 1993). The Anna Lake uranium deposit in Labrador contains 0.18 ppm Re at a cut-off grade of 0.015% U_3O_8 (Fraser and Giroux, 2009). The rhenium grade at the Palangana deposit is approximately 5 ppm Re, while selected samples from the Sun Valley mine contained as much as 1000 ppm Re.

Magmatic nickel–copper–platinum-group element (PGE) deposits

Magmatic nickel–copper–PGE deposits represent a potentially significant source of rhenium. According to Besser et al. (1997), as much as 400 kg of rhenium is concentrated annually in the flue dusts resulting from the smelter operations associated with the Noril'sk deposits in Russia, although apparently none of this rhenium is recovered at the present time.

Rhenium has many properties in common with PGE, in particular a strong affinity for metallic or sulfide phases, and is present at elevated levels in many nickel–copper–PGE deposits. Data on grades of rhenium in nickel-copper-PGE are generally unavailable. However, rhenium contents of typical ore samples analysed for Re–Os dating ranged from 0.01 to 0.52 ppm Re (average 0.12 ppm Re) for the Noril'sk-Talnakh deposits in Russia (data in Walker et al., 1994a) and from 0.07 to 0.31 ppm Re (average 0.19 ppm Re) for Sudbury, Ontario deposits (data in Walker et al., 1994b). In some deposits, the rhenium occurs as tarkianite and unnamed rhenium-bearing sulfide minerals (Dare et al., 2010); in others it is concentrated mainly in base-metal sulfides such as pyrrhotite, pentlandite and pyrite (Dare et al., 2011).

World resources and production

The location of significant rhenium-bearing deposits in the world is shown in Figure 14.3.

World mine production of rhenium from the main known sources in 2012 has been estimated at about 44.8 tonnes (MMTA, 2012a), most of which was produced as a by-product of molybdenum production associated with porphyry copper and copper–gold deposits. The largest portion (28 tonnes Re) was produced by Molymet in Chile, primarily from large porphyry deposits such as El Teniente, Escondida, Los Pelambres and Collahuasi and, to a lesser extent, Chuquicamata (Figure 14.5). However, Molymet also sourced molybdenite concentrates from deposits in other countries, including Mexico, Peru and the United States. The United States was the second largest producer (7 tonnes Re), with production primarily from Sierrita-Esperanza and Bagdad, although the facility at Sierrita also treated other molybdenite concentrates on a toll basis. Other porphyry deposits from which rhenium was recovered include Kadzharan and Agarak in

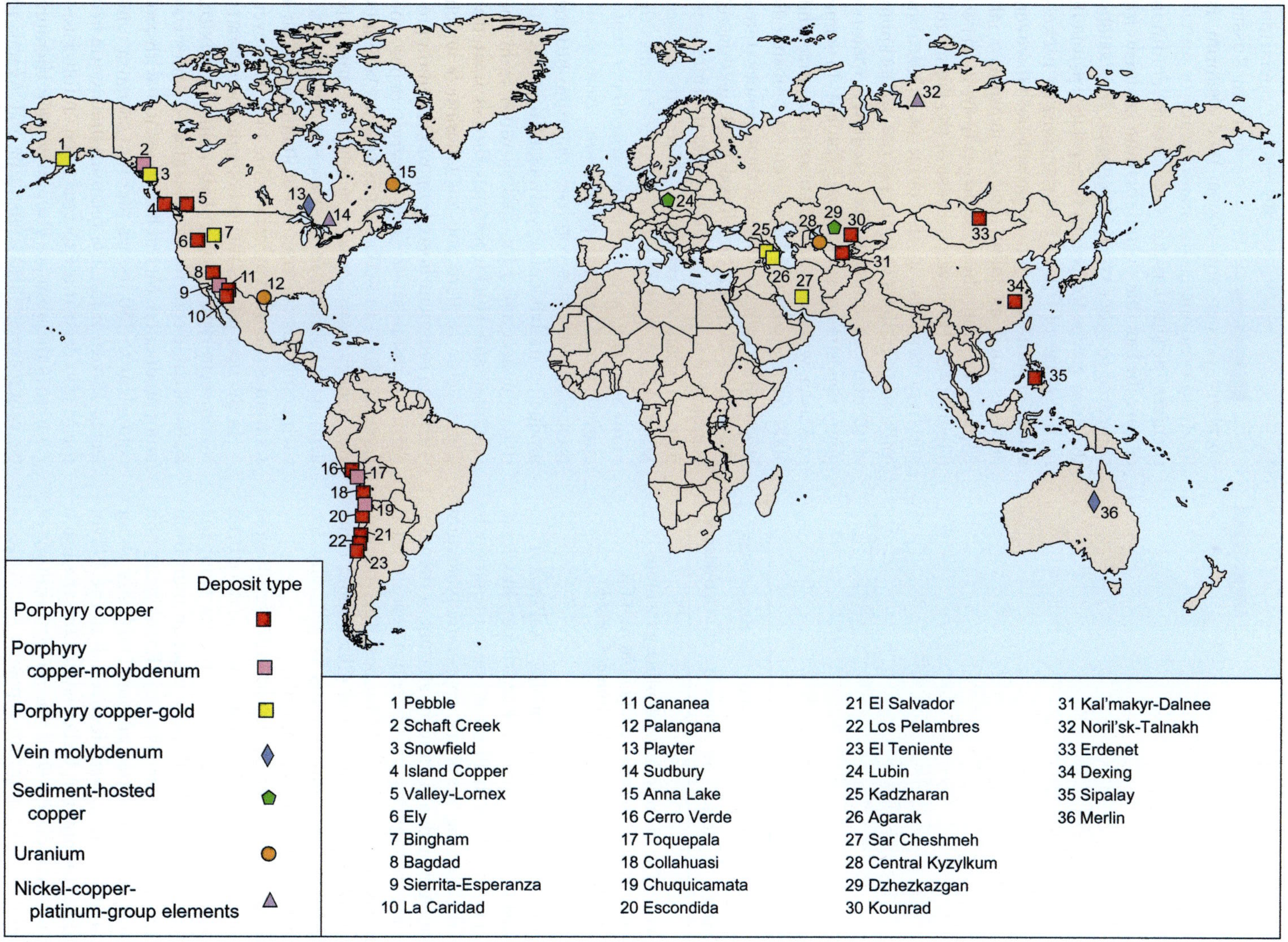

Figure 14.3 The global distribution of significant rhenium-bearing deposits.

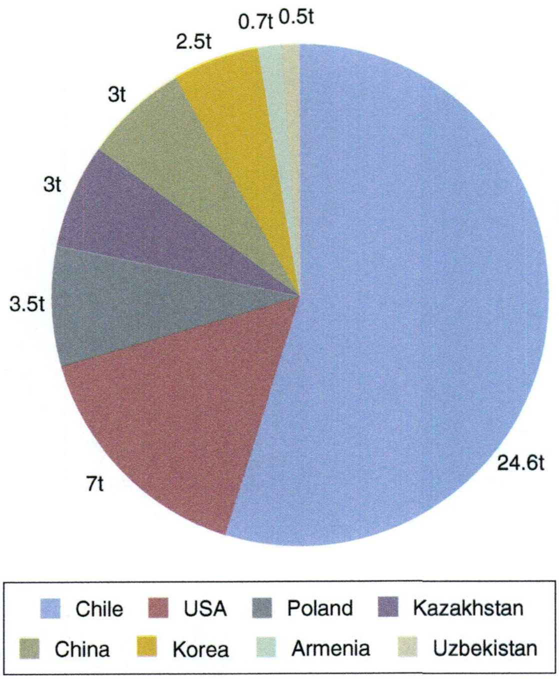

Figure 14.4 The main sources of primary rhenium production in 2012 (tonnes). (Data from MMTA, 2012a, courtesy of Lipmann Walton & Co Ltd.)

Armenia, Erdenet in Mongolia, Kal'makyr in Uzbekistan and Kounrad in Kazakhstan. In China, Jiangxi Copper lists rhenium as one of the products from the Dexing porphyry copper–molybdenum deposit. Although a significant amount of rhenium is present in porphyry molybdenum deposits such as Climax in Colorado (127 tonnes), the rhenium content of the molybdenite at Climax and other porphyry molybdenum deposits is generally too low to be recovered economically.

Rhenium production from sediment-hosted copper deposits in 2011 is estimated to have been six to seven tonnes, about 3.5 tonnes from the Lubin deposit in Poland and a lesser amount from the Dzhezkazgan deposit in Kazakhstan. In addition, an unspecified amount, probably about 0.5 tonnes or less, was recovered at the Navoi metallurgical plant in Uzbekistan from the leach solutions associated with the in situ mining of sandstone-hosted uranium deposits.

Future supplies

Porphyry copper and copper–gold deposits are likely to continue as the main source of primary rhenium production for the foreseeable future, despite the fact that overall rhenium grades are low compared to other types of deposits. For example, rhenium grades for porphyry deposits fall mainly in the range of 0.1 to 0.5 ppm Re, compared to about 1 ppm Re for sediment-hosted copper deposits such as Lubin and Dzhezkazgan, and 23 ppm Re for the Merlin vein-type deposit (Figure 14.5). However, because of the large size of porphyry deposits, in the range of hundreds of millions to billions of tonnes of ore, the resources of rhenium in molybdenite associated with porphyry deposits are substantial. The Pebble deposit in Alaska, with an estimated resource of more than 4000 tonnes contained rhenium, is probably the largest currently unexploited rhenium resource in the world. When it is developed, the Pebble deposit is projected to produce 1200 tonnes rhenium in the first 45 years of operation (Northern Dynasty Minerals Ltd, 2011), equivalent to 26.7 tonnes rhenium annually, which is more than 50 per cent of current world annual production. Additionally, there are numerous porphyry deposits currently producing molybdenite with significant rhenium that is unaccounted for, including Valley-Lornex, Gibraltar and Huckleberry in British Columbia, Cerro Verde in Peru and Sar Cheshmeh in Iran. Although sediment-hosted copper deposits contain rhenium resources as large as those in porphyry deposits (3000 to 4000 tonnes each in the Lubin and Dzhezkazgan deposits), production from these deposits is unlikely to increase significantly from their combined current production of about five to six tonnes rhenium per year.

In the near future, rhenium production from the Merlin vein-type deposit, if successfully commercialised, could contribute substantially to world rhenium production. Merlin has an identified resource equivalent to 150 tonnes rhenium and over a projected 15-year mine life, rhenium production from the Merlin deposit is expected to be on the order of 7.5 tonnes rhenium per year, allowing for about 75 per cent recovery (SRK Consulting, 2010). However,

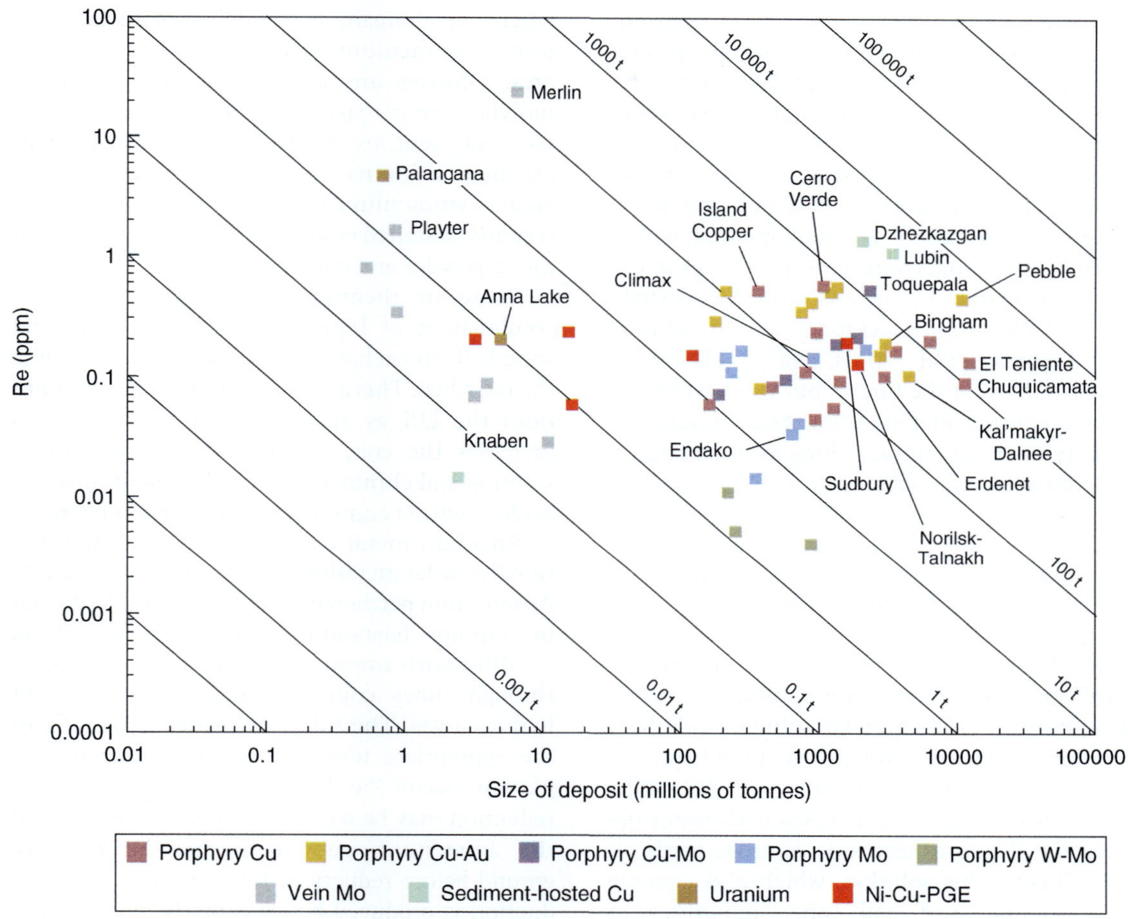

Figure 14.5 Rhenium grade versus tonnage of significant rhenium-bearing deposits; diagonal lines represent tonnes (t) of contained rhenium. (Revised after Sinclair et al., 2009.) (Cu, copper; Mo, molybdenum; Au, gold; W, tungsten; Ni, nickel; PGE, platinum-group elements)

Merlin is an unusual molybdenum and rhenium-rich vein-type deposit. Rhenium resources in other vein deposits, such as the Playter deposit in Ontario (about 1.5 tonnes rhenium), are relatively small and much lower in grade, and are unlikely to contribute significantly to world rhenium production except, possibly, on a relatively short-term basis.

Although rhenium production from uranium mining is limited at the present time, rhenium is present in many sediment-hosted uranium deposits, although probably unrecognised (not analysed) in many cases. It typically occurs in jordisite, the amorphous variety of molybdenite, which is generally soluble in the leach solutions used for in situ mining. This rhenium is readily recoverable using existing technology and represents a potentially important source of rhenium in the future, even though the rhenium content of individual deposits is relatively small (Figure 14.5).

Rhenium resources in other types of deposits, such as nickel–copper–PGE deposits, is significant, although no rhenium production from them has been reported. For example, the Sudbury nickel–copper–PGE deposits originally contained in the order of 300 tonnes rhenium, those in the Noril'sk-Talnakh district are estimated to have contained about 200 tonnes rhenium (Sinclair et al., in preparation). However, the technology to recover the rhenium in these deposits, from the flue dust associated with the smelting and refining of the PGE-rich ores, exists and presumably could be implemented if warranted by economic conditions. Resources in carbonaceous materials, such as coal and oil shale, may be substantial (Chekmarev and Troshkina, 1997; Lippmaa et al., 2011), and possibly represent resources for the future, but recovery of rhenium (and other metals) from these materials is unlikely to be economic as long as other sources of rhenium are available.

Extraction methods, processing and beneficiation

When rhenium-bearing molybdenite is roasted, converting the molybdenum sulfide to molybdenum trioxide (MoO_3) and to sulfur dioxide (SO_2), rhenium is oxidised to volatile rhenium heptoxide (Re_2O_7) which exits the roaster with the sulfur dioxide. Scrubbing the exit gases with water dissolves the rhenium heptoxide as crude perrhenic acid, $HReO_4$. This solution, which also contains some sulfuric acid and other impurities, is treated to prepare for rhenium recovery by either solvent extraction (liquid ion exchange) or solid-bed ion exchange (Figure 14.6). From either method, rhenium is stripped and crystallised as ammonium perrhenate, NH_4ReO_4. Generally, repeated recrystallisation is required to achieve the required ammonium perrhenate purity of 99.95 per cent, metal basis.

New methods of production include the new Kennecott MAP (Molybdenite Autoclave Process) facility which involves high-pressure oxidation of molybdenite to achieve better recovery and more pure products of molybdic oxide and ammonium perrhenate. This has been underway since 2010 according to numerous reports and press releases. Since 2007, rhenium has also been produced in Poland by KGHM Ecoren from copper operations where the company has installed plant facilities to recover rhenium from the smelting of copper ores. Offgases are scrubbed as is done where molybdenite is roasted. The sulfuric acid liquids are then sent to filtration and solid-bed ion-exchange columns where the rhenium is recovered as ammonium perrhenate. KGHM has also recently installed equipment to produce rhenium metal powder and pellets.

There are rhenium recovery plants where a combination of liquid ion exchange (LIX), and solid-bed ion exchange is used with LIX preceding the resin bed. This allows stripping of the rhenium from the LIX as an ammonium solution. This increases the concentration of the rhenium in solution and eliminates the need to neutralise the acidic solution coming from the wet scrubbers.

Rhenium metal powder is produced by traditional powder metallurgy techniques (Figure 14.7). Ammonium perrhenate is reduced using hydrogen in common boats-in-tubes type furnaces. Boats are filled with ammonium perrhenate and pushed through tubes counter-currently to a flow of hydrogen gas. The tubes are heated externally to the appropriate temperature. Depending on the particle size of the rhenium powder required, the reduction may be a one- or two-stage process and the ammonium perrhenate may be specially ground before reduction. Rhenium for alloy production is produced by pressing the powder into pellets of varying sizes, some as large as 15–20 mm in diameter and 8–10 mm thick, and others 5–8 mm in diameter and 3–5 mm in thickness. The pellets then are sintered to improve physical integrity as well as to reduce gases further, especially oxygen. Metal products such as wire and plate are made by pressing rhenium powder into bars or rods followed by resistance heating to sinter. The sintered rods or bars then are drawn into wire or rolled into sheets or plates. Rhenium also can be deposited on various types of parts and moulds using chemical vapour deposition. The production of catalysts may require the use of perrhenic acid, $HReO_4$. This is a solution of rhenium heptoxide (Re_2O_7) in water with a rhenium concentration of 35–50 weight per cent rhenium contained.

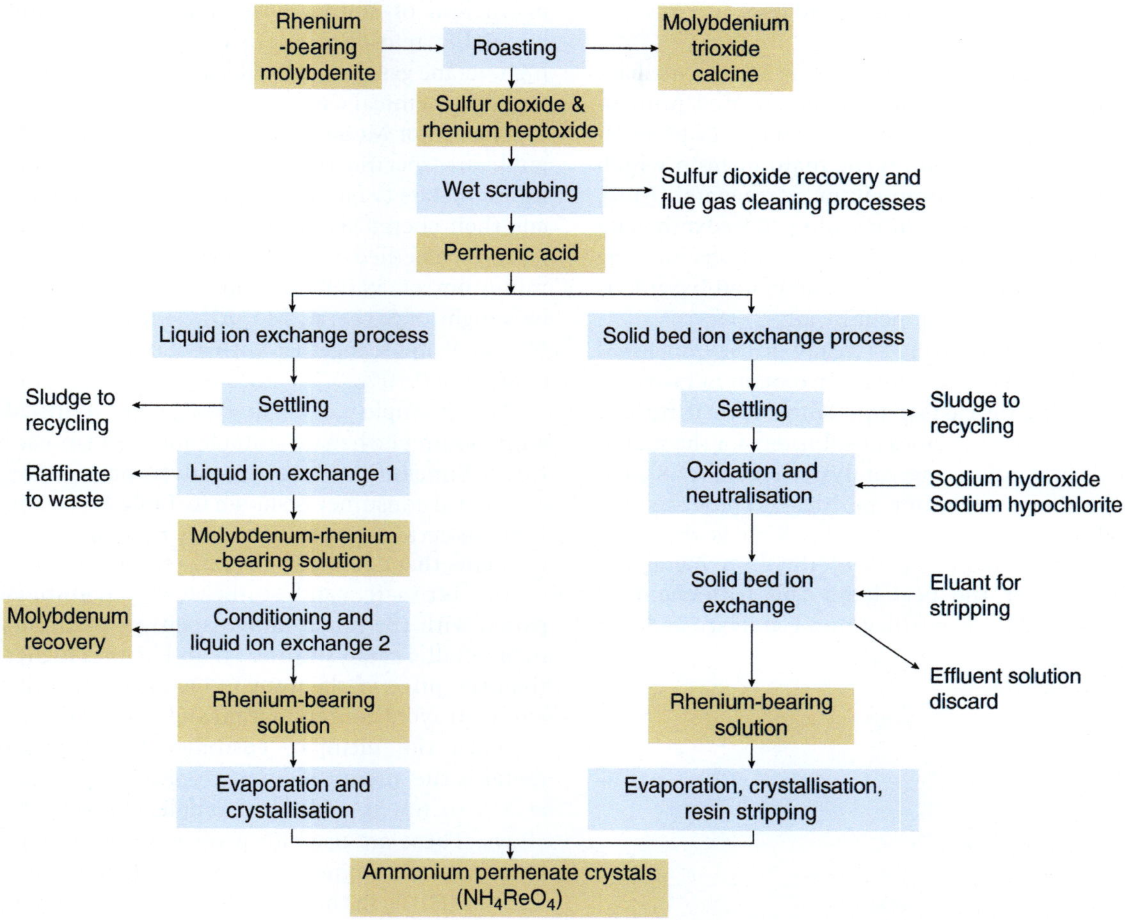

Figure 14.6 Typical rhenium recovery process by liquid ion exchange and solid-bed ion exchange.

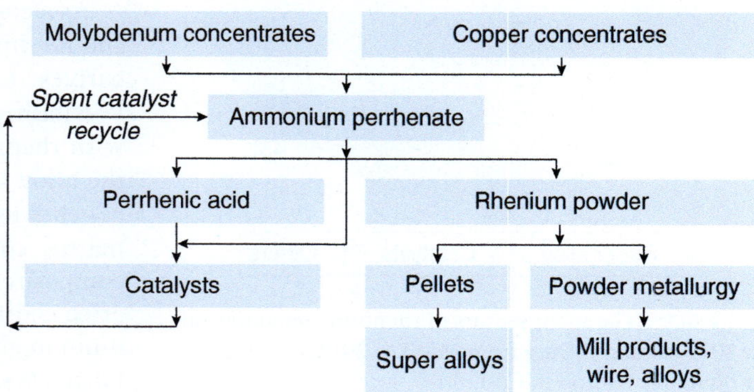

Figure 14.7 Summary of the production of rhenium and products.

Specifications and uses

Minor metals are often traded as an intermediate product, which is then further treated prior to incorporation into a final component or product. In the case of rhenium the main form in which it is traded is ammonium perrhenate (APR), sometimes called ammonium tetraoxorhenate (NH_4ReO_4). In this form, a salt of free-flowing white crystals, rhenium is stable, non-hazardous and suitable for shipment by air.

Rhenium in APR is ready to be processed further for use in one of its two main industries – superalloy for casting into single-crystal turbine blades for aero-engines or solutions for the manufacture of reforming catalysts (Figure 14.8). For superalloys, rhenium powder is produced by hydrogen reduction of APR which is then sintered into small pellets of rhenium metal (Re 99.9%). For the petroleum and petrochemical industry, APR is purified into Catalyst Grade for

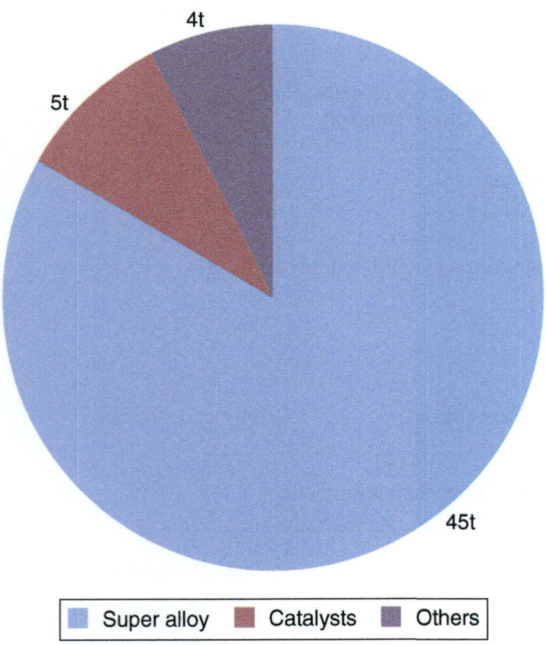

Figure 14.8 The main sectors of rhenium consumption in 2012 (tonnes). (Data from MMTA, 2012a, courtesy of Lipmann Walton & Co Ltd.)

production of alumina base platinum–rhenium bimetallic reforming catalysts used to produce high-octane gasoline, and added to other catalysts for petrochemical catalysis.

The Minor Metals Trade Association (MMTA) publishes specifications for numerous rare and minor metals to facilitate trading in those metals and their chemicals (MMTA, 2012b). Basic-grade APR is specified by the MMTA to contain a minimum of 69.00% Re and less than 0.05% by weight of Si, Fe, K, Mo, Al, Ca, Cu, Mg, Mn, Ni, Na, P and S, together with a maximum water content of 0.10%.

An example of 'catalyst'-grade purified ammonium perrhenate suitable for the manufacture of bimetallic reforming catalysts provided by a principal consumer is shown in Table 14.4a.The limit of certain impurities is set to avoid those elements that could be poisonous to the function of the ultimate catalyst. Because rhenium is paired with the much more expensive platinum in bimetallic catalysts (more than 13 times higher than the price of rhenium in October 2012) the aim is to avoid contamination and costly failure.

When the ultimate customer for rhenium metal is the aircraft industry, typically, rhenium at 3% to 6% is added to complex nickel-base alloys. These are used for precision casting into single-crystal turbine blades for the High-Pressure Turbine (HPT). In this use, the control of contaminants is also critical and a typical specification is given in Table 14.4b.

An example of the typical composition of a rhenium-bearing superalloy is Cannon Muskegon's CMSX-4 (Table 14.5).

As a consequence of the high price of rhenium and its limited production and availability, alternatives to it are continually being sought. Possibilities include coating blades and end tips with rhenium (to avoid the need for rhenium in the alloy itself), fibre-reinforced superalloy components using rhenium–tungsten fibres, ceramic matrix composites (CMCs) and metal matrix composites (MMCs). It has been suggested that a price ceiling for rhenium, at which users of rhenium might cease to use rhenium, could be in the range US$5000 to 6000/kg. However, even when rhenium did reach US$12,000/kg in August 2008, usage was not reduced. It should be noted, however, that GE Aviation has been working for 10 years to lessen its dependence on rare metals including rhenium. This is being accomplished through a combination of innovative component designs, advanced manufacturing processes and the development of new alloys, as well as recycling (General Electric, 2012.)

Rhenium has proven successful in aircraft gas turbines, accordingly manufacturers of land-based gas turbines used for primary, as well as emergency, power generation, are looking closely at the use of rhenium in turbine blades for those

Table 14.4 (a) Specification for catalyst grade ammonium perrhenate. (Data courtesy of MMTA, 2012b.)

Element	Maximum permitted content (ppm, unless otherwise stated)
Re	68.5–69.8% minimum
Sb	30
As	30
Ca	50
Cl_2	100
Cr	50
Co	50
Cu	50
Ir	75
Fe	50
Pb	30
Mg	50
Mn	50
Mo	100
Ni	50
Pd	50
P	200
K	200
Rh	50
Si	50
Ag	50
Na	50
Sn	50
Zn	50
As, Ca, Fe, K, Mg, Na, Pb, Sn	200 total
As, Co, Cr, Cu, Mn, Ni, Sb, Si, Sn	200 total
H_2O	0.10%

Table 14.4 (b) Specification for rhenium metal pellets, superalloy grade. (Data courtesy of MMTA, 2012b.)

Element	Maximum permitted content (ppm, unless otherwise stated)
Re	99.9% minimum
Se	5
Te	5
Tl	5
H	50
Mg	100
Ca	5
Na	10
Ag	5
As	10
Bi	1
Cd	50
Ga	50
In	50
Pb	5
Sb	50
Th	20
Sn	10
Zn	50
Cl	50
K	10
C	30
Si	20
S	20
O	300
N	50
Fe	100
Mo	500
W	500
Mn	20
P	200
Al	30
B	20
Co	100
Cr	100
Cu	70
Hf	200
Nb	200
Ni	100
Ta	200
Ti	200
V	200
Zr	200

Table 14.5 Composition of Cannon Muskegon's CMSX-4 superalloy. (Data from C-M Group, 2012.)

Element	Content (wt per cent)
Cr	6.5
Co	9.0
Mo	0.6
W	6.0
Ta	6.5
Re	3.0
Al	5.6
Ti	1.0
Hf	0.1
Ni	61.7

applications. There have been reports that rhenium is already in use in those turbine blades. However, a significant problem with these blades is their size and the quantity of rhenium that would be consumed. The present limited supply of rhenium could limit this use.

Other recent developments include new catalysts in petrochemical production. One has been the addition of rhenium to the silver catalyst in ethylene oxide production or even replacing the silver in some cases. Several companies have developed new catalysts which contain rhenium for the production of intermediates and precursors for several types of elastomeric materials for a variety of consumer products.

Rhenium has also been investigated for use in catalysts for the gas to liquids, GTL, process. The term gas to liquids is applied to various processes which have been developed to convert natural gas to liquid fuels (National Petroleum Council, 2007). These are all based on the Fischer–Tropsch process developed by Fischer and Tropsch in Germany during the 1920s to convert low-grade coal to liquid fuels. These processes all make use of catalysts consisting of iron, cobalt or nickel, although it was determined recently that rhenium added to the catalyst improved the efficiency. Another potential application of rhenium is that of rhenium as rhenium boride, ReB_2, as an abrasive tool for cutting or polishing. Although this use was discovered many years ago it has apparently not developed due to the price and limited availability of rhenium.

Based on published compositions of some alloys and recent rhenium prices, one can see the price sensitivity of those superalloys to the price of rhenium (Table 14.6). When the rhenium price exceeds about US$2200 per kg the rhenium content of these alloys accounts for at least two thirds of the price of the alloy.

Recycling and re-use

Rhenium provides a fascinating study in regard to re-use and recycling. Here, perhaps unsurprisingly, the determining factor has been value rather than concern about the environment, although a positive environmental outcome has resulted. Rhenium's twin markets of catalysts and superalloys provide two different case histories.

Catalysts

In the case of catalysts it was, and still is, the value of platinum with which rhenium is allied at the rate of Pt 0.3% and Re 0.3% which provides the incentive for recycling. A recycling industry developed originally to recover the valuable platinum. Recovery of rhenium developed later as rhenium began to be used in the reforming catalysts. Those that recover platinum and rhenium from spent reforming catalysts are Heraeus Precious Metals GmbH & Co. KG, who recycle Pt–Re catalysts at their plants in Germany and the USA, and Gemini in the USA. It is thought that about 15 tonnes of rhenium is recovered per year by this method, which remains in a continuous closed loop within the catalyst industry sector (and therefore does not appear in supply/demand figures). In Europe, the recovery of rhenium did not begin as soon as it did in the United States. In Europe it was regarded as a nuisance during platinum recovery while in the US rhenium was recovered along with the platinum by the spent catalyst processor and stockpiled for later sale as long as the owner of the spent catalyst did not want the rhenium returned or held for them. There was a time when the catalyst user could purchase virgin rhenium

Table 14.6 The influence of the price of rhenium on the cost of superalloys as a percentage of the total cost of input raw materials. (The source of the prices used in these calculations is Metal Prices.com on 28 September 2011. The rhenium contents of the superalloys are from various published sources.)

			Rhenium price (US$/kg)					
			$1215		$2205		$5952	
Alloy	Rhenium content (%)	Alloy cost (US$/kg)	Alloy cost (US$/kg)	% alloy cost due to Re content	Alloy cost (US$/kg)	% alloy cost due to Re content	Alloy cost (US$/kg)	% alloy cost due to Re content
CMSX-3	0	5.15						
CMSX-4	3		66.95	55	95.99	69	208.42	86
CMSX-10	6		103.24	72	162.63	81	387.51	92
Rene N5	3		66.95	54	96.65	68	209.09	85

at a lower price than for the cost of recovery from the spent catalyst. That has not been the case for quite some time.

Superalloys

In the case of complex superalloys, with typical compositions as mentioned in Table 14.5, the growth of a recycling industry has been slow and spasmodic. The main reason for this was that during the first two decades of rhenium's commercialisation in single-crystal turbine blades from the 1980s, the rhenium price was relatively low – as low as US$ 300/kg 1994 and rarely higher than US$1500/kg between 1980 and 2000. In other words, prices were mostly below the cost of recovery. During this period rhenium was regularly wasted when the alloy blades, containing 60 per cent nickel, were sold for their nickel value only and disposed of into the stainless steel industry as a nickel addition.

The incentive for greater efficiency in recycling rhenium-bearing alloys came when rhenium reached US$12,000/kg in August 2008. At a stroke, recovery became economically viable and two approaches were employed by recyclers: a) chemical decomposition of the complex alloy followed by an ion-exchange process to recover rhenium in the form of APR, and b) a unique pyrophoric process where rhenium is evaporated as the heptoxide and subsequently captured and then precipitated as APR. (The processes and those using them are confidential and require Non-Disclosure Agreements for detailed information.)

Feed required by recyclers for the various processes includes used blades, casting scrap (runners and risers) as well as grindings generated either by cutting virgin alloy ingots or surface-finishing blades. At the time of writing, it is thought that as much as (but not more than) five tonnes of rhenium is recovered by this method per year, while another five tonnes of rhenium returns into new alloy via internal and external 'revert', i.e. clean alloy that may be re-melted without the need for recycling.

Substitution

Presently, rhenium has no substitute in its main use as a three per cent constituent in complex nickel-based alloys for single-crystal turbine blades. Here, the main purpose of rhenium is to increase the presence of the gamma prime (γ^1) phase which increases creep resistance. With turbine inlet temperatures estimated at 1600 °C or above in the current generation of large gas-turbine engines (e.g. Rolls-Royce Trent 1000, lead engine for the Boeing 787 Dreamliner), no other technology has yet been commercialised that has similar defect tolerance and at the same time meets the requirements of temperature ranges from sub-zero to 1600 °C with minimal deformation. Substitute materials that have been considered include ceramic matrix composites and metal matrix composites, which are being studied at leading institutes such as Imperial College London and at leading engine makers such as General Electric, Rolls-Royce and Pratt & Whitney. It is thought that these composites may one day be applied for static parts but are unlikely to be approved for rotating sections.

In the catalyst industry, mono-metallic catalysts, consisting solely of platinum as the precious metal, are regularly used but, in most applications, rhenium is preferred alongside platinum because it increases efficiency. The price of rhenium compared to platinum is so much cheaper that removal of rhenium has little cost benefit.

Environmental issues

Rhenium is radioactive in all forms because of the content of the isotope ^{187}Re. However, ^{187}Re decays by emission of very low energy beta particles (electrons) that pose minimal risk to human health. There are few if any concerns related to radioactivity during the mining of copper–molybdenum ore or in the processing of the concentrates as the concentration of rhenium in these processes is extremely low. Only after the rhenium is recovered could there be any potentially harmful exposure. However, even at this stage, huge quantities of rhenium, of the order of several tonnes per annum, would need to be ingested for an occupational worker to exceed recommended annual limits.

The quest for fuel efficiency affects many critical metals, none more than rhenium. In so far as rhenium contributes to high operating temperatures, increased fuel efficiency of gas turbines, reduction of nitrous oxide emissions to the upper atmosphere, and blade longevity, it may be said that rhenium is a green metal. With law-makers ever more concerned to reduce emissions, aero-engineering and material science tends towards the use of elements and techniques which will serve this purpose. With forecasts of 27,000 aircraft deliveries between 2009–2018, and a further 38,000 between 2019–2028 (Rolls Royce, 2010), airlines are competing with each other to commission the most efficient gas turbines. In order to save cost from expensive fuel and also comply with ever more stringent laws this trend looks likely to continue. As an example, the EU's latest law directed towards the airline industry seeks to set carbon limits per airline that, if exceeded, will be penalised via the enforced purchase of carbon credits. The development of more efficient engines, which is reflected in rhenium demand, will continue to drive this market. With rhenium's value now recognised, due to the attention it received when prices peaked in August 2008, it now means that a strong incentive exists to recover rhenium from end-of-life blades and casting scrap. The recovery of rhenium units from the aerospace industry will never be as efficient as the catalyst industry due to the wide dispersal of rhenium units in end-of-life blades, the cost of recovery and the length of time (sometimes up to 12 months) from the recycling of used blade to production of new rhenium pellet.

World trade

The international trade in rhenium is partly determined by the geological/geographical locations in which it arises and by the development of the industries that have commercialised its use. During the cold war in the 1950s, when rhenium was just beginning to be commercialised in catalysts, rhenium produced in Kazakhstan and Armenia was used to supply the USSR and no exports were made to the West. During the same period, the first commercial production was at Shattuck Chemical in Denver, Colorado and Kennecott Copper, in Garfield, Utah and enough was produced to supply the oil-refining industry in the western world.

The key change to the structure of the trade occurred in the late 1970s when, due to a large build-up of stock at that time, (as much as 12 tonnes in 1974), producers were discouraged out of the business and ceased production. The well-known trading company, Philipp Brothers, then bought the technology for the recovery of rhenium from flue dusts from Shattuck Chemical and transported the system to Santiago in Chile. Here, at the heart of the copper industry, where by-product rhenium-bearing molybdenite arose, it was logical for rhenium to be recovered. This was the origin of Carburo y Metalurgia (now known as Molybdenos y Metales, Molymet), today's leading producer of rhenium.

Today, as rhenium's rarity value, use and unsubstitutability has become more recognised, Molymet's dominance has emerged by virtue of this strategic position, controlling the main world supply of rhenium via the recovery of rhenium units from the flues at their large molybdenite roasters. Production of rhenium here in 2012 was 24.68 tonnes (more than 55 per cent of world primary supply, Figure 14.4). However, in keeping with the history of initial over-production in the 1970s, Molymet, in the 1980s, pursued a policy of long-term fixed price contracts in order to encourage the commercialisation of rhenium in the aerospace industry. During this period Molymet supplied more than 70 per cent of the world's rhenium and the free trade in rhenium was extremely limited.

All this changed in the early 1990s with a coincidence of two factors, a) the break-up of the Soviet Union, and b) the rapid development of rhenium's use in modern jet engines. For the first time, Soviet stockpiles of rhenium flowed to the west from Kazakhstan, which gave life-blood to the free market, allowing companies in Europe such as HC Starck GmbH and WC Heraeus GmbH (now called Heraeus Precious Metals GmbH & Co. KG) to manufacture rhenium metal pellets suitable for the superalloy industry using non-South American rhenium raw material.

At the time of writing (August 2011) there is much greater plurality in the trade of rhenium, led by more availability of information and reporting of prices within numerous publications. The need to price rhenium content within scrap at a discount to a published base price of APR, has also given life to these quotations.

Finally, with the maturing of this market, it is now thought that Molymet will in future construct sales contracts based on objective published prices rather than the previous fixed-price system. One of the motives for this change is pressure from mining companies to receive a credit for the rhenium content within their molybdenite. A number of trading parties and stockholders, who are knowledgeable about rhenium, are to be found amongst members of the Minor Metals Trade Association (MMTA).

Prices

At present, daily and weekly prices for rhenium (APR and metal pellets) are available by subscription to several publications, such as Metals Week, Metal-Pages and Metal Bulletin.

From the 1970s, at the inception of the wider trade in rhenium, prices ranged from US$3000/kg Re contained, descending to a low of about US$300/kg in the mid-1990s, then peaking at US$12,000/kg in August 2008 (Figure 14.9).

The prices in the 1970s reflected first-time buying of rhenium as raw material for bimetallic catalysts, prior to the recycling that made the industry efficient. The next peak came in the era following the OPEC oil crisis of the mid-1970s. This resulted in wider use of small cars and worldwide laws to remove lead from petrol. The proliferation in demand for high-octane petroleum increased the demand for bi-metallic reforming catalysts. The trough in prices in the early 1990s, when prices sank to near US$300/kg for almost five years, resulted from the disposal of redundant Soviet/Kazakhstan stocks onto the Western market. The slow recovery from the late 1990s reflected the emerging use of rhenium in modern gas-turbine engines, while peak prices in 2008, as high as US$12,000/kg, were attributable to the fact that demand from the aerospace industry exceeded Molymet's ability to supply. The subsequent decline mirrored the Lehman Brothers financial crisis.

With growing demand in aerospace as the main driver for rhenium consumption, it is expected

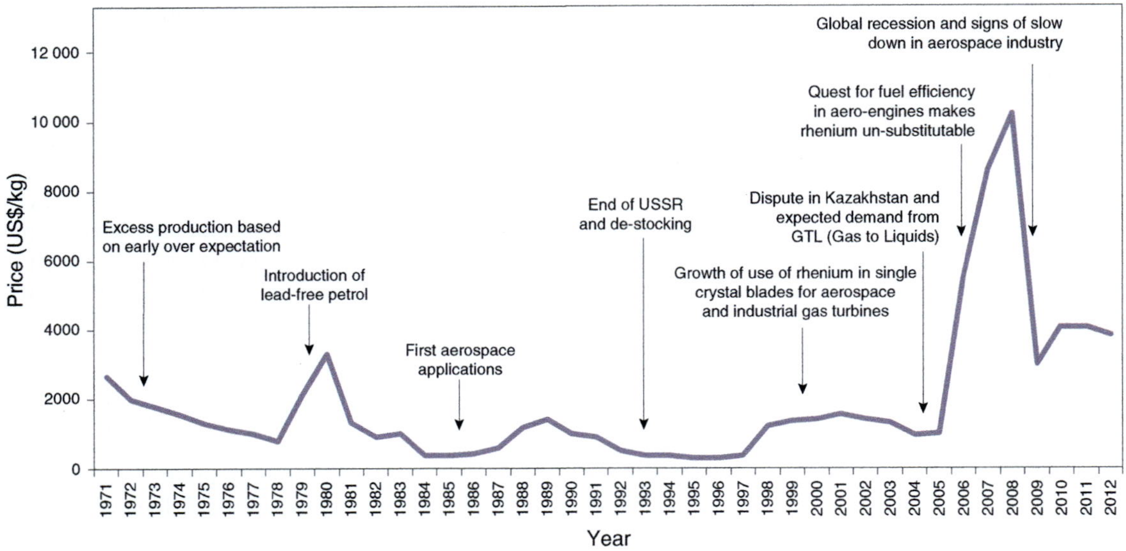

Figure 14.9 Rhenium price trend, 1971–2010. (Data from USGS up to 1993 and from Lipmann Walton & Co Ltd after 1993.)

that prices are unlikely to decline, as such an outcome would remove the incentive for recycling, thus cutting off much needed units to the market and sowing the seeds for further shortage.

Outlook

The outlook for the use of rhenium is mixed. In the 1970s it was appreciated that there were many potential uses for rhenium although their development and uptake were dependent on a reasonable price and adequate supply. Since then rhenium has been incorporated with silver in the ethylene oxide catalyst and also used in more than one proprietary petrochemical catalyst. Its application in the Fischer–Tropsch gas-to-liquids catalyst has also been investigated, but the scarcity of rhenium has curtailed that use.

There seems to be more rhenium coming available now with offers from such as Uzbekistan for 500 kg per month of APR, recovery from Mongolian ores being toll roasted, as well as the promise of more from deposits yet to be mined such as Merlin in Australia. There are doubts about how much rhenium might be recovered from this operation as it would be the first primary molybdenum ore body with significant levels of rhenium. Prior to the discovery of Merlin, primary deposits such as the Climax and Henderson mines in Colorado have contained only small amounts of rhenium, well under 100 ppm, which is insufficient to justify recovery efforts. It has been only the co- or by-product copper mines that had molybdenite concentrates with sufficient rhenium content, above 150 ppm, to justify rhenium recovery. The Pebble project in Alaska is an enormous deposit with much potential, but which faces very high levels of environmental opposition. It is likely that it will be many years before rhenium is produced from this project.

Meanwhile, the reduction in rhenium use by the aircraft engine producers has resulted in increased rhenium availability for some of the other applications such as land-based gas turbines for use in power generation. The use in X-ray tube target production seems to be one area where rhenium demand is on the increase, but there is little doubt that rhenium's use in nickel-based superalloys for the production of single-crystal turbine blades is set to remain rhenium's main market for many years to come.

References

Bernard, A. and Dumortier, P. (1986) Identification of natural rhenium sulfide (ReS_2) in volcanic fumaroles from the Usu volcano, Hokkaido, Japan. Proceedings of the XIth International Congress on Electron Microscopy (Kyoto), 1691–1692.

Bernard, A., Symonds, R.B. and Rose, W.I. (1990) Volatile transport and deposition of Mo, W and Re in high temperature magmatic fluids. Applied Geochemistry 5, 317–326.

Besser, A.D., Peredereev, A.V. and Tarasov, A.V. (1997) The perspectives of the development of the rhenium's production in Russia. In: Bryskin, D.D. (ed.) Rhenium and Rhenium Alloys. The Minerals, Metals and Materials Society, Warrendale, PA, USA, 59–65.

Chekmarev, A.M. and Troshkina, I.D. (1997) Rhenium recovery from non-traditional resource. In: Bryskin, D.D. (ed.) Rhenium and Rhenium Alloys. The Minerals, Metals and Materials Society, Warrendale, PA, USA, 99–107.

Chekmarev, A.M., Troshkina, I.D., Nesterov, Y.V., Maiboroda, A.B., Ushanova, O.N. and Smirnov, N.S. (2004) Associated rhenium extraction in complex processing of productive solutions of underground uranium leaching. Chemistry for Sustainable Development 12, 113–117.

C-M Group (2012) Vacuum melt alloy index, Nickel Base Single Crystal. http://www.c-mgroup.com/vacuum_melt_index/nickel_base_sx.htm

Dare, S.A.S., Barnes, S.-J., Prichard, H.M. and Fisher, P.C. (2010) The timing and formation of platinum-group minerals from the Creighton Ni-Cu-platinum-group element sulfide deposit, Sudbury, Canada: early crystallization of PGE-rich sulfarsenides. Economic Geology 105, 1071–1096.

Dare, S.A.S., Barnes, S.-J., Prichard, H.M. and Fisher, P.C. (2011) Chalcophile and platinum-group element (PGE) concentrations in the sulfide minerals from the McCreedy East deposit, Sudbury, Canada, and the origin of PGE in pyrite. Mineralium Deposita 46, 381–407.

Fraser, R.D. and Giroux, G.H. (2009) Form 43-101 Technical Report on the Anna Lake Uranium Project, Central Mineral Belt, Labrador, Canada; Prepared for Bayswater Uranium Corp., pp. 94.

General Electric (2012) Rhenium Reduction Program: using less of a rare material. http://citizenship.geblogs.com/rhenium-reduction-program-using-less-of-a-rare-mineral/#

Genkin, A.D., Poplavko, E.M., Gorshkov, A.I., Tsepin, A.I. and Sivtsov, A.V. (1994) New data on dzhezkazganite – rhenium-molybdenum-copper-lead sulfide – from the Dzhezkazgan deposit (Kazakhstan). Geology of Ore Deposits 36, 481–489.

Goddard, J.B. (1984) Recovery of rhenium from uranium in-situ leach liquor. Society of Mining Engineers of AIME Transactions 274, 1996–2000.

Habashi, F. (1996) Rhenium Seventy Years Old, In: Bryskin, B.D (ed.) Rhenium and Rhenium Alloys, TMS, Warrendale, PA, USA.

Hitzman, M., Kirkham, R., Broughton, D., Thorson, J. and Selley, D. (2005) The sediment-hosted stratiform copper ore system. In: Economic Geology 100th Anniversary Volume, 609–642.

Horton, J. (2010) Mount Dore-Merlin Deposit, NWQld, Australia. NI 43-101 Technical Report prepared for Ivanhoe Australia Ltd., 162.

Kojonen, K.K., Roberts, A.C., Isomäki, O.-P., Knauf, V.V., Johanson, B. and Pakkanen, L. (2004) Tarkianite, (Cu,Fe)$(Re,Mo)_4S_8$, a new mineral species from the Hitura mine, Nivala, Finland. The Canadian Mineralogist 42, 539–544.

Korzhinsky, M.A., Tkachenko, S.I., Shmulovich, K.I., Taran, Y.A. and Steinberg, G.S. (1994) Discovery of a pure rhenium mineral at Kudriavy Volcano. Nature 369, 51–52.

Kucha, H. (1990) Geochemistry of the Kupferschiefer, Poland. Geologische Rundschau 79, 387–389.

Leonhardt, T. (2009) Processing and Properties of Tungsten 25% Rhenium with and without Hafnium Carbide, W-Re 2009 Plansee, Seminar Proceedings, Reutt/Tirol.

Lipinski, M., Warning, B. and Brumsack, H.-J. (2003) Trace metal signatures of Jurassic/Cretaceous black shale from Norwegian Shelf and the Barents Sea. Palaeogeography, Palaeoclimatology, Palaeoecology 190, 459–475.

Lippmaa, E., Maremäe, E. and Pihlak, A.-T. (2011) Resources, production and processing of Baltoscandian multimetal black shales. Oil Shale 28, 68–77.

Maier, W. D., Barnes, S.-J., Chinyepi, G., Barton, J.M., Jr., Eglington, B. and Setshedi, I. (2008) The composition of magmatic Ni–Cu–(PGE) sulfide deposits in the Tati and Selebi-Phikwe belts of eastern Botswana. Mineralium Deposita 43, 37–60.

Millensifer, T.A. (1997) Rhenium Background and Markets. In: Bryskin, D.D. (ed.) Rhenium and

Rhenium Alloys. The Minerals, Metals and Materials Society, Warrendale, PA, USA, 37–47.

Min, M., Chen, J., Wang, J., Wei, G. and Fayek, M. (2005) Mineral paragenesis and textures associated with sandstone-hosted roll-front uranium deposits, NW China. Ore Geology Reviews 26, 51–69.

Minor Metals Trade Association, MMTA (2012a) Rhenium Supply and Demand. http://www.mmta.co.uk/rhenium-market-overview

Minor Metals Trade Association, MMTA (2012b) Metal and assaying norms. http://www.mmta.co.uk/metals/Re/

Morris, D.F.C. and Short, E.L. (1969) Rhenium. In: Wedepohl, K.H. (ed.) Handbook of Geochemistry, Vol. II/5. Springer-Verlag, Berlin and Heidelberg GmbH & Co. K.

National Petroleum Council (2007) Working Document of the NPC Global Oil and Gas Study, Topic Paper #9, Gas to Liquids.

Naumov, A.V. (2007) Rhythms of rhenium. Russian Journal of Non-Ferrous Metals 48, 418–423.

Noddack, I. and Noddack, W. (1931) Die Geochemie des Rheniums. Zeitschrift für Physikalische Chemie (Leipzig) 154, 207–244.

Northern Dynasty Minerals Ltd (2011) News Release, February 23, 2011. http://www.northerndynastyminerals.com/ndm/NewsReleases.asp?numCols=1andpage=2

Petersen, R.G., Hamilton, J.C. and Myers, A.T. (1959) An occurrence of rhenium associated with uraninite in Coconino County, Arizona. Economic Geology 54, 254–267.

Rolls-Royce (2010) Rolls-Royce Market Outlook, 2010. http://www.aarcorp.com/mro/files/Rolls_Royce_Forecast.pdf.

Sinclair, W.D., Jonasson, I.R., Kirkham, R.V. and Soregaroli, A.E. (2009) Rhenium and other platinum-group metals in porphyry deposits. Geological Survey of Canada, Open File 6181, 1 sheet.

Sinclair, W.D., Jonasson, I.R., Kirkham, R.V., and Soregaroli, A.E. (in prep.) Rhenium in Canadian mineral deposits. Geological Survey of Canada, Open File.

SRK Consulting (2010) Independent Technical Report on the Merlin Project, Queensland. 43–101 Report on Merlin. SRK Consulting (Australasia) Pty Ltd., pp. 168.

Treichel, P.M. (2000) Rhenium and rhenium compounds. In: Encyclopaedia of Chemical Technology, posted online December 4, 2000.

Venatovskij, I.V. (1993) Study of the geotechnical conditions of uranium deposits of Uzbekistan during exploration work. In: Uranium in situ leaching. Proceedings of a Technical Committee Meeting held in Vienna, 5–8 October, 1992. International Atomic Energy Agency, TECDOC-720, 95–103.

Voudouris, P.C., Melfos, V., Spry, P.G. et al. (2009) Rhenium-rich molybdenite and rheniite in the Pagoni Rachi Mo-Cu-Te-Ag-Au prospect, northern Greece: implications for the Re geochemistry of porphyry-style Cu-Mo and Mo mineralization. The Canadian Mineralogist 47, 1013–1036.

Walker, R.J., Morgan, J.W., Horan, M.F., Czamanske, G.K., Krogstad, E.J., Fedorenko, V.A. and Kunilov, V.E. (1994a) Re-Os isotopic evidence for an enriched-mantle source for the Norils'k-type, ore-bearing intrusions, Siberia. Geochemica et Cosmochimica Acta 58 (19), 4179–4197.

Walker, R.J., Morgan, J.W., Hanski, E. and Smolkin, V.F. (1994b) The role of the Re-Os system in deciphering the origin of magmatic sulfide ores: A tale of three ores. In: Lightfoot, P.C. and Naldrett, A.J. (eds.) Proceedings of the Sudbury-Norils'k Symposium. Ontario Geological Survey, Special Volume 5, 343–355.

Wooten, D.J.R. and Lanshaw, P.T. (1989) Proceedings Joint Army-Navy-NASA/Air Force Publication (JANNAP) Propulsion Meeting, Laurel, Maryland.

15. Tantalum and niobium

ROBERT LINNEN[1], DAVID L. TRUEMAN[2] AND RICHARD BURT[3]

[1]*Robert W. Hodder Chair in Economic Geology, Department of Earth Sciences, University of Western Ontario, London, Ontario, Canada*
[2]*Consulting Geologist, Richmond, British Columbia, Canada*
[3]*GraviTa Inc., Elora, Ontario, Canada*

Introduction

Tantalum derives its name from king Tantalus in Greek mythology. He stood in a pool of water beneath a fruit tree. When he tried to grasp some fruit the branches moved and when he tried to drink the water receded, which gives rise to the word *tantalise*. His daughter was Niobe, after which the element niobium is named. In the geological literature, these elements are not referred to as father and daughter, but rather as 'geochemical twins' because their behaviour is very similar. Charles Hatchett discovered niobium, first called columbium, in 1801 from a sample that was sent from Connecticut, USA. Anders Gustaf Ekeberg first discovered tantalum in 1802 but it was difficult to distinguish tantalum from niobium. This issue was not resolved until Heinrich Rose in 1844 and Jean Charles Galissard de Marignac in 1866 were able to demonstrate that niobium and tantalum were two different elements (Winter, 2011).

Physical and chemical properties

Niobium sits above tantalum in the Periodic Table and both have very high melting temperatures, 2468 and 2996 °C, respectively. A summary of their properties is given in Table 15.1. Niobium has only one stable isotope ^{93}Nb, but has eight radiogenic isotopes that range in mass from ^{89}Nb to ^{97}Nb. Tantalum has two stable isotopes: 99.988 per cent occurs as ^{181}Ta and 0.012 per cent as ^{180}Ta. It also has six radiogenic isotopes that range in mass from ^{177}Ta to ^{183}Ta. Niobium and tantalum both have a valency of +5 under most natural redox conditions within the Earth and have nearly the same ionic radius. These elements have a high charge to ionic radius and, because of this, they are insoluble in most geological fluids, are strong Pearson acids that are only complexed by strong ligands such as O^{2-}, OH^- and F^- (Wood, 2005). However, these elements are soluble at weight per cent levels in silicate melts, particularly alkaline melts (Linnen and Cuney, 2005), and can attain even higher solubilities in carbonatite melts (Mitchell, 2005).

Distribution and abundance in the Earth

The estimated abundances of niobium and tantalum in the upper continental crust are 12 and 0.9 ppm, respectively, which is enriched relative to the bulk continental crust, 8 and 0.7 ppm, respectively (Rudnick and Gao, 2004). These values are much higher than the estimated concentrations in primitive mantle, 548 ppb Nb and 40 ppb Ta (Palme and O'Neill, 2004). Both

Critical Metals Handbook, First Edition. Edited by Gus Gunn.

Table 15.1 Selected properties of niobium and tantalum.

Property	Value		Units
Name	Niobium	Tantalum	
Symbol	Nb	Ta	
Atomic number	41	73	
Atomic weight	92.91	180.95	
Density at 25 °C	8578	16670	kg/m^3
Melting point	2468	2996	°C
Boiling point	4930	5425	°C
Hardness (Mohs scale)	6.0	6.5	
Electrical resistivity at 25 °C	144	134	nΩ m
Crystal structure	Body-centred cubic	Body-centred cubic	
Ionic radius (six-fold coordination)	64	64	pm

Table 15.2 Selected niobium and tantalum minerals and indicative contents of Nb_2O_5 and Ta_2O_5.

Mineral name	Formula	Nb_2O_5 (%)	Ta_2O_5 (%)
Columbite	$(Fe,Mn)(Nb,Ta)_2O_6$	78.72	na
Tantalite	$(Fe,Mn)(Ta,Nb)_2O_6$	na	86.17
Pyrochlore	$(Na,Ca)_2Nb_2O_6(O,OH,F)$	75.12	na
Microlite	$(Na,Ca)_2Ta_2O_6(O,OH,F)$	na	83.53
Tapiolite	$(Fe,Mn)\ (Ta,Nb)_2O_6$	1.33	83.96
Ixiolite	$(Ta,Nb,Sn,Mn,Fe)_4O_8$	8.30	68.96
Wodginite	$(Ta,Nb,Sn,Mn,Fe)O_2$	8.37	69.58
Loparite	$(Ce,La,Na,Ca,Sr)(Ti,Nb)O_3$	16.15	na
Lueshite	$NaNbO_3$	81.09	na
Euxenite	$(Y, Ca, Ce, U, Th)(Nb, Ti, Ta)_2O_6$	47.43	22.53
Strüverite	$(Ti,Ta,Fe)O_2$	11.32	37.65
Ilmenorutile	$Fe_x(Nb,Ta)_{2x}4Ti_{1-x}O_2$	27.9	na

elements are highly incompatible (partition in favour of the melt over minerals), are enriched in alkaline magmas and are characteristically depleted in calc-alkaline magmas (e.g. Winter, 2010). Apart from minerals of niobium, tantalum and tin, the highest concentrations of niobium and tantalum are observed in titanium-bearing minerals, notably rutile and titanite.

Mineralogy

Niobium and tantalum do not occur naturally as free metals. The charge to ionic ratio of tantalum and niobium results in strong bonds being formed with oxygen; in fact, most niobium–tantalum minerals are oxides, rather than silicates, sulfides or phosphates. The most abundant niobium–tantalum minerals are listed in Table 15.2. The two most important groups of minerals by far are columbotantalite and pyrochlore. Columbotantalite has an orthorhombic crystal structure, with the end-members being columbite-(Fe), columbite-(Mn), tantalite-(Fe) and tantalite-(Mn). However, tantalite-(Fe) is very rare because the Fe–Ta end-member is typically tapiolite, which has a tetragonal crystal structure. Columbotantalite group minerals also range from completely ordered to more disordered, and ixiolite is the completely disordered oxide phase. The pyrochlore group of minerals has numerous end-members because of the large number of potential substitutions into its crystal structure (cubic). In terms of economic importance the most important are pyrochlore, the Nb end-member and microlite, the Ta end-member. There is considerable overlap in the mineralogy of niobium and tantalum deposits, although each commodity has its own characteristic ore mineralogy.

Niobium deposits are predominantly hosted by carbonatites, where pyrochlore group minerals are by far the most important ore minerals. The world's largest niobium deposit is Araxá, which contains primary pyrochlore. However, most of the deposit consists of enriched, secondary laterite ore (Pell, 1996) with the main niobium mineral being pyrochlore-(Ba). Other niobium minerals present in carbonatites include columbite-(Fe), perovskite group minerals (loparite and lueshite) and rare silicate minerals, such as niocalite. However, these minerals generally are not economically important. The niobium mineralogy of alkaline intrusions is similar to that of carbonatites, being dominated by pyrochlore and perovskite group minerals. However, a wide range of 'exotic' minerals is also present (Salvi and Williams-Jones, 2005).

Tantalum deposits are hosted by peraluminous pegmatites and granites, where three mineral groups, columbotantalite, wodginite and microlite, dominate the ore mineralogy. The columbotantalite group minerals are the most common and contain both tantalum and niobium. Because of this, ore, particularly from central Africa, is commonly referred to as 'coltan'. The crystallisation of granite-pegmatite melts results in an evolution from niobium-rich columbite to tantalum-rich tantalite; deposits of the latter are associated with the most evolved melts. Typically, there is also a progression from iron-rich columbite to manganese-rich tantalite, although a number of different iron–manganese crystallisation paths have been recognised (Černý et al., 1986). Wodginite is also an important tantalum ore mineral and tapiolite is less abundant, but is nevertheless present in many tantalum deposits. Microlite is a common tantalum mineral that occurs in many ore deposits, but it is rarely the dominant mineral in an ore body.

Another source of tantalum is as a by-product of tin mining. Cassiterite can contain weight per cent levels of tantalum and niobium and also commonly has inclusions of columbotantalite and rutile–strüverite–ilmenorutile.

Deposit types

All primary niobium and tantalum deposits are associated with igneous rocks and can be classified on the basis of the associated igneous rocks. Three types of deposits are recognised:

1. carbonatite-hosted deposits (niobium and industrial minerals);
2. alkaline to peralkaline granites and syenites (niobium, yttrium, rare earths, zirconium, tin, tantalum);
3. peraluminous pegmatites and granites (tantalum, niobium, tin, tungsten, caesium, lithium).

Tantalum and niobium minerals are also resistant to mechanical and chemical weathering and have high specific gravities. These properties favour the accumulation of these minerals as placer and alluvial deposits, which provide an important component of global production, especially during periods of tight supply from primary mines. A number of tin deposits also contain niobium and tantalum, particularly in south-east Asia. Processing of tin has led to accumulation of tin slags in the smelting process, which can be re-processed and thus are another important source of tantalum. The global distribution of these deposit types is shown in Figure 15.1.

Carbonatite deposits

Carbonatites are igneous rocks that contain more than 50 per cent carbonate minerals (calcite, dolomite or ankerite). Most carbonatites occur in rift settings, but it is clear that there are several different types of carbonatites, many of which are unmineralised and do not contain anomalous niobium concentrations. Carbonatites are genetically related to strongly alkaline silicate rocks, but it is

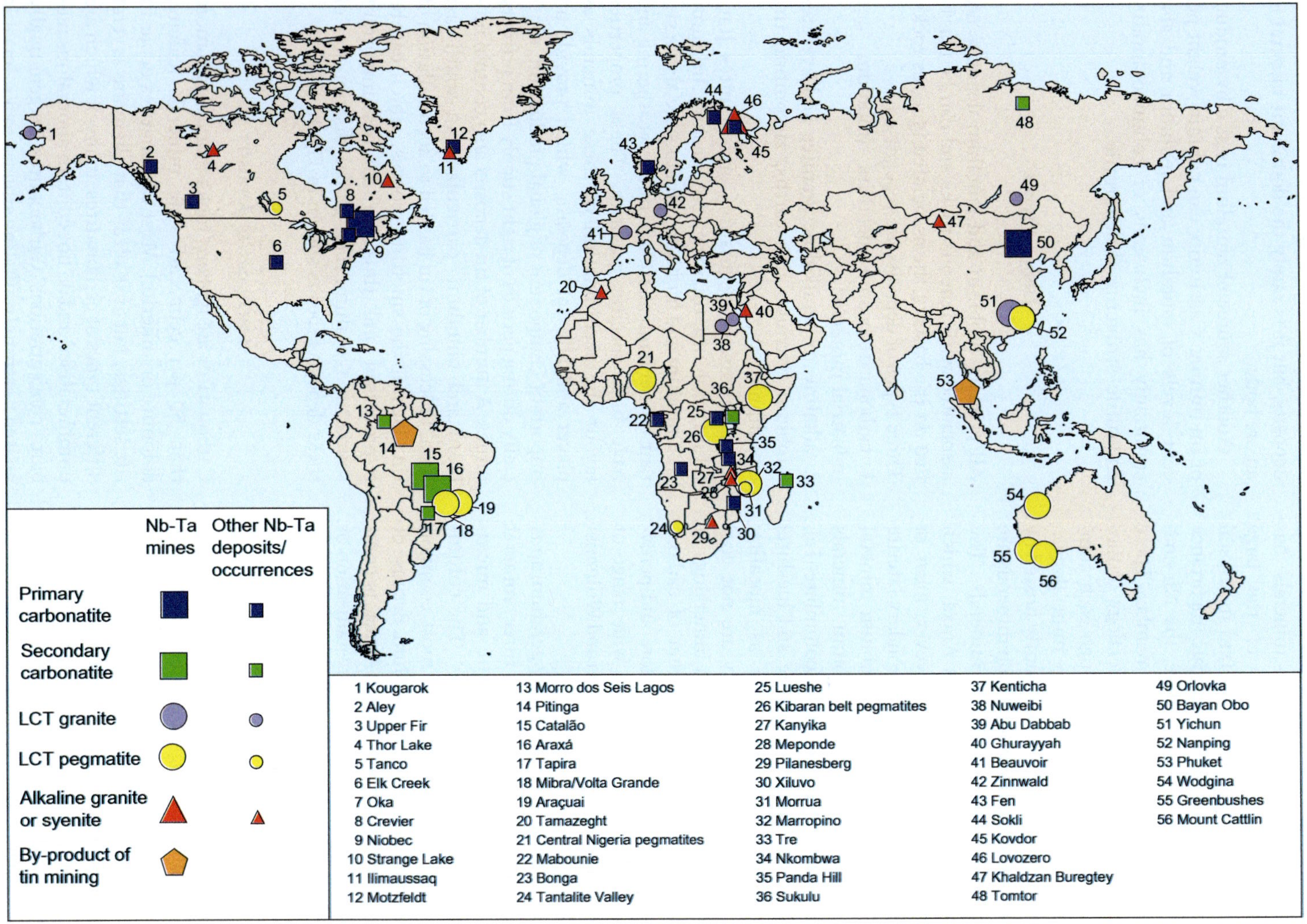

Figure 15.1 Global distribution of niobium and tantalum mines, selected deposits and occurrences. (Modified after Shaw and Goodenough, 2011.)

not clear whether these are related by fractional crystallisation or by silicate–carbonatite melt immiscibility (most likely both are important, depending on the individual carbonatite). However, both the alkaline silicate and carbonatite melts are interpreted to be the product of melting of a metasomatised mantle. According to Mitchell (2005), niobium mineralisation is associated with the melilitite family of carbonatites.

The most important carbonatite-hosted niobium deposits in the world are in Brazil, which produce approximately 92 per cent of the world's niobium. The mines currently producing niobium are the Araxá and Catalão I. Both of these deposits occur in Late Cretaceous alkaline ultramafic–carbonatite complexes of the Alto Paranaiba igneous province.

The Araxá deposit originally contained approximately 808 million tonnes of ore with an average grade of about 2.3% Nb_2O_5 (CBMM, personal communication, 2013). It lies within the Barreiro carbonatite complex, a circular intrusion approximately 4.5 km across (Figure 15.2). The surrounding Proterozoic quartzites have been affected by fenite alteration (consisting of alkali feldspar, eckermanite, arfvedsonite, aegirine–augite and dolomite) for a distance of 2.5 km from the intrusive contact. In

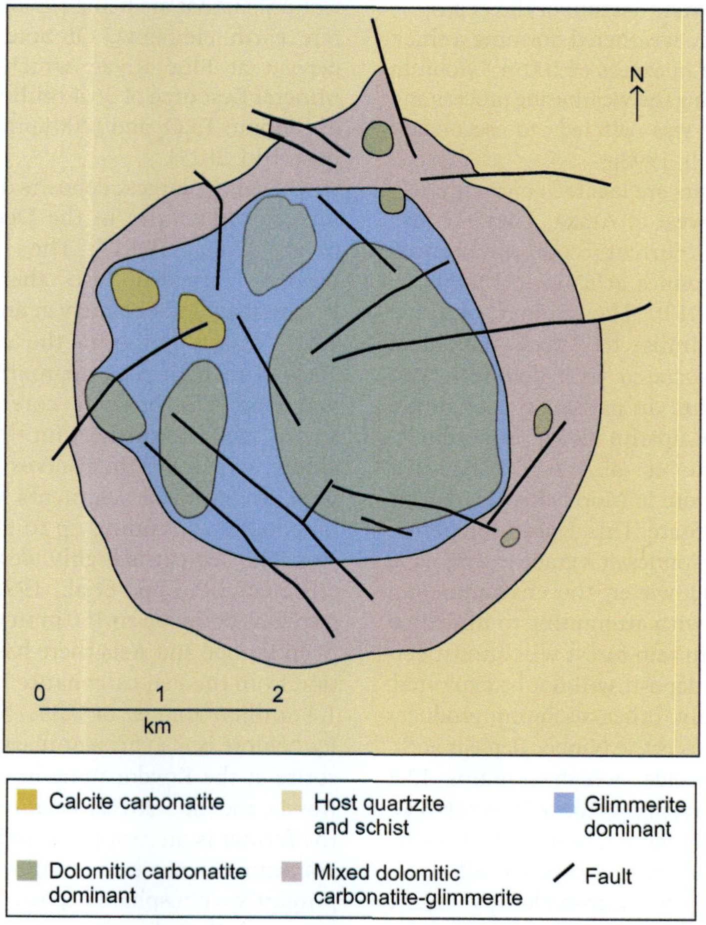

Figure 15.2 Geology of the Araxá carbonatite complex. (Modified after Pell, 1996.)

addition to these metasomatic minerals, the wallrock also contains hematite, goethite, anatase, rare earth phosphate, titanite and pyrochlore (Nasraoui and Waerenborgh, 2001). The pyrochlore deposit is approximately 1.8 km across, and is associated with dolomitic carbonatite that is surrounded by glimmerite, a rock that consists dominantly of phlogopite, with dolomite and accessory magnetite and apatite. Dolomite is the dominant mineral in the carbonatite with subordinate calcite and ankerite and accessory barite, apatite, magnetite, perovskite, quartz, pyrite, phlogopite and sodic amphibole. It is also noteworthy that the highest-grade niobium mineralisation is associated with phoscorite, a magnetite–apatite–phlogopite–carbonate rock. An important feature of the deposit is that it has been deeply weathered, forming a thick lateritic cover, locally in excess of 100 m. Niobium was concentrated during the weathering process and primary pyrochlore was altered to secondary barium-pyrochlore (Pell, 1996).

The Catalão deposits are located approximately 200 km north-north-west of Araxá. They are also lateritic deposits, with current resources of approximately 58 million tonnes at about 1.1% Nb_2O_5 (Anglo American, 2013). The geology of these deposits is broadly similar to Araxá; pyrochlore mineralisation is associated with dolomitic carbonatite and phoscorite (in particular a nelsonite phase of magnetite–apatite with subordinate phlogopite (Cordeiro et al., 2010). Another Brazilian deposit of note is Morro dos Seis Lagos, located in Amazonas State. This deposit is reported to contain 2.9 billion tonnes at a grade of 2.85 wt% Nb_2O_5 (Pell, 1996). However, the environmental problems associated with attempting to mine in a national park of virgin rain forest will almost certainly mean that the deposit will not be exploited.

Currently the only other niobium producer exploiting carbonatites is the Niobec deposit at St. Honoré, Quebec, Canada, which contains 12.2 million tonnes at an average grade of 0.66% Nb_2O_5 (Pell, 1996). The St. Honoré Complex is a 6.5 by 8 km elliptical intrusion that consists of dolomitic and ankeritic carbonatite surrounded by calcite carbonatites with magnetite and locally pyroxene or micas. An outer ring of syenite, ijolite, urtite and carbonatite partially encircles the inner carbonatites (Pell, 1996). The ore consists of both pyrochlore and columbite. Separate rare earth element (REE) zones with bastnäsite and monazite are also present. A past niobium producer is also located in Quebec at Oka. Between 1961 and 1976 this deposit produced 2000 tonnes of Nb_2O_5 from 6.3 million tonnes of ore. At the time of the mine closure proven reserves of 23 million tonnes of 0.44% Nb_2O_5 remained (Pell, 1996). The Oka complex is a Cretaceous intrusion six km long and two km wide which contains two intrusive carbonatite centres that alternate with alkaline silicate rocks (Pell, 1996). There are several other carbonatite complexes in Canada, particularly in British Columbia, that are being explored for niobium and rare earth elements. Of note is the Upper Fir deposit at Blue River, which has an Indicated Mineral Resource of 36.4 million tonnes at a grade of 195 ppm Ta_2O_5 and 1700 ppm Nb_2O_5 (Chong and Postolski, 2011).

One of the largest deposits outside Brazil is the Lueshe carbonatite in the Democratic Republic of the Congo (DRC). The 120 million tonne deposit, intruded into the late Proterozoic Bugandian period country at around 500 Ma (Pohl, 1994), stands out from the country rocks as a 200–300 m high plug approximately two km in diameter. The busorite core is surrounded by sovite, which is the host for the apatite and pyrochlore, which are themselves surrounded on the west by fenitised sediments with an outlier of rauhaugite containing up to 90 per cent magnesite. The deposit is highly weathered, with a lateritic cap (Philippo et al., 1997), with the sovite horizons being up to 150 m in thickness.

In Europe and Asia there has been past production from the Fen carbonatite in Norway (roughly 1.4 million tonnes of 0.3% Nb_2O_5 (Pell, 1996)). Pyrochlore is also present in carbonatite and phoscorite at the Kovdor mine in the Kola Peninsula, Russia, and the Sokli carbonatite, Finland. However, the former is an iron mine and the niobium mineralisation is not economic, whereas the latter is primarily a phosphate deposit. Lastly, Tomtor is a large, weathered carbonatite in northern Siberia (Kravchenko and Pokrovsky, 1995).

Alkaline to peralkaline granites and syenites

Significant concentrations of niobium and tantalum occur in alkaline to peralkaline granites and syenites. These intrusions also contain high concentrations of zirconium, yttrium and rare earth elements and generally occur in rift or failed rift tectonic settings, although there are exceptions such as the Lovozero and Khibiny intrusions in the Kola Peninsula of Russia. This style of mineralisation has received considerable interest recently because of related high yttrium and rare earth element contents. One of the characteristics of this deposit type is the wide range of zirconium, niobium and rare earth element minerals that are present, including hydrous or anhydrous silicate, phosphate, oxide and mixed silicate–phosphate mineral phases (Salvi and Williams-Jones, 2005). The dominant mineral of niobium and tantalum is pyrochlore, but other complex phases such as eudialyte and secondary fergusonite are also important carriers of niobium. There are no current producing mines of this type, but they are potential sources of a variety of metals in the future.

An important peralkaline granite-hosted deposit is the Pitinga tin–niobium–tantalum deposit in Amazonas State, Brazil. The deposit is associated with the ~1820 Ma[1] Madeira A-type granite, emplaced into a post-collisional extensional setting (Bastos Neto et al., 2009). This deposit contains 424 million tonnes of measured and indicated resources (Salles, 2000) at a mining grade of 0.17% Sn (as cassiterite) with niobium and tantalum (pyrochlore and columbotantalite) recovered as by-products. The granite is unusual in that it has a massive cryolite deposit in its core, ten million tonnes of 31.9% Na_3AlF_6. The granite consists of four facies, a metaluminous biotite–hornblende syenogranite, a peraluminous biotite–alkali feldspar granite and younger albite-enriched and border albite-enriched granites. The albite-enriched granites contain riebeckite, cryolite, and polylithionite, which indicate a peralkaline composition. Cassiterite in the main ore is disseminated and is interpreted to be magmatic (Bastos Neto et al., 2009). However, the granite is also altered and albite and greisen alteration are observed.

The Strange Lake deposit, along the Quebec–Labrador boundary in Canada, is a well-studied example of a peralkaline granite-hosted deposit. Current grade and tonnage estimates of the Indicated Resource at the B-Zone of this deposit are of 140.3 million tonnes grading 0.933% total REE, 1.93% ZrO_2, 0.18% Nb_2O_5, 0.05% HfO_2 and 0.08% BeO (Quest Rare Minerals Ltd. 2011). There is little doubt that magmatic processes concentrated zirconium, niobium and REE, but ore grade increases with the extent of hydrothermal alteration and Salvi and Williams-Jones (2005) propose that ore metals were enriched during these hydrothermal events.

The Meponde alkaline complex, just north of the Mozambique–Malawi border and 3 km east of Lake Niassa, primarily consists of leuocratic syenites and nepheline syenite. The complex is deformed, and appears to be a synform with limbs that dip generally westwards in conformity with the enclosing rocks. Pyrochlore, including uraniferous pyrochlore (hatchettolite), is the most abundant niobium-bearing mineral phase, with the ore zone containing up to 8800 ppm Nb_2O_5. In addition, a columbite zone contains grades generally exceeding 1% Nb_2O_5 and 500 ppm Ta_2O_5. Other minerals include zircon, fergusonite, monazite and other rare earth element minerals (Mroz, 1983). This 50 million tonne deposit has not been developed to date.

There are also several large peralkaline granites in the Arabian Shield, the most important of which is the Ghurayyah deposit. The Ghurayyah stock is approximately 900 m across and contains fine-grained disseminations of pyrochlore and Y-columbite. It contains an Inferred Resource of 385 million tonnes of 245 g/t Ta_2O_5, 2840 g/t Nb_2O_5, 8915 g/t ZrO_2, 1270 g/t Y_2O_5 and 140 g/t U_3O_8 (Küster, 2009). Rare earth element concentrations are also high, but grades have not been reported. In terms of contained metal, this is the largest tantalum deposit in the world, but it is not clear whether the tantalum can be economically separated from niobium (which is an order of magnitude higher in concentration) or from the other heavy minerals. The Ghurayyah has not been dated, but its age is likely around 580 Ma[1] (Küster, 2009). It contains arfvedsonite and aegirine as well as a wide range of accessory minerals including pyrochlore, zircon, smarskite, aeschynite-(Y), columbotantalite and cassiterite.

The second type of deposit hosted by peralkaline rocks are those hosted by syenites. There are a variety of different types of nepheline syenites that are mineralised and commonly these intrusions are well layered. Deposits of note are the Khibiny and Lovozero intrusions of the Kola Peninsula in Russia, the Khaldzan-Buregtey deposit in Mongolia, the Ilímaussaq and Motzfeldt Complexes in Greenland and the Nechalacho deposit at Thor Lake in the Northwest Territories of Canada. These deposits are primarily of interest for their rare earth element mineralisation and will not be discussed in detail. The most common niobium mineral in these deposits is pyrochlore, although at Nechalacho the dominant niobium mineral is fergusonite (see Salvi and Williams-Jones, 2005, for a review). The Motzfeldt Complex in Greenland is an exception, and here, niobium and tantalum are the primary commodities. Little has been published on this deposit, but according to the Ram Resources website (2011), the Geological Survey of Greenland estimated a mineralised zone of 500 million tonnes with average grades between 1320–1480 ppm Nb and 110–130 ppm Ta, whereas Tukiainen (1988) concluded that the nepheline syenite could be host to 50 million tonnes grading between 300–1000 ppm Ta_2O_5. Pyrochlore is the primary ore mineral for both niobium and tantalum.

Peraluminous pegmatites

Peraluminous pegmatites have historically been the most important source of tantalum. Many of these pegmatites have been mined intermittently and may or may not be in current production. They are described here because they are well documented and may resume tantalum production in the future. They are commonly associated with peraluminous S-type granites and occur as late syn- to post-tectonic intrusions in collision belts, typically in association with shear zones. Other elements of economic significance include lithium, caesium, beryllium, rubidium, niobium and tin. These pegmatites are assigned to the lithium–caesium–tantalum (LCT) family of pegmatites (Černý and Ercit, 2005). A complete discussion of pegmatite classification is beyond the scope of this article, but tantalum pegmatite deposits are most commonly Complex-Type pegmatites, although Albite and Albite–Spodumene-Type pegmatites, e.g., Wodgina, Mount Cassiterite, respectively, also host economic tantalum mineralisation (Sweetapple and Collins, 2002). The Complex Type is further subdivided into spodumene, petalite, lepidolite, elbaite and amblygonite subtypes on the basis of their dominant lithium, boron and phosphorus mineral species.

The best studied of these is the Tanco deposit, which is a Petalite Subtype. This pegmatite is approximately 1600 × 820 m and in the central portion is greater than 100 m thick. It intruded metavolcanic rocks and has been dated at 2640 Ma[1] (Černý, 2005). The complex zoning consists of eight major zones (Figure 15.3) with the economic mineralisation, Li, Cs and Ta, occurring in different parts of the pegmatite. The pre-production grades at Tanco were 2.07 million tons of 2160 ppm Ta_2O_5, 7.3 million tons of 2.76% Li_2O and 0.35 million tons of 23.3% Cs_2O (Černý et al., 1996). Tantalum mineralisation is dominantly hosted by aplitic and metasomatised (muscovite replacement) K-feldspar pegmatite (Van Lichtervelde et al., 2007). Figure 15.4 shows mineralised aplite from the Tanco deposit, where tantalum–niobium oxide minerals occur in primary magmatic layers. By contrast, lithium mineralisation is in Intermediate Zones and the caesium mineralisation is within a separate Pollucite Zone (Černý, 2005).

Two of the largest historic tantalum producers are the Greenbushes and Wodgina deposits in Western Australia. The Greenbushes pegmatite, in the Yilgarn Craton, contains Measured, Indicated and Inferred Resources of 135.1 million tonnes at 220 ppm Ta_2O_5. It is a Spodumene Subtype pegmatite that intruded into a shear

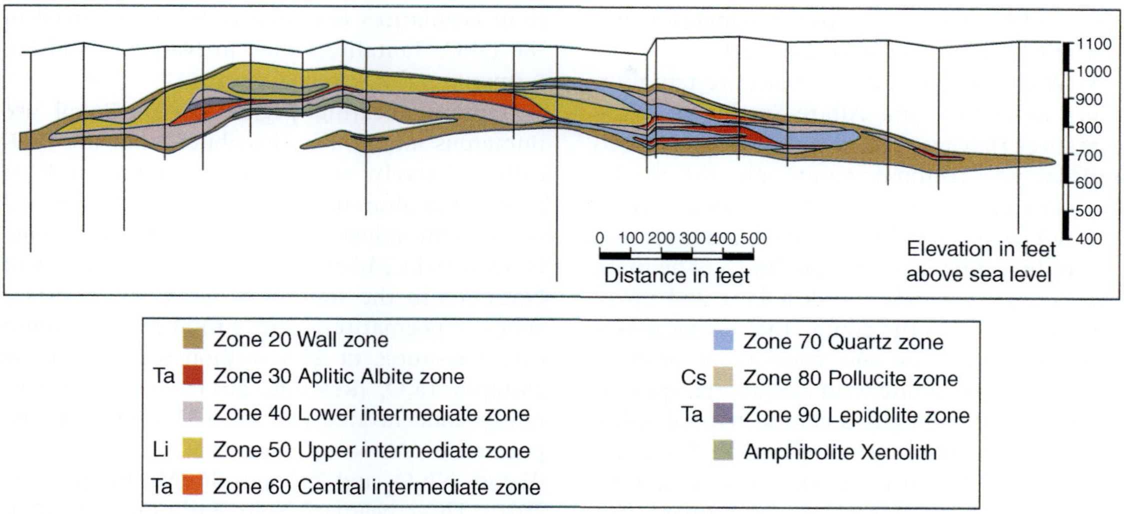

Figure 15.3 Longitudinal east–west section through the Tanco pegmatite. (Modified after Černý et al., 1996). (Cs, caesium; Li, lithium; Ta, tantalum.)

Figure 15.4 Banded aplite ore from the Tanco deposit. Tantalum mineralisation is contained in the banded aplite (red and blue-grey at the bottom). A massive quartz zone (dark grey) is at the top and coarse, crystalline white beryl separates the aplite and quartz. The scale is approximately 2 m across. (Courtesy of R. Linnen.)

zone and consists of four layers, a Li-Zone, K-Zone, Na-Zone and Border Zone. The tantalum mineralisation is associated with a massive albite–quartz-rich unit, and like Tanco, the lithium is mined from a different unit. Cassiterite is also an important ore mineral and early formed tantalum minerals occur as inclusions in cassiterite and tourmaline (Partington et al., 1995 and Fetherston, 2004). This pegmatite crystallised at relatively high temperatures and pressures, from

750°C and five kbar to a greisen and metasomatic stage at 620 °C and five kbar (op. cit.).

At Wodgina and Mount Cassiterite tantalum occurs in Albite- and Albite–Spodumene-Type pegmatites that are related to a suite of 2890 to 2830 Ma[1] post-tectonic granite plutons in the North Pilbara Craton. The last published reserves (Sons of Gwalia, 2002) for the Wodgina mine was 63.5 million tonnes of 370 ppm Ta_2O_5. The main dyke is approximately one km long and up to 40 m wide. It lacks the zoning that characterises Complex pegmatites and consists of primary layers of massive albite with rare quartz, spessartine and muscovite, and a central banded aplite that is largely granitic in composition. Tantalum minerals are primarily in the massive albite, where they are interpreted to be primary, but metasomatic styles of mineralisation are also observed (Sweetapple and Collins, 2002). The Mount Cattlin deposit, also in Western Australia, is a lithium deposit that produces a minor amount of by-product tantalum (Fetherston, 2004).

Current tantalum production elsewhere is primarily from pegmatites in Africa and Brazil, but less information on these pegmatites has been published, at least in recent years. In Africa, most of the tantalum pegmatite deposits are associated with the Pan-African orogeny. One of the most important tantalum pegmatites in Africa is Kenticha in Ethiopia. This is a Spodumene-Type pegmatite that occurs as a dyke over two km long and 400–700 m wide that intruded at 550 Ma[1] (Küster, 2009). It is complexly zoned but Küster et al. (2009) simplifies the zones into three units. Most of the tantalum mineralisation occurs in the Upper Zone, which also contains most of the spodumene mineralisation and is thought to represent the most evolved unit (bottom-to-top crystallisation; Küster et al. (2009). Other pegmatites in Africa include Morrua and Marrapino in Mozambique, the latter of which is deeply weathered. In Central Africa, particularly the DRC, the Manono deposit in Katanga was the largest historical producer. Here, most production is from shallow deposits and from the many eluvial and alluvial placer deposits that are pegmatite derived (Fetherston, 2004), although some primary ore from pegmatites is currently being produced in the DRC. Tantalite Valley in Namibia is also a former tantalum-producing area.

Tantalum-bearing pegmatites in Brazil are numerous and widely distributed, but are typically relatively small (less than one million pounds tantalum in size) and are generally mined on a semi-industrial or even artisanal scale. However, the Mibra (Volta) mine, near the city of Nazareno in the state of Minas Gerais, hosts a series of pegmatites, with a total reported (indicated) resource of 25.5 million tonnes grading 290 ppm Ta_2O_5 (Resende, 2011). Mining is currently concentrated on the 6.32 million tonne pegmatite 'A' with grades of 375 ppm Ta_2O_5, 92 ppm Nb_2O_5 and 283 ppm Sn (Mining Journal, 2010). These pegmatites are Early Proterozoic in age and are hosted by an Archean greenstone belt at the southern border of the São Francisco Craton (Lagache and Quéméneur, 1997). There are a number of flat, subhorizontal sills, the largest of which measured 1000 by 150 by 20 m before exploitation. The pegmatites show internal zoning with an aplitic wall zone, an intermediate zone that contains 20 to 30 per cent spodumene and a core zone that is comprised of 60 to 80 per cent spodumene crystals (Lagache and Quéméneur, 1997).

Peraluminous granites

The last class of niobium–tantalum deposits considered here are those hosted by peraluminous granites. They account for only a minor amount of tantalum production, but geochemically are similar to LCT pegmatites, being enriched in lithium, fluorine and phosphorus. The Yichun deposit, China, reported to contain approximately 175 million tonnes (personal communication, 1996), is a vertically zoned granite occupying the top of the south-eastern section of the Yashan granite batholith. The upper portion of the granite, which carries the majority of the tantalum values, is an albitic granite high in lepidolite and topaz, relatively low in quartz, and depleted in dark minerals. The lower portion, which has sub-economic tantalum values, is high in biotite mica rather than lepidolite (Yin et al., 1995).

The granites at Beauvoir, France (Raimbault et al., 1995) and in Egypt (Abu Dabbab and Nuweibi) contain disseminated cassiterite–columbotantalite mineralisation that is apparently magmatic, as well as peripheral tungsten–tin mineralisation that is hydrothermal. These deposits have not produced, but Abu Dabbab contains approximately 40 million tonnes of 243 g/t Ta_2O_5 and Nuweibi 98 million tonnes of 146 g/t Ta_2O_5 (Küster, 2009). Granites which have formerly produced minor quantities of tantalum are the Orlovka deposit in Russia (Badanina et al., 2004), and at Podlesí and Cínovec in the Zinnwald area of the Czech Republic (Breiter et al., 2007 and Rub et al., 1998). These deposits have several features in common: most are associated with late, multi-phase intrusions; magmatic-style columbotantalite is disseminated in the granite; and there is also a hydrothermal stage, typically with cassiterite–wolframite. The granites are also vertically zoned, in particular with respect to the variety of lithium mica present. The Kougarok tin–tantalum–niobium deposit in the Seward Peninsula of Alaska is another granite-hosted deposit of note (Puchner, 1986).

Extraction methods and processing

Tantalum and niobium minerals are recovered through industry-standard open pit, underground and placer mining methods. In the case of primary sources or lode deposits, ores are usually mined by standard methods, e.g. drill, blast, and muck cycles, prior to further comminution and concentration. In deposits where weathering processes are advanced, initial processing may be facilitated through simple screening.

Pyrochlore ores are concentrated by flotation, preceded by removal of irrecoverable ultrafines to achieve acceptable concentrate grades – generally in excess of 50% Nb_2O_5 – and rejection of unwanted contaminants. Several flotation stages as well as electrical separation are required (Figure 15.5).

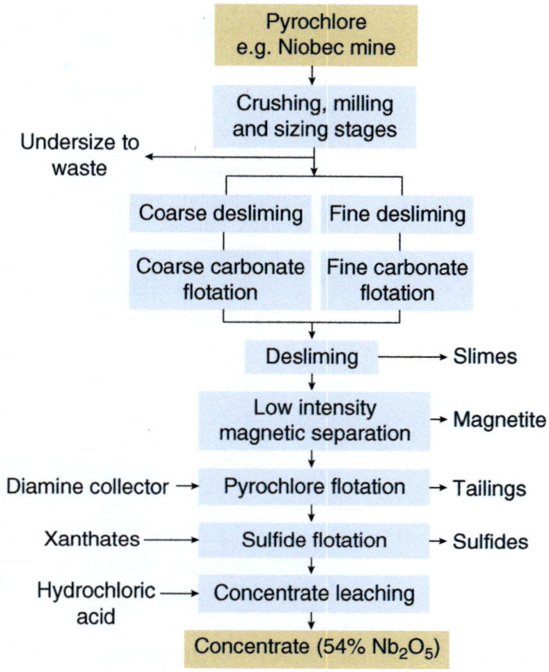

Figure 15.5 Niobium production flowsheet for the Niobec mine. (Modified after Shaw and Goodenough, 2011.)

Columbotantalite ores almost exclusively utilise standard wet gravity concentration equipment including jigs, spirals and tables, at least for production of a low- to medium-grade 'heavy mineral' concentrate. While the trend will be toward ever-finer recovery by gravity using centrifugal-type concentrators, there is no reason to assume that any other technology will take over, although tantalum flotation has been used commercially in at least one operation. The heavy mineral concentrates are often further concentrated in 'dry' plants consisting of such units as dry gravity concentrators, magnetic and electrostatic concentrators and occasionally removal of sulfides by flotation (Figure 15.6).

Artisanal mining of both alluvial and heavily weathered ores utilises the very simplest techniques. Primary concentration generally consists of nothing more than a 'ground sluice' and possibly winnowing. The primary concentrates are

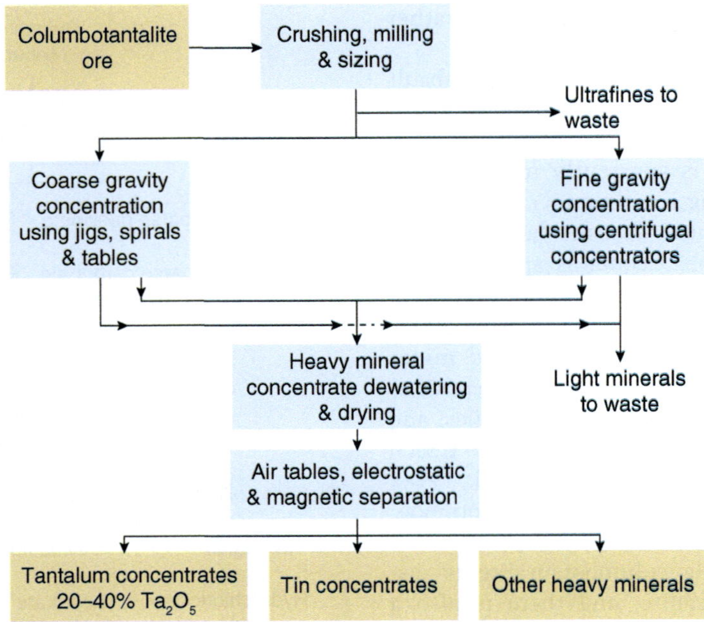

Figure 15.6 Schematic summary of the processes involved in the preparation of tantalum concentrates.

sold to central secondary processing plants similar to those used in larger-scale industrial plants.

Pyrochlore concentrates are converted by aluminothermic reduction, followed by electron beam refining, to ferro-niobium, which is the product that dominates the niobium market.

Tantalum concentrates generally contain between 25–35% Ta_2O_5, much reduced from the 1970s 'norm' of 45–50% Ta_2O_5. While lower grades can be treated, the relative chemical processing cost is usually regarded as prohibitive. Those concentrates from central Africa commonly contain a similar or higher content of Nb_2O_5, whereas those from Canada, Brazil and Australia have a significantly lower Nb_2O_5 content (Burt, 2011).

Tin slags, primarily from south-east Asia, are the other main feedstock to tantalum processors, essentially in two ways. Higher-grade tin slags (those containing more than 2% Ta_2O_5) are suitable for processing directly, whilst lower grade slags (commonly with 1–1.5% Ta_2O_5) require treatment – generally by an aluminothermic process – to increase the tantalum content to a level suitable for chemical processing. The latter are generally referred to as 'syncons'. While the former are generally sold direct from tin smelter to processor, the greatest source of the latter are 'old' slags dumped over the decades, and later dug up and stockpiled, for treatment when tantalum prices peak.

Production of tantalum metals and chemicals is a multi-stage process (Figure 15.7), ideally suited to batch or semi-batch style production, as some of the procedures are quite time consuming. Many of the products are made to order, which can be as long as six months from receipt of ore to final dispatch of a specific product.

The first stage is to convert tantalum concentrates to an intermediate chemical – generally potassium tantalum fluoride (K_2TaF_7) or tantalum oxide (Ta_2O_5). The 'standard' method, which is essentially unchanged for decades, is to digest the ore at high temperature in a sulfuric acid – hydrofluoric acid mix (or 'neat' HF) and, after filtering out the insoluble minerals, further processing via solvent extraction using methyl

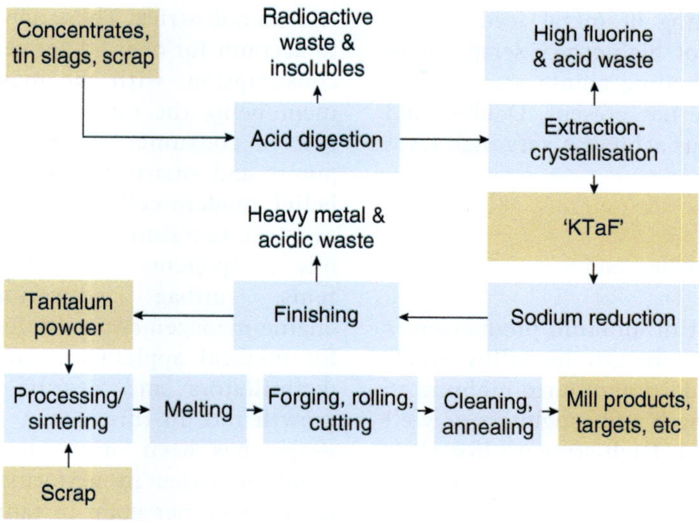

Figure 15.7 Schematic flowsheet for the production of tantalum from concentrates, tin slags and scrap. ('KTaF', potassium fluortantalate).

isobutyl ketone (MIBK) or liquid ion exchange using an amine extractant in kerosene. This produces highly purified solutions of tantalum and niobium, from which tantalum is crystallised, as potassium fluortantalate ('KTaF') by reaction with potassium fluoride. Where the level of niobium in the feed warrants, it is recovered as niobium oxide (Nb_2O_5) via neutralisation of the niobium-fluoride complex with ammonia to form the hydroxide, followed by calcination to the oxide.

Alternative methods do exist and are used when they are better suited to particular local conditions. One used for a titanium–niobium–tantalum–rare earth mineral concentrate involves blending the crushed concentrate with coke and passing this through a chlorination stage which separates out the rare earths and other elements including most of the thorium. The temperature of the resulting titanium–niobium–tantalum oxichloride gas is then decreased, which causes the iron, thorium and alkali metals to precipitate out. The cleaned titanium–niobium–tantalum oxichloride gas is next cooled to a liquid and distilled to separate out low-boiling titanium chloride gas. The niobium–tantalum oxichloride gas is then further chlorinated to produce niobium and tantalum chlorides, $NbCl_5$ and $TaCl_5$. These are fractionally distilled and the niobium chloride subsequently reacted with steam to produce the hydroxide, which is calcined to oxide. The tantalum chloride is reacted with ammonium hydroxide to produce the oxide (Tantalum-Niobium International Study Center (T.I.C.), 2011).

Metal powder, including the precursor to capacitor-grade powder, is produced by sodium reduction of the potassium tantalum fluoride in a molten-salt system at high temperature. The metal can also be produced by the carbon or aluminium reduction of the oxide or the hydrogen or alkaline-earth reduction of tantalum chloride. The choice of process is based on the specific application and whether the resultant tantalum will be further consolidated by processing into ingot, sheet, rod, tubing, wire and other fabricated articles.

The consolidation of metal powder for ingot and processing into various metallurgical products begins with either vacuum arc melting or electron beam melting of metal feed stocks, comprising powder or high-purity scrap where the elements with boiling points greater than that of tantalum are not present. Double- and triple-melt ingots thus achieve a very high level of purification.

Specifications and uses

Close to 90 per cent of niobium production is used to make high-strength low-alloy steels (T.I.C., 2011). The addition of ferro-niobium to steel increases strength and toughness as well as reduces the weight. High-strength low-alloy steels are used to manufacture oil and gas pipelines, car and truck bodies, tool steels, ship hulls and railroad tracks (T.I.C., 2011). The electrical resistance of niobium–titanium and niobium–tin alloy wire drops to virtually zero at or below temperature of liquid helium (–268.8 °C). Consequently, these alloys are utilised in superconducting magnetic coils in magnetic resonance imagery (MRI), magnetoencephalography, magnetic levitation transport systems, and particle physics experiments. Elsewhere, niobium is used for alloys and chemicals including carbides. Niobium oxide (Nb_2O_5) is used to manufacture lithium niobate ($LiNbO_3$) for surface acoustic wave filters, camera lenses, coating on glass for computer screens and ceramic capacitors. Niobium carbide is used for cutting tools and niobium metal and alloys have various speciality applications (T.I.C., 2011).

The physical and chemical properties of tantalum have been long known; such properties having made its discovery difficult in the first instance. Its use in light-bulb filaments hails from over 100 years ago, while it has been used to improve the performance of optical glasses for more than 75 years. Its widespread usage, however, arose with the advent of the transistor and solid-state avionics in the Korean War, in a complementary role with the selenium rectifier and transistor. By the 1980s close to 70 per cent of tantalum was destined for the electronics industries. These applications continue to account for over 50 per cent of all tantalum consumption, with the most significant segment being the tantalum capacitor, used not just in consumer electronics, such as computers and smart-phones (contrary to popular belief, modern cellphones utilise very little, or even no, tantalum), but also in many automobile components (e.g. anti-lock braking systems, airbag activation, GPS and engine-management modules), in aircraft and for medical appliances, such as pacemakers, defibrillators and hearing aids. While the growth rate in this sector, in terms of metal usage, has been minimal, or even negative, rapid increases in performance (in terms of capacitance per gram of tantalum) and consequent miniaturisation means the growth in number of units used has continued to grow. Elsewhere in electronics, tantalum metal is used in hard-disk drives and ink-jet printer heads. A significant growth area is for sputtering targets, with tantalum being used as the diffusion barrier between interconnects on copper-based semiconductors. Lithium tantalate is used in surface acoustic wave filters where the electronic signal wave dampening results in better audio and video output in mobile phones, audio systems and televisions. Tantalum oxide has a high refractive index and hence lenses in spectacles can be made lighter and brighter, as can lenses in mobile phones and the high-end digital camera market. Tantalum carbide-based cutting tools are now a relatively minor segment of the tantalum industry.

Other uses for tantalum include corrosion-resistant objects, prosthetic devices and high-temperature alloys for turbines. Single-crystal alloys, containing three to eleven per cent tantalum, are used in jet (and some land-based) engine turbines where they offer greater resistance to hot gases than other alloys, allowing higher operating temperatures and hence improved fuel efficiency (T.I.C., 2011).

Overall, there was a rapid increase in tantalum demand (as opposed to supply) in all its forms

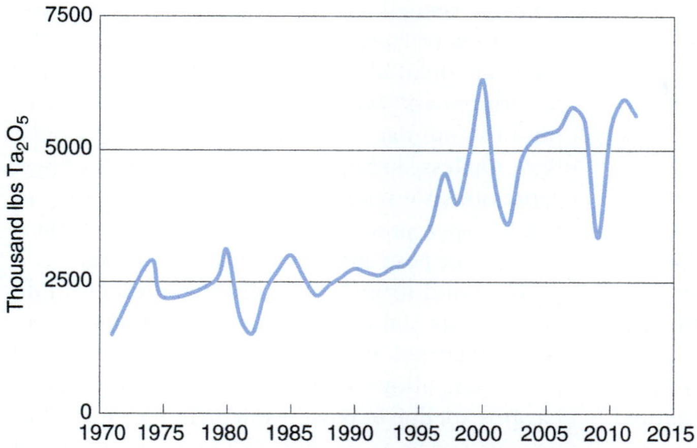

Figure 15.8 Tantalum demand, shown as tantalum pentoxide (Ta_2O_5) equivalent, based upon tantalum processor shipments, 1971–2012. (Data from various T.I.C. bulletins, trade magazines, government statistics and industry observers.)

during the 1990s, but the growth rate has slowed since 2000 to about a one per cent Compound Growth Rate (Figure 15.8).

Recycling, re-use and resource efficiency

According to the USGS (2011a), niobium is recycled when niobium-bearing steels and superalloys are recycled, but scrap recovery specifically for niobium is negligible. Statistics for the amount of recycled niobium are not available, but this could be as high as 20 per cent of apparent consumption. Tantalum is recycled from electronic components and from tantalum-bearing cemented carbide and superalloy scrap – about 25 per cent originates from material recycled from the manufacturing stage of these components (i.e. 'new scrap'). (T.I.C., 2011). However, statistics are not available for the amount of post-consumer tantalum that is recycled and much of that of electronic origin is ignored because of the small amounts involved in miniaturised capacitors.

Substitution

As tantalum prices rose in the early 2000s, niobium was more widely substituted into electronic capacitors. Niobium, having a lower dielectric capacity, resulted in a physically larger but lower-cost capacitor, that only operated in lower peak-temperature ranges. Accordingly, it failed to make military specifications in many critical end uses.

Tantalum capacitors have a relatively small, but specific portion of the overall capacitor market, with aluminium, ceramic and wet electrolytic capacitors having a substantially greater market share. The tantalum industry regards that, where substitution of tantalum with an alternative is possible, it has already occurred, and where tantalum capacitors are currently used, their potential for substitution is low. Likewise, it is expected that other energy-storage devices will continue to have a greater role than tantalum in future green technologies – for example, in electric or hybrid vehicle charging systems where volume is of lesser consequence and electrical storage capability is paramount. However, in airbag triggers, the tantalum capacitor is a must – it is the only capacitor, to date, that can operate at both high temperatures and sub-zero temperatures with essentially no variation in performance.

Tantalum-carbide machining, cutting and grinding tools have longer operating lives and higher operating temperatures, which allow faster machining times and finer finishing. These advantages over tungsten carbides are surrendered as tantalum prices rise, and industry falls back on convention until tungsten prices rise.

Environmental aspects of niobium and tantalum

Niobium and tantalum ores do not pose any special environmental problems, and standard industry-wide safety and environmental precautions during mining and mineral concentration are satisfactory. Flotation reagents, where used, are generally biodegradable and, because of a lack of sulfide minerals associated with the deposits, the tailing impoundments pose little environmental hazard such as acid mine drainage.

Tantalum concentrates produced from pegmatites generally contain minute quantities of natural thorium and uranium, and, as a consequence, the concentrates are described as naturally occurring radioactive materials (NORMs). Some concentrates, from Mozambique, central Africa and Brazil, have higher levels and are classified as 'Class 7' under 'Dangerous Goods' under the International Maritime Dangerous Goods (IMDG) Code and other modal regulations (IMO, 2011). Consequently, minor requirements in terms of packaging, labelling and staff training have to be met. NORMs pose a very low radiological risk during transport and, further, the regulations are designed so that safety is provided by passive safety inherent in the package. Next-generation concentrates, especially those produced from alkaline and peralkaline deposits, will almost certainly have higher levels of these radioactive elements, and some form of on-site processing prior to shipment may well be required. This could include by-product production of uranium chemicals (uranium concentrates are reportedly produced from ores containing as little as 0.05% U), or simply suitable impoundment at site.

Likewise, conversion of niobium concentrates to ferro-niobium poses no special environmental problem, as long as standard safety procedures are in place. The pyrochlore concentrates used to produce ferro-niobium also contain thorium and uranium, which report in the ferro-niobium slag. This slag contains elevated levels of thorium and uranium and is generally stored on site.

The first step in treatment of tantalum concentrates requires their digestion in strong acids – generally sulfuric acid with at least stoichiometric quantities of hydrofluoric acid. The acids are neutralised with alkali salts and the resulting mass has thorium and uranium concentrations similar to the feed raw material. Nevertheless, appropriate, fully licensed, discharge containment is required. Later stages of the process require appropriate safety procedures but otherwise do not pose any environmental hazards.

The solid forms of tantalum and niobium do not pose special environmental problems. There is no reported information on toxicity of the metals and alloys, and the only associated health hazards are due to powders, which, like any other powder, can be an irritant. As the key to increased tantalum capacitor efficiency is the fineness of the powder, care must be taken to ensure a static- and ignition-free environment as, like many other very fine powders, they are pyrophoric and may explode if handled incorrectly.

Comprehensive regulation of materials is now in place in the European Union, called REACH (Registration, Evaluation, Authorisation and Restriction of Chemicals). All companies wishing to produce substances in the EU or import them into the EU must ensure they meet their registration obligations if they are to continue with their activities; further guidance can be obtained from the European Chemical Agency website (ECHA, 2011) or their national competent authority. Dossiers for some of the many niobium and tantalum substances are being developed. Along similar lines to REACH, the US government is drafting legislation to have more comprehensive regulation of chemicals.

Geopolitical aspects

From 1998 to 2003 central Africa was embroiled in what has become known as 'Africa's World War'. The underlying reasons for this conflict are outside the scope of this chapter, but one undoubted effect of it was the involvement of illegal forces in the mining and transportation of minerals, especially from eastern DRC, enabling them to continue funding their war – and subsequent rebel activities.

While tantalum has had only a very minor role (generally less than five per cent of all such funding was from tantalum) it rapidly became the cause célèbre for many activist groups. The tantalum industry may have appeared slow to react, but several of the larger processors had disengaged by 2003. However, it was the 2008 report by the United Nations Group of Experts that really challenged the tantalum, and other, industries to act (Stearns et al., 2008). By mid-2009, a tin industry-led Due Diligence initiative (the 'iTSCi' Programme) was being put in place, with the tantalum industry becoming partners in early 2010. By mid-2011 it had been activated throughout Rwanda and the Katanga province in the DRC. Concurrently, the electronics and telecommunications industries, through the Electronic Industry Citizenship Coalition, Global e-Sustainability Initiative (EICC/GeSI), developed the 'Conflict Free Smelter' programme which, since 2010, requires processors to provide credible evidence of their conflict-free sourcing (Electronic Industry Citizenship Coalition, 2011).

Governments and intergovernmental agencies also reacted, and in early 2012 the Organisation for Economic Cooperation and Development (OECD) published a set of 'guidelines' that had been endorsed by governments, NGOs and industry, and these are being piloted in 2011–12 (OECD, 2011). The United Nations have also published guidelines, these closely following the OECD guidelines.

At the same time, two Senators sponsored legislation that was eventually included in the US Financial Stability Act (commonly known as the 'Dodd Frank Act') promulgated in July 2010. The Securities and Exchange Commission (SEC) published the appropriate regulations in mid 2012. Essentially the law places a legal obligation on all US companies that report to the SEC to declare the use of 'conflict minerals' (cassiterite, wolframite, columbotantalite and gold) and metals originating from the DRC or adjoining countries. The law does not prohibit the use of these 'conflict minerals': it states that '… a product may be labeled as 'DRC conflict free' if the product does not contain conflict minerals that directly or indirectly finance or benefit armed groups in the DRC or an adjoining country'. Other countries, including Canada as well as the EU, are considering similar legislation.

In addition, the International Conference on the Great Lakes Region (ICGLR), which is composed of eleven GLR states, has developed regulations to be put in place by all member states. While embracing the iTSCi Programme as one programme, the ICGLR is also supporting the development of a 'forensic' fingerprinting method for columbite–tantalite concentrates, so that the origin may be traced. While not, as yet, in commercial use, mineral chemistry data is available for a number of African pegmatites through the efforts of the German Federal Institute for Geosciences and Natural Resources (BGR) (Melcher et al., 2008 and Graupner et al., 2010). The fingerprinting combines a number of methods including age dating, mineral liberation analysis and major- and trace-element analysis of tantalum minerals.

World resources and production

Current significant resources and supplies of niobium and tantalum are shown in Figure 15.1 and quantified in Tables 15.3 and 15.4, respectively.

Although niobium deposits are globally widely dispersed, in excess of 90 per cent of the world's niobium is produced from two mines in Brazil: CBMM's operations at Araxá, and Anglo American's Catalão operations. Canada's Niobec Mine, operated by Camet Metallurgy, produces between five and ten per cent of global supply and the remainder comes from loparite concentrates from the Karnasursk Mine in the Russian Kola Peninsula. In the ten-year period from 1997 to 2007 production increased rapidly, from just 20,000 tonnes of niobium in concentrates in 1997 to the level of approximately 100,000 tonnes from 2007 onward, apart from the decline due to the 2008 economic recession (Figure 15.9).

Tantalum deposits are similarly widely dispersed, and, like niobium, there are few mines in production. Annual tantalum concentrate production climbed at an eight per cent annual

Table 15.3 Estimated global reserves and resources of niobium pentoxide, Nb_2O_5. (Data from numerous company websites and other published sources.)

Source	Contained Nb_2O_5 in proven and probable reserves (thousand tonnes)	Contained Nb_2O_5 in measured and indicated resources (thousand tonnes)
Australia	165	164
Brazil	44	78,133
Canada	1810	3005
China	-	2200
Egypt	-	4
Malawi	-	174
Mozambique	-	52
USA	-	129
Total	2019	81,662

[1] Inferred resources are also reported in Brazil, Gabon, Kenya, Canada, Tanzania, Ethiopia, Saudi Arabia, Spain, Angola, Mozambique and USA.
[2] Some deposits are omitted because no reliable reserve or resource data are available.

Table 15.4 Estimated global tantalum reserves and resources of tantalum. (Data from USGS, 2011b; Burt, 2010; DNPM, 2011.)

Source	Most likely resource base (tonnes Ta_2O_5)	Percentage of resources	Reserves (tonnes Ta)
Brazil	129,274	40	87,360
Australia	65,771	21	40,560
China and Southeast Asia	33,112	10	7800
Russia and Middle East	31,298	10	–
Central Africa	28,576	9	3120
Other Africa	21,318	7	12,480
North America	5443	2	1500*
Europe	2268	1	–
Total	317,060	100	152,820

*Deemed uneconomic at 2010 prices (USGS, 2011b).

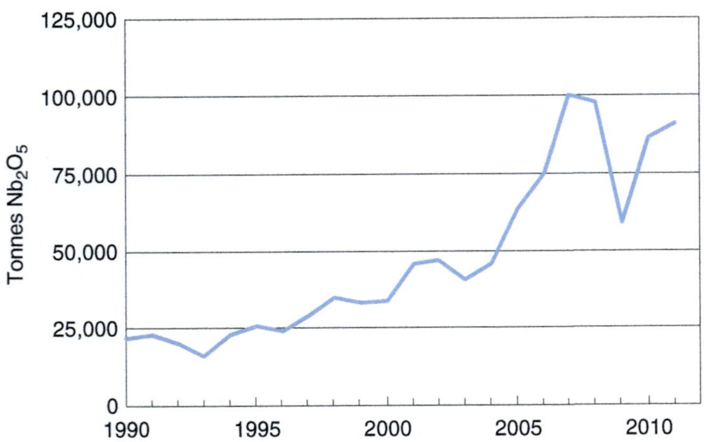

Figure 15.9 Global niobium production, 1990–2011. (After Schwela, 2011, with additional data from T.I.C. Bulletins.)

growth rate through the later 1990s, resulting in major expansion at both Greenbushes and Wodgina in Australia, and the development of other, smaller, mines in Africa and Australia. This new production peaked in 2002, at approximately 2300 tonnes, prior to stabilising at approximately 2100 tonnes until the 2008/9 recession. Mine production declined dramatically to a little over 1000 tonnes due to the temporary closure of several of the larger industrial mines. The downstream industry recovered rapidly, mined supplies tightened, prices rose accordingly and dormant supplies are returning to production and mined production is returning toward pre-2008 levels (Figure 15.10). This tightening was partially exacerbated by governments', and industry's, efforts to break the link between mining and the financing of illegal forces in central Africa.

Future supplies

The Araxá niobium mine has a history of increasing production to meet long-term market demand and it will therefore remain the major producer of niobium. Nevertheless, new or alternative supplies do exist and could be brought on stream if required. Worldwide known resources of niobium are very large in comparison with current global consumption.

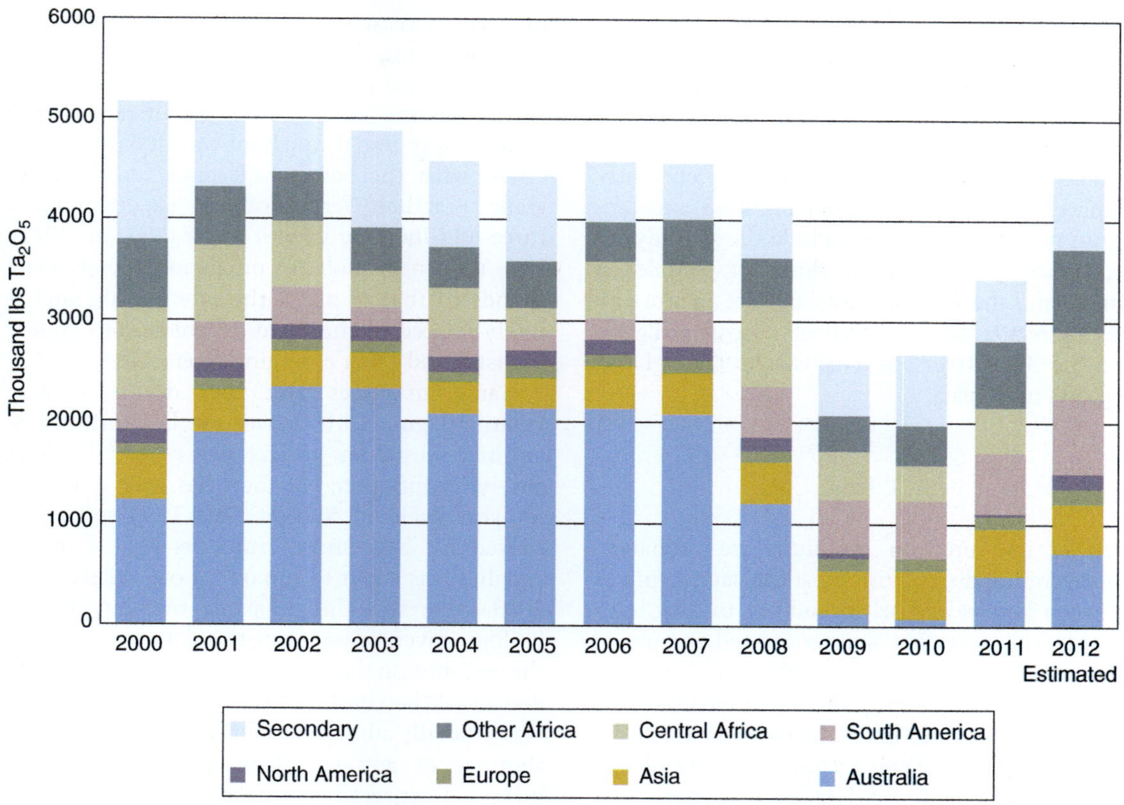

Figure 15.10 Global production of tantalum from concentrates and tin slags, 2000–2012. (Data from various trade documents, government statistics, and industry sources and estimates.) The 'Secondary' category includes low-grade tin slags and shipments of old stocks from the US Defence Logistics Agency 'stockpile'.

With high tantalum prices in 2011, some of the idle mine capacity in Australia (the Wodgina mine), Canada and Mozambique came back on stream. The Mibra Mine, in Brazil, having captured market share during the economic downturn, may continue to produce in excess of 20 per cent of the world's demand. Other mines that will continue to produce in excess of 100,000 lbs Ta_2O_5 per annum include Marrapino (Mozambique), Kenticha (Ethiopia), Pitinga (Brazil) and Yichun (China), while small, industrialised mines in Asia and South America will continue to contribute to overall supply.

Artisanal production in South America, southeast Asia, Nigeria and elsewhere in Africa will continue. Although production in individual mines is minor, in total it accounts for approximately 20 per cent of global production. As industry incorporates appropriate due diligence into its central African supply chain, 'conflict-free' production will return to, and may even exceed, historic levels in the Great Lakes Region, especially in the DRC.

There are several large deposits currently being examined for potential niobium and tantalum production. These include the Ghurayyah deposit in Saudi Arabia, the Motzfeldt Complex in Greenland, the Zashikhinsky deposit in Russia's Irkutsk region, the Abu Dabbab deposit in Egypt, and the Blue River and Nechalacho (Thor Lake) deposits in Canada.

Prices

Neither niobium nor tantalum are exchange-traded commodities, with most transaction prices between buyers and sellers subject to long-term confidential contracts. Only a limited amount of concentrate is traded against the 'spot' market, and even here prices are little more than 'best estimates' that are available through trade-journal subscription (e.g. Metal Pages, Asian Metals) or from annually compiled data, such as that of the US Geological Survey or the Brazilian government. The industry association, the Tantalum-Niobium International Study Center, by mandate, cannot collect or publish price information.

Price data are usually shown in a form other than that of elemental metal. For example, tantalum concentrates are quoted in the oxide form Ta_2O_5 – there are no data available for 'downstream' products, i.e. those from the processors to manufacturers. Of this, the tantalum oxide, approximately 82 per cent is tantalum metal. The contained metal price is therefore about 122 per cent of that shown in the oxide form. Niobium prices were historically quoted in terms of pyrochlore or columbite concentrate. However, since the early 1990s when pyrochlore was no longer sold on the open market, they are now normally quoted in terms of ferro-niobium – either as 'ferro-niobium' itself or based upon the niobium content.

Historically, niobium and ferro-niobium prices had held essentially constant, or declined once inflation had been taken into account. Production capacity increased to meet demand and with rapidly increasing demand, a price correction was inevitable and occurred in 2007 (Figure 15.11). Ferro-niobium prices rose significantly and continued to rise throughout the global financial crisis, with the benchmark price for standard-grade Brazilian ferro-niobium experiencing a three-fold increase in four years (Naidoo, 2011). This upturn in niobium prices in general corresponds to that of all of the rare metals, and is likely related to the rapid expansion of Chinese industry and other emerging economies.

Tantalum prices have been decidedly more volatile (Figure 15.12). Over the last forty years, 'normal' tantalum pricing has essentially kept pace with inflation, but there has been a roughly ten-year cycle of 'spikes'. The 1980 spike was caused by (unfounded) rumours that a major supplier was about to run out of ore, whereas the 2000 spike was due to panic over-buying and building inventories during the 'dot.com boom' on the assumption that mines could not keep up with demand. When clearer heads prevailed in 2002, it was generally admitted that there was never a true shortage of any significance, partially due to increased artisanal mining especially in central Africa (this is now accepted as being 'conflict mining' and was a result of – rather than the cause of – the civil war in the DRC). While the 'spot' prices in 2000 were far in excess of the 1980s peak,

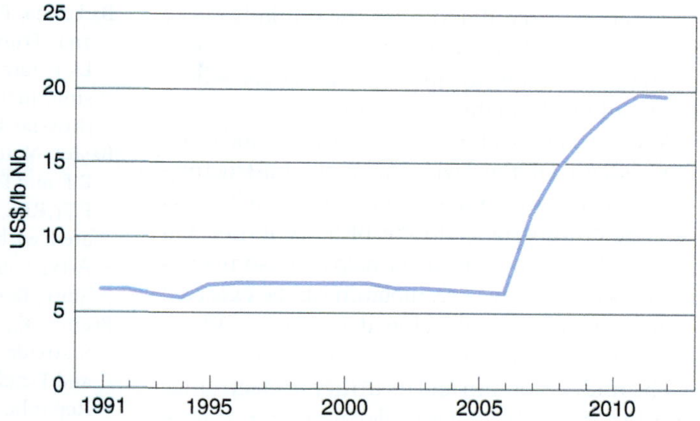

Figure 15.11 Average annual ferro-niobium price per pound of contained niobium, 1991–2012. (Based upon various technical and commercial sources.)

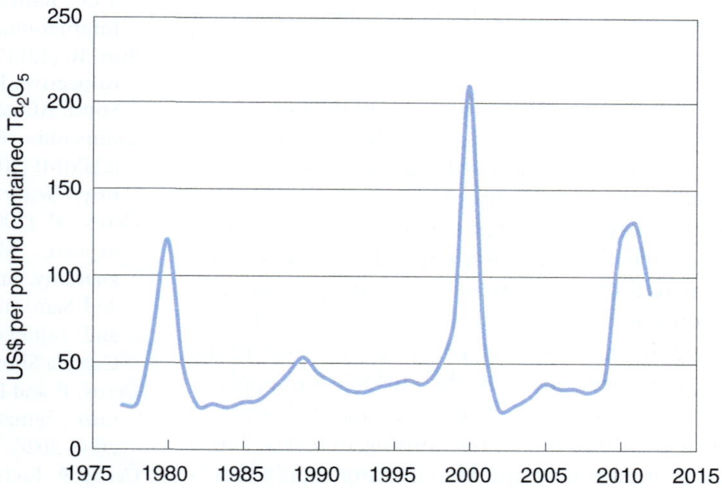

Figure 15.12 Average annual price for tantalum pentoxide contained in 30% tantalum pentoxide (Ta_2O_5) concentrate, 1977–2012. (Based upon various technical and commercial sources.)

it actually reflected far less of the market, and the spot + contract price peak was substantially lower. The 2010/11 spike is the most complex: its cause actually started as early 2003 when mine supply outstripped demand, and inventories throughout the supply chain grew to excessive levels, forcing prices down and the closure of several big, higher cost, mines by 2007/8. The dramatic, short-term reduction in demand in 2008/9 masked this, but while demand could and did rapidly return to normal levels, mine supply could not. As a result inventories declined, and prices once again spiked. While prices are declining again in late 2011, it is too early to determine whether they will return to the historic trend, or will remain at a level (and at a steady enough level) to satisfy the industrial miners while at the same time not resulting in equipment re-design to use alternatives without such price volatility as tantalum, even if they are less technically effective.

Outlook

The known global resources of niobium are exceptionally large and geological availability is, therefore, unlikely to be a significant problem in the foreseeable future. More problematic, however, is a lack of suppliers and, at present, consumers are essentially dependent on the policies of a single country, Brazil. If, in the unlikely event that the Brazilian sources were to become restricted, there are adequate resources elsewhere that would be brought on stream.

A major end use of niobium is in steels and other alloys essential to the development of infrastructure, i.e. railways, pipelines and automobiles. Niobium is therefore of importance to the BRIC countries and they can be expected to be principal consumers as growth occurs. Elsewhere, niobium can be expected to be consumed in developed countries where 'ageing' infrastructure has become a concern.

The role of niobium and its utility in electronics is still in the early stages of development. It now finds its way into computer chips and superconductors for medical technologies. Research is not for a replacement metal, but for expansion, and novel functions, of the metal.

Tantalum demand, after a rapid increase in the late 1990s, has slowed to about a 1 per cent compound growth rate throughout the last decade (ignoring the 2008/9 dip). As noted earlier, demand in the historically largest market sector – capacitors – has declined, while in other sectors it has increased. It will remain a critical component in electronic end uses, and there is no reason to presume the 'new' uses will decline. However, in the short to medium term, the outlook for supply is somewhat unclear. The tantalum industry, as with most sectors of the mining industry, has a habit of relatively quickly ensuring supply meets (and often exceeds) demand. The large tonnage low-grade deposits, identified globally, have yet to play a role in the tantalum supply chain, and these may await demand for their potential co-products, or new beneficiation techniques, to economically capture ultrafine tantalum mineralisation.

Note

1. Ma, million years before present

References

Anglo American (2013) Fact Book 2012/13. http://media.investis.com/A/Anglo-American-plc/2013/AA-Factbook-2012-2013.pdf

Badanina, E.V., Vekslerb, I.V., Thomas, R. Syritsoc, L.F. and Trumbull, R.B. (2004) Magmatic evolution of Li–F, rare-metal granites: a case study of melt inclusions in the Khangilay complex, Eastern Transbaikalia (Russia). Chemical Geology 210, 113–133.

Bastos Neto, A.C., Pereira, V.P., Ronci, L.H., de Lima, E.F. and Frantz, J.C. (2009) The world-class Sn, Nb, Ta F (Y,REE,Li) deposit and the massive cryolite associated with the albite-enriched facies of the Madeira A-type granite, Pitinga mining district, Amazonas State, Brazil. Canadian Mineralogist 47, 1329–1357.

Breiter, K., Škoda, R. and Uher, P. (2007) Nb-Ta-Ti-W-Sn-oxide minerals as indicators of a peraluminous P- and F-rich granitic system evolution: Podlesí, Czech Republic. Mineralogy and Petrology 91, 225–248.

Burt, R. (2010) Tantalum – a rare metal in abundance? T.I.C. Bulletin 141, March 2010, Tantalum-Niobium International Study Center, Brussels.

Burt, R. (2011) Trends in the tantalum industry – a retrospective. Keynote address at ZrTa 2011 conference. South Africa.

Companhia Brasileira de Metalurgia e Mineração (CBMM) (2011). Mines and operations – production. http://www.cbmm.com/

Černý, P. (2005) The Tanco rare-element pegmatite deposit, Manitoba: Regional context, internal anatomy, and global comparisons. In: Linnen, R.L. and Samson, I.M. (eds.) Rare element geochemistry and mineral deposits. Geological Association of Canada Short Course Notes, 17, 127–158.

Černý, P. and Ercit, T.S. (2005) The classification of granitic pegmatites revisited. Canadian Mineralogist 43, 2005–2026.

Černý, P., Ercit, T.S. and Vanstone, P.T. (1996) Petrology and mineralisation of the Tanco rare-element pegmatite, southeastern Manitoba. Geological Association of Canada Fieldtrip Guidebook.

Černý, P., Goad, B.E., Hawthorne, F.C. and Chapman, R. (1986) Fractionation trends of the Nb- and Ta-bearing oxide minerals in the Greer Lake pegmatitic granite and its pegmatite aureole, southeastern Manitoba. American Mineralogist 71, 501–517.

Chong, A. and Postolski, T. (2011) Blue River Ta-Nb Project NI 43–101 Technical Report.

Cordeiro, P.F.O., Brod, J.A., Dantas, E.L. and Barbosa, E.S.R. (2010) Mineral chemistry, isotope geochemistry and petrogenesis of niobium-rich rocks from the Catalão I carbonatite-phoscorite complex, Central Brazil. Lithos 118, 223–237.

DNPM. (2011) Sumário Mineral Brasiliero 2011. Tântalo. https://sistemas.dnpm.gov.br/publicacao/mostra_imagem.asp?IDBancoArquivoArquivo=6394

European Chemical Agency, ECHA. (2011) REACH: Regulations. http://echa.europa.eu/web/guest/regulations/reach.

Electronic Industry Citizenship Coalition, EICC. (2011) www.eicc.info/

Fetherston, J. M. (2004) Tantalum resources in Western Australia. Geological Survey of Western Australia, Mineral Resources Bulletin 22.

Graupner, T., Melcher, F., Gäbler, H.-E., Sitnikova, M., Brätz, H. and Bahr, A. (2010) Rare earth element geochemistry of columbite-group minerals: LA-ICP-MS data. Mineralogical Magazine 74, 691–713.

International Maritime Organisation, IMO. (2011) www.imo.org

Kravchenko, S.M. and Pokrovsky, B.G. (1995) The Tomtor alkaline ultrabasic massif and related REE-Nb deposits, northern Siberia. Economic Geology 90, 676–689.

Küster, D. (2009) Granitoid-hosted Ta mineralization in the Arabian-Nubian Shield: Ore deposit types, tectono-metallogenetic setting and petrogenetic framework. Ore Geology Reviews 35, 68–86.

Küster, D., Romer, R.L., Tolessa, D., Bheemalingeswara, D.Z.K., Melcher, F. and Oberthür, T. (2009) The Kenticha rare-element pegmatite, Ethiopia: internal differentiation, U–Pb age and Ta mineralization. Mineralium Deposita 44, (7), 723–750.

Lagache, M. and Quéméneur, J. (1997) The Volta Grande pegmatites, Minas Gerais, Brazil: An example of rare-element granitic pegmatites exceptionally enriched in lithium and rubidium. Canadian Mineralogist 35, 153–165.

Linnen, R.L. and Cuney, M. (2005) Granite-related rare-element deposits and experimental constraints on Ta-Nb-W-Sn-Zr-Hf mineralization. In: Linnen, R.L. and Samson, I.M. (eds.) Rare element geochemistry and mineral deposits. Geological Association of Canada Short Course Notes, 17, 45–68.

Melcher, F., Sitnikova, M.A., Graupner, T. et al. (2008) Fingerprinting of conflict minerals: columbite-tantalite ("coltan") ores. SGA News 23, 1–14.

Mining Journal (2010) Tantalum Supplement.

Mitchell, R.H. (2005) Mineralogical and experimental constraints on the origins of niobium mineralization in carbonatites. In: Linnen, R.L. and Samson, I.M. (eds.) Rare element geochemistry and mineral deposits. Geological Association of Canada Short Course Notes, 17, 201–215.

Mroz, J.P. (1983) Inventory of minerals in the extreme north of Mozambique IN Library report #1205 (extended extracts in Hunting 1988).

Naidoo, B. (2011) http://www.miningweekly.com/print-version/ferro-niobium-price-stability-benefits-malawi-project-says-globe-2011-06-07.

Nasraoui, M. and Waerenborgh, J.C. (2001) Fe speciation in weathered pyrochlore-group minerals from the Lueshe and Araxá (Barreiro) carbonatites by ^{57}Fe Mössbauer spectroscopy. Canadian Mineralogist 39, 1073–1080.

Organisation for Economic Co-operation and Development, OECD. (2011) OECD Due Diligence Guidance for Responsible Supply Chains of Minerals from Conflict-Affected and High-Risk Areas, OECD Publishing. http://dx.doi.org/10.1787/9789264111110-en

Palme, H. and O'Neill, H.St.C. (2004) Cosmochemical estimates of mantle composition. In: Carlson, R.W., Holland, H.D. and Turekian, K.K. (eds.) Treatise on Geochemistry 2, 1–38.

Partington, G.A., McNaughton, N.J. and Williams, I.S. (1995) A review of the geology, mineralization, and geochronology of the Greenbushes pegmatite, Western Australia. Economic Geology 90, 616–635.

Pell, J. (1996) Mineral deposits associated with carbonatites and related alkaline igneous rocks. In: Mitchell, R.H. (ed.) Mineralogical Association of Canada Short Course Volume 24, 271–312.

Philippo, S., Naud, J. and Verkaeren, J. (1997) Geochemical evaluation of the Lueshe niobium deposit (Zaire) by Rietveld quantitative X-ray diffraction. Applied Geochemistry 12, 175–180.

Pohl, W. (1994) Metallogeny of the northeastern Kibara belt, central Africa – recent perspectives. Ore Geology Reviews 9, 105–130.

Puchner, C.C. (1986) Geology, alteration and mineralization of the Kougarok Sn deposit, Seward Peninsula, Alaska. Economic Geology 81, 1775–1794.

Quest Rare Minerals Ltd. (2011) www.questrareminerals.com.

Raimbault, L., Cuney, M., Azencott, C., Duthou, J.L. and Joron, J.L. (1995) Geochemical evidence for a multistage magmatic genesis of Ta-Sn-Li mineralization in the granite at Beauvoir, French Massif Central. Economic Geology 90, 548–576.

Ram Resources Inc. (2011) www.ramresources.com.au.

Resende, I. (2011) AMG Mining Mibra Mine Update. Presentation at 52nd Annual Assembly of the Tantalum-niobium International Study Center, Almaty, Kazakhstan, October 2011.

Rub, A.K., Štemprok, M. and Rub, M.G. (1998) Tantalum mineralization in the apical part of the Cínovec (Zinnwald) granite stock. Mineralogy and Petrology 63, 199–222.

Rudnick, R.L. and Gao, S. (2004) Composition of the continental crust. In: Rucnick, R.L., Holland, H.D. and Turekian, K.K. (eds.) Treatise on Geochemistry 3, 1–64.

Salvi, S. and Williams-Jones, A.E. (2005) Alkaline granite-syenite deposits. In: Linnen, R.L. and Samson, I.M. (eds.) Rare element geochemistry and mineral deposits. Geological Association of Canada, Short Course Notes, 17, 315–341.

Salles, J.J. (2000) Production of niobium and tantalum from the Pitinga hard rock tin mine. T.I.C. Bulletin No. 101, March 2000.

Schwela, U. (2011) Tantalum-Niobium International Study Center (T.I.C.), Fifty-second General Assembly, Almaty, 16–19 October 2011, http://tanb.org/webfm_send/166.

Shaw, R.A. and Goodenough, K.M. (2011) Niobium-tantalum. Mineral Profile. British Geological Survey. http://www.bgs.ac.uk/mineralsuk/statistics/mineralProfiles.html

Sons of Gwalia (2002) Sons of Gwalia Annual Report, 28–35.

Stearns, J., Mahtani, D., Diallo, M.K., Danssaert, P. and Finardi, S. (2008) Final report of the Group of Experts on the Democratic Republic of Congo S/2008/773. United Nations Security Council.

Sweetapple, M.T. and Collins, P.L.F. (2002) Genetic Framework for the Classification and Distribution of Archean Rare Metal Pegmatites in the North Pilbara Craton, Western Australia. Economic Geology 97, 873–895.

Tantalum-Niobium International Study Center (T.I.C.). (2011) www.tanb.org

Tukiainen, T. (1988) Niobium-tantalum mineralisation in the Motzfeldt Centre of the Igaliko Complex, South Greenland. Mineral Deposits within the European Community. In: Boissonnas, J. and Omenetto, P. (eds.) Springer-Verlag, Berlin. 230–246.

U.S. Geological Survey, USGS. (2011a) Mineral Commodity Summaries: Niobium. http://minerals.usgs.gov/minerals/pubs/commodity/niobium/mcs-2011-niobi.pdf

U.S. Geological Survey, USGS. (2011b) Mineral Commodity Summaries: Tantalum. http://minerals.usgs.gov/minerals/pubs/commodity/niobium/mcs-2011-tanta.pdf

Van Lichtervelde, M., Linnen, R.L., Salvi, S. and Didier, B. (2007) Textures and chemical evolutions in tantalum oxides: a discussion of magmatic versus metasomatic origins for Ta mineralization in the Tanco Lower Pegmatite, Manitoba, Canada. Economic Geology 102, 257–276.

Winter, M. (2011) WebElements: the periodic table on the web: Home of the periodic table. WebElements, The University of Sheffield and WebElements Ltd, UK. www.webelements.com.

Winter, J.D. (2010) Principles of Igneous and Metamorphic Petrology. Prentice Hall, Upper Saddle River, New Jersey, USA.

Wood, S.A. (2005) The aqueous geochemistry of zirconium, hafnium, niobium and tantalum. In: Linnen, R.L. and Samson, I.M. (eds.) Rare element geochemistry and mineral deposits. Geological Association of Canada Short Course Notes, 17, 217–268.

Yin, L., Pollard, P.J., Shouxi, H. and Taylor, R.G. (1995) Geologic and geochemical characteristics of the Yichun Ta-Nb-Li deposit, Jiangxi Province, South China. Economic Geology 90, 577–585.

16. Tungsten

TERESA BROWN AND PETER PITFIELD

British Geological Survey, Keyworth, Nottingham, UK

Introduction

Tungsten, also known as wolfram, was discovered in 1781 when Carl Wilhelm Scheele produced tungstic acid from scheelite. In 1783 brothers Juan José and Fausto de Elhuyar discovered that the same acid could also be produced from the mineral wolframite and subsequently they managed to isolate the metal by reducing this acid with charcoal. The name tungsten comes from the Nordic words '*tung*' and '*sten*' meaning 'heavy' and 'stone'. The name 'wolfram' has older roots and is believed to have been used because the yield of tin during smelting was reduced if tungsten minerals were present (for more details see Schubert and Lassner, 2009).

The main applications for tungsten are in 'hard metals', i.e. tungsten carbide and cemented carbides, used for cutting, drilling and wear-resistant parts or coatings. Tungsten is also used to add hardness and strength to steel alloys, particularly where heat resistance is also required.

Physical and chemical properties

Tungsten is a hard, very dense, steel-grey to greyish-white metal. It has the highest melting point of all non-alloyed metals and the second highest of all elements behind carbon. Of all pure metals, tungsten has the lowest coefficient of expansion and the highest tensile strength at temperatures over 1650 °C (Christie and Brathwaite, 1996). Tungsten is also known for its high density, which is similar to gold, and its high thermal and electrical conductivities. It has excellent corrosion resistance, does not react with air or water at room temperature (although fresh surfaces will oxidise) and is largely unaffected by most acids. Key properties are summarised in Table 16.1.

Distribution and abundance in the Earth's crust

The average abundance of tungsten in the Earth's continental crust is estimated to be 1.0 parts per million (ppm) (Rudnick and Gao, 2003). The upper crust contains about 1.9 ppm tungsten, whereas the middle and lower crust is estimated to contain 0.6 ppm. The abundance of tungsten in the oceans is estimated to be 0.1 parts per billion (ppb). The average concentration in workable ores is usually between 0.1 and 1.0 per cent tungsten trioxide (WO_3).

Critical Metals Handbook, First Edition. Edited by Gus Gunn.

Mineralogy

Tungsten does not occur in nature as a free metal, but only in the form of chemical compounds with other elements. Although several tungsten-bearing minerals are known, most are rare or very rare. Only scheelite and the wolframite group are abundant enough to be considered ores (Table 16.2).

Scheelite, a calcium tungstate ($CaWO_4$), is typically white to yellowish in colour, and has blue-white fluorescence in ultraviolet light; a property which is especially utilised in exploration and mining. The colour of the fluorescent light is influenced by the molybdate content and changes from blue to cream and then to pale yellow and orange with increasing molybdenum content (Lassner and Schubert, 1999).

Wolframite is a general term for iron–manganese tungstate and is a solid-solution series between two end members: ferberite ($FeWO_4$, with less than 20 per cent manganese) and hübnerite ($MnWO_4$, with less than 20 per cent iron). In practice, the name wolframite is often used for the intermediate mineral between these two end members. The wolframite group exhibit typically tabular morphology, and are usually black, dark grey or reddish-brown in colour (Lassner and Schubert, 1999).

Secondary tungsten minerals, such as hydrotungstite ($H_2WO_4 \cdot H_2O$) or cerotungstite ($CeW_2O_6(OH)_3$), can be produced by alteration processes or weathering and may cause problems during processing leading to reduced recovery of tungsten (Schmidt, 2012).

Table 16.1 Selected properties of tungsten.

Property	Value	Units
Symbol	W	
Atomic number	74	
Atomic weight	183.84	
Density at 25 °C	19254	kg/m^3
Melting point	3422	°C
Boiling point	5555	°C
Hardness (Mohs scale)	7.5	
Specific heat capacity at 25 °C	0.13	J/(g °C)
Electrical conductivity	18.2×10^6	S/m
Coefficient of linear thermal expansion	4.5×10^{-6}	/°C
Tensile strength at 20 °C	1000	MPa
Tensile strength at 1650 °C	approx 100	MPa
Thermal conductivity	174	W/(m °C)

Deposit types

Tungsten deposits usually occur within, or near to, orogenic belts resulting from subduction-related plate tectonics. All major deposit types are associated with granitic intrusions or with

Table 16.2 Properties of the most common tungsten minerals.

		Wolframite Group		
	Scheelite	Ferberite	Wolframite	Hübnerite
Chemical formula	$CaWO_4$	$FeWO_4$	$(Fe,Mn)WO_4$	$MnWO_4$
Tungsten trioxide content (WO_3 %)	80.6	76.3	76.5	76.6
Specific gravity (g/cm^3)	5.4–6.1	7.5	7.1–7.5	7.2–7.3
Colour	Pale yellow to orange, green to dark brown, pinkish-tan, dark blue to black, white or colourless	Black	Dark grey to black	Red-brown to black
Lustre	Vitreous or resinous	Submetallic to metallic	Submetallic to metallic	Submetallic to adamantine
Hardness (Mohs scale)	4.5–5.0	5.0	5.0–5.5	5.0
Crystal structure	Tetragonal	Monoclinic	Monoclinic	Monoclinic

medium- to high-grade metamorphic rocks. The locations of selected major tungsten mines and deposits are shown in Figure 16.1.

Werner et al. (1998) classified major tungsten deposits into seven types: vein/stockwork, skarn, disseminated, porphyry, stratabound, placer, and brine/evaporite. Four additional types of relatively minor economic interest were also identified as: pegmatite, breccia, pipe, and hot-spring deposits. In reality the categorisation of individual localities can be complicated by multiple stages of formation creating deposits that could be ascribed to more than one of these types. Table 16.3 contains a summary of typical sizes and grades for the major producing deposit types.

Vein and stockwork deposits

Vein and stockwork deposits are genetically related to the development of fractures that occur in or near granitic intrusions during emplacement and crystallisation. These fissures are frequently filled with quartz and can be up to several metres in width. Large vein deposits may contain several individual veins, while stockworks comprise swarms of parallel, or near parallel, veins with interconnecting veinlets (Werner et al., 1998).

The veining is commonly bordered by greisen (a form of endoskarn alteration) and is often spatially associated with disseminated greisen and porphyry-style tungsten mineralisation.

The mineralogy of vein deposits range from the simple, consisting almost entirely of quartz and wolframite, to the complex, as at Pasto Bueno in Peru (Landis and Rye, 1974) or Panasqueira in Portugal (Kelly and Rye, 1979) where more than 50 vein-forming minerals have been identified. The wolframite series is the main tungsten-bearing mineral but scheelite also occurs in some deposits of this type. Tin, copper, molybdenum, bismuth and gold may also be present in economic quantities. In addition, uranium, thorium, rare earth elements and phosphate minerals may also occur (Elliott et al., 1995).

In general, vein and stockwork deposits tend to be low grade. However, even with grades as low as 0.1 per cent tungsten trioxide they can still be exploited economically by bulk mining methods, as demonstrated at Mount Carbine mine in Australia (De Roo, 1988), which operated successfully at this low grade between 1973 and 1986. Other notable examples of vein/stockwork deposits are found at Verkhne-Kayrakty, Kazakhstan (Rubinstein and Barsky, 2002); Xihuashan in the Nanling tungsten–tin province, China (Elliott, 1992; Guiliani et al., 1988; Wang et al., 2011); Bolsa Negra and Chicote Grande, Bolivia (Cox and Bagby, 1986) and Hemerdon, United Kingdom (Mining Magazine, 1979; Keats, 1981). The tungsten mineralisation at Hemerdon in Devon, is hosted in sheeted greisens-bordered vein systems and stockworks in the apex of a steeply dipping, dyke-like, granite body (known as Hemerdon Ball) on the south-west side of the early Permian-age Dartmoor pluton. Mineralisation extends to depths of at least 400 metres below surface.

The Ta'ergou tungsten deposit, located in the western part of the Qilian orogen in Gansu Province, north-western China, consists of scheelite skarn bodies and wolframite–quartz veins. The deposit is genetically related to the Caledonian Teniutan granodiorite emplaced within Proterozoic rocks bordering the Archaean North Chine Platform. The tungsten veining overprints the calcic skarns (Zhang et al., 2003).

The tungsten vein system in the Nanling Range, South China, has been described using a so-called "five-floor building" model (Gu, 1982; Liu and Ma, 1993; Li et al., 2011). Within this model there are broadly five vertical zones: a thread or stringer zone at the top, over a veinlet zone, thin vein or mixed zone, large vein zone and finally a thin-out or extinction zone (Figure 16.2). Large veins can also be sparsely developed in the underlying granite where typically they are weakly mineralised. The large-vein zone and the thin-vein or mixed-vein zones are the most economically valuable.

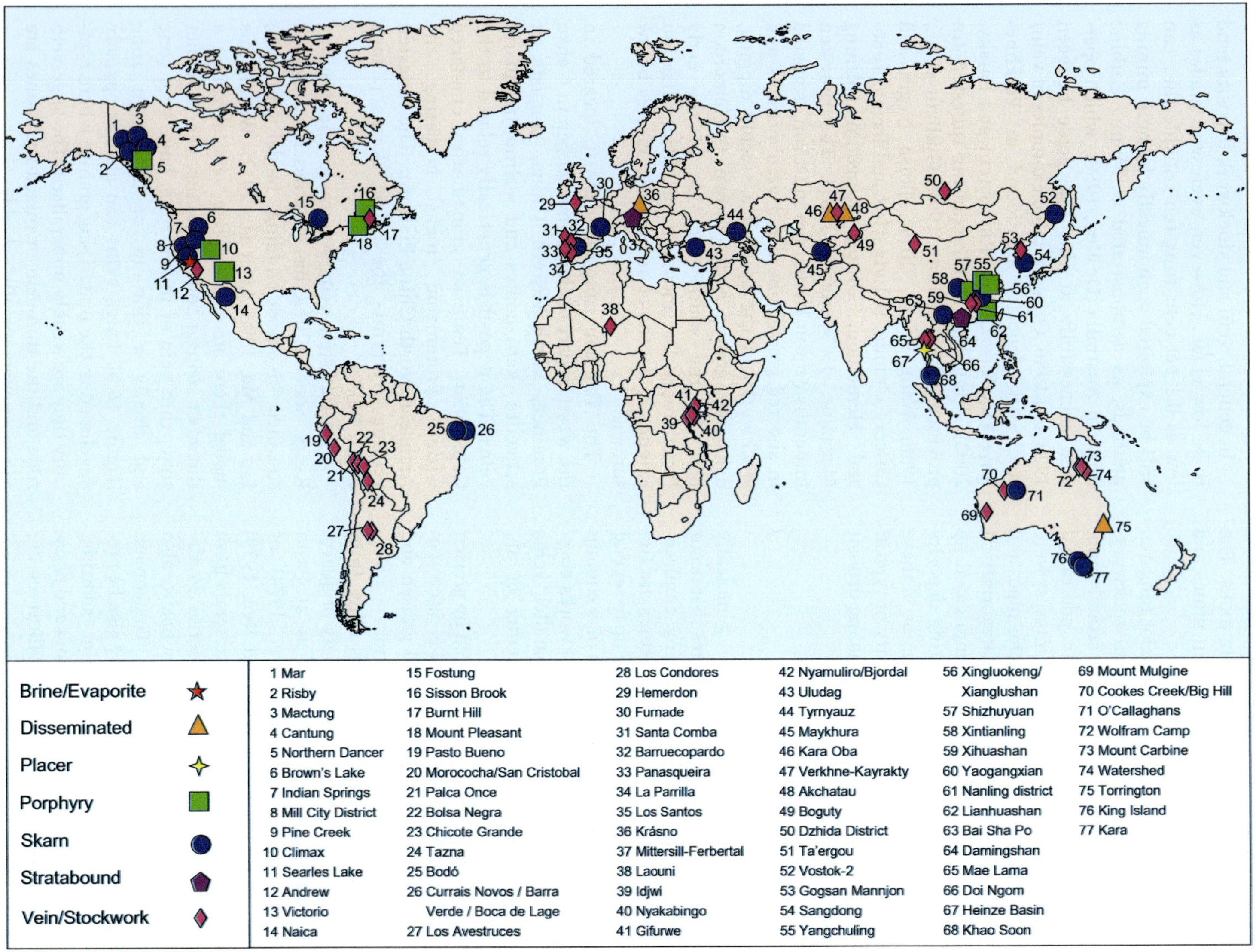

Figure 16.1 The location of selected major tungsten mines and deposits. Note: In China there are many other deposits noted in the literature and it is not possible at this scale to include them all.

Table 16.3 Typical size and grade of major producing tungsten deposit types.

Deposit type	Deposit size range (metric tonnes)	Typical grade (WO_3 %)	Examples
Vein/stockwork	<10^5 to 10^8	0.1 to 0.8	Panasqueira (Portugal), Pasto Bueno (Peru), Hemerdon (UK)
Skarn	<10^4 to 10^7	0.1 to 1.5	Mactung and Cantung (Canada), Xintianling and Yaogangxian (China)
Disseminated	<10^7 to 10^8	0.1 to 0.5	Akchatau (Kazakhstan), Krásno (Czech Republic)
Porphyry	<10^7 to 10^8	0.08 to 0.4	Xingluokeng and Yangchulin (China), Northern Dancer (Canada)
Stratabound	<10^6 to 10^7	0.2 to 1.0	Mittersill-Ferbertal (Austria), Damingshan (China)

	Morphological zones	Zones thickness (metres)	Average vein width (metres)	Density (veins per metre)	Mineral association
I	Stringer or thread zone	100–200	0.001–0.01	0.05–5	muscovite, cassiterite, wolframite, tourmaline
II	Veinlet zone	50–150	0.02–0.1	5–20	cassiterite, wolframite, chalcopyrite, bismuthine, beryl, muscovite, pyrite
III	Thin vein or mixed zone	150–250	0.05–0.5	1–8	wolframite, cassiterite, scheelite, chalcopyrite, pyrite, molybdenite, bismuthine, beryl, galena, sphalerite
IV	Large vein zone	200–450	0.2–2	0.03–2	wolframite, chalcopyrite, molybdenite, potash, feldspar, pyrite, galena
V	Extinction or thin-out zone	>50	0.05–2	0.05–0.1	wolframite, molybdenite, chalcopyrite, pyrite, potash, feldspar

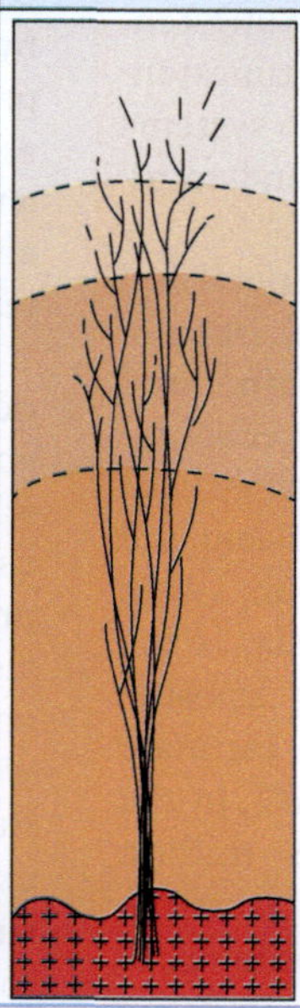

Figure 16.2 Schematic model of 'five-floor building' vein-type tungsten deposit. (Data sourced from: Gu, 1982; Huang and Xiao, 1986.)

Skarn deposits

Skarns are coarse-grained rocks dominated by calc-silicate minerals that have formed by metasomatic processes in sequences containing carbonate-bearing rocks such as limestone (Einaudi et al., 1981). Most are found adjacent to plutons and are often associated with hornfels, skarnoid, marble and other similar rocks in the thermal aureoles. They can also occur along faults, major shear zones, in shallow geothermal systems and in metamorphic rocks at lower crustal depths (Meinert et al., 2005). Less common types of skarns form in contact with sulfidic or carbonaceous rocks such as banded iron formations or black shales. Calcic skarns are characterised by calcium- and iron-rich silicates (andradite, hedenbergite, wollastonite); magnesian skarns by calcium- and magnesium-rich silicates (forsterite, diopside, serpentine); and aluminous skarns by aluminium- and magnesium-rich calc-silicates (grossularite, vesuvianite, epidote). Dolomitic rocks tend to inhibit the development of tungsten-bearing skarns; consequently magnesian tungsten-bearing skarns are uncommon (Ray, 1995).

Newberry and Einaudi (1981) distinguished two types of skarn on the basis of host-rock composition and relative depth: reduced skarns such as the Cantung and Mactung deposits in the North West Territories, Canada, and oxidised skarns, such as King Island, Tasmania, Australia (Kwak and Tan, 1981; Kwak, 1987). In general, oxidised tungsten-bearing skarns are smaller than reduced tungsten-bearing skarns. The highest grades in both systems are associated with hydrous minerals and retrograde alteration (Meinert et al., 2005).

Scheelite is the principal tungsten mineral and this may occur as disseminated grains or fracture fillings. Copper, molybdenum, tin, zinc and bismuth may also be present and can be economically recoverable. Economically exploitable skarn deposits usually contain between 0.1 per cent and 1.5 per cent tungsten trioxide (Werner et al., 1998). Mactung in the Yukon Territory, Canada, is the one of the largest tungsten-bearing skarn deposits, with a NI43-101 compliant resource estimate including 33 million tonnes @ 0.88% WO_3 in the indicated category and a further 11.9 million tonnes @ 0.78% WO_3 in the inferred resource category (Narciso et al., 2009). The deposits occur in the thermal aureole of a late Cretaceous felsic intrusion, which was emplaced into a dominantly pelitic, Lower Paleozoic sequence along the eastern margin of the Selwyn basin. The mineralisation is stratabound and confined to four individual beds (Dick and Hodgson, 1982).

Other tungsten-bearing skarn deposits include Los Santos in Spain (Tornos et al., 2007), Tyrnyauz and Vostock-2 in Russia (Soloviev and Krivoschchekov, 2011) and Xintianling and Yaogangxian in China (Zhao et al., 1990; Chang, 2005). The Xintianling scheelite deposit, located on the north-east side of the Qitianling batholith in Hunan Province, is one of the largest amongst several tungsten-bearing skarn deposits in China. It is spatially associated with two-stage granite emplacement into Carboniferous dolomitic limestones over a 10 Ma interval in the mid–late Jurassic. Skarn ore formation is related to the older stage granite. It is a typical example of the Jurassic tungsten–tin ore-forming event in the Nanling Range of south China (Zhang et al., 2011).

Disseminated or greisen deposits

In disseminated or greisen deposits wolframite or scheelite are disseminated in highly altered (greisenised) granite or granitic pegmatite. Greisen comprises mainly quartz and mica and is formed by post-magmatic metasomatic replacement of the primary granite minerals. Disseminated deposits are distinguished from the greisen-bordered veins and stockworks by the pervasive nature of the alteration and the absence of fluid pathways. In reality, these deposit types commonly coexist.

Disseminated greisen deposits usually occur near to the upper parts of intrusions that are emplaced at depths of between 0.5 and 5 km, where fluids can boil but are prevented from escaping to the surface. Tungsten is usually present as wolframite although some deposits also contain scheelite. Tin, molybdenum, bismuth and other base metals may also be present, along with quartz, topaz, white mica, tourmaline and fluorite. Tungsten grades are generally low but exploitation can be economic as a by-product of tin extraction. Examples of disseminated deposits include the Akchatau and Kara Oba deposits in central Kazakhstan (Zaraisky and Dubinina, 2001), the Torrington district of New South Wales, Australia and the Hub stock at Krásno in the Czech Republic (Jarchovsky, 2006).

Porphyry deposits

Porphyry deposits are extensive, low-grade deposits formed following the separation of metal-rich fluids from a crystallising wet magma. Tungsten tends to be concentrated in stockwork zones and fractures either in or near to the upper parts of granitic intrusions emplaced at shallow depths. Mineralised breccia zones may also be present.

Tungsten occurs either as wolframite or scheelite, and sometimes both are present. Molybdenum, bismuth and tin often occur and may represent an opportunity for co-production. Tungsten-bearing porphyry deposits tend to be large in size but may not be economic due to their low grade. Nevertheless, important examples include Northern Dancer (formerly known as Logtung), Sisson Brook and Mount Pleasant in Canada (Brand, 2008; Snow and Coker, 1986; Kooiman et al., 1986) and the Xingluokeng, Shizhuyan, Lianhuashan and Yangchulin deposits in China (Liu, 1980; Zhaolin and Zhongfang, 1996; Werner et al., 1998).

The Northern Dancer porphyry of the Western Cordillera, which straddles the boundary between Yukon Territory and British Columbia Province, comprises multiple, mid-Cretaceous felsic intrusions hosting four vein systems with different tungsten/molybdenum ratios (Noble et al., 1984). Mineralisation is centred on a felsic porphyry dyke complex and includes stockworks, sheeted vein systems, disseminations and skarns but mostly comprises typical porphyry-style crackle breccias showing many similarities with porphyry molybdenum deposits. Resources, which are NI43-101 compliant, are estimated to include 30.8 million tonnes @ 0.114% WO_3 in the measured category, 192.6 million tonnes @ 0.1% WO_3 in the indicated category and a further 201.2 million tonnes @ 0.089% WO_3 in the inferred resource category (Molavi et al., 2011).

The Xingluokeng tungsten–molybdenum deposit of the Fujian province is hosted in a late Jurassic Yanshanian granite porphyry stock in the Wuyishan metallogenic belt. The central zone underwent strong silicic and potassic alteration and is enriched in rare earth elements (Zhang et al., 2008).

The Shizhuyuan deposit in the Dongpo orefield of Hunan Province is a world-class polymetallic tungsten deposit. The mineralisation is diverse in character and has a complex origin related to multiple phases of granite intrusion (Lu et al., 2003). It comprises dominantly W–Mo–Bi–Sn–F calcic skarn-greisen zones developed around the late Jurassic Qianlishan granite complex of the Yanshanian granitoid province. The highest tungsten grades occur in vein/stockwork W–Sn–Mo–Bi–Be mineralisation associated with a later granite phase which is superimposed on the early stage massive skarn-greisen zone. On this basis some researchers have described this deposit as porphyry in style (Li et al., 2004).

Breccia deposits

Breccia deposits are composed of angular, broken fragments of rock located within, above or marginal to the apex of an intrusion. They are formed either by magmatic/hydrothermal hydraulic fracturing or by explosive interactions between water and magma. Many vein/stockwork and porphyry deposits have breccia zones associated with them. However, some tungsten-bearing breccia bodies appear to have formed independently of other deposit types (Werner et al., 1998). An example is the Washington copper–molybdenum–tungsten breccia pipe in Sonora, Mexico (Simmons and Sawkins, 1983).

Stratabound deposits

Tungsten mineralisation in stratabound deposits is confined to a single stratigraphic unit, although they may not be strictly conformable to bedding, i.e. mineralisation may cross bedding planes. Stratabound deposits occur in volcano-sedimentary sequences and are considered to be syngenetic in origin. They can be distinguished from skarn deposits which are largely controlled by the composition of the host rock lithology and are assumed to be epigenetic. Stratabound tungsten mineralisation occurs within iron–magnesite and dolomitic marbles in the Eastern Alps (Neinavale et al., 1989). Many stratabound tungsten occurrences appear to have been affected by later mobilisation and reconcentration and therefore their syngenetic origin is questionable (Werner et al., 1998).

Examples of this type of tungsten deposit include the Mittersill-Ferbertal deposits in the Salzburg province of Austria and Damingshan in China (Ma, 1982). The Cambrian-age Mittersill-Ferbertal orefield comprises several lenses of scheelite-rich quartzite, an underlying vein-stockwork zone, an eruption breccia and quartz-rich aureole to a granitoid intrusion. Further scheelite enrichments occur along shear zones. Geochronological and geochemical data on the Ferbertal deposit indicate a genetic link with mantle-dominated granitic melts (Eichhorn et al., 1999). Subsequent metamorphic events and granitic intrusions have remobilised the scheelite (Höll and Eichhorn, 2000). The Cambrian-Ordovician Damingshan tungsten deposit in the Danchi metallogenic belt in Guangxi Province includes Late Cretaceous vein and stockwork wolframite mineralisation as well as the massive stratiform types (Li et al., 2008).

Pegmatite deposits

Pegmatites are coarse-grained igneous rocks, generally of granitic composition, formed by the late-stage crystallisation of magma and containing many incompatible elements such as lithium, beryllium, niobium, tantalum, tin and uranium. Tungsten is not a common constituent of pegmatites and tungsten-bearing pegmatite deposits are therefore rare. Grades for tungsten tend to be low but it can be extracted as a by-product. For example, tungsten occurs in the pegmatite which is worked primarily for tantalum at the Wodgina mine in Western Australia, and it is also found in the Okbang deposit in South Korea (Chung, 1975).

Pipe deposits

Pipe deposits can be cylindrical or irregular, elongated or bulbous masses or quartz that occur at the margins of granitic intrusions. Mineralisation, most frequently wolframite, is often erratically distributed in high-grade shoots or pockets containing up to 20 per cent wolframite, but deposits tend to be small. Examples include the Wolfram Camp deposit in Queensland, Australia (Plimer, 1975).

Hot-spring deposits

These deposits are probably derived from bedrock tungsten-bearing deposits and are formed by circulating hot ground water. Deposits of calcareous tuffs or travertine are formed by precipitation as this hot ground water cools and tungsten mineralisation has been found in selected locations. Examples, where these deposits have been worked in the past, include Golconda in Nevada, USA (Kerr, 1940; Marsh and Erickson, 1975) and Uncia in Bolivia (Werner et al., 1998).

Placer deposits

Placer deposits are concentrations of heavy and chemically resistant minerals that occur in sediments. Wolframite and scheelite, although heavy, will eventually decompose during weathering and therefore tend not to be preserved long enough to form widespread placer deposits. However, they do occur, in both alluvial and coastal sediments, albeit they are usually small in size. Typically, they are located very close to the bedrock deposit from which they were originally derived (Werner et al., 1998). A few tungsten-bearing placer deposits have been worked on an industrial scale, for example in the Heinze Basin in Burma (Myanmar) (Goosens, 1978) and in the Dzhida district of eastern Siberia.

Brine and evaporite deposits

Tungsten-bearing brines and evaporite deposits occur in recent lakes and/or palaeolake settings in arid regions of Asia and North America. The tungsten is thought to have been leached from bedrock deposits by hot fluids. The most significant example of this type of deposit is the Searles Lake deposit in California, USA (Guerenko and Schmincke, 2002; Altringer, 1985). Stratabound scheelite at Halls Creek in Western Australia is of evaporitic origin analogous to continental-sabkha playa basins of the Mojave desert (Todd, 1989).

Extraction methods, processing and beneficiation

Extraction

Tungsten mining techniques are similar to many other metals of similar occurrence. Most tungsten is mined from sub-surface (or underground) mines. A few tungsten mines have used surface (open-pit) methods, but in most cases, these mines were later converted to underground mines to access deeper ores.

Surface mining

Open-pit mining is generally cheaper and safer to operate than underground mining and is normally used if a near-surface ore body occurs. Open-pit mining can reach depths of several hundred metres but rarely exceeds 100 metres in tungsten mine operations before going underground. Some placer deposits are amenable to strip mining or dredging operations.

Underground mining

Underground mining is preferred when surface mining is, or becomes, prohibitively expensive. A major factor in the decision to mine underground is the ratio of waste to ore (strip ratio). Once this ratio becomes large, surface mining is no longer economic. Ore is mined in stopes on a number of roughly horizontal levels at various depths below the surface. The particular method used for a mine will depend on the size, shape and grade of the ore body, its depth below the surface and the competence of the ore zone and wall rocks (Shedd, 2011).

Processing

The first phase of processing the ore is beneficiation at the mine site to increase the tungsten content. The resulting concentrate containing more than 65 per cent tungsten trioxide can either be used directly for production of ferrotungsten and steel manufacture, converted to a number of intermediate tungsten compounds by hydrometallurgical processes or further refined to pure tungsten using pyrometallurgical techniques (Figure 16.3).

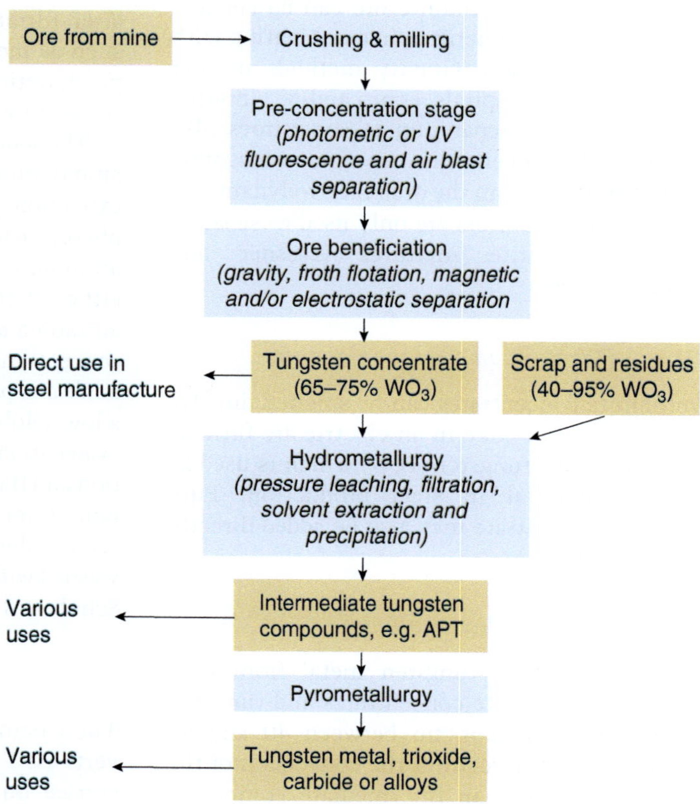

Figure 16.3 Simplified flow diagram illustrating the generic steps in processing tungsten.

Ore benefication

The tungsten ore is first crushed and milled to liberate the tungsten mineral grains. The slurry containing tungsten minerals and waste rock is then concentrated using several methods (e.g. gravity, froth flotation, magnetic and electrostatic separation) depending on the characteristics and composition of the ore. Some plants incorporate a pre-concentration stage before other conventional processing methods. This usually involves an ore-sorting process whereby a continuous stream of rock particles is subjected to a 'sensing' stage and are subsequently separated using an air blast. The sensing stage can consist of a photometric method if the dark tungsten ore minerals are contained within white quartz or a UV-fluorescence process if the ore is scheelite (Lassner and Schubert, 1999).

After that, scheelite ore can be concentrated by gravity methods, often combined with froth flotation, whilst wolframite ore can be concentrated by gravity, sometimes in combination with magnetic separation. Gravity methods usually involve the use of spirals, cones and/or vibrating tables. Magnetic separation is sometimes also used to clean scheelite (i.e. to remove magnetic gangue minerals from the ore). Electrodynamic or electrostatic separators are only used to separate scheelite–cassiterite mixtures (Lassner and Schubert, 1999).

Direct use in steel manufacture

Wolframite concentrates can be smelted directly with charcoal or coke in an electric arc furnace to produce ferrotungsten (FeW) which is used as alloying material in steel production. Pure scheelite concentrate may also be added directly to molten steel.

Hydrometallurgy

The production of tungsten metal from wolframite and scheelite concentrates and tungsten scrap (which can contain between 40 and 95 per cent tungsten) requires the conversion of the feed material into an intermediate compound that can be readily purified and then reduced. In modern plants most tungsten concentrates are processed chemically to an ammonium paratungstate (APT) intermediate.

In the hydrometallurgical processes, scheelite is first digested by sodium carbonate pressure leaching and wolframite by sodium hydroxide. Wolframite–scheelite mixtures can be successfully digested by a pressure leach with sodium hydroxide under simultaneous mechanical activation.

Tungsten scrap with only a few exceptions is easier to convert to APT than ore concentrates because it does not contain detrimental elements like phosphorus, arsenic and silicon. The tungsten scrap is first oxidised to facilitate dissolution in the alkaline leach process. Dissolution is achieved either by sodium hydroxide pressure digestion or an oxidising alkaline melt or by electrolysis.

Following the dissolution process the undissolved digestion residues are removed by filtration. However, the sodium tungstate solution after filtration may still contain contaminants such as molybdenum. These can be removed by precipitation under carefully controlled conditions or by ion exchange and adsorption.

The sodium tungstate solution is converted into ammonium isopolytungstate by either solvent extraction (liquid ion exchange using an organic phase) or ion exchange by resins. The resultant ammonium isopolytungstate solution from either of these processes is evaporated whereby ammonia and water are volatilised. The ammonia concentration decreases relative to tungsten trioxide and paratungstate, as the ammonium salt has a low solubility, and crystallises out. The degree of evaporation depends on the purity of the feed solution and the required purity of the APT, which may range from 90 to 99 per cent. The tungsten-bearing crystal slurry is filtered from the mother liquor, washed with deionised water and dried (Lassner and Schubert, 1999).

Pyrometallurgy

The calcination under oxidising condition converts APT to tungsten trioxide. This is usually carried out in a rotary furnace at 500–700 °C.

Calcination under slightly reducing conditions results in tungsten blue oxide. The latter can be carried out in either a push-type or rotary furnace at temperatures between 400 °C and 900 °C.

The conversion of either oxide into tungsten metal powder requires further treatment in a furnace, almost exclusively under hydrogen reducing conditions. This is carried out either in push-type or rotary furnaces with carefully controlled temperature zones varying between 600 °C and 1100 °C. The hydrogen not only provides a reducing environment, it also acts to remove any water vapour which forms during the process (Lassner and Schubert, 1999).

Most uses of tungsten powder, including in the production of tungsten carbide or the various tungsten alloys, involves further heat treatment. Unlike other metals, compaction and sintering processes are normally used, rather than smelting technology, because tungsten has an exceptionally high melting temperature. Compaction involves pressing tungsten powder in rigid dies or flexible moulds and these are then sintered at high temperatures to increase their strength and density, often in a reducing environment created by the presence of hydrogen (Lassner and Schubert, 1999).

Specifications and uses

Specifications

Tungsten is used and traded in a variety of forms (Lassner and Schubert, 1999; International Tungsten Industry Association (ITIA), 2012), the most important of which are:

Ammonium paratungstate (APT) is a white free-flowing crystalline salt with the chemical formula $(NH_4)_{10}(W_{12}O_{41}){\cdot}5H_2O$. It is the main intermediate and also the main tungsten raw material traded in the market. APT is usually calcined to yellow trioxide (WO_3) or blue oxide ($W_{20}O_{58}$).

Ammonium metatungstate (AMT) with the formula $(NH_4)_6H_2W_{12}O_{40}{\cdot}xH_2O$, is a highly soluble white crystalline powder and a source of high-purity, water-soluble tungsten, which is essentially free of alkali and other metallic impurities. It is produced from APT in an electrolytic cell.

Tungsten trioxide (WO_3) or tungstic oxide, is a light yellow powder synthesised by oxidative calcination of ammonium paratungstate. It can also be produced by calcination of tungstic acid.

Tungsten blue oxide (TBO) is a dark blue or black powder with a nominal chemical formula of $WO_{2.97}$. It is not a well-defined chemical compound but a mixture of constituents. TBO is formed by calcination of APT under slightly reducing conditions.

Tungsten metal powder (W) is produced from yellow or blue tungsten oxide in hydrogen reduction furnaces. Tungsten is sometimes used in powder form but commonly must be consolidated into solid form. The metal powder is not smelted directly because tungsten has the highest melting point of any metal. The pure tungsten powder is pressed into bars, and then sintered at high temperature (1800–2500 °C). These are worked and rolled at lower temperatures to increase density and ductility, usually under a hydrogen atmosphere to prevent metal oxidation.

Tungsten–heavy-metal alloy (WHAs) are a category of tungsten alloy that typically contain 90–98 per cent by weight of tungsten in combination with a mix of nickel, iron, copper, and/or cobalt. These metals serve as a binder to hold the tungsten particles in place after sintering. These heavy metal alloys do not require millwork to enhance their properties and are machined much more easily than pure tungsten. The most common forms are tungsten–nickel–iron alloys, although the addition of cobalt enhances both strength and ductility. The tungsten–molybdenum-nickel–iron alloy has greater strength but reduced ductility.

Tungsten carbide (WC) and semi-carbide (W_2C) are produced by the reaction of tungsten metal powder with pure carbon powder at 900–2200 °C in a furnace, a process called carburisation. Other methods include a fluid-bed process that react either tungsten metal or TBO with a carbon monoxide/dioxide mixture and hydrogen at 900–1200 °C. By melting tungsten metal and WC together, a eutectic composition of WC and W_2C is formed. This melt is cast and rapidly quenched to form extremely hard solid particles.

The solids are crushed and classified to various mesh sizes.

Tunsgten carbide is, quantitatively, the most important tungsten compound. In its most basic form, it is a fine grey powder, but it can be pressed and formed into shapes for a multitude of uses. Tungsten carbide is approximately three times stiffer than steel and is much denser than steel or titanium. It has a high melting point (2870 °C), is extremely hard (8.5–9.0 on Mohs scale, comparable with corundum or sapphire) and can only be polished and finished with abrasives of superior hardness such as silicon carbide, cubic boron nitride and diamond. It also has a low electrical resistivity, comparable with some metals (e.g. vanadium).

Ferrotungsten (FeW and Fe_2W) is an alloy with a steel-grey appearance formed by combining iron and tungsten. There are two grades with tungsten contents in the range 75–82 per cent and 70–75 per cent. Ferrotungsten is a remarkably robust alloy because of its high melting point. By combining iron with tungsten, the brittleness of metallic tungsten is much reduced.

Non-ferrous tungsten alloys in current use include composites with copper, silver, nickel and rare metals. The combination of tungsten (or tungsten carbide) with these metals produces relatively hard, heat-resistant materials with superior wear resistance, robust physical properties at elevated temperatures, and good electrical and thermal conductivity. The tungsten component imparts a high density, high vibration and damping capacity, excellent radiation shielding, higher temperature strength and thermal shock resistance and improved corrosion resistance.

Superalloys containing tungsten are high-performance alloys that exhibit excellent mechanical strength and creep resistance at high temperatures, good surface stability, and corrosion and oxidation resistance. A superalloy's base alloying element is usually nickel, cobalt, or nickel–iron.

Uses

The unique range of properties of tungsten makes it an essential component in a wide range of products for many applications. These properties include:

- very high melting point;
- very high density;
- extreme strength;
- high wear resistance;
- high tensile strength;
- low coefficient of expansion;
- high thermal and electrical conductivity.

The most important uses are shown in Figure 16.4.

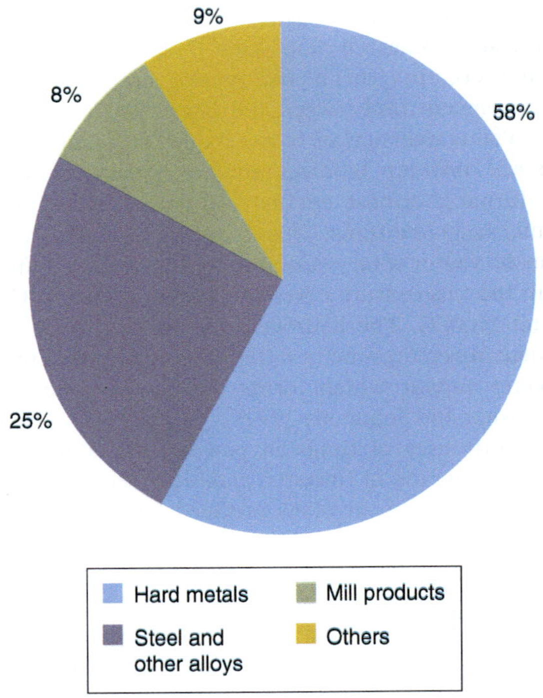

Figure 16.4 The main global uses of tungsten. (Data from Wolf Minerals, 2012.)

Hard metals

Hard metals consist of tungsten carbide and cemented carbides, which are formed from tungsten carbides and cobalt, or occasionally other metals such as titanium, tantalum and niobium. They are very hard materials that are used for cutting, drilling and wear-resistant parts or coatings.

Tungsten carbide (WC) is important in the metal-working (40 per cent of the total WC market), mining (30 per cent), and petroleum (20 per cent) industries. Its main applications are in machine tools especially where they need to withstand higher temperatures (e.g. punches, drill tips, tile and glass cutters), in armour-piercing ammunition, sports equipment and domestic items such as scratch-resistant jewellery and the rotating ball in the tips of ballpoint pens. Fine- and ultra-fine-grained WC hard metals are becoming increasingly important because they last even longer than conventional WC and can be used for specialist applications where very fine tools are required, e.g. the drilling of holes in electronic circuit boards.

Steel and other alloys

Tungsten was one of the first elements to be systematically alloyed with steel to improve its properties. Special steel alloys, such as high-speed steels, heat-resistant steel and tool steels, are largely utilised in metal cutting and specialist engineering applications where hardness and strength are required, particularly over a wide temperature range.

High-speed steel (HSS) is a variety of tool steel in which tungsten is a common heat-resisting alloying element. It is superior to the older high-carbon steel in that it can withstand higher temperatures without losing its hardness. This property allows HSS to cut faster than high-carbon steel, which gave rise to the name. There are many different types and designations of high-speed steel, each with its own combination of added metals.

Superhigh-speed steels have very high wear resistance, improved temperature resistance or both. These steels are used for high cutting speeds, dry or semi-dry cutting or simply to maximise tool life. Because of the high hardness and wear resistance some grades are considered to provide a bridge between HSS and hard metals (Tarney, 2004). Heat treatment (usually with a laser or electron beam), surface finish and coatings can provide significantly enhanced HSS tool performance.

Superalloys containing tungsten have particular applications in the aerospace, industrial gas turbine and marine turbine industries due to their very high corrosion and wear resistance in a variety of challenging environments. Examples of applications include flue-gas desulfurisation systems, chemical processing components such as heat exchangers, industrial furnaces, erosion shields and jet-engine combustion chambers.

The unique properties of conventional WHAs – high density, high strength, high ductility, good corrosion resistance, high radiation adsorption capability, and reasonably high toughness – make them well suited for a variety of defence and civilian applications including X-ray and radiation shields, counter weights, kinetic-energy penetrators, vibration dampening devices, medical devices for radioactive isotope containment, heavy-duty electrical contact materials and gyroscopes. Due to eco-toxicity concerns WHAs have been used as weights in fishing to replace lead. A tungsten alloy including a proportion of tin, copper or nylon is also used as a substitute for lead shot in shotgun pellets.

Tungsten-copper (WCu) alloys are dense, high-performance, easily machinable materials with excellent thermal and electrical conductivity. WCu alloys are widely used in the electrical contacts, high-voltages switches, electrodes, circuit breakers, refractory parts, self-cooling heat sinks, etc. As electrodes they exhibit low wear and maintain high contour sharpness.

Nickel–tungsten (WNi) alloys remain stable indefinitely at room temperature and are highly resistant to decomposition when heated. WNi electroplating is harder and longer lasting than chrome. Furthermore, the electroplating process is more efficient than that for chrome, because multiple layers can be applied in one step, and it is potentially safer and less environmentally polluting to produce. This is a relatively new technology with potential application in coating shock absorbers, print rolls and connectors for portable electronics (Nanomaterials, 2009).

Several other tungsten alloys exist including molybdenum–tungsten (MoW), tungsten–rhenium (WRe), tantalum–tungsten (TaW) and niobium–tungsten (NbW). These have superior properties in terms of strength, corrosion resistance, temperature stability, etc. and are used for specialist applications in the glass, aerospace and electrical industries.

Mill products

The term 'mill products' is used in the industry to refer to tungsten wire, sheets or rods. The combination of extremely high melting point, conductivity and ductility of pure metallic tungsten makes it ideal for electrical and electronic applications, most notably in incandescent lightbulb filaments, discharge lighting electrodes and vacuum tubes, and heating elements. In medical X-ray tubes both the filament and target is usually tungsten or tungsten alloy. Unalloyed tungsten is also used in electronic circuit interconnectors, filaments in vacuum-metallising furnaces, and for electrical contacts such as the distributor points in automotive ignition systems. Its high melting point is the basis for the use of tungsten welding electrodes in tungsten inert (TIG) welding, where it is often doped (0.5–2% weight/weight) with lanthanum, cerium, thorium, or zirconium. Water-cooled tungsten tips are used for non-consumable electrode vacuum-arc melting of alloys.

Chemical and other specialist applications

Many of the intermediate tungsten products have uses in their own right. For example, ammonium paratungstate (APT) can be used as a colouring agent in the porcelain industry or in catalysts, phosphors and absorbent gels; ammonium metatungstate (AMT) is used as a reagent for chemical analysis during medical diagnosis or as a corrosion inhibitor; and tungsten trioxide (WO_3) can be used as a catalyst or as a pigment in ceramics and paints.

Other substances containing tungsten have a wide range of applications. For example, sodium tungstate is a widely used tungsten chemical with applications in the manufacture of organic dyes and pigments, catalysts, fireproofing of textiles and hard surfacing for graphite crucibles.

Tungsten hexafluoride (WF_6) gas is most commonly used in the production of semiconductor circuits and circuit boards through the process of chemical vapour deposition. The semiconductor industry consumes around 200 tonnes of WF_6 per year worldwide. It is also used to deposit films of tungsten on ceramics or metal products such as bearings.

Recycling, re-use and resource efficiency

As the price of tungsten has increased substantially in the last decade there are economic and environmental imperatives to use tungsten more efficiently, to seek substitutes for less-demanding applications and to recycle it more effectively.

Recycling is an important element of the world's tungsten supply. It is estimated that more than 30 per cent of total world supply is from recycled sources (EC, 2010). In some countries this figure is much higher, for example in the USA 46 per cent of tungsten supply is from recycled material (Shedd, 2011). The tungsten processing industry is able to treat almost every kind of tungsten-containing scrap and waste to recover tungsten and, if present, other valuable constituents.

Old scrap

Old scrap, also referred to as end-of-life scrap, consists of tungsten-bearing products that are no longer being used. These include:

- cemented carbide parts;
- tungsten metal and tungsten alloy parts;
- old superalloy scrap;
- tungsten-bearing catalysts.

It is estimated that on average 10–25 per cent of old scrap is recycled globally (UNEP, 2011). However, this figure will vary greatly between individual countries, for example in the USA it is as high as 66 per cent (Shedd, 2011).

New scrap

This is generated during the processing of tungsten concentrates, scrap, and chemicals to make metal powder, carbide powder, or chemicals, and during the fabrication of tungsten products from these materials. Recovery rates of new scrap are high, and metal losses are kept to a minimum by internal recycling of the scrap generated.

Alloy industries that produce parts by cast or wrought processes typically generate large quantities of new scrap. During the casting of corrosion-resistant alloys 40 to 60 per cent of the melt becomes new scrap (Shedd, 2011), which is either recycled in-house or sold to be recycled elsewhere.

Unrecovered scrap

In various applications scrap tungsten is not recovered, either due to the lack of an efficient collection system, or because it is not economic to recycle owing to contamination. Most uses of tungsten in chemical applications, except for catalysts, are dissipative. Tungsten is also lost in use due to wear of cemented carbide parts and hard-faced products, arc erosion of electrical contacts and electrodes, and the oxidation of alloys under high-temperature conditions.

Recycling methods

The tungsten industry uses a variety of recycling methods to optimise recovery and reduce costs. Metallurgical and hydrometallurgical techniques have the ability to treat all scrap types and remove impurities.

Contaminated cement carbide scrap, tungsten grindings and powder scrap can be oxidised and chemically processed to APT in a similar way to that used for the processing of tungsten ores. If present, cobalt, tantalum and niobium are also recovered separately. Other tungsten-containing scrap might require a modified process (EC, 2010; Roskill, 2010).

Tungsten in high-speed steel is frequently recycled, and a typical melt contains 60 to 70 per cent scrap, including internally generated scrap (EC, 2010; Roskill, 2010).

Substitution

Because of the unique combination of properties of tungsten, there are limited options for substitution in many applications, especially where optimum performance is required at high temperatures. Unlike many other metals, there are no substitutes for tungsten that do not involve a considerable cost increase and compromise in product performance. In the aerospace and defence industries where product performance is paramount, substitution is generally avoided (Pitfield and Brown, 2011).

Potential substitutes for cemented tungsten carbides or hard metals include cemented carbides based on molybdenum carbide and titanium carbide, ceramics, ceramic–metallic composites (cermets), diamond tools, and tool steels (EC, 2010). Potential substitutes for other applications are:

- molybdenum can replace tungsten in certain mill products;
- molybdenum steels can substitute for tungsten steels for some applications;
- lighting based on compact fluorescent lamps, low-energy halogen light bulbs and light-emitting diodes (LEDs) are gradually replacing the traditional use of tungsten in light bulb filaments as inefficient incandescent light bulbs are being phased out;
- depleted uranium can be used in weights and counter-weights instead of tungsten alloys or unalloyed tungsten, but generally it has been tungsten that has substituted for depleted uranium for health and environmental reasons;
- depleted uranium alloys can also be used in armour-piercing projectiles instead of cemented tungsten carbides or tungsten alloys.

Because of its perceived benign impacts on human and environmental health, tungsten alloys have been the preferred substitute for lead-based munitions since the mid-1930s.

Environmental aspects of the life cycle of the metal and its products

Tungsten is much less toxic than other metals, such as lead or mercury, for which it is sometimes used as a substitute, e.g. in ammunition. However, there is growing evidence that tungsten should not be considered as completely 'non-toxic' and that it is not as environmentally inert as previously thought (Petkewich, 2009).

Concern regarding the health impacts of tungsten in the environment has been expressed following the identification of three clusters of leukaemia in the USA, which are associated with elevated tungsten levels. It has not yet been proven that tungsten is the cause of these clusters and research is continuing (Koutsospyros et al., 2006; and several others as listed in Strigul, 2010). Tungsten levels in water have been subject to regulation in Russia for some years, but similar regulations do not yet exist in the USA nor in the EU (Strigul, 2010).

A recent study by Clauson and Korte (2009) found that particles of tungsten metal could oxidise in air and then dissolve into soil through the action of rainwater. The tungsten, in this more soluble form, became mobile in the soil and elevated levels were detected in local groundwater. Elevated levels of tungsten have been found in soils, trees and plant tissue, aquatic systems, the atmosphere, animals and humans in various locations (see Koutsospyros et al., 2006 for a discussion of the locations and values reported). Often these appear to be associated with natural tungsten deposits but also occur in the vicinity of anthropogenic activities such as the processing of tungsten ore or metal, military firing ranges, the use of agrichemicals and the disposal of waste in various forms.

The dissolution of tungsten powder appears to cause soil acidification with associated adverse consequences on the micro-organisms, invertebrates and plant communities (Strigul et al., 2005). However, the geochemistry of tungsten is complicated and its mobility, bioavailability and toxicity in the environment depends on the exact form of the tungsten compounds (Strigul et al., 2009), the environmental conditions (e.g. composition of chemical substrates, pH, oxidation states) and the nature of the exposure pathway (Bednar et al., 2008; Strigul et al., 2005).

Many studies have been conducted in recent years (Strigul, 2010) but the environmental and health impacts of tungsten are not yet fully understood.

World resources and production

Resources and reserves

World tungsten resources have been estimated at seven million tonnes (contained tungsten metal) including deposits that have so far not been proven to be economically workable (Hinde, 2008). It is believed that 30 per cent of the resources are wolframite (76.5 per cent tungsten trioxide) and 70 per cent are scheelite (80.5 per cent tungsten trioxide) ores (Hinde, 2008).

Werner et al., (1998) noted that the ten largest known deposits at that time were located in Kazakhstan, Canada, China and Russia. However, since then new deposits have been discovered, for example O'Callaghans in Australia, and previously identified occurrences have been found to contain more tungsten than previously thought, for example Hemerdon in the United Kingdom. It is clear that China, Kazakhstan and Russia have considerable resources but detailed information is difficult to obtain. Resources in Canada are known to exceed one million tonnes of contained tungsten, whereas resources in Australia are believed to be nearly 0.5 million tonnes of contained tungsten.

The USGS (Shedd, 2012a) estimated in January 2012 that reserves stood at 3.1 million tonnes (contained tungsten metal) with more than 60 per cent of these located in China (Figure 16.5).

The National Bureau of Statistics of China reported that their reserves of tungsten in 2010 were approximately 1.75 million tonnes of contained metal (NBSC, 2012) and it is believed this is concentrated in the provinces of Hunan and Jiangxi (Pitfield et al., 2010). These two provinces also receive the highest proportion of China's production quota (Shedd, 2012b). Deposits of tungsten are known to exist in many other Chinese provinces but it is not known whether they are categorised formally as resources or reserves.

In Russia, reserves of tungsten are believed to be mainly located in the North Caucases area and the Far East region. Canadian reserves are dominated by the huge deposit at Mactung and the nearby operating mine at Cantung, which are located in the Yukon and North West Territories, respectively.

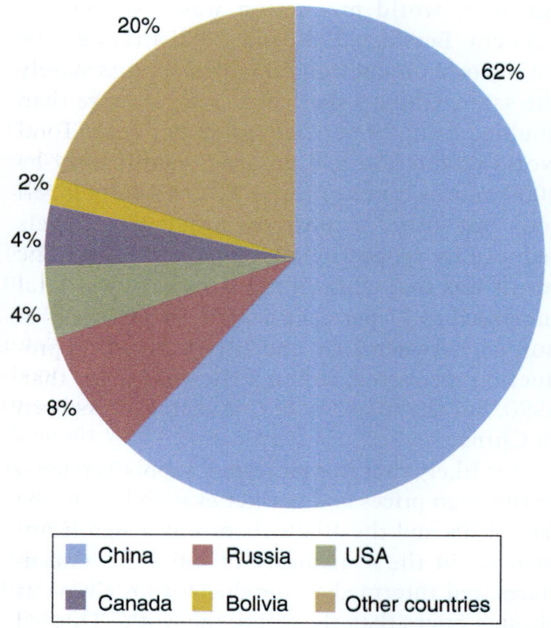

Figure 16.5 Location of the estimated world reserves of tungsten, as at December 2011 (Shedd, 2012a).

Production

In 2011, total world production was nearly 73,000 tonnes of tungsten (metal content of concentrates). This was a six per cent increase compared to 2010 and 173 per cent higher than the low of 1999 shown in Figure 16.6 (BGS, 2013).

Tungsten is currently produced in approximately 20 countries. China has been the world's leading tungsten producer for many years. In 1989 it accounted for 58 per cent of the world total but this rose to reach a peak of 89 per cent in 2004. In recent years this proportion has fallen slightly but it was still 82 per cent in 2011 (Figure 16.7).

China's output was produced mainly in Jiangxi and Hunan provinces, accounting for 44 per cent and 24 per cent of the total respectively (Research in China, 2011). Its major operating mines are at Shizhuyuan in Hunan Province and Yangchulin, Xingluokeng and Xianglushan in Jiangxi Province. There are further mines, in these and other provinces, and new mines are believed to have opened but accurate information is difficult to obtain. Russia's output is believed to come from Tyrnyauz in the North Caucases and the Vostok-2 area of the Far East region, but Russian mines were reported to be struggling in the economic crisis (Levine, 2011). Historically, Bolivia has had

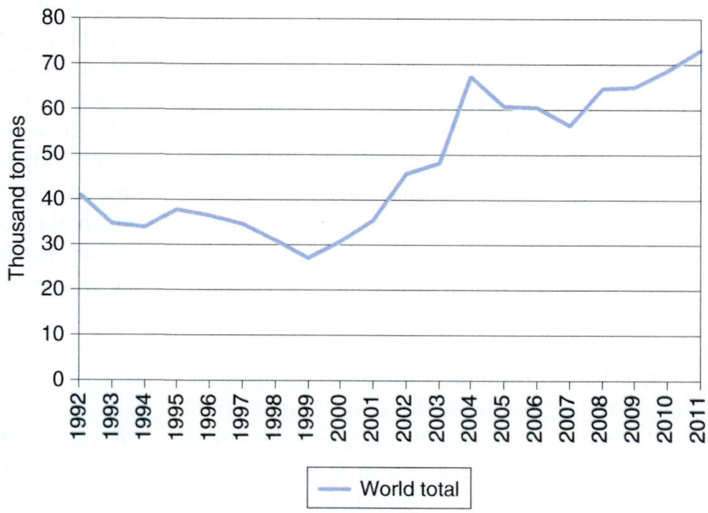

Figure 16.6 Mine production of tungsten, 1992–2011. (Data from British Geological Survey World Mineral Statistics database.)

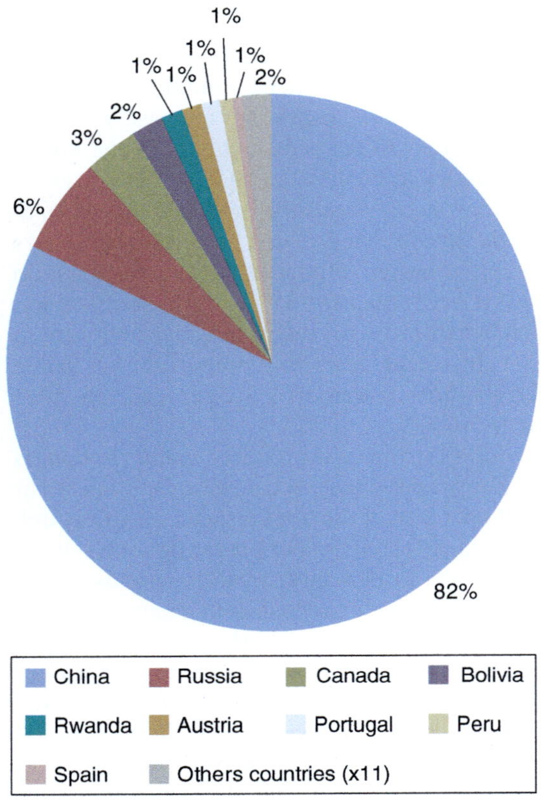

Figure 16.7 Mine production of tungsten in 2011. (Data from British Geological Survey, 2013.)

numerous small-scale mining operations extracting tungsten and this is thought to still be the case (Anderson, 2011). However, output from Bolivia has increased significantly in recent years from approximately 400 tonnes of contained tungsten in 2004 to 1400 tonnes in 2011 (BGS, 2013). Austria's output comes from the Mittersill mine, operated by Wolfram Bergbau and Hütten AG, while Canada's is from the Cantung mine in the Yukon Territory, which is operated by North American Tungsten Corp.

Other countries producing tungsten in 2011 included: Portugal (mainly from the Panasqueira mine), Peru (mainly from the Pasto Bueno mine), Rwanda, Spain, Uzbekistan, Brazil, Thailand, Burma (Myanmar), North Korea, Burundi, Kyrgyzstan, Democratic Republic of Congo, Uganda, Australia (mainly from the Kara mine on Tasmania) and Mongolia (BGS, 2013).

The concentration of production in China is a relatively recent occurrence, as illustrated by Figure 16.8. Prior to the 1990s, China's share of the total world production was less than 35 per cent. Between 1980 and 1990, although the total world production remained approximately the same, China's share of that total more than doubled from 29 per cent to 62 per cent. Total world production had dropped significantly by 2000 and in tonnage terms China's output fell too, but other countries' production levels reduced by proportionally greater extents. The result was that China's share of the world total increased to 77 per cent despite the decrease in output. Between 2000 and 2010 total world production recovered, and in 2010 was higher than 1990, but the majority of this increase has been in China.

It is likely that this pattern of supply is related to tungsten prices and market availability. In the late 1970s and the 1980s there was a significant increase in the availability of tungsten concentrates and intermediate products from China at cheaper prices than the rest of the world. This led to oversupply and a significant fall in tungsten prices, with the result that many mines could not sustain economic production and world output decreased. The cost of production was probably lower in China than other countries and consequently output there fell less. As demand and prices recovered, Chinese producers were able to react more quickly and therefore China's output has grown much more rapidly than that in the rest of the world. The subsequent imposition of production and export quotas in China has pushed prices up further and encouraged both the exploration for new deposits and the development of new mines outside of China.

Future supplies

A recent increase in prices for most forms of tungsten, and concerns over security of supply, particularly from China, have caused a revival in

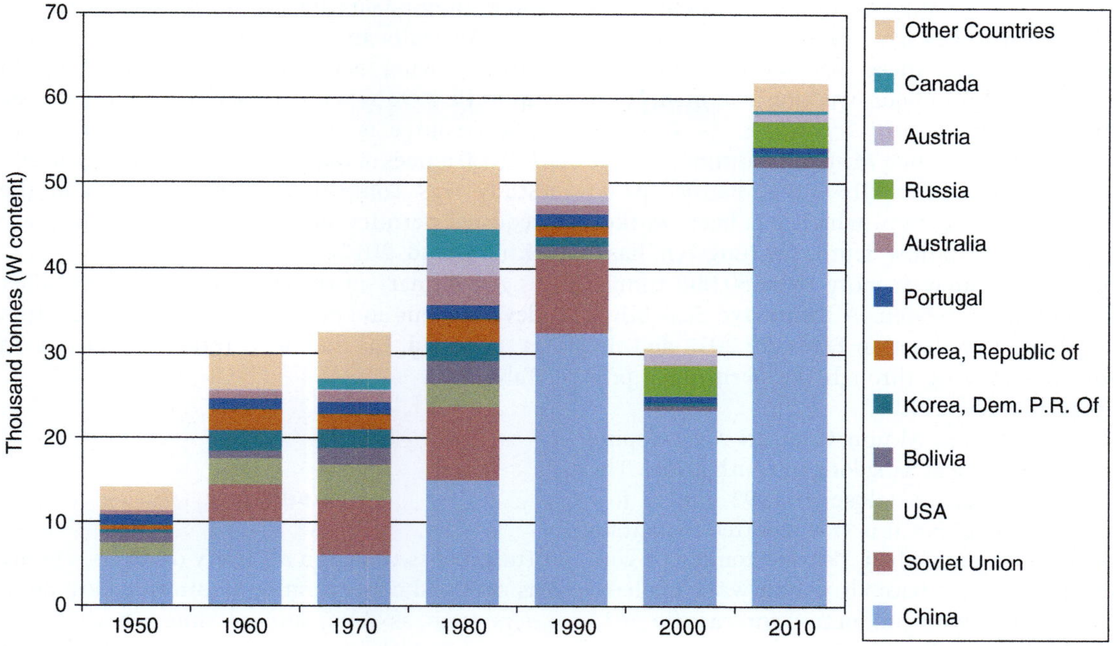

Figure 16.8 Evolution of tungsten production concentration, output by country for 1950, 1960, 1970, 1980, 1990, 2000 and 2010. (Data from British Geological Survey World Mineral Statistics database.)

the development of new mines and the reopening of inactive mines in other parts of the world. New production has been recorded in Australia (2005), Peru (2006) and Spain (2008).

In North America, the Andrew mine in California, USA was reopened in 2007. The Cantung mine in Canada was reopened, again, in October 2010 following a short suspension in 2009. Resources at Cantung are reported to contain more than 40,000 tonnes of tungsten. Its owners, North American Tungsten Corp., also own the nearby Mactung tungsten deposit and completed a feasibility study there in 2009, which reported a contained tungsten resource of more than 370,000 tonnes. The Company is currently evaluating the options for the deposit and seeking the necessary environmental permits (North American Tungsten Corp., 2011 and 2012).

An even larger resource of more than 390,000 tonnes of contained tungsten has been reported for the Northern Dancer deposit in Canada's Yukon Territory. Its owner, Largo Resources, completed a scoping study in 2008 and a Preliminary Economic Assessment in 2011 (Largo, 2011a).

In New Brunswick, eastern Canada, Northcliff Resources completed a full feasibility study in 2013 and is conducting an environmental impact assessment at its Sisson Brook project. The project has a resource of more than 270,000 tonnes of contained tungsten (Northcliff, 2012).

In South America, Largo Resources announced in December 2011 that it had shipped the first tungsten concentrates from its Currais Novas project in Brazil, which is reworking tailings from past mining operations at Barra Verde and Boca de Lage (Largo, 2011b).

In Europe, Wolf Minerals have completed the construction of a new link road at the Hemerdon project in the United Kingdom. A favourable

Definitive Feasibility Study was completed in 2011 and finances have been put in place. The resource contains more than 460,000 tonnes of tungsten and mine construction commenced in 2013 (Wolf, 2012).

In Spain, Ormonde Mining is aiming to start production in 2014 or 2015 at their Barruecopardo project. This location, which has been worked in the past to shallow depths for tungsten, has a reported resource of more than 50,000 tonnes of contained tungsten. A Definitive Feasibility Study was completed in February 2012 and the project is moving through the permitting process (Ormonde, 2012).

In Asia, Woulfe Mining Corp. are redeveloping the former mine at Sangdong in South Korea. The mine was forced to close in 1992 due to low tungsten prices, but it is still reported to contain resources of more than 280,000 tonnes of contained tungsten. A feasibility study was completed in April 2012 with production forecast for 2014 or 2015 (Woulfe, 2012).

In Australia, Newcrest Mining Ltd discovered the O'Callaghan's deposit in 2008 near to the Telfer gold mine in the Pilbara region of Western Australia. The deposit is reported to contain more than 200,000 tonnes of tungsten (Newcrest, 2012) and work towards a prefeasibility study continued in 2012. Also in Western Australia, Hazelwood Resources is conducting a Definitive Feasibility Study on its Cookes Creek tungsten deposit, which has a reported resource of more than 40,000 tonnes of contained tungsten (Hazelwood, 2012).

Developments are also taking place at the historical Mount Carbine mine in Queensland, Australia, where Carbine Tungsten Ltd commenced reprocessing of tailings in 2012. The company proposes to follow this by processing low-grade stockpiles on the site before progressing on to mining of the hard rock resource. A feasibility study for the latter was completed in 2012. It is reported that there is more than 40,000 tonnes of contained tungsten remaining at the mine (Carbine, 2012). King Island Scheelite Ltd is also planning to reopen the former Dolphin mine on King Island (between the mainland Australia and Tasmania). The underground mine on King Island was operated for tungsten up until 1990 when it closed due to low prices. The resource is reported to still contain over 190,000 tonnes of tungsten. A Definitive Feasibility Study was completed early in 2012 and the required permits and approvals are all in place (King Island, 2012).

A summary of selected major deposits under development and projects where new production is expected in the near future is shown in Table 16.4.

World trade

Tungsten is traded in a variety of forms, chiefly as ores and concentrates, as intermediate products such as APT, and as tungsten oxide (or trioxide). Although China is by far the largest producer of tungsten ores and concentrates, it does not export much material in these forms, preferring to process the raw material into either APT or other products. The main exporting countries in 2011 are shown in Figure 16.9a. China is the largest importer of tungsten ores and concentrates, followed by the USA and Austria (Figure 16.9b).

Due to its high economic importance, together with concerns over access to raw tungsten materials and uncertainty over the ability of Western producers to compete with China in the sale of tungsten products, the European Commission (EC) has identified tungsten as one of its 14 'critical' raw materials (EC, 2010). This means that they consider there to be a risk of supply shortage with an associated potential impact on the European Union economy. Strategies to reduce this risk are being considered and these include the diversification of supply sources, resource efficiency and recycling. Various countries, including China, the USA, Japan and Russia, maintain stockpiles of critical raw materials, including tungsten.

Table 16.4 Selected major tungsten deposits under development and those where production is expected in the near future.

Name	Country	Current Status (as at January 2013)	Possible Production	Resources (tonnes contained tungsten)
Hemerdon	United Kingdom	Feasibility study completed May 2011, mine construction expected 2013	2014	>460,000
Barruecopardo	Spain	Feasibility study completed Feb 2012, commissioning expected 2014	2014 or 2015	>50,000
Sangdong	South Korea	Feasibility study completed April 2012, commissioning expected 2014	2014 or 2015	>280,000
King Island	Australia	Feasibility study completed Feb 2012, all permits in place, raising funding	2015 ?	>190,000
Sisson Brook	Canada	Feasibility study completed 2013. Environmental impact assessment ongoing	2015 ?	>270,000
Cookes Creek	Australia	Feasibility study ongoing	Unknown	>40,000
Mactung	Canada	Feasibility study completed 2009, some environmental permitting in 2012	Unknown	>370,000
Northern Dancer	Canada	Preliminary economic assessment completed in 2011	Unknown	>390,000
O'Callaghans	Australia	Prefeasibility study ongoing	Unknown	>200,000

Note: Resources are from all categories and in some cases include reserves.

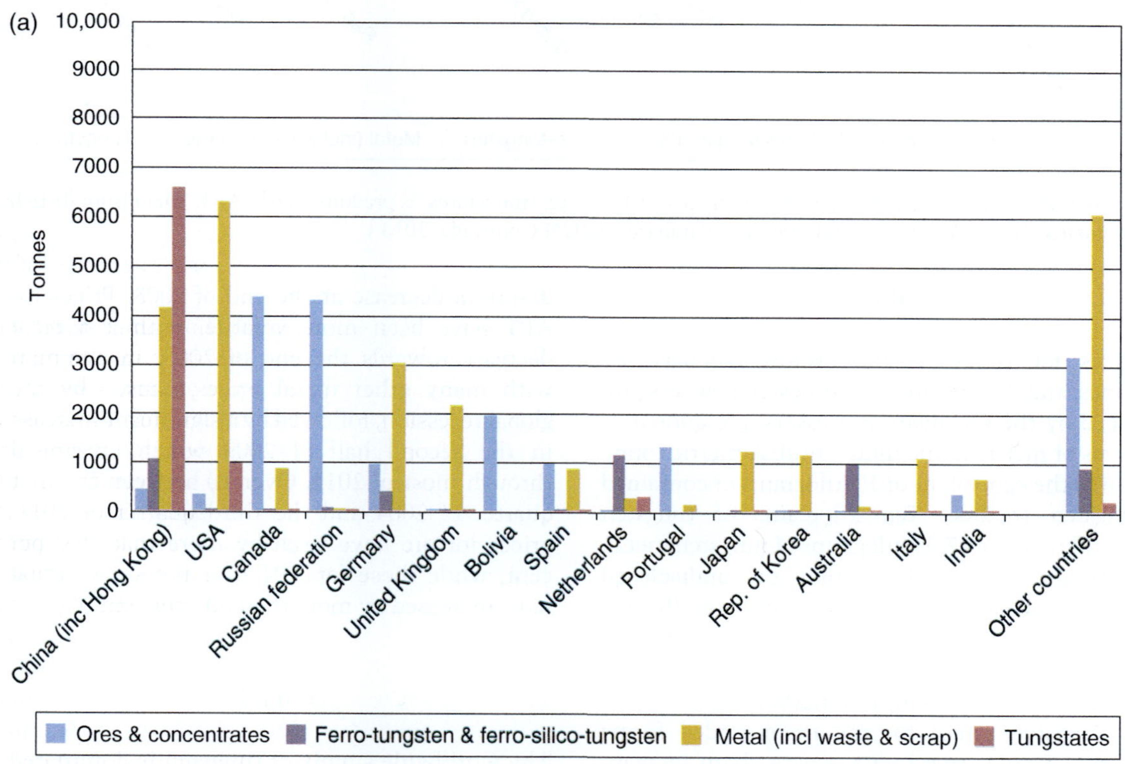

Figure 16.9a Main tungsten exporting countries, 2011. Note: 'tungstates' is predominantly APT. (Data from British Geological Survey World Mineral Statistics database and UN Comtrade, 2013.)

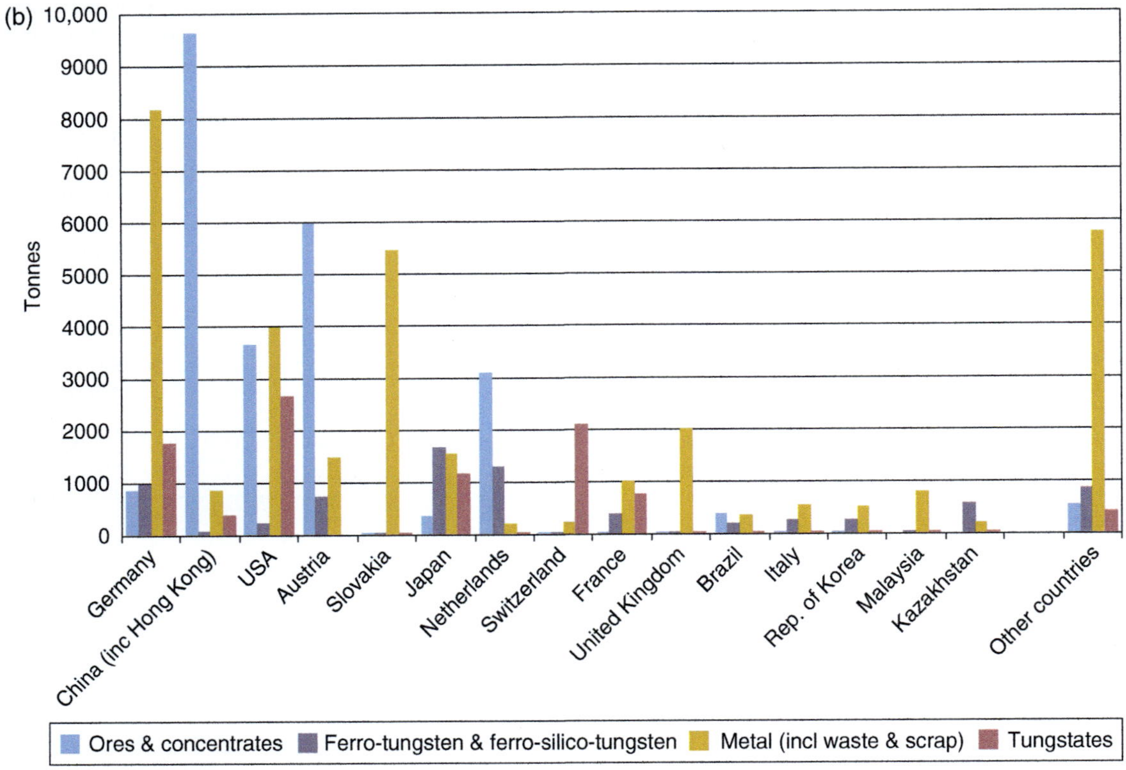

Figure 16.9b Main tungsten importing countries, 2011. Note: 'tungstates' is predominantly APT. (Data from British Geological Survey World Mineral Statistics database and UN Comtrade, 2013.)

Prices

Prices for tungsten concentrates produced by mines, and the intermediate tungsten powders produced by the secondary processors, are quoted in terms of metric tonne units (mtu). A metric tonne unit is the equivalent of 10 kilograms of contained tungsten trioxide. Ten kilograms of tungsten trioxide contains 7.93 kilograms of tungsten metal.

Tungsten is traded either on undisclosed supply contracts between the primary producers, secondary processors and tertiary manufacturers, or via traders. The Chinese price for APT is the benchmark in the tungsten market.

Tungsten prices for the main traded forms show a similar trend with a very sharp increase during the later part of 2004 through to the early part of 2006 (Figure 16.10). Since then prices for ores and concentrates have generally stabilised, despite a decrease at the end of 2008. Prices for APT have been more volatile, with a notable decrease towards the end of 2008, in common with many other metal prices, caused by the global recession, followed by a significant increase in the second half of 2009 which continued through most of 2011. Overall, between the first quarter of 2004 and the third quarter of 2011, prices for ore have risen by more than 200 per cent, while those for APT over the same period have increased by more than 500 per cent.

Outlook

The worldwide supply of tungsten is dominated by China, with more than 70 per cent of total world production in every year since 1994 and more than 60 per cent of estimated world

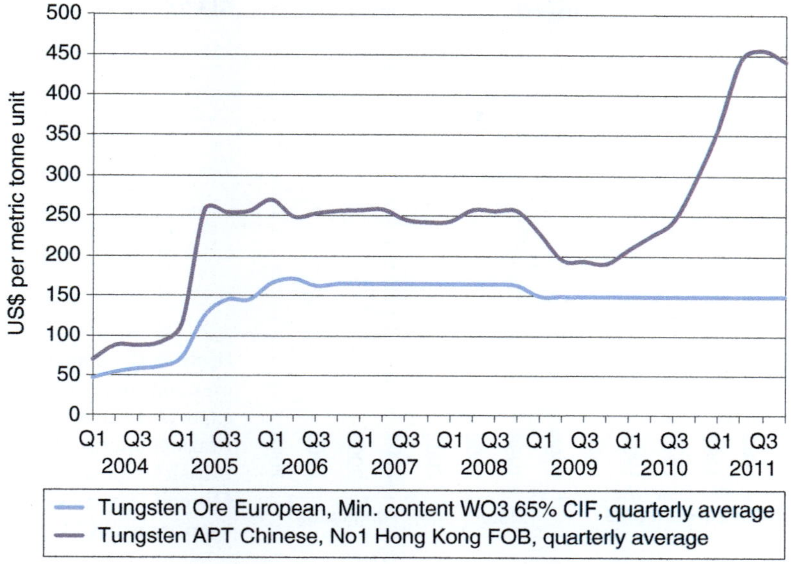

Figure 16.10 Tungsten ore and APT price trend 2004–2011 (Data from Metal Bulletin.)

reserves in 2011. China imposes strict controls on its tungsten industry by limiting the number of exploration and mining licences issued within its borders, applying production quotas and imposing export taxes and quotas (Shedd, 2012b). Production quotas have been increased in recent years. However, export quotas have been reduced from 18,075 tonnes in 2003 to 15,400 tonnes in 2012 (Hayes, 2011; Smith, 2011). It is believed that the higher prices for tungsten ore and APT shown in Figure 16.10 are principally due to these export restrictions. Further reductions in the export quota are likely, although China's production output is expected to continue to increase.

Whereas in the past China has exported considerable quantities of tungsten ores, in recent years its exports have been concentrated on the higher-value intermediates, such as APT, and to a lesser extent on tungsten metal products (Figure 16.11). This reflects the development of the domestic tungsten industry in China, and may also be the underlying reason for the export quotas, i.e. that China is seeking to preserve its resources and support its domestic industry. The development of this domestic industry has resulted in China becoming the world's largest consumer of tungsten, as well as the largest producer, and this is reflected in the sharp increase in China's imports of tungsten ores (Figure 16.12). It was reported in 2010 that China accounted for 37 per cent of the world's consumption of tungsten (Toovey, 2010) and the main consuming industries within the country have continued to grow. It is likely that China's tungsten industry will remain dominant and that these trends will continue. China's exports of 'tungstates' (which includes APT) have fallen steadily since their peak in 1995 and it is possible that supplies of Chinese APT to the rest of the world may be further restricted in future if the downstream tungsten metal industries in China continue to develop.

Concerns have been raised that China's export controls disproportionately disadvantages non-Chinese consumers (Smith, 2011), and in March 2012 the United States of America, European Union and Japan all requested formal 'consultations' with China though the World Trade Organisation (WTO, 2012a). These

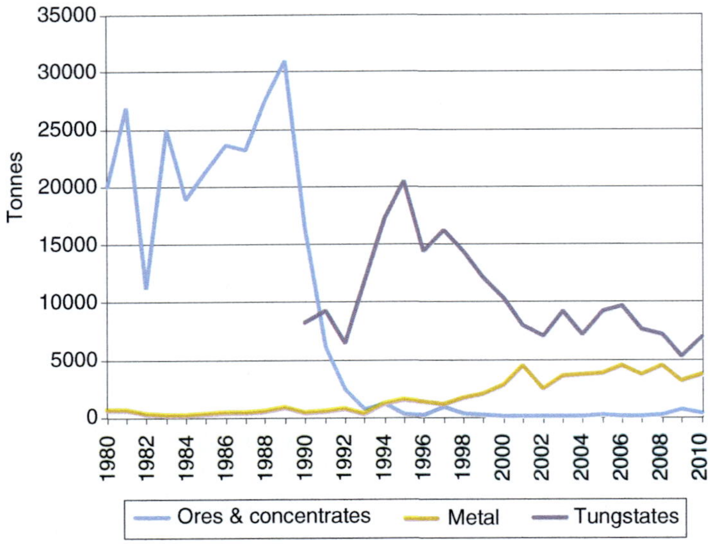

Figure 16.11 China exports of tungsten ores and concentrates compared to exports of tungstates (mainly APT) and tungsten metal, 1980–2010. (Data from British Geological Survey World Mineral Statistics database and UN Comtrade, 2013.)

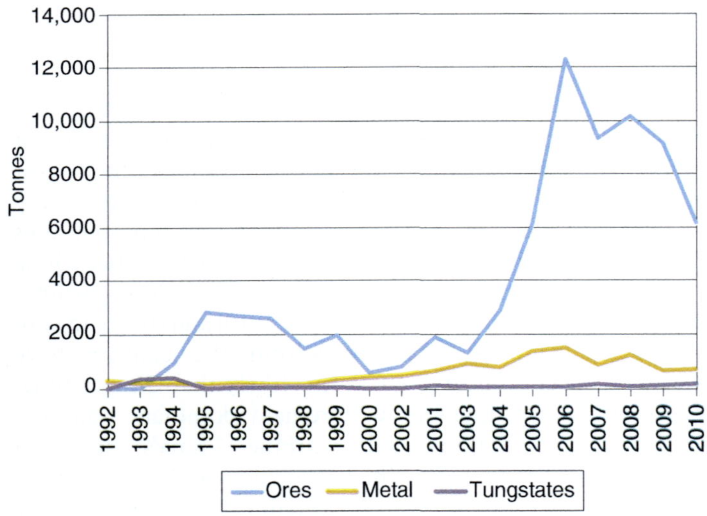

Figure 16.12 China imports of tungsten ores and concentrates, tungstates (mainly APT) and tungsten metal, 1992–2010. (Data from British Geological Survey World Mineral Statistics database and UN Comtrade, 2013.)

requests are the first stage in disputes through the WTO and enable discussions between parties to take place in an attempt to find solutions. However, these discussions did not resolve the concerns and in July 2012 the Dispute Settlement Body within the WTO formally agreed to establish a panel to examine the issues (WTO, 2012b). A report and WTO ruling is expected in 2013.

Higher prices, and the recognition of the risks associated with production being concentrated in a few producing countries (BGS, 2012), has encouraged the development of tungsten mines and projects outside of China. Whilst it is inevitable that not all proposals will come to fruition, several new non-Chinese mines are likely to start producing in the near future, as shown in Table 16.4.

Tungsten use is largely influenced by conditions in the global economy. As a consequence, the consumption of tungsten in most Western countries declined significantly in 2009 as a result of the global recession but subsequently recovered. Future consumption of tungsten will mainly depend on the performance of sectors that use hard metals (tungsten carbide and cemented carbide), particularly in machine tools, which include the manufacturing, metal working, mining, construction and petroleum industries. Further development of specialist steel alloys and 'superalloys' for a variety of applications requiring high degrees of wear and temperature resistance are also likely to require significant quantities of tungsten. Consumption in this area will depend on developments in the aerospace, industrial turbines and similar industries. Future developments of new technologies, such as the current designs for fusion reactors, may also require significant quantities of tungsten.

Acknowledgements

The authors publish with the permission of the Executive Director of the British Geological Survey.

References

Altringer, P.B. (1985) Recovery of Tungsten from Searles Lake Brines by an Ion-exchange Process. Bull. US Dept of the Interior, Bureau of Mines, 682, pp. 47.

Anderson, S.T. (2011) Mineral Industry of Bolivia. 2009 Minerals Yearbook, United States Geological Survey, pp. 17.

Bednar, A.J., Jones, W.T., Boyd, R.E., Rindelberg, D.B. and Larson, S.L. (2008) Geochemical parameters influencing tungsten mobility in soils. Technical reports: Heath metals in the environment. American Society of Agronomy, Crop Science Society of America, and Soil Science Society of America.

Brand, A.A. (2008) Mineralogy, geochemistry and geochronology of the Northern Dancer tungsten-molybdenum deposit, Yukon and British Columbia. MSc thesis, University of British Columbia.

British Geological Survey (BGS). (2012) The risk list. British Geological Survey, Keyworth, Nottingham. 9 pp. http://www.bgs.ac.uk/mineralsuk/statistics/riskList.html

British Geological Survey (BGS). (2013) World Mineral Production 2007–2011, British Geological Survey, Keyworth, Nottingham.

Carbine Tungsten Ltd. (2012) Quarterly Activities Report for quarter ended 30 September 2012. http://www.carbinetungsten.com.au

Chang, Z. (2005) World Skarn Deposits - China. Appendix 2 of Mernert et al. (2005) World Skarn Deposits. In: Hedenquist, J.W., Thompson, J.F.H., Goldfarb R.J. and Richards J.R. (eds.) Society of Economic Geologists One hundredth Anniversary Volume (1905–2005).

Christie, T and Brathwaite, B. (1996) Mineral commodity report 12 - Tungsten. Ministry of Economic Development, New Zealand. http://www.nzpam.govt.nz/cms/pdf-library/minerals/publications/Commodity%20Reports

Chung, C.K. (1975) Tungsten and molybdenum deposits in South Korea. Korean Institute of Mineral and Mining Engineers Journal 12, 192–199.

Clauson, J.L. and Korte, N. (2009) Environmental fate of tungsten from military use. Science of the Total Environment 407, 2887–2893.

Cox, D.P. and Bagby, W.C. (1986) Descriptive model of W veins. In: Cox, D.P. and Singer, D.A. (eds.) Mineral deposit models: U.S. Geological Survey Bulletin 1693, p. 64.

De Roo, J.A. (1988) Structural controls on the emplacement of the vein-type tungsten-tin ore at Mount Carbine, Queensland, Australia. Economic Geology 83, 1170–1180.

Dick, L.A. and Hodgson, C.J. (1982) The Mactung tungsten-copper-zinc contact metasomatic and related deposits of the northeastern Canadian cordillera. Economic Geology 77, 845–867.

Eichhorn, R., Höll, R., Loth, G. and Kennedy, A. (1999) Implications of U-Pb SHRIMP zircon data on the age and evolution of the Ferbertal tungsten deposit (Tauern Window, Austria). Geologische Rundschau 88, 496–512.

Einaudi, M.T., Meinert, L.D. and Newberry, R.J. (1981) Skarn deposits. Economic Geology 75th Anniversary Volume, 317–391.

Elliott, J.E. (1992) Tungsten - geology and resources of deposits in southeastern China. In: DeYoung Jr, J.J. and Hammarstrom J.M. (eds.) Contributions to commodity geology research. U.S. Geological Survey Bulletin 1877, I1–I10.

Elliott, J.E., Kamilli, R.J., Miller, W.R. and Livo, K.E. (1995) Vein and greisen Sn and W deposits. In: Du

Bray, E.A. (ed.) Preliminary compilation of descriptive geoenvironmental mineral deposit models. U.S. Geological Survey Open-File Report 95-831, 62–69.

European Commission (EC). (2010) Critical raw materials for the EU. Report of the Ad-hoc Working Group on defining critical raw materials. European Commission Enterprise and Industry. http://ec.europe.eu/enterprise/policies/raw-materials/critical/index_en.htm

Goossens, P.J. (1978) The metallogenic provinces of Burma—their definitions, geologic relationships, and extension into China, India, and Thailand. In: Nutalaya, P. (ed.) Proceedings of the Third Regional Conference on Geology and Mineral Resources of Southeast Asia, Bangkok, Thailand, 14–18 November 1978, 431–492.

Gu, J Y. (1982) Morphological zoning of the vein-type tungsten deposits in southern China. In: Hepworth, J.V. and Zhang, Y.H. (eds.) Proceedings of Symposium on Tungsten Ore deposits, Geological Publishing House, Beijing, 269–278.

Guerenko, A. and Schmincke, H.U. (2002) Tungsten in Searles Lake. Mining Engineering 11, 301–303.

Guiliani, G., Li, Y.D. and Sheng, T.F. (1988) Fluid inclusion study of Zihuashan tungsten deposit in southern Jiangxi province, China. Mineralium Deposita 23, 24–33.

Hayes, M. (2011) China's dominance in tungsten to continue as exports decrease. Fast Markets Ltd, 15 November 2011.

Hazelwood Resources Ltd. (2012) Cookes Creek Tungsten - overview. http://www.hazelwood.com.au/Projects/Cookes_Creek_Tungsten/

Hinde, C. (ed.). (2008) Tungsten. Mining Journal Special Publication (June 2008). p16 http://www.mining-journal.com/_data/assets/supplement_file_attachment/0014/ 123314/tungsten-scr.pdf

Höll, R. and Eichhorn, R. (2000) Tungsten mineralisation and metamorphic remobilisation in the Ferbertal scheelite deposit (Central Alps, Austria). In: Spry, P.G., Longstaff, F.J. and Vokes F.M. (eds.) Metamorphosed and metamorphogenic Ore Deposits: Rev. Economic Geology 11, 233–264.

Huang, C. and Xiao, Z. (1986) A general review of the tungsten deposits in China. In Beus, A.A. (ed.) 1986 Geology of Tungsten. International Geological Correlation Programme Project 26 'MAWAM' (Mineralisation Associated with Acid Magmatism). United Nations Educational, Scientific and Cultural Organisation, Paris, France.

International Tungsten Industry Association (ITIA) (2012). Tungsten Products. http://www.itia.info/tungsten-products.html

Jarchovsky, T. (2006) The nature and genesis of greisens stocks at Krásno, Slavkovský les area – western Bohemia, Czech Republic. Journal of the Czech Geological Society 51, 201–215.

Keats, W. (1981) Alteration and replacement of wolframite in the Hemerdon tungsten deposit, Devon. PhD thesis, Open University.

Kelly, W.C. and Rye, R.O. (1979) Geologic, fluid inclusion, and stable isotope studies of the tin-tungsten deposits of Panasqueira, Portugal. Economic Geology 74, 1721–1822.

Kerr, P.F. (1940) Tungsten-bearing manganese deposits at Galconda, Nevada. Geological Society of America Bulletin 51, 1359–1389.

King Island Scheelite Ltd. (2012) Dolphin Project – overview and press releases. http://www.kingisland-scheelite.com.au/index.php?id=18

Kooiman, G.J.A., McLeod, M.J. and Sinclair, W.D. (1986) Porphyry tungsten-molybdenum orebodies, polymetallic veins and replacement bodies and tin-bearing greisens zones in the Five Tower zone, Mount Pleasant, New Brunswick. Economic Geology 81, 1356–1373.

Koutsospyros, A., Braida, W., Christodoulatos, C., Dermatas, D. and Strigul, N. (2006) A review of tungsten: From environmental obscurity to scrutiny. Journal of Hazardous Materials 136, 1–19.

Kwak, T.A.P. (1987) W-Sn skarn deposits and related metamorphic skarns and granitoids. Developments in Economic Geology, 24, Elsevier Publ. Co., pp. 451.

Kwak, T.A.P., and Tan, T.H. (1981) The geochemistry of zoning in the skarn minerals at King Island (Dolphin) mine. Economic Geology 76, 468–497.

Landis, Gary P. and Rye, R.O. (1974), Geologic fluid inclusion and stable isotope studies of the Pasto Bueno tungsten-base metal ore deposit, northern Peru. Economic Geology 69, 1025–1059.

Largo Resources Ltd. (2011a) Largo begins drill program for current prefeasibility study at Northern Dancer tungsten-molybdenum project, Yukon. Press release 6 July 2011. http://www.largoresources.com/Investors/Press-Releases/

Largo Resources Ltd. (2011b) Largo Resources announces first commercial shipment from Currais Novos tungsten project. Press release 14 December 2011. http://www.largoresources.com/Investors/Press-Releases/

Lassner, E. and Schubert, W-D. (1999) Tungsten: Properties, Chemistry, Technology of the Element,

Alloys and Chemical Compounds. Kluwer Academic/Plenum Publishers, New York, USA pp. 422.

Levine, R.M. (2011) The Mineral Industry of Russia. 2009 Minerals Yearbook. United States Geological Survey, pp. 26.

Li, G.L. Hua, R.M. Zhang, W.L. et al. (2011) He-Ar isotope composition of pyrite and wolframite in the Tieshanlong tungsten deposit, Jiangxi, China: Implications for fluid evolution. Resource Geology 61, 356–366.

Li, S.R., Wang, D.H., Liang, T., Qu, W.J. and Ying, L.J. (2008) Metallogenic epochs of the Damingshan tungsten deposit in Guangxi and its prospecting potential. Acta Geologica Sinica 82 873–879 (in Chinese with English abstract).

Li, X-H., Liu, D., Sun, M., Li, W-X., Liang, X-R. and Liu Y. (2004) Precise Sm-Nd and U-Pb isotopic dating of the supergiant Shizhuyuan polymetallic deposit and its host granite, SE China. Geological Magazine 141 (2), 225–231.

Liu, W. (1980) Geological Features of Mineralisation of the Xingluokeng Tungsten (Molybdenum) Deposit, Fujian Province. In: Hepworth, J.V. and Lu, H.Z. (eds.) Tungsten Geology, China, ESCAP/RMRDC, Bandung, Indonesia, 338–348.

Liu, Y.J and Ma, D.S. (1993) Vein-type tungsten deposits of China and adjoining regions. Ore Geology Reviews 8, 233–246.

Lu, H-Z., Liu, Y., Wang, C., Xu, Y. and Li, H. (2003) Mineralisation and fluid inclusion study of the Shizhuyuan W-Sn-Bi-Mo-F skarn deposit, Hunan Province, China. Economic Geology 98, 955–974.

Ma, L. (1982) The geological characteristics of Damingshan sedimentary-magmatic- hydrothermal wolframite deposit, Guangxi Zhuang Autonomous Region, China. In: Hepworth, J.V., and Yu, H.Z., (eds.) Symposium on Tungsten Geology [Jiangxi, China, October 1981]: Bandung [Indonesia], ESCAP/RMRDC and Beijing, Geological Publishing House, 374–384.

Marsh, S.P. and Erickson, R.L. (1975) Integrated geologic and geochemical studies, Edna Mountain, Nevada. In: Elliot, I.L. and Fletcher, W.K. (eds.) Geochemical Exploration 1974. Elsevier Publ. Co., Amsterdam, Netherlands, 237–250.

Meinert, L.D., Dipple, G.M. and Nicolescu, S. (2005) World Skarn Deposits. In: Hedenquist, J.W., Thompson, J.F.H., Goldfarb, R.J. and Richards, J.R. (eds.) Society of Economic Geologists One hundredth Anniversary Volume (1905–2005), 299–336.

Metal Bulletin Online (2012) Price information. www.metalbulletin.com

Mining Magazine (1979) Hemerdon – Britain's largest tungsten deposit. Mining Magazine October 1979, 342–351.

Molavi, M., Hollett, G., Board, W.S., Bolu, M., Smith, P., Lemieux, J., Netherton, D.E., Nyland, E. and Weston, S. (2011) Preliminary Economic Assessment for the Northern Dancer Project, Yukon, Canada. Prepared by AMC for Largo Resources Ltd. Filed on www.sedar.com on 7 April 2011.

Nanomaterials (2009) New nickel-tungsten alloy that is not only safer chrome but also more durable. News article posted on the A-Z of Nanotechnology webpage. 19 May 2009. http://www.azonano.com/news.asp?newsID=11591.

Narciso, H., Iakovlev, I., de Ruijter, M.A. et al.(2009) Amended Technical Report on the Mactung Property, prepared by Wardrop et al. for North American Tungsten Corp. Ltd. www.sedar.com on 18 May 2010.

National Bureau of Statistics of China (NBSC). (2012) Statistical Yearbook of China 2011. http://www.stats.gov.cn/english/statisticaldata/yearlydata/.

Neinavale, H., Thalmann, F., Ataii, B. and Beran, A. (1989) Wolframite and scheelite-bearing carbonate rocks of the Nock Mountains, Austria: A new type of tungsten mineralisation in the Eastern Alps. Mineralium Deposita 24, 14–18.

Newberry, R.J. and Einaudi, M.T. (1981) Tectonic and geochemical setting of tungsten skarn mineralization in the Cordillera. Arizona Geological Society Digest 14, 99–111.

Newcrest Mining Ltd. (2012) December 2011 Resources and Reserves Statement. Market Release 10 February 2012. http://www.newcrest.com.au//resources.asp.

Noble, S.R., Spooner, E.T.C. and Harris, F.R. (1984) The Logtung large tonnage, low-grade W (scheelite)-Mo porphyry deposit, south-central Yukon Territory. Economic Geology 79, 848–868.

North American Tungsten Corp. (2011) North American Tungsten establishes special committee for Mactung, releases second quarter financial results and provides an update on Cantung operations. Press release 30 May 2011. http://www.natungsten.com/.

North American Tungsten Corp. (2012) North American Tungsten Achieves Major Permitting Milestone for its Mactung Project, Yukon. Press release 16 October 2012. http://www.natungsten.com/.

Northcliff Resources Ltd. (2012) Management Discussion and Analysis report for period ended 31 October 2012 released on http://www.sedar.com

Ormonde Mining Plc. (2012) Barruecopardo tungsten project summary. http://www.ormondemining.com/

en/projects/barruecopardo_tungsten_project/summary_btp#Summary
Petkewich, R A. (2009) Unease over tungsten. Science and Technology, 19 January 2009, 87, 3, 63–65. http://pubs.acs.org/cen/science/87/ 8703sci2.html
Pitfield, P.E.J. and Brown, T.J. (2011) Tungsten. Mineral Profile series, British Geological Survey. http://www.bgs.ac.uk/mineralsuk
Pitfield, P.E.J. Brown, T J and Idoine, N E. (2010) Mineral Information and Statistics for the BRIC countries 1999–2008. British Geological Survey. http://www.bgs.uk/mineralsuk
Plimer, I. 1975. Wolfram Camp wolframite-molybdenite-bismuth quartz pipes, North Queensland. In: Knight, C.L. (ed.) Economic Geology of Australasia and Papua New Guinea v.1 – Metals. Australasian Institute of Mining and Metallurgy Monograph Series 5, 760–762.
Ray, G.E. (1995) Tungsten Skarns, in Selected British Columbia Mineral Deposit Profiles, Volume 1 - Metallics and Coal, Lefebure, D.V. and Ray, G.E., Editors, British Columbia Ministry of Employment and Investment, Open File 1995-20, 71–74. http://www.empr.gov.bc.ca/Mining/Geoscience/MineralDepositProfiles/ListbyDepositGroup/Pages/KSkarn.aspx#K05
Research in China (2011) China Tungsten Industry Report 2010-11, Abstract. http://www.researchinchina.com/Htmls/Report/2011/6076.html
Roskill Consulting Group. (2010) A custom study of the market for tungsten. Commissioned by Wolf Minerals Ltd. Roskill Consulting Group, London, UK.
Rubinstein, J.P. and Barsky, L.A. (2002) Mineral Deposit and Resources of the Former Soviet Union, Vol 1: Non-Ferrous Metal Ores: Deposits, Minerals and Plants. Taylor and Francis, pp. 411.
Rudnick, R L and Gao, S. (2003) The Composition of the Continental Crust. In: The Crust Vol. 3, Treatise on Geochemistry (eds. Holland, H.D. and Turekian, K.K.). Elsevier-Pergamon, Oxford, United Kingdom. 1–64.
Schmidt, S. (2012) From deposit to concentrate: The basics of tungsten mining. Part 1: Project generation and project development. International Tungsten Industry Association newsletter, June 2012.
Schubert, W.D. and Lassner, E. (2009) Tungsten. International Tungsten Industry Association. http://www.itia.info/news-2-title.html
Shedd, K.B. (2011) Tungsten recycling in the United States in 2000, chap. R of Sibley, S.F., Flow studies for recycling metal commodities in the United States: U.S. Geological Survey. Circular 1196–R, p. R1–R19, http://pubs.usgs.gov/circ/circ1196-R/. (Supersedes U.S. Geological Survey Open-File Report 2005–1028.)
Shedd, K B. (2012a) Tungsten. Mineral Commodity Summaries, United States Geological Survey. http://minerals.usgs.gov/minerals/pubs/commodity/tungsten/680400.pdf
Shedd, K.B. (2012b) Tungsten. 2010 Minerals Yearbook. United States Geological Survey. http://minerals.usgs.gov/minerals/pubs/commodity/tungsten/myb1-2010-tungs.pdf
Simmons, S.F. and Sawkins, F.J. (1983) Mineralogic and fluid inclusion studies of the Washington Cu-Mo-W-bearing breccia pipe, Sonora, Mexico. Economic Geology 78, 521–526.
Smith, M. (2011) China cuts 2012 export quotas for tungsten. Tungsten Investing News, 9 November 2011.
Snow, R.J. and Coker, W.B. (1986) Overburden geochemistry related to W-Cu-Mo mineralisation at Sisson Brook, New Brunswick, Canada: an example of short- and long-distance glacial dispersal. Journal of Geochemical Exploration 28 (1987), 353–368.
Soloviev, S.G. and Krivoschchekov, N.N. (2011) Vostok-2 gold-base-metal-tungsten skarn deposit, Central Sikhote-Alin, Russia. Geology of Ore Deposits 53, 478–500.
Strigul, N., Koutsospyros, A., Arienti, P., Christodoulatos, C., Dermatas, D. and Braida, W. (2005) Effects of tungsten on environmental systems. Chemosphere, 61, 248–258.
Strigul, N., Galdun, C., Vaccari, L., Ryan, T., Braida, W. and Christodoulatos, C. (2009) Influence of speciation on tungsten toxicity. Desalination 248, 869–879.
Strigul, N. (2010) Does speciation matter for tungsten ecotoxicology? Ecotoxicology and Environmental Safety 73, 1099–1113.
Tarney, E. (2004) High speed steel properties: different grades for different requirements. Article for Gear Technology Journal, Randall Publishing, Elk Grove Village, IL60009, USA. http://www.technology.com/pa/members/sepoct04/section9. pdf
Todd, R.G. (1989) Evaporites and strata-bound tungsten mineralisation. Geology 17, 139–143.
Toovey, L.M. (2010) China's tungsten demand to grow 8 per cent this year. Tungsten Investing news, 22 July 2010.
Tornos, F., Galindo, C., Crespo, J. L., and Spiro, B. F. (2007) Geochemistry and origin of calcic tungsten-bearing skarns, Los Santos, Central Iberian Zone, Spain. Canadian Mineralogist 46, 87–109.
United Nations (UN) Comtrade (2013) Commodity Trade Statistics Database. http://comtrade.un.org/db.

United Nations Environment Programme (UNEP). (2011) Recycling Rates of Metals, A Status Report. International Resource Panel. Working Group on the Global Metal Flows.

Wang, F.Y., Li, C.Y., Ling, M.X., Zhang, H., Sun, Y.L. and Sun, W. (2011) Geochronology of the Xihuashan tungsten deposit in southeastern China: constraints from RE-Os and U-Pb dating. Resource Geology 61, 414–423.

Werner, A B T., Sinclair, W.D. and Amey, E.B. (1998) International Strategic Mineral Issues Summary Report – Tungsten. U.S. Geological Survey Circular 930-0. http://pubs.usgs.gov/pdf/circular/c930-0.html

Wolf Minerals Ltd. (2012) Hemerdon tungsten and tin project. http://www.wolfminerals.com.au/investor-relations/press-releases

World Trade Organisation (WTO) (2012a) US, EU and Japan file disputes against China. WTO 2012 News Items, 13 March 2012. http://www.wto.org/english/news_e/news12_e/dsrfc_13mar12_e.htm

World Trade Organisation (WTO) (2012b) Panel established on China's rare earth exports. WTO News Items, 23 July 2012. http://www.wto.org/english/news_e/news12_e/dsb_23jul12_e.htm

Woulfe Mining Corp. (2012) Sangdong – Overview. http://www.woulfemining.com/s/Sangdong.asp

Zaraisky, G.I. and Dubinina, E.O. (2001) On duration of the magmatic and hydrothermal processes at Akchatau greisen W-Mo deposit in central Kazakhstan. In: Piestrzynski et al. (eds.) Mineral Deposits at the Beginning of the 21st Century, Swets and Zeitlinger B.V., Lisse, Netherlands, 505–510.

Zhang, J., Chen, Z., Wang, D., Chen, Z., Liu, S. and Wang, C. (2008) Geological characteristics and metallogenic epoch of the Xungluokeng tungsten deposit, Fujian Province. Geotectonica et Metallogenia 32, 92–97.

Zhang, R., Lu, J., Zhu, J., Wang, R., Chen, J. and Gao, J. (2011) Re-Os and U-Pb geochronology of large Xintianling skarn-type scheelite deposit, Nanling Range, China. 11th Biennial Meeting SGA 2011. https://sga2011conference services.net/ resources/

Zhang, Z., Mao, J., Yang, J., Wang, Z. and Zhang, Z. (2003) Geology and genesis of the Ta'ergou skarn-quartz vein tungsten deposit in north Qilian Caledonian orogenic belt, northwest China. Resource Geology 53, 101–114.

Zhao, Y., Wenwei, L., Chengsi, B., Daxin, L. and Chongjun, J. (1990) Skarn Ore Deposits in China. Beijing, Geological Press, pp. 354.

Zhaolin, L. and Zhongfang, Y. (1996) Rock forming and ore-forming temperatures of Lianhuashan tungsten deposit, Guangdong Province. Chinese Journal of Geochemistry 15, 239–248.

Appendix 1 Units and symbols used in this volume

Throughout this publication units are, in general, expressed in SI units (i.e. according to the International System of Units or *Système international d'unités*). The main exception to this is the unit for temperature where the more commonly recognised degrees Celsius (°C) is used in preference to Kelvin (K). The intervals between degree Celsius and Kelvin are identical; the only difference is the position of zero. The abbreviations or symbols used in this book are shown in Table A.1.1 together with the unit name in full and the property for which they are used. In some instances the internationally recognised Standard Prefixes are used alongside the SI units and these are also included in Table A1.1.

There are several acceptable methods of displaying compound units, for example speed which has the unit of 'metres per second' can be shown as 'm/s' or 'm s^{-1}'. In this volume, the symbol ' / ' has been used to represent the word 'per', which represents the action of dividing one unit by another. Symbols for compound units involving multiplication (rather than division) can be either joined using a raised dot or a non-breaking space, for example 'N·m' or 'N m'. In this volume the non-breaking space has been used.

For some properties the commonly used units are not included within the SI system. In these instances the values for the property have been converted into SI units. Similarly, within the SI system certain properties can be described using more than one compound unit. The conversion factors between certain units are shown in Table A1.2.

Table A1.1 Units used in this volume

Symbol or abbreviation	Unit name	Property
/°C	per degree Celsius	coefficient of thermal linear expansion
°C	degree Celsius	temperature
μΩ m	micro-ohm metre (10^{-6} ohm metre)	electrical resistivity
GPa	gigapascal (10^{9} pascal)	pressure or stress (including Young's modulus)
J/(g °C)	joule per gram degree Celsius	specific heat capacity
J/g	joule per gram	latent heat of fusion or vaporisation
kg	kilogram	mass

(*continued*)

Critical Metals Handbook, First Edition. Edited by Gus Gunn.

Table A1.1 *(cont'd)*

Symbol or abbreviation	Unit name	Property
kg/m^3	kilogram per cubic metre	density
kJ/kg	kilojoule per kilogram	latent heat of fusion or vaporisation
km	kilometre	length
m	metre	length
m^3/kg	cubic metre per kilogram	mass magnetic susceptibility
MN/m^2	meganewton per square metre	Brinell hardness
MPa	megapascal (10^6 pascal)	pressure or stress (including tensile strength)
nΩ m	nano-ohm metre (10^{-9} ohm metre)	electrical resistivity
pm	picometre (10^{-12} metre)	length
S/m	siemen per metre	electrical conductivity
V	volt	electric potential difference or electromotive force
W/(m °C)	watt per metre degree Celsius	thermal conductivity
Ω m	ohm metre	electrical resistivity

Table A1.2 Conversion factors for units

Property name	Convert from unit	Convert to unit	Conversion required
Brinell hardness	meganewton per square metre MN/m^2	megapascal MPa	no conversion necessary because 1 pascal is equal to 1 newton per square metre
Density	gram per cubic centimetre g/cm^3	kilogram per cubic metre kg/m^3	multiply by 1000
Latent heat of fusion or vaporisation	joule per gram J/g	kilojoule per kilogram kJ/kg	no conversion necessary
Radius of atoms or cations	Ångstrom Å	picometre pm	multiply by 100
Specific heat capacity	joule per kilogram degree Celsius J/(kg °C)	joule per gram degree Celsius J/(g °C)	divide by 1000
Specific heat capacity	joule per gram degree Celsius J/(g °C)	joule per gram kelvin J/(g K)	no conversion necessary
Temperature	degree Celsius °C	kelvin K	add 273
Thermal conductivity	watt per metre degree Celsius W/(m °C)	watt per centimetre degree Celsius W/(cm °C)	divide by 100
Thermal conductivity	watt per metre degree Celsius W/(m °C)	watt per metre kelvin W/(m K)	no conversion necessary

Appendix 2 Geological time periods (simplified)

Eon	Era	Period	Age (Ma, million years before present)	
Phanerozoic	Cenozoic	Quaternary	2.6	
		Neogene	23	also known as Tertiary
		Palaeogene	65	
	Mesozoic	Cretaceous	145	
		Jurassic	199	
		Triassic	251	
	Palaeozoic	Permian	299	
		Carboniferous	359	
		Devonian	416	
		Silurian	443	
		Ordovician	488	
		Cambrian	542	
Proterozoic			2500	Precambrian
Archaean				

Critical Metals Handbook, First Edition. Edited by Gus Gunn.

Appendix 3 List of the elements in the Periodic Table sorted alphabetically by element symbol

Chemical symbol	Element name	Atomic number	Chemical symbol	Element name	Atomic number	Chemical symbol	Element name	Atomic number
Ac	Actinium	89	Db	Dubnium	105	Lr	Lawrencium	103
Ag	Silver	47	Ds	Darmstadtium	110	Lu	Lutetium	71
Al	Aluminium	13	Dy	Dysprosium	66	Lv	Livermorium	116
Am	Americium	95	Er	Erbium	68	Md	Mendelevium	101
Ar	Argon	18	Es	Einsteinium	99	Mg	Magnesium	12
As	Arsenic	33	Eu	Europium	63	Mn	Manganese	25
At	Astatine	85	F	Fluorine	9	Mo	Molybdenum	42
Au	Gold	79	Fe	Iron	26	Mt	Meitnerium	109
B	Boron	5	Fl	Flerovium	114	N	Nitrogen	7
Ba	Barium	56	Fm	Fermium	100	Na	Sodium	11
Be	Beryllium	4	Fr	Francium	87	Nb	Niobium	41
Bh	Bohrium	107	Ga	Gallium	31	Nd	Neodymium	60
Bi	Bismuth	83	Gd	Gadolinium	64	Ne	Neon	10
Bk	Berkelium	97	Ge	Germanium	32	Ni	Nickel	28
Br	Bromine	35	H	Hydrogen	1	No	Nobelium	102
C	Carbon	6	He	Helium	2	Np	Neptunium	93
Ca	Calcium	20	Hf	Hafnium	72	O	Oxygen	8
Cd	Cadmium	48	Hg	Mercury	80	Os	Osmium	76
Ce	Cerium	58	Ho	Holmium	67	P	Phosphorus	15
Cf	Californium	98	Hs	Hassium	108	Pa	Protactinium	91
Cl	Chlorine	17	I	Iodine	53	Pb	Lead	82
Cm	Curium	96	In	Indium	49	Pd	Palladium	46
Cn	Copernicium	112	Ir	Iridium	77	Pm	Promethium	61
Co	Cobalt	27	K	Potassium	19	Po	Polonium	84
Cr	Chromium	24	Kr	Krypton	36	Pr	Praseodymium	59
Cs	Caesium	55	La	Lanthanum	57	Pt	Platinum	78
Cu	Copper	29	Li	Lithium	3	Pu	Plutonium	94

(continued)

Critical Metals Handbook, First Edition. Edited by Gus Gunn.

(continued)

Chemical symbol	Element name	Atomic number	Chemical symbol	Element name	Atomic number	Chemical symbol	Element name	Atomic number
Ra	Radium	88	Si	Silicon	14	Uuo	Ununoctium	118
Rb	Rubidium	37	Sm	Samarium	62	Uup	Ununpentium	115
Re	Rhenium	75	Sn	Tin	50	Uus	Ununseptium	117
Rf	Rutherfordium	104	Sr	Strontium	38	Uut	Ununtrium	113
Rg	Roentgenium	111	Ta	Tantalum	73	V	Vanadium	23
Rh	Rhodium	45	Tb	Terbium	65	W	Tungsten	74
Rn	Radon	86	Tc	Technetium	43	Xe	Xenon	54
Ru	Ruthenium	44	Te	Tellurium	52	Y	Yttrium	39
S	Sulfur	16	Th	Thorium	90	Yb	Ytterbium	70
Sb	Antimony	51	Ti	Titanium	22	Zn	Zinc	30
Sc	Scandium	21	Tl	Thallium	81	Zr	Zirconium	40
Se	Selenium	34	Tm	Thulium	69			
Sg	Seaborgium	106	U	Uranium	92			

Glossary of technical terms

A

absorption spectrum when electromagnetic radiation is passed through a substance, certain wavelengths are absorbed by that substance; the variation in absorption intensity is the absorption spectrum.

accessory mineral a mineral that is present in small amounts in a rock but is not considered to be characteristic of the rock.

accretionary terrane an assemblage of fragments of different tectonic plates joined together along a convergent plate boundary. See also: plate, convergent plate boundary.

actively spreading ridge a ridge created by the forces of plate tectonics at divergent plate margins. Typically occur under the oceans, where it is known as 'oceanic spreading,' or 'mid-ocean spreading', but also on land at 'rift zones'.

albitite a rock composed almost totally of the feldspar mineral albite.

alkaline igneous rock rock formed from magmas enriched in the alkalis, sodium and potassium; they are characterised by the presence of alkali feldspar and/or feldspathoids.

alluvial placer deposit a type of mineral deposit laid down in streams or rivers.

alpha particle a positively charged particle containing two protons and two neutrons, identical to the nucleus of a helium atom, which may be emitted during radioactive decay.

aluminosilicate mineral a mineral composed primarily of aluminium, silicon and oxygen, together with other cations.

aluminothermic reduction an exothermic reaction involving aluminium oxide and iron oxide as reducing agents.

amorphous lacking a clear shape; non-crystalline.

amphoteric a molecule that can react as an acid as well as a base (alkali).

anhydrous containing no water.

anode slimes material collected at the bottom of electrolytic cells during the refining of copper metal.

anorthosite a type of igneous rock largely composed of plagioclase feldspar.

anthroposphere that part of the biosphere that is made or modified by humans for use in human activities; sometimes referred to as technosphere.

Critical Metals Handbook, First Edition. Edited by Gus Gunn.

argillic alteration hydrothermal alteration of rocks, generally at low temperature, that results in the formation of clay minerals such as kaolinite, smectite or illite.

A-type granite (or granitoid) derived from mantle depths and emplaced in zones of large-scale extension of the Earth's crust.

B

back-arc rift zone a rift zone that sometimes occurs near to convergent plate boundaries as a result of the subduction processes, particularly if the converging plates are both oceanic. They are often associated with island arcs.

Basel Convention on the Control of Transboundary Movements of Hazardous Wastes and their Disposal a convention adopted in Basel, Switzerland in 1989 with the objectives of protecting human health and the environment from the adverse effects of hazardous wastes.

basement terranes underlying, generally older and deformed, rock units.

Bayer liquor the fluid produced by digesting bauxite in the Bayer process from which the alumina is precipitated.

Bayer process the main industrial process used to refine bauxite into alumina (aluminium oxide).

beach placer deposit mineral deposit formed along coastlines by the action of the sea.

Beijing consensus a term used to describe alternative plans for economic development compared to the Washington consensus; as the name suggests the model for these alternatives is China.

beneficiation the treatment of ore to separate the desired mineral from the gangue. May involve a range of processes such as crushing, washing, sizing of particles and concentration of the desired mineral by various methods.

bioleaching a type of leaching that utilises bacteria to extract the metals. See also: leaching.

biosphere all life on Earth.

blister an intermediate stage in copper metal production, with a copper content of 97–99%, before refining into pure metal.

breakdown voltage the minimum voltage required to cause part of an insulator to begin conducting electricity.

breccia a rock composed of angular fragments of one or more rock types cemented by a fine-grained groundmass.

C

calc-alkaline a suite of igneous rocks which evolves by fractional crystallisation from basic rock types, such as basalt or gabbro, to acid rock types, such as rhyolite or granite, with little or no associated iron enrichment. See also: tholeiitic.

calcine to heat a substance to a temperature at which it is oxidised or reduced, or loses water or at which carbonate compounds are decomposed. The term is also used to describe the product of this process.

capacitor an electrical component used to store energy in an electric field.

carbonaceous a substance rich in carbon.

carbonate a compound containing the carbon-bearing anion CO_3^{2-}.

carbonatisation the entire or partial replacement of rocks by carbonate minerals, such as calcite or dolomite.

carbonatite igneous rock composed of more than 50% carbonate minerals.

catalyst a substance that changes the rate of a chemical reaction whilst not being consumed by the reaction itself.

cathode depolariser a substance that removes hydrogen build up on the cathode of a battery.

cathodic protection a technique used to control corrosion of a metal surface by making it the cathode of an electrochemical cell.

chalcogenide glass a glass containing one of the elements from Group 16 of the Periodic Table.

chalcophile a group of elements that have a strong affinity with sulfur.

chloritisation the alteration of ferromagnesian minerals to chlorite.

chromatic dispersion an effect that results from a variation in propagation delay with wavelength; the most familiar example is the rainbow, where white light is split into its component wavelengths (as evidenced by different colours) as a result of dispersion.

chromitite a rock containing a high concentration of the mineral chromite (an iron chromium oxide, $FeCr_2O_4$).

circular economy a circular economy is a model which aims to reduce the generation of waste, re-use materials for a new purpose or treat waste materials as a resource and recycle them into new products.

cleavage cleavage planes are planes of relative weakness as a result of the positioning of atoms within a crystal and allow the crystal to be more easily split.

coefficient of expansion the degree of expansion of a substance as temperature increases; a low coefficient means that an element expands little as temperature rises.

collision tectonics convergent plate margins. See also: orogeny, orogenic belts, subduction and obduction.

conchoidal the description of fractures in brittle materials that do not follow any natural planes of weakness. Often results in curved breakage surfaces.

concordant intrusive igneous rock bodies which are parallel to bedding planes. The opposite is described as 'discordant'.

conflict metals or minerals metals or minerals which are extracted in areas experiencing conflict and which are sold to provide funds for armed groups.

continental brines fluids containing a high level of dissolved solids that occur close to the surface of continental crust.

continental plate plate of the Earth's crust forming the land areas; typically 35 to 50 km thick and more variable in composition, but higher in silica, than oceanic crust.

convergent plate boundary the boundary between two plates which are moving towards each other. The crust is destroyed at these boundaries. Also known as 'destructive plate boundary'.

cosmogenic produced by cosmic rays.

Cottrell ash (or dust) ash or dust collected in a type of electrostatic precipitator which was invented by Frederick Cottrell and designed to remove dust from industrial exhaust fumes.

covalent bond a form of chemical bond which is characterised by the sharing of pairs of electrons between two atoms.

crystalline structure the arrangement of atoms in a crystal; there are seven basic 3-dimensional patterns that can be formed: triclinic, monoclinic, orthorhombic, rhombohedral, tetragonal, hexagonal and cubic.

D

decrepitation the breaking apart of a substance on heating.

diadochic replacement the replacement of one element in a crystal structure with another.

die casting a metal casting process that involves forcing metal into a mould cavity under high pressure.

dielectric capacity the ability to hold an electric charge.

differentiates different types of igneous rocks formed by the process of magmatic differentiation in which a single magma body evolves over time to produce rocks of varying composition.

digestion the dissolving of a substance in a hot fluid.

diode semiconductor device which allows current to flow in one direction but not the other.

disseminated when a substance is evenly distributed throughout another substance. Commonly used in describing the distribution of ore minerals.

divergent plate boundary the boundary between two plates which are moving away from each other. New crust is generated at these boundaries. Also known as a 'constructive plate boundary'.

dopant a trace impurity element that is inserted into a substance to confer particular, usually electrical or optical, properties.

ductility the ability of a solid material to deform under tensile stress.

E

elasticity the ability of a material to return to its original shape after deformation.

electrical conductivity a measure of how easily an electrical current will pass through a material; the opposite of electrical resistivity.

electrical resistance furnace a furnace that uses the electrical resistance of a substance as the means for creating heat.

electrical resistivity a measure of how strongly a material opposes the flow of electricity; the opposite of electrical conductivity.

electrochemical potential see electrochemical series.

electrochemical series a list of chemical elements in order of their standard potentials to a hydrogen electrode, (also known as their electrochemical potential). The more negative the value the stronger reducing agent the metal is, and the more positive the stronger the oxidising agent.

electrodeposition an industrial process whereby particles are deposited from a liquid on an electrode using the influence of an electrical field.

electroluminescent display material that emits light in response to the passage of an electrical current.

electrolysis a method of using a direct electrical current to drive a chemical reaction.

electron beam refining a process that uses a high-energy electron beam as a heat source.

electron shell a grouping of electrons around an atom's nucleus. The configuration of electrons in the outermost shells determines the chemical properties of the atom.

electronegative elements elements that have the potential to attract electrons to themselves.

electronegativity the tendency of an atom to attract electrons towards itself.

electrophoretic display an electronic display that is designed to mimic the appearance of ink on paper.

electrostatic concentrator a process that uses an electrical charge to separate minerals.

electrostatic discharge the sudden flow of electricity between two objects; it can be caused by a build-up of static electricity or by electrostatic induction.

electrowinning an electrodeposition process whereby an electrical current is passed from an inert anode through the liquid leach solution containing the metal to a cathode; the metal is extracted from the liquid and deposited on the cathode.

eluvial placer deposit a residual type of placer deposit formed by the removal of lighter material.

emission spectrum the spectrum of frequencies of electromagnetic radiation (including light) emitted by an element's atoms or a compound's molecules.

end-of-life recycling rate the proportion of a metal that is discarded at the end of its life which is actually recycled.

epigenetic deposition or alteration by external agents after the rocks have formed.

epithermal formed by warm waters at shallow depths.

evaporite sedimentary rocks produced by the crystallisation of minerals from evaporating water bodies, either the sea or a lake.

exhalative processes the venting of hydrothermal fluids on the sea floor.

exothermic reaction a reaction that releases energy; the opposite of endothermic reaction.

F

felsic light-coloured minerals, such as quartz, feldspar and feldspathoids, found in igneous rocks. Also used to describe rocks which are rich in these minerals.

fenite high-temperature metasomatic rocks composed mainly of potassium and sodium feldspars.

ferromagnetic having a high susceptibility to magnetisation, the strength of which depends on that of the applied magnetising field, and that may persist after removal of the applied field. This is the kind of magnetism displayed by iron, nickel and cobalt and many of their alloys.

flood basalt suite areas of basalt resulting from extremely large volcanic eruptions which have covered the land or sea floor on a regional scale.

flotation, also known as 'froth flotation' a process for separating hydrophobic and hydrophilic materials in a mineral concentrate. The concentrate and reagents are mixed in water to form a slurry which is aerated to create bubbles; the hydrophobic materials attach to the bubbles and rise to the surface in a froth which can be skimmed off.

footwall rocks located immediately below a mineral deposit.

fractional crystallisation the precipitation and segregation of minerals from magma.

fractional distillation the separation of a mixture into its component parts by heating to a temperature at which one or more fractions will vaporise.

fumarole precipitates fumaroles are openings in the earth's crust, usually in the vicinity of volcanoes, which emit steam and other gases; on cooling, contained elements, such as sulfur, may be deposited to form new rock.

G

gabbroic relating to a gabbro, a coarse-grained igneous rock composed of the minerals plagioclase feldspar, pyroxene and, commonly, olivine.

geosphere the solid parts of the Earth.

geothermal field an area of the Earth's crust that is unusually hot.

geothermal gradient the rate at which temperature increases with depth in the Earth's crust.

grade the concentration of a mineral or metal in its ore; expressed as a percentage content, or as 'parts per million' (ppm) or 'grams per tonne' (g/t).

gravity concentration the process of separating two mineral types based on their specific gravity. It can be performed wet or dry, depending on the particular method used.

greisen deposit mineral deposit comprising quartz and mica formed by post-magmatic metasomatic alteration of granitic rocks.

greisenisation the formation of greisen, a highly altered granitic rock composed of quartz and muscovite mica, often rich in fluorine.

H

Hall effect the production of a voltage difference across a conductor, transverse to an electric current in the conductor and a magnetic field perpendicular to the current.

Hall–Héroult process the main industrial process used for smelting alumina (aluminium oxide) to produce aluminium metal.

halogens the elements fluorine, chlorine, bromine, iodine and astatine.

hard metal a mixture of cemented carbides containing one or more of tungsten, titanium, tantalum or vanadium embedded in a matrix of cobalt or nickel by sintering; widely used for cutting tools.

heap leaching metal extraction in which metals are leached from a heap of crushed ore by a percolating reagent. See also: leaching.

humic acid one of the principal components of soil, peat and coal produced by the biodegradation of dead organic matter.

humin chemical compounds in soil, peat and coal which do not breakdown when treated with diluted alkali solutions.

hydrogen reduction a process that involves a molecule gaining hydrogen atoms.

hydrolysis the decomposition of compounds by interaction with water; can also be carried out in the presence of acids or alkalis.

hydrometallurgy processes that use aqueous chemistry to recover metals.

hydrosphere the combined mass of water found on Earth.

hydrothermal related to processes involving hot fluids.

hydroxide a chemical compound containing an oxygen atom and a hydrogen atom bound together.

hypabyssal, also known as 'sub-volcanic' igneous rocks that originate from shallow or medium depths within the crust.

I

igneous rock rock types formed from the cooling and solidification of magma.

immiscible liquids that cannot mix with each other.

incompatible element an element that because of its size or charge does not fit into the crystalline structure of minerals.

in situ leaching the extraction of metal from an ore without the need for conventional mining. A leach solution is pumped via boreholes into the ore and the solution containing dissolved metal is pumped back to surface. See also: leaching.

interstitial lung disease a group of diseases affecting the tissue and space around the air sacs of the lungs.

intracontinental extensional basin depression in a continental plate where the forces of plate tectonics are divergent; may lead to formation of a new rift system.

ion an atom with either a positive electrical charge (a 'cation') or negative electrical charge (an 'anion').

ion adsorption deposit a type of residual deposit in which elements released during the weathering of granites are adsorbed by kaolin and other clay minerals.

ion exchange reversible exchange of ions contained in a crystal for different ions in solution without destruction of the crystal structure. It is sometimes carried out using resins which are small beads fabricated from an inorganic polymer.

ionic radius the radius of an ion.

island arc typically a curved line of volcanoes forming islands near a convergent plate boundary where both plates are oceanic.

isotope forms of a particular chemical element differing in the number of neutrons contained in their nucleus.

I-type granite (or granitoid) derived from the melting of mainly igneous rocks.

K

komatiitic basalt a type of ultramafic mantle-derived volcanic rock containing high levels of magnesium but low levels of silicon, potassium and aluminium.

L

late magmatic fluids fluids remaining in final stages of crystallisation of a magma.

laterite deposits formed by intensive weathering, usually in tropical settings, resulting in the enrichment of some elements and the depletion of others.

leaching a widely used technique for extracting metals from ores by converting the metal into salts soluble in an aqueous liquid; it can be carried out in a number of ways, using either acid or alkaline liquids. See also: heap leaching; bioleaching; in situ leaching; pressure leaching.

leukocyte scintigraphy leukocytes are white blood cells which form the body's immune system; scintigraphy is a form of diagnostic test in which radioisotopes are used to 'tag' cells so they can be detected using scanning detectors.

ligand molecular groups bonded around a central metal atom.

lignite a soft brown coal with characteristics intermediate between those of peat and hard coal; it has a low calorific value and high content of water and volatile components.

liquefaction the process of converting a solid into a liquid by heating or lowering pressure, or of converting a gas to a liquid by cooling or raising pressure.

lithology a description of the physical characteristics of a rock unit.

lithophile a group of elements that combine readily with oxygen.

lixiviant the liquid medium used in hydrometallurgy to selectively extract the desired metal from an ore or mineral.

M

mafic dark-coloured minerals such as olivine, pyroxene and amphibole found in igneous rocks, which are rich in iron and magnesium. The term can also be used to describe rocks which are rich in these minerals.

magmatic related to magma, molten rock and fluid originating deep within or below the Earth's crust.

magnetic concentration a process of separating two mineral types based on their magnetic properties.

magnetoencephalography a technique for mapping brain activity by recording magnetic fields produced by electrical currents occurring naturally in the brain.

malleability the ability of a solid material to deform under compressive stress.

matte molten metal formed during smelting processes.

mesothermal hydrothermal deposits formed at intermediate temperature and pressure.

metalloids also known as semi-metals; elements which are in between metals and non-metals in the Periodic Table and display properties similar to both. Generally these include boron, silicon, germanium, arsenic, antimony and tellurium.

metaluminous a type of igneous rock where the aluminium oxide content is greater than the content of sodium oxide and potassium oxide combined, but less than the content of calcium oxide combined with sodium and potassium oxides.

metamorphic rock rock types formed by the alteration of existing rock by heat and/or pressure.

metamorphogenic deposits formed during metamorphic processes as a result of high pressure and/or temperature.

metasedimentary sedimentary rock that shows evidence of metamorphism.

metasomatic the process of chemical alteration of a rock by interaction with hydrothermal or other fluids and replacement of one mineral by another without melting.

metastable a state which is not the most stable (full equilibrium) and yet might be long lived.

meteoric water water derived from rain or surface water courses that circulates through the rocks or is stored in pore spaces.

meteoritic abundance the abundance of elements in meteorites; for many elements this is commonly used as a best estimate of the abundance of the elements in the entire solar system.

mid-ocean spreading see actively spreading ridges.

mineral deposit an accumulation of a mineral, or group of minerals, of sufficient size and quality that may allow extraction under favourable economic conditions.

mineral deposit model systematically arranged information describing the essential attributes, both descriptive and genetic, of a class of mineral deposits.

mineral occurrence the presence in the Earth's crust of a mineral, or group of minerals, which may indicate a concentration of economic, scientific or technical interest.

monochromator an optical device that transmits only a narrow band of wavelengths of light or other radiation.

N

non-stoichiometric chemical compounds with an elemental composition that cannot be represented by a ratio of whole numbers.

noritic relating to a norite, an igneous rock in which orthopyroxene is the dominant mafic mineral.

O

obduction the overthrusting of continental crust by oceanic crust or mantle rocks at a convergent plate boundary. See also: plate, plate margins, plate tectonics.

oceanic plate plate of the Earth's crust forming the ocean floor; generally less than 10 km thick and composed largely of igneous rocks in layered form.

oceanic spreading see actively spreading ridges.

offgas stream / offgas exhaust gases emitted during industrial processes.

ophiolite a section of the Earth's oceanic crust and underlying upper mantle that has been emplaced above sea level or onto continental crust by plate tectonic processes.

ore the part of mineral deposit comprising the metallic minerals that can be economically extracted.

organolithium molecules with a direct bond between carbon and lithium.

organophile elements that have an affinity for organic compounds.

orogenic belt a structurally complex tract of land comprising a mountain chain, or former mountain chain, created during an orogeny.

orogeny a period of mountain building; they occur at convergent plate boundaries where one or both plates are continental.

oxidation see oxidation state

oxidation state the degree of oxidation of an atom in a chemical compound. An increase in oxidation state through a chemical reaction is known as 'oxidation'; a decrease is known as a 'reduction'. Oxidation state is recorded as an integer and can be either positive, negative or zero. See also valency.

oxide a chemical compound containing at least one atom of oxygen and one other element.

oxyhydroxide a mixed oxide and hydroxide.

P

para-autochthonous rocks found close to the location of their formation.

paramagnetic a form of magnetism that only occurs in the presence of an externally applied magnetic field.

pegmatite very coarse grained igneous rock, commonly of granitic composition.

peralkaline igneous rocks which have a higher molecular proportion of combined sodium and potassium than aluminium.

peraluminous a type of igneous rock where the aluminium oxide content is greater than the content of calcium oxide, sodium oxide and potassium oxide combined.

permanent magnet an object that is magnetised and creates its own persistent magnetic field.

photometric sorting a method of concentrating minerals using optical properties to distinguish between the required mineral and gangue.

photovoltaic the method of generating electrical power by converting solar radiation into direct electrical current, as used in photovoltaic cells.

piezoelectricity electricity that builds up in solid substances, such as certain crystals and ceramics, as a result of pressure.

placer deposit an accumulation of valuable minerals formed as a result of differences in specific gravity. They can be alluvial, eluvial or beach types.

plate the Earth's crust is divided into large slabs of solid rock, known as plates, which are constantly moving in relation to each other. See also: convergent plate boundary, divergent plate boundary, transform plate boundary.

plate margin the edge of a plate, also known as a plate boundary.

plate tectonics collective term for the science related to the moving of Earth's plates.

pneumoconiosis a restrictive lung disease often caused by the inhalation of dust during mining.

porphyritic an igneous rock texture comprising some large crystals surrounded by a finer-grained groundmass.

porphyry deposit orebody associated with porphyritic intrusive rocks and the fluids associated with them. See also: porphyritic.

powder metallurgy processes involving the blending of finely ground materials, pressing them into a desired form and then heating them in a controlled atmosphere to bond the material together.

power density the rate of energy transfer over a given time per unit volume of material.

power-added efficiency a metric for rating the efficiency of a power amplifier.

precipitation the formation of a solid from a fluid because of a change in solubility, either due to temperature variation or the introduction of another substance.

pressure leaching a form of leaching carried out under increased pressure.

primary raw material material extracted from the geosphere.

primitive magma primary melt from the mantle, unaffected by later processes such fractional crystallisation, magma mixing or contamination by crustal rocks.

protolith the original unmetamorphosed rock from which a given metamorphic rock is formed.

pyroclastic deposits deposits of rock material that has been expelled aerially from a volcanic vent, such as agglomerate, tuff and ash.

pyrometallurgy processes that use heat to bring about physical or chemical transformations that enable the recovery of metals.

pyrophoric capable of igniting spontaneously in air.

pyroxenite a rock containing a high concentration of the mineral pyroxene.

R

redox reduction-oxidation. See: oxidation states.

reducing agent the element or compound in a reduction-oxidation reaction that donates an electron.

reduction see oxidation state

refining the purification of an impure metal, commonly by hydrometallurgy or pyrometallurgy.

refractive index a number that describes how light, or other radiation, will pass through a particular substance.

refractory metals a class of metals which are highly resistant to heat and wear.

replacement deposit mineral deposit formed by chemical processes that remove certain elements and leave others in their place.

reserves the part of a resource which has been fully geologically evaluated and is commercially and legally mineable using current technology.

resource nationalism the tendency of people and governments to assert control over natural resources located in their territory. The control may be exerted in many ways, including nationalisation or partial state ownership, imposition of trade restrictions through taxation, quotas, levies, etc.

resources a concentration of minerals or a body of rock that is, or may become, of potential economic interest for the extraction of a mineral commodity.

reverberatory furnace metallurgical furnace in which the material being processed is kept separate from the fuel but is exposed to the combustion gases and radiant heat.

rift zone a region of the crust at a divergent plate boundary associated with a linear zone of faulting and volcanic activity.

S

secondary material material sourced from recycling.

sedimentary-exhalative (SEDEX) mineralisation formed at or below the palaeo-seafloor by fluids exhaling from underneath.

sedimentary basin a geologically depressed area containing sedimentary rocks.

sedimentary rock rock types formed from the settling and accumulation of particles or the precipitation of material from solution.

semiconductor a material with electrical conductivity intermediate between that of a conductor and an insulator. The conductivity may be affected by temperature, light, electric and magnetic fields or by the addition of small amounts of other materials ('dopants').

sericitisation a type of alteration which leads to the formation of sericite, a form of white mica.

serpentinisation the conversion of ultramafic rock into serpentinite by low-temperature metamorphism.

shale a fine-grained, finely laminated, clastic sedimentary rock composed of clay minerals and tiny grains of other minerals, most commonly quartz.

siderophile a group of elements that have a strong affinity with iron.

silicate a compound with structure based on anion of SiO_4^{4-}.

siliciclastic rock silica-rich fragmentary rock.

silicification the entire or partial replacement of rocks with silica, in the form of quartz, chalcedony or opal.

sill a tabular sheet of igneous rock that has been emplaced between layers of older rock.

skarn deposit mineral deposit consisting of metamorphic rocks formed as a result of chemical alteration by hydrothermal or other fluids.

smelting the production of metal from its ore, normally by heating with a chemical reducing agent, commonly carbon.

solid solution series a continuum of solid crystals containing two or more chemical elements where each element can substitute for the other within the crystal structure.

solvent extraction a method used to separate compounds based on their relative solubilities in two immiscible liquids.

sputtering a process used in thin-film deposition whereby atoms are ejected from a target material by bombarding them with particles; these atoms are deposited in a thin layer across a required surface.

stockwork an irregular network of many cross-cutting veins.

stratabound deposits where the mineralisation is confined to a single stratigraphic unit.

stratiform bedded or layered.

stripping ratio the ratio of the amount of waste rock removed to ore mined.

structurally controlled the localisation of a mineral deposit as a result of geological structure, such as a fold or fault.

S-type granite (or granitoid) derived from the melting of mainly sedimentary rocks.

subduction processes that occur at convergent plate margins where one plate (usually an oceanic plate) is forced below another. May result in formation of an oceanic trench and either an island arc (if both plates are oceanic) or orogenic belt (if one plate is continental).

sulfate a compound containing the sulfur-bearing anion SO_4^{2-}.

sulfosalts a complex set of minerals which generally contain a metal, semi-metal and sulfur.

superalloy high-performance alloy with particular applications in aerospace and gas turbines due to their strength, resistance to creep, and resistance to wear and corrosion.

superconductor a material with zero electrical resistivity; this happens to most metals when cooled below a certain, very low, temperature.

supply chain the system of organisations, people, resources and activities that are required to move a product from the raw materials stage to a finished product delivered to a customer.

supracrustal deposited on existing basement rocks.

syenite a coarse-grained alkaline igneous rock with the same composition as granite except that quartz is either absent or present in only small quantities.

synform a downward bending fold; generally the term 'syncline' is only used when it can be demonstrated that the strata in the centre of the fold are younger.

syngenetic formed contemporaneously with rock formation.

synorogenic contemporaneous with orogenic processes.

synsedimentary deposit mineral deposit formed at the same time as the host sedimentary rock.

syntectonic metamorphism metamorphism synchronous with, and resulting from, tectonic activity.

T

technosphere see anthroposphere.

tensile strength the maximum load a material can withstand when being stretched.

teratogenic a substance that may cause malfunctions of an embryo or foetus.

thermal aureole a zone in the country rock surrounding an igneous intrusion within which metamorphic changes, mainly thermal, have occurred.

thermal conductivity a measure of how easily heat is transferred through a material.

thermoelectric device device that converts temperature differences to electrical voltage, and vice versa.

tholeiitic a suite of igneous rocks which evolves by fractional crystallisation from basic rock types, such as basalt or gabbro, to acid rock types, such as rhyolite or granite, with marked enrichment in iron. See also: calc-alkaline.

toll refining the refining of metal-bearing concentrates, or scrap, for which the refinery is paid a fee but the ownership of the metal recovered remains with the owner of the feed material.

transform plate boundary the boundary between two plates which are moving sideways past each other along 'transform' faults. Crust is neither destroyed not created at these boundaries. Also known as a 'conservative plate boundary'.

transition metal elements in Groups 3 to 12 of the Periodic Table.

U

ultrabasic an igneous rock composed of less than 45% silica (SiO_2).

ultramafic igneous rocks containing more than 90% mafic minerals.

V

vacuum arc melting a form of pyrometallurgy carried out in a vacuum arc furnace.

valency the number of chemical bonds formed by the atoms of a given element; monovalent (single bond), divalent (two bonds), trivalent (three bonds), etc.

value chain the series of processes through which a product passes on the way to a customer; at each stage additional value is added to the product.

vapour pressure the pressure exerted by a vapour when it is in equilibrium with its solid or liquid condensed phases at a given temperature in a closed container.

vein deposit mineral deposit in narrow, elongate cavities of the parent rock such as cracks or fissures.

volatile in a chemical context, the tendency of a substance to vaporise.

volcano-sedimentary sequence a mixture of rocks of volcanic and sedimentary origin laid down in layers.

W

wafer also known as 'substrate'; a thin slice of semiconductor material used in the fabrication of integrated circuits or other electronic devices.

Washington consensus a term used since 1989 to describe 10 relatively specific economic policy prescriptions considered as 'standard' by Washington D.C.-based institutions such as the International Monetary Fund and the World Bank.

X

xenolith a fragment of one type of rock enveloped by a different rock type during the latter's formation.

Index

Alaskan/Alaskan-Ural type deposits 289
alkaline granites and syenites 363, 367–8
alkaline igneous rock deposits 319, 323–4
Alpine-type (APT) deposits 181–2, 184
aluminium
  beryllium alloys **102**, **103**, **113**, **116**
  gallium recovery 151, 153
  magnesium alloys 269, 280
  mining industry **21**
  recycling and reuse **52**
  resource criticality **6**, **8**
aluminium-doped zinc oxide (AZO) 224
ammonium metatungstate (AMT) 395, 398
ammonium paratungstate (APT) 394, 395, 398–9, 406–7
ammonium perrhenate (APR) 352
ammonothermal 160
AMT *see* ammonium metatungstate
anode slime residues 212
antimony 70–98
  abundance in the Earth **71**
  carbonate replacement deposits **72**, **76–8**
  contemporary uses **70**, **82–5**
  definitions and characteristics **70**
  deposit characteristics **76**
  environmental factors **94–5**
  epithermal deposits **72–7**
  future developments and needs **96**
  global consumption **84**
  global distribution **73**, **75**, **86**
  global production **87–90**
  gold–antimony ore **72–7**
  greenstone-hosted quartz-carbonate vein deposits **76–8**
  historical uses **70**
  hot spring deposits **74**, **76–7**
  major deposit classes 72–8
  mineralogy 70–72
  mining methods 78–9
  polymetallic base metal vein deposits 74, 76
  prices 92–4
  processing, beneficiation and conversion to metal 79–82
  projects under development 90–91
  recycling and reuse 57, 85, 88
  reduced magmatic deposits 74, 76, 78
  resources and reserves 86–7, 90–91
  size and grade of deposits 74
  specifications **82**
  substitution 86
  world trade 91–3
antimony trioxide (ATO) 82, 84, 95
apartheid 34
apatites 317, 324
Apex Mine, Utah 152
APR *see* ammonium perrhenate
APT *see* Alpine-type (APT) deposits; ammonium paratungstate
Araxá deposit 363, 365–6, 379
arsenide ores
  antimony deposits 77
  cobalt 122–3, 125, 138
artisanal mining 22, 371–2
ATO *see* antimony trioxide
autocatalysts 297–9, 307, 309
AZO *see* aluminium-doped zinc oxide

basalt deposits 288, 292–3
Basel Convention 60
bastnäsite 315, 327
batteries
  antimony 85–6
  cobalt 140, 142–3, 145–6
  gallium 155
  lithium 239–41, 244, 255–8
  rare earth elements 330
bauxite 151, 153
Bayer process 151, 153, 155
BEE *see* black economic empowerment
Beja process 153
bertrandite 106, 109–111
beryl 106, 107, 110
beryllia ceramics 100, 103
beryllium 99–121
  beryllia ceramics 100, 103
  beryllium oxide production from beryllium hydroxide 113–15
  contemporary uses 100–103, 118–19
  deposit characteristics 107–10
  discovery and isolation 99
  distribution and abundance in Earth's crust 100
  environmental factors 116–18
  future developments and needs 118–19
  global distribution 107, 108
  global production 103–5
  high-beryllium alloys 102–3
  hydrothermal deposits 110

Critical Metals Handbook, First Edition. Edited by Gus Gunn.

beryllium (*cont'd*)
low-beryllium alloys 100, 102
metal and alloy production from beryllium hydroxide 113–15
metal preparation 269
mineralogy 106–7
mining methods 110–11
pegmatite deposits 107–10
prices 118
processing of ores to beryllium hydroxide 111–12
properties 99–100, 102–3
pure beryllium metal 102–3
rare element zoning 107, 109
recycling and reuse 57, 113–16
resources and reserves 106–7
substitution 116
world trade 105
BGS *see* British Geological Survey
Bielfeld and Laspeyres process 153
bismuth alloys 215–16
black economic empowerment (BEE) 28
blast furnaces 85–6
Bou Azer, Morocco 125
breccia deposits 391
British Geological Survey (BGS)
beryllium 103–5
cobalt 138, 140
lithium 247
rare earth elements 333–4
Bushveld Complex 286, 290–292
by-product metals
cobalt 123, 125
germanium 193–4
indium 226–7
mining industry 28–9
rare earth elements 326
recycling and reuse 46
resource criticality 6–8

cadmium
alloys 215–16
cadmium telluride 163
mining industry 28
Canadian Institute of Mining, Metallurgy and Petroleum (CIM) 4
capacitors 375, 382
capital investment 24, 30–31
carbon nanotubes 224
carbonate-hosted zinc-lead deposits 182–4
carbonate ores 316
carbonate replacement deposits 74, 76–8
carbonatite deposits
niobium and tantalum 363–6
platinum-group metals 289
rare-earth elements 319–23
carbo-thermic reduction 266–7
carcinogenicity 117
cars 43–4, 58–60
casting cover-gases 273–4
catalysts
cobalt 142
germanium 189–91
platinum-group metals 297–9, 307, 309
rare earth elements 330
recycling and reuse 43
rhenium 340–341, 352–6, 358
CBD *see* chronic beryllium disease
CCZ *see* Clarion–Clipperton Zone
Central African Copperbelt 126–8, 135–40, 142, 198
CFL *see* compact fluorescent lightbulbs
chalcopyrite deposits 206
Chromitite reef type deposits 286, 288, 291
chromium 34
chronic beryllium disease (CBD) **116–18**
CIGS *see* copper indium gallium selenium
CIM *see* Canadian Institute of Mining, Metallurgy and Petroleum
CIS *see* copper indium selenide
civil war *see* geopolitical risk
Clarion–Clipperton Zone (CCZ) 132
closed-loop recycling systems 58–60, 62, 66, 300–301, 309
coal deposits 185–6, 188, 194, 197, 199
cobalamin 143
cobalt 122–49
contemporary uses 122, 140–142
deposit characteristics 123–34
distribution and abundance in the Earth 122
environmental factors 143–4
extraction, processing and refining 134–8
ferromanganese oxyhydroxide crusts 123, 132–4, 138
future developments and needs 144–6
global distribution 123–4
global production 138–40
historical uses 122
hydrothermal deposits 123–9, 135–6
iron-oxide–copper–gold deposits 125, 129
laterite deposits 130–132, 134–5
magmatic deposits 129–30, 134
manganese nodules 132, 140
mineralogy 122–3
mining industry 27, 28, 34
physical and chemical properties 122–3
prices 144–5
recycling and reuse 43, 57, 138
resources and reserves 139–40
specification 140–142
substitution 142–3, 146
world trade 139, 141
cobalt–chromium alloys 122
collection schemes 47–8, 52–4
columbotantalite 362–3, 371
combustion 81
comminution 294–5
compact fluorescent lightbulbs (CFL) 161
computers *see* Waste Electrical and Electronic Equipment
concentrator photovoltaics (CPV) 162, 170
ConRoast process 296
construction minerals 1
contact type deposits 288, 291
continental brines 232–4, 236–8, 241–2, 249–51, 392
Copaux-Kawecki process 111
copper and copper ores
beryllium alloys 100, 101, 113, 115, 118–19
cobalt recovery 125–7, 129–30, 135–6
germanium recovery 178–9, 181–3
indium recovery 206, 208–14, 217
mining industry 26–31, 35
recycling and reuse 41–2, 52
resource criticality 6–7
rhenium recovery 342–50
tungsten alloys 397
copper–indium–gallium–selenide (CIGS) 155, 163–4, 166–7, 171, 216
copper–indium–selenide (CIS) 163, 216
CPV *see* concentrator photovoltaics
criticality *see* resource criticality
cyanidation 81
Czochralski (CZ) method 155–6, 188–9

de la Bretèque process 153
Defence Advanced Research Projects Agency (DARPA) 170
demand rationing 26
depleted uranium 399
deuterium-tritium (DT) reaction 257–8
didymium 312
dietary ingestion 143–4

dismantling processes 47–8, 51–4
disseminated deposits 390
dolomite deposits 366
DT *see* deuterium-tritium (DT) reaction
ductile-to-brittle transition
temperature 341
dunite pipes 288, 291–2

economic forecasting 24, 25, 39
economic viability 3–5
electric vehicles (EV) 240, 244, 255–7
electrical insulation 103
electrodeposition 81–2
electrolytic processes
indium 212, 214
lithium 239–40
magnesium 263–4, 267, 272–3, 279–80
End-of-life Vehicles (ELV) Directive 47, 60
energy minerals 2
environmental factors
antimony 94–5
beryllium 116–18
cobalt 143–4
gallium 163–4
germanium 192
indium 225–6
lithium 241
magnesium 272
mining industry 25
niobium and tantalum 376
platinum-group metals 301–2
rare earth elements 330–331, 336–7
recycling and reuse 42, 43
resource criticality 14
rhenium 356
tungsten 399–400
epitaxial layers 156–7, 160
epithermal deposits 72–7, 210–211
equity financing 22–3
essential nutrients 143–4
ETFs *see* exchange-traded funds
European Commission (EC) 12–13, 35
eutectic alloys 157
EV *see* electric vehicles
evaporite deposits 392
exchange-traded funds (ETFs) 298, 307
exploration companies **22–3**

ferberite 386
ferromanganese oxyhydroxide crusts 123, 132–4, 138
ferro-niobium 376, 380
ferrotungsten 396, 406
fibre glass 308
fibre optics 189–91, 199
Financial Stability Act (2010) 377
Fischer–Tropsch process 354, 358
flame retardants 83, 85
flash smelting 134–5
flotation processes
germanium 186, 188
platinum-group metals 294–5
rare earth elements 326
fluorcarbonate ores 316
fluoride ores 316
fluorite 317
fluorspar 37
fluxes 239, 271
fly ash 152
forensic fingerprinting 377
Frary and Pechiney process 153
fuel cells 307–8, 328
fusion reactors **103**

gallium 150–176
bauxite deposits 151
contemporary uses 153–63
deposit characteristics 151–2
discovery 150
environmental factors 163–4
future developments and needs 166–7, 170–172
gallium antimonide 157
gallium arsenide 153–9, 163, 166, 224–5
gallium chemicals 159–60
gallium metal 157
gallium nitride 153, 160–163, 170
gallium phosphide 162
global distribution 150–151
global resources and production 164–7
IREDs, LDs and LEDs 158–64, 170–172
mineralogy 150–151
mining industry 29–30, 37
photovoltaics 162–4, 170–171
physical and chemical properties 150–151
prices 167–9
primary production 164–5
primary recovery 152–3
recycling and reuse 41, 57, 153–5, 165–7
refining and purification 155
resource criticality 6
secondary production 165–6
secondary recovery 153–5
semiconductors 153–60, 163
specifications 157–63
sphalerite deposits 151–2
substitution 163
world trade 167–9
gallium-doped zinc oxide (GZO) 170–171
gallium–indium–tin alloys 216–17
galvanic corrosion 271–2
gas turbines 353–4, 356–8
geopolitical risk
cobalt 142, 144
magnesium 261
mining industry 26, 28, 32–4, 37
niobium and tantalum 376–7, 380–381
platinum-group metals 308–9
policy attractiveness 34–6
recycling and reuse 43
geothermal brines 234, 251
germanium 177–204
accumulation in sulfide deposits 181–5
contemporary uses 189–91, 199–200
deposit characteristics 179–86
discovery 177
distribution and abundance in the Earth 177–8
enrichment in lignite and coal 185–6, 188, 194, 197–9
environmental factors 192
extraction methods 186
first reduction metal 188
future developments and needs 196–7, 199–200
germanium dioxide 188, 197
germanium tetrachloride 188
global distribution 179–81
global production 194–6
mineralogy 178–9
mining industry 28, 37–8
physical and chemical properties 177–8
prices 197–8
processing and beneficiation 186–8
recycling and reuse 47, 57, 189–91, 199
resources and reserves 192–4
single crystals 188–9
specifications 188–9
substitution 191–2
supply and demand 198–200
world trade 197
zone-refined metal 188
GHG *see* greenhouse gases
global diversified miners 21
Global Mineral Resource Assessment Project 6

gold
antimony deposits 73–80, 90
antimony ores 73–7
cobalt ores 125, 129
mining industry 21, 24
recycling and reuse 41–2, 47–8
granite deposits 363, 367–8, 370–371
granitic pegmatites 107–10
gravity methods 394
greenhouse gases (GHG) 272–5, 302
greenstone-hosted quartz-carbonate vein deposits 76–8
greisen deposits 390
Grignard reagents 269
GZO *see* gallium-doped zinc oxide

Harris Process 213
HB *see* horizontal Bridgman (HB) method
hectorite clays 234–5, 248–9, 252–3
high nitrogen pressure solution (HNPS) 160
high-pressure acid leach (HPAL) technology 135, 138
high-speed steel (HSS) 397
HNPS *see* high nitrogen pressure solution
Hoboken universal process 54–5
horizontal Bridgman (HB) method 155–6
hot spring deposits
antimony 72, 76–7
indium 211
tungsten 392
HPAL *see* high-pressure acid leach
HSS *see* high-speed steel
hübnerite 386
hydride vapour phase epitaxy (HVPE) 160
hydrometallurgical processes
antimony 80, 81
cobalt 134
germanium 187
lithium 240
platinum-group metals 296
recycling and reuse 47–8, 54, 56
tungsten 394
hydrothermal deposits
antimony 72, 74, 76, 77
beryllium 110
cobalt 123–9, 135–6
platinum-group metals 289
rare earth elements 317, 319–24
hypothetical resources **4**

ICGLR *see* International Conference on the Great Lakes Region
Idaho Cobalt Belt (ICB) 125, 138
IMA *see* International Mineralogical Association
IMDG *see* International Maritime Dangerous Goods
imperial smelting process (ISP) 212
indigenisation 32
indium 204–30
abandoned production 221–2
abundance in Earth's crust 205
alloys and solders 215–16
base-metal-rich epithermal deposits 210–211
base-metal-rich tin–tungsten and skarn deposits 210
base-metal sulfide deposits 209
contemporary uses 204, 214–17
deposit characteristics 206–10
discovery 204
environmental factors 225–6
extraction methods and processing 210–214
future developments and needs 226–7
global distribution 207
global production 218–20
historical uses 204
mineralogy 205–6
mining industry 28, 37
mining methods 210–212
physical and chemical properties 204–5
polymetallic vein-type deposits 209, 211
prices 223–4
processing, beneficiation and conversion to metal 212–14
production from copper ores 213–14
production from tin ores 214
projects under development 221
recovery from secondary sources 214
recycling and reuse 41, 42, 46, 51, 57, 220–221, 224
resource criticality 5, 8
resources and reserves 217–18
semiconductors 215, 216
specifications 214–15
substitution 224–5
world trade 222–3
indium–boron oxide 216
indium gallium aluminium phosphide 162
indium gallium phosphide 162
indium–tin oxide (ITO) 46, 51, 54, 171, 204, 215, 223–6
industrial minerals 1–2
infrared emitting diodes (IREDs) 158–9
infrared optics 189–91, 199
inhalation exposure 117
intermetallic compounds **101**
International Conference on the Great Lakes Region (ICGLR) 377
International Maritime Dangerous Goods (IMDG) Code 376
International Mineralogical Association (IMA) 178
international reporting codes 4
IOCG *see* iron-oxide–copper–gold
ion adsorption deposits 322–4
ion-exchange processes 238
IREDs *see* infrared emitting diodes
iridium *see* platinum-group metals
Irish-type (IRT) deposits 181–2, 184
iron and iron ores
germanium recovery 179, 182–3
indium recovery 206
mining industry 36
rare earth elements 325
recycling and reuse 52
resource criticality 6, 8
iron-oxide–copper–gold (IOCG) deposits 125, 129, 324
IRT *see* Irish-type (IRT) deposits
isotropy of beryllium 102
ISP *see* imperial smelting process
ITO *see* indium–tin oxide

jadarite 235–6, 244, 253
jarosite 152, 212
jewellery 298
Joint Ore Reserves Committee (JORC) 4
Joint Research Council (JRC) **13**

Kambalda Dome 129
Kipushi-type (KPT) deposits 181–2, 184, 196–7
Kjellgren–Sawyer process 111
Komatiite related deposits 288, 293
KPT *see* Kipushi-type (KPT) deposits
Kroll process 269

Landsat images 235
lanthanides *see* rare earth elements
laser diodes (LDs) 158–61
laterite deposits 130–132, 134–5, 289
LCA *see* life-cycle analysis
LCD *see* liquid crystal displays
LCEs *see* lithium carbonate equivalents
LCI *see* life-cycle inventories
LCT *see* Lithium-Caesium-Tantalum

LDs *see* laser diodes
leaching
antimony 81
beryllium 111, 112
cobalt 134–8
germanium 186
lead and lead ores
germanium 181–4
indium alloys 215–16
recycling and reuse 52
resource criticality 8
LEC *see* liquid encapsulated Czochralski
LEDs *see* light emitting diodes
lepidolite 232, 247
life-cycle analysis (LCA) 274–5, 399–400
life-cycle inventories (LCI) 273
light emitting diodes (LEDs)
gallium 155, 158–9, 161–4, 170–172
germanium 199
indium 216
lignite deposits 185–6, 194
limonite 152
liquid crystal displays (LCD)
gallium 161–2, 170
indium 215, 221, 223–7
recycling and reuse 46, 51
liquid encapsulated Czochralski (LEC) method 155–6
liquid ion exchange (LIX) 350–351
liquid phase epitaxy (LPE) 157, 166
lithium 230–258
abundance in the Earth 230
chemical and non-chemical demand 238
continental brines 232–8, 241–2, 249–51
current producers 245–8
discovery 230
environmental factors 241
extraction methods and processing 236–8
future developments and needs 249–53, 255–8
geothermal brines 234, 251
global distribution 231
global production 244–5
hectorite clays 234–5, 248–9, 252–3
jadarite 235–6, 244, 253
lithium carbonate 239–40, 253–4
mineralogy and deposit characteristics 230–232
mining industry 24
oilfield brines 234, 251–2
pegmatites 232, 236–8, 241–9, 255
prices 254–5
production costs 248–9
properties 230, 232
recycling and reuse 240–241
resources and reserves 241–5
salars 231–8, 241–4, 249–51
specifications and uses 238–40, 254–8
substitution 240–241
terminology 230
world trade 253–4
lithium carbonate equivalents (LCEs) 230
lithium niobate 374
lithium tantalate 374
Lithium-Caesium-Tantalum (LCT) association 107
LIX *see* liquid ion exchange
London Metal Exchange (LME) 144
LPE *see* liquid phase epitaxy
lubricants 239

magmatic deposits
cobalt 129–30, 134
platinum-group metals 292
rhenium 346, 349–50
magnesium 261–83
casting cover-gas issues 273–4
contemporary uses 261, 267–9, 280–281
deposit characteristics 263
distribution and abundance in the Earth 262
electrolytic processes 263–4, 267, 272–3, 279–80
environmental factors 272
extraction methods, processing and beneficiation 263–7
future developments and needs 277, 279–81
global production 275–7
life-cycle analysis 274–5
metallothermic reduction 263, 265–7, 273–5, 279
mineralogy 262–3
nineteenth century production processes 266
physical and chemical properties 261–2
prices 277, 279
recycling and reuse 269–71
resources and reserves 275–7
specifications 267–8
substitution 271–2
supply security 261
twentieth century production processes 266–7
world trade 277–8
magnet alloys 140–142, 157
magnet ores 328–9, 334
magnetic separation 394
magnetite 179
manganese
ferromanganese oxyhydroxide crusts 123, 132–3, 138
steel alloys 34
manganese nodules
cobalt 133, 140
germanium 179
rare earth elements 325
market capitalisation 21, 22
matte-smelting process 296
MBE *see* molecular beam epitaxy
medical applications
indium 217
platinum-group metals 299–300, 308
MEG *see* Metals Economics Group
memory and storage applications 200, 299–300
mercury 78
Merensky Reef type deposits 286, 288, 290–291, 295
metal organic vapour phase epitaxy (MOVPE) 157, 160
metallothermic reduction 263, 265–7, 273–5, 279
metallurgical metals recovery *see* hydrometallurgical processes; pyrometallurgical processes
Metallurgical Miners' Association of China (MMAC) 36
Metals Economics Group (MEG) 23
metals life cycle 45–6, 62, 63
meteorite impacts 292, 293
microlite 363
military applications 158–9
mineral commodities, definition 2
mineral occurrences, definition 2
mineral sands 324, 326
mining industry **20–39**
artisanal mining 22
capital and operating costs 24–5, 30
constraints on mineral supply response 27–36
demand rationing 26
economic constraints 29–31
economic forecasting 24, 25, 39
environmental factors 25
geopolitical risk and policy attractiveness 33–5, 38
industry dynamics 23–7

mining industry (*cont'd*)
institutional constraints 31–4
market capitalisation 21–2
market size and market share 36
miners and explorers 21–3
nationalisation 32
natural constraints 27–8
policy issues 38–9
prospect development 24
resource criticality 20, 24, 34–8
role of China 34–8
securitisation 20, 32
supply and demand 20–21, 26–7
Minor Metals Trade Association (MMTA) 82, 352, 357
Mississippi Valley-type (MVT) deposits 181–4, 196
MLCC *see* multi-layer ceramic capacitors
MMAC *see* Metallurgical Miners' Association of China
MMIC *see* monolithic microwave integrated circuits
MMTA *see* Minor Metals Trade Association
mobile phones 158, 239 *see also* Waste Electrical and Electronic Equipment
modifying factors 5
molecular beam epitaxy (MBE) 157, 160
molybdenites 342–5, 349–50
molybdenum 28–29, 342–5, 347–50
monazite 315–17, 323, 326–7
monolithic microwave integrated circuits (MMIC) 158–9
MOVPE *see* metal organic vapour phase epitaxy
multi-layer ceramic capacitors (MLCC) 299, 301
multi-stage solvent extraction 327–8
MVT *see* Mississippi Valley-type (MVT) deposits

National Bureau of Statistics of China (NBSC) 400
National Institute of Occupational Safety and Health (NIOSH) 95
National Research Council (NRC)
lithium 240–241
mining industry 35
resource criticality 12, 14
nationalisation 32
naturally occurring radioactive materials (NORMs) 376
NBSC *see* National Bureau of Statistics of China
neodymium–iron–boron magnets 328, 334
neutron moderation 102
nickel
beryllium alloys 113, 115
cobalt ores 129–35
mining industry 26, 27, 36
platinum-group metal alloys 308
recycling and reuse 52
resource criticality 7, 8
tungsten alloys 397
nickel–copper-dominant deposits 288, 292–7, 309
niobium 361–84
alkaline to peralkaline granites and syenites 363, 367–8
carbonatite deposits 363–6
deposit characteristics 363–71
discovery 361
distribution and abundance in the Earth 361–2
environmental factors 376
extraction methods and processing 371–4
future developments and needs 379–82
geopolitical risk 376–7, 380–381
global distribution 364
global production 377–80
mineralogy 362–3
mining industry 28
peraluminous granites 363, 370–371
peraluminous pegmatites 363, 368–70
physical and chemical properties 361–2
prices 380–381
recycling and reuse 43, 57, 375
resources and reserves 377–80
specifications and uses 374–5
substitution 375
niobium-yttrium-fluorine (NYF) association 109
NIOSH *see* National Institute of Occupational Safety and Health
non-ferrous tungsten alloys 396
Norilsk region 130
NORMs *see* naturally occurring radioactive materials
nuclear weapon pits 157
NYF *see* niobium-yttrium-fluorine (NYF) association

occupational health
antimony 95
beryllium 117
indium 226
platinum-group metals 301
Oddo Harkins effect 314
OECD *see* Organisation for Economic Cooperation and Development
oilfield brines 234, 251–2
ongonites 110
open-loop recycling systems 58–9, 61–2, 300, 309
operating costs 25–6, 30–31
ophiolites 289
Organisation for Economic Cooperation and Development (OECD) 377
osmium *see* platinum-group metals
oxide ores
antimony 79, 81
germanium 179, 181
rare earth elements 316

palladium *see* platinum-group metals
peak production concept 9
pegmatite deposits
beryllium 107–10
lithium 232, 236, 241–9, 255
niobium and tantalum 363
tungsten 392
peralkaline granites and syenites 363, 367–8
peraluminous granites 363, 370–371
peraluminous pegmatites 363, 368–70
PET *see* polyethylene terephthalate
petalite 232
PGM *see* platinum-group metals
phosphate ores 316
photovoltaics
gallium 162–4, 170–171
germanium 190–192
indium 221, 223
platinum-group metals 308
Pidgeon process 263, 265–7, 273–5, 279
pipe deposits 392
placer deposits 289, 324, 325, 392
platinum-group metals (PGM) 284–309
cobalt 129–30
contemporary uses 284, 297–300, 307–9
deposit characteristics 285–94
environmental factors 301–2
extraction methods 294
future developments and needs 306–9
global distribution 287
global production 302–4
mineralogy 285–6
mining industry 27–8, 35
nickel–copper-dominant deposits 288, 292–7, 309

nomenclature 284
PGM-dominant deposits 286–92, 294–6, 309
prices 306–7
processing methods 294–7
properties and abundance in the Earth 284–5
recycling and reuse 41–4, 46, 49, 51, 54, 57–62, 300–301, 309
resource criticality 5, 8, 13, 14, 16
resources and reserves 302
specifications 297
substitution 301
supply security 308–9
world trade 304–6
policy attractiveness 33–5
polyethylene terephthalate (PET)
antimony 83–5, 95
germanium 189, 191, 199
polymetallic vein deposits
antimony 72, 76
indium 209, 211
porphyry deposits
germanium 182–3
indium 208, 211
platinum-group metals 289
rhenium 342–8
tungsten 390–391
pre-processing 47–8, 51–4
primary resources 9–10
prospect development 24
pyrochlore 362–3, 366–7, 371–2
pyrometallurgical processes
antimony 79, 81
cobalt 134
germanium 187–8
lithium 239
recycling and reuse 47–8, 54–7
tungsten 394–5

quantum dots 224

rare earth elements (REE) 312–39
alkaline igneous rock deposits 319, 323–4
by-products, co-products and waste products 325
carbonatite-related deposits 317–23
deposit characteristics 317–25
didymium 312
distribution and abundance in the Earth's crust 313–15
environmental factors 330–331, 336–7
extraction and separation 327–8
future developments and needs 329, 332–3, 336–7
global distribution 318
global production 331–3
inclusion of yttrium and scandium 312
ion adsorption deposits 319, 324
iron-oxide-apatite deposits 316–17, 324
IUPAC definition 312
lanthanide contraction 313
mineralogy 315–17
mines and advanced projects 317, 319–22
mining industry 24, 25
mining methods 325
physical and chemical properties 312–5
placer deposits 324, 326
prices 334–6
processing and beneficiation 325–8
recycling and reuse 41, 54, 58, 328–31
resource criticality 5–8, 14, 16
resources and reserves 331–3
seafloor deposits 325
separation from each other 312
size and grade of ore deposits 317, 319
specifications and uses 328–9
substitution 330
world trade 333–4
rare element zoning 107, 109
Raw Materials Group 36
Raw Materials Initiative 62
REACH guidelines 376
recycling and reuse **41–67**
antimony 58, 85, 88
automotive PGM applications 60
beryllium 57, 113–16
cars 43
cobalt 142
dedicated processes for metal combinations 56–7
electronic PGM applications 60
EU critical metal recycling status 57–8
future developments and needs 62–6
gallium 153–5, 165–7
germanium 189–91, 198–9
global end-of-life recycling rates 49
improvements in primary production and wastes 45
indium 220–221, 224
industrial PGM applications 59–60
innovation needs 62–4
lifecycle structures 58–62
lithium 239–40
magnesium 269–71
metallurgical metals recovery 47–8, 54–7
metals life cycle 45–6, 62
mining industry 26
mining and recycling as complementary systems 64–6
mobile phones 42–3
niobium and tantalum 375
platinum-group metals 300–301, 309
rare earth elements 328–31
rationale and benefits 41–3
recyclability factors 48–50, 62, 63
recycling challenges 48–50
recycling rates and improvement pathways 61–2
recycling technologies 51–8
recycling value chain 47–8, 51–7
resource criticality 9–10, 16
resource security as societal driver 64
rhenium 354–5
seven conditions of effective recycling 50–51
status and challenges 45–51
tungsten 398–9
urban mine concept 41–3, 66
waste and resource legislation 47
reduced magmatic deposits 74, 76, 78
REE *see* rare earth elements
reserve base, definition 5
reserves, definition 5
resource classification 4
resource criticality **1–18**
assessments of criticality 11–15
construction minerals 2
criticality concept 10–18
crustal abundances 2
definitions and terminology 3–4
demand and usage patterns 6
economic viability 4–6
energy minerals 2
geological assessment 6
geology and technology of metals 1–5
improving criticality assessment 14, 15
industrial minerals 1–2
key concepts 1–4
metals 2
mining industry 20, 24, 34–8
peak production concept 9
policy implications 16
recycling and reuse of metals 9–10, 16

resource criticality (*cont'd*)
resources and reserves 4–5
supply and demand 6–9
Resource Super Profits Tax 32
reverberatory furnaces 85
rheniite 341–2
rhenium 340–360
contemporary uses 340, 352–4, 356–9
deposit characteristics 342–6
discovery 340
distribution and abundance in the Earth's crust 341
environmental factors 356
extraction methods, processing and beneficiation 350–351
future developments and needs 348–9, 358–9
global distribution 347
global production 340, 346–50
magmatic nickel–copper–PGE deposits 346, 349–50
mineralogy 341–2
mining industry 28, 35
physical and chemical properties 340–341
porphyry deposits 342–8
prices 352–3, 355, 357–8
recycling and reuse 354–5
resources and reserves 346–50
sediment-hosted copper deposits 345, 348
specifications 352–4
substitution 355–6
uranium deposits 346, 349
vein molybdenum deposits 343, 345
world trade 356–7
rhodium *see* platinum-group metals
rift-and continental-flood basalts 288, 292–3
ruthenium *see* platinum-group metals

salars 232–5, 237, 238, 241–4, 249–51
samarium–cobalt magnets 328
scandium *see* rare earth elements
scheelite 386, 390–392
Schwenzfeir process 113, 114
SEC *see* Securities and Exchange Commission
secondary resources 9
Securities and Exchange Commission (SEC) 377
securitisation 20, 32
sediment-hosted copper deposits 345, 348
sediment-hosted massive sulfides (SMS)
germanium 181–2, 184, 196
indium 208–9, 211, 217–18, 222
selenium 28
semiconductors
gallium 153–60, 163
germanium 189, 199
indium 215, 216
tungsten 398
shales 289
SIC *see* Sudbury Igneous Complex
silicates 178–9, 316
silico-thermic reduction 267
silver 42–3, 49
skarn deposits 210, 389–90
skutterudite 123
slag/tailings recovery
antimony 79–81, 85
cobalt 136–7
germanium 185, 192
indium 213
niobium and tantalum 372–3, 379
recycling and reuse 45, 57
SMS *see* sediment-hosted massive sulfides
sodium antimonate 84–5
solid-bed ion exchange 350–351
solid-state lighting (SSL) 161–2
solution-annealing 113
solvent extraction 111–12, 327–8
speculative resources 4
sphalerite deposits 151–2, 183, 205–6
spodumene 232, 236, 243–5, 254
SSL *see* solid-state lighting
State Reserves Bureau (SRB) 35–6
static life index 27
stibnite 70–73, 78–81
stock market financing 23
stockpiling
antimony 89–90
beryllium 99, 101–3, 111
China and criticality 34–6
cobalt 137
germanium 185
indium 223, 227
lithium 246
niobium and tantalum 372, 379
policy issues 39
rare earth elements 331–2
recycling and reuse 62
resource criticality 11, 14, 34–7, 39
rhenium 354, 357
tungsten 404
stockwork deposits 387, 389
stratabound deposits 391–2
structural scarcity 7
Sudbury Igneous Complex (SIC) 129–30, 292, 293
sulfide ores
antimony 79–81
cobalt 122–3, 125, 134–6
gallium 151–2
germanium 178–9, 181–5, 193, 197
indium 205–6, 209
platinum-group metals 285–6
rhenium 341–2
sulfur hexafluoride 273–4
superalloys
cobalt 140–142, 145–6
platinum-group metals 308
rhenium 352–5, 359
tungsten 396, 397, 409
supply risk 12–14
surface-coating applications 216
syelite deposits 363, 367–8
synorogenic hydrothermal deposits **77–8**

tailings *see* slag/tailings recovery
Tanco deposit 368–70
tantalum 361–84
alkaline to peralkaline granites and syenites 363, 367–8
carbonatite deposits 363–6
deposit characteristics 363–71
discovery 361
distribution and abundance in the Earth 361–2
environmental factors 376
extraction methods and processing 371–4
future developments and needs 379–82
geopolitical risk 376–7
global distribution 364
global production 377–80
mineralogy 362–3
mining industry 28
peraluminous granites 363, 370–371
peraluminous pegmatites 363, 368–70
physical and chemical properties 361–2
prices 380–381
recycling and reuse 57, 375
resources and reserves 377–80
specifications and uses 374–5
substitution 375
taxation 34
TCOs *see* transparent conducting oxides
technical scarcity 7

tellurium 7–8, 28, 35
TFSC *see* thin-film solar cells
TFT *see* thin-film transistor
thermometers 157
thin-film deposition 157
thin-film solar cells (TFSC) 163, 167, 223
thin-film transistor (TFT) LCDs 161–2
tin and tin ores. *see also* indium–tin oxide
  indium alloys 215–16
  indium recovery 210, 214
  niobium and tantalum 363, 372–3, 379
  resource criticality 8
titanium 269
TMI *see* trimethylindium
total leaching 296
transparent conducting oxides (TCOs) 170
trimethylindium (TMI) 216
tungsten 385–413
  antimony deposits 78
  breccia deposits 391
  brine/evaporite deposits 392
  contemporary uses 385, 396–8
  deposit characteristics 386–92
  discovery 385
  disseminated or greisen deposits 390
  distribution and abundance in the Earth's crust 385
  environmental factors 399–400
  extraction methods 392–3
  future developments and needs 402–9
  global distribution 388
  global production 401–5
  hot spring deposits 392
  indium recovery 210
  life-cycle analysis 399–400
  mill products 398
  mineralogy 386
  mining industry 36
  pegmatite deposits 392
  physical and chemical properties 385–6
  pipe deposits 392
  placer deposits 392
  porphyry deposits 390–391
  prices 406–9
  processing and beneficiation 393–5
  recycling and reuse 57, 398–9
  resources and reserves 400–405
  skarn deposits 389–90
  specifications 395–6
  steel and other alloys 397–8, 409
  stratabound deposits 391–2
  substitution 399
  vein and stockwork deposits 387, 389
  world trade 404–8
tungsten carbides 375, 395–7
tungsten–heavy-metal alloys (WHAs) 395, 397

UBC *see* Used Beverage Containers
unconformity-related deposits 289
undiscovered resources 4
United Nations Environment Programme (UNEP), beryllium 115
United States Bureau of Mines (USBM) 5
United States Environmental Protection Agency (USEPA)
  beryllium 117
  magnesium 273–4
  rare earth elements 331
United States Geological Survey (USGS) **5**, **7**
  antimony 87
  beryllium 118
  cobalt 139–40
  indium 217–18, 222
  lithium 241
  mining industry 27–8
  niobium and tantalum 380
  rare earth elements 312
  tungsten 400
uranium and uranium ores
  metal preparation 269
  mining industry 33
  rhenium 346, 349
  tungsten alloys 399
urban mine concept 41–3, 66
USBM *see* United States Bureau of Mines
Used Beverage Containers (UBC) 270
USEPA *see* United States Environmental Protection Agency
USGS *see* United States Geological Survey

vacuum melting 113, 114
vanadium 34
VB *see* vertical Bridgman (VB) method
vein deposits
  antimony 74, 76–8
  indium 209, 211
  rhenium 343, 345
  tungsten 387, 389
vertical Bridgman (VB) method 155–6
vertical gradient freeze (VGF) method 155–6
vitamins 143–4
VMS *see* volcanogenic massive sulfide
Voisey's Bay 130
volatilisation, antimony 79–82
volcanogenic massive sulfide (VMS) deposits
  germanium 181, 184, 197
  indium 206, 208–9, 211, 217–18, 222
vulnerability to supply restriction (VSR) **14**

Waste Electrical and Electronic Equipment (WEEE)
  germanium 191
  lithium 239–40
  platinum-group metals 300
  recycling and reuse 42–3, **47**, **50**, **52–3**, **59–60**, **64**
WHAs *see* tungsten–heavy-metal alloys
whole ore leach (WOL) method 136–7
wireless communications 159
wodginite 363
WOL *see* whole ore leach
wolframite 386, 392, 394
World Trade Organisation (WTO)
  magnesium 279
  mining industry 32
  tungsten 407–8

X-ray transparency 102
xenotime 315–16, 326–8

yttrium *see* rare earth elements
yttrium aluminium garnet (YAG) lasers 328

zinc and zinc ores
  gallium recovery 151–3
  germanium recovery 178–9, 181–4, 195
  indium recovery 206, 212, 217–20, 226
  mining industry 28
  resource criticality 8
zircon 317
zirconium 269
zirconium–indium-based fluoride glass 191

Printed and bound by CPI Group (UK) Ltd, Croydon, CR0 4YY
16/09/2026
14963384-0003